Thermodynamics in Nuclear Power Plant Systems

7

Bahman Zohuri • Patrick McDaniel

Thermodynamics in Nuclear Power Plant Systems

 Springer

Bahman Zohuri
Department of Nuclear Engineering
University of New Mexico
Albuquerque
USA

Patrick McDaniel
Department of Chemical and Nuclear
Engineering
University of New Mexico
Albuquerque
USA

A solution manual for this book is available on Springer.com.

ISBN 978-3-319-13418-5   ISBN 978-3-319-13419-2 (eBook)
DOI 10.1007/978-3-319-13419-2

Library of Congress Control Number: 2015935571

Springer Cham Heidelberg New York Dordrecht London

Springer International Publishing AG Switzerland is part of Springer Science+Business Media (www.springer.com)

This book is dedicated to my Parents Marzieh and Akbar Zohuri
Bahman Zohuri

This book is dedicated to Colonel Ben Pollard
Patrick McDaniel

Preface

This text covers the fundamentals of thermodynamics required to understand electrical power generation systems. It then covers the application of these principles to nuclear reactor power systems. It is not a general thermodynamics text, but is a thermodynamics text aimed at explaining the fundamentals and applying them to the challenges facing actual nuclear power systems. It is written at an undergraduate level, but should also be useful to practicing engineers.

It starts with the fundamental definitions of thermodynamic variables such as temperature, pressure and specific volume. It defines the Zeroth Law of Thermodynamics. It then explains open and closed systems. The Ideal Gas law is introduced along with some of its limitations for real gases. Gas kinetic theory is then introduced to provide a background for the Ideal Gas Law and a foundation for understanding for the theory of specific heats. Then it moves on to the First Law of Thermodynamics and its realization in the internal energy and enthalpy potentials. After addressing several applications, it moves on to the Second Law of Thermodynamics and the concept of entropy. It then approaches entropy from the statistical mechanics viewpoint to validate that it truly is a measurable physical quantity. It concludes the fundamental theory portion of the book by discussing irreversibility, availability, and the Maxwell relations, touching slightly on the Third Law of Thermodynamics.

The second portion of the book is devoted to specific applications of the fundamentals to Brayton and Rankine cycles for power generation. Brayton cycle compressors, turbines, and recuperators are covered, along with the fundamentals of heat exchanger design. Rankine steam generators, turbines, condensers, and pumps are discussed. Reheaters and feed water heaters are also covered. Ultimate heat rejections by circulating water systems are also discussed.

The third part of the book covers current and projected reactor systems and how the thermodynamic principles are applied to their design, operation and safety analyses.

Detailed appendices cover metric and English system units and conversions, detailed steam and gas tables, heat transfer properties, and nuclear reactor system descriptions.

Acknowledgments

The authors would like to acknowledge all the individuals for their help, encouragement, and support. We have decided not to name them all since some of them may not be around to see the end result of their encouragement, but we hope they can at least read this acknowledgment wherever they may be.

Last but not least, special thanks to our parents, wives, children and friends for providing constant encouragement, without which this book could not have been written. We especially appreciate their patience with our frequent absence from home and long hours in front of the computer during the preparation of this book.

Contents

About the Authors

Dr. Bahman Zohuri is currently at the Galaxy Advanced Engineering, Inc. a consulting company that he started himself in 1991 when he left both semiconductor and defense industries after many years working as a chief scientist. After graduating from University of Illinois in field of Physics and Applied Mathematics, he joined the Westinghouse Electric Corporation where he performed thermal hydraulic analysis and natural circulation for the inherent shutdown heat removal system (ISHRS) in the core of a liquid metal fast breeder reactor (LMFBR) as a secondary fully inherent shut system for secondary loop heat exchange. All these designs were used for Nuclear Safety and Reliability Engineering for Self-Actuated Shutdown System. He designed the Mercury Heat Pipe and Electromagnetic Pumps for Large Pool Concepts of LMFBR for heat rejection purpose for this reactor around 1978 where he received a patent for it. He later on was transferred to defense division of Westinghouse where he was responsible for the dynamic analysis and method of launch and handling of MX missile out of canister. The results are applied to MX launch seal performance and muzzle blast phenomena analysis (i.e. missile vibration and hydrodynamic shock formation). He also was involved in analytical calculation and computation in the study of Nonlinear Ion Wave in Rarefying Plasma. The results are applied to the propagation of "Soliton Wave" and the resulting charge collector traces, in the rarefactions characteristic of the corona of the a laser irradiated target pellet. As part of his graduate research work at Argonne National Laboratory, he performed computation and programming of multi-exchange integral in surface physics and solid state physics. He holds different patent in areas such as diffusion processes and design of diffusion furnace while he was senior process engineer working for different semiconductor industries such as Intel, Varian, and National Semiconductor corporations. Later on, he joined Lockheed Missile and Aerospace Corporation as Senior Chief Scientist. At this position, he was responsible for Senior in R&D and the study of vulnerability, survivability and both radiation and laser hardening of different components of payload (i.e. IR Sensor) for Defense Support Program (DSP), Boost Surveillance and Tracking Satellite (BSTS) and Space Surveillance and Tracking Satellite (SSTS) against laser or nuclear threat. While in there, he also studied and performed the analysis of characteristics of laser beam and nuclear radiation interaction with materials, Transient Radiation

Effects in Electronics (TREE), Electromagnetic Pulse (EMP), System Generated Electromagnetic Pulse (SGEMP), Single-Event Upset (SEU), Blast and, Thermo-mechanical, hardness assurance, maintenance, device technology.

He did few years of consulting under his company Galaxy Advanced Engineering with Sandia National Laboratories (SNL), where he was supporting development of operational hazard assessments for the Air Force Safety Center (AFSC) in connection with other interest parties. Intended use of the results was their eventual inclusion in Air Force Instructions (AFIs) specifically issued for Directed Energy Weapons (DEW) operational safety. He completed the first version of a comprehensive library of detailed laser tools for Airborne Laser (ABL), Advanced Tactical Laser (ATL), Tactical High Energy Laser (THEL), Mobile/Tactical High Energy Laser (M-THEL), etc.

He also was responsible on SDI computer programs involved with Battle Management C3I and artificial Intelligent, and autonomous system. He is author few publications and holds various patents such as Laser Activated Radioactive Decay and Results of Thru-Bulkhead Initiation.

Recently he has published two other books with CRC and Francis Taylor on the subject of;

1. Heat Pipe Design and Technology: A Practical Approach
2. Dimensional Analysis and Self-Similarity Methods
3. Directed Energy Weapons Technologies

Dr. Patrick McDaniel is currently research professor at Department of Chemical and Nuclear Engineering, University of New Mexico. Patrick began his career as a pilot and maintenance officer in the USAF. After leaving the Air Force and obtaining his doctorate at Purdue University, he worked at Sandia National Laboratories in fast reactor safety, integral cross section measurements, nuclear weapons vulnerability, space nuclear power, and nuclear propulsion. He left Sandia to become the technical leader for Phillips Laboratory's (became part of Air Force Research Laboratory) Satellite Assessment Center. After 10 years at PL/AFRL, he returned to Sandia to lead and manage DARPA's Stimulated Isomer Energy Release project, a $ 10 M per year effort. While at Sandia, he worked on the Yucca Mountain Project and DARPA's classified UER-X program. Having taught at the University of New Mexico in the Graduate Nuclear engineering program for 25 years, when he retired from Sandia in early 2009, he joined the faculty at the University of New Mexico full time. He has worked on multiple classified and unclassified projects in the application of nuclear engineering to high energy systems. Dr. McDaniel holds PhD in nuclear engineering from Purdue University.

Disclaimer

Chapter 1
Definitions and Basic Principles

Nuclear power plants currently generate better than 20 % of the central station electricity produced in the United States. The United States currently has 104 operating power producing reactors, with 9 more planned. France has 58 with 1 more planned. China has 13 with 43 planned. Japan has 54 with 3 more planned. In addition, Russia has 32 with 12 more planned. Nuclear generated electricity has certainly come into its own existent and is the safest, cleanest and greenest form of electricity currently is in produced on this planet. However, many current thermodynamics texts ignore nuclear energy and use few examples of nuclear power systems. Nuclear energy presents some interesting thermodynamic challenges and it helps to introduce them at the fundamental level. Our goal here will be to introduce thermodynamics as the energy conversion science that it is and apply it to nuclear systems. Certainly, there will be many aspects of thermodynamics that are given little or no coverage. However, that is true for any textual introduction to this science; however by considering concrete systems, it is easier to give insight into the fundamental laws of the science and to provide an intuitive feeling for further study. Although brief summary of definition and basic principles of thermodynamic are touched up in this chapter for the purpose of this book, we encourage the readers to refer themselves to references [1–6] provided at the end of this chapter.

1.1 Typical Pressurized Water Reactor

By far the most widely built nuclear system is the Pressurized Water Reactor (PWR). There are a number of reasons for this. Steam turbines have for many decades been the dominant means of generating mechanical energy to turn electrical generators. The temperatures reached in the thermodynamic cycle of a PWR are within the range of fairly, common engineering materials. They were the first system built and operated reliably to produce electricity. A typical PWR system is described in Fig. 1.1.

© Springer International Publishing Switzerland 2015
B. Zohuri, P. McDaniel, *Thermodynamics In Nuclear Power Plant Systems*,
DOI 10.1007/978-3-319-13419-2_1

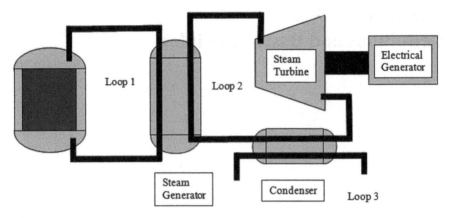

Fig. 1.1 Pressurized water reactor schematic

   The basic PWR consists of five major components, the reactor core, steam generator(s), steam turbine, condenser, and electrical generator and three water/steam loops. Each loop requires a pump that is not shown to keep the diagram cleaner. The nuclear energy is converted to thermal energy in the reactor core. This thermal energy is then transported via the first loop to the steam generator where it is passed to the water in the second loop. The water in the second loop enters as a liquid and is turned to steam. The steam then passes to the turbine where the thermal energy is converted to mechanical energy to rotate the electrical generator. After the thermal energy has been converted to mechanical energy in the steam turbine, the low-pressure steam passes to the condenser to be cooled by the water in the third loop. The second law of thermodynamics tells us that we cannot simply expand the steam to a low enough energy state that it can return to the steam generator in its original liquid state. Therefore, we must extract more thermal energy from the low-pressure steam to return it to its liquid state where it can be pumped back into the steam generator. The third loop is called the circulating water system and it is open to the environment. There are multiple ways of providing this cooling water including intake and return to a river, or the ocean, intake and return to a cooling pond, or intake from a river and exhaust through a cooling tower. However, we are getting ahead of ourselves.
   Consider for a minute why nuclear energy is so useful. A great deal of energy is produced by a very little mass.
   Example Calculation: Calculate the U-235 consumed to produce 1 MW of thermal energy for 1 day. Note that a Megawatt is a unit of power, or energy per unit time,

$1 \text{ MW} = 10^6 \text{ W} = 10^6 \text{ joules/s}$ $1 \text{ day} = 24 \text{ h} = 24 * 3600 \text{ s}$

The energy released in fission of a U-235 atom is ~200 Mev

$1 \text{ ev} = 1.6 \times 10^{-19} \text{ J}$ $1 \text{ Mev} = 1.6 \times 10^{-13} \text{ J}$ $200 \text{ Mev} = 32 \text{ pJ}$

Fissioning 1 atom of U-235 produces 3.2×10^{-11} J

To produce 10^6 J requires $106/3.2 \times 10^{-11}$ atoms $= 3.125 \times 10^{16}$ atoms

And for a duration of 8.64×10^4 s

The total number of atoms consumed will be $3.125 \times 8.64 \times 10^{20}$ atoms

Therefore 2.7×10^{21} atoms will be consumed

A gram mole of U-235 is 6.022×10^{23} atoms

So a gram is $6.022 \times 10^{23}/235 = 2.563 \times 10^{21}$ atoms/gram

Therefore 1 Megawatt-Day of nuclear energy consumes 1.05 g of U-235

The fundamental thing to understand is that a PWR converts nuclear energy to electrical energy and it does this by converting the nuclear energy first to thermal energy and then converting the thermal energy to mechanical energy, which is finally converted to electrical energy. The science of thermodynamics deals with each of these conversion processes. To quantify how each of these processes takes place we must understand and apply the laws of thermodynamics.

1.2   Scope of Thermodynamics

Thermodynamics is the science that deals with energy production, storage, transfer and conversion. It is a very broad subject affects most fields of science including biology and microelectronics. The primary forms of energy considered in this text will be nuclear, thermal, chemical, mechanical and electrical. Each of these can be converted to a different form with widely varying efficiencies. Predominantly thermodynamics is most interested in the conversion of energy from one form to another via thermal means. However, before addressing the details of thermal energy conversion, consider a more familiar example. Newtonian mechanics defines work as force acting through a distance on an object. Performing work is a way of generating mechanical energy. Work itself is not a form of energy, but a way of transferring energy to a mass. So when one mass gains energy, another mass, or field, must lose that energy.

Consider a simple example. A 65-kg woman decides to go over Niagara Falls in a 25-kg wooden barrel. (The first person to go over the fall in a barrel was a woman, Annie Taylor.) Niagara Falls has a vertical drop of 50 m and has the highest flow rate of any waterfall in the world. The force acting on the woman and barrel is the force of gravity, which at the surface of the earth produces a force of 9.8 Newtons for every kilogram of matter that it acts on. So we have

$$W = F \times D \quad F = (65+25) \times 9.8 = 882.0 \text{ N} \quad D = 50 \text{ m}$$

$$W = 882.0 \times 50.0 = 44,100 \text{ N m} = 44.1 \text{ k J}$$

A Newton meter is a joule and 1000 J is a kilojoule. Therefore, when the woman and barrel went over the falls, by the time they had reached the bottom, the force of gravity had performed 44.1 kilojoules (kJ) of work on them. The gravitational field had 44.1 kJ of potential energy stored in it, when the woman and the barrel were at the top of the falls. This potential energy was converted to kinetic energy by the time the barrel reached the bottom of the falls. Kinetic energy is also measured in Joules, as with all other forms of energy. However, we are usually most interested in velocities when we talk about kinetic energies, so let us extract the velocity with which she hit the waters of the inlet to Lake Ontario.

$$\Delta KE = \Delta PE = 44.1 \text{ kJ} = 1/2mV^2 = (90/2) \text{ kg} \times V^2 \quad V^2 = 44.1 \text{ kJ}/(90/2) \text{ kg}$$

Now it is a matter of converting units. A Joule is a Newton-meter. 1 Newton is defined as 1 kg accelerated at the rate of 1 m/second/second. So

$$44.1 \text{ kJ} = 44,100 \text{ N m}$$
$$= 44,100 \text{ kg m/s/s m}$$
$$= 44,100 \text{ kg } (\text{m/s})^2$$

$$V^2 = 44,100 \text{ kg}(\text{m/s})^2/(90/2) \text{ kg}$$
$$= 490/(1/2) = 980 \ (\text{m/s})^2$$
$$V = 31.3 \text{ m/s} \ (\sim 70 \text{ mph})$$

Needless to say she recommended that no one ever try that again. Of course, others have, some have made it, and some have drowned.

Before leaving this example, it is worth pointing out that when we went to calculate the velocity, it was unaffected by the mass of the object that had dropped the 50 m. So one-half the velocity squared represents what we will call a specific energy, or energy per kilogram. In addition, the potential energy at the top of the falls could be expressed as a specific potential energy relative to the waters below. The potential energy per pound mass would just be the acceleration of gravity times the height of the falls. Typically, we will use lower case letters to represent specific quantities and upper case letters to represent extensive quantities. Extensive quantities are dependent upon the amount of mass present. Specific quantities are also referred to as intensive variables, though there are some intensive variables that have no extensive counterpart, such as pressure or temperature.

$$p.e. = mgh/m = gh = 9.8 \times 50 = 0.49 \text{ kJ/kg}$$

It is also worth pointing out that Newton's law of gravity states that

$$F = G\frac{m_1 M_2}{R^2} \tag{1.1}$$

where m_1 is the smaller mass and M_2 is the mass of the Earth. We can find the specific force on an object by dividing the gravitational force by the mass of the object. For distances like 50 m on the surface of the Earth ($R=6,378,140$ m) we can treat R as constant, but if the distance the gravitational force acts through is comparable to the radius of the Earth, an integration would be required. Even on the top of Mount Everest, the gravitational potential is within 0.25% of that at Sea Level, so gravity is essentially constant for all systems operating on the face of the Earth.

1.3   Units

In this section, we will discuss the System International (SI) and English (E) Systems

1.3.1   Fundamental Units

The Before going further it will be a very good idea to discuss units for physical quantities and the conversion of units from one system to another. Unfortunately, the field of thermodynamics is beset with two popular systems of units. One is the System International (SI) system consisting of the kilogram, meter, and second. The other is the English (E) system consisting of the pound-mass, foot, and second.

Starting with the SI system, the unit of force is the Newton. The unit of work or energy is the Joule and the unit of pressure is the Pascal. We have,

$$1\,N = 1\,kg\ m/s^2$$

$$1\,J = 1\,N\ m$$

$$1\,Pa = 1\,N/m^2$$

Now the acceleration of gravity at Sea Level on Earth is 9.8066 m/s², so a 100 kg mass will weight 980.66 Newton. Also when we want avoid spelling out very large or small quantities we will usually use the standard abbreviations for powers of ten in units of 1000. We have,

$$kilo = 10^3$$

$$mega = 10^6$$

$$giga = 10^9$$

$$deci = 10^{-1}$$
$$centi = 10^{-2}$$
$$milli = 10^{-3}$$
$$micro = 10^{-6}$$
$$nano = 10^{-9}$$

For the English system, we have

$$lbm => 1 \, lbf \, (at \, Sea \, Level)$$
$$1 \, ft\text{-}lbf = 1 \, lbf \times 1 \, ft$$
$$1 \, British \, Thermal \, Unit \, (BTU) = 778 \, ft\text{-}lbf$$
$$1 \, psi = 1 \, lbf/in^2$$

Note that the fact that 1 lbf=1 lbm at Sea Level on Earth, means that a mass of 100 lbm will weigh 100 lbf at Sea Level on Earth. The acceleration of gravity at Sea Level on Earth is 32.174 ft/s². Thus we have 1 lbf/(1 lbm-ft/s²)=32.174. If we move to another planet where the acceleration of gravity is different, the statement that 1 lbm=>1 lbf doesn't hold.

Consider comparative weights on Mars. The acceleration of gravity on Mars is 38.5% of the acceleration of gravity on Earth. So in the SI system we have

$$W = 0.385 * 9.8066 \, m/s^2 \times 100 \, kg = 377.7 \, N$$

In the English system, we have,

$$W = 0.385 * 100 \, lbm = 38.5 \, lbf$$

1.3.2   Thermal Energy Units

The British thermal unit (Btu) is defined to be the amount of heat that must be absorbed by a 1 lb-mass to raise its temperature 1 °F. The calorie is the SI unit that is defined in a similar way. It is the amount of heat that must be absorbed by 1 g of water to raise its temperature 1 °C. This raises the question as to how a calorie compares with a joule since both appear to be measures of energy in the SI system. James Prescott Joule spent a major part of his life proving that thermal energy was simply another form of energy like mechanical kinetic or potential energy. Eventually his hypothesis was accepted and the conversion factor between the calorie and joule has been defined by,

1 calorie=4.1868 J

The constant 4.1868 is called the mechanical equivalent of heat.

1.3.3 Unit Conversion

As long as one remains in either the SI system or the English system, calculations and designs are simple. However, that is no longer possible as different organizations and different individuals usually think and work in their favorite system. In order to communicate with an audience that uses both SI and English systems it is important to be able to convert back and forth between the two systems. The basic conversion factors are,

$$1 \, kg = 2.20462 \, lbm$$
$$1 \, lbm = 0.45359 \, kg$$

$$1 \, m = 3.2808 \, ft$$
$$1 \, ft = 0.3048 \, m$$

$$1 \, J = 0.00094805 \, Btu$$
$$1 \, Btu = 1055 \, J$$

$$1 \, atm = 14.696 \, psi$$
$$1 \, atm = 101325 \, Pa$$
$$1 \, psi = 6894.7 \, Pa$$
$$1 \, bar = 100000.0 \, Pa$$
$$1 \, bar = 14.504 \, psi$$

The bar unit is simply defined by rounding off Sea Level atmospheric pressure to the nearest 100 kPa. There are many more conversion factors defined in the Appendix, but they are all derived from this basic few.

1.4 Classical Thermodynamics

Classical thermodynamics was developed long before the atomic theory of matter was accepted. Therefore, it treats all materials as continuous and all derivatives well defined by a limiting process. Steam power and an ability to analyze it and optimize it was one of the main drivers for the development of thermodynamic theory. The fluids involved always looked continuous. A typical example would be the definition of the density of a substance at a point. We have,

$$\rho = \lim_{\Delta V \to 0} \frac{\Delta m}{\Delta V} \qquad (1.2)$$

As long as ΔV does not get down to the size of an atom, this works. Since classical thermodynamics was developed, however, we have come to understand that all gases and liquids are composed of very small atoms or molecules and a limiting process that gets down to the atomic or molecular level will eventually become discontinuous and chaotic. Nevertheless, the continuous model still works well for the macroscopic systems that will be discussed in this text and Classical Thermodynamics is based on it.

At times, we will refer to an atomistic description of materials in order to develop a method of predicting specific thermodynamic variables that classical thermodynamics cannot predict. A typical example is the derivative that is called the constant volume specific heat. This variable is defined as the rate of change of the internal energy stored in a substance as a function of changes in its temperature. Classical thermodynamics demonstrates that this variable has to exist and makes great use of it, but it has no theory for calculating it from first principles. An atomistic view will allow us to make some theoretical estimates of its value. Therefore, at times we will deviate from the classical model and adopt an atomistic view that will improve our understanding of the subject.

Classical thermodynamics is also an equilibrium science. The laws of thermodynamics apply to objects or systems in equilibrium with themselves and their surroundings. By definition, a system in equilibrium is not likely to change. However, we are generally interested in how systems change as thermal energy is converted to and from other forms of energy. This presents a bit of a dilemma in that the fundamental laws are only good for a system in equilibrium and the parameters we want to predict are a result of thermal energy changes in the system. To get around this dilemma, we define what is called a quasi-equilibrium process. A quasi-equilibrium process is one that moves from one system state to another so slowly and so incrementally, that it looks like a series of equilibrium states. This is a concept that classical thermodynamics had a great deal of difficulty clarifying and quantifying. Basically, a process was a quasi-equilibrium process if the laws of equilibrium thermodynamics could characterize it. This is sort of a circular definition, but once again, we will find that the atomistic view allows us to make some predictions and quantifications that identify a quasi-equilibrium process. Quasi-equilibrium processes can occur very rapidly on time scales typical of human observation. For example, the expansion of the hot gases out the nozzle of a rocket engine can be well described as a quasi-equilibrium process with classical thermodynamics.

1.5   Open and Closed Systems

In the transfer and conversion of thermal energy, we will be interested in separating the *entire universe* into a *system* and its *environment*. We will mainly be interested in the energy transfers and conversions that go on within the *system*, but in many cases, we will need to consider its interactions with the rest of the world or its *environment*. Systems that consist of a *fixed amount of mass* that is contained within fixed boundaries are called *closed systems*. Systems that *pass the mass back and forth* to the environment will be called *open systems*. Both *open* and *closed systems*

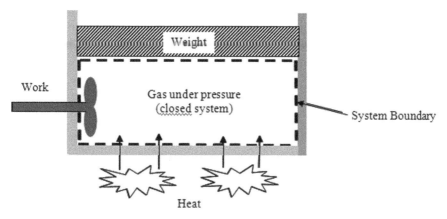

Fig. 1.2 A closed system

allow energy to flow across their borders, but the flow of mass determines whether they are *open* or *closed systems*. *Open systems* will also carry energy across their borders with the mass as it moves. Consider the simple compressed gas in the piston below as a *closed* system (Fig. 1.2).

In analyzing the closed system, we will be concerned about the changes in the internal energy of the compressed gas as it interacts with its environment and the transfers of mechanical and thermal energies across its boundary.

In analyzing open systems, the concept of a ***control volume*** comes into play. The ***control volume*** is the boundary for the open system where the energy changes that we are interested in takes place. The thing separates the open system from its environment. Consider the following open system where we have now allowed mass to flow in and out of the piston of our closed system above (Fig. 1.3).

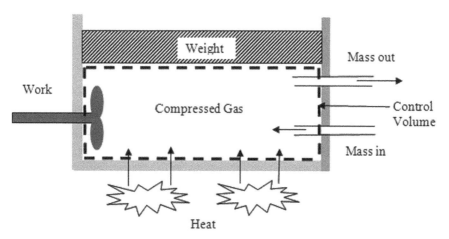

Fig. 1.3 An open system

The *control volume* looks a lot like our system boundary from before, and it is. The only difference is that we now allow mass to flow in and out of our *control volume*. Thermal and mechanical energy can still flow across the boundary, or in and out of the *control volume*. The mass flowing in and out can also carry energy with it either way.

1.6   System Properties

In order to characterize a system we will have to identify its properties. Initially there are three main properties that we will be concerned with—density, pressure and temperature all of which are *intensive* variables. We will use intensive properties to characterize the equilibrium states of a system. Systems will be composed of *pure substances* and *mixtures of pure substances*. A *pure substance* is a material that consists of only one type of atom, or one type of molecule. A *pure substance* can exist in multiple phases. Normally the phases of concern will be gas, liquid, and solid, though for many pure substances there can be several solid phases. Water is an example of a pure substance that can readily be observed in any of its three phases.

A solid phase is typically characterized as having a fixed volume and fixed shape. A solid is rigid and incompressible. A liquid has a fixed volume but no fixed shape. It deforms to fit the shape of the container that is in it. It is not rigid but is still relatively incompressible. A gas has no fixed shape and no fixed volume. It expands to fit the container that is in it. To characterize a system composed of one or more pure components and one or more phases we will need to specify the correct number of intensive variables required to define a state. Gibbs Phase Rule named after J. Willard Gibbs who first derived it gives the correct number of intensive variables required to completely define an equilibrium state in a mixture of pure substances. It is

$$V = C - P + 2 \qquad (1.3)$$

V = Number of variables required to define an equilibrium state.
C = The number of pure components (substances) present.
P = The number of phases present.

So for pure steam at Sea Level and above 100 °C, we have one component and one phase so the number of variables required to specify an equilibrium state is 2, typically temperature and pressure. However, temperature and density would also work. If we have a mixture of steam and liquid water in the system, we have one component and two phases, so only one variable is required to specify the state, either pressure or temperature would work. If we have a mixture like air that is composed of oxygen, nitrogen, and argon, we have three components and three phases (the gas phase for each component), we are back to requiring two variables. As we progress, we will introduce additional intensive variables that can be used to characterize the equilibrium states of a system in addition to density, pressure, and temperature.

1.6.1 Density

Density is defined as the mass per unit volume. The standard SI unit is kilograms per cubic meter (kg/m^3). The Standard English unit is pounds mass per cubic foot (lbm/ft^3). If the mass per unit volume is not constant in a system, it can be defined at a point by a suitable limiting process that converges for engineering purposes long before we get to the atomistic level. The inverse of density is specific volume. Specific volume is an intensive variable, whereas volume is an extensive variable. The standard unit for specific volume in the SI system is cubic meters per kilogram (m^3/kg). The standard unit in the English system is cubic feet per pound mass (ft^3/lbm).

1.6.2 Pressure

Pressure is defined as force per unit area. The standard unit for pressure in the SI system is the Newton per square meter or Pascal (Pa). This unit is fairly small for most engineering problems so pressures are more commonly expressed in kilo-Pascals (kPa) or mega-Pascals (MPa). The standard unit in the English system really does not exist. The most common unit is pounds force per square inch (psi). However, many other units exist and the appropriate conversion factors are provided in the Appendix.

Pressure as an intensive variable is constant in a closed system. It really is only relevant in liquid or gaseous systems. The force per unit area acts equally in all directions and on all surfaces for these phases. It acts normal to all surfaces that contain or exclude the fluid. (The term fluid includes both gases and liquids). The same pressure is transmitted throughout the entire volume of liquid or gas at equilibrium (Pascal's law). This allows the amplification of force by a hydraulic piston. Consider the system in the following Figure. In the Fig. 1.4, the force on the piston at B is greater than the force on the piston at A because the pressure on both is the same and the area of piston B is much larger.

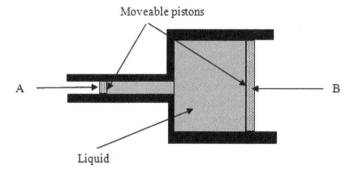

Fig. 1.4 A hydraulic amplifier

Fig. 1.5 Pressure in a liquid
column

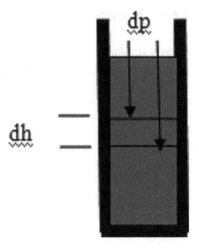

In a gravity field, the pressure in a gas or liquid increases with the height of a column of the fluid. For instance, in a tube containing a liquid held vertically, the weight of all of the liquid above a point in the tube is pressing down on the liquid at that point. Consider Fig. 1.5, then:

$$dp = \rho g dh$$

$$p(0) = P(H) + \int_0^H \rho g dh \qquad (1.4)$$

Thus, the pressure at the bottom of the container is equal to the pressure on the top of the fluid in the container plus the integral of the weight of the fluid per unit area in the container.

This raises an interesting concept. Often it will be important to distinguish between **absolute pressure** and **gage pressure**. The preceding equation calculates the **absolute pressure**. The **gage pressure** is simply the pressure exerted by the weight of the column without the external pressure on the top surface of the liquid. It is certainly possible to have a negative gage pressure, but not possible to have a negative absolute pressure. A **vacuum pressure** occurs when the absolute pressure in a system is less than the pressure in the environment surrounding the system.

Using the setup in Fig. 1.6, a very common way of measuring pressure is an instrument called a manometer. A manometer works by measuring the difference in height of a fluid in contact with two different pressures. A manometer can measure absolute pressure by filling a closed end tube with the liquid and then inverting it into a reservoir of liquid that is open to the pressure that is to be measured. Manometers can also measure a vacuum gage pressure. Consider Fig. 1.6 as below:

The tall tubes on the right in each system are open to the atmosphere. System A is operating at a small negative pressure, or vacuum, relative to the atmosphere. System B is operating at a positive pressure relative to the atmosphere. The magnitude

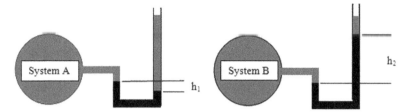

Fig. 1.6 Pressure measurement with manometers

of the pressure in each case can be calculated by measuring the height difference between the fluids in the two sides of the U-tube and calculating its weight per unit area. This is the difference in the pressures inside the Systems A or B and the atmospheric pressure pushing down on the open columns on the right.

1.6.3   Temperature

The other intensive variable to be considered at this point is the temperature. Most everyone is familiar with temperature as a measure of coldness or hotness of a substance. As we continue our study of thermodynamics, we will greatly refine our concept of temperature but for now it is useful to discuss how a temperature scale is constructed. Traditionally the Fahrenheit scale was established by defining the freezing point of water at Sea Level pressure to be 32 °F and the boiling point of water to be 212 °F under the same conditions. A thermometer containing a fluid that expands readily as a function of temperature could be placed in contact with a system that contained ice and water vapor saturated air. The height of the fluid in the thermometer would be recorded as the 32 °F height. Then the same thermometer would be placed in a water container that was boiling and the height of the fluid in the thermometer marked as the 212 °F point. The difference in height between the two points would then be marked off in 180 divisions with each division representing 1 °F. The Celsius scale was defined in the same way by setting the freezing point of water at 0 °C and the boiling point at 100 °C. Water was chosen as the reference material because it was always available in most laboratories around the world.

When it became apparent that absolute temperatures were possibly more important than simply temperatures in the normal range of human experience, absolute temperature scales were defined. The freezing point of water was defined as 273.15 K and the boiling point was defined as 373.15 K, to match up with the Celsius scale. Note that the unit on the absolute scale is Kelvins, not degrees Kelvin. It was named in honor of Lord Kelvin who had a great deal to do with the development of temperature measurement and thermodynamics. The freezing point of water was further defined as the equilibrium of pure ice and air saturated water. However, it was difficult to attain this point because as ice melts it forms a layer of pure water around itself, which prevents direct contact of pure ice, and air-saturated water. Therefore, in 1954, the two-point method was abandoned and the triple point of water was chosen as a single standard. The triple point of water is 273.16 K, 0.01 K above the ice point for water at Sea Level pressure. A single point can be used to define the temperature scale if temperatures are measured with a constant volume, ideal gas thermometer. Basically, the ideal gas thermometer can measure the pressure exerted by a constant volume of gas in contact with the system to be measured. It can also measure the pressure exerted by the gas when in contact with a system at the triple point of water. The ratio of the two pressures gives the ratio of the measured absolute temperature to the absolute temperature of the triple point of water.

However, additional secondary standards are defined to simplify calibration over a broad range of temperatures. The International Practical Temperature Scale is defined by

Triple point of equilibrium hydrogen	13.81 K
Boiling point of hydrogen at 33.33 kPa	17.042 K
Boiling point of hydrogen at 1 atm	20.28 K
Boiling point of neon	27.102 K
Triple point of oxygen	54.361 K
Boiling point of oxygen	90.188 K
Triple point of water	273.16 K
Boiling point of water	373.15 K
Freezing point of zinc	692.73 K
Freezing point of silver	1235.08 K
Freezing point of gold	1337.58 K

Once the absolute temperature scale in Kelvins was defined it became part of the SI system. An absolute scale matching the Fahrenheit scale between the freezing point of water and its boiling point has been defined for the English system. Since there are 180° between the freezing and boiling points in the Fahrenheit scale and 100° over the same range in the Kelvin scale, the absolute scale for the English system, where the unit of measurement is called a degree Rankine, is simply 1.8 times the number of Kelvins. So the freezing point of water on the Rankine scale is 491.67 °R and the boiling point is 671.67 °R. Absolute zero on the Rankine scale is −459.67 °F. To convert back and forth the following formulas apply.

$$T_K = T_C + 273$$
$$T_C = T_K - 273$$
$$T_R = T_F + 460$$
$$T_F = T_R - 460$$

(1.5)

$$T_R = 1.8 T_K$$
$$T_K = \frac{5}{9} T_R$$
$$T_F = 1.8 T_C + 32$$
$$T_C = \frac{5}{9}(T_F - 32)$$

(1.6)

1.7   Properties of the Atmosphere

Before going further, it will be useful to have a model for the atmosphere that can be used for calculations. This is important to realize that the atmosphere at Sea Level supports a column of air that extends upwards of 50 miles. Given the equation derived earlier for the pressure in a column of fluid, we have as always to begin at Sea Level.

$$dp = -\rho g dh$$

$$Let\ \rho = {}^{p}\!/_{RT} \tag{1.7a}$$

Then

$$dp = -p\frac{g}{RT}dh$$

Or integration the last term of Eq. 1.7a, we obtain

$$p = p_{SL}e^{-\frac{g}{RT}h} \tag{1.7b}$$

To perform the integration, the above temperature has been assumed constant. This is not quite true as the standard lapse rate for the Troposphere up to about 40,000 ft is approximately $2\,°C/1000$ ft or $3.6\,°F/1000$ ft. This means that the air is denser than the exponential model predicts. However, it is approximately correct for the Troposphere particularly if only a limited range of elevations is considered and the average temperature is used. The initial values at Sea Level for the standard atmosphere are,

Pressure: 14.696 psi   101.325 kPa
Temperature   59 °F (519 °R)   15 °C (288 K)
Density   076474 lbm/ft^3 1.225 kg/m^3

Composition	Mole fraction (%)
Nitrogen	78.08
Oxygen	20.95
Argon	0.93
Carbon dioxide	0.03
Ne, He, CH^4 et al.	0.01

A more extensive model of the atmosphere as a function of altitude is provided in the Appendix. The relative composition is essentially constant up to the top of the Troposphere.

1.8   The Laws of Thermodynamics

It is useful at this time to state the Laws of Thermodynamics. Later chapters will expand on them greatly, but realizing that there are four simple laws that all of the analysis is built around will provide some structure to guide the way forward.

Zeroth Law of Thermodynamics: *Two bodies in thermal contact with a third body will be at the same temperature.*

This provides a definition and method of defining temperatures, perhaps the most important intensive property of a system when dealing with thermal energy conversion problems.

First Law of Thermodynamics: *Energy is always conserved when it is transformed from one form to another.*

This is the most important law for analysis of most systems and the one that quantifies how thermal energy is transformed to other forms of energy.

Second Law of Thermodynamics: *It is impossible to construct a device that operates on a cycle and whose sole effect is the transfer of heat from a cooler body to a hotter body.*

Basically, this law states that it is impossible for heat to spontaneously flow from a cold body to a hot body. If heat could spontaneously flow from a cold body to a hot body, we could still conserve energy, so the First Law would hold. But every experiment that has ever been performed indicates that thermal energy always flows the other way. This law seems obvious enough but the implications are very significant, as we will see.

Third Law of Thermodynamics: *It is impossible by means of any process, no matter how idealized, to reduce the temperature of a system to absolute zero in a finite number of steps.*

This allows us to define a zero point for the thermal energy of a body.

To be taken under consideration and subject of this matter is beyond the scope of this book.

Problems

Problem 1.1: A bell jar 60 cm in diameter is made to rest on a flat plate and is evacuated with the help of a vacuum pump until the pressure inside the jar reduces to 35 Pa. If the atmospheric pressure is 101.335 kPa, determine the force required to lift the bell jar off the plate.

Problem 1.2: Atmospheric pressure is usually measured with the help of a barometer shown in Fig. 1.7 here. On a particular day at a particular location where $g = 9.7$ m/s^2, if a barometer reads 735 mmHg, determine the atmospheric pressure in kPa and in bars.

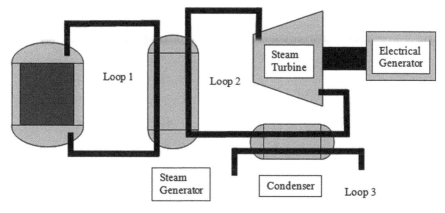

Fig. 1.7 Schematic Diagram of a Barometer for Problem 1.2

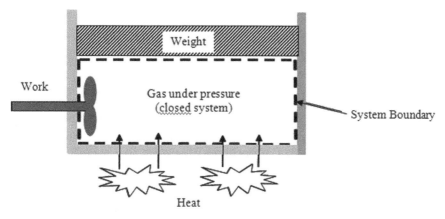

Fig. 1.8 Sketch for Problem 1.4

Problem 1.3: The pressure gages, in common use, are usually calibrated in terms of kg/cm² (the pressure exerted by 1 kg mass on an area of 1 cm²). If a pressure gage connected to a gas chamber reads 5 kg/cm², what is the absolute pressure (in bars) of the gas in the chamber? Assume that $g = 9.78$ m/s².

Problem 1.4: The flow rate of water through a pipe is correlated with the pressure drop across a special length of the pipe. In one such measurement, a U-tube manometer filled with mercury of density 13.6×10^3 kg/m³ shows a deflection of 20 cm. Determine the pressure drop if the density of water is 1000 kg/m³.

Problem 1.5: Newton's second law, $F = ma$, relates a net force acting on a body to its mass acceleration. If a force of 1 N accelerates a mass of 1 kg at on m/s²; or, a force of 1 lbf accelerates 32.2 lbm (1 slug) at a rate 1 ft/s², how are the units related?

Problem 1.6: Newton's second law also defines *Weight* is the force of gravity and can be written as $W = mg$. How does weight change with elevation?

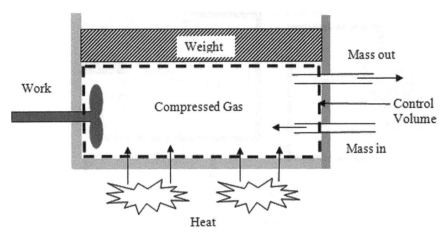

Fig. 1.9 Sketch for Problem 1.8

Problem 1.7: Express the energy unit **J** (joules) in term of **SI** base units: *mass, length,* and *time* (i.e. these units are the bases for *Dimensional Analysis* subject in **SI** form).

Problem 1.8: An assembly of cylinder-piston shown in Fig. 1.9, which contains 0.1 m³ of a gas at a given pressure of 101.325 kPa. At this stage, the spring is touching the piston but applies no force on it. The gas is heated until the volume is doubled. During this process, the force exerted by the spring is proportional to the displacement of the piston. If the spring constant is 50 k N/m and the cross-sectional area of the piston is 0.05 m², then calculate the final pressure of the gas in the cylinder.

Problem 1.9: Assume that the atmosphere is locally isothermal, that is, the variation of the pressure with the specific volume of the atmospheric air follows the relation $P\upsilon = P_0\upsilon_0$ where the subscript zero denotes the conditions at the surface of the earth. Show that the pressure variation with the height in such an atmosphere is given by:

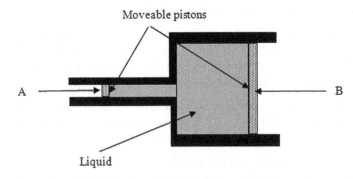

Fig. 1.10 Sketch for Problem 1.9

Fig. 1.11 Sketch for Problem 1.11

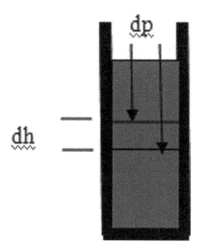

$$P = P_0 \exp\left(\frac{-gh}{P_0 \upsilon_0}\right)$$

where h is the height above the earth's surface and g the acceleration due to gravity.

Problem 1.10: A mixture of nitrogen and hydrogen in the mole ratio of 1:3 enters an ammonia synthesis reactor at the rate of 100 kg/min. Express the flow rate in terms of kmol/min.

Problem 1.11: A container is filled with oil whose density is $\rho = 800 \text{kg}/\text{m}^3$. If the volume of the tank is $V = 2m^3$, determine the amount of mass m in the container (See Fig. 1.11 below).

Problem 1.12: A vacuum gage connected to a chamber reads 5.8 psi at a location where the atmospheric pressure is 14.5 psi. Determine the absolute pressure in the chamber.

Problem 1.13: A spring is stretched a distance of 0.9 m and attached to paddle wheel (See Fig. 1.13). The paddle wheel then rotates until spring is unstretched. Determine the heat transfer necessary to return the system to its initial state.

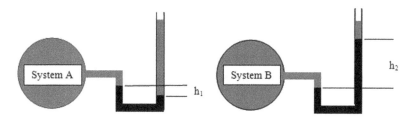

Fig. 1.12 Sketch for Problem 1.2

Fig. 1.13 Sketch for Problem
1.13

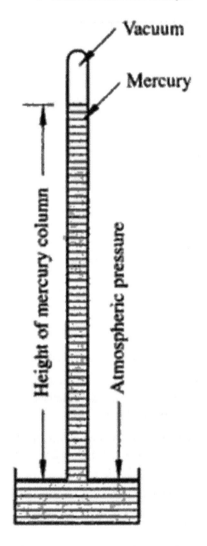

Problem 1.14: A manometer is used to measure the pressure in a tank. The fluid used has a specific gravity of 0.85, and the manometer column height is 55 cm, as shown in Fig. 1.14. If the local atmospheric pressure is 96 kPa, determine the absolute pressure within the tank.

Problem 1.15: A vacuum gage connected to a chamber reads 35 kPa at a location where the atmospheric pressure is 92 kPa. Determine the absolute pressure in the chamber. Use Fig. 1.15 for your analysis

Problem 1.16: If a temperature given in Celsius is equal to 27 °C, then express it in absolute temperature oK?

Problem 1.17: If a Celsius temperature is equal to 40 °C, then express it in °K, °F and °R.

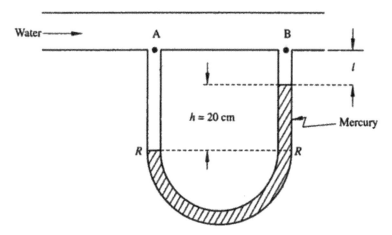

Fig. 1.14 Sketch for Problem 1.14

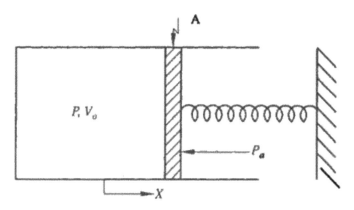

Fig. 1.15 Sketch for Problem 1.15

Problem 1.18: The temperature of a system drops by 30 °F during a cooling process. Express this drop in temperature in Kelvin (°K), Rankin (°R) and Celsius (°C).

Problem 1.19: Consider two closed system A and B. System A contains 1000 kJ of thermal energy at 10 °C whereas system B contains 100 kJ of thermal energy at 60 °C. Now systems are brought into contact with each other. Determine the direction of any heat transfer between the two systems

Problem 1.20: A 250-pound man has a total foot imprint area of 70 in². Determine the pressure this man exerts on the ground if

   a.   he stands on both feet and
   b.   he stands on one foot.

Assume that the weight of the person is distributed uniformly on foot imprint area
Problem 1.21: Consider a 70-kg woman who has a total foot imprint area of 400 cm². She wishes to walk on the snow, but the snow cannot withstand pres-

Fig. 1.16 Sketch for Problem
1.21

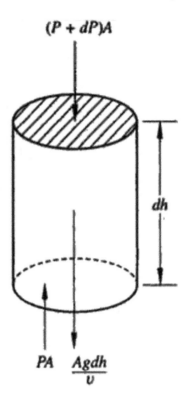

$(P + dP)A$

dh

PA $\dfrac{Agdh}{v}$

sures greater than 0.5 kPa. Determine the minimum size of the snowshoes needed
(imprint area per shoe) to enable her to walk on the snow without sinking (See
Fig. 1.16). Assume that,

1. The weight of the person is distributed uniformly on the imprint area of the
   shoes.
2. One foot carries the entire weight of a person during walking, and the shoe is
   sized for walking conditions (rather than standing).
3. The weight of the shoes is negligible

Problem 1.22: The absolute pressure in water at a depth of 5 m is read to be
145 kPa. Determine (a) the local atmospheric pressure, and (b) the absolute pressure at a depth of 5 m in a liquid whose specific gravity is 0.85 at the same location.
(See Fig. 1.17) and assume that the liquid and water are incompressible

Fig. 1.17 Sketch for Problem
1.32

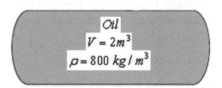

Oil
$V = 2m^3$
$\rho = 800 \ kg/m^3$

Problem 1.23: Find the mass and weight of the air in a living room with a 4.0×5.0 m floor and a ceiling 3.0 m high. What is the mass and weight of an equal volume of water? Assume that air is homogeneous, so that the density is the same throughout the room.

Problem 1.24: In the room described in Problem 1.33, what is the total downward force on the surface of the floor due to air pressure of 1.00 atm? Assume the pressure is uniform, se we use relationship between pressure P on surface A and force F as $P = F/A$.

Problem 1.25: A solar water-heating system uses solar panels on the roof, 12.0 m above the storage tank. The water pressure at the level of the panels is one atmosphere. What is the absolute pressure in the tanks? The gage pressure?

Problem 1.26: A 150-lbm astronaut takes his bathroom scale (a spring scale) and a beam/weight scale (that compares masses) to the Moon where the local gravity is $g = 5.48$ ft/s^2. Determine how much he will weigh: (a) on the spring scale; and (b) on the beam scale. Use English units throughout and convert your final results to SI units.

Problem 1.27: Consider a nuclear power plant that produces 1000 MW of electrical power and has a thermal conversion efficiency of 30 % (that is, for each unit of nuclear fuel energy used, the plant pressure 0.3 units of electrical energy). Assuming continuous operation, determine the amount of nuclear fuel (kilograms of U-235) consumed by this plant per year. Assume that only 180 MeV of the energy released by the fission of U-235 atom is recoverable thermally in the nuclear reactor and the plant.

Problem 1.28: Repeat problem 2 for a coal power plant that burns coal with a heating values of 28,000 kJ/kg.

Problem 1.29: The barometer of a mountain hiker reads 930 mbars at the beginning of a hiking trip, and 780 mbars at the end. Neglecting the effect of altitude on the local gravitational acceleration, determine the vertical distance climbed by the hiker. Assume an average air density of 1.20 kg/m^3 and take $g = 9.7$ m/s^2.

References

1. Cengel YA, Boles MA (2008) Thermodynamics an engineering approach, 6th edn. McGraw Hill, Boston
2. Elliott JR, Lira CT (1999) Introductory chemical engineering thermodynamics. Prentice Hall, Upper Saddle River
3. Hseih JS (1975) Principles of thermodynamics. McGraw Hill, New York
4. Moran MJ, Shapiro HN (2008) Fundamentals of engineering thermodynamics, 6th edn. Wiley, New York
5. Van Wylen GJ, Sonntag RE (1978) Fundamentals of classical thermodynamics, SI Version 2e. Wiley, New York
6. Images are taken with permission form http://www.nasa.gov

Chapter 2
Properties of Pure Substances

2.1 Introduction

A pure substance is a material with a constant chemical composition throughout its entire mass. A pure substance can exist in one or more physical phases such as a solid, liquid or vapor. Each phase will have homogeneous physical characteristics, but all three phases could be different physical forms of the same pure substance. The temperature and pressure boundaries between phases are well defined and it usually requires an input or extraction of thermal energy to change from one phase to another. Most pure substances have a well defined Triple Point where all three phases exist in equilibrium.

In general matter can be classified into two broad categories:

1. **Pure-Substances.**
2. **Mixture.**

Each of these categories can be described as.

1. **Pure substance:** A pure substance is defined as a substance having a constant and uniform chemical composition. Typically, it can be divided in two groups as:

    1. **Elements**—all the same type of atom.
    2. **Compounds**—substances made from two or more different kinds of atoms.

2. **Mixture:** The thermodynamic properties of a mixture of substances can be determined in the same way as for a single substance. The most common example of this is dry air, which is a mixture of oxygen, nitrogen, a small percentage of argon, and traces of other gases. The properties of air are well determined and it is often considered as a single substance. Mixtures can be categorized as two general types:

This chapter deals with the relationship between pressure, specific volume, and temperature for a pure substance.

© Springer International Publishing Switzerland 2015
B. Zohuri, P. McDaniel, *Thermodynamics In Nuclear Power Plant Systems,*
DOI 10.1007/978-3-319-13419-2_2

1. **Homogeneous**—A substance that has uniform thermodynamic properties throughout is said to be homogeneous. The characteristics of a homogeneous mixture are;

   a. Mixtures, which are the same throughout with identical properties everywhere in the mixture.
   b. Not easily separated.
   c. This type of mixture is called a solution. A good example would be sugar dissolved in water or some type of metal alloy like the CROmium-MOLYbdenum steel used in many bike frames.

2. **Heterogeneous**—A heterogeneous mixture is a type of mixture in which the composition can easily be identified. Often there is two or more phases present. Each substance retains its own identifying properties (e.g., granite) and it includes.

   a. Mixtures, which have different properties when sampled from different areas. An example of this would be sand mixed with water.
   b. A mixture in which the individual components can be seen with the naked eye.
   c. A mixture that can be easily separated.

Air is a homogeneous mixture of the gases nitrogen, oxygen, and other minor gases. Here are some other examples of homogeneous mixtures

- Salt water
- Brewed tea or coffee
- Soapy water
- A dilute solution of hydrochloric acid
- Hard alcohol
- Wine

Here are some examples of heterogeneous mixtures

- Sandy water.
- Cake mix and cookie dough.
- Salad.
- Trail mix.
- Granite.
- Sodium chloride (table salt) stirred up with iron filings.
- Sugar and salt mixed in a bowl.
- Carbonated beverage or beer (the CO_2 gas is mixed with the liquid).
- Orange juice with pulp in it
- Water with ice cubes in it.
- Chicken noodle soup.

A pure substance normally requires only two independent properties to specify its state. If pressure and specific volume, for example, are fixed, then all the other properties become fixed. The equation relating pressure, volume, and temperature to each other is called an *Equation of State*. However, a more fundamental

Classification of Matter

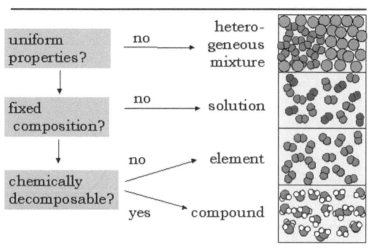

Fig. 2.1 Classification of matter. (Courtesy of NASA)

equation is required to specify all thermodynamic variables as a function of only two properties. These fundamental equations will be called Thermodynamic Potentials (Fig. 2.1).

An example of a simple equation of state which is satisfactory for most dilute gases is the Ideal Gas Law $-pV=nRT$.

2.2 Properties of Pure Substances—Phase Changes

Now consider how a pure substance changes phases. the most common pure substance that is available around the world is water in its three phases—ice, liquid water and steam.

Start with a solid body like ice and add heat. At first the temperature of the body increases proportional to the amount of heat that is added. However, at some point continued addition of heat will cause the body to start to melt. Once it starts to melt the temperature stops increasing and remains constant while the solid is melting. The amount of heat that is added to complete the melting is called the ***Heat of Fusion*** and is normally expressed on per unit mass or per unit mole basis. Once the entire solid is melted the temperature increases again in proportion to the amount of heat input. Note that the increase in temperature per unit heat input for the solid and liquid are not usually equal. As the substance continues to heat up, at some point the liquid will start to vaporize. Once it starts to vaporize, the temperature remains constant until all of the liquid is vaporized. The heat input per unit mass or unit mole required to change the substance from a liquid to a vapor is called the ***Heat of Vaporization***. Once all of the liquid is vaporized, the temperature of the substance starts to increase again proportional to the heat input. This sequence of events is illustrated in Fig. 2.2 below, which is called Temperature-Specific Volume or T-υ Diagram.

A three dimensional view of these processes is presented in the Fig. 2.3. Note that the surface has the following regions; Solid, Liquid, Vapor, Solid-Liquid,

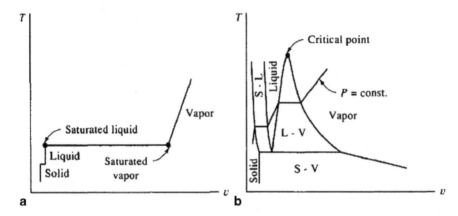

Fig. 2.2 The T-υ diagram

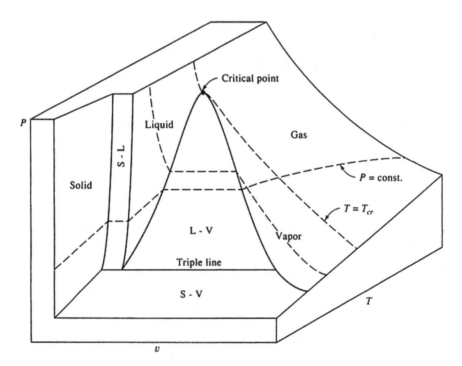

Fig. 2.3 The P-υ-T rendering of a substance that contract on freezing

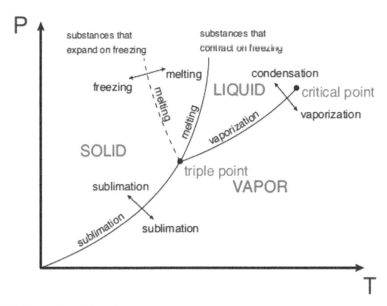

Fig. 2.4 Illustration of phase diagram

Solid-Vapor, and Liquid-Vapor. It also has a line where all three phases can coexist called the *Triple Line* with an interest point that called Triple Point and depicted in Fig. 2.4 as well. At the top if the Liquid-Vapor region, a point exists called the *critical point*. Above the Critical Point, in either pressure or temperature, the fluid cannot be identified as either Liquid or Vapor. In the Liquid-Vapor region called the **Vapor Dome**, the following definition in Sect. 2.2.1 (next section) applies.

2.2.1 Phases of Pure Substances

A pure substance may exist in different phases, where a phase is considered to be a physically uniform form of the substance. The three principle phases are as solid, liquid and gas.

Figure 2.4 shows the typical behavior of pure substances. It is called a *"phase diagram"* because three lines separate all three phases from each other.

2.2.2 Equations of State

Consider a *closed system*, in a vessel so equipped that the pressure, volume, and temperature may be easily measured. If the volume is set at some arbitrary value and the temperature is maintained at a specific value, then the pressure will be fixed at a definite value. Once the V and T are chosen, then the value of P at equilibrium is fixed. That is, of the three thermodynamic coordinates P, V, and T, only two are independent variables. There exists an equation of equilibrium which connects the thermodynamic coordinates and which robs one of them of its independence. Such an equation, called an equation of state, is a mathematical function relating the appropriate thermodynamic coordinates of a system in equilibrium. Every thermodynamic system has its own equation of state, although in some cases the relation may be so complicated that it cannot be expressed in terms of simple mathematical functions. For a closed system, the equation of state relates the temperature to two other thermodynamic variables.

An equation of state expresses the individual peculiarities of one system as compared with another system and must, therefore, be determined either by experiment or by molecular theory. A general theory like thermodynamics, based on general laws of nature, is incapable of generating an equation of state for any substance. An equation of state is not a theoretical deduction from the theory of thermodynamics, but is an experimentally derived law of behavior for any given pure substance. It expresses the results of experiments in which the thermodynamic coordinates of a system were measured as accurately as possible, over a range of values. An equation of state is only as accurate as the experiments that led to its formulation, and holds only within the range of values measured. As soon as this range is exceeded, a different form of equation of state may be required.

Note that in any of the three homogeneous phases discussed in Sect. 2.2.1 in above, a relationship exists that gives $P = P(V,T)$. Or any of the variables can be expressed in terms of the other two. These equations are called *Equations of State*. In the two-phase regions, including their borders, specifying temperature alone will set the pressure and vice versa. Giving both pressure and temperature will not define the volume because we will need to know the relative proportion of the two phases present. The mass fraction of the vapor in a two-phase liquid-vapor region is called the quality.

2.3   Ideal Gas

Any equation that relates the pressure, temperature, and specific volume of a substance is called an equation of state. There are many equations of state, some simple and others very complex. The simplest and best-known equation of state for substances in the gas phase is the Ideal Gas equation of state. This equation predicts the *p-v-T* behavior of a gas quite accurately for dilute or low-pressure gases. Probably the definition of a low pressure or dilute gas is that it obeys the Ideal Gas Law. *It is based on the two modeling assumptions that (1) each molecule is a point mass with no volume, and (2) they only interact by billiard ball-like collision conserving energy and momentum of the colliding particles.* The Ideal Gas equation of state was formulated long before the atomic hypothesis was demonstrated, but these two assumptions quickly lead to the properties of the Ideal Gas equation of state.

An *Ideal Gas* is one that obeys the following *Equation of State*.

$$pV = n\Re T \tag{2.1}$$

p Absolute pressure
V Volume of gas
n Number of moles of the gas
$\Re$ Universal Gas Constant $=8314$ J/kmol/K$=1545$ ft-lbf/lbmol/°R$=1.986$ Btu/lbmol/°R
T Absolute temperature in degrees Rankine or Kelvins

Note that $\Re$ is the Universal Gas Constant. A gas constant for a particular gas can be obtained by dividing the universal constant by the molar mass to obtain the following equation.

$$R = \Re/M \tag{2.2}$$

where M is molecular weight of gas. If we identify m as the mass of gas in kg or lbm, then another form of the Ideal Gas Law can be written as:

$$pV = m\Re T \tag{2.3}$$

Identifying $\rho = m/V$ as the gas density, then another form of the Ideal Gas Law is:

$$p = \rho\Re T \tag{2.4}$$

Normally an Ideal Gas must be a pure substance. However, air is a mixture that obeys the Ideal Gas Equation over a broad range of values for temperature and pressure. Most gases obey the Ideal Gas Equation of State if the pressure is not too high or the temperature too low.

The Ideal Gas law gives is a simple enough equation that given any two of the thermodynamic variables, p, v, and T, the third can easily be found. Consider 2 kg-moles of H_2 at 1000 °K and 0.2 MPa. Calculate the volume required to store the gas at this temperature and pressure. The required volume is,

$$V = \frac{n\Re T}{p} = 2.0\text{kg-moles}*8314.47 \text{ J/kg-mole}/\text{K}*1000\,^{\circ}\text{K}/200,000\,\text{nt/m}^2 = 83.1\text{m}^3$$

Obviously, given temperature and density, or specific volume, the pressure could be found in a similar manner. Given pressure and density, or specific volume, the temperature is easily found from the same equation. For this reason, applying the Ideal Gas Law is usually a good first guess when trying to solve for pressure, density, or temperature.

2.4   Real Gases and Vapors

In this section, the behavior and properties of real gases and vapors are described and equations of state are identified.

An ideal gas is made up of particles that do not attract or repel one another. Real gases are made up of atoms or molecules that may attract one another strongly, like ammonia, water vapor, or sulfur dioxide. On the other hand, they may attract one another hardly at all, like helium.

Real gases behave like ideal gases at "ordinary" temperatures and pressures. However, if you heat them up and compress them to high pressure, then their behavior departs from ideal. If the molecules attract one another, a molecule in the center of the gas is attracted equally on all sides and its motion is not affected. For a molecule, which is very close to the wall of container, exerts lees force on the wall, due to the intermolecular attractive forces with other molecules.

2.4.1   Simple Real Gas Equations of State

At higher pressures or lower temperatures, the equation of state becomes more complicated. The volume taken up by the molecules of the gas must be considered and the attraction of the molecules for each other lessens the pressure they exert on their container. The first Equation of State to take these two effects into account was the Van der Waals Equation of State given by,

$$p = \frac{\Re T}{(\upsilon - b)} - \frac{a}{\upsilon^2} \tag{2.5}$$

where a and b are constants appropriate to the specific gas. As far as thermodynamics is concerned, the important idea is that an equation of state exists, not whether it can be written down in a simple mathematical form. Also there exists no equation of state for the states traversed by a system that is not in mechanical and thermal equilibrium, since such states cannot be described in terms of thermodynamic coordinates referring to the system as a whole.

It is generally impossible to express the complete behavior of a substance over the whole range of measured values of p, v, and T by means of one simple equation with two adjustable parameters (a and b). Several equations of state, such as the Ideal Gas Law and those found below can be used to characterize the gas or vapor phase. Several Equations of State that have found utility in thermodynamic analysis are listed here.

a. $p = \frac{\Re T}{(\upsilon - b)} - \frac{a}{\upsilon^2}$ Van der Waals equation of state

b. $p = \frac{\Re T}{v - b} - \frac{a}{T^{1/2} v(v + b)}$ Redlich-Kwong equation of state

c. $p = \dfrac{\Re T}{v-b} - \dfrac{\alpha a}{v^2 + 2bv - b^2}$ Peng-Robinson equation of state

d. $pv = \Re T(1 + BP + CP^2 + \cdots)$ Virial Expansion

e. $(pe^{a/\Re T v})(v - b) = \Re T$ Dieterici equation of state

f. $\left(p + \dfrac{a}{v^2 T} \right)(v - b) = \Re T$ Berthelot equation of state

g. $\left(p + \dfrac{a}{(v+c)^2 T} \right)(v - b) = \Re T$ Clausius equation of state.

h. $pv = \Re T \left(1 + \dfrac{B'}{v} + \dfrac{C'}{v^2} + \cdots \right)$ Another type of virial expansion

2.4.2   Determining the Adjustable Parameters

Every equation of state must satisfy a number of conditions.

1. It must reduce to the Ideal Gas Law as the pressure approaches zero or the temperature increases without bound.
2. The critical isotherm must show a point of inflection at the critical point.
3. The isometrics on a p-T diagram should approach straight lines with either decreasing density or increasing temperature. The critical isometric should be a straight line.

Since the critical point is the limiting position on a p-v diagram (see Fig. 2.5 below) as the two end-points (saturated liquid and saturated vapor) on the same isotherm approach each other, it follows that the slope of the isotherm passing through the critical point (the critical isotherm) is zero, or stated mathematically as;

$$\left(\frac{\partial P}{\partial V} \right)_{T=T_c} = 0 \qquad\qquad (2.6a)$$

Also, the critical point is a point of inflection on the critical isotherm, because the isotherm is concave upward at volumes less than the critical volume and concave downward at specific volumes more than the critical volume; hence

$$\left(\frac{\partial^2 P}{\partial V^2} \right)_{T=T_c} = 0 \qquad\qquad (2.6b)$$

Equations 2.6a and 2.6b, along with the equation of state itself, enable one to evaluate the constants in any two parameter equation of state based on the critical values P_C, V_C, and T_C. Consider, for example, the Van der Waals equation of state, which can be written:

Fig. 2.5 *p-v* diagram for pure
substance showing isotherms
in the region of critical point.
Solid lines represent the
values predicted by the Van
der Waals equation of state.
Points represents the experimental values

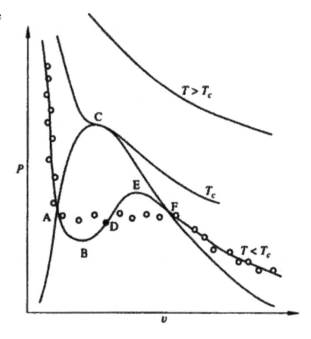

$$p = \frac{\Re T}{\upsilon - b} - \frac{a}{\upsilon^2} \tag{2.7}$$

where $\upsilon = V / n$ is the molar volume. This equation holds fairly well in the vapor region near and above the critical point. Equations 2.6a and 2.6b for molar volume yield, respectively;

$$\left(\frac{\partial P}{\partial \upsilon}\right)_{T=T_C} = -\frac{\Re T}{(\upsilon - b)^2} + \frac{2a}{\upsilon^3} = 0 \tag{2.8a}$$

and

$$\left(\frac{\partial^2 P}{\partial \upsilon^2}\right)_{T=T_C} = \frac{2\Re T}{(\upsilon - b)^3} - \frac{6a}{\upsilon^4} = 0 \tag{2.8b}$$

Equations 2.8a and 2.8b can be rewritten as;

$$\frac{2a}{\upsilon^3} = \frac{\Re T}{(\upsilon - b)^2} \tag{2.9}$$

and

$$\frac{3a}{\upsilon^4} - \frac{\Re T}{(\upsilon - b)^3} \qquad (2.10)$$

Dividing the first equation by the second to obtain the critical molar volume as;

$$\upsilon_C = 3b \qquad (2.11)$$

Substituting this value for υ in the first of the two equations, we obtain a relationship for the critical temperature as;

$$T_C = \frac{8a}{27b\Re} \qquad (2.12)$$

and finally, substituting these two values in the Van der Waals equation to obtain the critical pressure as

$$P_C = \frac{a}{27b^2} \qquad (2.13)$$

At the critical point these equations can be written as follows;

$$\left(\frac{\partial P}{\partial \upsilon}\right)_{T=T_C} = -\frac{\Re T_C}{(\upsilon_C - b)^2} + \frac{2a}{\upsilon_C^3} = 0 \qquad (2.14)$$

and

$$\left(\frac{\partial^2 P}{\partial \upsilon^2}\right)_{T=T_C} = \frac{2\Re T_C}{(\upsilon_C - b)^3} - \frac{6a}{\upsilon_C^4} = 0 \qquad (2.15)$$

so that at the critical point, Van der Waals equation is given by

$$\left(P_C + \frac{a}{\upsilon_C^2}\right)(\upsilon_C - b) = \Re T_C \qquad (2.16)$$

Based on the critical point data then, we can calculate the Van der Waals constants a and b in terms of the critical constants. Since it is possible to experimentally measure the critical temperature and critical pressure, a and b can be evaluated from knowledge of P_C and T_C through the following relations

$$a = \frac{27\Re^2 T_C^2}{64P_C} \quad \text{and} \quad b = \frac{\Re T_C}{8P_C} \qquad (2.17)$$

Substance	Z_c
Water	0.230
Ammonia	0.242
Carbon dioxide	0.275
Nitrogen	0.287
Helium	0.291
Hydrogen	0.307
Van der Waals gas	0.375
Ideal gas	1.00

Table 2.1 Calculated values of Zc

The values of a and b are provided for a number of gases in the Appendix.

It follows for the Van der Waals equation of state at the critical point,

$$Z_c = \frac{P_C \upsilon_C}{\Re T_C} = \frac{\frac{a}{27b^2} \cdot 3b}{\Re \cdot \frac{8a}{27b\Re}} = \frac{3}{8} = 0.375 \qquad (2.18)$$

where Zc is the critical compressibility factor. If a substance behaved like an Ideal Gas at the critical point, then Z_c would equal 1.0. If it obeys the Van der Waals equation, then this ratio should equal 0.375, which would be a measure of the departure of the Van der Waals gas from an Ideal Gas. In Table 2.1 below, the calculated values of Zc are listed for a number of interesting gases, and in no case is this ratio equal to 0.375, or even close. Above the critical point, at higher pressure, the Van der Waals equation is fairly satisfactory and is useful in many cases. Other equations of state give better values of Zc, but no two parameters Equation of State adequately describes all properties of pure substances near the vapor dome.

2.4.3   Other Useful Two Parameter Equations of State

Many equations of state have been proposed to represent $P-V-T$ data more accurately than the Ideal Gas Law for those regions where it does not apply. Most of the equations of state that have been proposed are empirical and only a few of them are in wide use in thermodynamics and related engineering and physics fields. Two other equations of state, commonly used in engineering analysis, are presented below.

2.4.3.1   Redlich-Kwong Equation of State

The Redlich-Kwong (RK) equation of state is an empirical equation that is widely used for engineering calculations.

$$P = \frac{\Re T}{\upsilon - b} - \frac{a}{T^{0.5}\upsilon(\upsilon + b)} \qquad (2.19)$$

The constants a and b of the Redlich-Kwong equation of state can be estimated from the critical constants by the following relations. It is generally thought to provide satisfactory results above the critical temperature for any pressure.

$$a = \frac{0.42748\Re^2 T_c^{2.5}}{P_c} \qquad (2.20a)$$

$$b = \frac{0.08679\Re T_c}{P_c} \qquad (2.20b)$$

This gives $Z_C = 0.333$ which is significantly closer to the range of interest for most gases. The constants a and b are evaluated in the Appendix for a number of gases.

2.4.3.2   Peng-Robinson Equation of State

The Peng-Robinson equation of state gives a slightly better approximation below the critical temperature by adding another parameter, ω, the acentricity factor given by,

$$\omega = -1 - \log_{10}\left(\frac{p^{sat}}{p_C}\right)_{T/T_C = 0.7} \qquad \kappa = 0.37464 + 1.54226\omega - 0.26993\omega^2 \qquad (2.21a)$$

$$\alpha = \left[1 + \kappa\left(1 - \sqrt{T/T_C}\right)\right]^2 \qquad P = \frac{\Re T}{v - b} - \frac{\alpha a}{v^2 + 2bv - b^2} \qquad (2.21b)$$

The Peng-Robinson constants are determined by

$$a = 0.45723553\frac{\Re^2 T_C^2}{P_C} \qquad b = 0.07779607\frac{\Re T_C}{P_C} \qquad (2.21c)$$

It gives a $Z_C = 0.307$, closer to the range of a number of gases. The constants for the Peng-Robinson Equation of State are provided for a number of gasses in the Appendix.

2.4.4   Common Equations of State with Additional Parameters

Equations of state play an important role in chemical engineering design and they
have assumed an expanding role in the study of the phase equilibrium of fluids and
fluid mixtures. Originally, equations of state were used mainly for pure components. Many equations of state have been proposed in the literature with either an
empirical, semi-empirical or a theoretical basis. Brief reviews can be found in the
following sections.

2.4.4.1   Beattie-Bridgeman Equation of State

The Beattie-Bridgemen equation of state is given by

$$Pv^2 = \Re T\left(1 - \frac{c}{vT^3}\right)\left(v - B_0 - \frac{bB_0}{v}\right) - A_0\left(1 - \frac{a}{v}\right) \tag{2.22}$$

The constants A_0, B_0, a, b, and c are characteristic of a gas. These constants for
some substances are given in Table 2.2.

Table 2.2 Constants of Beattie-Bridgeman equation of state

Gas	$A_0\left(\dfrac{\mathrm{Pam}^3}{\mathrm{mol}^2}\right)$	$B_0 \times 10^6 \left(\dfrac{\mathrm{m}^3}{\mathrm{mol}}\right)$	$a \times 10^6 \left(\dfrac{\mathrm{m}^3}{\mathrm{mol}}\right)$	$b \times 10^6 \left(\dfrac{\mathrm{m}^3}{\mathrm{mol}}\right)$	$c \left(\dfrac{\mathrm{m}^3\mathrm{K}^3}{\mathrm{mol}}\right)$
Air	0.1318	46.11	19.31	-11.010	43.40
Ammonia	0.2425	34.15	170.31	19.112	4768.70
n-Butane	1.8030	246.20	121.61	94.230	3500.00
Carbon dioxide	0.5073	104.76	71.32	72.350	660.00
Ethane	0.5958	94.00	58.61	19.150	900.00
Ethylene	0.6234	121.56	49.64	35.970	226.80
Helium	0.00219	14.00	59.84	0.000	0.04
Hydrogen	0.0200	20.96	-5.06	-43.590	5.04
Methane	0.2307	55.87	18.55	-158.700	128.30
Neon	0.0215	20.60	21.96	0.000	1.01
Nitrogen	0.1362	50.46	26.17	-6.910	42.00
Oxygen	0.1511	46.24	25.62	4.208	48.00
n-Pentane	2.8634	394.00	150.99	139.600	4000.00
Propane	1.2078	181.00	73.21	42.930	1200.00

Table 2.3 Source Kenneth Wark, thermodynamics, 4th ed., p. 141

Gas	a	A_0	B	B_0	$c \times 10^{-4}$	$C_0 \times 10^{-5}$	$\alpha \times 10^5$	Γ
n-Butane, C_4H_{10}	190.68	1021.6	0.039998	0.12436	3205	1006	110.1	0.0340
Carbon dioxide, CO_2	13.86	277.30	0.007210	0.04991	151.1	140.4	8.470	0.0054
Carbon monoxide, CO	3.71	135.87	0.002632	0.05454	10.54	8.673	13.50	0.0060
Methane, CH_4	5.00	187.91	0.003380	0.04260	25.78	22.86	12.44	0.0060
Nitrogen, N_2	2.54	106.73	0.002328	0.04074	7.379	8.164	12.72	0.0053

2.4.4.2 Benedict-Webb-Rubin Equation of State

The Benedict-Webb-Rubin (BWR) equation of state is given by;

$$P = \frac{\Re T}{v} + \frac{1}{v^2}\left(\Re T\left(B_0 + \frac{b}{v}\right) - \left(A_0 + \frac{a}{v} + \frac{a\alpha}{v^4}\right) - \frac{1}{T^2}\left(C_0 - \frac{c}{v}\left(1 + \frac{\gamma}{v^2}\right)\exp\left(\frac{-\gamma}{v^2}\right)\right)\right)$$

(2.23)

where A_0, B_0, C_0, a, b, c, α and γ are constants for a given fluid. The BWR constants for a few selected gases can be found in Table 2.3 above or in Perry's Chemical Engineer's Handbook. This equation of state is quite complex and contains eight constants, and is able to predict the $p-v-T$ data with higher accuracy compared to many other equations of state.

The Equations of State used to calculate the steam properties in the Appendix was broken down into five regions. Each region required between 10 and 43 constants to adequately represent the data.

2.4.4.3 Virial Equation of State

The word virial comes from the Latin meaning force, thus it refers to the interaction forces between molecules. In 1901 Kamerlingh Onnes suggested the virial equation of state expressed as a power series in reciprocal volume; it is given by

$$\frac{pv}{\Re T} = 1 + \frac{B}{v} + \frac{C}{v^2} + \frac{D}{v^3}$$

(2.24a)

where B, C, D, etc. are known as second virial coefficient, third virial coefficient etc. Virial coefficients express the deviations from the Ideal Gas Law due to intermolecular forces. These virial coefficients are functions of temperature only. The advantage of the virial equation of state is that it may be made to represent the experimental $p-\upsilon-T$ data as accurately as required by increasing the number of constants. The values of the second virial coefficients have been determined experimentally for a number of gases. The third virial coefficients are not known for many substances and much less information is available beyond the third virial coefficient. Moreover, the virial equation of state with more than three terms is difficult to handle. The virial equation of state and the ideal gas law have a strong theoretical base. They have been derived through *statistical mechanical methods*. All other equations of state are empirical or semi-empirical. The virial equation of state is sometimes written as a power series in the pressure as;

$$\frac{p\upsilon}{\Re T} = 1 + B'P + C'P^2 + D'P^3 + \ldots \tag{2.24b}$$

where the coefficients B', C', D', etc. are functions of temperature only. The coefficients B', C', D', etc. are related to the virial coefficients B, C, D, etc. by the following relations:

$$B' = \frac{B}{\Re T} \tag{2.25a}$$

$$C' = \frac{C - B^2}{(\Re T)^2} \tag{2.25b}$$

$$D' = \frac{D - 3BC + 3B^3}{(\Re T)^3} \tag{2.25c}$$

It has been found that the virial Eq. 2.24a adequately represents the experimental data over a wide range of pressure, compared to the virial Eq. 2.24b when both these equations are truncated after the third term [1]. The general form of Eq. 2.24a can be written as;

$$\frac{p\upsilon}{\Re T} = 1 + \frac{B}{\upsilon} + \frac{C}{\upsilon^2} + \frac{D}{\upsilon^3} + \ldots = \sum_{i=0}^{R} \frac{c_i}{\upsilon^i} \tag{2.26}$$

The parameters in the equation $(B, C, D = c_i)$ are again called "virial coefficients". If $c_i = 0$ for $i > 0$, the virial equation reduces to the ideal gas equation. The accuracy required determines the number of terms that are kept—more terms makes the equation more accurate, but also more complicated to work with. Virial coefficients are different for each gas, but other than that are functions of temperature only.

Coefficients are normally obtained by making measurements of p, v, and T, and fitting the equation. These values are then published so that others may use them.

Many forms of the virial equation exist. Truncating this equation after one coefficient gives a quadratic equation in v. Thus, it retains some of the simplicity of the Ideal Gas law allowing quick analytic solutions for v given p and T.

$$\frac{P\upsilon}{\Re T} = 1 + \frac{B}{\upsilon} \tag{2.27a}$$

A number of methods (correlations, etc.) are available to determine B. In order to improve accuracy and capture more behaviors, additional parameters are sometimes added. One example is the Benedict-Webb-Rubin (BWR) Equation of State Eq. 2.23.

This equation provides a first order correction to the Ideal Gas Law for non-polar species. It should not be attempted for polar compounds such as water that have a non-zero dipole moment [1]. The following procedure may be used to estimate υ or P for a given T for a non-polar species, one with a dipole moment close to zero, such as hydrogen or oxygen and all other symmetrical molecules.

To use the truncated virial Equation of State proceeds in the following manner.

• Look up the critical temperature and pressure (T_c and P_c) for the species of interest in Appendix. Also, look up the acentric factor, ω, a parameter that reflects the geometry and polarity of a molecule, in the constants table for the Peng-Robinson Equation of State in the Appendix. (A more complete list can be found in Reid et al. [2].)
• Calculate the reduced temperature T_r using the relationship $T_r = T / T_c$.
• Calculate the following coefficients:

$$B_0 = 0.083 - \frac{0.422}{T_r^{1.6}} \tag{2.27b}$$

$$B_1 = 0.139 - \frac{0.172}{T_r^{4.2}} \tag{2.27c}$$

$$B = \frac{\Re T_c}{P_c}(B_0 + \omega B_1) \tag{2.27d}$$

• Substitute into Eq. 2.27 the value of B and whichever of the variables p and υ is known and solve for the other variable. Solution for p is straightforward. If υ is to be determined, the equation can be rearranged into a quadratic and solved using the quadratic formula.

$$v^2 - \frac{\Re T}{p}v - \frac{\Re T}{p}B = 0$$

• Normally one of the two solutions is reasonable and the other is not and should be discarded; if there is any doubt, estimate υ from the ideal gas equation of state and accept the virial equation solution that comes closest to υ_{ideal}.

2.4.4.4   Equation of State Comparison

Virial equations with one coefficient cannot represent thermodynamic systems where both liquid and vapor are present. A *"cubic"* equation of state is needed to do this. We have identified three two-parameter equations of state above for which data is presented in the Appendix. The most sophisticated of these is the Peng-Robinson equation because it corrects the "a" coefficient for the acentric factor.

$$P = \frac{\Re T}{(v-b)} - \frac{\alpha a}{v^2 + 2bv - b^2} \tag{2.28a}$$

where the constants are given by;

$$a = 0.45723553 \frac{\Re^2 T_c^2}{P_c} \tag{2.28b}$$

$$b = 0.07779607 \frac{\Re T_c}{P_c} \tag{2.28c}$$

$$\kappa = 0.37464 + 1.54226\omega - 0.26993\omega^2 \tag{2.28d}$$

$$\alpha = \left(1 + \kappa\left(1 - \sqrt{\frac{T}{T_c}}\right)\right)^2 \tag{2.28e}$$

In this equation, the b term is a volume correction, while a is a molecular interaction parameter. The constants all depend on the critical temperature and pressure of the gas. These can be looked up easily in a data table.

The "*acentric factor*", omega ω, is also easily looked up. It is related to the geometry of the gas molecule.

To use the Peng-Robinson equation:

1. Look up T_c, P_c, and the *acentric factor* for the species of interest in the Appendix.
2. Plug in and find a, b, and alpha α.
3. Plug these into the Peng-Robinson equation; the result will be a cubic equation in v depending on p and T.
4. Solve for the unknown you seek.

Solving the cubic equation can be accomplished with a binary search using the computer or by analytically solving the cubic equation. The equation can be transformed to

$$v^3 + \left(b - \frac{\Re T}{p}\right)v^2 + \left(\frac{\alpha a}{p} - 3b^2 - 2\frac{\Re T}{p}b\right)v + \left(b^3 + \frac{\Re T}{p}b^2 - \frac{\alpha a}{p}b\right) = 0$$

$$v^3 + a_1 v^2 + a_2 v + a_3 = 0$$

The analytic solution is given by,

$$v^3 + a_1 v^2 + a_2 v + a_3 = 0$$

Transform to

$$x^3 + b_1 x + b_2 = 0 \quad v = x - \frac{a_1}{3}$$

$$b_1 = \frac{3a_2 - a_1^2}{3} \quad b_2 = \frac{2a_1^3 - 9a_1 a_2 + 27a_3}{27}$$

$$\frac{b_2^2}{4} + \frac{b_1^3}{27} > 0, \ 1 \ \text{real, 2 imaginary,} \quad \frac{b_2^2}{4} + \frac{b_1^3}{27} = 0, \ 3 \ \text{real, 2 equal,}$$

$$\frac{b_2^2}{4} + \frac{b_1^3}{27} < 0, \ 3 \ \text{real \& distinct}$$

For the first case

$$C = \sqrt[3]{-\frac{b_2}{2} + \sqrt{\frac{b_2^2}{4} + \frac{b_1^3}{27}}} \quad D = \sqrt[3]{-\frac{b_2}{2} - \sqrt{\frac{b_2^2}{4} + \frac{b_1^3}{27}}}$$

$$x = C + D \quad x = -\frac{C+D}{2} + \frac{C-D}{2}\sqrt{-3} \quad x = -\frac{C+D}{2} - \frac{C-D}{2}\sqrt{-3}$$

$$v = x - \frac{a_1}{3}$$

For the third case of three real unequal roots let,

$$\cos\phi = \frac{-a_2/2}{\sqrt{-a_1^3/27}}$$

$$x_1 = 2\sqrt{-b_1/3}\cos\left(\phi/3\right) \quad x_2 = 2\sqrt{-b_1/3}\cos\left(\phi/3 + 2\pi/3\right) \quad x_3 = 2\sqrt{-b_1/3}\cos\left(\phi/3 + 4\pi/3\right)$$

Example 2.1 Carbon dioxide at 500 K and 6.5 MPa flows at 100 kg/h. Use the one parameter Viral Equation of State and the Peng-Robinson Equation of State to determine the volumetric flow.

Solution The pressure and temperature are known, so look up the critical properties, the acentric factor, and the Peng-Robinson constants in the Appendix.

The critical properties are $T_c = 304.2\,°K$, $p_c = 7.39$ MPa, and the acentric factor is 0.225.

Evaluating the B coefficients,

$$T_r = 500.0/304.2 = 1.64365 \quad B_0 = 0.083 - 0.422/T_r^{1.6} = -0.1076$$
$$B_1 = 0.139 - 0.422/T_r^{4.2} = 0.0867$$

$$B = 8314.47 * 304.2/7.39E+6 * (-0.1076 + 0.225 * 0.0867) = -0.03014$$

$$v = \frac{\Re T}{2p} \pm \sqrt{\left(\frac{\Re T}{2p}\right)^2 + \frac{\Re TB}{p}} = (0.31978 + 0.081651)/44 = 0.01382\, \text{m}^3/\text{kg}$$

The Peng-Robinson coefficients are

$$a = 0.39576 \text{ MPa-m}^3/\text{kgmol}^2 b = 0.02662 \text{ m}^3/\text{kgmol}$$

$$\kappa = 0.37464 + 1.5422 * 0.225 - 0.26993 * 0.225^2 = 0.70797$$

$$\alpha = \left(1.0 + 0.70797\left\{1 - \sqrt{T/T_c}\right\}\right)^2 = 0.6357$$

$$p = \frac{\Re T}{v-b} - \frac{\alpha a}{v^2 + 2bv - b^2}$$

$$v^3 + \left(b - \frac{\Re T}{p}\right)v^2 + \left(\frac{\alpha a}{p} - 3b^2 - 2\frac{\Re T}{p}b\right)v + b^3 + \frac{\Re T}{p}b^2 - \frac{\alpha a}{p}b = 0$$

Applying the cubic formula gives

$$a_1 = -0.61295, \ a_2 = 0.002526, \ a_3 = -0.00055844, \ b_1 = -0.12271, \ b_2 = -0.017101$$

$$\frac{b_2^2}{4} + \frac{b_1^3}{27} > 0, \ 1 \text{ real, 2 imaginary}$$

$$C = 0.22044 \quad D = 0.18555$$

$$v = 0.5 * (0.22044 + 0.18555)/44.0 = 0.01387 \text{ m}^3/\text{kg}$$

It is worth noting that the Ideal Gas solution is

$$v = \frac{\Re T}{pAM} = \frac{8314.47 * 500}{6500000.0 * 44} = 0.01454 \text{ m}^3/\text{kg and the tables give } 0.01389 \text{ m}^3/\text{kg}$$

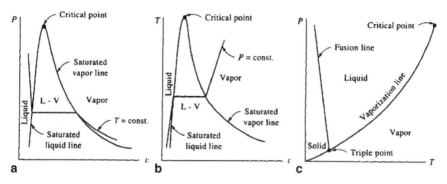

Fig. 2.6 The $P-V$, $T-V$, and $P-T$ diagrams

So the Ideal Gas solution is high by 4.65%. The virial solution is low by 0.54%, and the Peng-Robinson solution is low by 0.14%.

The volumetric flow rate with the viral solution will be $100*0.01382 = 1.382$ m³/h

The volumetric flow rate with the Peng-Robinson solution would be $100*0.01387 = 1.387$ m³/h

2.4.5   The Liquid-Vapor Region

Appling Fig. 2.6 shows that at any given (T, υ) between saturated points **1** and **2**, liquid and vapor exist as a mixture in equilibrium. Let υ_f and υ_g to represent the specific volumes of the *saturated liquid* and the *saturated vapor*, respectively, while m is the total mass of the system that is shown in Fig. 2.6 and m_f the mass amount of mass in the liquid phase, and m_g the amount of mass in the vapor phase, then for a state of the system represented by (T, υ) the total volume of the mixture is the same of the volume occupied by the liquid and the occupied by the vapor as [3]:

$$m\upsilon = m_f \upsilon_f + m_g \upsilon_g \tag{2.29}$$

$$m = m_f + m_g \tag{2.30}$$

or dividing both side of Eq. 2.29 by m, then utilizing Eq. 2.30, we have;

Fig. 2.7 The $T-V$ diagram showing the saturated liquid and saturated vapor points

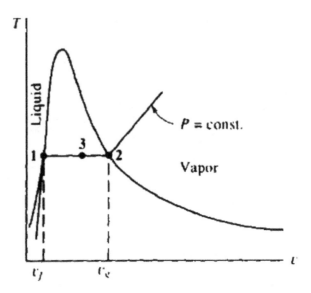

(2.31)

$$v = \left(\frac{m_f}{m}\right)v_f + \left(\frac{m_g}{m}\right)v_g$$

$$= \left(\frac{m - m_g}{m}\right)v_f + \left(\frac{m_g}{m}\right)v_g$$

$$= v_f + \left(\frac{m_g}{m}\right)v_g - \left(\frac{m_g}{m}\right)v_f$$

$$= v_f + \left(\frac{m_g}{m}\right)(v_g - v_f)$$

$$= v_f + x(v_g - v_f)$$

$$= v_f + xv_{fg}$$

The ratio $x = \dfrac{m_g}{m}$ is called quality because steam that has a larger proportion of vapor is considered "higher quality" than steam with a lesser mass of vapor. $v_{fg} = v_g - v_f$ is the heat of vaporization. If we take a slice through the 3-D plot to form the P-T plane and include the **Critical Point** we will obtain the plot that is shown in Fig. 2.4.

Note that the percentage liquid by mass in a mixture is $1000(1 - x)$ and the percentage vapor is $100x$. See Fig. 2.7.

For most substances, the relationships among thermodynamic properties are too complex to be expressed by simple equations. Therefore, properties are frequently presented in the form of tables. Some thermodynamic properties can be measured easily and those that can't are calculated by using the thermodynamic relations that they must satisfy and the measurable properties.

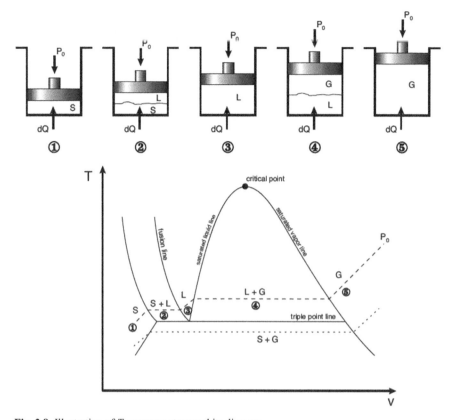

Fig. 2.8 Illustration of *T-v* process steps and its diagram

The working fluid of most interest to engineer's and by far the fluid most studied is water. Its properties have been tabulated for years in what are called Steam Tables. A set of *Steam Tables* are provided in the back of the book, within the Appendix. The Tables are Appendix 14.1–14.7.

2.5   *T – V* Diagram for a Simple Compressible Substance

Consider an experiment in which a substance starts as a solid and is heated up at constant pressure until it becomes a gas. The process is depicted in Fig. 2.8.

As heat is applied to the solid, the temperature increases and the volume increases slightly. When the melt temperature is reached, the temperature remains constant but the volume continues to increase as the solid is converted to a liquid. Once all of the material has been converted to a liquid, the temperature begins to increase again as more heat is added. When the vaporization temperature is reached, the liquid begins to be converted to a vapor and the temperature remains constant as more heat is added. Once all of the liquid has been converted to vapor, adding more heat will once again cause the temperature to rise.[3–7].

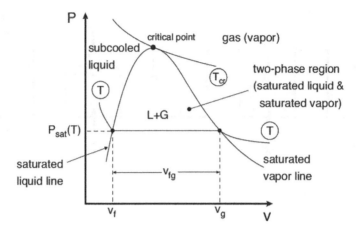

Fig. 2.9 Illustration of $P-V$ diagram

2.6 $P-V$ Diagram for a Simple Compressible Substance

The general shape of a $p-V$ diagram for a pure substance is very similar to that of a $T-V$ diagram and its representation by the **vapor dome** as discussed before. Figure 2.9 is a presentation of a $p-V$ diagram.

On this diagram the subscript f denotes a saturated liquid (fluid) and g denotes a saturated vapor (gas).

2.7 $P-V-T$ Diagram for a Simple Compressible Substance

All the data that are represented on both the p-v and p-T diagrams can be shown one diagram if the three coordinates p, v, and T are plotted along orthogonal axes. The result is called the p-v-T surface and two such surfaces are shown in Figs. 2.10 and 2.11, the first for a kilogram of an unusual substance like water that contracts upon melting, and the second for a kilogram of a typical substance like carbon dioxide that expands upon melting.

Where the critical point is denoted by the letter C and the triple point by TP. The critical isotherm is marked T_C. Every point on the $P-V-T$ surface represents a state of equilibrium for the substance. If the p-v-T surface is projected on the p-v plane, then the usual p-v diagram is seen and upon projecting the p-v-T surface onto the p-T plane, the entire solid-vapor region projects into the sublimation curve, the entire liquid-vapor region projects into the vaporization curve, the entire solid-liquid region projects into the fusion curve, and, finally, the *Triple-Point Line* projects into the triple point on the phase diagram [4].

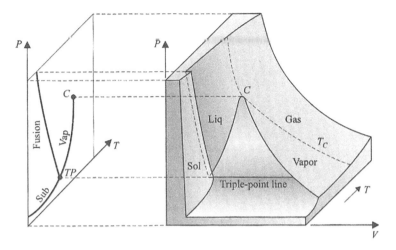

Fig. 2.10 $P-V-T$ surface for H_2O, which contracts while melting

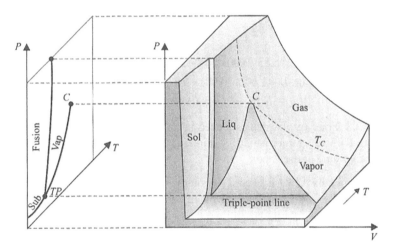

Fig. 2.11 $P-V-T$ surface for CO_2, which expands while Melting

The $P-V-T$ surfaces present a great deal of information at once, but in typical thermodynamic analysis it is more convenient to work with two-dimensional diagrams, such as the p-v and T-v diagrams (Fig. 2.12).

Example 2.2 Determine the volume change when 1 kg of saturated water is completely vaporized at a pressure of (a) 1 kPa, (b) 100 kPa, and (c) 10000 kPa.

Solution Appendix A14.2 provides the necessary values. The quantity being sought is $v_{fg} = v_g - v_f$. Note that p is given in MPa

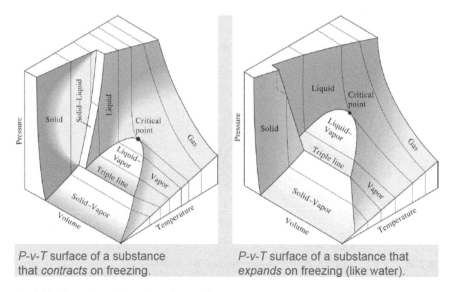

P-v-T surface of a substance   P-v-T surface of a substance that
that *contracts* on freezing.   *expands* on freezing (like water).

Fig. 2.12 Illustration of $P-V-T$ diagram for two cases

a. 1 kPa. Thus, $\upsilon_{fg} = 129.183 - 0.001 = 129.182$ m³/kg.
b. 100 kPa MPa. Again $\upsilon = 1.673 - 0.001 = 1.672$ m³ / kg.
c. 10,000 kPa = 10 MPa. Finally, $\upsilon_{fg} = 0.018034 - 0.001453 = 0.016581$ m³ / kg.

Example 2.3 Four kg of water is placed in an enclosed volume of 1 m³. Heat is added until the temperature is 420 °K. Find (a) the pressure, (b) the mass of vapor, and (c) the volume of the vapor.

Solution Appendix A14.1 is used. The volume of 4 kg of saturated vapor at 420 °K is (0.425255) (4) = 1.701 m³. Since the given volume is less than this, we assume the state to be in the quality region.

a. In the quality region the pressure is given as $p = 437.24$ kPa.
b. To find the mass of the vapor we must determine the quality. It is found from Eq. 2.3, using the actual $v = 1/4$ m³/kg, as;

$$0.25 = 0.001087 + (0.425255 - 0.001087)$$

thus $x = 0.2489/0.425255 = 0.5853$. Using the relationship of $x = \dfrac{m_g}{m}$, the vapor mass is

$$m_g = mx = (4)(0.5853) = 2.341 \text{ kg}$$

c. Finally, the volume of the vapor is found from

$$V_g - \upsilon_g m_g - (0.4253)(2.341) = 0.9956 \text{ m}^3$$

Note that in a mixture where the quality is not very close to zero, the vapor phase occupies most of the volume. In this example, with a quality of 58.53% it occupies 99.56% of the volume.

Example 2.4 Four kg of water is heated at a pressure of 220 kPa to produce a mixture with quality $x = 0.8$. Determine the final volume occupied by the mixture.

Solution Using Appendix A14.2 to determine the appropriate number at 220 kPa we linearly interpolate between 0.2 and 0.3 MPa. This provides, at 220 kPa.

$$\upsilon_g = \left(\frac{220-200}{250-200}\right)(0.718697 - 0.885735) + 0.885735 = 0.8189 \text{ m}^3/\text{kg}$$

Note that no interpolation is necessary for υ_f, since for both pressures υ_f is the same to four decimal places. Using Eq. 2.6, we now find

$$\upsilon = \upsilon_f + x(\upsilon_g - \upsilon_f) = 0.0011 + (0.8)(0.8189 - 0.001) = 0.6554 \text{ m}^3/\text{kg}$$

The total volume occupied by 4 kg is $V = m\upsilon = (4 \text{ kg})(0.6640 \text{ m}^3/\text{kg}) = 2.621 \text{ m}^3$.

Example 2.5 Two lb of water is contained in a constant-pressure container held at 540 psia. Heat is added until the temperature reaches 1100 °R. Determine the final volume of the container.

Solution: Use Appendix 14.7. Since 540 psia lies between the table entry values, the specific volume is simply

$$\upsilon = 1.2223 + (0.4)(1.0017 - 1.2223) = 1.1341 \text{ ft}^3 / \text{lbm}$$

The final volume is then $V = m\upsilon = (2)(1.2115) = 2.2681 \text{ ft}^3$

Example 2–6 Calculate the pressure of steam at a temperature of 500 °C and a density of 24 kg/m³ using (a) the ideal-gas equation, (b) the van der Waals equation, (c) the Redlich-Kwong equation, (d) the Peng-Robinson equation, and (e) the steam table.

Solution
a. Using the ideal-gas equation, $P = \rho RT = (24/18)(8.31447)(773) = 8569.4$ kpa.
b. Using values for a and b from the Appendix for the Van der Waals equation provides;

$$P = \frac{RT}{\upsilon - b} - \frac{a}{\upsilon^2} = \frac{8.31447(773)}{\dfrac{18}{24} - 0.03084} - \frac{553.04}{\left(\dfrac{18}{24}\right)^2} = 7954 \text{ kpa}$$

c. Using values for a and b from the Appendix for the Redlich-Kwong equation gives;

$$P = \frac{RT}{\upsilon - b} - \frac{a}{\upsilon(\upsilon + b)\sqrt{T}} = \frac{(8.31447)(773)}{\frac{18}{24} - 0.02110} - \frac{14258.5}{\left(\frac{18}{24}\right)\left(\frac{18}{24} + 0.02110\right)\sqrt{773}} = 7931 \text{ kpa}$$

d. For the Peng-Robinson equation the acentric factor for water is 0.3437.

$$\kappa = 0.37464 + 1.54226\omega - 0.26993\omega^2$$

$$\alpha = \left(1 + \kappa\left(1 - \sqrt{\frac{T}{T_c}}\right)\right)^2$$

$$\alpha = 0.8447 :$$

$$p = \frac{8.31447 * 773}{0.75 - 0.01895} - \frac{0.8447 * 599.4}{(0.75^2 + 2 * 0.01895 * 0.75 - 0.01895^2)} = 7934.24 \text{ kPa}$$

e. The steam table provides the most precise value for the pressure. Using $T = 500\,^{\circ}\text{C}$ and $\upsilon = 1/24 = 0.04166$ m³/kg, we find $P = 8141$ kPa. Note that the ideal-gas law has an error of 5.3 %, and the errors of each of the other three equations are Van der Waals $= -2.29\%$, Redlich Kwong $= -2.58\%$, Peng-Robinson $= -2.54\%$

Problems

Problem 2.1: It is necessary to store one kmol of Methane at $300\,^{\circ}\text{K}$ and 60 MPa. Determine the volume of the cylinder that is required for storage by each of the following:

a. Ideal gas law.
b. Van der Waals equation and
c. Redlich-Kwong equation

Problem 2.2: One kmol of Ethylene is contained in a 0.6 m³ steel vessel immersed in a constant temperature bath at $200\,^{\circ}\text{C}$. Calculate the pressure developed by the gas by each of the following:

a. Ideal gas law
b. Van der Waals equations, and
c. Redlich-Kwong

Problem 2.3: Expand the following equations in the form

$$P\upsilon = RT(1 + BP + CP^2 + \cdots)$$

and determine the second virial coefficient B in each case:

$$\left(P + \frac{a}{\upsilon^2}\right)(\upsilon - b) = RT \qquad\qquad \text{(van der Waals equation of state).}$$

$$\left(Pe^{a/RT\upsilon}\right)(\upsilon - b) = RT \qquad\qquad \text{(Dieterici equation of state).}$$

$$\left(P + \frac{a}{\upsilon^2 T}\right)(\upsilon - b) = RT \qquad\qquad \text{(Berthelot equation of stat).}$$

$$\left(P + \frac{a}{(\upsilon + c)^2 T}\right)(\upsilon - b) = RT \quad \text{(Clausius equation of state).}$$

$$P\upsilon = RT\left(1 + \frac{B'}{\upsilon} + \frac{C'}{\upsilon^2} + \cdots\right) \qquad \text{(another type of virial expansion)}$$

Problem 2.4: Two gram-moles of nitrogen is placed in a three-liter tank at $-150.8\,^\circ$C. Estimate the tank pressure using the ideal gas equation of state and then using the virial equation of state truncated (Eq. 2.27) after the second term. Taking the second estimate to be correct, calculate the percentage error that results from the use of the ideal gas equation at the system conditions.

Problem 2.5: A gas cylinder with a volume of $2.50\,\text{m}^3$ contains $1.00\,\text{kmol}$ of carbon dioxide at $T = 300^\circ$K. Use the Peng-Robinson equation of state to estimate the gas pressure in atm.

Problem 2.6: A rigid tank contains 20 lbm of air at 20 psia and 70°F. More air is added to the tank until the pressure and temperature rise to 35 psia and 90°F, respectively. Determine the mass of air added to the tank.

Problem 2.7: A perfectly fitting pot and its lid often stick after cooking, and it becomes very difficult to open the lid when the pot cools down. Explain why this happens and what you would do to open the lid.

Problem 2.8: Estimate the error in using the Ideal Gas law to calculate the specific volume of nitrogen at 20 Mpa and 300 K. Use either the Van der Waals or the Redlich-Kwong real gas models for the more correct answer.

References

1. Rao YVC (2004) An introduction to thermodynamics, Rev edn. Sangam Books Ltd, Hyderabad
2. Reid RC, Prausnitz JM, Poling BE (1986) The properties of gases and liquids, 4th edn. McGraw-Hill, New York
3. Potter MC, Somerton CW (2006) Thermodynamics for engineers, 2nd edn. McGraw-Hill, New York
4. Zemansky MW, Dittman RH (1997) Heat and thermodynamic, 7th edn. McGraw-Hill, New York
5. Nag PK (2002) Basic and applied thermodynamics, 2nd edn. McGraw-Hill, New York
6. Mooney DA (1965) Thermodynamics and heat transfer, 7th edn. Prentice-Hall, New York
7. Cengel YA, Boles MA (2005) Thermodynamics, an engineering approach, 5th edn. McGraw-Hill, New York

Chapter 3
Mixture

Not all thermodynamic systems contain only pure substances. Many systems of interest are composed of mixtures of pure substances. It is important to be able to analyze these systems as well as those containing only pure substances. Therefore, an understanding of mixtures is essential to the study of thermodynamics.

3.1 Ideal Gas Mixtures

In the following section we will be studying gas mixtures and introduce certain conceptual framework and properties of substances that are mixed. Rederas can also refer to references at the end of this chapter for further information on the subject of mixture.[1–3].

3.1.1 Avogadro's Number

In order to provide a conceptual framework for understanding mixtures it is easiest to start with Ideal Gas Mixtures and address the fundamentals. Then real gas effects and liquid effects can be added in. The Ideal Gas is composed of point molecules that have no volume and only interact with billiard ball-like collisions. This description does not distinguish between one type of molecule or another. So the only thing that matters when combining two or more ideal gases is how many molecules of each gas are present. To get the number of molecules of a gas present it is a simple matter to divide the mass of the gas by the molecular weight of the gas and multiply by a constant known as *Avogadro's Number*. A kilogram-mole of any pure substance contains 6.022×10^{26} molecules or atoms. Avogadro's number gives the number of molecules in a mole of a pure substance. To get the number of

© Springer International Publishing Switzerland 2015
B. Zohuri, P. McDaniel, *Thermodynamics In Nuclear Power Plant Systems,*
DOI 10.1007/978-3-319-13419-2_3

kilogram-moles in a given mass, the mass of the substance is divided by the molecular mass in kilograms. For instance,

$$5 \text{ kilograms of He} = 5 \text{ kg}/4 \text{ kg per kg-mole}$$
$$= 1.25 \text{ kg-moles He} = 7.528 \text{ x } 10^{26} \text{ atoms He}$$
$$5 \text{ kilograms of N}_2 = 5 \text{ kg}/28 \text{ kg per kg-mole}$$
$$= 0.1786 \text{ kg-moles N}_2 = 1.075 \text{ x } 10^{26} \text{ molecules N}_2$$

It is generally not important to know the number of atoms or molecules present, but
in most cases, it is important to know the number of moles present. So it is useful
to remember that when quantities are measured in moles, it is the same as if they
were measured in atoms or molecules. Classical thermodynamics was developed
long before the atomic hypothesis was demonstrated but the concept of a mole of
material superseded classical thermodynamics.

3.1.2   Mass Fractions

When pure substances are mixed, they are typically quantified by the amount of
mass of each substance present. The mass fraction for a component of a mixture is
the mass of that component divided the total mass of the mixture. Knowing the mass
fractions for a mixture is useful if one wants to know the recipe for putting the mixture together, but they are generally not usually useful for predicting the thermodynamic characteristics of the mixture. Consider a mixture of 5 kg N_2 an 15 kg of CO_2.

$$\text{Total Mass} = 5 \text{ kg} + 15 \text{ kg} = 20 \text{ kg}$$
$$\text{Mass fraction N2} = 5/20 = 0.25$$
$$\text{Mass Fraction CO2} = 15/20 = 0.75$$

Unfortunately, the mass fractions tell very little about the thermodynamic characteristics of the mixture other than the recipe for putting it together.

3.1.3   Mole Fractions

A more useful characterization of the mixture is the mole fraction of the components. In order to calculate mole fractions, the moles of each component present
must be calculated first. To get the moles of a component present the components
mass must be divided by its molecular weight. For instance

$$5 \text{ kg N}_2 = 5 \text{ kg}/28 \text{ kg per kg-mole} = 0.1786 \text{ kg-moles of N}_2$$
$$15 \text{ kg CO}_2 = 15 \text{ kg}/44 \text{ kg per kg-mole} = 0.3409 \text{ kg-moles of CO}_2$$

$$\text{Total kg-moles} = 0.1786 + 0.3409 = 0.5195 \text{ kg-moles}$$
$$\text{Mole Fraction N}_2 - 0.1786/0.5195 = 0.3438$$
$$\text{Mole Fraction CO}_2 = 0.3409/0.5195 = 0.6562$$

Now that we have the mole fractions present, we have the relative numbers of molecules of each gas present. These fractions will be far more useful for determining the properties of the gas mixture than the relative masses.

Example 3.1 Consider a mixture of O_2 and H_2 that is 5 wt % H_2. If it is burned, which component will be consumed entirely, and which component will be left over.

Solution To solve this problem we must find the mole fractions of the two gases and consider that 1 mol of O_2 combines with 2 moles of H_2 to form water.

$$n_{H_2} = 0.1/2 = 0.05$$
$$n_{O_2} = 0.9/32 = 0.0281$$
$$n_{total} = 0.0781$$
$$n_{f,H_2} = 0.05/0.0781 = 0.64$$
$$n_{f,O_2} = 0.0281/0.0781 = 0.36$$

0.32 *moles of* O_2 *will combine with* 0.64 *moles of* H_2 *leaving* 0.04 *moles of excess* O_2
The final mixture will contain 0.32 *moles of* H_2O *and* 0.04 *moles of* O_2

$$n_{f,H_2O} = \frac{0.32}{0.32 + 0.04} = 0.889 \quad n_{f,O_2} = \frac{0.04}{0.32 + 0.04} = 0.111$$

3.1.4   Dalton's Law and Partial Pressures

Ideal gases when combined to form a mixture can be modeled fairly easily. Basically each if the gases expand to fill the volume of the mixture. Each gas exerts a pressure proportional to the number of atoms or molecules present. So each gas behaves as it is an Ideal Gas ignoring the other gases present. Dalton' s Law states that *the total pressure exerted by a mixture is simply the sum of the partial pressures of the gases in the mixture.*

$$p_{Total} = \sum_i p_i = \sum_i \frac{n_i \Re T}{V}$$

$$\frac{p_i}{p_{Total}} = \frac{n_i}{\sum_i n_i} = n_{f,i}$$

Thus, the partial pressure of an Ideal Gas is equal to the mole fraction of the gas in the mixture.

3.1.5   Amagat's Law and Partial Volumes

Amagat's law is similar in that it states that *the partial volumes occupied by a mixture add up to the total volume of the mixture at the system pressure.*

$$V_i = \frac{n_i \Re T}{p}$$

$$\frac{V_i}{V} = \frac{n_i}{\sum_i n_i} = n_{f,i}$$

It is a little harder to measure partial volumes but the concept has utility in some cases.

Example 3.2 A rigid tank contains 2 kg of N^2 and 4 kg of CO^2 at a temperature of 25 °C and 2 MPa. Find the partial pressures of the two gases and the gas constant of the mixture.

Solution To find the partial pressures we need the mole fractions. The moles of N_2 and CO_2 are, respectively as follows;

$$\left. \begin{array}{l} N_1 = \dfrac{m_1}{M_1} = \dfrac{2}{28} = 0.0714 \text{ mol} \\[3mm] N_2 = \dfrac{m_2}{M_2} = \dfrac{4}{44} = 0.0909 \text{ mol} \end{array} \right\} \quad \text{Therefore} \quad N_m = N_1 + N_2 = 0.1623 \text{ mol}$$

The mole fractions are

$$\left. \begin{array}{l} n_{f,1} = \dfrac{N_1}{N_m} = \dfrac{0.0714 \text{ mol}}{0.1623 \text{ mol}} = 0.440 \\[3mm] n_{f,2} = \dfrac{N_2}{N_m} = \dfrac{0.0909 \text{ mol}}{0.1623 \text{ mol}} = 0.560 \end{array} \right\}$$

The partial pressures are then;

$$P_1 = n_{f,1}P = (0.44)(2) = 0.88 \text{ MPa} \quad \text{and} \quad P_2 = n_{f,2}P = (0.56)(2) = 0.1.12 \text{ MPa}$$

The molecular weight is $M_m = M_1 n_{f,1} + M_2 n_{f,2} = (28)(0.44) + (44)(0.56) = 36.96$ kg/kmol. The gas constant of the mixture is then given by;

$$R_m = \frac{R_u}{M_m} = \frac{8.314}{36.96} = 0.225 \text{ kJ/kg.K}$$

3.2 Real Gas Mixtures

Real gas mixtures are more complicated than Ideal Gas Mixtures because the molecules have volume and they are attracted to each other in different ways. The best that can be done is to develop a real gas model based on equivalent critical temperatures and pressures.

3.2.1 Pseudo Critical States for Mixtures—Kay's Rule

Kay's rule states that mixtures of real gases can be approximately modeled by calculating a psuedo-critical state for the mixture based on a mole fraction weighted critical temperature and mole fraction weighted critical pressure. We have,

$$T_{c,equiv} = \sum_i n_{f,i} T_{c,i}$$

$$P_{c,equiv} = \sum_i n_{f,i} P_{c,i}$$

This will not work very well if the conditions of interest are too close to the highest critical temperature of one of the components.

3.2.2 Real Gas Equations of State

Once the pseudo-critical temperature and pressure have been determined, either the Van der Waals or Redlich-Kwong equation of state can be used to estimate properties. The virial and Peng-Robinson equations present the added difficulty of trying to estimate the acentric factor, which adds an additional source of uncertainty.

Example 3.3 Estimate the pressure exerted on a 3 m³ tank used to store 100 kg of air at 200 K. The mole fractions for air are 0.78 N_2, 0.21 O_2, and 0.01 Ar.

Solution First calculate the pseudo-critical state for the mixture.

$$T_{c,equiv} = 0.78*126.2 + 0.21*154.8 + 0.01*151.0 = 132.45\,K$$

$$P_{c,equiv} = 0.78*3.39 + 0.21*5.08 + 0.01*4.86 = 3.76\,MPa$$

$$MW = 0.78*28 + 0.21*32 + 0.01*39.95 = 28.96$$

Choose to model the air with the Redlich-Kwong model. Note,

$$v = 0.03\,m^3/kg = 0.03*28.96 = 0.8687\,m^3/kgmol$$

The Redlich-Kwong constants become

$$a = \frac{0.42748 * \Re^2 T_c^{2.5}}{p_c} = \frac{0.42748 * 8314.47^2 132.45^{2.5}}{3760000} = 1.5868x10^6$$

$$b = \frac{0.0867 * \Re * T_c}{p_c} = \frac{0.0867 * 8314.47 * 132.45}{3760000} = 0.02539$$

Then calculating the required pressure,

$$p = \frac{\Re T}{v - b} - \frac{a}{T^{0.5} v(v - b)}$$

$$= \frac{8314.47 * 200}{0.8687 - 0.02539} - \frac{1.5868x10^6}{200^{0.5}(0.8687)(0.8687 - 0.02539)}$$

$$= 1.819 \, MPa$$

For the Ideal Gas model the pressure would have been 1.914 MPa or about 5.2% more.

3.3   Liquid Mixtures

Liquid mixtures can be very simple or quite complicated.

3.3.1   Conservation of Volumes

Typically gases are assumed to have an indefinite shape and an indefinite volume, whereas liquids are assumed to have an indefinite shape but a definite volume, and solids have a definite shape and definite volume. When this simple model works, combining two volumes of different liquids will produce a volume that is simply the sum of the volumes of the components. If the volumes of the molecules of the two liquids are similar, this is a good approximation.

3.3.2   Non-Conservation of Volumes and Molecular Packing

However, if one of the components of the mixture has a large molecular structure and the other component is a fairly small molecule like H_2O, it is possible for the smaller molecules to take up space between the large molecules and the net volume of the mixture to be significantly smaller than the simple sum of the volumes of the two components. Quantifying this effect is beyond the level of this text.

Problems

Problem 3.1: The effect of high pressure on organisms, including humans, is studied to gain information about deep-sea diving and anaesthesia. A sample of air occupies 1.00 L at 25 °C and 1.00 atm. What pressure is needed to compress it to 100 cm³ at this temperature?

Problem 3.2: You are warned not to dispose of pressurized cans by throwing them on to a fire. The gas in an aerosol container exerts a pressure of 125 kPa at 18 °C. The container is thrown on a fire, and its temperature rises to 700 °C. What is the pressure at this temperature?

Problem 3.3: At sea level, where the pressure was 104 kPa and the temperature 21.1 °C, a certain mass of air occupied 2.0 m³. To what volume will the region expand when it has risen to an altitude where the pressure and temperature are

a. ·52 kPa, −5.0 °C,
b. 880 Pa, −52.0 °C?

Problem 3.4: Consider a gas mixture that consists of 3 kg of O^2, 5 kg of N^2, and 12 kg of CH^4, as shown in Figure below. Determine (a) the mass fraction of each component, (b) the mole fraction of each component, and (c) the average molar mass and gas constant of the mixture (Fig. 3.1).

Problem 3.5: The molar analysis of a gaseous fuel indicates that it contains 40% CH_4; 20% C_2H_6; 25% H_2 and 15% N_2. Determine;

c. The molar mass of the fuel and,
d. The gravimetric analysis.

Problem 3.6: A vessel of volume 0.4 m³ contains 0.45 kg of carbon monoxide and 1 kg of air, at 15 °C. Calculate the partial pressure of each constituent and the total pressure in the vessel. The gravimetric analysis of air is to be taken as 23.3% oxygen (O_2) and 76.7% nitrogen (N_2). Take the molar masses of carbon monoxide, oxygen and nitrogen as 28, 32 and 28 kg/kmol.

Problem 3.7: A particular gas mixture has the following analysis by volume. $O_2=20\%$, $N_2=50\%$, $H_2O=10\%$ and $CO_2=20\%$. If the gas mixture is available at

Fig. 3.1 Schematic for Problem **3.2**

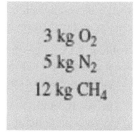

3 kg O_2
5 kg N_2
12 kg CH_4

300 K and 100 kPa, calculate the partial pressures of each constituent and express the composition in mass fractions. If the gas mixture held in a closed vessel is treated with diethylamine solution, which absorbs the CO_2, and the gases are maintained at constant temperature, calculate the final pressure of the gas and express its composition by volume as well as by mass fraction.

Problem 3.8: A mixture of gases has the following gravimetric analysis
  Constituent CO_2 O_2 N_2 H_2 H_2O
  Mass % 25 20 25 20 10
  Determine:

a. The composition of the mixture on a volume basis,
b. The mole numbers of the mixture containing 100 kg of mixture at 100 kPa and 400 K,
c. The specific heat at constant pressure for the mixture.

Problem 3.9: A mixture of carbon monoxide and oxygen is to be prepared in the proportion of 7–4 kg in a vessel of 0.3 m³ capacity. If the temperature of the mixture is 15 °C, determine the pressure to which the vessel is subject. If the temperature is raised to 40 °C, what will then be the pressure in the vessel?

Problem 3.10: For the mixture of Problem **3.**9 calculate the volumetric analysis, the molar mass and the characteristic gas constant. Calculate also the total amount of substance in the mixture.

Problem 3.11: An exhaust is analyzed and is found to contain, by volume, 78 % N_2, 12 % CO_2, and 10 % O_2. What is the corresponding gravimetric analysis? Calculate the molar mass of the mixture, and the density if the temperature is 550 °C and the total pressure is 1 bar.

Problem 3.12: A vessel of 3 m³ capacities contains a mixture of nitrogen and carbon dioxide, the analysis by volume showing equal quantities of each. The temperature is 15 °C and the total pressure is 3.5 bar. Determine the mass of each constituent.

Problem 3.13: The mixture of Problem **3.**12 is to be changed so that it is 70 % CO_2 and 30 % N_2 by volume. Calculate the mass of mixture to be removed and the mass of CO_2 to be added to give the required mixture at the same temperature and pressure as before.

Problem 3.14: A 5 m³ tank contains 60 % H_2 and 40 % methane by volume at 100 kPa and 300 K. Determine the amount of methane to be added at 300 K to change the composition to 50 % methane by volume. Also determine the final pressure of the mixture in the tank

Problem 3.15: A gas mixture being used to simulate the atmosphere of another planet consists of 320 mg of methane, 175 mg of argon, and 225 mg of nitrogen. The partial pressure of nitrogen at 300 K is 15.2 kPa. Calculate

d. The volume and
e. The total pressure of the mixture.

Problem 3.16: A combustion gas mixture consists of 2 moles of H_2O and 2 moles of CO_2 in a 10 L tank. It is cooled to 700 K. What is the tank pressure? Develop a real gas model for the mixture and estimate the pressure base on a Redlich-Kwong equation of state.

References

1. Potter MC, Somerton CW (2006) Thermodynamics for engineers Schaum's outlines series, 2nd edn. McGraw-Hill, Boston
2. Rao YVC (2004) An introduction to thermodynamics, Rev edn. Sangam Books Ltd, Hyderabad
3. Eastop TD, McConkey A (1993) Applied thermodynamic for engineering technologists, 5th edn. Pearson, Prentice Hall

Chapter 4
Work and Heat

This chapter deals with two quantities that affect the thermal energy stored in a system. Work and heat represent the transfer of energy to or from a system, but they are not in any way stored in the system. They represent energy in transition and must carefully defined to quantify their effect on the thermal energy stored in a system. Once they are quantified, they can be related to the conservation of energy principle known as the First Law of Thermodynamics.

4.1 Introduction of the Work and Heat

A closed system can interact with its surroundings in two ways, either by;

a. Work Transfer.
b. Heat Transfer.

These may be called *energy transfer* or *energy interactions* and they bring about changes in the properties of the system. Positive work occurs when the system transfers energy to its surroundings by some mechanical or electrical process. Positive heat transfer occurs when the surroundings transfer thermal energy to the system. Normally a temperature difference is the driving potential that moves thermal energy into or out of a system.

4.2 Definition of Work

The formal definition of work is "a force acting through a distance". When a system undergoes a displacement due to the action of a force, *work* is taking place and the amount of work is equal to the product of the force and the displacement in the direction of the force. The term *work* is so common with many meanings in the English language that it is important to be very specific in its thermodynamic definition.

© Springer International Publishing Switzerland 2015
B. Zohuri, P. McDaniel, *Thermodynamics In Nuclear Power Plant Systems,*
DOI 10.1007/978-3-319-13419-2_4

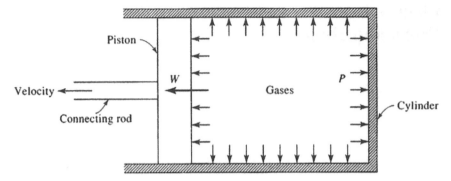

Fig. 4.1 Work being done by expanding gases in a cylinder

Work is done by a force as it acts upon a body moving in the direction of the force.

If the force acts, but no movement takes place, no work is done. Work is performed by the expanding exhaust gases after combustion occurs in a cylinder of an automobile engine as shown in Fig. 4.1. In this case the energy produced by the combustion process can be transferred to the crankshaft by means of the connecting rod, in the form of work. Therefore, the work can be thought of as energy being transferred across the boundary of a system, the system being the gases in the cylinder.

A similar concept is the work done in the turbine to generate electricity in a nuclear power plant. The gas pressure rotates the turbine blades producing a torque that turns generator. Thermal energy is transferred from the reactor core to the steam generator in the first loop. The second loop then uses this steam to drive the turbine. See Fig. 4.2 for the basic configuration of the loops.

Work is done by a system, if the sole *external* effect on the surroundings would be the raising of a weight [1]. The work done, however, by one part of a system on another part is called *internal work*. *Internal* work is not discussed in *macroscopic* thermodynamics. Only the work that involves an interaction between a system and its surroundings can be analyzed. When a system does *external* work, the changes that take place can be described by means of macroscopic quantities referring to the system as a whole, in which case the change may be imagined as the raising or lowering of a suspended weight, the winding or unwinding of a spring, or more generally the alteration of the position or *configuration* of some external mechanical device.

The magnitude of the work is the product of the weight and the distance that the weight is lifted. Figures 4.3a, b shows that the battery cell is connected to an external circuit through which charge flows. The current may be imagined to produce rotation of the armature of a motor, thereby lifting a weight or winding a spring. For an electrochemical cell to do work, it must be connected to an external circuit. Figure 4.3b is the interaction for Fig. 4.3a that qualifies as work in the thermodynamic sense.

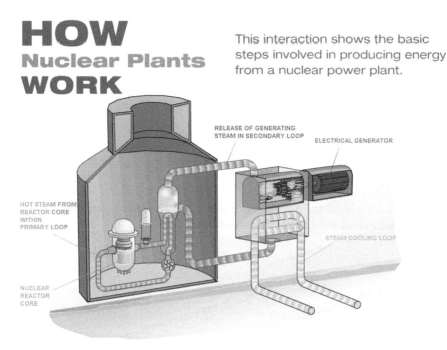

HOW
Nuclear Plants
WORK

This interaction shows the basic steps involved in producing energy from a nuclear power plant.

RELEASE OF GENERATING
STEAM IN SECONDARY LOOP
ELECTRICAL GENERATOR

HOT STEAM FROM
REACTOR CORE
WITHIN
PRIMARY LOOP

STEAM COOLING LOOP

NUCLEAR
REACTOR
CORE

Fig. 4.2 Basic schematic of nuclear power plants and steam loops

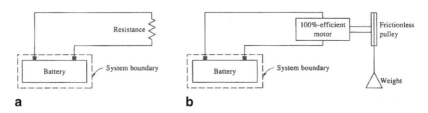

Fig. 4.3 Work being done by electrical means

The thermodynamic convention defines *positive work* as that done by the system on its surroundings. *Negative work* is defined as work is done on the system. Figure 4.4 is a simple presentation of positive and negative work W for interactions between a system and its surroundings.

The units of work in the SI system are Newton-meters. A Newton–meter is also defined as a Joule. In the English system, the basic unit is foot-pound force. There is no other name. A new quantity defined as *power* can be introduced as the rate of doing work W. In the SI system, the unit for power is Joules per second (J/s) or Watts (W), while in English system the unit is **ft-lbf/sec**. An additional English system unit is the horsepower (**hp**) which is defined as 550 ft-lbf/sec. Note that 1 hp = 746 W.

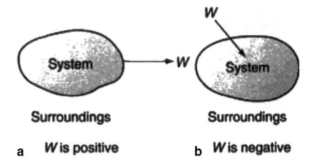

Fig. 4.4 Work interaction between a system and its surroundings

The work associated with a unit mass will be designated as w or specific work. It should not be confused with specific weight as is given by:

$$w = \frac{W}{m} \tag{4.1}$$

4.3   Quasi-Static Processes

Before going further, it is important to note that thermodynamics can only be used to describe equilibrium states. A system in thermodynamic equilibrium will satisfy the following requirements:

1. *Mechanical equilibrium*: There are no unbalanced forces or torques acting on any part of the system or on the system as a whole.
2. *Thermal equilibrium*: There are no temperature differences between parts of the system or between the system and its surroundings.
3. *Chemical equilibrium*: There are no chemical reactions within the system and no motion of any chemical constituent from one part of a system to another part.

A system in thermodynamic equilibrium with its surroundings will have no motion take place and as result, no work will be done, since there is no displacement of any kind.

If the sum of the external forces is changed so that there is a finite unbalanced force acting on the system, then the condition for mechanical equilibrium is no longer satisfied and the following situations will arise:

1. Unbalanced forces or torques will be created within the system; resulting in turbulence, waves, etc.. The system as a whole may execute some sort of accelerated motion.
2. As a result of this turbulence, acceleration, etc., a non-uniform temperature distribution may be brought about, as well as a finite difference of temperature between the system and its surroundings. The sudden change in the forces and in the temperature may produce a chemical reaction or the motion of a chemical constituent.

The above finite unbalanced force may cause the system to pass through *non-equi-librium* states. If it is desired, during a process, to describe every state of a system by means of system-wide thermodynamic coordinate, then the process must *not* be performed using a finite unbalanced force or torque. Under these circumstances, the external forces acting on a system are varied only slightly so that the unbalanced force is infinitesimal, and the process proceeds infinitesimally slowly. A process performed in this mode is said to be *quasi-static.*

If all of the states through which the system passes can be described by means of thermodynamic coordinates referring to the system as a whole, and an equation of state for all these states is valid, the process is called quasi-static. A quasi-static process is an idealization that is applicable to any thermodynamic system, including electric and magnetic ones. The conditions for such a process can never be achieved in the real world, but can often be approached with almost any degree of accuracy.

Classical thermodynamics does not quantify how infinitesimally slowly a process must take place to be considered quasi-static. Molecular gas kinetics requires only that the process proceed slowly compared to the speed of the molecules in the gas. This allows system properties to be equilibrated across the system faster than the system configuration changes. Examples of processes that seem rapid, but can be treated as quasi-static are the expansion of combustion products in a gasoline engine, or the expansion of the exhaust gases of a chemical rocket.

The reason for the introduction of a quasi-static process is to allow calculations without addressing the complications of friction within the system. This approach is no different from that of Newtonian's mechanics with its mass-less springs and ideal pulleys, or that of circuit theory with wires with no resistance, or batteries with constant voltage. Later reversible processes will be considered that are synonymous quasi-static processes because dissipative processes are ignored.

4.4   Quasi-Equilibrium Work Due to Moving Boundary

Consider the piston-cylinder arrangement with the included gas in it as shown in Fig. 4.5. The expanding gas can be treated as instantaneously in equilibrium at a given pressure p and volume V. Initially the system is characterized by the pressure p_1 and volume V_1. If we let the piston move out to the new equilibrium state at position 2 that is specified by pressure p_2 and volume V_2 via a quasi-static process, all intermediate points in the travel path of the piston can be characterized by the pressure p and the volume V at those points. This is required because the macroscopic properties p and V are significant only for equilibrium states.

If A is the area of the piston and the piston moves an infinitesimal distance dl, the force F acting on the piston is $F = pA$. The infinitesimal amount of work done by the gas on the piston is:

$$\delta W = F \cdot dl = pAdl \qquad (4.2)$$

Fig. 4.5 *pdV* work

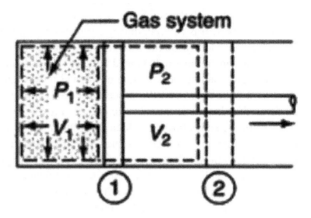

Fig. 4.5 *pdV* work

where $dV = Adl$ = infinitesimal displacement volume. The small delta (δ) sign in δW represents an *inexact differential*. When the piston moves out from position **1** to position **2** with the volume changing from V_1 to V_2, the amount of work W done by the system will be (Fig. 4.6);

$$W_{1-2} = \int_{V_1}^{V_2} pdV \qquad (4.3)$$

In Fig. 3.6, the magnitude of the work done is given by the area under path 1–2. Since p is at all times a thermodynamic coordinate, all the states passed through by the system as the volume changes from V_1 to V_2 must be equilibrium states, and

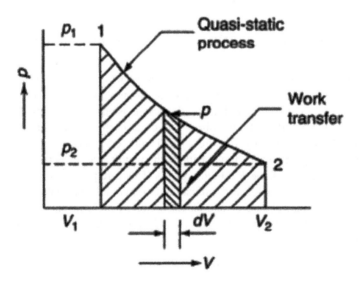

Fig. 4.6 Quasi-state *pdV* work

the path 1–2 must be *quasi-static*. The piston moves infinitely slowly so that every state passed through is an equilibrium state.

The integration $\int pdV$ can be performed only on a quasi-state path.

The significant key in Eq. 4.3 is that we assume the pressure is known for each position as the piston moves from volume V_1 to volume V_2 and typical pressure-volume (P-V) diagrams are shown in Fig. 4.7 below. The work W_{1-2} is the cross-hatched area under the P-V Curve from the definition of the integration process. The integration process highlights two very important features.

First, as work is performed from state 1 to state 2 by the moving piston of Fig. 4.7, pressure and volume changes of a gas during expansion may be indicated by the area under the curve of Fig. 4.8a. However, the expansion of the gas could be represented by the area under curve of Fig. 4.8b. The area under curve of Fig. 4.8b work is significantly larger than the work under the curve of Fig. 4.8a.

The end states 1 and 2 are identical, yet the areas under the P-V curves are very different. In addition to being dependent on the end points, work depends on the actual path that connects the two end points.

Thus, work is a path function, as contrasted with a point, or state, function, which is dependent only on the end-points. The differential of a path function is called an

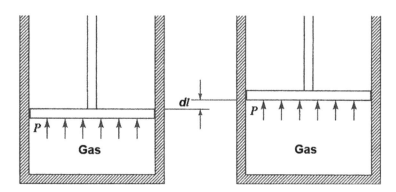

Fig. 4.7 Work due to a moving boundary

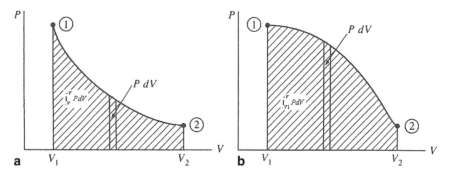

Fig. 4.8 Work depends on the path between two states

inexact differential, whereas the differential of a point or state, function is an *exact differential*. An inexact differential will be denoted with the symbol δ and the integral of δW is W_{1-2} as shown in Eq. 4.2 in above, where the subscript emphasizes that the work is associated with the path as the process passes from state 1 to state 2, however the subscript may be omitted when the work done is written simply as W. In this case, we would never write W_1 or W_2, since work is not associated with a state but with a process. Work is not a property.

The integral of an exact differential, for example, dT, would be

$$\int_{T_1}^{T_2} dT = T_2 - T_1 \tag{4.4}$$

where T_1 is the temperature at state 1 and T_2 is the temperature at state 2.

The second observation to be made from Eq. 4.3 is that the pressure is assumed to be constant throughout the volume at each intermediate position. The system passes through each equilibrium state shown in the *P-V diagrams* of Fig. 4.8a and b. An equilibrium state can usually be assumed even though the variables may appear to be changing quite rapidly. Combustion is a very rapid process that cannot be modeled as a quasi-static process. The other processes in the internal combustion engine (*expansion*, exhaust, intake, and *compression*) can be assumed to be quasi-static processes, as they occur at a slow rate, thermodynamically.

As a final comment regarding work we may now discuss what is meant by a simple system. For a system free of surface, magnetic, and electrical effects, the only work mode is that due to pressure acting on a moving boundary. Such simple systems require only two independent variables to establish an equilibrium state of the system composed of a homogeneous substance. If other work modes were present, such as an electric field, additional independent variables would be necessary, such as the electric field intensity.

On the P-V Diagram depicted in Fig. 4.9, an initial equilibrium state characterized by the coordinates P_i, V_i, and T_i, as well as a final equilibrium state coordinates by P_f, V_f and T_f of a hydrostatic system are represented by two the points i and f, respectively.

There are many ways in which the system may expand from i to f. For example, using Fig. 4.9 the pressure may be kept constant from i to a (*isobaric process*) and then the volume kept constant from a to f (*isochoric processes*), in which case the work done is equal to the area under the line ia, $W = 2P_0V_0$ and positive, because work is being done *by* the system. Another possibility is the path ibf, in which case the work is the area under the line bf, or P_0V_0. The straight line from i to f represents another path, where the work is $\frac{3}{2}P_0V_0$. The most work is done by system traversing path iaf, which does more work than traversing path if, which does more work than traversing path ibf. We can see, that the *work done by a system depends not only on the initial and final states but also on the intermediate*

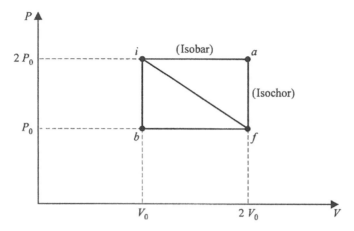

Fig. 4.9 Work depends on the path of integration from initial equilibrium state i to the final equilibrium state f

states, namely, on the path of integration. This basically another way of saying that, for a quasi-static process, the expression

$$W = \int_{V_i}^{V_f} P dV \qquad (4.5)$$

cannot be integrated until P is specified as a function of V using an appropriate equation of state.

The expression $P dV$ is an infinitesimal amount of work and is represented by the symbol of δW. There is, however, an important distinction between an infinitesimal amount of work and the other infinitesimals, such as dP or dV. An infinitesimal amount of hydrostatic work is an *inexact differential*, that is, δW is *not* the differential of an actual function of the thermodynamic coordinates. To indicate that an infinitesimal amount of work is *not* a mathematical differential of a function W and to emphasize at all times that it is an inexact differential, it gets denoted by δW.

4.5   Definition of a Cycle in Thermodynamic

Any process or series of processes whose end states are identical is termed a **cycle**. The processes through which the system has passed can be shown on a state diagram, but a complete description of the path also requires a statement of the heat and work crossing the boundary of the system. Consider Fig. 4.10. It shows such a cycle in which a system starts at state '1' and changes pressure and volume through a path 1–2–3 to return to its initial state '1'.

Fig. 4.10 Cycle of
operations

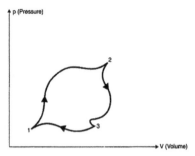

With this definition of a cycle, consider the following p-V diagrams. For curve I in Fig. 4.11a an expansion, the volume increases, dV is positive, and the integral Eq. 4.3 is positive. For curve II in Fig. 4.11b, where the gas is being compressed, the volume decreases, so the same integral is negative. According to the sign convention for work, work is done *by* the system in the process represented by curve I, and work is done *on* the system in the process represented by curve II. In Fig. 4.11c, curve I and II are drawn together so that they constitute two processes that bring the gas back to its initial state. The net work for the cycle is positive as represented by the area enclosed between the two curves.

Such a series of two or more processes, represented by a closed figure is called a *cycle*. The area within the closed figure in Fig. 4.11c is obviously the difference between the areas curves I and II and, therefore, represents the *net* work done in the cycle. Notice that the cycle is traversed in a direction such that the net work is positive, and the net work is done *by* the system. If the direction of the cycle were reversed, then the net work would be negative as the net work is done *on* the system [2].

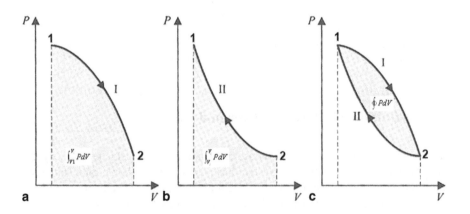

Fig. 4.11 P-V diagram of a gas with shaded area to show work done by the system or work done on the system. **a** curve I, expansion; **b** curve II, compression; **c** curves I and II together constitute a cycle

4.6   Path Functions and Point or State Functions

Further interpretation of Eq. 4.3 and with reference to Fig. 4.12, it is possible to take a system from state 1 to state 2 along many quasi-static paths as mentioned in above, such as A, B or C. Since the area under each curve represents the work for each process, the amount of work involved in each case is *not a function of the states* of the process, and it depends on the path the system follows in going from state 1 to state 2. For this reason, work is called a *path function, and* δW *is an inexact or imperfect differential.*

Thermodynamic properties are *point or state functions*, since for a given state; there is a definite value for each property. The change in a thermodynamic property of a system when changing states is independent of the path the system follows during the change of state, and depends only on the initial and final states of the system. The differentials of point, or state, functions are *exact or perfect differentials,* and the integration is simply

$$\int_{V_1}^{V_2} dV = V_2 - V_1 \tag{4.6}$$

The change in volume thus depends only the end states of the system irrespective of the path the system of follows. On the other hand, work done in a quasi-static process between two given states depends on the path followed and will be expressed as;

$$\int_1^2 \delta W \neq W_2 - W_{11} \tag{4.7a}$$

Or a more simple form will be written as

$$\int_1^2 \delta W \neq W_{1-2} =_1 W_2 \tag{4.7b}$$

Path Functions and **Point or State Functions** can be expressed as:

Fig. 4.12 Work-a *path function*

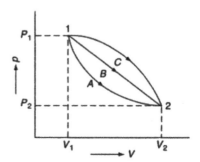

- **Path functions**: Magnitudes depend on the path followed during a process as well as the end states. Work (W), heat (Q) are considered as path functions. Work and heat are examples of path functions. Heat and work are *inexact differentials*. Their change cannot be written as difference between their end states. Thus;

In the case of Work,

$$\int_1^2 \delta W \neq W_2 - W_1 \quad \text{abbreviated as } W_{1-2} \text{ or } {}_1W_2 \tag{4.8a}$$

In case of Heat we will have;

$$\int_1^2 \delta Q \neq Q_2 - Q_1 \quad \text{abbreviated as } Q_{1-2} \text{ or } {}_1Q_2 \tag{4.8b}$$

- **Point or State Functions**: Depend on the state only, and not on how a system reaches that state. Properties are point functions, (i.e. pressure, volume, temperature, etc) and they are exact differentials. For example *Temperature* and *Volume* can be expressed as,

$$\int_{T_1}^{T_2} dT = T_2 - T_1 \tag{4.9a}$$

$$\int_{V_1}^{V_2} dV = V_2 - V_1 \tag{4.9b}$$

In addition, to distinguish an inexact differential δW from an exact differential dV or dP as we explained in Sect. 4.4 of this chapter the δ symbol is used.

From Eq. 4.2b, we can write the following expression.

$$dV = \frac{1}{p} \delta W \tag{4.10}$$

Here, $1/p$ is called the *integration factor*. Therefore, an inexact differential δW when multiplied by an integrating factor $1/p$ becomes an exact differential dV.

For a cyclic process, the initial and final states of the system are the same, and hence, the change in any property is zero, i.e.

$$\oint dV = 0 \cdot \oint d\varphi = 0 \cdot \oint dT = 0 \tag{4.11}$$

where the symbol $\oint$ denotes the cyclic integral around a closed path. Therefore, *the cyclic integral of a property is always zero.*

4.7   *PdV* **Work for Quasi-Static Process**

It must be emphasized that the area on a *P-V diagram* represents the work for a qua-si-static process only. For non-equilibrium processes the work cannot be calculated using $\int PdV$. Either p must be given as a function of V or it must be determined by some other means. Consider the following examples in which integration of $\int PdV$ can be carried out, because the path of integration is provided by an equation of state or a state function.

1. Constant pressure expansion presented by Fig. 3.13 process 1–2, which is depicts an *isobaric process* (Fig. 4.13).

$$W_{1-2} = \int_{V_1}^{V_2} PdV = P(V_2 - V_1) \tag{4.12}$$

2. Constant volume process represented in Fig. 4.14 as the process 1–2, depicting an *isochoric process*.

$$W_{1-2} = \int_{V_1}^{V_2} PdV = 0 \tag{4.13}$$

Fig. 4.13 Constant pressure process

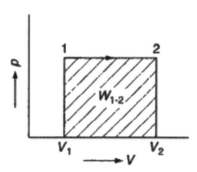

Fig. 4.14 Constant volume process

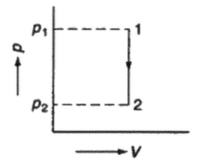

Fig. 4.15 Process in which
PV is constant

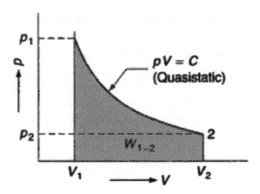

3. A process in which $PV = \text{Constant}$ as shown in Fig. 4.15.

$$W_{1-2} = \int_{V_1}^{V_2} PdV, PV = P_1V_1 = C \text{ , where } C \text{ is a constant presentation.}$$

$$P = \frac{(P_1V_1)}{V}$$

Substitution gives;

$$W_{1-2} = P_1V_1 \int_{V_1}^{V_2} \frac{dV}{V} = P_1V_1 \ln \frac{V_2}{V_1}$$

$$= P_1V_1 \ln \frac{P_1}{P_2}$$
(4.14)

4. A process in which $PV^n = C$ as shown in Fig. 4.16, where both n and C are constant.

Fig. 4.16 Process in which
PV^n is constant

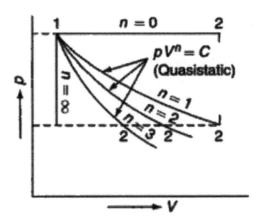

$$PV^n = P_1V_1^n = P_2V_2^n = C$$

$$P = \frac{P_1V_1^n}{V^n}$$

$$W_{1-2} = \int_{V_1}^{V_2} PdV = \int_{V_1}^{V_2} \frac{P_1V_1^n}{V^n} \cdot dV = \left(P_1V_1^n\right)\left[\frac{V^{-n+1}}{-n+}\right]_{V_1}^{V_2}$$

$$= \frac{P_1V_1^n}{1-n}\left(V_2^{1-n} - V_1^{1-n}\right) = \frac{P_2V_2^n \times V_2^{1-n} - P_1V_1^n \times V_1^{1-n}}{1-n} \qquad (4.15)$$

$$= \frac{P_1V_1 - P_2V_2}{n-1} = \frac{P_1V_1}{n-1}\left[1 - \left(\frac{P_2}{P}\right)^{n-1/n}\right]$$

For each of these processes the inexact differential can be converted to an exact differential so that the integration can easily be performed. The states that the quasi-static process passes through are defined by the function that performs this conversion

4.8   Non-equilibrium Work

In order to have a concept for a non-equilibrium work process, we consider a system to be formed by the gas in Fig. 4.17. In part (a) work is crossing the boundary of the system by means of the rotating shaft and the volume doesn't change. We calculate the work input, neglecting any friction in the pulley system, by multiplying the distance the weight drops by its weight. This action does not mean the work is equal to $W = \int PdV$, which is zero. The paddle wheel provides a non-equilibrium work mode.

Suppose the membrane in Fig. 4.17b ruptures, allowing the gas to be expanded and fill the evacuated volume. There is no resistance to the expansion of the gas at

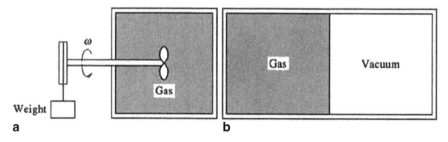

Fig. 4.17 A system with rotating shaft paddle and weight attached to a pulley

Fig. 4.18 Illustration of
example 4.1

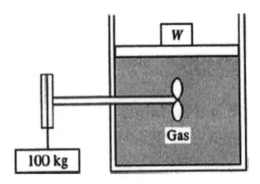

the moving boundary as the gas fills the volume; hence, there is no work done. Yet there is a change in volume by the gas's expansion to fill the entire container. The sudden expansion is a non-equilibrium process, and again the $W = \int PdV$ relationship cannot be used to calculate the work.

Example 4.1 A 100-kg mass drops 3 m, resulting in an increased volume in the cylinder of 0.002 of Fig. 4.18. The piston maintains a constant gage pressure of 200 kPa. Determine the net work done by the gas on the surroundings.

Solution Assessing the problem and considering Fig. 3.18, we see that, the paddle wheel does work on the system, the gas, due to the 100 kg mass dropping 3 m, consequently the work done is negative.

$$W = -(F)(d) = -(mg)(d) = -(100)(9.8)(3) = -2940 \text{ Joules}$$

The work done by the system on this frictionless piston is positive since the system is doing the work. It is

$$W = (PA)(h) = (P)(Ah) = PV = (200000)(0.002) = 400 \text{ Joules}$$

Therefore, the net work done is;

$$W_{net} = -2940 + 400 = -2540 \text{ Joules}$$

4.9   Other Work Modes

There are many other forms of work than $W = \int PdV$ or simple displacement work on a straight line. Some additional types of work are,

a. **Electrical Work:** When a current flows through a resistor that is shown in Fig. 4.19, taken as a system, there is work transfer into the system. This is because the current can drive a motor, the motor can drive a pulley and the pulley can raise a weight.

Fig. 4.19 Electrical work illustration

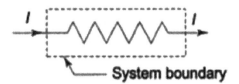

Work and Heat Transfer

System boundary

The current flow is $I = \dfrac{dC}{d\tau}$, and C is the charge in coulombs and τ is the time in seconds. Thus dC is the charge crossing a boundary during time $d\tau$. If E is the voltage potential, the work is given by:

$$\delta W = E.dC$$
$$= EId\tau$$

Therefore

$$W = \int_1^2 EId\tau \qquad (4.16)$$

The electrical power will be;

$$\dot{W} = \lim_{d\tau \to \infty} \frac{dW}{d\tau} = EI \qquad (4.17)$$

Work is transferred at this rate.

b. **Shaft Work:** When a shaft, taken as the system according to Fig. 4.10, is rotated by a motor, there is work transfer into the system. This is because the shaft can rotate a pulley which can raise a weight. If T is the torque applied to the shaft and $d\theta$ is the angular displacement of the shaft, the shaft work is then given by:

$$W = \int_1^2 Td\theta \qquad (4.18)$$

and the shaft power is

$$\dot{W} = \int_1^2 T \frac{d\theta}{d\tau} = T\omega \qquad (4.19)$$

where ω is the angular velocity and T is applied torque (Fig. 4.20).

c. **Paddle-Wheel Work or Stirring Work:** As the weight is lowered, and the paddle wheel turns as shown in Fig. 4.21, there is work transfer into the fluid system which gets stirred. Since the volume of the system remains constant, $\int PdV = 0$.

Fig. 4.20 Shaft work
illustration

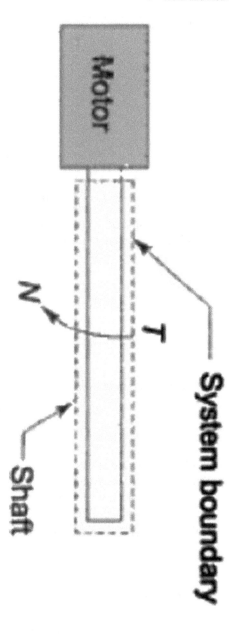

If m is the mass of the weight lowered through a distance dz and T is the torque
transmitted by the shaft in rotating through an angle $d\theta$, the differential work
transfer to the fluid is given by;

$$dW = mgdz = Td\theta$$

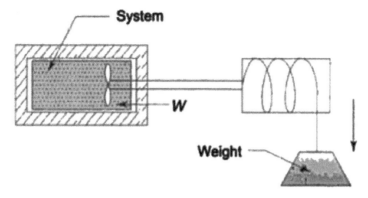

Fig. 4.21 Paddle-wheel work illustration

and the total work transfer is presented by.

$$W = \int_1^2 mgdz = \int_1^2 Td\theta \qquad (4.20)$$

d. **Flow Work:** The flow work, significant only in a flow process or an open system, represents the energy transferred across the system boundary because of the
energy imparted to the fluid by a pump, blower or compressor to make the fluid
flow across the control volume. Flow work is analogous to displacement work.
Let P be the fluid pressure in the plane of the imaginary piston, which acts in a
direction normal to it as it can be seen in Fig. 4.22. The work done on this imaginary piston by the external pressure as the piston moves forward is given by:

$$\delta W_{\text{flow}} = pdV \qquad (4.21)$$

Fig. 4.22 Flow work
illustration

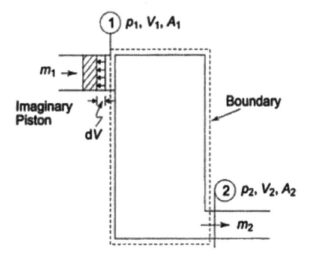

where dV is the volume of fluid element about to enter the system.

$$\delta W_{flow} = p\upsilon dm \tag{4.22}$$

where $dV = \upsilon dm$. Therefore, flow work at inlet Fig 4.22 is given by:

$$(\delta W_{flow})_{in} = p_1\upsilon_1 dm_1 \tag{4.23}$$

Equation 4.23 can also be derived in a slightly different manner. If the normal pressure P_1 is exerted against the area A_1, giving a total force $(P_1 A_1)$ against the piston, in time $d\tau$, this force moves a distance a distance $V_1 d\tau$, where V_1 is the velocity of flow (piston). The work in time $d\tau$ is $P_1 A_1 V_1 d\tau$, or the work per unit time is $p_1 A_1 V_1$. Since the flow rate

$$w_1 = \frac{A_1 V_1}{\upsilon_1} = \frac{dm_1}{d\tau} \tag{4.24}$$

The work done in time $d\tau$ becomes;

$$(\delta W_{flow})_{in} = p_1\upsilon_1 dm_1 \tag{4.25}$$

Similarly, flow work of the fluid element leaving the system is;

$$(\delta W_{flow})_{out} = p_2\upsilon_2 dm_2 \tag{4.26}$$

The flow work per unit mass is thus

$$W_{flow} = p\upsilon \tag{4.27}$$

It is the displacement work done by mass moving across the system boundary.

e. **Work Done in Stretching a Wire:** Let us consider a wire as the system. If the length of the wire is changed from L to $L + dL$ by the tension f_T the infinitesimal amount of work done is equal to,

$$\delta W = -f_T dL$$

The minus sign is used because a positive value of dL means an expansion of the wire, for which work must be done on the wire, i.e., negative work. For a finite change of length,

$$W = -\int_1^2 f_T dL \tag{4.28}$$

If the stretching is kept within the elastic limit, where E is the modulus of elasticity, s is the stress, ε is the strain, and A is the cross-sectional area, then;

$$f_T = sA = E\varepsilon A \qquad \text{since } \frac{s}{\varepsilon} = E$$

$$d\varepsilon = \frac{dL}{L}$$

$$\delta W = -f_T dL = -E\varepsilon ALd\varepsilon$$

Therefore;

$$W = -A\varepsilon L \int_1^2 \varepsilon d\varepsilon = -\frac{AEL}{2}\left(\varepsilon_2^2 - \varepsilon_1^2\right) \qquad (4.29)$$

f. **Work Done in Changing the Area of a Surface Film:** A film on the surface of a liquid has a surface tension, which is a property of the liquid and the surroundings. The surface tension acts to make the surface area of the liquid a minimum. It has units of force per unit length. The work done on a homogeneous liquid film in changing its surface area by an infinitesimal amount dA is;

$$\delta W = -\sigma dA$$

where σ is the surface tension (N/m). Therefore.

$$W = -\int_1^2 \sigma dA \qquad (4.30)$$

g. **Magnetization of a Paramagnetic Solid:** The work done per unit volume on a magnetic material through which the magnetic and magnetization fields are uniform is;

$$\delta W = -HdI$$

and

$$W_1 - W_2 = -\int_{I_1}^{I_2} HdI \qquad (4.31)$$

where H is the field strength, and I is the component of the magnetization field in the direction of the field. The minus sign provides that an increase in magnetization (positive dI) involves negative work.

The following equations summarize the different forms of work transfer:

Displacement Work:

Compressible Fluid $W = \int_1^2 pdV$

Electrical Work $W = \int_1^2 E dC = \int_1^2 E I d\tau$

Shaft Work $W = \int_1^2 T d\theta$

Surface Film $W = \int_1^2 \sigma dA$

Stretched Wire $W = \int_1^2 f_T dL$

Magnetized Solid $W = -\int_1^2 H dI$

It may be noted in the above expressions that the work is equal to the integral of the product of an intensive property and the change in its related extensive property. These expressions are valid only for quasi-static processes.

There are some other forms of work, which can be identified in processes that are not quasi-static, for example, the work done by shearing forces in a process involving friction in a viscous fluid.

Example 4.2 One kg of steam with a quality of 20% is heated at a constant pressure of 200 kPa until the temperature reaches 700 K. Calculate the work done by the steam.

Solution The work is given by;

$$W = \int P dV = P(V_2 - V_1) = mP(\upsilon_2 - \upsilon_1)$$

To evaluate the work we must determine υ_1 and υ_2. Using Steam Tables, we find

$$\upsilon_f + x(\upsilon_g - \upsilon_f) = 0.001061 + (0.2)(0.8857 - 0.001061) = 0.1780 \, \text{m}^3/\text{kg}$$

From the superheat table we locate state 2 at $T_2 = 400 \,^{\circ}\text{C}$ and $P_2 = 0.2$ MPa as:

$$\upsilon_2 = 1.61172 \, \text{m}^3/\text{kg}$$

The work is then

$$W = (1)(200)(1.61172 - 0.1780) = 286.7 \, \text{kJ}$$

Note: With the pressure having units of kPa, the result is in kJ

Example 4.3 A 110-mm-diameter cylinder contains 100 cm³ of water at 330 K. A 50-kg piston sits on top of the water. If heat is added until the temperature is 500 K, find the work done.

Solution The pressure in the cylinder is due to the weight of the piston and remains constant. Assuming a frictionless seal (this is always done unless information is given to the contrary), a force balance provides

$$mg = pA - p_{atm} A \quad p\frac{\pi(0.110)^2}{4} = (50)(9.81) + 101325.0\frac{\pi(0.110)^2}{4}$$

$$\therefore p = 152938.6 \, \text{Pa}$$

The atmospheric pressure is included so that absolute pressure results. The volume at the initial state 1 is given as

$$V_1 = 100 \text{ x } 10^{-6} = 10^{-4} \text{ m}^3$$

Using υ_1 at 330 K, the mass is calculated to be

$$m = \frac{V_1}{\upsilon_1} = \frac{10^{-4}}{0.001015} = 0.09852\text{kg}$$

At state 2 the temperature is 500 K and the pressure is 0.1516 MPa. Interpolating to find the specific volume gives,

$$v_2 = 3.18482 + \frac{(1.61172 - 3.18482)}{(0.2 - 0.101325)}(0.152938.6 - 0.101325) = 2.362m^3/kg$$

$$V_2 = m\upsilon_2 = (0.09852)(2.362) = 0.2327\text{ m}^3$$

Finally, the work is calculated to be

$$W = P(V_2 - V_1) = 152938.6 * (0.2327 - 0.0001) = 35573.5\text{J or } 35.6\text{kJ}$$

Example 4.4 Energy is added to a piston-cylinder arrangement, and the piston is withdrawn in such a way that the quantity PV remains constant. The initial pressure and volume are 200 kPa and 2 m³, respectively. If the final pressure is 100 kPa, calculate the work done by the gas on the piston.

Solution The work is found from Eq. 3.3 to be

$$W_{1-2} = \int_2^{V_2} PdV = \int_2^{V_2} \frac{C}{V}dV$$

Where we have used $PV = C$. To calculate the work we must find C and V_2. The constant C is found from

$$C = P_1V_1 = (200)(2) = 400\text{kJ}$$

To find V_2 we use $P_1V_1 = P_2V_2$, which is, of course, the equation that would result from an isothermal process (constant temperature) involving an ideal gas. This can be written as

$$\frac{P_1V_1}{P_2} = \frac{(200)(2)}{100} = 4 \text{ m}^3$$

Finally,

$$W_{1-2} = \int_2^4 \frac{400}{V}dV = 400\ln\frac{4}{2} = 277\text{kJ}$$

This is positive, since work is done during the expansion process by the system (the gas contained in the cylinder).

Example 4.5 Determine the horsepower required to overcome the wind drag on a modern car traveling 90 km/h if the drag coefficient C_D .is 0.2. The drag force is given by $F_D = \frac{1}{2}\rho V^2 A C_D$, where A is the projected area of the car and V is the velocity. The density ρ of air is 1.23 kg/m³. Use $A = 2.3$ m².

Solution To find the drag force on a car we must express the velocity in m/s:
$V = (90)(1000/3600) = 25$ m/s. The drag force is then
The drag force is then

$$F_D = \frac{1}{2}\rho V^2 A C_D = \left(\frac{1}{2}\right)(1.23)(25^2)(2.3)(0.2) = 177 \text{ N}$$

To move this drag force at 25 m/s the engine must do work at the rate

$$W = F_D V = (177)(25) = 4425 \text{ W}$$

The horsepower is then

$$H_P = \frac{4425 \text{ W}}{746 \text{ W/hp}} = 5.93 \text{ hp}$$

Example 4.6 The drive shaft in an automobile delivers 100 N-m of torque as it rotates at 3000 rpm. Calculate the horsepower delivered.

Solution The power is found by using $\dot{W} = T\omega$. This requires ω to be expressed in rad/s.

$$\omega = (3000)(2\pi)\left(\frac{1}{60}\right) = 314.2 \text{ rad}/s$$

Hence $\dot{W} = (100)(314.2) = 31420$ Watt or $H_p = \dfrac{31420}{746} = 42.1$ hp

Example 4.7 The air in a circular cylinder of Fig. 4.23 is heated until the spring is compressed 50 mm. Find the work done by the air on the frictionless piston. The spring is initially unstretched, as shown in the figure.

Solution The pressure in the cylinder is initially found from a force balance:

$$P_1 A_1 = P_{atm} A + W \quad P_1 \frac{\pi (0.1)^2}{4} = (101325)\frac{\pi (0.1)^2}{4} + (50)(9.81) \quad \therefore \ p_1 = 163777 \text{Pa}$$

Fig. 4.23 For example 4.7

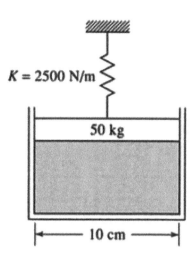

To raise the piston a distance of 50 mm, without the spring, the pressure would be constant required would be force times distance:

$$W = PA \times d = (163777)\frac{\pi(0.1)^2}{4}(0.05) = 64.32 \text{ Joules}$$

For the additional work performed to compress the spring, with spring constant K and the compression from a length x_1 to x_2 the force doing the work given by $F = Kx$. This becomes,

$$W = \int_{x_1}^{x_2} Fdx = \int_{x_1}^{x_2} Kxdx = \frac{1}{2}K\left(x_2^2 - x_1^2\right) = \left(\frac{1}{2}\right)(2500)(0.05)^2 = 3.125 \text{ Joules}$$

The total work is then, found by summing the above two values:

$$W_{Total} = 64.32 + 3.125 = 67.45 \text{ Joules}$$

4.10 Reversible and Irreversible Process

Reversible Process: A *reversible process (also sometimes called a quasi-static process) is one that can be stopped at any stage and reversed so that the system and surroundings are exactly restored to their initial stage.* See Fig. 4.24

This Process has the following *characteristics*:

1. It must pass through the same states on the reversed path as were initially visited on the forward path.
2. This process when undone will leave no history of events in the surroundings.
3. It must pass through a continuous series of equilibrium states.

Fig. 4.24 Reversible process

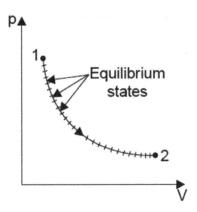

No real process is truly reversible but some processes may approach reversibility, to a close approximation.

Example Some examples of nearly reversible processes are:

1. Frictionless relative motion
2. Expansion and compression of a spring
3. Frictionless adiabatic expansion or compression of a fluid
4. Polytropic expansion or compression of a fluid
5. Isothermal expansion or compression of a fluid
6. Electrolysis

Irreversible Process An *irreversible process* is one *in which there are frictions-like losses, usually involving heat transfer, such that it cannot be reversed.* See Fig. 4.25.

An irreversible process is usually represented by a dotted (or discontinuous) line joining the end states to indicate that the intermediate states are indeterminate.

Irreversibilities are of two types:

Fig. 4.25 Irreversible process

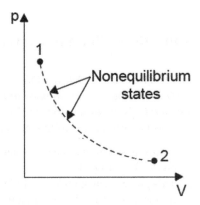

1. **External Irreversibilities:** These are associated with *dissipating effects outside the working fluids.* An example is mechanical friction occurring during due to some external movement.
2. **Internal Irreversibilities:** These are associated with *dissipating effects within the working fluid.* An example is unrestricted expansion of a gas.

Example: Some examples of irreversible processes are:

1. Relative motion with friction
2. Combustion
3. Diffusion
4. Free expansion
5. Throttling
6. Electricity flow through a resistance
7. Heat Transfer
8. Plastic deformation

4.11   Definition of Energy (Thermal Energy or Internal Energy)

Energy is generally defined as the potential to do work. Mechanical energy is classically divided into kinetic and potential energy. Kinetic mechanical energy is related to the velocity that a mass possesses. Potential mechanical energy is related to the distance a mass is above some neutral reference plane. Tension in a spring or surface film tension are other forms of potential mechanical energy. There are many other forms of energy including electrical, chemical, and nuclear. Thermodynamics deals with another type of energy called "thermal energy" or "internal energy". If energy were not such a broad term, it might simply be possible to call the internal energy stored in a fluid simply "energy". But to distinguish the thermal energy stored in a fluid as a result of the temperature of the fluid, from other forms of energy the normal thermodynamic term is **Internal Energy**. Internal Energy, or thermal energy, is generally represented by a capital **U** (sometimes and E). Normally the two ways that the Internal Energy of a system is changed is by Work or Heat Transfer. Both Work and Heat Transfer affect the Internal Energy of a system. When a system does Work on its environment, it gives up Internal Energy. When Heat is transferred to a system, its Internal Energy increases. Heat Transfer, like Work, is a path dependent process. Both Heat transfer and Work represent energy in transition. These are the only forms in which energy can cross the boundaries of a system. *Neither heat nor work can exist as stored energy.*

4.12   Definition of Heat

Detailed analysis of *heat* and *heat transfer* will be addressed in Chapter twelve of the text, but it will be discussed conceptually at this time to relate it to changes in Internal Energy of systems. In the preceding section, several work modes by which energy is transferred macroscopically to or from a system were addressed. Energy can also be transferred microscopically to or from a system by means of interactions between the molecules that form the surface of the system and those that form the surface of the surroundings. If the molecules of the system boundary are more active than those of the surrounding boundary, they will transfer energy from the system to the surroundings, with the faster molecules transferring energy to the slower molecules. On this microscopic scale the energy is transferred by a work mode: collisions between particles. A force occurs over an extremely short time span, with work transferring energy from the faster molecules to the slower ones. The problem is that this microscopic transfer of energy is not observable macroscopically as any of the work modes addressed so far and a means must be developed to account for it.

System temperature is a property, which increases with increased molecular activity. Thus, it is not surprising that microscopic energy transfer can be related to the macroscopic property temperature. This macroscopic transfer of energy that we cannot account for by any of the classic macroscopic work modes will be called heat. Heat is energy transferred across the boundary of a system due to a difference in temperature between the system and its surroundings. A system does not contain heat, it contains energy, and heat is energy in transit.

Heat, like work, is something that *crosses a boundary*. Because a system does not contain heat, heat is not a property. Thus, its differential is *inexact* and is written as δQ where Q is the heat transfer. For a particular process between state 1 and state 2 the heat transfer could be written as Q_{1-2} (or $_1Q_2$), but it will generally be denoted by Q. The *rate of heat transfer* will be denoted by $\dot{Q}$.

By convention, if heat is transferred *to a system* it is considered *positive*. If it is transferred *from a system,* it is *negative*. This is opposite from the convention chosen for work; if a system performs work on the surroundings it is positive. Positive heat transfer adds energy to a system, whereas positive work subtracts energy from a system. A process in which there is zero heat transfer is called an *adiabatic process*. Such a process is approximated experimentally by insulating the system so that negligible heat is transferred.

Heat denoted by the symbol of Q, and it may be defined in an analogous way to work as follows:

"*Heat is something which appears at the boundary when a system changes its state due to a difference in temperature between the system and its surroundings*".

Heat, like work, is a transient quantity, which only appears at the boundary while a change is taking place within the system.

It is apparent that neither δW or δQ are *exact differentials* and therefore any integration of the elemental quantities of work or heat which appear during a change from state 1 to state 2 must be written as;

$$\int_1^2 \delta W = W_1 - W_2 = W_{1-2} =_1 W_2 \text{ or simply } W \qquad (4.32a)$$

and

$$\int_1^2 \delta Q = Q_1 - Q_2 = Q_{1-2} =_1 Q_2 \qquad (4.32\text{b})$$

Sign Convention   If the heat flows *into* a system *from* the surroundings, the quantity is said to be *positive* and, conversely, if heat flows *from* the system to the surroundings it is said to be *negative*. In other words, we can say:

- *Heat received by the system= +Q*
- *Heat rejected or given up by the system= −Q*

4.13   Comparison of Work and Heat

There certain *Similarities* and *Dissimilarities* between Work and Heat,

1. *Similarities*:

    1. Both are *path functions* and *inexact differentials*.
    2. *Both are boundary phenomenon i.e., both are recognized at the boundaries of the system as they cross them.*
    3. Both are associated with a process, not a state. Unlike properties, work or heat has no meaning at a state.
    4. Systems possess energy, but not work.

2. *Dissimilarities:*

    1. In heat transfer temperature difference is required.
    2. In a stable system there cannot be work transfer, however, there is no restriction for the transfer of heat.
    3. The sole effect external to the system from work could be reduced to rise of a weight but in the case of a heat transfer other effects are also observed.

In case of heat and heat transfer process, it is sometimes convenient to refer to heat transfer per unit mass. Heat transfer per unit mass will be designated by letter q and defined by the following expression as;

$$q = \frac{Q}{m} \qquad (4.33)$$

There are three modes of heat transfer,

1. Conduction
2. Convection
3. Radiation

Briefly, each of these modes can be explained as follows,

1. Conduction

This results from the collision of neighboring molecules in which the kinetic energy of vibration of one molecule is transferred to its nearest neighbor. Thermal energy is thus spread by conduction even if the molecules themselves do not move their location appreciably. The mathematical expression of this process is given by *Fourier's Law of heat transfer*, which for a one-dimensional plane wall takes the form

$$\dot{Q} = -kA \frac{\Delta T}{\Delta L} \tag{4.34}$$

where k is the thermal conductivity with units of W/m.K (Btu/sec-ft-^{0}R), ΔL is the thickness of the wall, ΔT is the temperature difference, and A is the wall area. Often, the heat transfer is related to the common R-factor, resistivity, given by $R_{mat} = \Delta L / k$. Note that heat flows in the opposite direction of the temperature gradient.

2. Convection

In addition to conduction, when a vibrating molecule moves from one region to another, it takes its thermal energy with it. This type of movement of thermal energy is called convection. Convection is expressed in terms of the temperature difference between the bulk temperature of a fluid T_∞ and the temperature of the surface T_s. *Newton's law of cooling* expresses this as;

$$\dot{Q} = h_c A(T_s - T_\infty) \tag{4.35}$$

where h_c is the *convective heat transfer coefficient*, with units of W/m^2 K (Btu/sec-ft$^{2-0}$R), and depends on the properties of the fluid including its velocity and the wall geometry. Free convection occurs due to the temperature difference only, whereas forced convection results from the fluid being forced, as with a fan. Convection is once again in the direction opposite the temperature difference. Heat must move from high to low temperature.

3. Radiation

Radiation is energy that is transmitted by electromagnetic radiation. All bodies emit electromagnetic radiation as a result of electron transitions within their atoms. The electrons are moved to excited states via collisions with other atoms. When they return to their normal state, they emit radiation. At absolute zero, temperature there is no atomic motion, so there is no exciting of the electrons and bodies do not radiate. Radiation heat transfer is calculated using the *Stefan-Boltzmann law* and accounts for the energy emitted and the energy absorbed from the surroundings.

$$\dot{Q} = \varepsilon \sigma A \left(T^4 - T_{surr}^4 \right) \tag{4.36}$$

where σ is the *Stefan-Boltzmann constant* ($\sigma = 5.67 \times 1^{-8}$ W/m^2.K^4), ε is the emissivity which is a number within the interval of $0 < \varepsilon < 1$ where $\varepsilon = 1$ is for a blackbody, a body that emits the maximum amount of radiation, and T_{surr} is the uniform temperature of the surroundings. The temperatures must be absolute temperatures and the area A must be the area over which the systems in question face each other.

Example: 4.8 A paddle wheel adds work to a rigid container by dropping a 50 kg weight a distance of 2 m from a pulley. How much heat must be transferred to result in an equivalent effect?

Solution For this non-equilibrium process the work is given by;

$$W = (mg)(d) = (50)(9.8)(2) = 980 \text{ Joules}$$

The heat Q that must be transferred equals the work, 980 J.

Example: 4.9 A 10-m long by 3-m high wall is composed of an insulation layer with $R=2$ m² K/W and a wood layer with $R=0.5$ m² K/W. Estimate the heat transfer rate through the wall if the temperature difference is 40 °C.

Solution The total resistance to heat flow through the wall is

$$R_{Total} = R_{Insulation} + R_{Wood} = 2 + 0.5 = 2.5 \text{ m}^2 \cdot \text{K/W}$$

The heat transfer rate is then;

$$\dot{Q} = \frac{A}{R_{Total}} \Delta T = \frac{10 \times 3}{2.5} \times 40 = 480 \text{ W}$$

Note that ΔT measured in °C is the same as ΔT measured in Kelvin

Example: 4.10 The heat transfer from a 2-m-diameter sphere to a 25 °C air stream over a time interval of 1 h is 3000 kJ. Estimate the surface temperature of the sphere if the heat transfer coefficient is 10 W/m²K. Note that the surface area of a sphere is $4\pi r^2$.

Solution The heat transfer is

$$Q = h_c A(T_s - T_\infty) \text{ or } 3 \times 10^6 = 10 \text{ x } 4\pi \times 1^2 (T_s - 25) \times 3600$$

The surface temperature is calculated to be

$$T_s = 31.6 \text{ °C}$$

Example: 4.11 Estimate the rate of heat transfer from 200 °C sphere which has an emissivity of 0.8 if it is suspended in a cold room maintained at -20 °C. The sphere has a diameter of 20 cm.

Solution The rate of heat transfer is given by;

$$\dot{Q} = \varepsilon \sigma A \left(T^4 - T_{surr}^4 \right) = 0.8 \text{ x } 5.67 \times 10^{-8} \times 4\pi \times 0.1^2 (473^3 - 253^4) = 262 \text{ J/s}$$

Problems

Problem 4.1: Energy is added to a piston-cylinder arrangement, and the piston is withdrawn in such a way that the quantity $PV = C = \text{constant}$. The initial pressure and volume are 400 kPa and 2 m³, respectively. If the initial pressure is 200 kPa, calculate the work done by the gas on the piston.

Problem 4.2: Calculate the work done due to a volume change using Eq. 4.5 and knowing the pressure as function of volume, using ideal-gas relationship $PV = nRT$. Further we assume temperature is constant (i.e. isothermal condition)

Problem 4.3: Consider a gas enclosed in a piston-cylinder assembly as the system. The gas is initially at a pressure of 500 kPa and occupies a volume of 0.2 m³. The gas is taken to the final state where $P_2 = 100$ kPa by the following two different processes. Calculate the work done by the gas in each case.

a. The volume of the gas is inversely proportional to the pressure.
b. The process follows the path $PV^\gamma = \text{constant}$, where $\gamma = 1.4$.

Problem 4.4: Determine the horsepower required to overcome the wind drag on a modern car traveling 90 km/h if the drag coefficient $C_D = 0.2$. The drag force F_D is given by $F_D = \dfrac{1}{2}\rho V^2 A C_D$, where A cross section area (projected area) of the car and V is the velocity. The density ρ of air is 1.23 kg/m³. Use $A = 2.3$ m².

Problem 4.5: A scoter of mass 90 kg is moving at a speed of 60 km/h. Estimate the kinetic energy of the scooter.

Problem 4.6: An aircraft with a mass of 25,000 kg flies at a speed of 1000 km/h at an altitude of 10 km. Calculate the kinetic and potential energy of the aircraft.

Problem 4.7: A hollow sphere of mass m and volume V is immersed in a liquid of density ρ. If the sphere is raised through a distance of h in the liquid by an external agent, determine the work done by the external agent. Is there any energy transfer as work between the sphere and the liquid? What happens to the energy of the fluid? Use the Fig. 4.26 to solve this problem.

Fig. 4.26 Schematic for
Problem 4.7

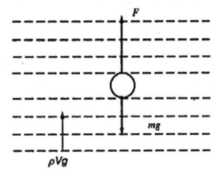

Problem 4.8: A gas is compressed reversibly from the initial state of P_1, V_1 to final state of P_2, V_2. During the compression process the pressure and volume are related by the equation $PV^\gamma = C$ (constant). Calculate the work done by the gas.

Problem 4.9: A balloon, which is initially collapsed and flat, is slowly filled with helium from a cylinder, forming the balloon into a sphere of 5 m in diameter. The ambient pressure is 100 kPa. During the filling process the temperature of helium inside the cylinder remains constant at 300 K. Determine the work done by the cylinder balloon system.

Problem 4.10: Three thermodynamic quantities A, B, and C are defined as $dA = Pd\upsilon$, $dB = \upsilon dP$ and $dC = Pd\upsilon - \upsilon dP$, where $P\upsilon = RT$. Which of the quantities can be used as properties?

Problem 4.11: A particular gas obeys the Van der Waals equation of state. Calculate the work done per kmol of the gas if the gas is compressed reversibly at constant temperature from the initial volume υ_1 to the final υ_2. The Van der Waals equation is give by $P = \dfrac{RT}{\upsilon - b} - \dfrac{a}{\upsilon^2}$.

Problem 4.12: Consider the system shown in Fig. 4.27. Initially the gas is at 500 kPa and occupies a volume of 0.2 m³. The spring exerts a force, which is proportional to the displacement from its equilibrium position. The ambient is 100 kPa. The gas is heated until the volume is doubled at which point the pressure of the gas is 1 MPa. Calculate the work done by the gas.

Problem 4.13: A gas is compressed reversibly in a piston-cylinder assembly, from the initial state of 2 bar and 0.2 m³ to a final state of 10 bar and 0.04 m³. The pressure-volume relationship during the compression process is $P = a + bV$. Use Fig. 4.28 and show the path followed by the gas on a pressure versus volume diagram and calculate the work done on the gas.

Problem 4.14: Consider the system shown in Fig. 4.29. Initially the gas is at 200 kPa and occupies a volume of 0.1 m³. The spring exerts a force which is proportional to the displacement from its equilibrium position. The atmospheric pressure

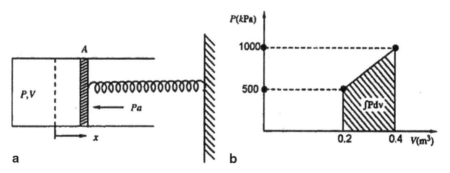

Fig. 4.27 Schematic for Problem 4.12

Fig. 4.28 Schematic for
Problem 4.13

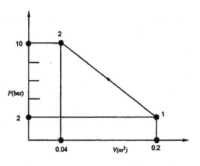

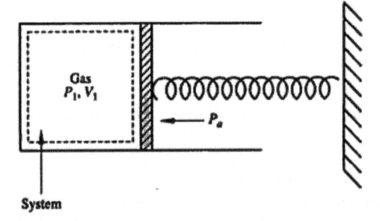

Fig. 4.29 Schematic for Problem 5.14

of 100 kPa acts on the other side of the position. The gas is heated until the volume is doubled and the final pressure is 500 kPa. Calculate the work done by the gas.

Problem 4.15: A spherical balloon of 1 m diameter contains a gas at 150 kPa. The gas inside the balloon is heated until the pressure reaches 450 kPa. During the pressure of heating, the pressure of the gas inside the balloon is propersional to the cube of the diameter of the balloon. Determine the work done by the gas inside the balloon.

References

1. Potter MC, Somerton CW (2006) Thermodynamics for engineers Schuam's outlines series, 2nd edn. McGraw-Hill, Boston
2. Zemansky MW Dittman RH (1997) Heat and thermodynamic, 7th edn. McGraw-Hill, New York

Chapter 5
First Law of Thermodynamics

The first law of thermodynamics states that the total energy of a system remains constant, even if it is converted from one form to another.

5.1 Introduction

The first law of thermodynamics is generally thought to be the least demanding to grasp, for it is an extension of the law of conservation of energy, meaning that energy can be neither created nor destroyed. However much energy there was at the start of the universe, there will be that amount at the end. However, thermodynamics is a subtle subject, and the first law is much more interesting than this remark might suggest. Moreover, like the zeroth law, which provided an impetus for the introduction of the property 'temperature' and its clarification, the first law motivates the introduction and helps to clarify the meaning of the elusive concept of 'energy'.

Energy balance, based on the First Law of Thermodynamics, is developed to better understand any process, to facilitate design and control, to point at the needs for process improvement, and to enable eventual optimization. The degree of perfection in the energy utilization of the process, or its particular parts, allows comparison with the degree of perfection, and the related process parameters, to those in other similar processes. Comparison with the currently achievable values in the most efficient systems is especially important. Priorities for the required optimization attempts for a system, or its components, can be established. Such priorities can be carried out based either on the excessive energy consumption, or on the particularly low degree of perfection.

However, the energy approach has some deficiencies. Generally, energy exchange is not sensitive to the assumed direction of the process, e.g., energy analysis allows heat to be transferred spontaneously in the direction of the increasing temperature. Energy also does not distinguish its quality, e.g., 1 Watt of heat equals 1 Watt of work or electricity.

The first law of thermodynamics states that the total energy of a system remains constant, even if it is converted from one form to another. For example, kinetic

© Springer International Publishing Switzerland 2015
B. Zohuri, P. McDaniel, *Thermodynamics In Nuclear Power Plant Systems,*
DOI 10.1007/978-3-319-13419-2_5

energy—the energy that an object possesses when it moves—is converted to heat energy when a driver presses the brakes on the car to slow it down. The first law of thermodynamics relates the various forms of kinetic and potential energy in a system to the work, which a system can perform, and to the transfer of heat. This law is sometimes taken as the definition of **internal energy**, and also introduces an additional state variable, **enthalpy.** The first law of thermodynamics allows for many possible states of a system to exist. However, experience indicates that only certain states occur. This eventually leads to the second law of thermodynamics and the definition of another state variable called **entropy**.

Work is motion against an opposing force. Raising a weight against the opposing force of gravity requires work. The magnitude of the work depends on the mass of the object, the strength of the gravitational pull on it, and the height through which it is raised. Work is the primary foundation of thermodynamics and in particular of the first law. Any system has the capacity to do work. For instance, a compressed or extended spring can do work such as that can be used to bring about the raising of a weight. An electric battery has the capacity to do work, for it can be connected to an electric motor, which in turn can be used to raise a weight. It is not an entirely obvious point, but when an electric current passes through a heater, it is doing work on the heater, for the same current could be used to raise a weight by passing it through an electric motor rather than the heater. Then why a heater is called a '*heater*' and not a '*worker*' is obvious from the concept of heat that was defined in Chap. 4.

The first law of thermodynamics is commonly called the conservation of energy. In elementary physics courses, the study of the conservation of energy emphasizes changes in mechanical kinetic and potential energy and their relationship to work. A more general form of conservation of energy includes the effects of heat transfer and internal energy changes. This more general form is usually called the *first law of thermodynamics*. Other forms of energy may also be included, such as electrostatic, magnetic, strain, and surface energy.

To understand and have better concept of work from thermodynamics point of view, a term is needed to denote the capacity of a system to do work. That term is *energy*. A fully stretched spring has a greater capacity to do work than the same spring only slightly stretched. A liter of hot water has a greater energy than a liter of cold water. Therefore concept of energy is *just a measure of the capacity of a system to do work.*

The First Law of Thermodynamics states that e*nergy can neither be created nor destroyed only altered in form*. For any system, energy transfer is associated with mass crossing the control boundary, external work, or heat transfer across the boundary. These produce a change of stored energy within the control volume. The mass flow of a fluid is associated with the kinetic, potential, internal, and "flow" energies that affect the overall energy balance of the system. The exchanges of external work and heat complete the energy balance. That is why *The First law of Thermodynamics is referred to as the Conservation of Energy principle*, meaning that *energy can neither be created nor destroyed*, but rather transformed into various forms as the fluid within the control volume changes. A system is a region in space (control volume) through which a working fluid may or may not pass. The various energies associated with the fluid are then observed as they cross the boundaries of the system and the balance is made. As discussed in Chap. 1, a system may be one of three types:

1. **Isolated System**
2. **Closed System**
3. **Open System**

The open system, the most general of the three, allows mass, heat, and external work to cross the control boundary. The balance is expressed in work, as all energies into the system are equal to all energies leaving the system plus the change in storage of energies within the system.

The system might be a mechanical device, a biological organism, or a specified quantity of material such as the refrigerant in an air conditioner, or the steam expanding in a turbine. A thermodynamic system is a system that can interact (and exchange energy) with its surroundings, or environment, in at least two ways, one of which is heat transfer. A familiar example is a quantity of popcorn kernels in a pot with a lid. When the pot is placed on a stove, energy is added to the popcorn by conduction of heat; as the popcorn pops and expands, it does work as it exerts an upward force on the lid and displaces it (Fig. 5.1).

The state of the popcorn changes in this process, since the volume, temperature, and pressure of the popcorn all change as it pops. A process such as this one, in which there are changes in the state of a thermodynamic system, is called a thermodynamic process. With thermodynamic systems, it is essential to define clearly at the start exactly what is and is not included in the system. Only then can the energy transfers be unambiguously described. For instance, in the popcorn example, the system was defined to include the popcorn, but not the pot, lid, or stove.

Fig. 5.1 The popcorn in the pot is a thermodynamic system. In the thermodynamic process shown here, heat is added to the system, and the system does work on its surroundings to lift the lid of the pot

5.2   System and Surroundings

The First Law of Thermodynamics tells us that energy is neither created nor destroyed, thus the energy of the universe is a constant. However, energy can certainly be transferred from one part of the universe to another. To work out thermodynamic problems we will need to isolate a certain portion of the universe (the system) from the remainder of the universe (the surroundings).

For example, consider the pendulum example given in the last section. In real life, there is friction and the pendulum will gradually slow down until it comes to rest. We can define the pendulum as the system and everything else as the surroundings. Due to friction there is a small but steady transfer of heat energy from the system (pendulum) to the surroundings (the air and the bearing upon which the pendulum swings). Due to the first law of thermodynamics the energy of the system must decrease to compensate for the energy lost as heat until the pendulum comes to rest. [Remember though the total energy of the universe remains constant as required by the First Law.]

When it comes time to work homework, quiz and exam problems not to mention to design a power plant the first Law of Thermodynamics will be much more useful if we can express it as an equation.

$\Delta E = Q + W$ (First Law of Thermodynamicsa chemical reaction the energy)

- ΔE = The change internal energy of the system.
- Q = The heat Transferred into/out of the system.
- W = The work done by/on the system.

This reformulation of the First Law tells us that once we define a system (remember we can define the system in any way that is convenient) the energy of the system will remain constant unless there heat added or taken away from the system, or some work takes place.

5.2.1   Internal Energy

We have already discussed work and heat extensively, but a few comments are in order regarding internal energy. The internal energy encompasses many different things, including:

- The kinetic energy associated with the motions of the atoms,
- The potential energy stored in the chemical bonds of the molecules,
- The gravitational energy of the system.

It is nearly impossible to sum all of these contributions up to determine the absolute energy of the system. That is why we only worry about ΔE, the change in the energy of the system. This saves all of us a lot of work, for example:

- If the temperature doesn't change we can ignore the kinetic energy of the atoms,
- If no bonds are broken or destroyed we can ignore the chemical energy of the system,

- If the height of the system does not change then we can ignore gravitational potential energy of the system

Our convention for ΔE is to subtract the initial energy of the system from the final energy of the system.

$$\Delta E = E(final) - E(initial) = Q + W$$

In a chemical reaction the energy of the reactants is E (initial) and the heat of the products is E (final). Now if we ask the question whether the nuclear power plant follow the first law of thermodynamics, we can positive say yes to the question and the reason behind is that because a Nuclear Power plant does not create or destroy energy, and they convert energy, and the amount of energy in the system and the surroundings remains constant.

5.2.2 Heat Engines

The work-producing device that best fits into the definition of a heat engine is the *steam power plant,* which is an **external-combustion engine**. That is, the combustion process takes place outside the engine, and the thermal energy released during this process is transferred to the steam as heat.

Heat engines, technically speaking, are continuously operating thermodynamic systems at the boundary of where there are heat and work interactions. Simply, a heat engine converts heat to work energy or vice versa (Fig. 5.2).

An example of a common heat engine is a power plant. It consists of four main elements, a boiler, turbine, condenser, and feed pump, and the main circulating heat transfer entity is water. If we consider the power plant to be a closed system with its boundary enclosing the operating components, we can apply the First Law of Thermodynamics. The boiler burns a fuel source, causing a transfer of $\mathbf{Q}_{combustion}$ heat to water inside, vaporizing it. The high-pressure vapor enters the turbine, resulting in a work output of $\mathbf{W}_{turbine}$, and then leaves still as steam but at lower pressure and temperature. The vapor moves through the condenser where it condenses back into water, losing $\mathbf{Q}_{condensation}$ heat to the surroundings. The water is pumped back into the boiler, requiring $\mathbf{W}_{pump}$ work.

Since $\Delta E = Q + W$, and assuming a steady state of operation ($\Delta E = 0$),

$$(Q_{combustion} - Q_{condensation}) + (W_{pump} - W_{turbine}) = 0$$

or

$$Q_{combustion} - Q_{condensation} = W_{turbine} - W_{pump}$$

Generally, $\mathbf{W}_{pump}$ is significantly less than the $\mathbf{W}_{turbine}$ attained. However, $\mathbf{Q}_{condensation}$ may be even more than two-thirds the magnitude of $\mathbf{Q}_{combustion}$, meaning that the total useful work obtained from the combustion of fuel is less than one third of

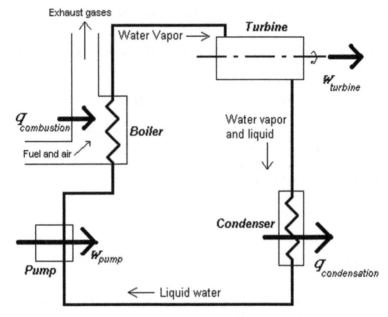

Fig. 5.2 Diagram of steam power plant

the total work theoretically possible from a complete conversion of $Q_{combustion}$. The second law of thermodynamics embodies the fact that no engine can be constructed that is 100 % efficient.

5.3   Signs for Heat and Work in Thermodynamics

As noted in Chap. 4 energy transfers in any thermodynamic process are measured in terms of the quantity of heat Q added to the system and the work W done by the system. Both Q and W may be positive, negative, or zero (Fig. 5.3).

A positive value of Q represents heat flow *into* the system, with a corresponding input of energy to the system. A negative value of Q represents heat flow *out* of the system. A positive value of W represents work done *by* the system against its surroundings, such as work done by an expanding gas, and hence corresponds to energy *leaving* the system. Negative W, such as work done during compression of a gas by its surroundings, represents energy *entering* the system.

5.4   Work Done During Volume Changes

A gas in a cylinder with a movable piston is a simple example of a thermodynamic system. Internal-combustion engines, steam engines, and compressors in refrigerators and air conditioners all use some version of such a system.

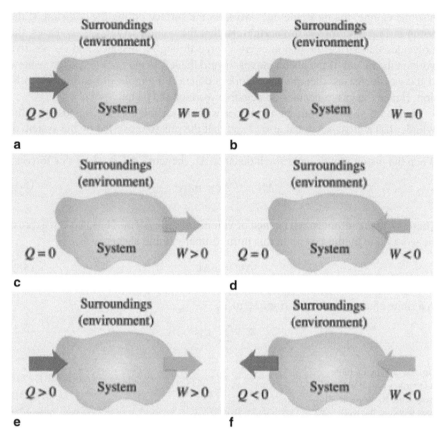

Fig. 5.3 A thermodynamic system may exchange energy with its surroundings (environment) by means of heat and work. **a** When heat is added to the system, Q is positive. **b** When heat is transferred out of the system, Q is negative. **c** When work is done by the system, W is positive. **d** When work is done on the system, W is negative. Energy transfer by both heat and work can occur simultaneously; in **e** heat is added to the system and work is done by the system, and in **f** heat is transferred out of the system and work is done on the system ([1])

Consider a microscopic viewpoint, based on the kinetic and potential energies of individual molecules in a material, to develop intuition about thermodynamic quantities. (It is important to understand that the central principles of thermodynamics can be treated in a completely macroscopic way, without reference to microscopic models. Indeed, part of the great power and generality of thermodynamics is that it does not depend on details of the structure of matter.) First, consider the work done by the system during a volume change. When a gas expands, it pushes outward on its boundary surface as it moves outward. Hence an expanding gas always does positive work. The same thing is true of any solid or fluid material that expands under pressure, such as the popcorn in Fig. 4.1.

The work done by a gas in a volume change can be understood by considering the molecules that make up the gas. When one such molecule collides with a stationary surface, it exerts a momentary force on the wall but does no work because the wall does not move. However, if the surface is moving, such as a piston in a gasoline engine, the molecule does work on the surface during the collision. If the piston in Fig. 5.4a moves to the right, so that the volume of the gas increases, the molecules that strike the piston exert a force through a distance and do positive work on the piston. If the piston moves toward the left as in Fig. 5.4b, so the volume of the gas decreases, then positive work is done on the molecule during the collision. Hence the gas molecules do negative work on the piston.

Figure 5.5 shows a fluid in a cylinder with a movable piston. Suppose that the cylinder has a cross-sectional area A and that the pressure exerted by the system at the piston face is p. The total force F exerted by the system on the piston is $F = pA$. When the piston moves out a small distance Δx, the work ΔW done by this force is

$$\Delta W = F\Delta x = pA\Delta x \qquad (5.1)$$

where ΔV is the infinitesimal change of volume of the system. Thus, we can express the work done by the system in this infinitesimal volume change as

$$\Delta W = p\Delta V \qquad (5.2)$$

In a finite change of volume from V_1 to V_2

$$W = \sum p\Delta V \qquad (5.3)$$

Fig. 5.4 a When a molecule strikes a wall moving away from it, the molecule does work on the wall—the molecule's speed and kinetic energy decrease. The gas does positive work on the piston. **b** When a molecule strikes a wall moving toward it, the wall does work on the molecule—the molecule's speed and kinetic energy increase. The gas does negative work on the piston ([1])

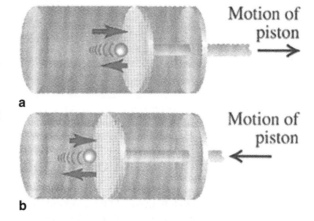

Fig. 5.5 The infinitesimal work done by the system during the small expansion Δx is $\Delta W = pA\Delta x$

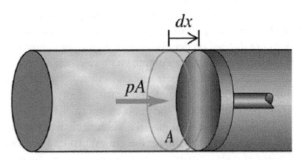

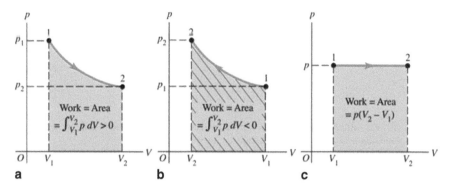

Fig. 5.6 The work done equals the area under the curve on a pV-diagram. **a** In this change from state 1 to state 2 the volume increases and the work and area are positive. **b** In this change from state 1 to state 2 the volume decreases; the area is taken to be negative to agrees with the sign of W. The two states are reversed from part (**a**). **c** The area of the rectangle gives the work done for a constant-pressure process. In this case, volume increases and $W > 0$

This equation says the total work is equal to the sum of all the infinitesimally small volume changes time the pressure at that volume.

In general the pressure of the system may vary during the volume change. To evaluate Eq. 5.3, we have to know how the pressure varies as a function of volume. We can represent this relationship as a graph of p as a function of V (called a pV-diagram). Figure 5.5a shows a simple example. In the figure, Eq. 5.3 is represented graphically as the area under the curve of p versus V between the limits V_1 and V_2.

According to the rule we stated for the sign of work, work is *positive* when a system *expands*. In an expansion from state 1 to state 2 in Fig. 5.6a the area under the curve and the work are positive. A *compression* from 1 to 2 in Fig. 5.6b gives a *negative* area; when a system is compressed, its volume decreases and it does *negative* work on its surroundings.

If the pressure p remains constant while the volume changes from V_1 to V_2 (Fig. 5.5c), the work done by the system is

$$W = p(V_2 - V_1) \tag{5.4}$$

In any process in which the volume is constant, the system does no work because there is no displacement.

5.5 Paths Between Thermodynamic States

We have seen that if a thermodynamic process involves a change in volume, the system undergoing the process does work (either *positive* or *negative*) on its surroundings. Heat also flows into or out of the system during the process, if there is a temperature difference between the system and its surroundings. Consider

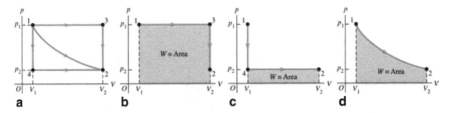

Fig. 5.7 a Three different paths between state 1 and state 2. **b–d**. The work done by the system during a transition between two states depends on the path chosen [1]

now how the work done by, and the heat added to, the system during a thermodynamic process depend on the details of how the process takes place.

When a thermodynamic system changes from an initial state to a final state, it passes through a series of intermediate states. We call this series of states a ***path***. There are always infinitely many different possibilities for these intermediate states. When they are all equilibrium states, (a quasi-static process) the path can be plotted on a pV -diagram (Fig. 5.7a). Point 1 represents an initial state with pressure p_1 and volume V_1. Point 2 represents a final state with pressure p_2 and volume V_2. To pass from state 1 to state 2, we could keep the pressure constant at p_1 while the system expands to volume V_2 (point 3 in Fig. 5.7b.), then reduce the pressure to p_1 (probably by decreasing the temperature) while keeping the volume constant at V_2 (to point 2 on the diagram). The work done by the system during this process is the area under the line $1 \rightarrow 3$; no work is done during the constant-volume process $3 \rightarrow 2$. Or the system might traverse the path $1 \rightarrow 4 \rightarrow 2$ (Fig. 5.7c); in that case the work is the area under the line $4 \rightarrow 2$, since no work is done during the constant-volume process $1 \rightarrow 4$. The smooth curve from 1 to 2 is another possibility (Fig. 5.7d), and the work for this path is different from that for either of the other paths.

The work done by the system depends not only on the initial and final states, but also on the intermediate states, that is, on the path. Furthermore, the system can move through a series of states forming a closed loop, such as $1 \rightarrow 3 \rightarrow 2 \rightarrow 4 \rightarrow 1$. In this case, the final state is the same as the initial state, but the total work done by the system is not zero. (In fact, it is represented on the graph by the area enclosed by the loop). It follows that it does not make sense to talk about the amount of work contained in a system. In a particular state, a system may have definite values of the state coordinates p, V, and T, but it wouldn't make sense to say that it has a definite value of W.

Example 5.1 A gas is taken through the cycle illustrated below. During one cycle, how much work is done by an engine operating on this cycle? (Fig. 5.8)

Solution Start in the lower left hand corner of the rectangle and work clockwise. Call the lower left corner point 1, the upper left corner point 2, the upper right corner point 3 and the lower right corner point 4.

$$W = W_{1\rightarrow2} + W_{2\rightarrow3} + W_{3\rightarrow4} + W_{4\rightarrow1}$$

Fig. 5.8 Sketch of a full
cycle in Example 5.1

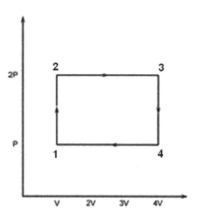

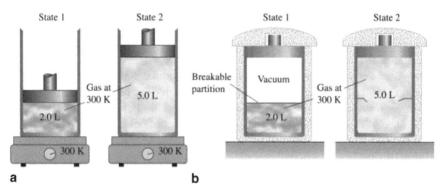

Fig. 5.9 **a** Slow, controlled isothermal expansion of a gas from an initial state 1 to a final state 2 with the same temperature but lower pressure. **b** Rapid, uncontrolled expansion of the same has starting at the same state 1 and ending at the same state 2

$$W = (p\Delta V)_{1\to 2} + (p\Delta V)_{2\to 3} + (p\Delta V)_{3\to 4} + (p\Delta V)_{4\to 1}$$

From 1 to 2, $\Delta V = 0$, so $(p\Delta V)_{1\to 2} = 0$. From 2 to 3, $(p\Delta V)_{2\to 3} = 2P \times 3V = 6PV$. From 3 to 4, $(p\Delta V)_{3\to 4} = 0$, and from 4 to 1, $(p\Delta V)_{4\to 1} = -P \times 3V = -3PV$. Adding these up gives

$$W = 0 + 6PV + 0 - 3PV = 3PV$$

Therefore, the work done by an engine operating on this cycle is a positive $3PV$ on each cycle.

Like work, the heat added to a thermodynamic system when it undergoes a change of state depends on the path from the initial state to the final state. Here is an example. Consider changing the volume of a certain quantity of an ideal gas from 2.0 to 5.0 L while keeping the temperature constant at $T = 300$ K. Figure 5.9 shows two different ways to do this. In Fig. 5.9a the gas is contained in a cylinder with a piston, with an initial volume of 2.0 L. The gas expands slowly, supplying heat from the electric heater to keep the temperature at 300 K. After expanding in this slow, controlled, isothermal manner, the gas reaches its final volume of 5.0 L; it absorbs a definite amount of heat in the process.

Figure 5.9b shows a different process leading to the same final state. The container is surrounded by insulating walls and is divided by a thin, breakable partition into two compartments. The lower part has volume 2.0 L, and the upper part has volume 3.0 L. In the lower compartment, place the same amount of the same gas as in Fig. 5.9a, again at $T = 300$ K. The initial state is the same as before. Now break the partition. The gas undergoes a rapid, uncontrolled expansion, with no heat passing through the insulating walls. The final volume is 5.0 L, the same as in Fig. 5.9a. The gas does not do any work during this expansion because it does not push against anything that moves. The uncontrolled expansion of a gas into vacuum is called a **free expansion.**

Experiments have shown that when an ideal gas undergoes a free expansion, there is no temperature change. Therefore the final state of the gas is the same as in Fig. 5.9a. The intermediate states (pressures and volumes) during the transition from state 1 to state 2 are entirely different in the two cases; Fig. 5.9a and b represent *two different paths* connecting the *same states* 1 and 2. For the path in Fig. 5.9b, no heat is transferred into the system and the system does not do any work. Like work, *heat depends not only on the initial and final states but also on the path.*

Because of this path dependence, it would not make sense to say that a system "contains" a certain quantity of heat. To see this, suppose that an arbitrary value is assigned to "the heat in a body" in some reference state. Then presumably the "heat in the body" in some other state would equal the "heat in the body" in the reference state and the heat added when the body goes to the second state. However, that is ambiguous. The heat added depends on the *path* taken from the reference state to the second state. The obvious conclusion is that there is *no* consistent way to define "heat in a body". It is not a useful concept.

While it does not make sense to talk about "work in a body" or "heat in a body," it *does* make sense to speak of the amount of *internal energy* in a body.

5.6   Path Independence

Consider a system enclosed in adiabatic (thermally non-conducting) walls. In practice, 'adiabatic' means a thermally insulated container, like a well-insulated vacuum flask. The temperature of the contents of the flask can be monitored using a thermometer, a concept introduced by the *zeroth law*. Now do some experiments.

First, churn the contents of the flask (that is, the system) with paddles driven by a falling weight, and note the change in temperature this churning brings about. J. P. Joule (1818–1889), one of the fathers of thermodynamics, performed exactly this type of experiment in the years following 1843. The work accomplished can be calculated based on heaviness of the weight and the distance through which it

Fig. 5.10 The observation
that different ways of doing
work on a system and thereby
changing its state between
fixed endpoints required
the same amount of work is
analogous to different paths
on a mountain resulting in
the same change of altitude
leads to the recognition of the
existence of a property called
the internal energy

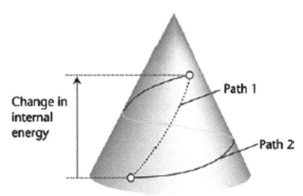

fell. Then remove the insulation and let the system return to its original state. After
replacing the insulation, put a heater into the system and pass an electric current
through it for a time that result in the same work being done by the heater as was
done by the falling weight. The conclusion arrived at in this pair of experiments and
in a multitude of others of a similar kind is that *the same amount of work, however
it is performed, brings about the same change of state of the system.*

This is analogous to climbing a mountain by a variety of different paths, each
path corresponding to a different method of doing work (see Fig. 5.10). Provided
we start at the same base camp and arrive at the same destination, we shall have
climbed through the same height regardless of the path we took between them. That
is, we can attach a number (the 'altitude') to every point on the mountain, and calculate the height we have climbed, regardless of the path, by taking the difference
of the initial and final altitudes for our climb. Exactly the same process, applies to
our system. The fact that the change of state is path-independent means that we can
associate a number, which we shall call the *internal energy* (symbol U) with each
state of the system. Then we can calculate the work needed to travel between any
two states by taking the difference of the initial and final values of the internal energy, and using Fig. 5.9, then we can write;

$$\text{Work required } (W) = U_f \text{ (final)} - U_i \text{ (initial)} \tag{5.5}$$

The observation of the path-independence of the work required to go between two
specified states in an *adiabatic* system (remember, at this stage the system is *adiabatic*) has motivated the recognition that there is a property of the system that is a
measure of its capacity to do work. In thermodynamics, a property that depends
only on the current state of the system and is independent of how that state was
prepared (like altitude in geography) is called a *state function*. The state function
for the energy stored in a closed system is called internal energy. It has a definite
parallel with the state function called temperature that was defined by the Zeroth
Law of Thermodynamics.

5.7   Heat and Work

Internal energy can change by an agency other than by doing work. One way of regarding this additional change in internal energy is to interpret it as arising from the transfer of energy from the system into the surroundings due to the difference in temperature caused by the work that has been done. This transfer of energy because of a temperature difference is called *heat*.

A simple model gives *"work as the transfer of energy that makes use of the uniform motion of atoms in the surroundings"* as depicted in Fig. 5.11. Moreover, *"heat is the transfer of energy that makes use of the random motion of atoms in the surroundings"*.

In more pictorial terms, using Fig. 5.11 and noting that temperature is a property that tells us the relative numbers of atoms in the allowed energy states, with the higher energy states progressively more populated as the temperature is increases. A block of iron at high temperature consists of atoms that are oscillating vigorously around their average positions. At low temperatures, the atoms continue to oscillate, but with less vigor. If a hot block of iron is put in contact with a cooler block, the vigorously oscillating atoms at the edge of the hot block jostle the less vigorously oscillating atoms at the edge of the cool block into more vigorous motion, and they pass on their energy by jostling their neighbors. There is no net motion of either block, but energy is transferred from the hotter to the cooler block by this random jostling where the two blocks are in contact. That is why the above statement that was made, and is repeated here again, is a valid statement. That is, *"heat is the transfer of energy that makes use of the random motion of atoms in the surroundings"* [2].

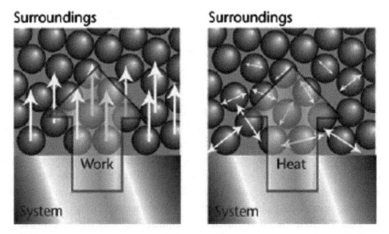

Fig. 5.11 The molecular distinction between the transfer of energy as work (*left*) and heat (*right*). Doing work results in the uniform motion of atoms in the surroundings; heating stimulates their disorderly motion

5.8   Heat as Energy in Transition

The amount of energy that is transferred as heat into or out of the system can be measured very simply: we measure the work required to bring about a given change in an adiabatic system, and then the work required to bring about the same change of state in a non-adiabatic system (the one with thermal insulation removed), and take the difference of the two values.

That difference is the energy transferred as heat. A point to note is that this type of the measurement of the rather elusive concept of 'heat' has been put on a purely mechanical foundation as the difference in the heights through which a weight falls to bring about a given change of state under two different conditions (See Fig. 5.12 below).

Once the energy is inside the system, either by making use of the uniform motion of atoms in the surroundings (a falling weight) or of randomly oscillating atoms (a hotter object, such as a flame), there is no memory of how it was transferred. Once inside, the energy is stored as the kinetic energy (the energy due to motion) and the potential energy (the energy due to position) of the constituent atoms and that energy can be withdrawn either as heat or as work. The distinction between work and heat is made in the surroundings: the system has no memory of the mode of transfer nor is it concerned about how its store of energy will be used.

Consider a closed system and use it to do some work or allow a release of energy as heat. Its internal energy falls. Then leave the system isolated from its surroundings for an indefinite amount of time. Return to it and measure its capacity to do work. Invariably it's capacity to do work—its internal energy—is the same as when it was first isolated. In other words,

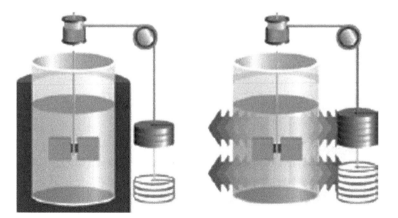

Fig. 5.12 When a system is adiabatic (*left*), a given change of state is brought about by doing a certain amount of work. When the same system undergoes the same change of state in a non-adiabatic container (*right*), more work has to be done. The difference is equal to the energy lost as heat

The internal energy of an isolated system is constant.

That is the first law of thermodynamics, or at least one statement of it, for the law comes in many equivalent forms.

5.9   The First Law of Thermodynamics Applied to a Cycle

Having discussed the concepts of work and heat completely in Chap. 3 and briefly in Sect. 5.7 of this chapter again, we are now ready to present the first law of thermodynamics. Recall that a law is not derived or proved from basic principles but is simply a statement that we write based on our observations of many experiments. If an experiment shows a law to be violated, either the law must be revised or additional conditions must be placed on the applicability of the law. Historically, *the first law of thermodynamics* was stated for a cycle and the net heat transfer is equal to the net work done for a system undergoing a cycle. This is expressed in equation form by Eq. 5.6 as follow;

$$\sum W = \sum Q \qquad (5.6)$$

or

$$\oint \delta W = \oint \delta Q \qquad (5.7)$$

where the symbol $\oint$ implies an integration around a complete cycle. Therefore, the first law of thermodynamics can be stated as: *whenever a system undergoes a cyclic change, however complex the cycle may be, the algebraic sum of the work transfer is equal to the algebraic sum of energy transfer as heat*. Another way of interpreting the first law of thermodynamics is the conclusion that Joule made from his experiment which is depicted in Fig. 5.9a, b and that is, the net work done on the system is always proportional to the net energy removed from the system as heat, irrespective of the type of work interaction, the rate at which work was done on the system, and the method employed for transferring the energy in the form of work into thermal energy and it can be shown as the following equation

$$\oint dQ_{(from\,system)} = \oint dW_{(on\,system)} \qquad (5.8)$$

The first law can be illustrated by considering the following experiment. Let a weight be attached to a pulley/paddle-wheel setup, such as that shown in Fig. 5.9a. Let the weight fall certain distance thereby doing work on the system, contained in the insulated (adiabatic) tank shown, equal to the multiplied by the distance dropped. The temperature of the system (the fluid in the tank) will rise an amount of ΔT. Now, the system is returned to its initial state (the completion of the cycle) by transferring heat to the surrounding, as implied by the Q in Fig. 5.9b. This reduces the temperature of the system to its initial temperature. The first law states that this heat transfer will be exactly equal to the work, which was done by the falling weight [3].

5.10   Sign Convention

To avoid confusion as to whether the work is being done on the system by the surrounding, or by the system on the surroundings and whether energy is transferred as heat from the system, or to the system, the following sign convention is defined with reference to Fig. 5.13. The work done by a system on its surroundings is treated as a positive quantity. Similarly, the energy transfer as heat to a system from its surroundings is also treated as a positive quantity. With this convention, Eq. 5.8 becomes,

$$-\oint dQ - \oint (-dW) = \oint dQ - \oint (dW) = 0 \tag{5.9}$$

or

$$\oint (dQ - dW) = 0 \tag{5.10}$$

It should be noted that there is no restriction on the type of process the system has undergone. Therefore, the first law of thermodynamics is applicable to reversible as well as irreversible processes.

The first law of thermodynamics, as stated above, has a number of important consequences.

5.11   Heat is a Path Function

Suppose a system is taken from an initial state 1 to state 2 by the path 1a2 and it is restored to the initial state through the 2b1 according to Fig. 5.14. Then, the system has undergone a cyclic change. If the system is restored to the initial state by following path 1a2c1 it has experienced a different cycle. Apply the first law of thermodynamics to the two cycles 1a2b1 and 1a2c1.

$$\int_{1a2} dQ + \int_{2b1} dQ - \int_{1a2} dW - \int_{2b1} dW = 0 \tag{5.11}$$

$$\int_{1a2} dQ + \int_{2c1} dQ - \int_{1a2} dW - \int_{2c1} dW = 0 \tag{5.12}$$

Fig. 5.13 Sign convention
for heat and work interactions

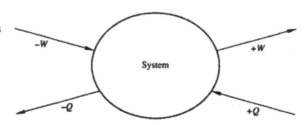

Fig. 5.14 Reversible cycles
on $P-V$ diagram

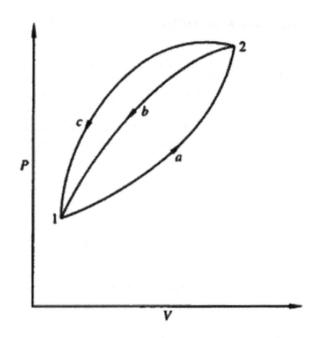

Subtracting Eq. 5.12 from 5.11, gives,

$$\int_{2b1} dQ - \int_{2c1} dQ - \left(\int_{2b1} dW - \int_{2c1} dW\right) = 0 \qquad (5.13)$$

Since work depends on the path, we have

$$\int_{2b1} dW - \int_{2c1} dW \neq 0 \qquad (5.14)$$

Therefore

$$\int_{2b1} dQ \neq \int_{2c1} dQ \qquad (5.15)$$

That is, energy transfer as heat is not a point function, neither is it a property of the system. Therefore, heat interaction is a path function and its differential is not exact.

Example 5.2 A spring is stretched a distance of 0.8 m and attached to a paddle using the figure below. The paddle wheel then rotates until the spring is unstretched. Calculate the heat transfer necessary to return the system to its initial state (Fig. 5.15).

Solution The work done by spring on the system is given by;

$$W_{1-2} = \int_0^{0.8} F dx = \int_0^{0.8} 100x dx = (100)\left[\frac{(0.8)^2}{2}\right] = 32N$$

Fig. 5.15 Example 5.2

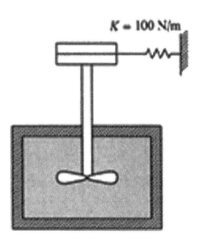

$K = 100 \text{ N/m}$

Since the heat transfer returns the system to its initial state, a cycle results. The first law then states that $Q_{1-2} = W_{1-2} = 32$ J.

5.12 Energy is a Property of System

Quite often changes in the state of a system when it undergoes a process are important, rather than when it passes through a cycle. Evaluating the energy changes of the system while it interacts with its surrounding for a specific process is necessary. Rearranging Eq. 5.13 gives;

$$\int_{2b1} (dQ - dW) = \int_{2c1} (dQ - dW) \tag{5.16}$$

This shows that while $\int dQ$ and $\int dW$ depend on the path followed by the system, the quantity $\int (dQ - dW)$ is the same for both the processes $2b1$ and $2c1$, connecting states 2 and 1. By considering several other cycles, it can be observed that $\int (dQ - dW)$ is the same for all the processes connecting states 1 and 2. Therefore $\int (dQ - dW)$ does not depend on the path followed by the system, but depends only on the initial and final states of the system. Hence, the quantity $\int (dQ - dW)$ is an exact differential. Therefore it is the differential of a property of the system. This property is the energy of the system and is represented by E. The differential change in the energy of the system is given by,

$$dE = dQ - dW \tag{5.17}$$

The energy of a system is the sum of the macroscopic and microscopic modes of energy, i.e.

$$E = KE + PE + U \tag{5.18}$$

and

$$dE = d(KE) + d(PE) + dU = dQ - dW \qquad (5.19)$$

Therefore, whenever a system undergoes a change of state, energy may cross the boundary as either heat or as work, and the change in the energy of the system is equal to the total energy that crosses the boundary. The energy of the system may change from one form to the other. For example, kinetic energy may change into either potential energy or internal energy. Equation 5.19 simply accounts for the energy when the system interacts with the surroundings. The quantity $[d(KE) + d(PE) + dU]$ represents the net change in the energy of the system while dQ and dW represent the energy transfer across the boundary of the system in the form of heat and work, respectively. The net change in the energy of the system is exactly equal to the net energy transfer across the boundary of the system.

5.13   Energy of an Isolated System is Conserved

A system, which does not exchange energy with the surroundings, in the form of either heat or work, is called an isolated system. During any process in such a system $dQ = 0$ and $dW = 0$. The first law of thermodynamics then reduces to

$$dE = 0 \text{ or } E_2 = E_1 \qquad (5.20)$$

for a reversible or an irreversible process. Therefore, the energy of an isolated system remains constant.

Example 5.3 A radiator of a heating system with a volume of 0.1 m³ contains saturated steam at 0.2 MPa. The inlet and outlet valve of the radiator are closed. Due to the energy transfer as heat to the surroundings, the pressure drops to 0.15 MPa. Determine the amount of steam and water at the final pressure.

Solution Steam contained in the radiator.

The process followed by the system is shown on a $P - \upsilon$ diagram in Fig. 5.14 below (Fig. 5.16).

Specific volume of the saturated steam at 0.2 MPa, $\upsilon_1 = 0.885$ m³/kg.

Volume of the radiator, $V = 0.1$ m³

Total mass of the steam, $m = \dfrac{V}{\upsilon_1} = \dfrac{0.1}{0.8854} = 0.113$ kg.

Since the quantity of steam and the volume of the radiator are constant, the steam undergoes a constant volume process. Therefore, $\upsilon = $ constant; or $\upsilon_1 = \upsilon_2$

At $P_2 = 0.15$ MPa $\upsilon_f = 0.001053$ m³/kg $\upsilon_g = 1.159$ m³/kg $T = 111.37$ °C

Fig. 5.16 Sketch for
Example 5.3 on a
$P-\upsilon$ diagram

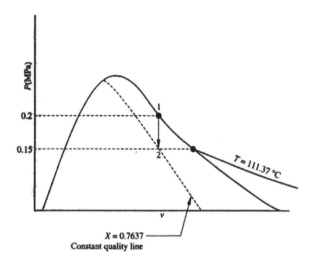

$$X = 0.7637$$
Constant quality line

Since $\upsilon_f < \upsilon < \upsilon_g$, the final state is wet steam of quality X_2.

$$\upsilon_2 = X_2\upsilon_g + (1-X_2)\upsilon_f \quad 0.8854 = X_2(1.159) + (1-X_2)(0.001053)$$

Or $X_2 = 0.7637$

The saturation temperature at $P_2 = 0.15$ MPa, $T_2 = 111.37$ °C. Quality of the steam in the final state, $X_2 = 0.7637$. Mass of vapor$=0.7637(0.1138)=0.86$ kg and mass of liquid$=0.027$ kg.

Example 5.4 An insulated and rigid container of 1 m³ volume contains helium at 20 °C and 100 kPa pressure. A paddle wheel is rotated inside the container raising the temperature of helium to 60 °C. Assume that helium obeys the relation $PV = NRT$ and $du = C_v dT$ with $C_v = 12.4717$ kJ/kmol K. Determine the work done on helium and the final pressure of helium.

Solution

$$N = \frac{P_1 V_1}{RT_1} = \frac{100 \times 10^3 \times 1}{8.314 \times 10^3 \times 293.15} = 0.041 \text{ kmol}$$

$$\Delta u = C_v \Delta T$$

or

$$\Delta U = NC_v\Delta T = 0.041 \times 12.4717(40) = 20.4685 \text{ kJ}$$

$$\Delta U = -W \text{ or } W = -20.4685 \text{ kJ}$$

$$\frac{P_2}{P_1} = \frac{T_2}{T_1} \quad \text{or} \quad P_2 = \frac{P_1 T_2}{T_1} = \frac{100 \times 10^3 \times 313.15}{293.15} = 106.82 \text{ kPa}$$

5.14   Internal Energy and the First Law of Thermodynamics

Internal energy is one of the most important concepts in thermodynamics. What is internal energy? It can be approached in various ways. Start with one based on the ideas of mechanics. Matter consists of atoms and molecules, and these are made up of particles having kinetic and potential energies. Tentatively define the internal energy of a system as the sum of the kinetic energies of all its constituent particles, plus the sum of all the potential energies of interaction among these particles.

Note that internal energy does not include potential energy arising from the interaction between the system and its surroundings. If the system is a glass of water, placing it on a high shelf increases the gravitational potential energy arising from the interaction between the glass and the earth. But this has no effect on the interaction between the molecules of the water, and so the internal energy of the water does not change.

The symbol U represents internal energy. During a change of state of the system the internal energy may change from an initial value U_1 to a final value U_2. Denote the change in internal energy as $\Delta U = U_2 - U_1$.

Heat transfer is energy transfer. Adding a quantity of heat Q to a system, without doing any work during the process, will increase the internal energy by an amount equal to Q. That is, $\Delta U = Q$. When a system does work W by expanding against its surroundings and no heat is added during the process, energy leaves the system and the internal energy decreases. That is, when W is positive, ΔU is negative, and conversely. So $\Delta U = -W$. When *both* heat transfer and work occur, the total change in internal energy is

$$U_2 - U_1 = \Delta U = Q - W \quad \text{(First law of thermodynamics)} \tag{5.21}$$

This can be rearranged to

$$Q = \Delta U + W \tag{5.22}$$

The message of Eq. 5.5.22 is that in general when heat Q is added to a system, some of this added energy remains within the system, changing its internal energy by an amount ΔU. The remainder leaves the system again as the system does work W against its surroundings. Because W and Q may be positive, negative, or zero, U can be positive, negative, or zero for different processes (Fig. 5.17).

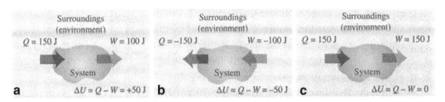

Fig. 5.17 In a thermodynamic process, the internal energy of a system may increase, decrease, or stay the same. **a** If more heat is added to the system than the system does work, ΔU is positive and the internal energy increases. **b** If more heat flows out of the system than work is done on the system, ΔU is negative and the internal energy decreases. **c** If the heat added to the system equals the work done by the system, $\Delta U = 0$ and the internal energy is unchanged

Equations (5.21) or (5.22) is **The First Law of Thermodynamics**. It is a generalization of the principle of conservation of energy to include energy transfer through heat as well as mechanical work. This principle can be extended to ever-broader classes of phenomena by identifying additional forms of energy and energy transfer. In every situation in which it seems that the total energy in all known forms is not conserved, it has been possible to identify a new form of energy such that the total energy, including the new form, *is* conserved. There is energy associated with electric fields, with magnetic fields, and, according to the theory of relativity, even with mass itself.

At the beginning of this discussion, internal energy was tentatively defined in terms of microscopic kinetic and potential energies. This has drawbacks, however. Actually *calculating* internal energy in this way for any real system would be hopelessly complicated. Furthermore, this definition isn't an *operational* one because it doesn't describe how to determine internal energy from physical quantities that can be measured directly.

So look at internal energy in another way. Starting over, define the *change* in internal energy ΔU during any change of a system as the quantity given by Eq. 5.21, $\Delta U = Q - W$. This is an operational definition, because Q and W can be measured. It does not define U itself, only ΔU. This is not a shortcoming, because the value of the internal energy of a system can be defined to have a specified value in some reference state, and then use Eq. 5.21 to define the internal energy in any other state.

This new definition trades one difficulty for another. If U is defined by Eq. 5.21, then when the system goes from state 1 to state 2 by two different paths, is ΔU the same for both paths? Q and W are, in general, not the same for different paths. If ΔU, which equals $Q - W$, is also path-dependent, then ΔU is ambiguous. If so, the concept of internal energy of a system is subject to the same criticism as the erroneous concept of quantity of heat in a system, as discussed earlier in the chapter.

The only way to answer this question is through experiment. For various materials and processes, Q and W have been measured for various changes of state and various paths to learn whether ΔU is, or is not, path-dependent. The results of many such investigations are clear and unambiguous: While Q and W depend on the path, $\Delta U = Q - W$ *is independent of path. The change in internal energy of a system during any thermodynamic process depends only on the initial and final states, not on the path leading from one to the other.*

Experiment, then, is the ultimate justification for believing that a thermodynamic system in a specific state has a unique internal energy that depends only on that state. An equivalent statement is that the internal energy U of a system is a function of the state coordinates p, V, and T (actually, any two of these, since the three variables are related by the equation of state).

To say that the first law of thermodynamics, given by Eqs. 5.21 or 5.22, represents conservation of energy for a thermodynamic process is correct, as far as it goes. However, an important *additional* aspect of the first law is the fact that internal energy depends only on the state of a system. In changes of state, the change in internal energy is path-independent.

All this may seem a little abstract if one wants to think of internal energy as microscopic mechanical energy. But remember classical thermodynamics and the concept of internal energy were developed long before atoms and molecules were proven to exist. There's nothing wrong with the microscopic view. But in the interest of precise *operational* definitions, internal energy, like heat, can and must be defined in a way that is independent of the detailed microscopic structure of the material.

Two special cases of the first law of thermodynamics are worth mentioning. A process that eventually returns a system to its initial state is called a *cyclic* process. For such a process, the final state is the same as the initial state, and so the *total* internal energy change must be zero. Then;

$$U_2 = U_1 \quad \text{and} \quad Q = W \tag{5.23}$$

If a net quantity of work W is done by the system during this process, an equal amount of energy must have flowed into the system as heat Q. But there is no reason why either Q or W individually has to be zero.

Another special case occurs in an isolated system, one that does not do any work on its surroundings and has no heat flow to or from its surroundings. For any process taking place in an isolated system,

$$W = Q = 0 \tag{5.24}$$

and therefore

$$U_2 - U_1 = \Delta U = 0 \tag{5.25}$$

In other words, *the internal energy of an isolated system is constant.*

Example 5.5 You propose to eat a 900 Calorie hot fudge sundae (with whipped cream) and then run up several flights of stairs to work off the energy you have taken in. How high do you have to climb? Assume that your mass is 60 kg.

Solution The system consists of you and the earth. Remember that one food-value calorie is 1 kcal = 1000 cal = 4190 J. The energy intake is;

$$Q = 900 \text{ kcal}(4190 \text{ J/kcal}) = 3.77 \times 10^6 \text{ J}$$

The potential energy output required to climb a height h is;

$$W = mgh = (60\text{kg})(9.8\text{m/sec}^2)h = (588\text{N})h$$

If the final state of the system is the same as the initial state (that is, no fatter, no leaner), these two energy quantities must be equal: $Q = W$. Then

$$h = \frac{Q}{mg} = \frac{3.77 \times 10^6 \text{ J}}{588 \text{ N}} = 6410 \text{ m about } 21{,}000 \text{ ft) Good luck!}$$

Example 5.6 Figure 5.18 below shows a PV-diagram for a cyclic process, one in which the initial and final states are the same. It starts at point a and proceeds counterclockwise in the PV-diagram to point b, then back to a, and the total work is $W = -500$ J. (a) Why is the work negative? (b) Find the change in internal energy and the heat added during this process.

Solution

a. The work done equals the area under the curve, with the area taken as positive for increasing volume and negative for decreasing volume. The area under the lower curve from a to b is positive, but it is smaller than the absolute value of the negative area under the upper curve from b back to a. Therefore, the net area (the area enclosed by the path, shown with red stripes) and the work are negative. In other words, 500 more joules of work are done on the system than by the system.

b. For this and any other cyclic process (in which the beginning and end points are the same), $\Delta U = 0$, so $W = Q = -500$ J. That is, 500 J of heat must come out of the system.

Fig. 5.18 The net work done by the system in the process *aba* is -500 J. Would it have been if the process had proceeded clockwise in this *PV*-diagram?

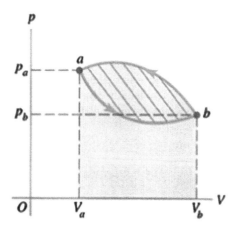

Example 5.7 A rigid volume tank contains 6 ft^3 of steam originally at a pressure of 400 psia and a temperature of 1300 °R. Estimate the final temperature if 800 Btu of heat is added.

Solution The first law of thermodynamics, with $\Delta KE = \Delta PE = 0$, is $Q - W = \Delta U$. For a rigid container the work is zero. Thus

$$Q = \Delta U = m(u_2 - u_1)$$

From Steam tables on page A-80 first column and second row group, we read $u_1 = 1302$ Btu/lbm and $v_1 = 1.8871$ ft^3/lbm for a temperature of 1300 °R. The mass is then;

$$m_1 = \frac{V_1}{v_1} = \frac{6\,[\text{ft}^3]}{1.8871\,[\text{ft}^3/\text{lbm}]} = 3.179\text{lbm}$$

Then using the First Law for the energy transferred to the volume by heat is given, thus we + have

$$800 = 3.179\,(u_2 - 1302) \quad \therefore u_2 = 1553.7 \text{ Btu/lbm}$$

So the state is specified by $v_2 = 1.8871$ ft^3/lbm and $u_2 = 1553.7$ Btu/lbm. u and v are two thermodynamic parameters which are enough to define a state. However no tables exist for temperature as a function of u and v. So a double linear interpolation is required. First four pressure-temperature combinations must be identified that bound both the desired u and v. For this case these combinations are (from Table A14.7. pages A-86 and A-87)

$$P = 550 \text{ psi} \quad T = 1800 \text{ R} \quad v = 1.9358 \text{ ft}^3/\text{lbm} \quad u = 1513.8 \text{ Btu/lbm}$$

$$P = 600 \text{ psi} \quad T = 1800 \text{ R} \quad v = 1.7729 \text{ ft}^3/\text{lbm} \quad u = 1513.1 \text{ Btu/lbm}$$

$$P = 550 \text{ psi} \quad T = 1900 \text{ R} \quad v = 2.0479 \text{ ft}^3/\text{lbm} \quad u = 1558.8 \text{ Btu/lbm}$$

$$P = 600 \text{ psi} \quad T = 1900 \text{ R} \quad v = 1.8760 \text{ ft}^3/\text{lbm} \quad u = 1558.3 \text{ Btu/lbm}$$

A double linear interpolation is required. The first interpolation can be done in pressure or temperature, but then the second one must be performed in the other variable. Consider interpolating in pressure first. At 1800 R this gives

$$U_{1800} = 1513.8 + (1.9358 - 1.8871)/(1.9358 - 1.7729) * (1513.1 - 1513.8)$$

$$U_{1800} = 1513.8 + 0.298 * (-0.7) = 1513.6 \; Btu/lbm$$

$$U_{1900} = 1558.8 + (2.0479 - 1.8871) / (2.0479 - 1.8760) * (1558.3 - 1558.8)$$

$$U_{1900} = 1558.8 + 0.936 * (-0.5) = 1558.33$$

$$T(1553.7) = 1800 + (1553.7 - 1513.6) / (1558.33 - 1513.6) * 100$$

$$T(1553.7) = 1800 + 0.896 * 100 = 1889.6R$$

Example 5.8 A frictionless piston is used to provide a constant pressure of 400 kPa in a cylinder containing steam originally at 500 K with a volume of 2 m^3. Calculate the final temperature if 3500 kJ of heat is added.

Solution The first law of thermodynamics gives $\Delta U = Q - W$. The work done is $W = p(V_2 - V_1)$. The mass remains unchanged $m_1 = m_2$.

$$m_1 = \frac{V_1}{v_1} = \frac{2}{0.56722} = 3.526 \text{ kg}$$

Then the First Law gives

$$3500 = 400 * (3.526 v_2 - 2) + 3.526 * (u_2 - 2690)$$

This requires iteration on v_2 and u_2 to find the final temperature. After some effort this gives,

$$T_f = 965 \text{ K}$$

5.15   Internal Energy of an Ideal Gas

For an ideal gas, the internal energy U depends only on temperature, not on pressure or volume. Consider again the free-expansion experiment described earlier. A thermally insulated container with rigid walls is divided into two compartments by a partition (Fig. 5.19). One compartment has a quantity of an ideal gas, and the other is evacuated.

When the partition is removed or broken, the gas expands to fill both parts of the container. The gas does not do any work on its surroundings because the walls of the container don't move, and there is no heat flow through the insulation. So both Q and W are zero, and the internal energy U is constant. This is true of any substance, whether it is an ideal gas or not.

Does the *temperature* change during a free expansion?. Suppose it *does* change, while the internal energy stays the same. In that case we have to conclude that the internal energy depends on both the temperature and the volume or on both the temperature and the pressure, but certainly not on the temperature alone. But if T is

Fig. 5.19 The partition is
broken (or removed) to start
the free expansion of gas into
the vacuum region

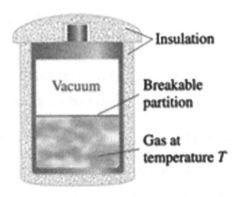

constant during a free expansion, for which we know that U is constant even though
both P and V change, then we have to conclude that U depends only on T, not on
P or V.

Many experiments have shown that when a low-density gas undergoes a free
expansion, its temperature *does not* change. Such a gas is essentially an ideal
gas. The conclusion is: **The internal energy of an ideal gas depends only on
its temperature, not on its pressure or volume**. This property, in addition to the
ideal-gas equation of state, is part of the ideal-gas model. Both properties will be
used frequently in later systems calculations.

5.16   Introduction to Enthalpy

In the solution of problems involving systems, certain products or sums of properties
occur with regularity. One such combination of properties can be demonstrated by
considering the addition of heat to the constant-pressure situation shown in Fig. 5.20.
Heat is added slowly to the system (the gas in the cylinder), which is maintained
at constant pressure by assuming a frictionless seal between the piston and the
cylinder. If the kinetic energy changes and potential energy changes of the system
are neglected and all other work modes are absent, the first law of thermodynamics
requires that Eq. 5.21 apply,

$$Q - W = U_2 - U_1 \tag{5.26}$$

The work done using the weight for the constant pressure process is given by;

$$W = P(V_2 - V_1) \tag{5.27}$$

The first law can then be written as;

$$Q = (U + PV)_2 - (U + PV)_1 \tag{5.28}$$

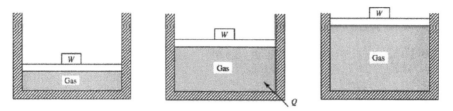

Fig. 5.20 Constant pressure heat addition

The quantity in parentheses $U+pV$ is a combination of properties, and it is thus a property itself. It is called the **enthalpy** H of the system,

$$H = U + PV \tag{5.29}$$

The specific enthalpy h is found by dividing by the mass.

$$h = u + P\upsilon \tag{5.30}$$

Enthalpy is a property of a system. It is so useful that it is tabulated in the steam tables along with specific volume and specific internal energy. The energy equation can now be written for a constant pressure process as

$$Q_{1-2} = H_2 - H_1 \tag{5.31}$$

The enthalpy has been defined assuming a *constant-pressure system* with difference in enthalpies between two states being the heat transfer. For a *variable-pressure process*, the difference in enthalpy is not quite as obvious. However, enthalpy is still of use in many engineering problems and it remains a property as defined by Eq. 5.29. In a non-equilibrium *constant-pressure process* ΔH would not equal the heat transfer.

Because only *changes* in the enthalpy or the internal energy are important, the datum for each can be chosen arbitrarily. Normally the saturated liquid at $0\,^{\circ}\text{C}$ is chosen as the datum point for water.

Example 5.9 A frictionless piston is used to provide a constant pressure of 400 kPa in a cylinder containing steam originally at 500 K with a volume of 2 m³. Calculate the final temperature if 3500 kJ of heat is added.

Solution This is the same problem as Example 5.8. However using the concept of enthalpy the solution is simpler, $Q = H_2 - H_1$. Then

$$3500 = (h_2 - 2916.2) * 3.526$$

$$h_2 = \frac{3500 + 2916.2 * 3.526}{3.526} = 3908.8$$

$$T_f = 964.8K$$

5.17   Latent Heat

When a substance changes phase from either a solid to a liquid or a liquid to a gas, it requires an input of energy to do so. The potential energy stored in the inter-atomic forces between molecules needs to be overcome by the kinetic energy of the motion of the particles before the substance can change phase.

Starting with a substance that is initially solid, as it is heated, its temperature will rise as depicted in the graph of Fig. 5.21.

Starting a point **A**, the substance is in its solid phase, heating it brings the temperature up to its melting point but the material is still a solid at point **B**. As it is heated further under constant pressure, the energy from the heat source goes into breaking the bonds holding the atoms in place. This takes place from **B** to **C**. At point **C** all of the solid phase has been transformed into the liquid phase. Once again, as energy is added the energy goes into the kinetic energy of the particles raising the temperature, (**C** to **D**). At point **D** the temperature has reached its boiling point but it is still in the liquid phase. From points, **D** to **E** thermal energy overcomes the bonds holding the particles in the liquid state and the particles have enough kinetic energy to escape to the vapor state. Then substance enters the gas phase and its temperature continues to rise with added heat inputs. Beyond **E**, further heating under pressure can raise the temperature still further.

Fig. 5.21 Temperature change with time. Phase changes are indicated by flat regions where heat energy used to overcome attractive forces between molecules

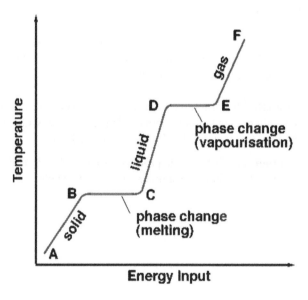

Table 5.1 Latent heat of fusion and vaporization for certain substances

Substance	Specific latent heat of fusion kJ/kg^{-1}	°C	Specific latent heat of vaporization kJ/kg^{-1}	°C
Water	334	0	2258	100
Ethanol	109	−114	838	78
Ethanoic acid	192	17	395	118
Chloroform	74	−64	254	62
Mercury	11	−39	294	357
Sulphur	54	115	1406	445
Hydrogen	60	−259	449	−253
Oxygen	14	−219	213	−183
Nitrogen	25	−210	199	−196

The Latent Heats of Fusion and Vaporization are the heat inputs required to produce the phase changes. The word latent means hidden, when the phase change is from solid to liquid it is the latent heat of fusion, and when the phase change is from liquid to a gas, it is the latent heat of vaporization.

The energy required is $Q = mL$, where m is the mass of the substance and L is the specific latent heat of fusion or vaporization required to produced the phase change.

Table 5.1 lists the Latent Heat of Fusion and Vaporization for certain substances.

The amount of energy that must be transferred in the form of heat to a substance held at constant pressure in order that phase changes occur is called the *latent heat.* It is the change in enthalpy of the substance at the saturated conditions of the two phases. The heat that is necessary to melt (or freeze) a unit mass at the substance at constant pressure is the *heat of fusion* and is equal to $h_{if} = h_f - h_i$, where h_i is the enthalpy of saturated solid and h_f is the enthalpy of saturated liquid. The *heat of vaporization* is the heat required to completely vaporize a unit of saturated liquid (or condense a unit mass of saturated vapor) and it equal to $h_{fg} = h_g - h_f$. When a solid changes phases directly to a gas, sublimation occurs and the *heat of sublimation* is equal to $h_{ig} = h_g - h_i$.

The heat of fusion and the heat of sublimation are relatively insensitive to pressure or temperature changes. For ice the heat of fusion is approximately 320 kJ/kg (140 Btu/lbm) and the heat of sublimation is about 2040 kJ/kg (880 Btu/lbm). The heat of vaporization of water is identified as h_{fg} in Appendices 14.1, 14.2, 14.5 and 14.6.

5.18 Specific Heats

For a simple system, only two independent variables are necessary to establish the state of the system as specified by Gibbs Phase Rule. This means that properties like specific internal energy u can be tabulated as a function of two variables, In the case of u, it is particularly useful to choose T and v. Or,

$$u = u(T, \upsilon) \tag{5.32}$$

Using the calculus chain rule, we express the differential in terms of the partial derivative as follow using the function u and its independent variables T and υ in Eq. 5.32;

$$du = \left.\frac{\partial u}{\partial T}\right|_{\upsilon} dT + \left.\frac{\partial u}{\partial \upsilon}\right|_{T} d\upsilon \tag{5.33}$$

The Chain Rule for Functions of Two Variables

The Chain Rule

    If $\quad x = x(t) \quad$ and $\quad y = y(t) \quad$ are differentiable at $\quad t \quad$ and $z = f(x, y) = f(x(t), y(t))$ is differentiable at $(x(t), y(t))$, then $z = f(x(t), y(t))$ is differentiable at t and is written as;

$$\frac{dz}{dt} = \frac{\partial z}{\partial x}\frac{dx}{dt} + \frac{\partial z}{\partial y}\frac{dy}{dt} = \left.\frac{\partial z}{\partial x}\right|_{y} dx + \left.\frac{\partial z}{\partial y}\right|_{x} dy$$

This can be proved directly from the definitions of z being differentiable at $(x(t), y(t))$ and $x(t)$ and $y(t)$ being differentiable at t.

Example of Chain Rule

In physics and chemistry, the pressure P of a gas is related to the volume V, the number of moles of gas n, and temperature T of the gas by the following equation:

$$P = \frac{nRT}{V}$$

where R is a constant of proportionality. We can easily find how the pressure changes with volume and temperature by finding the *partial derivatives* of P with respect to V and P, respectively. But, now suppose volume and temperature are functions of variable time t (with n constant): $V = V(t)$ and $T = T(t)$. We wish to know how the pressure P is changing with time. To do this we need a chain rule for functions of more than one variable. We will find that the chain rule is an essential part of the solution of any related rate problem, then we can write dP / dt as;

$$\frac{dP}{dt} = \frac{\partial P}{\partial T}\frac{dT}{dt} + \frac{\partial P}{\partial V}\frac{dV}{dt} = \frac{nR}{V}\frac{dT}{dt} - \frac{nRT}{V^2}\frac{dV}{dt}$$

Since u, v, and T are all properties, the partial derivative is also a property and is called the *constant volume specific heat*, C_v and that is

$$C_v = \left.\frac{\partial u}{\partial T}\right|_v \qquad (5.34)$$

Since many experiments have shown that when a low-density gas undergoes a free expansion, its temperature does not change. Such a gas is by definition an Ideal Gas. The conclusion is that, **the internal energy of an Ideal Gas depends only on its temperature, not on its pressure or volume**. This property, in addition to the Ideal Gas equation of state, is part of the Ideal Gas model. Because there is no change in temperature, there is no net heat transfer to the substance under experiment. Obviously since no work is involved, the first law requires that the internal energy of an ideal gas does not depend on volume. Thus the second term in Eq. 5.33 is zero.

$$\left.\frac{\partial u}{\partial v}\right|_T = 0 \qquad (5.35)$$

Combining Eqs. 5.33, 5.34 and 5.35 we have;

$$du = C_v dT \qquad (5.36)$$

This can be integrated to give;

$$u_2 - u_1 = \int_{T_1}^{T_2} C_v dT \qquad (5.37)$$

For a known $C_v(T)$ the Eq. 5.37 can be integrated to find the change in internal energy over any temperature interval for an ideal gas. By a similar argument, considering specific enthalpy to be dependent on the two variables temperature T and pressure P, we have;

$$dh = \left.\frac{\partial h}{\partial T}\right|_P dT + \left.\frac{\partial h}{\partial P}\right|_T dP \qquad (5.38)$$

The *constant-pressure specific heat* C_p is defined as the partial,

$$C_p = \left.\frac{\partial h}{\partial T}\right|_P \qquad (5.39)$$

For an Ideal Gas and the definition of enthalpy,

$$h = u + Pv = u + RT \qquad (5.40)$$

Where we have used the ideal-gas equation of state. Since u is only a function of T per Eq. 5.5.32, h also is only a function of T for an ideal gas. Hence, for an ideal gas;

$$\left.\frac{\partial h}{\partial P}\right|_T = 0 \tag{5.41}$$

Then from Eq. 5.38, we have;

$$dh = C_p dT \tag{5.42}$$

and integrating 5.5.5.42 over the temperature range T_1 to T_2 will give the following result.

$$h_2 - h_1 = \int_{T_1}^{T_2} C_p dT \tag{5.43}$$

Often it is convenient to denote the specific heat on a per-mole rather than a per kilogram basis and the specific heats can be written as:

$$\bar{C}_v = \bar{M} C_v \quad \text{and} \quad \bar{C}_p = \bar{M} C_p 3 \tag{5.44a}$$

where M is the molar mass. The value of $\bar{C}_v$ and $\bar{C}_p$ have been fit very accurately with a five parameter fit whose coefficients are given in Appendix A11 for a number of gases and combustion products. C_v and C_p can be derived from the specific heats per mole by dividing by the Molar mass $\bar{M}$ defined as

$$\bar{M} = \frac{m}{n} \tag{5.44b}$$

where m is the mass and n is the number of moles of the substance. The normal units used for m and n are kg and kgmol. Therefore the normal unit for $\bar{M}$ is kg/kgmol.

The enthalpy Eqs. 5.30 or 5.40 for an ideal gas can written as,

$$dh = du + d(Pv) \tag{5.45}$$

Introducing the specific heat relations and the ideal-gas equation, gives,

$$C_p dT = C_v dT + R dT \tag{5.46}$$

Dividing both sides by dT results in the following relation for an Ideal Gas:

$$C_p = C_v + R \tag{5.47}$$

The same relation holds for the molar specific heats and the universal gas constant $\bar{C}_p = \bar{C}_v + \bar{R}$. Note that the difference between C_p and C_v for an ideal gas is always a constant, even though both are functions of temperature. Defining the ratio of specific heats γ, we can see this ratio is also property of interest and is written as,

$$\gamma = \frac{C_p}{C_v} \qquad\qquad (5.48)$$

Substitution of Eq. 5.48 into Eq. 5.47, results in the following useful relationships.

$$C_p = \left(\frac{\gamma}{\gamma-1}\right)R \qquad\qquad (5.49a)$$

or

$$C_v = \left(\frac{1}{\gamma-1}\right)R \qquad\qquad (5.49b)$$

Since R for an ideal gas is constant, the specific heat ratio γ just depends on temperature T.

For gases, the specific heat slowly increases with increasing temperature. Since they do not vary significantly over fairly large temperature differences, it is often acceptable to treat C_p and C_v as constants. In this case, the integration is simple and the internal energy and enthalpy can be expressed as,

$$u_2 - u_1 = C_v(T_2 - T_1) \qquad\qquad (5.50)$$

$$h_2 - h_1 = C_p(T_2 - T_1) \qquad\qquad (5.51)$$

Example 5.10 The specific heat of superheated steam at 200 kPa can be determined by the equation

$$C_p = 2.0 + \frac{T - 700}{2500} \quad \text{kJ/kg}^\circ\text{K}$$

a. What is the enthalpy change between 500 and 1000 K for 5 kg of steam? Compare with the steam tables.
b. What is the average Cp between 500 and 1000 °K based on the equation and based on the tabulated data?

Solution

a. The enthalpy change is found to be

$$\Delta H = m\int_{T_1}^{T_2} C_p dT = 5\int_{500}^{1000}\left(2.0 + \frac{T - 700}{2500}\right)dT = 5323 \text{ kJ}$$

From Table 14.3 at page A-53 and A-59 we find, using $P=200$ kpa

$$\Delta H = m(3990.1 - 2924.8) = 5(1065.3) = 5326.5 \text{ kJ}$$

b. The average value $C_{P,\text{Avg}}$, is found by using the relation

$$mC_{P,\text{Avg}}\Delta T = m\int_{T_1}^{T_2} C_p dT$$

or

$$(5)(1000 - 500)C_{P,\text{Avg}} = 5\int_{500}^{1000}\left(2.0 + \frac{T - 700}{2500}\right)dT$$

The integral on right-hand side of above relation was evaluated in part (a) of the problem; hence, we have

$$\begin{cases} (5)(500)C_{P,\text{Avg}} = 5323 \\ C_{P,\text{Avg}} = \dfrac{5323}{2500} = 2.1292 = 2.13 \text{ kJ/kg.}^\circ K \end{cases}$$

Using the values from the steam table, we have

$$C_{P,\text{Avg}} \approx \frac{\Delta h}{\Delta T} = \frac{(3990.1 - 2924.8)}{500} = \frac{1065.3}{500} = 2.1306 \approx 2.13 \text{ kJ/kg.}^\circ K$$

Because the steam tables give the same values as the linear equation of this example, we can safely assume that the $C_P(T)$ relationship for steam over this temperature range is closely approximated by a linear relation. This linear relation would change, however, for each pressure chosen; hence, the steam tables are essential.

Example 5.11 Determine the value of C_p for steam at $T = 1500\ ^\circ R$ and $P = 800$ psia (See Appendix 14.6 page A-87).

Solution To determine C_P we use a finite-difference approximation to Eq. 5.39. We use the entries at $T = 1600\ ^\circ R$ and $T = 1400\ ^\circ R$ are used (a forward difference), C_P is too low. If values at $T = 1500\ ^\circ R$ and $T = 1300\ ^\circ R$ are used (a backward difference), C_p is too high. Thus, both a forward and a backward value (a central difference) should be used resulting in a more accurate of the slope.

Example 5.12 Calculate the change in enthalpy between 300 and 1200 K for nitrogen gas three ways:
1. Look up the change in the enthalpy given in the tables in Appendix A13.7.

2. Calculate the constant pressure specific heat at 300 K based on the fitted coefficients in Appendix A11. Multiply this specific heat value times the temperature difference to get the change in enthalpy over this temperature range.

3. Integrate the fitted expansion from 300 to 1200 K to get the enthalpy change.

Solution

1. From Appendix A13.7

$$h(300\ K) = 7754.4\ kj\,/\,kgmol\ h(1200\ K) = 35806.4\ kJ\,/\,kgmol$$

$$\Delta h = 35806.4 - 7754.4 = 28,052\ \text{kJ/kgmol}$$

2. From Appendix A11

$$\theta = \frac{T}{100}\quad C_p = c_0\left\{\frac{\theta^2}{\theta^2+\tau}\right\} + c_1\theta^{1q\ 2} + c_2\theta + c_3\theta^{3q\ 2} + c_4\theta^2$$

$$\theta = 3C_p = 12.2568(1) - 7.0276*3^{1q\ 2} + 3.1766*3 - 0.5465*3^{3q\ 2} + 0.0328*3^2$$

$$C_p = 12.2568 - 12.1722 + 9.5298 - 2.8396 + 0.2952 = 7.070\frac{kcal}{kgmol-K}$$

$$C_p = 7.07*4.1868 = 29.60\ \text{kJ/kgmol}$$

$$\Delta h = 29.6*(1200-300) = 26,640\ \text{kJ/kgmol}(-5\%)$$

3. The following steps is used to calculate the step 3 of this problem

$$\theta_1 = \frac{300}{100} = 3 \qquad \theta_2 = \frac{1200}{100} = 12$$

$$\Delta h = \int_{T_1}^{T_2} C_p dT = 100\int_{\theta_1}^{\theta_2} C_p d\theta = 100\int_{\theta_1}^{\theta_2}(c_0 + c_1\theta^{1/2} + c_2\theta + c_3\theta^{3/2} + c_4\theta^2)d\theta$$

$$\Delta h = 100 * \left\{ \begin{array}{l} 12.2568*(12-3) - 7.0276*\dfrac{(12^{3/2}-3^{3/2})}{3/2} \\[2ex] +3.1766*\dfrac{(12^2-3^2)}{2} - 0.5465*\dfrac{(12^{5/2}-3^{5/2})}{5/2} + 0.0328*\dfrac{(12^3-3^3)}{3} \end{array} \right\}$$

$$\Delta h = 6728.15\,\text{kcal/kgmol} = 28,169.4\ \text{kJ/kgmol}(+0.4\%)$$

5.19   Heat Capacities of an Ideal Gas

It is usually easiest to measure the heat capacity of a gas in a closed container under constant-volume conditions. The corresponding heat capacity is called the **specific heat at constant volume** and is designated C_v. Heat capacity measurements for solids and liquids are usually carried out in the atmosphere under constant atmospheric pressure, and we call the corresponding heat capacity the **specific heat at constant pressure**, designated C_p. If neither p nor V is constant, we have an infinite number of possible heat capacities.

Consider C_v and C_p for an ideal gas. To measure C_v, an ideal gas is heated in a rigid container at constant volume. To measure C_p, the gas is allowed expand just enough to keep the pressure constant as the temperature rises.

Why should these two molar heat capacities be different? The answer lies in the first law of thermodynamics. In a constant-volume temperature increase, the system does not do any work, and the change in internal energy ΔU equals the heat added Q. In a constant-pressure temperature increase, on the other hand, the volume *must* increase; otherwise, the pressure (given by the ideal-gas equation of state) could not remain constant. As the material expands, it does an amount of work W. According to the first law,

$$Q = \Delta U + W \qquad\qquad (5.52)$$

For a given temperature increase, the internal energy change U of an ideal gas has the same value no matter what the process (remember that the internal energy of an ideal gas depends only on temperature, not pressure or volume). Equation 5.52 then shows that the heat input for a constant-pressure process must be greater than that for a constant-volume process because additional energy must be supplied to account for the work done during the expansion. So C_p is greater than C_v for an ideal gas. The pV-diagram in Fig. 5.22 shows this relationship. For air, C_p is 40 % greater than C_v at temperatures near room temperature.

The ratio of specific heats then is a dimensionless parameter that will be useful for predicting many Ideal Gas processes. It is denoted by, denoted by the Greek letter gamma,

$$\gamma = \frac{C_p}{C_v} \qquad\qquad (5.53)$$

For gases, C_p is always greater than C_v and γ is always greater than unity. This quantity plays an important role in *adiabatic* presses for an ideal gas. Below is a table with specific heats and ratios of specific heats for various temperatures for air. Note that in most gas dynamics text books γ also is known as Adiabatic Index (Table 5.2).

Here's a final reminder: For an Ideal Gas the internal energy change in *any* process is given by $\Delta U = nC_v\Delta T$, *whether the volume is constant or not.*

Fig. 5.22 Raising the temperature of an ideal gas from T_1
to T_2 by a constant-volume or
a constant-pressure process.
For an ideal gas, U depends
only on T, so ΔU is the same
for both processes. In the
constant-volume process, no
work is done, so $Q = \Delta U$.
But for the constant-pressure
process, Q is greater, since
it must include both ΔU
and $W = p_1(V_2 - V_1)$. Thus
$C_p > C_v$

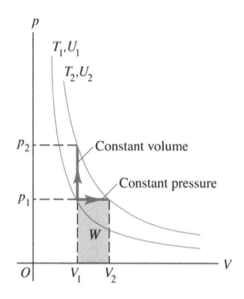

Table 5.2 Table of ratio of
specific heat for air

Temperature K	C_p kJ/kg K	C_v kJ/kg K	γ
200	1.002	0.715	1.401
300	1.005	0.718	1.400
500	1.029	0.742	1.387
1000	1.140	0.853	1.336
2000	1.249	0.962	1.298

Example 4.13 Cooling a Dorm Room.
    A typical dorm room or bedroom is about 8 m × 4 m × 2.5 m. The density of air on
a typical day is about 1.225 kg/m^{-3}. First find the mass of the air in the room. Then
find the change in the internal energy of this much air when it is cooled from 23.9 to
11.6 °C at an constant pressure of 1.00 atm. Treat the air as an ideal gas.

Solution This is a constant pressure process. One way to do this is to find Q from
$Q = nC_p\Delta T$, then find the volume change. After finding the volume change, then
find the work done by the gas from $W = p\Delta V$. Finally, use the first law to find ΔU.
This would be perfectly correct, but there's a much easier way. For an ideal gas the
internal-energy change is $\Delta U = nC_v\Delta T$ for *every* process, *whether the volume is
constant or not*. So all that is needed is C_v. The average of 23.9°C and 11.6 °C is
about 18 °C or about 291 K. We'll use the value for C_v from the table at 300 K which
is 0.718 kJ/kg K. We also need the mass of the air.

$$\text{mass} = \rho \times V = 1.225 \text{ kg/m}^{-3} \times 8 \times 4 \times 2.5 \text{ m} = 98 \text{ kg}$$

Then

$$\Delta U = 98 \text{ kg} \times 0.718 \text{ kJ / kg K} \times (11.6^\circ\text{C} - 23.9^\circ\text{C})$$

$$\Delta U = -865 \text{ kJ} = -8.65 \times 10^5 \text{ J}$$

A room air conditioner must extract this much internal energy from the air in the room and transfer it to the air outside.

5.20   Adiabatic Processes for an Ideal Gas

An adiabatic process is a process in which no heat transfer takes place between a system and its surroundings. Zero heat transfer is an idealization, but a process is approximately adiabatic if the system is well insulated or if the process takes place so quickly that, there is not enough time for appreciable heat flow to occur.

In an adiabatic process, $Q = 0$, so from the first law, $\Delta U = -W$. An adiabatic process for an ideal gas is shown in the pV-diagram in Fig. 5.23 below.

As the gas expands from volume V_a to V_b, it does positive work, so its internal energy decreases and its temperature drops. If point a, representing the initial state, lies on an isotherm at temperature $T + \Delta T$, then point b for the final state is on a different isotherm at a lower temperature T. For an ideal gas, an adiabatic curve at any point is always steeper than the isotherm passing through the same point. For an adiabatic compression from V_b to V_a, the situation is reversed and the temperature rises accordingly.

Fig. 5.23 A pV-diagram of an adiabatic process for an ideal gas. As the gas expands from V_a to V_b, its temperature drops from $T + \Delta T$ to T, corresponding to the decrease in internal energy due to the work W done by the gas (indicated by the shaded area). For an ideal gas, when an isotherm and an adiabatic pass through the same point on a pV-diagram, the adiabatic is always steeper

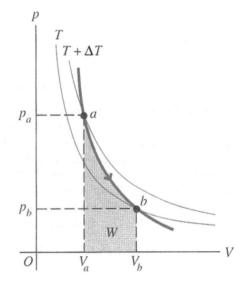

The air in the output pipes of air compressors used in gasoline stations and in paint-spraying equipment is always warmer than the air entering the compressor; this is because the compression is rapid and hence approximately adiabatic. Adiabatic *cooling* occurs when you open a bottle of your favorite carbonated beverage. The gas just above the beverage surface expands rapidly in a nearly adiabatic process; the temperature of the gas drops so much that water vapor in the gas condenses, forming a miniature cloud.

Caution Keep in mind that "adiabatic heating" and "adiabatic cooling," really mean "raising the temperature" and "lowering the temperature," in an adiabatic process respectively. In an adiabatic process, the temperature change is due to work done by, or on the system; there is *no* heat flow at all.

A relation can be derived between volume and temperature changes for an infinitesimal adiabatic process in an ideal gas. Equation 5.55 gives the internal energy change ΔU for any process for an ideal gas, adiabatic or not, so we have $\Delta U = nC_v\Delta T$. Also, the work done by the gas during the process is given by F. Then, since $\Delta U = -\Delta W$ for an adiabatic process, we have

$$nC_v\Delta T = -p\Delta V \qquad (5.54)$$

To obtain a relation containing only the volume V and temperature T, eliminate p using the ideal-gas equation in the form $p = \rho RT$. Since $m = \rho V$, write the ideal gas equation as $p = (n/V)\cdot RT$. Substituting this into Eq. 4.63, gives

$$nC_v\Delta T = -\frac{n}{V}RT\Delta V$$

Rearranging, gives

$$\frac{\Delta T}{T} + \frac{R}{C_v}\frac{\Delta V}{V} = 0$$

The coefficient R/C_v can be expressed in terms of $\gamma = C_p/C_v$. We have

$$\frac{R}{C_v} = \frac{C_p - C_v}{C_v} = \frac{C_p}{C_v} - 1 = \gamma - 1$$

$$\frac{\Delta T}{T} + (\gamma - 1)\frac{\Delta V}{V} = 0 \qquad (5.55)$$

Because γ is always greater than unity for a gas, $(\gamma - 1)$ is always positive. This means that in Eq. 5.55, ΔV and ΔT always have opposite signs. An adiabatic *expansion* of an ideal gas $(\Delta V > 0)$ always occurs with a *drop* in temperature $(\Delta T < 0)$, and an adiabatic compression $(\Delta V < 0)$ always occurs with a rise in temperature $(\Delta T > 0)$; this confirms the earlier prediction.

Equation 5.5.55 does not help much in its current form for solving problems. To get it to a form that will help requires a little calculus. Taking differentials and integrating gives,

$$\ln T + (\gamma - 1)\ln V = \text{constant}$$

$$\ln(TV^{\gamma-1}) = \text{constant} \tag{5.56}$$

Thus, for an initial state (T_1, V_1) and a final state (T_2, V_2),

$$T_1 V_1^{\gamma-1} = T_2 V_2^{\gamma-1} \text{(Adiabatic process, ideal gas)} \tag{5.57}$$

The T's must always be *absolute* (Kelvin) temperatures.

Equation 5.56 can be converted into a relation between pressure and volume by eliminating T, using the ideal-gas equation in the form $T = p/R\rho = pV/nR$. Substituting this into Eq. 5.56, gives

$$\frac{pV}{nR}V^{\gamma-1} = \text{constant}$$

or because n and R are constant

$$pV^{\gamma} = \text{constant} \tag{5.58}$$

For an initial state (p_1, V_1) and a final state (p_2, V_2), Eq. 5.58 becomes

$$p_1 V_1^{\gamma} = p_2 V_2^{\gamma} \text{(Adiabatic process, ideal gas)} \tag{5.59}$$

V_1 and V_2 can be eliminated between Eqs. 5.58 and 5.59 to obtain

$$\frac{p_1}{p_2} = \left(\frac{T_1}{T_2}\right)^{\frac{\gamma}{\gamma-1}} \tag{5.60}$$

It is also possible to calculate the work done by an ideal gas during an adiabatic process. In this case $Q = 0$ and $W = -\Delta U$ which holds true for *any* adiabatic process. For an ideal gas, $\Delta U = nC_v(T_2 - T_1)$. If the mass n and the initial and final temperatures T_1 and T_2 are known, then

$$W = nC_v(T_1 - T_2) \tag{5.61}$$

Using the Ideal Gas law $p = \rho RT = (m/V)\cdot RT$ in this equation to obtain

$$W = \frac{C_v}{R}(p_1 V_1 - p_2 V_2) = \frac{1}{\gamma - 1}(p_1 V_1 - p_2 V_2) \tag{5.62}$$

Fig. 5.24 Hot steam pressure cooker

where $C_v = R/(\gamma - 1)$. If the process is an expansion, the temperature drops. T_1 is greater than T_2. p_1V_1 is greater than p_2V_2, and the work is positive, as expected. If the process is a compression, the work is negative.

Throughout this analysis of adiabatic processes assumed the Ideal Gas equation of state, which is valid only for equilibrium states. Strictly speaking, these results are valid only for processes that are fast enough to prevent appreciable heat exchange with the surroundings (so that $Q = 0$ and the process is adiabatic), yet slow enough that the system does not depart very much from thermal and mechanical equilibrium. Even when these conditions are not strictly satisfied, though, Eq. 5.59 through 5.61 gives useful approximate results.

Hot steam escapes from the top of this pressure cooker at high speed. Hence, it has no time to exchange heat with its surroundings, and its expansion is nearly adiabatic. As the steam's volume increases, its temperature drops so much (See Eq. 5.59) that it feels cool on this chef's hand (See Fig. 5.24 above).

Example 5.14 Adiabatic Compression in a Diesel Engine

The compression ratio of a diesel engine is 15 to 1; this means that air in the cylinders is compressed to 1/15 of its initial value (Fig. 5.25). If the initial pressure is 1.10×10^5 Pa and the initial temperature is $27\,^\circ$C (300 K), find the final pressure and the temperature after compression. Air is mostly a mixture of diatomic oxygen and nitrogen; treat it as an ideal gas with $\gamma = 1.4$.

Solution We have $p_1 = 1.10 \times 105$ Pa, $T_1 = 300$K , and $V_1 / V_2 = 15$. From Eq. 5.57,

$$T_2 = T_1 \left(\frac{V_1}{V_2} \right)^{\gamma-1} = (300 \text{ K})(15)^{0.4} = 886 \text{ K} = 613\,^\circ\text{C}$$

From Eq. 5.59,

$$p_2 = p_1 \left(\frac{V_1}{V_2} \right)^{\gamma} = (1.01x10^5 \, pa)(15)^{1.4}$$

$$= 44.8x10^5 \, pa = 44 atm$$

Fig. 5.25 Adiabatic compression of air in a cylinder of a diesel engine

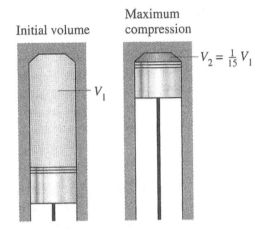

If the compression had been isothermal, the final pressure would have been 15 atm, but because the temperature also increases during an adiabatic compression, the final pressure is much greater.

When fuel is injected into the cylinders near the end of the compression stroke, the high temperature of the air attained during compression causes the fuel to ignite spontaneously without the need for spark plugs.

Example 5.15 Work done in an adiabatic process

In Example 5.12, how much work does the gas do during the compression if the initial volume of the cylinder is 1.00 L$= 1.00 \times 10^{-3}$ m^3? Assume that C_v for air is 0.718 kJ/kg K and $\gamma = 14$.

Solution Determine the mass using the Ideal Gas equation of state

$$p = \rho RT = (m/V)RT.$$

$$m = \frac{p_1 V_1}{RT_1} = \frac{(1.01 \times 10^5 \, Pa)(1.00 \times 10^{-3} \, m^3)}{(287.05^{Nm} q_{ kgK})(300 \text{ K})}$$

$$m = 1.173 \times 10^{-3} \, \text{kg}$$

and Eq. 5.26 gives

$$W = nC_v(T_1 - T_2)$$

$$= 1.173 \times 10^{-3} \text{ kg}(0.718 \, kJ/kg \, K)(300 \, K - 886 \, K)$$

$$W = -0.494 \text{ kJ} = -494 \text{ J}$$

With the second method,

$$W = \frac{1}{\gamma - 1}(P_1 V_1 - P_2 V_2)$$

$$W = \frac{1}{1.40 - 1}\left[(1.01 \times 10^5\, Pa)(1.00 \times 10^{-3}\, m^3) - (44.8 \times 10^5\, Pa)\left(\frac{1.00 \times 10^{-3}\, m^3}{15}\right)\right]$$

$$W = -494\ kJ = -494\ J$$

The work is negative because the gas is compressed.

5.21 Summary

This has been a long Chapter but its fundamentals can be summarized with the following main points and equations.

- A thermodynamic system can exchange energy with its surroundings by heat transfer or by mechanical work and in some cases by other mechanisms. When a system at pressure p expands from volume V_1 to V_2, it does an amount of work W given by

$$W = \sum p\Delta V \tag{5.63}$$

If the pressure is constant during the expansion

$$W = p(V_2 - V_1) \tag{5.64}$$

- In any thermodynamic process, the heat added to the system and the work done by the system depends not only on the initial and final states, but also on the path (the series of intermediate states through which the system passes).
- The first law of thermodynamics states that when heat Q is added to a system while it does work W, the internal energy U changes by an amount

$$U_2 - U_1 = \Delta U = Q - W \text{(First Law of Thermodynamics)} \tag{5.65}$$

- The internal energy of any thermodynamic system depends only on its state. The change in internal energy during a process depends only on the initial and final states, not on the path. The internal energy of an isolated system is constant.
- An adiabatic process requires no heat transfer into or out of a system, $Q = 0$.
- An isochoric process implies constant volume, $\Delta V = 0, \Rightarrow W = 0$.
- An isobaric process implies constant pressure, $\Delta p = 0 \Rightarrow W = p(V_2 - V_1)$.

- An isothermal process implies constant temperature.
- The heat capacities C_υ and C_p of an ideal gas are related by

$$C_p = C_\upsilon + R \qquad (5.67)$$

The ratio of heat capacities, C_p/C_υ, is denoted by γ

$$\gamma = \frac{C_p}{C_\upsilon} \qquad (5.68)$$

- For an adiabatic process for an ideal gas the quantities $TV^{\gamma-1}$ and pV^γ are constant. For an initial state (p_1, V_1, T_1) and a final state (p_2, V_2, T_2),

$$T_1 V_1^{\gamma-1} = T_2 V_2^{\gamma-1} \qquad (5.67)$$

$$p_1 V_1^\gamma = p_2 V^\gamma \qquad (5.68)$$

Or eliminating V_1 and V_2 between these equations gives,

$$\frac{p_1}{p_2} = \left(\frac{T_1}{T_2}\right)^{\frac{\gamma}{\gamma-1}} \qquad (5.69)$$

The work done by an ideal gas during an adiabatic expansion is

$$W = nC_\upsilon(T_1 - T_2) = \frac{C_\upsilon}{R}(p_1 V_1 - p_2 V_2) = \frac{1}{\gamma-1}(p_1 V_1 - p_2 V_2) \qquad (5.70)$$

Problems

Problem 5.1: For circulation of air in a large room, an 8-hp fan is used. If we assume the room is fully insulated, then determine the internal energy increase after 1 h of fan operation.

Problem 5.2: If your mass body is 60 kg and you consume a 900 kcal hot fudge with whipped cream, then how much work you should do in order to run up several flights of stairs in order to burn off all those calories.

Problem 5.3: A rigid volume contains 6 ft^3 of steam originally at a pressure of 400 psi and a temperature of 900 °F. Estimate the final temperature if 800 Btu of heat is added.

Problem 5.4: A frictionless piston is used to provide a constant pressure of 400 kPa in a cylinder containing steam originally at 200 °C with a volume of 2 m^3. Calculate the final temperature if 3500 kJ of heat is added.

Problem 5.5: A mixture of 25% nitrogen and 75% hydrogen by volume is compressed isontropically from 300 K and 100–500 kPa in the first stage of a multistage compressor in a fertilizer plant. The compression to still higher pressure is achieved in subsequent stages after the gas mixture is passed through intercoolers. It is desired to predict the temperature of the gas mixture after compression as well as the work required per unit mass of the mixture. Also, evaluate the entropy change for each gas. Assume that the mixture behaves like an ideal gas.

Problem 5.6: A rigid and insulated tank of volume 2 m³ contains 50% helium (by volume) at 100 kPa and 300 K. The tank is connected to high pressure line carrying helium at 4 MPa and 600 K. The valve is opened and the helium enters the tank until the pressure inside the tank is 4 MPa. Then, the valve is closed and the tank is isolated. Determine:

a. The final temperature of the gas mixture in the tank,
b. The composition of the mixture in the tank, and
c. The amount of helium that enters the tank (Fig. 5.26).

Problem 5.7: In a certain steam plant the turbine develops 1000 kW. The heat supplied to the steam in the boiler is 2800 kJ/kg, the heat rejected by the steam to the cooling water in the condenser is 2100 kJ/kg and the feed-pump work required to pump the condensate back into the boiler is 5 kW. Calculate the steam flow rate (Fig. 5.27).

Problem 5.8: Using the Fig. 5.28 below, determine the maximum pressure increase across 10-hp. The inlet velocity of the water is 30 ft/s.

Problem 5.9: In turbine of a gas turbine unit the gases flow through the turbine at 17 kg/s and the power developed by turbine is 14,000 kW. The specific enthalpies of the gases at inlet and outlet are 1200 and 360 kJ/kg respectively, and the velocities of the gases at inlet and outlet are 60 and 150 m/s respectively. Calculate the rate at which heat is rejected from the turbine. Find also the area of the inlet pipe given that the specific volume of the gases at inlet is 0.5 m³/kg (Fig. 5.29).

Problem 5.10: Air flows steadily at the rate of 0.4 kg/s through an air compressor, entering at 6 m/s with a pressure of 1 bar and a specific volume of 0.85 m³/kg, and leaving at 4.5 m/s with a pressure of 6.9 bar and a specific volume of 0.16 m³/kg.

Fig. 5.26 Schematic for
Problem 5.6

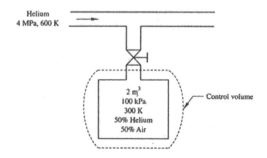

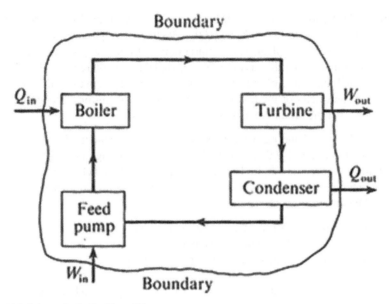

Fig. 5.27 Schematic for Problem 5.7

Fig. 5.28 Schematic for
Problem 5.8

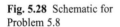

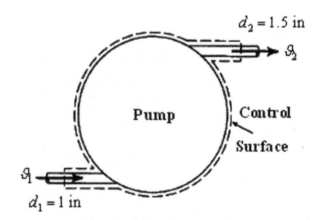

The specific internal energy of the air leaving is 88 kJ/kg greater than that of the air entering. Cooling water in a jacket surrounding the cylinder absorbs heat from the air at the rate of 59 kW. Calculate the power required to drive the compressor and the inlet and outlet pipe cross-sectional areas (Fig. 5.30).

Problem 5.11: A steam receives a steam flow of 1.3 kg/s and the power output is 500 kW. The heat loss from the casing is negligible. Calculate:

1. The change of specific enthalpy across the turbine when the velocities at entrance and exit and the difference in elevation are negligible,

Fig. 5.29 Schematic for
Problem 5.9

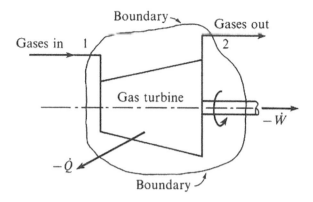

Fig. 5.30 Schematic for
Problem 5.10

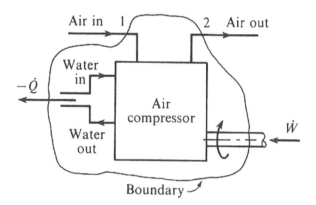

2. The change of specific enthalpy across the turbine when the velocity at entrance is 60 m/s, the velocity at exit is 360 m/s, and the inlet pipe is 3 m above the exhaust pipe.

Problem 5.12: A steady flow of steam enters a condenser with a specific enthalpy of 2300 kJ/kg and a velocity of 350 m/s. The condensate leaves the condenser with a specific enthalpy of 160 kJ/kg and a velocity of 70 m/s. Calculate the heat transfer to the cooling fluid per kilogram of steam condensed.

Problem 5.13: A turbine in a steam power plant operating under steady state conditions receives 1 kg/s. of superheated steam at 3 MPa and 350 °C with a velocity of 50 m/s at an elevation of 2 m above the ground level. The steam leaves the turbine at 10 kPa with a quality of 0.95 at an elevation of 5 m above the ground level. The exit velocity of the steam is 120 m/s. The energy losses as heat from the turbine are estimated at 5 kJ/s. Using Fig. 6.6, calculate:

Figure 5.31 Schematic for
Problem 5-13

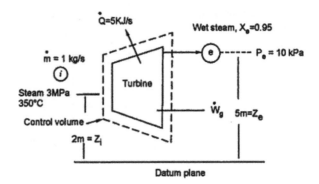

1. The power output of the turbine.
2. How much error will be introduced if the kinetic energy and the potential energy terms are ignored? (Fig. 5.31)

Problem 5.14: Consider the steady state operation of a compressor. The fluid enters the compressor at P_1 and υ_1 then leaves the compressor at P_2 and υ_2. Show that the work done on the compressor is given by $W = -\int_{\upsilon_1}^{\upsilon_2} \upsilon \, dP$. Sketch the $P-\upsilon$ diagram to represent the work done on the compressor and compare it with the work done if the compression is carried out in a piston-cylinder assembly under non-flow conditions.

Problem 5.15: In a steam power plant, saturated liquid water at 10 kPa enters a feed pump at the rate of 1 kg/s. The feed pump delivers the water to the boiler at a pressure of 3 MPa. Assuming that the pump is adiabatic, estimate the power input to the pump.

Problem 5.16: In an air-conditioning plant, saturated Freon-12 at $-20\,°C$ with a quality of 0.8 enters an adiabatic compressor and leaves as saturated at $40\,°C$. If the flow rate of Freon through the compressor is 1 kg/s, estimate the power input to the compressor.

Problem 5.17: A turbine operating under steady-flow conditions receive steam at the following state: pressure, 13.8 bar, specific volume 0.143 m³/kg, specific internal energy 2590 kJ/kg, velocity 30 m/s. The state of the steam leaving internal energy 2360 kJ/kg, velocity 90 m/s. Heat is rejected to the surrounding at the rate of 0.25 kW and the rate of steam flow through the turbine is 0.38 kg/s. Calculate the power developed by the turbine.

Problem 5.18: A nozzle is a device for increasing the velocity of a steady flowing fluid. At the inlet to a certain nozzle the specific enthalpy of the fluid is 3025 kJ/kg and the velocity is 60 m/s. At the exit from the nozzle, the specific enthalpy is 2790 kJ/kg. The nozzle is horizontal and there is a negligible heat loss from it. Calculate:

1. The velocity of the fluid at exit,
2. The rate of flow of fluid when the inlet area is 0.1 m² and the specific volume at inlet is 0.19 m³/kg,

3. The exit area of the nozzle when the specific volume at the nozzle exit is 0 5 m³/kg.

Problem 5.19: Steam enters an adiabatic nozzle operating at steady state at 4 bar and 200 °C with negligible velocity, and exits at 2 bar with a velocity of 300 m/s. Determine the temperature of the steam leaving the nozzle.

References

1. Atkins P (2010) The laws of thermodynamics, a very short introduction. Oxford University Press, Oxford
2. Young HD, Freedman RA (2003) University physics with modern physics with mastering physics, 11th edn. Addison Wesley, New York
3. Potter MC, Somerton CW (2006) Thermodynamics for engineers Schaum's outlines series, 2nd edn. McGraw-Hill, Boston

Chapter 6
The Kinetic Theory of Gases

As stated previously Classical Thermodynamics is very much a mathematical discipline. Given that the defining equations are known, the theory is developed around multi-variable calculus. The theory is actually quite elegant, but it does not predict how to estimate or calculate the fundamental quantities or the properties that characterize them. For this, a transition to Statistical Thermodynamics is required. Statistical Thermodynamics starts with the kinetic theory of gases and treats fluids as made up of large assemblages of atoms or molecules. It can be a very detailed and extensive theory that extends well beyond the subjects of interest to this text. However, a smattering of Statistical Thermodynamics, including the kinetic theory of gases, will be useful for understanding a number of Classical Thermodynamics phenomena. A brief sojourn into the kinetic theory of gases is useful.

6.1 Kinetic Theory Basis for the Ideal Gas Law

Classical thermodynamics is based on treating all fluids (gases and liquids) as a continuum. This means there is no macroscopic structure to the fluid. However, liquids and gases in particular are made up of individual atoms or molecules. To extend Classical Thermodynamics to be able to predict some of the parameters that show up in the classical equations, molecular structure must be considered. Begin by considering the size of some molecules of interest. Though there is some internal structure to each of these molecules, they can be approximated as spheres to first order (Table 6.1).

Generally of more interest to the analysis are the number of molecules per cubic millimeter and the fraction of available volume they take up. For the above molecules, this data is presented in Table 6.2.

Example 6.1 Estimate the molecules per cubic millimeter for CO_2 and the fraction of volume they occupy at STP. The average radius for the CO_2 molecule is 0.226 nm.

© Springer International Publishing Switzerland 2015
B. Zohuri, P. McDaniel, *Thermodynamics In Nuclear Power Plant Systems*,
DOI 10.1007/978-3-319-13419-2_6

Table 6.1 Effective molecular radii and volumes per molecule

Gas	Effective atom/molecule radius (nm)	Volume per atom/molecule (cubic mm)
Helium	0.152	1.46E-20
Argon	0.181	2.50E-20
Nitrogen	0.175	2.23E-20
Oxygen	0.161	1.76E-20
Hydrogen	0.155	1.57E-20
Water	0.157	1.62E-20

Table 6.2 Volume fraction occupied at STP

Gas	Fraction of volume at STP
Helium	3.93E-5
Argon	6.71E-5
Nitrogen	5.99E-5
Oxygen	4.73E-5
Hydrogen	4.21E-5
Water	4.34E-5

STP standard temperature and pressure, 273.15 K and 1 atmosphere pressure (101.3 k Pa)

Solution The effective volume of the CO_2 molecule is

$$V = \frac{4\pi}{3}r^3 = \frac{4\pi}{3}(0.226x10^{-9})^3 = 4.835x10^{-20}\, nm^3$$

And the number of molecules per cubic meter for any gas at STP is given by

$$N = N_a\frac{\rho}{M} = \frac{N_a \, P/RT}{M} = \frac{N_a P}{M\frac{\Re}{M}T} = \frac{N_a P}{\Re T} = \frac{6.022x10^{26} * 101325}{8314.47 * 273.15} = 2.687x10^{25}\, \frac{molecules}{m^3}$$

$N_a = Avogadro's\ Number$

Then the volume per cubic millimeter occupied by the molecules is,

$$V_{molecules} = V * N = 4.835x10^{-20} * 2.687x10^{25}/1.0x10^9 = 1.299x10^{-3}\, mm^3$$

It is clear that in a volume as small as a cubic millimeter there are a huge number of molecules and they fill only a very small fraction of the volume.

Fig. 6.1 Column volume for
molecules impacting the wall

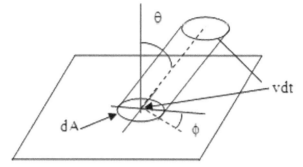

Now consider these molecules as they hit a stationary wall that confines them. Assume that they are moving equally in all directions and have a distribution of speeds. Thus at each speed the number that pass through a small area on the surface of a sphere that is characterized by the polar and azimuthal angles θ and ϕ is given by

$$dn_{v\theta\phi} = dn_v \frac{\sin\theta d\theta d\phi}{4\pi} \tag{6.1}$$

Consider a cylinder that ends on the surface of the wall and intercepts a small area dA on the wall. The cylinder is of length vdt. (Only consider molecules close enough to the wall that they do not collide with each other before they hit the wall.) (Fig. 6.1).

Then the molecules with a given speed that will intercept the surface during dt are those that are contained in the volume dV given by,

$$dV = dA(vdt\cos\theta) \tag{6.2}$$

The number that are headed in the right direction and can intercept the surface in the time dt are given by,

$$dn_{colliding} = dn_{v\theta\phi}dV = dAdtdn_v \frac{v\sin\theta\cos\theta d\theta d\phi}{4\pi} \tag{6.3}$$

So the number colliding per $dAdt$ is given by,

$$\frac{dn_{colliding}}{dAdt} = vdn_v \frac{\sin\theta\cos\theta d\theta d\phi}{4\pi} \tag{6.4}$$

Now for every collision in which the molecule is scattered elastically, each molecule has a change in momentum corresponding to twice its momentum perpendicular to the fixed surface.

$$\Delta mv = mv\cos\theta - (-mv\cos\theta) = 2mv\cos\theta \tag{6.5}$$

The total momentum change for those colliding in $dAdt$ is

$$\Delta m v dn_{colliding} = m v^2 dn_v \frac{\sin\theta\cos^2\theta d\theta d\phi}{2\pi} dAdt \tag{6.6}$$

Integrating over θ from 0 to $\pi/2$ and ϕ from 0 to 2π gives,

$$\Delta m v dn_{colliding} = \frac{1}{3} m v^2 dn_v dAdt \tag{6.7}$$

for the total momentum change due to molecules at velocity v. Integrating over all velocities to get the total momentum change and setting that equal to the total impulse gives,

$$dFdt = \frac{1}{3} m \int_0^\infty v^2 dn_v dAdt \tag{6.8}$$

or

$$\frac{dFdt}{dAdt} = \frac{1}{3} m \left[\int_0^\infty v^2 dn_v \right] = \frac{dF}{dA} = p \tag{6.9}$$

Then defining

$$n\langle v^2 \rangle = \int_0^\infty v^2 dn_v \tag{6.10}$$

Thus

$$p = \frac{1}{3} nm \langle v^2 \rangle \tag{6.11}$$

Remembering that $n=N/V$, gives,

$$pV = \frac{1}{3} Nm \langle v^2 \rangle \tag{6.12}$$

This looks like the Ideal Gas law,

$$pV = nRT \tag{6.13}$$

Note that $n=N/N_a$ where N_a is Avogadro's number 6.022E+26 molecules per kilogram-mole. So the Ideal Gas law is

$$pV = N(R/N_a)T = NkT$$
$$k = R/Na = 8314.47 J/kmol/K/6.022E + 26 molecules/kmol$$
$$k = 1.381E - 23 J/molecule/K - Boltzmanns\ constant$$

and,

$$pV = NkT = \frac{1}{3}Nm\langle v^2 \rangle \qquad \langle v^2 \rangle = \frac{3kT}{m} \qquad v_{rms} = \sqrt{\langle v^2 \rangle} = \sqrt{\frac{3kT}{m}} \qquad (6.14)$$

Example 6.2 Estimate V_{rms} for a nitrogen molecule in a gas at 300 K. Compare this with the speed of sound in nitrogen at this temperature given by $V_{sound} = \sqrt{\gamma RT}$.

Solution So estimating the **Vrms** for nitrogen at 300 K gives

$$V_{rms} = \sqrt{\frac{3kT}{m}} = \sqrt{\frac{3*1.381x10^{-23}\frac{nt-m}{K}*300K}{28*1.661x10^{-27}kg}} = 516.9m/\sec$$

The speed of sound in nitrogen is

$$V_{sound} = \sqrt{\gamma RT} = \sqrt{1.4*8314.47/28*300} = 353.2m/s$$

So the molecules, on average, are traveling significantly faster than the speed of sound. Thus if the gas is pushing a piston at speeds below 0.1 the speed of sound, the gas molecules hardly know the piston is moving and the process of moving the piston looks like a quasi-static one.

Perhaps more important is the following relationship.

$$E_{KE} = \frac{1}{2}m\langle v^2 \rangle = \frac{3}{2}kT \qquad (6.15)$$

which relates the average kinetic energy of a gas molecule to the absolute temperature.

6.2 Collisions with a Moving Wall

Now using the same type of analysis to evaluate collisions with a moving wall, consider a wall moving relatively slowly compared to the RMS velocity of the molecules in the gas. The change in kinetic energy of a molecule striking a wall moving away from it is given by,

$$\Delta KE_{A,v,\theta,\phi} = \frac{1}{2}m\left[(v\cos\theta)^2 - (v\cos\theta - 2u)^2\right]$$

$$\Delta KE_{A,v,\theta,\phi} = \frac{1}{2}m\left[v^2\cos^2\theta - v^2\cos^2\theta + 4uv\cos\theta - 4u^2\right]$$

$$\Delta KE_{A,v,\theta,\phi} = 2muv\cos\theta \qquad (6.16)$$

Multiplying this by the number of molecules striking the wall with velocity v gives,

$$\Delta KE_{A,v} = 2mvu \int_0^{\frac{\pi}{2}} \cos\theta v dn_v \frac{\sin\theta\cos\theta d\theta \int_0^{2\pi} d\phi}{4\pi}$$

$$\Delta KE_{A,v} = muv^2 dn_v \int_0^{\frac{\pi}{2}} \sin\theta\cos^2\theta d\theta$$

$$\Delta KE_{A,v} = \frac{1}{3} muv^2 dn_v \tag{6.17}$$

Then integrating over velocity gives,

$$\Delta KE_A = \frac{1}{3} nmu \left\langle v^2 \right\rangle = pu \tag{6.18}$$

Multiplying by the area of the moving wall gives

$$pAu = pA\frac{dx}{dt} = p\frac{dV}{dt} = \frac{\Delta KE_{gas}}{dt}$$

$$pdV = \Delta KE_{gas} \tag{6.19}$$

6.3   Real Gas Effects and Equations of State

The Ideal Gas assumes that the molecules are small spherical masses of negligible volume that only interact with each other by bouncing off each other elastically. This is not true because as the temperature goes down and the pressure increases, eventually each gas will condense into a liquid. So as a minimum, there are two corrections that must be made to the Ideal Gas law as an Ideal Gas becomes a 'real gas' or vapor. The first correction is for the volume that the molecules take up as they get in the way of each other. Rudolf Clausius proposed that the actual volume V could be corrected to a volume V' such that $V' = V - b$. He estimated that the excluded volume should be four times the volume of each molecule. So he calculated b' as $\frac{16}{3}N\pi r^2$ where r is the molecular radius. Van der Waals in 1873 included a second correction factor to take into account the forces of attraction that molecules exert on each other. These forces typically fall off very rapidly with distance between molecules $\sim 1/r^6$, so they are appreciable only between one molecule and its nearest neighbor. Since the molecules are attracted to each other one would expect that the actual pressure would be reduced from the pressure calculated by the Ideal Gas Law. If p' is the pressure based only on instantaneous collisions, the actual pressure would be given by,

$$p = p' - a'\left(\frac{N}{V}\right)^2 \tag{6.20}$$

Combining this effect with the Clausius correction gives the Van der Waals equation of state.

$$\left[p+a'\left(\frac{N}{V}\right)^2\right](V-b') = m\Re T \qquad (6.21)$$

Redefining

$$a = a'Na^2 \qquad b = b'Na/N \qquad (6.22)$$

gives Van der Waal's equation of state,

$$(p+a/v^2)(v-b) = RT \quad (Van\ der\ Waals) \qquad (6.23)$$

6.4   Principle of Corresponding States

Van der Waals also noticed that if the actual temperature and pressure for a gas are normalized by the pressure and temperature at the critical point (the top of the vapor dome) to produce a reduced temperature and pressure. The compressibility factor, $Z=pv/RT$, for most gases could then be plotted on a small number of charts with the value of Z_C being the distinguishing factor among the charts. A typical generalized compressibility chart is given below for $Z_C=0.27$ (Fig. 6.2).

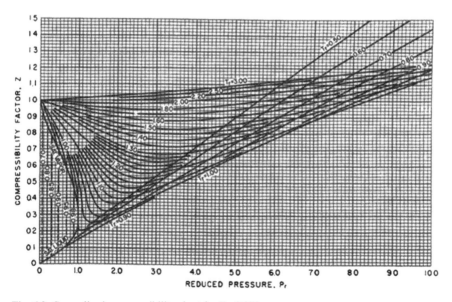

Fig. 6.2 Generalized compressibility chart for $Z=0.270$

As shown previously the Van der Waals coefficients a and b are found in terms of the critical state values. The coefficients are given by

$$a = \frac{27\Re^2 T_{cr}^2}{64 p_{cr}} \qquad b = \frac{\Re T_{cr}}{8 p_{cr}} \qquad\qquad (6.24)$$

when these values are substituted into the Van der Waals equation, it can be expressed in terms of the reduced pressure, temperature and specific volume.

$$\left(P_r + \frac{3}{v_r^2}\right)(3v_r - 1) = 8T_r \quad P_r = \frac{P}{P_C}, \quad T_r = \frac{T}{T_C}, \quad v_r = \frac{v}{v_C} \qquad (6.25)$$

For the Redlich-Kwong model this becomes,

$$P_r = \frac{T_r}{v_r - 0.0867} - \frac{0.42748}{T_r^{1/2} v_r (v_r + 0.0867)} \qquad\qquad (6.26)$$

and for the Peng-Robinson model; this becomes

$$P_r = \frac{T_r}{v_r - 0.077796} - \frac{0.45724\alpha}{v_r^2 + 0.1556 v_r - 0.00605} \qquad\qquad (6.27)$$

Note that based on the corresponding states theory, the Ideal Gas Constant does not show up and units depend only on what is used for critical state values.

6.5   Kinetic Theory of Specific Heats

The relationship for point molecules was derived above

$$\frac{1}{2} m \langle v^2 \rangle = \frac{3}{2} kT \qquad\qquad (6.28)$$

where the average kinetic energy of a point molecule is equal to $3/2kT$. Had the same derivation been made for point molecules in a plane, the average kinetic energy would equal to kT, and if the molecules were confined to a line, the average kinetic energy would be equal to $1/2kT$. Since the molecules are bouncing around in three dimensions, they essentially have three degrees of freedom. The Equi-Partition of Energy theorem states that each degree of freedom contributes $1/2kT$ to the thermal energy of the gas. Admittedly at this point this is somewhat of a hand waving argument, but it will be proven more rigorously later on.

Now when dealing with diatomic and polyatomic molecules, they can store energy in two additional modes. Since they have a non-spherical shape, they can store energy as a rotation about their center of mass. In addition, since they are not rigidly connected, they can store energy because of vibration between the atoms.

For any three-dimensional structure, there are potentially three different axes about which energy can be stored in the rotation of the structure. However, for a diatomic molecule there are really only two degrees of freedom associated with the rotation of the molecule. Rotation about the axis joining the two atoms stores an insignificant amount of energy and its contribution to stored energy is negligible. Therefore, for a diatomic molecule, gas rotation can contribute only kT to the stored energy per molecule. Of course, for polyatomic molecules, the full three degrees of freedom associated with rotation are possible and rotation can potentially contribute $3/2kT$ to the stored energy per molecule.

A diatomic molecule stores the energy of vibration as the kinetic energy of separation and the potential energy of separation. Therefore, vibration contributes two degrees of freedom to the energy storage in a diatomic molecule. The relationship for polyatomic molecules depends on the vibration modes available and excited, but certainly more than two degrees of freedom are possible.

Based on this model a diatomic gas should store internal energy in the following manner.

$$U = U_{trans} + U_{rot} + U_{vib}$$
$$U = \frac{3}{2}NkT + NkT + NkT \qquad (6.29)$$

Or factoring out the number of molecules,

$$U = N\left(\frac{3}{2}kT + kT + kT\right)$$
$$U = n\left(\frac{3}{2}\Re T + \Re T + \Re T\right) \qquad (6.30)$$

Then to calculate the specific heat we have

$$u = \frac{U}{n} = \left(\frac{3}{2}\Re T + \Re T + \Re T\right)$$
$$C_v = \frac{du}{dT}\bigg|_v = \frac{3}{2}\Re + \Re + \Re = \frac{7}{2}\Re \qquad (6.31)$$
$$C_p = C_v + \Re = \frac{9}{2}\Re$$

This implies that for a monatomic Ideal Gas we should have,

$$C_v = \frac{3}{2}\Re = 1.5*8.3143kj/(kgmole-K) = 12.47kj/(kgmole-K)$$

$$C_p = \frac{5}{2}\Re = 20.79kj/(kgmole-K) \hspace{2cm} (6.32)$$

$$\gamma = 5/3 = 1.667$$

All of which are in excellent agreement with experiment. However, for a diatomic gas like hydrogen this implies that

$$C_v = \frac{7}{2}\Re = 3.5*8.3143kj/(kgmole-K) = 29.1kj/(kgmole-K)$$

$$C_p = \frac{9}{2}\Re = 37.4kj/(kgmole-K) \hspace{2cm} (6.33)$$

$$\gamma = 9/7 = 1.286$$

which is not in good agreement with experiments. These are over estimates of the specific heats for diatomic gases at normal temperatures below the onset of disassociation.

Consider the following chart for hydrogen. Basically at low temperatures, C_p takes the value of a monatomic gas. As the temperature increases, the specific heat increases in what appears to be two steps when plotted against the temperature on a logarithmic scale. The first step corresponds to the activation of rotational degrees of freedom. The second step corresponds to the activation of the vibration degrees of freedom (Fig. 6.3).

The transitions between the levels are functions of characteristic temperatures for rotation and vibration. For hydrogen these temperatures are $T_{rot}=87.5$ K and $T_{vib}=6382$ K. For other diatomic gases the transition temperatures are listed in Table 6.3.

Perhaps the most useful thing to observe from this table is that for temperatures of interest with air as the working fluid, both oxygen and nitrogen are in the region where the rotational degrees of freedom have been activated and the vibration degrees of freedom are still dormant. Thus for both gases $\mathbf{C_v}$ should be **5/2 R** and $\mathbf{C_p}$ should be **7/2 R**. This gives a value for γ of **7/5** or **1.4** which matches the experimental value quite well near room temperature.

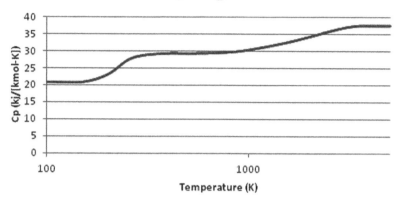

Fig. 6.3 Constant pressure specific heat for H_2 vs. temperature

Table 6.3 Characteristic rotation & vibration temperatures for diatomic molecules (Reference: [1])

Gas	Characteristic rotation temp (K)	Characteristic vibration temp (K)
H_2	87.5	6382
OH	27.2	5411
HCl	15.2	4331
CH	20.8	4145
N_2	2.88	3416
HBr	12.2	3839
CO	2.78	3143
NO	2.45	2758
O_2	2.08	2289
Cl2	0.346	813
Br_2	0.116	468
Na_2	0.223	231
K_2	0.081 134	

6.6 Specific Heats for Solids

Based on the simple models above, its worth commenting on the specific heat of a solid. The atoms vibrating in a solid at room temperature store energy in two modes—the kinetic energy of movement about the neutral position, and the potential energy of the field trying to restore the atoms to the neutral position. So for

Table 6.4 Debye temperatures for selected materials

Substance	Debye temperature (K)
Lead	88
Mercury	97
Iodine	106
Sodium	172
Silver	215
Calcium	226
Zinc	235
Copper	315
Aluminum	398
Iron	453
Diamond	1860

each possible direction of motion, there is kT energy stored. (A solid is essentially a constant volume phase so the constant pressure and constant volume specific heats are equal.) This gives

$$U = 3NkT = 3RT \quad C_v = 3R \tag{6.34}$$

This is known as the law of Dulong and Petit. It is only valid if the temperature of the solid is well above the Debye temperature for the solid. Below the Debye temperature of a solid the constant volume specific heat is given by

$$C_v = 3\Re\left[\frac{12}{x_o^3}\int_0^{x_o}\frac{x^3 dx}{e^x - 1} - \frac{3x_o}{e^{x_o} - 1}\right], x_o = \frac{\Theta_D}{T}, \Theta_D = Debye Temperature \tag{6.35}$$

Some typical values of Debye temperatures are given in Table 6.4.

This ends the excursion to solids. Unfortunately, liquids are a bit more complicated and cannot easily be addressed here.

6.7   Mean Free Path of Molecules in a Gas

So far, the collisions between gas molecules have been neglected, which is the equivalent of treating the molecules as point masses. Collisions actually do not affect the velocities of the molecules if the gas is in an equilibrium state. When two molecules collide their energies change, but because the gas is in equilibrium, for every molecule whose energy increases another's energy must decrease. Now consider some properties that can apply to non-equilibrium states. In order to do this, collisions and the size of the molecules must be taken into account.

Fig. 6.4 A track of a
molecule in a gas

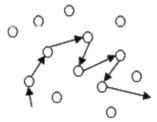

The first thing that must be considered is how far molecules travel between collisions. Begin by assuming that one molecule of interest is moving and all of the other molecules are frozen in place. The moving molecule moves from place to place colliding with the other molecules as it travels with a track like that described in Fig. 6.4.

Treating the molecules as hard spheres, when the moving molecule approaches another molecule, the closest its center can come to the other molecule's center is two times their radii. Now in many cases the two molecules will not meet with a collision velocity vector aligned with their centers. Often there will be skimming collisions. So the area that a stationary molecule presents to the moving molecule looks like a circular target with a radius twice that of one of the molecules. The area presented to the moving molecule that will produce a collision will be called its cross section and will be symbolized by σ. Thus,

$$\sigma = \pi(2r)^2 = 4\pi r^2 \tag{6.36}$$

Now if the molecule is moving with velocity v along this zigzag path, in a time t, it will sweep out a volume given by

$$V = vt\sigma$$

The number of non-moving molecules per unit volume can be represented by N, measured in molecules per cm^3.

$$N = N_a \frac{\rho}{M}$$
$$\rho = density\ in\ kilograms/m^3$$
$$M = molecular\ mass \tag{6.37}$$
$$N_a = Avogadro's\ Number$$

Then the number of collisions occurring in the volume swept out by the moving molecules path will be

$$C = N\sigma vt$$
$$\frac{C}{t} = N\sigma v \tag{6.38}$$

The collisions per unit time will depend on the number of atoms per unit volume, the collision cross section, and the speed of the colliding molecule. The average distance between collisions will be the distance traveled divided by the number of collisions.

$$\lambda = \frac{vt}{N\sigma vt} = \frac{1}{N\sigma} \tag{6.39}$$

The average distance between collisions is called the **Mean Free Path** and is usually identified with the symbol λ.

Example 6.3 Calculate the scattering cross section for CO_2 and the mean free path at STP.

Solution The effective molecular radius for CO_2 is 0.226 nm so,

$$\sigma = 4\pi r^2 = 4*3.1415926*(0.226x10^{-9})^2 = 6.41x10^{-19}m^2$$

Then the mean free path is

$$\lambda = \frac{1}{2.687x10^{25}*6.41x10^{-19}} = 5.8x10^{-8}m = 58nm$$

Therefore, a CO_2 molecule will normally average traveling about 128 diameters before having a collision at STP.

When the assumption that only one molecule is moving and all of the others are stationary is relaxed the equation for the mean free path must be modified slightly. If all of the molecules are assumed to have the same speed, the equation for the mean free path is given by,

$$\lambda = \frac{0.75}{N\sigma} \tag{6.40}$$

If the molecules have a distribution of speeds typical of a gas at temperature, (Maxwell-Boltzmann distribution) the mean free path is modified to,

$$\lambda = \frac{0.707}{N\sigma} \tag{6.41}$$

6.8   Distribution of Mean Free Paths

Not all collisions occur after a molecule travels one mean free path. Some travel farther and some travel less than a mean free path. Consider starting with a group of molecules n_0. In traveling a distance dx, some of them dn will collide with atoms in the gas. There will be $n_0 - dn$ left. In the next distance dx, some more will collide, but the number that collide will be proportional to the number that are traveling, so a balance equation can be written as,

$$dn = -p_c n dx \qquad (6.42)$$

where p_c is simply the probability of a collision. The number of collisions per unit distance is simply the number of collisions per unit time divided by the velocity of the molecules. Therefore, from Eq. 6.38 the collisions per unit distance are given by,

$$\frac{C}{vt} = N\sigma \qquad (6.43)$$

The inverse of collisions per unit distance is simply distance per collision, which is the probability of a collision in a unit distance. Thus

$$p_c = N\sigma \qquad (6.44)$$

Then separating variables and integrating Eq. 6.42 gives,

$$\frac{dn}{n} = -N\sigma dx$$
$$\ln(n) = -N\sigma x + \text{constant} \qquad (6.45)$$
$$n = n_0 e^{-N\sigma x}$$

Given this distribution of free paths, now calculate the average path, or mean free path.

$$\lambda = \frac{\int_0^\infty x dn}{n_0} = \frac{\int_0^\infty x p_c n_0 e^{-p_c x} dx}{n_0} = \frac{1}{p_c} \int_0^\infty (p_c x) e^{-p_c x} d(p_c x) = \frac{1}{p_c} = \frac{1}{N\sigma} \qquad (6.46)$$

as expected. So Eq. 6.45 can be written as

$$n = n_0 e^{-x/\lambda} \qquad (6.47)$$

Fig. 6.5 A gas between a
moving and a stationary plate

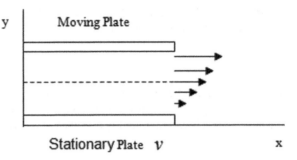

6.9   Coefficient of Viscosity

Consider Fig. 6.5. It represents two plates that are moving relative to each other
with a gas in between. The lower plate is stationary and the upper plate is moving to
the right. To keep the upper plate moving a force is required to overcome the friction
of the gas between the two plates. The friction force is defined by

$$F = \mu A \frac{dv}{dy}$$
(6.48)

μ = coefficient of viscosity

The coefficient of viscosity establishes the relationship between the friction force
and the velocity gradient in the fluid. The gas must have zero velocity in the x direction at the bottom plate and it must have the velocity of the upper plate at the point
in meets the upper plate.

The dashed line represents a surface with in the gas at an arbitrary distance from
the bottom plate. The macroscopic velocity of the gas, v, is much less than the
rms velocity of the gas molecules due to thermal excitation, so the results for an
equilibrium gas can be used. The gas molecules above the dashed line possess a
greater momentum than the gas molecules below the dashed line. As a result when
the molecules cross the dashed line there is a net rate of transport of momentum and
by Newton's second law there must be a force to sustain this momentum transport.

A macroscopic analogy would be two pickup trucks passing each other on the
road with one going faster than the other. If a rider in the bed of the faster pickup
truck tosses a bale of hay to the slower pickup truck at right angles to the trucks, the
slower truck will feel a jump in momentum because the bale of hay is traveling at
a faster speed when it arrives. In the reverse situation, if a rider in the back of the
slower pickup truck tosses a bale of hay to the faster truck, the faster truck will feel
a loss of momentum because the hay is traveling at a slower speed when it lands in
the faster truck.

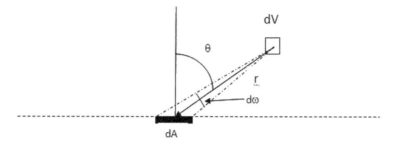

Fig. 6.6 Demonstration of small volume at distance r

Now consider Fig. 6.6. Compute the average height above the dashed line plane that a molecule has its last collision (or change in momentum) prior to crossing the plane.

In Fig. 6.6 dV is a small volume at a distance r from the differential dA in the crossing plane. The vector r makes an angle with the normal to the plane of θ and dA fills a solid angle $d\omega$ with respect to dV. Now if C_f is the collision frequency in dV, and the total number of molecules in dV is NdV the the total number of collisions in dV in time dt will be,

$$n_c = \frac{1}{2}C_f NdVdt \qquad\qquad (6.49)$$

The factor ½ appears so as to not count collisions twice. At each collision, two new free paths originate and they start off uniformly in all directions. The number headed toward dA is given by,

$$dn_{fp} = \frac{d\omega}{4\pi}C_f NdVdt \quad d\omega = \frac{dA\cos\theta}{r^2} \qquad (6.50)$$

The fraction that reach dA is given by Eq. 6.43 and writing dV in spherical coordinates as $dV = r^2\sin\theta d\theta d\phi dr$ gives

$$dn_{fp} = \frac{1}{4\pi}C_f NdAdt\sin\theta\cos\theta e^{-\frac{r}{\lambda}}d\theta d\phi dr \qquad (6.51)$$

Then

$$n_{fp} = \frac{1}{4\pi}C_f N\int_0^{\pi/2}\int_0^{2\pi}\int_0^{\infty} dAdt\sin\theta\cos\theta e^{-\frac{r}{\lambda}}d\theta d\phi dr = \frac{1}{4}C_f N\lambda dAdt \qquad (6.52)$$

And since

$$C_f = \frac{\bar{v}}{\lambda} \mathrm{n}_{fp} = \frac{1}{4} N \bar{v} dA dt \quad \bar{v} = average molecule speed \tag{6.53}$$

Now to get the average height of the last collision before crossing the plane, multiply by the height $r \cos \theta$ and integrate over r, θ, and ϕ. This gives,

$$n_y = \frac{1}{4\pi} C_f dA dt \int_0^{\pi/2} \sin\theta \cos^2\theta \, d\theta \int_0^{2\pi} d\phi \int_0^{\infty} r e^{-\frac{r}{\lambda}} dr = \frac{C_f}{6} N \lambda^2 dA dt \tag{6.54}$$

And dividing by the total collision rate in dV gives

$$y_{ave} = \frac{\frac{1}{6} C_f N \lambda^2 dA dt}{\frac{1}{4} C_f N \lambda dA dt} = \frac{2}{3} \lambda \tag{6.55}$$

Now at a height above the plane the flow velocity is

$$v = v_{plane} + \frac{2}{3} \lambda \frac{dv}{dy} \tag{6.56}$$

And the momentum carried by a single molecule is

$$p = m \left(v + \frac{2}{3} \lambda \frac{dv}{dy} \right) \tag{6.57}$$

So the total net momentum carried across the plane by molecules crossing from above is given by,

$$P_{upper} = \frac{1}{4} N \bar{v} m \left(v + \frac{2}{3} \lambda \frac{dv}{dy} \right)$$

And the total net momentum by those crossing from below is

$$P_{lower} = \frac{1}{4} N \bar{v} m \left(v - \frac{2}{3} \lambda \frac{dv}{dy} \right)$$

The net rate then is the difference of these two quantities, or,

$$P_{net} = \frac{1}{3} N m \bar{v} \lambda \frac{dv}{dy} \tag{6.58}$$

From Newton's second law this is equal to the viscous force per unit area. Therefore, the coefficient of viscosity is given by

$$\mu = \frac{1}{3} Nm\bar{v}\lambda \text{ or } \mu = \frac{1}{3}\frac{m\bar{v}}{\sigma} \tag{6.59}$$

The substituting for the mean velocity, $\bar{v}$

$$\mu = \frac{2}{3\sigma}\left(\frac{2mkT}{\pi}\right)^{1/2} \tag{6.60}$$

Once again, more advanced analysis taking into account the motion of all of the molecules in the gas will give

$$\mu = \frac{5\pi}{16\sqrt{2}\sigma}\left(\frac{2mkT}{\pi}\right)^{1/2} \tag{6.61}$$

Note that these two equations only differ by 3 %. An unexpected conclusion of this equation is that the viscosity of the gas only depends on the temperature and is independent of the pressure or density. Experiment bears this out except at very low pressures where the mean free path becomes of the order of the distance between the two plates. In addition, this relation was can be used to determine the size of molecules. Consider

$$\mu = \frac{5\pi}{16\sqrt{2}\sigma}\left(\frac{2mkT}{\pi}\right)^{1/2} = \frac{C}{4\pi r^2}$$

and for the liquid state

$$\frac{V}{m} = \frac{N_a V_{molecule}}{M} = \frac{N_a \, {}^{4\pi}/_3 r^3}{M} = \frac{1}{\rho} = \frac{4\pi N_a}{3M}r^3 = Br^3$$

Combining

$$r^3 = \frac{1}{B\rho} \qquad r^2 = \frac{C}{4\pi\mu} \qquad r = \frac{4\pi\mu}{CB\rho} \tag{6.62}$$

This was one of the earliest methods developed for determining the size of molecules.

6.10   Thermal Conductivity

The analysis for thermal conductivity proceeds in the same fashion as viscosity where the velocity gradient in Eq. 6.48 is replaced by the temperature gradient and both planes are stationary. This gives,

$$Q = k_T A \frac{dT}{dy} \qquad q = \frac{Q}{A} = k_T \frac{dT}{dy} \tag{6.63}$$

The mean energy of a molecule at a temperature T is $\frac{f}{2} k_B T$, where f is the number of degrees of freedom that are active. The energy carried across the arbitrary plane per unit area per unit time from a volume above the plane is

$$q_{upper} = \frac{1}{4} N \overline{v} \frac{f}{2} k_B \left(T + \frac{2}{3} \lambda \frac{dT}{dy} \right)$$

and the energy carried across the plane from a volume below the plane is

$$q_{lower} = \frac{1}{4} N \overline{v} \frac{f}{2} k_B \left(T - \frac{2}{3} \lambda \frac{dT}{dy} \right)$$

Differencing these two gives

$$q_{net} = \frac{1}{6} N \overline{v} f k_B \lambda \frac{dT}{dy} \tag{6.64}$$

Comparing Eq. 6.65 with Eq. 6.64 the thermal conductivity must be given by

$$k_T = \frac{1}{6} N \overline{v} f k_B \lambda = \frac{1}{6} \frac{\overline{v} f k_B}{\sigma} = \frac{1}{3} \frac{f k_B}{\sigma} \left(\frac{2 k_B T}{\pi m} \right)^{\!1/2} \tag{6.65}$$

Once again more exact theories give

$$k_T = \frac{25\pi}{64\sqrt{2}} \frac{f k_B}{\sigma} \left(\frac{2 k_B T}{\pi m} \right)^{\!1/2} \tag{6.66}$$

Thus, the thermal conductivity should only depend on temperature. This is in good agreement with experiment down to pressures where the mean free path becomes comparable to the separation between the plates.

Now the ratio of the viscosity to the thermal conductivity from Eqs. 6.57 and 6.62 is

$$\frac{\mu}{k_T} = \frac{4}{5} \frac{m}{f k_B}$$

Table 6.5 Comparison of reduced viscosity-thermal conductivity ratio for several gases

Gas (273 K)	$\dfrac{\alpha c_v}{Mk_{thermal}}$
Argon	0.405
Hydrogen	0.504
Oxygen	0.511
Nitrogen	0.512
Carbon dioxide	0.590

But

$$c_v = \frac{f}{2}\Re, \qquad k_B = \frac{\Re}{N_a}, \qquad m = \frac{M}{N_a}$$

This gives

$$\frac{\mu c_v}{Mk_{thermal}} = \frac{2}{5} \tag{6.67}$$

Consider Table 6.5 for some common gases.

The ratio is pretty accurate for Argon but the multi-atom molecules overshoot due to internal degrees of freedom for energy storage. The velocity does not carry all of the energy that is transported.

Problems

Problem 6.1: Estimate the rms velocity of a hydrogen molecule at STP and compare it with the speed of sound in hydrogen at STP.

Problem 6.2: Calculate the scattering cross section for the water molecule and the mean free path in water vapor at 500 K and 2 atmospheres.

Problem 6.3: Assume air to be composed of only nitrogen and calculate the mean free path for a nitrogen molecule in the standard atmosphere at 16 km altitude.

Problem 6.4: Given that the viscosity of air is 2.08E-05 N-s/m^2 at 350 K, and 1 atm, estimate its value at 10 atm and 1000 K.

Problem 6.5: Given that the thermal conductivity of water vapor at 1 atm and 400 K is 0.0261 W/m/K, estimate its value at 10 MPa and 700 K.

Reference

1. Lee JF, Sears FW, Turcotte Dl (1973) Statistical thermodynamics, 2nd edn. Addison-Wesley, Boston

Chapter 7
Second Law of Thermodynamics

The second law stipulates that the total entropy of a system plus its environment cannot decrease; it can remain constant for a reversible process but must always increase for an irreversible process.

7.1   Introduction

The First Law of Thermodynamics has been validated experimentally many times in many places. It is truly a law of physics. It always allows the conversion of energy from one form to another, but never allows energy to be produced or destroyed in the conversion process. But it is not a complete description of thermal energy conversion processes. The First Law would allow heat to be transferred from a cold body to a hot body as long as the amount of heat transferred decreased the internal energy of the cold body by the amount it increased the internal energy of the hot body. However, this never happens. Heat can only be transferred from a hot body to a cold body. Therefore, there is a requirement for a Law that explicitly states the direction of thermal energy transfer in addition to the conservation of energy expressed by the First Law. This is the Second Law of Thermodynamics. A simple statement of the Second Law would be that "heat can't be spontaneously transferred from a cold body to a hot body." More rigorous statements will be provided shortly.

7.2   Heat Engines, Heat Pumps, and Refrigerators

Consider Fig. 7.1. Heat is transferred from a heat reservoir to an ideal heat engine to produce work.

Unfortunately, it is virtually impossible to transfer heat to a heat engine without using some kind of a working fluid. Therefore, if the working fluid is passed to

© Springer International Publishing Switzerland 2015   173
B. Zohuri, P. McDaniel, *Thermodynamics In Nuclear Power Plant Systems*,
DOI 10.1007/978-3-319-13419-2_7

Fig. 7.1 Ideal heat engine

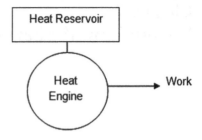

Fig. 7.2 Real heat engine

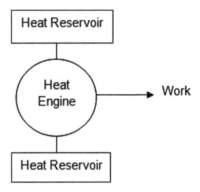

the heat engine and all of its contained heat is converted to work, it will have to be released to the environment with no heat or at zero absolute temperature. This is virtually impossible to do, so any realistic heat engine will need to operate between two heat reservoirs, one to provide the heat and another to absorb the working fluid after the work has been extracted from it. Therefore, a simple concept of a heat engine would have to look like Fig. 7.2. There must be a high temperature reservoir to provide heat and a low temperature reservoir to accept heat after the work has been performed. Then

$$W = Q_H - Q_L$$

Now there are basically three devices that operate in a cycle that can be represented by this model of a heat engine. These are a True Heat Engine, a Heat Pump, and a Refrigerator.

For the True Heat Engine, the objective is to produce work, so its efficiency is given by,

$$\eta = \frac{W}{Q_H} \tag{7.1}$$

Note that $Q_H = W + Q_L$ and h could be optimized by minimizing Q_L, The heat pump absorbs work and transfers heat to the upper reservoir. Its Coefficient of Performance is given by,

$$COP_{HP} = \frac{Q_H}{W} \qquad (7.2)$$

Once again $Q_H = W + Q_L$, but COP could be maximized by minimizing W. The refrigerator absorbs work and removes heat from the low temperature reservoir. Its Coefficient of Performance is given by,

$$COP_R = \frac{Q_L}{W} \qquad (7.3)$$

In this case COP could be maximized by minimizing W also.

7.3 Statements of the Second Law of Thermodynamics

The Second Law must apply to each of these devices, so consider two more elegant statements of the Second Law.

Clausius Statement *"It is impossible to construct a device which operates on a cycle and whose sole effect is the transfer of heat from a cooler body to a hotter body".*

Kelvin-Plank Statement *"It is impossible to construct a device which operates on a cycle and produces no other effect than the production of work and the transfer of heat from a single body".*

The two statements are equivalent. See Figs. 7.3a and b as follows;

Note that both statements are negations. They cannot be proven theoretically from some higher first principle argument, but have simply been determined to be correct by experiments similar to the First Law. They also rule out any heat engine that can be 100% efficient.

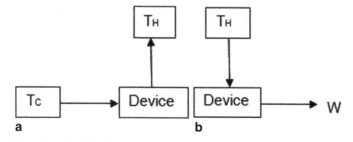

Fig. 7.3 a Not allowed by Kelvin-Plank statement. **b** Not allowed by Clausius statement

7.4 Reversibility

Now in order to find the most efficient engine possible given two heat reservoirs, the concept of reversibility must be quantified. Reversibility was touched on in the First Law chapter when quasi-static processes were discussed. A reversible process is simply one that once it has taken place it can be reversed back to the original state. It is the thermodynamic equivalent of a frictionless surface in mechanics or a resistance-less wire in an electrical circuit. No real process is reversible, but a process can be treated as reversible if the energy losses are so small as to be negligible. The energy losses that have to be considered are

1. Friction losses
2. Heat transfer across a finite temperature difference
3. Unrestrained expansion of a gas
4. Mixing of simple substances
5. Chemical or nuclear reactions

Friction losses mean that energy is lost when two bodies slide across each other. When the sliding is reversed, additional friction losses occur so a process involving friction cannot be reversed. Heat transfers across a finite temperature drop mean that when the process is reversed, the heat must transfer from a colder body to a hotter body. This is impossible. Unrestrained expansion of a gas means that the volume increased without work or heat being applied. A gas cannot be compressed without doing work. When substances are mixed, it requires work to separate them so this type of process cannot be reversed. Finally, once a chemical or nuclear reaction has occurred, the products cannot be converted back to the ingredients without the input of energy. Now if these processes are eliminated and only reversible processes considered, it is possible to talk about the Carnot engine.

7.5 The Carnot Engine

The heat engine that operates most efficiently between a given high temperature reservoir and a given low temperature reservoir is a Carnot engine named after Sadi Carnot who first proposed it. It uses reversible processes to complete a thermodynamic cycle and produce the most useful work possible. It also establishes a standard to which other heat engines can be compared to determine how well they use the thermal energy available in a high temperature reservoir. If their efficiency is significantly less than the Carnot efficiency, there may be improvements possible to boost their efficiency. The simplest model for a Carnot Engine is one based on an Ideal Gas cycle. The cycle consists of four processes and is described in Fig. 7.4, which the figure is the plot of the cycle in a T-v diagram.

Assume a piston cylinder arrangement that can alternately be placed in contact with hot and cold reservoirs and during the transition can be treated as adiabatic.

Fig. 7.4 T-v diagram for
Carnot cycle

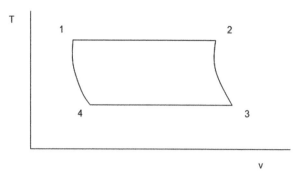

Process 1–2 is an isothermal expansion from point 1 to point 2. Heat is added reversibly at the high temperature reservoir and the volume contained by the piston expands.

$$Q_{H1-2} = W_{1-2} = \int p dV = mRT_H \ln\left(\frac{V_2}{V_1}\right) \tag{7.4}$$

Process 2–3 is an adiabatic expansion from point 2 to point 3. The volume continues to expand adiabatically.

$$Q = 0 \qquad \frac{V_2}{V_3} = \left(\frac{T_L}{T_H}\right)^{\frac{1}{\gamma-1}} \tag{7.5}$$

Process 3–4 is an isothermal compression from point 3 to point 4. The piston cylinder arrangement is placed in contact with the cold reservoir and the gas is compressed isothermally.

$$Q_{L,3-4} = W_{3-4} = \int p dV = mRT_L \ln\left(\frac{V_4}{V_3}\right) \tag{7.6}$$

Process 4–1 is an adiabatic compression from point 4 to point 1. The gas continues its compression back to its initial conditions.

$$Q = 0 \qquad \left(\frac{T_L}{T_H}\right)^{\frac{1}{\gamma-1}} = \frac{V_1}{V_2} \tag{7.7}$$

Then the efficiency for the cycle is given by

$$\eta = \frac{W}{Q_H} = \frac{Q_H - Q_L}{Q_H} = 1 - \frac{Q_L}{Q_H} = 1 + \frac{T_L \ln\left(\frac{V_4}{V_3}\right)}{T_H \ln\left(\frac{V_2}{V_1}\right)} \tag{7.8}$$

and for the adiabatic processes,

$$\frac{V_3}{V_2}=\frac{V_4}{V_1} \qquad V_3 V_1 = V_2 V_4 \qquad \frac{V_1}{V_2}=\frac{V_4}{V_3} \qquad \ln\left(\frac{V_4}{V_3}\right)=\ln\left(\frac{V_1}{V_2}\right) \tag{7.9}$$

Therefore

$$\eta = 1-\frac{T_L}{T_H} \tag{7.10}$$

which is the fundamental relationship for the Carnot efficiency of a reversible heat engine.

When the cycle is operated in reverse, the heat engine becomes a heat pump and this gives,

$$COP_{HP} = \frac{Q_h}{W_{net}} = \frac{Q_H}{Q_H - Q_L} = \frac{1}{1-\frac{Q_L}{Q_H}} = \frac{1}{1-\frac{T_L}{T_H}} \tag{7.11}$$

This is the fundamental relationship for a Carnot heat pump.

By the same analysis the COP for a Carnot refrigerator is given by

$$COP_R = \frac{Q_L}{W_{net}} = \frac{Q_L}{Q_H - Q_L} = \frac{1}{\frac{Q_H}{Q_L}-1} = \frac{1}{\frac{T_H}{T_L}-1} \tag{7.12}$$

and this is the fundamental relationship for a Carnot refrigerator.

Example 7.1: A Carnot Engine operates between two temperature reservoirs maintained at 240 °F and 40 °F. If the desired output from the heat engine is 15 hp what is the heat transfer from the high temperature reservoir and the heat transfer to the low temperature reservoir?

Solution: The efficiency of the Carnot Engine is given by

$$\eta = 1-\frac{T_L}{T_H} = 1-\frac{40+460}{240+460} = 1-\frac{500}{700} = \frac{2}{7} = 0.286$$

$$0.286 = \frac{Q_H}{W} \qquad Q_H = \frac{W}{0.286} = \frac{15*2545}{0.286} = 1.33x10^5 \; Btu \,/\, hr$$

$$Q_L = Q_H - W = 1.33x10^5 - 15*2545 = (1.33-0.38)x10^5 = 9.5x10^4 \; Btu \,/\, hr$$

Example 7.2: A Carnot engine operates with air on the following cycle.
Process 1–2 Constant temperature expansion at 327 °C

Process 2–3 Adiabatic expansion to 8 m³/kg
Process 3–4 Constant temperature compression at 77 °C
Process 4–1 Adiabatic compression starting at 80 kPa back to 327 °C
Determine the thermal efficiency and the work output per cycle per kg.

Solution: The thermal efficiency is

$$\eta = 1 - \frac{T_L}{T_H} = 1 - \frac{77 + 273}{327 + 273} = 1 - \frac{350}{600} = 0.417$$

At 4 the specific volume can be found from the Ideal Gas equation

$$v_4 = \frac{RT_4}{p_4} = \frac{8314.47/28.9669 * 350}{8.0x10^4} = 1.256 m^3 / kg$$

Then for an adiabatic compression from 4 to 1,

$$v_1 = v_4 \left(\frac{T_4}{T_1}\right)^{\frac{1}{\gamma-1}} = 1.256 \left(\frac{350}{600}\right)^{2.5} = 0.3264$$

Also for the adiabatic exp ansion from 2 to 3,

$$v_2 = v_3 \left(\frac{T_3}{T_2}\right)^{\frac{1}{\gamma-1}} = 8 \left(\frac{350}{600}\right)^{2.5} = 2.079$$

Then the work for the isothermal exp ansion from 1 to 2,

$$q_H = RT_H \ln\left(\frac{v_2}{v_1}\right) = 287 * 600 * \ln\left(\frac{2.079}{0.3264}\right) = 3.189x10^5 \, Joules$$

$$q_L = RT_L \ln\left(\frac{v_4}{v_3}\right) = 287 * 350 * \ln\left(\frac{1.256}{8}\right) = -1.860x10^5 \, Joules$$

$$w_{per \; cycle} = q_H - q_L = (3.189 - 1.860)x10^5 \, Joules = 1.329x10^5 \, Joules$$

Or

$$\eta = \frac{w}{q_H} = 0.417 = \frac{w}{3.189x10^5} \qquad w = 0.417 * 3.189x10^5 = 1.329x10^5 \, Joules$$

7.6   The Concept of Entropy

When the fundamental concepts of Classical Thermodynamics were being formulated the scientific community investigated a number of concepts that would formalize the Second Law analytically. The concept that was finally settled on was $\delta Q/T$. If a cyclic integral of this quantity is taken for the Carnot engine cycle it gives,

$$\oint \frac{\delta Q}{T} = \frac{Q_H}{T_H} - \frac{Q_L}{T_L} \qquad\qquad (7.13)$$

and for the Carnot cycle

$$\frac{Q_L}{Q_H} = \frac{T_L}{T_H} \quad \text{or} \quad \frac{Q_H}{T_H} = \frac{Q_L}{T_L} \tag{7.14}$$

This then leads to the conclusion that

$$\oint \frac{\delta Q}{T} = 0 \tag{7.15}$$

As a result $\delta Q/T$ is a perfect differential and it can define a new property of a thermodynamic system. This property is called entropy and given the symbol S. Its differential is given by,

$$dS = \frac{\delta Q}{T}\bigg|_{reversible} \tag{7.16}$$

$1/T$ serves as an integrating factor for dQ. It can be integrated for a process to give,

$$\Delta S = \int_1^2 \frac{\delta Q}{T} \tag{7.17}$$

Note that S is an extensive property and it can be converted to a specific entropy by dividing by the mass, $s = S/m$. In addition, an adiabatic-reversible process implies $\Delta S = 0$, the Carnot cycle can be plotted on a T-s diagram rather than a T-v diagram. Figure 7.5 is such a plot.
    Then,

Fig. 7.5 Temperature-entropy plot for a Carnot cycle

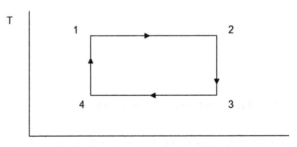

$$\Delta S_{1-2} = S_2 - S_1 = \int_1^2 \frac{\delta Q}{T} = \frac{Q_{H,1-2}}{T_H}$$

$$\Delta S_{2-3} = S_3 - S_2 = \int_2^3 \frac{\delta Q}{T} = 0$$

$$\Delta S_{3-4} = S_4 - S_3 = \int_3^4 \frac{\delta Q}{T} = -\frac{Q_{L,3-4}}{T_L} \qquad (7.18)$$

$$\Delta S_{4-1} = S_1 - S_4 = \int_4^1 \frac{\delta Q}{T} = 0$$

For a reversible process the heat transfer can be written as $dQ = TdS$. This allows the First Law for closed systems to be written as

$$dU = \delta Q - \delta W = TdS - pdV \qquad (7.19)$$

Or on a per unit mass basis it can be written,

$$du = Tds - pdv \qquad (7.20)$$

The First Law for a flow system using the enthalpy gives,

$$h = u + pv$$
$$dh = du + pdv + vdp \qquad (7.21)$$
$$dh = Tds + vdp$$

7.7   The Concept of Entropy

Now consider an Ideal Gas with constant specific heats. Dividing Eq. 7.20 by T and using the Ideal Gas law gives

$$ds = \frac{du}{T} + \frac{pdv}{T} = C_v \frac{dT}{T} + R \frac{dv}{v} \qquad (7.22)$$

This can be integrated to give,

$$s_2 - s_1 = C_v \ln\left(\frac{T_2}{T_1}\right) + R \ln\left(\frac{v_2}{v_1}\right) \qquad (7.23)$$

Equation 7.23 can also be evaluated as,

$$ds = C_p \frac{dT}{T} - \frac{v}{T} dp = C_p \frac{dT}{T} - R \frac{dp}{p}$$

$$s_2 - s_1 = C_p \ln\left(\frac{T_2}{T_1}\right) - R \ln\left(\frac{p_2}{p_1}\right)$$

(7.24)

Of course, these equations only apply to reversible processes. However since they relate changes in entropy to other thermodynamic properties at the two end states, the can be used for reversible and irreversible processes.

If the entropy change is zero, called an isentropic process, then Eq. 7.23 becomes

$$\frac{T_2}{T_1} = \left(\frac{v_1}{v_2}\right)^{\gamma-1} \qquad \frac{T_2}{T_1} = \left(\frac{p_2}{p_1}\right)^{\gamma-\frac{1}{\gamma}} \Rightarrow \frac{p_2}{p_1} = \left(\frac{v_1}{v_2}\right)^{\gamma}$$

(7.25)

Of course, these equations are exactly the ones obtained earlier for a quasi-static adiabatic process.

Example 7.3: Nitrogen is contained in a rigid volume at 27 °C and 100 kPa. A paddle wheel does 900 kJ of work on the nitrogen. If the volume is 3 m³, calculate the entropy increase assuming constant specific heats.

Solution: Treating the process as adiabatic

$$-W = \Delta U = m C_v \Delta T$$

Calculating the mass based on the Ideal Gas law

$$m = \frac{pV}{RT} = \frac{1.0x10^5 * 3}{8314.47 / 28 * 300} = 3.367 kg$$

$$C_v = \frac{R}{\gamma - 1} = \frac{8314.47 / 28}{1.4 - 1} = 742.4 J / kg / K$$

$$9.0x10^5 = 3.367 * 742.4 * (T_2 - 300)$$

$$T_2 = 300 + \frac{9.0x10^5}{3.367 * 742.4} = 300 + 360 = 660 K$$

$$\Delta S = m * C_v * \ln\left(\frac{T_2}{T_1}\right) = 3.367 * 742.4 * \ln\left(\frac{660}{300}\right) = 1.97 kJ / K$$

Example 7.4: Nitrogen is heated to 327 °C and a pressure of 1.5 MPa. It is then expanded to 200 kPa with a reversible adiabatic process. Calculate the work done by the nitrogen assuming it to be calorically perfect.

Solution: The first law gives

$$-W = mC_v(T_2 - T_1)$$

$$T_2 = T_1 \left(\frac{p_2}{p_1}\right)^{\gamma-1/\gamma} = 600 * \left(0.2/1.5\right)^{0.286} = 337K \tag{7.26}$$

$$-w = 742.4 * (337 - 600) = -195kJ / kg$$

7.8 Entropy for an Ideal Gas with Variable Specific Heats

If the Ideal Gas does not have constant specific heats over the temperature range of interest, the entropy change equations are very useful. Consider the constant pressure specific heat equation,

$$ds = \frac{dh}{T} - \frac{vdp}{T} = C_p \frac{dT}{T} - R\frac{dp}{p}$$

$$s_2 - s_1 = \int_{T_1}^{T_2} C_p \frac{dT}{T} - R\ln\left(\frac{p_2}{p_1}\right)$$

Now define

$$s_2^0 - s_1^0 = \int_{T_1}^{T_2} \frac{C_p}{T} dT \tag{7.27}$$

and

$$s_2 - s_1 = s_2^0 - s_1^0 - R\ln\left(\frac{p_2}{p_1}\right) \tag{7.28}$$

This can be rearranged as,

$$\frac{p_2}{p_1} = e^{\frac{s_2^0 - s_1^0}{R}} = \frac{e^{s_2^0/R}}{e^{s_1^0/R}} = \frac{f(T_2)}{f(T_1)} \tag{7.29}$$

This allows the definition of a relative pressure, p_r, as,

$$p_r = e^{s^0/R} \tag{7.30}$$

Thus for an isentropic process,

$$\frac{p_2}{p_1} = \frac{p_{r2}}{p_{r1}} \tag{7.31}$$

The volume ratio can be found by applying the Ideal Gas law to get

$$\frac{v_2}{v_1} = \frac{RT_2/p_2}{RT_1/p_1} = \frac{p_1}{p_2}\frac{T_2}{T_1} = \frac{T_2/p_2}{T_1/p_1} \tag{7.32}$$

Therefore a relative volume can be defined as, v_r, with

$$\frac{v_2}{v_1} = \frac{v_{r2}}{v_{r1}} \tag{7.33}$$

Both p_r and v_r are tabulated as a function of temperature for several gases in Appendix 13.

Example 7.5: Rework Example 7.3 assuming variable specific heats.

Solution: As before m=3.367 kg
    First law $-w = u_2 - u_1$

$$u_2 = -W/m + u_1 = -\frac{(-900)}{3.367} + 5551.2/28 = 465.56 kJ/kg = 13035.6 kJ/kmol$$

$$T(13035.6) = 650 + \frac{13035.6 - 13010.8}{13566.4 - 13010.8} * 25 = 650 + 0.0446 * 25 = 651.1K$$

$$s_1^0 = 186.3310$$
$$s_2^0 = 209.411 + 0.0446*(210.2928 - 209.1411) = 209.193$$

The Ideal Gas law gives the pressure at state 2

$$p_2 = p_1\left(\frac{T_2}{T_1}\right) = 100 * \left(\frac{651.1}{300}\right) = 217 kPa$$

$$\Delta S = m\left[(s_2^0 - s_1^0) - R\ln\left(p_2/p_1\right)\right]$$

$$= 3.367*\left[209.193 - 186.3310 - 8.31447 * \ln\left(217/100\right)\right]/28$$

$$= 1.97 kJ/K$$

Example 7.6: Rework Example 7.4 assuming variable specific heats.

Solution: Assume isentropic process,
    At $T = 600 K$, $P_{r1} = 16.13$

$$p_{,2} = p_{,1}\left(\frac{p_2}{p_1}\right) = 16.13*\left(\frac{0.2}{1.5}\right) = 2.15$$

$$T(2.15) = 325 + \frac{2.15 - 1.84}{2.38 - 1.84}*25 = 325 + 0.574*25 = 339.4K$$

$$-w = u_2 - u_1$$

$$u_1 = 11911.4 / 28 = 425.4$$

$$u_2 = \frac{6073.4 + 0.574*(6595.7 - 6073.4)}{28} = 227.6$$

$$w = 425.4 - 227.6 = 197.7kJ / kg$$

7.9   Entropy for Steam, Liquids and Solids

For pure substances like steam the specific entropy is included in the steam tables similar to specific volume, specific internal energy and specific enthalpy. In the mixed phase region, the specific entropy is calculated based on the steam quality similar to the other properties.

$$s = s_f + x s_{fg} \qquad (7.34)$$

For a compressed liquid or solid, the specific heat can be assumed constant and both assumed incompressible ($dv = 0$), so,

$$Tds = du = CdT$$

$$ds = C\frac{dT}{T} \qquad (7.35)$$

$$\Delta s_{1-2} = C\ln\left(\frac{T_2}{T_1}\right)$$

Certainly if the specific heat is known as a function of temperature, the functional dependence can be substituted in the above equations and the integration performed. Specific heats for liquids and metals are listed in Appendices A.8 and A.9.

Example 7.7: Steam is contained in a rigid container at 600 K and 0.8 MPa. The pressure is reduced to 60 kPa. Calculate the entropy change and heat transfer.

Solution: The specific volume must remain the same. The initial specific volume is taken from the superheated steam tables

$$v_1 = 0.34046 \, m^3/kg \quad u_1 = 2841.1 \, kJ/kg \quad s_1 = 7.33094 \, k/kg/K$$

Then at 60 kPa the fluid must be in the mixed liquid-vapor region, so from Table 14.2 of Appendix 14, the specific volumes are

$$v_f = 0.001033 \, m^3/kg \quad v_g = 2.73183 \, m^3/kg$$

$$0.34046 = 0.001033 + x * 2.73183 x = (0.34046 - 0.001033)/2.73183 = 0.124$$

So the final internal energy is

$$u_f = 359.8 + 0.124 * 2129.2 = 623.8 \, kJ/kg$$

and

$$s_f = 1.14524 + 0.124 * 6.38586 = 1.93709 \, kJ/kg/K$$

$$\Delta u = 2841.1 - 623.8 = 2217.3 \, kJ/kg$$

$$\Delta s = 7.330904 - 1.93709 = 5.38385 \, kJ/kg/K$$

7.10   The Inequality of Clausius

The Carnot cycle is a reversible cycle operating between two reservoirs. If an irreversible cycle operates between the same two reservoirs, the available work from the irreversible cycle will have to be less than the available work from the reversible cycle. That is

$$W_{irr} < W_{rev}$$

If the First Law is applied to a cycle and equal amounts of heat Q_H are transferred from the high temperature reservoir, more heat will be transferred to the low temperature reservoir for the irreversible cycle.

$$Q_H = W_{rev} + Q_{L,rev}$$

$$Q_H = W_{irr} + Q_{L,irr}$$

So when performing the cyclic integral for dS.

$$\oint \left(\frac{\delta Q}{T}\right)_{irr} \leq 0 \tag{7.36}$$

For a reversible process the above integral is 0, of course. If an irreversible refrigerator is being considered more work will be required that for a reversible refrigerator, the heat transferred at the high temperature reservoir will be greater so the cyclic integral will still be less than zero for an irreversible process. So the fundamental result can be written for any cycle,

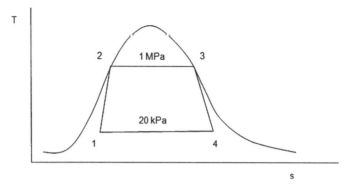

Fig. 7.6 Power plant cycle

$$\oint \left(\frac{\delta Q}{T} \right)_{irr} \leq 0 \tag{7.37}$$

This is known as the *inequality of Clausius*. It is a consequence of the Second Law of Thermodynamics.

Example 7.6: A steam power plant operates on the cycle shown in the Fig. 7.6 below
Process 1–2: Adiabatic compression in the pump
Process 2–3: Isothermal heat transfer in the boiler
Process 3–4: Adiabatic expansion in the turbine
Process 4–1: Isothermal heat transfer in the condenser
   The quality at point 4 after expansion through the turbine is 88%. The quality leaving the condenser and entering the pump at point 1 is 18%. Does this cycle satisfy the inequality of Clausius?

Solution: To check the inequality of Clausius the following integral must be evaluated.

$$\oint \frac{\delta Q}{T} = \frac{Q_{2-3}}{T_{2-3}} - \frac{Q_{4-1}}{T_{4-1}}$$

From the steam tables Appendix 14.2, the two temperatures are

$$T_{2-3} = 453.0 \, K \quad T_{4-1} = 333.2 \, K$$

The heat transfer in the boiler is $u_{2-3} = 1821.2 \, kJ/kg$
   The heat transfer in the condenser is $u_{4-1} = (0.88-0.18)*2204.6 = 1543.2 \, kJ/kg$
Then

$$\oint \frac{\delta Q}{T} = \frac{1821.2}{453.0} - \frac{1543.2}{333.2} = 4.02031 - 4.63145 = -0.61114$$

Therefore, the inequality of Clausius is satisfied.

7.11   Entropy Change for an Irreversible Process

Now consider a cycle composed of two reversible processes as shown in Fig. 7.7.

Process 1_{rev} goes from state A to state B as a reversible process. Process 2_{rev} returns from state B to state A as a reversible process. Process 3_{irr} also returns from state B to state A, but it is an irreversible process. The reversible processes give,

$$\int_A^B \frac{\delta Q}{T}\bigg)_{along\,1} + \int_B^A \frac{\delta Q}{T}\bigg)_{along\,2} = 0 \tag{7.38}$$

But for the cycle A-1-B-3-A,

$$\int_A^B \frac{\delta Q}{T}\bigg)_{along\,1} + \int_B^A \frac{\delta Q}{T}\bigg)_{along\,3} < 0 \tag{7.39}$$

Subtracting Eq. 7.38 from Eq. 7.39 gives

$$\int_B^A \frac{\delta Q}{T}\bigg)_{along\,2} > \int_B^A \frac{\delta Q}{T}\bigg)_{along\,3} \tag{7.40}$$

But along path 2,

$$\Delta S = \int_B^A \frac{\delta Q}{T}\bigg)_{Along\,2}$$

Thus for any path representing any process,

$$\Delta S \geq \int \frac{\delta Q}{T} \quad \text{or} \quad dS \geq \frac{\delta Q}{T} \tag{7.41}$$

The relationship expressed by Eq. 7.41 leads to a fundamental statement. Consider an infinitesimal heat transfer to a system at an absolute temperature T, if the process

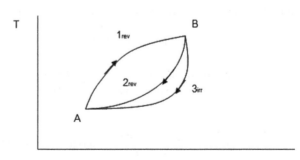

Fig. 7.7 A cycle with an irreversible process

Fig. 7.8 Insulated tank

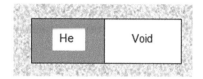

is reversible, the change in entropy is $\Delta Q/T$. If the process is irreversible the change in entropy is greater than $\Delta Q/T$. The effect of irreversibility, e.g. friction, is to increase the entropy of the system.

So to check to see if a process is possible, simply see if it satisfies Eq. 7.41. If it does not then the process is impossible. Entropy plays the same roll for the Second Law, as does energy for the First Law.

Now consider an isolated system. It exchanges no work or heat with its surroundings. The First Law requires that for any process $U_2 = U_1$. For this case Eq. 7.41 becomes $\Delta S \geq 0$. Therefore, the entropy of the isolated system must remain constant or increase. For any real process that must be irreversible to some extent, the entropy of the system increases.

It is possible to consider the universe as an isolated system and break it down to a subsystem and its environment. This would then give,

$$\Delta S_{universe} = \Delta S_{sub-system} + \Delta S_{environment} \geq 0 \qquad (7.42)$$

Once again the equality applies to a reversible process and the inequality applies to any real (irreversible) process. This is sometimes, referred to the mathematical statement of the Second Law.

Example 7.9: Helium is contained in one-half of an insulated tank as shown above, confined by a membrane. The membrane is ruptured and the helium expands to fill the tank. Calculate the specific entropy change for this system. See Fig. 7.8 above

Solution: There is no heat transfer so the final temperature is equal to the initial temperature. Using Eq. 7.23 for the entropy change gives,

$$\Delta s = R * ln(v_2/v_1) = 8.31447/4*ln2 = 1.44 \; kJ/kg/K$$

7.12   The Second Law Applied to a Control Volume

The Second Law can be applied to a control volume in the same way that the First Law was. It simply requires a balance equation that conserves, or increases, entropy. This can be written,

$$\Delta S_{In \; Control \; Volume} + \Delta S_{exiting} - \Delta S_{entering} + \Delta S_{surroundings} \geq 0 \qquad (7.43)$$

This can be refined to

$$\Delta S_{cv} + m_2 s_2 - m_1 s_1 + \frac{\delta Q_{surr}}{T_{surr}} \geq 0$$

Then considering changes as a function of time and using the dot notation to signify a rate this becomes,

$$\dot{S}_{cv} + \dot{m}_2 s_2 - \dot{m}_1 s_1 + \frac{\dot{Q}}{T} \geq 0 \tag{7.44}$$

For a steady flow process, the entropy in the control volume does not change and the mass entering is equal to the mass exiting so this becomes,

$$\dot{m}(s_2 - s_1) + \frac{\delta \dot{Q}_{surr}}{T_{surr}} \geq 0 \tag{7.45}$$

In all cases the equal sign applies to reversible processes and the greater than sign applies to all real (irreversible) processes. **If a process is adiabatic and reversible, $\Delta S = m\Delta s = 0$ and the process is called isentropic.**

Finally, for devices like turbines and compressors, an isentropic process defines the ideal process and the adiabatic efficiency of a device can be defined as,

$$\eta_{turbine} = \frac{W_{actual}}{W_{isentropic}} \qquad \eta_{compressor} = \frac{W_{isentropic}}{W_{actual}} \tag{7.46}$$

Note that since a turbine does work on its environment, the actual work is less than the isentropic work, so the efficiency requires the actual to be divided by the theoretically maximum possible, the isentropic work. On the other hand, for a compressor, the environment is doing work on the working fluid, so the work required will be greater that the theoretical minimum work, the isentropic work. The relationship is reversed.

Example 7.10: Superheated steam enters a turbine at 900 K and 1 MPa and exits at 10 kPa. If the mass flow is 2 kg/sec, determine the power output if the process is assumed to be isentropic.

Solution: The enthalpy for the high temperature steam is 3758.5 kJ/kg and the entropy is 8.09857 kJ/kg/K. For an isentropic expansion the entropy at 10 kPa must be the same. At 10 kPa the saturation entropies are $s_f = 0.64922$ kJ/kg/K and $s_g = 8.14889$ kJ/kg/K, so

$$8.09857 = 0.64922 + x * 7.49968 \; x = (8.09857 - 0.64933)/7.49968 = 0.9933$$

Therefore, the final enthalpy is

$$h = 191.8 + 0.9933 * 2392.1 = 2567.8 \, kJ/kg$$

$$\Delta h = 3758.5 - 2567.8 = 1190.7 \, kJ/kg$$

$$P = 2 * 1190.7 = 2381.4 \, kW$$

Example 7.11: Assume that the turbine in Example 7.10 is 80% efficient. Determine the entropy and temperature of the final state.

Solution: For 80% efficiency, the change in enthalpy must be 80% of the isentropic change in enthalpy.

$$h_{act} = 0.8 * 1190.7 = 952.6 \, kJ/kg$$

$$h_f = 3758.5 - 952.6 = 2805.9 \, kJ/kg$$

At 10 kPa, h=2805.9 kJ/kg gives a vapor in the super heated region between 400 K and 450 K.

$$T_f = 400 + (2805.9 - 2738.7)/(2834.7 - 2738.7) * 50 = 400 + 0.7 * 50 = 435 \, K$$
$$s_f = 8.58137 + 0.7 * (8.80763 - 8.58137) = 8.73975 \, kJ/kg/K$$

Problems

Problem 7.1: A Carnot heat engine produces 1 MW by transferring energy between two reservoirs at 100 and 5 °C. Calculate the rate of heat transfer from the high temperature reservoir and the rate of heat transfer to the low temperature reservoir.

Problem 7.2: A industrial plant wants to use hot groundwater from a hot spring to power a heat engine. The maximum temperature of the ground water is 200 °F, and the average atmospheric temperature is 60 °F. Assume that a supply of water at 1.0 lbm/s is available. What is the maximum power that can be generated?

Problem 7.3: An industrial proposer claims that he can extract 50 kw power by drawing 3000 kJ of heat per minute from a high temperature reservoir at 950 °C and dumping heat to a reservoir at 25 °C. Is this device feasible?

Problem 7.4: A Carnot engine rejects 100 MJ of heat every hour to a low temperature reservoir at 5 °C. If the high temperature reservoir provides 50 kW of heat, what is the power produced and the temperature of the high temperature reservoir?

Problem 7.5: Gas is stored in a rigid container at 7 °C. The volume of the container is 2 m³. The initial gage pressure is 0. The gas is heated and reaches a gage pressure of 0.9 MPa. Atmospheric pressure is 100 kPa. What is the entropy change of the gas if the gas is (a) helium, (b) hydrogen, (c) nitrogen, (d) air, (e) carbon dioxide?

Problem 7.6: The temperature of a gas changes from 17 to 487 °C while the pressure remains constant at 0.2 MPa. Compute the heat transfer and entropy change if the gas is (a) air, (b) helium, or (c) carbon dioxide.

Problem 7.7: A piston cylinder arrangement is used to compress 0.1 kg of air isentropically from initial conditions of 200 kPa and 17 °C to 3.0 MPa. Calculate the work necessary (a) assuming a constant specific heat, and (b) the gas table.

Problem 7.8: 3 kg of steam initially at a quality of 40 % and a pressure of 4 MPa is expanded in a cylinder at constant temperature until the pressure is halved. Determine the entropy change and the heat transfer.

Problem: 7.9: A Carnot engine using steam has a pressure of 50 kPa and a quality of 30 % at the beginning of the adiabatic compression process. If the thermal efficiency is 35 % and the adiabatic expansion process begins with a saturated vapor, determine the heat added.

Problem 7.10: A turbine accepts steam at 2.0 MPa and 627 °C and discharge at 20 kPa. Every second 4 kg of superheated steam passes through the turbine. Calculate the maximum power rating of the turbine.

Problem 7.11: A Carnot engine operates at 6000 cycles per minute with 0.5 lbm of steam per the diagram below. The quality of state 1 is 34.52 %. (A) What is the power output? (B) What is the quality at state 4?

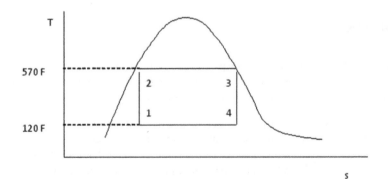

Problem 7.12: A 500 kW turbine works off steam entering at 627 °C and produces saturated steam at 50 kPa. What is the minimum mass flux of steam required?

Chapter 8
Reversible Work, Irreversibility, and Exergy (Availability)

From thermodynamic point of view, work is considered a macroscopic event, such as raising or lowering a weight or winding or unwinding of a spring. In this, chapter we talk about, which work can be reversible or irreversible and what do we mean by either of these processes.

8.1 Reversible Work, and Irreversibility

Second law efficiency is defined as the actual work divided by the reversible work.

$$\eta_{II} = \frac{W_{actual}}{W_{Reversible}} \tag{8.1}$$

It is easiest to explain in terms of a heat engine cycle. Let us consider two heat engines that operate between hot and cold temperature reservoirs. The first engine operates between 600 and 300 K. The second operates between 1200 and 300 K. Both have a thermal efficiency of 30%.

$$\eta_T = 0.30 = \frac{W}{Q_H}$$

Now the Carnot efficiency for a reversible engine is given by,

$$\eta_C = 1.0 - \frac{T_L}{T_H}$$

or

$$\eta_C = 1.0 - \frac{300}{600} = 50\% \text{ for the first}$$

© Springer International Publishing Switzerland 2015
B. Zohuri, P. McDaniel, *Thermodynamics In Nuclear Power Plant Systems*,
DOI 10.1007/978-3-319-13419-2_8

and $$\eta_C = 1.0 - \frac{300}{1200} = 75\% \text{ for the second.}$$

Since they are both operating at a thermal efficiency of 30%, the first is operating at a second law efficiency

$$\eta_{II,1} = \frac{0.3}{0.5} = 0.6 \text{ or } 60\%$$

and the second is operating at a second law efficiency

$$\eta_{II,2} = \frac{0.3}{0.75} = 0.4 \text{ or } 40\%$$

Therefore, the first is making better use of its heat source than the second is. This is the fundamental concept of a second law efficiency. It should be possible to improve the thermal efficiency of the second engine more than the first.

For individual processes, the concept becomes a little more difficult to define. For a turbine or engine the second law efficiency is given by

$$\eta_{II,t} = \frac{W_{Actual}}{W_{Reversible}} \tag{8.2}$$

but for a compressor or pump the second law efficiency is given by

$$\eta_{II,c} = \frac{W_{Reversible}}{W_{Actual}} \tag{8.3}$$

For an open system control volume the first law can be written as

$$\dot{Q} - \dot{W}_S = \dot{m}_2 h_2 - \dot{m}_1 h_1 + \dot{U}_{cv} \tag{8.4}$$

The second law for a control volume is

$$\dot{S}_{cv} + \dot{m}_2 s_2 - \dot{m}_1 s_1 - \frac{\dot{Q}}{T_o} - \dot{S}_{prod} = 0 \tag{8.5}$$

Eliminating $\dot{Q}$ between the two equations gives

$$\dot{W}_S = -\dot{U}_{cv} + T_o \dot{S}_{cv} - \dot{m}_2 (h_2 - T_o s_2) + \dot{m}_1 (h_1 - T_o s_1) - T_o \dot{S}_{prod} \tag{8.6}$$

$\dot{S}_{prod}$ represents the irreversibility. So setting it to zero gives the reversible shaft work. Then integrating over time gives,

$$W_{s,rev} = m_i (u_i - T_o s_i) - m_f (u_f - T_o s_f) + m_1 (h_1 - T_o s_1) - m_2 (h_2 - T_o s_2) \tag{8.7}$$

The actual work can be measured or determined from a First Law analysis

$$W_{s,act} = m_i u_i - m_f u_f + m_1 h_1 - m_2 h_2 + Q \tag{8.8}$$

The second law efficiency then is simply the ratio of the two. The irreversibility is given by

$$I = W_{s,rev} - W_{s,act} = T_o(m_f s_f - m_i s_i + m_2 s_2 - m_1 s_1) - Q \tag{8.9}$$

For a steady flow process

$$\dot{W}_{s,rev} = \dot{m}[h_1 - h_2 + T_o(s_2 - s_1)] \tag{8.10}$$

$$\dot{W}_{s,act} = \dot{m}(h_1 - h_2) + Q \tag{8.11}$$

$$\dot{I} = \dot{m}T_o(s_2 - s_1) - \dot{Q} \tag{8.12}$$

Example 8.1 An ideal steam turbine is supplied with steam at 14 MPa and 1100 K, and exhausts at 60 kPa. (a) Determine the reversible work and irreversibility. (b) If the turbine has an adiabatic efficiency of 0.88, what is the reversible work, irreversibility and second law efficiency?

Solution

a. From the steam tables at 14 MPa & 1100 K, h_1 = 4170 kJ/kg and s_1 = 7.37798 kJ/kg/K. From the saturated pressure table at 60 kPa, s_g = 7.53110 kJ/kg/K which is greater than s_1, so the final steam is in the quality region.

s_f = 1.14524 kJ/kg/K. s_{fg} = 6.38586 kJ/kg/K

h_f = 359.8 kJ/kg h_{fg} = 2293.0 kJ/kg

$$x = \frac{7.37798 - 1.14524}{6.38586} = 0.976$$

h_2 = 359.8 + 0.976 * 2293.0 = 2597.8 kJ/kg

Δh = 4170 − 2597.8 = 1572.2 kJ/kg

w_{rev} = 1572.2 kJ/kg $I = 0$

b. For an adiabatic efficiency of 0.88, $w_{act} = 0.88 * w_{rev} = 1383.5$ kJ/kg

$h_2 = h_1 - 1383.5 = 4170 - 1383.5 = 2786.5$ kJ/kg

Interpolating in the superheated steam tables at 60 kPa, $T_2 = 426.8$,

$s_2 = 7.86890$ kJ/kg/K

Assuming $T_0 = 298$ K,

$w_{rev} = h_1 - h_2 - T_0(s_1 - s_2) = 1383.5 - 298 * (7.37798 - 7.8689) = 1529.8$ kJ/kg

$w_{act} = h_1 - h_2 = 1383.5$ kJ/kg $\eta_{II} = 1383.5/1529.8 = 0.904$ $I = T_o(s_2 - s_1)$

= 146.3 kJ/kg

Example 8.2 Measurements are made on an adiabatic compressor with its supply air at 100 kPa and 300 °K. The exhaust air is measured at 600 kPa and 525 °K. Can these measurements be correct?

Solution At $300\,°K, p_{r1}=1.39$ so $p_{r2}=1.39*600/100=8.34$. This gives $T_{2,rev}=498.2°K$ so since the reversible temperature is less than 525 °K, the process is possible. The adiabatic efficiency can be calculated as $\eta=\Delta h_{rev}/\Delta h_{act}=5826.3/6627.0=0.879$.

8.2 Exergy

It is often important to know the maximum work available from a heat source or thermal body that must reject heat to the local atmosphere or environment. For energy production systems like geothermal or ocean gradients, the peak reservoir temperatures are specified. For combustion systems, the peak temperature of the working fluid is determined by the heating value of the fuel and its complete combustion in air. For nuclear systems and solar thermal systems, the peak temperature is essentially infinite. However, all systems are limited by the material that must contain them, so the peak temperature for combustion systems often does not reach that possible by complete combustion of all of the air available. Nuclear and solar thermal systems are even more limited by material properties because they must use a thermal gradient to transfer heat to the working fluid rather than heating it directly as in combustion systems. Therefore, one of the concepts useful for comparing thermal systems is a concept called *exergy*. It is defined as,

$$X=\left(W_{Reversible}\right)_{Max} \tag{8.13}$$

The maximum reversible work can be obtained from an energy source that must reject heat to the local environment. On the other hand, one can express that Exergy (also called Availability or Work Potential): the maximum useful work that can be obtained from a system at a given state in a given environment; in other words, the most work you can get out of a system. It is also sometimes called Availability and sometimes the symbol Ψ or Φ is used.

The local environment is called the "*dead state*" and symbolize it with a *o* because an energy source that is at the temperature and pressure of the local environment does not have a heat sink to which it can reject heat. To estimate the exergy available both the temperature and pressure of the local environment must be considered.

The work that can be done in expanding from the initial pressure of the fixed mass to the environmental pressure is,

$$W = pdV = (p-p_o)dV + p_o dV = \delta W_{useful} + p_o dV \tag{8.14}$$

For a reversible process, the only way that heat can be transferred is through a Carnot cycle. And since the energy will be removed from the mass with the excess energy above the dead state, the heat transfer will be negative. The work available from the heat engine is given by,

$$\delta W_{HE} = \left(1 - \frac{T_o}{T}\right)\delta Q = \delta Q - \frac{T_o}{T}\delta Q = \delta Q - T_o \frac{\delta Q}{T} \qquad (8.15)$$

In this case the $\dfrac{\delta Q}{T} = dS$ is negative because heat is being transferred from the system, so

$$\begin{aligned}
\delta W_{HE} &= \delta Q - (-T_o dS) = \delta Q + T_o dS \\
\delta Q &= \delta W_{HE} - T_o dS \qquad (8.16)
\end{aligned}$$

The First Law gives for a negative heat transfer

$$dU = -\delta Q - \delta W = -\delta W_{HE} + T_o dS - \delta W_{useful} - p_o dV \qquad (8.17)$$

So,

$$\delta W_{TotalUseful} = \delta W_{HE} + \delta W_{Useful} = -dU - p_o dV + T_o dS \qquad (8.18)$$

Performing the integration from the initial state to the dead state gives,

$$W_{TotalUseful} = X = (U - U_o) + p_o(V - V_o) - T_o(S - S_o) \qquad (8.19)$$

which gives the *exergy* for a closed system.

Now for an open system or control volume the First Law is,

$$\dot{Q} - \dot{W}_S = h_2 \dot{m}_2 - h_1 \dot{m}_1 + \dot{U}_{CV} \qquad (8.20)$$

and for the Second Law,

$$\dot{S}_{prod} = s_2 \dot{m}_2 - s_1 \dot{m}_1 + \dot{S}_{CV} - \frac{\dot{Q}}{T_o} \qquad (8.21)$$

Eliminating $\dot{Q}$ between the First and Second Law equations gives,

$$\dot{W}_s = -\dot{U}_{CV} + T_o \dot{S}_{CV} - (h_2 - T_o s_2)\dot{m}_2 + (h_1 - T_o s_1)\dot{m}_1 - T_o \dot{S}_{Prod} \qquad (8.22)$$

Now integrating over time and noting that S_{Prod} is due to irreversibilities, we have,

$$W_{Rev} = (u_i - T_o s_i)m_i - (u_f - T_o s_f)m_f + (h_1 - T_o s_1)m_1 - (h_2 - T_o s_2)m_2 \qquad (8.23)$$

The actual work can be determined from a First Law analysis as before,

$$W_{Actual} = m_i u_i - m_f u_f + m_1 h_1 - m_2 h_2 + Q \tag{8.24}$$

Then the irreversibility is given by,

$$I = T_o(s_f m_f - s_i m_i + s_2 m_2 - s_1 m_1) - Q \tag{8.25}$$

For steady flow,

$$\dot{W}_{Rev} = (h_1 - h_2)\dot{m} + T_o(s_2 - s_1) \text{ and } \dot{I} = \dot{m}T_o(s_2 - s_1) - \dot{Q} \tag{8.26}$$

Note that the reversible work is not necessarily the same as the isentropic work. If the entropy changes during the process, the reversible work does not equal to isentropic work. So a process must be both adiabatic and reversible to be isentropic.

In all of the above, the changes in kinetic or potential energy have been neglected. If the changes in kinetic or potential energy are significant, they can be included by noting that they will have zero values in the *"dead state"* so these energies are essentially completely recoverable in a reversible process.

If we include the kinetic and potential energies, we can define a state variable *exergy* by,

$$x = h + \frac{v^2}{2} + gz - T_o s \tag{8.27}$$

However, it is important to remember that x has a zero value at the appropriate *"dead state"* of interest for the particular process or engine.

The concept of second law efficiency can be generalized to that of second law effectiveness given by,

$$\varepsilon_{II} = \frac{(\text{exergy produced}) + (\text{work produced}) + (\text{adjusted heat produced})}{(\text{exergy supplied}) + (\text{work used}) + (\text{adjusted heat used})} \tag{8.28}$$

Remember that the adjusted heat produced or used by a device is based on the heat reservoir that interacts with the device and it must do so through a Carnot cycle, or the process would not be reversible.

$$Q_{adjusted} = \left(1 - \frac{T_o}{T_{hr}}\right)Q \tag{8.29}$$

Example 8.3 Which system can do more useful work, 1.0 kg of H_2O at1000 K and 500 kPa or 1.0 kg of Air at 1000 K and 500 kPa.

Solution Assume a dead state of 300 K and 101.325 kPa.

Then the exergy for the H_2O is given by

$$X = m\left[h - h_o - T_o\left(s_1^0 - s_2^0 - R\ln\left(\frac{p}{p_o}\right)\right)\right]$$

$$X = \frac{1}{18}\left[34790.1 - 8853.5 - 300\left(182.830 - 226.5235 - 8.31447\ln\left(\frac{500}{101.325}\right)\right)\right]$$

$$X = 2390.35\ \text{kJ}$$

And the exergy for the air is given by

$$X = m\left[h - h_o - T_o\left(s_1^0 - s_2^0 - R\ln\left(\frac{p}{p_o}\right)\right)\right]$$

$$X = \frac{1}{28.9669}\left[29016.1 - 7414.4 - 300\left(55.7943 - 92.4604 - 8.31447\ln\left(\frac{500}{101.325}\right)\right)\right]$$

$$X = 1262.93\ \text{kJ}$$

Therefore, for the same conditions, the kilogram of water has almost twice the exergy.

Example 8.4 How much useful work is wasted in the condenser of a power plant that takes in steam with a quality of 0.85 and 1 psi and delivers saturated liquid at the same pressure?

Solution Choose the dead state as 537 °R and 1 psi. Note the atmosphere is not the reference here.

The exergy at the inlet is $x_1 = h_1 - h_o - T_o(s_1 - s_o)$

The exergy at the outlet is $x_2 = h_2 - h_o - T_o(s_2 - s_o)$

The difference in exergy is the useful work wasted, so

$W_{wasted} = x_1 - x_2 = h_1 - h_2 - T_o(s_1 - s_2)$

$W_{wasted} = 69.9 + 0.85 * 1037.9 - 537(0.13292 + 0.85 * 1.84897 - 0.13292)$

$W_{wasted} = 0.85 * 1037.9 - 537 * 0.85 * 1.84897 = 38.3\ Btu/lbm$

Example 8.5 Calculate the *exergy* of steam at 1000 °R and 500 psia. The surroundings are at 520 °R and atmospheric pressure.

Solution

$$x = h - h_o - T_o(s - s_o) \quad \text{by definition } h_o = s_o = 0$$

$$x = h - T_o s = 1263.2 - 520 * 1.5253 = 470.0\ \text{kJ/kg}$$

Example 8.6 Determine the second law efficiency for an ideal isentropic nozzle. Hydrogen enters the nozzle at 3000 K and 7.0 MPa with negligible kinetic energy and exits to a pressure of 0.1 MPa. Assume the dead state is 300 K and 0.1 MPa.

Solution Since the process is isentropic, use the gas tables for hydrogen

$$s_2^0 = s_1^0 - R \ln \frac{p_1}{p_2} = 197.898 - 8.31447 * \ln \frac{7}{0.1} = 162.574 \text{ kJ/kmol}$$

$$T_2 = 1045.4K \quad h_1 = 96,299.9 \text{ kJ/kmol} \quad h_2 = 29,631.1 \text{ kJ/kmol}$$

$$h_1 = h_2 + \frac{MV^2}{2} \quad V = \sqrt{\frac{2*(h_1 - h_2)}{M}} = \sqrt{\frac{2*(96,299.9 - 29,631.1)*1000}{2}} = 8,165.1 \text{ m/s}$$

The exergies are

$$x_1 = [h_1 - h_o - T_o(s_1^0 - s_o^0 - R \ln \frac{p_1}{p_o})]/M$$

$$= [96,299.9 - 7,628.8 - 300(197.898 - 125.873 - 8.31447 * \ln \frac{7}{0.1})]/2 = 38,830.4 \text{ kJ/kg}$$

$$x_2 = [h_2 - h_0 + \frac{V^2}{2} - T_o(s_2^0 - s_o^0 - R \ln \frac{p_2}{p_o})]/M$$

$$= (36,259.5 - 7,628.8 + \frac{2*8165.1^2}{2*1000} - 300(168.337 - 125.873)]/2 = 38,830.4 \text{ kJ/kg}$$

And the second law effectiveness is

$$\varepsilon_{II} = \frac{x_2}{x_1} = \frac{38,830.4}{38,830.4} = 1.0$$

which is to be expected since it is an isentropic process.

Problems

Problem 8.1: A pump for a power plant takes in saturated water at 1.5 psi and boosts its pressure to 1500 psi. The pump has an adiabatic efficiency of 0.92. Calculate the irreversibility and the second law efficiency.

Problem 8.2: A power plant uses lake water in the circulating coolant loop. Water enters the loop at 280 K and 100 kPa and exits at 300 K and 90 kPa. If the heat transfer in the loop occurs at 325 K what is the irreversibility?

Problem 8.3: A feed water heater extracts steam from a turbine at 90 psi and 950 °R. It combines the steam with 0.8 lbm/s of liquid at 90 psi and 750 °R. The exhaust is saturated liquid at 90 psi. Determine the second law effectiveness of the heater.

Problem 8.4: A compressor with an adiabatic efficiency of 92 % takes in air at 280 K and 100 kPa and exhausts at 800 kPa. What is (a) the actual work and (b) the reversible work associated with this compressor?

Chapter 9
Gas Kinetic Theory of Entropy

This chapter will attempt to provide a physical understanding of the concept of entropy based on the kinetic theory of gases. Entropy in classical thermodynamics is a mathematical concept that is derived from a closed cycle on a reversible Carnot heat engine. For many students it lacks physical meaning. Most students have a physical understanding of variables like volume, temperature, and pressure. Internal energy and enthalpy are easy to understand, if not intuitive. However, entropy is a bit more difficult. The discussion that follows is an attempt to provide physical insight into the concept of entropy at the introductory level. This discussion closely follows the excellent text "Elements of Statistical Thermodynamics" by L. K. Nash, Dover 2006 [1–4].

In basic mechanics, the student learns that all two-body problems are solvable analytically, and some three-body problems are also solvable. However, beyond that, many-body problems are very complicated. However, they can theoretically be solved numerically for any given set of conditions that specify the initial position and momentum for all of the bodies involved, the problems become extraordinarily difficult for more than a handful of bodies. If we consider only the simplest of interactions, the collisions of hard spheres, for the molecules in a cubic millimeter of gas at standard temperature and pressure there is no hope of producing deterministic solutions for the $\sim 10^{20}$ molecules that could have any engineering meaning. However, we know that the behavior of a gas under these conditions is predictable and we need only measure a few parameters with relatively crude instruments to describe its condition. The analytic technique that allows us to explain how this is possible is statistical mechanics. We can characterize the hard sphere interactions of many molecules in a statistical manner and derive macroscopic properties, or relations between properties that allow us to predict gas behavior.

© Springer International Publishing Switzerland 2015
B. Zohuri, P. McDaniel, *Thermodynamics In Nuclear Power Plant Systems*,
DOI 10.1007/978-3-319-13419-2_9

9.1 Some Elementary Microstate and Macrostate Models

In order to develop this capability, let us start by analyzing some very simple experiments. The simplest experiment is the flipping of a fair coin multiple times. Consider the outcomes of flipping two coins. There are four possible outcomes. We have

First coin	H	H	T	T
Second coin	H	T	H	T

Now consider the total number of heads observed. One-fourth of the time, two heads will be observed. One-half of the time one head will be observed. In addition, one-fourth of the time no heads will be observed. Each tossing sequence is "independent and equally probable", but the "observation of one head" is most likely. Consider what happens when we toss four coins.

First coin	H	H	H	H	T	T	T	T	H	H	H	H	T	T	T	T
Second coin	H	H	H	H	T	T	T	T	T	T	T	T	H	H	H	H
Third coin	H	T	H	T	H	T	H	T	H	T	H	T	H	T	H	T
Fourth coin	H	H	T	T	H	H	T	T	H	H	T	T	H	H	T	T
Number of heads	4	3	3	2	2	1	1	0	3	2	2	1	3	2	2	1

Note that the distribution is,

One observation with zero heads,
Four observations with one head,
Six observations with two heads,
Four observations with three heads, and
One observation with four heads.

The number of observations W, with H heads, and T tails, out of N tosses, can be predicted by the formula

$$W = \frac{N!}{H!T!} \tag{9.1}$$

$$\text{Zero Heads}\quad W = \frac{4!}{0!4!} = 1$$

$$\text{One Head}\quad W = \frac{4!}{1!3!} = 4$$

$$\text{Two Heads}\quad W = \frac{4!}{2!2!} = \frac{4*3}{1*2} = 6$$

$$\text{Three Heads}\quad W = \frac{4!}{3!1!} = 4$$

$$\text{Four Heads}\quad W = \frac{4!}{4!0!} = 1$$

Now consider what happens with eight coins. We have

$$\text{Zero Heads} \quad W = \frac{8!}{0!8!} = 1$$

$$\text{One Head} \quad W = \frac{8!}{1!7!} = 8$$

$$\text{Two Heads} \quad W = \frac{8!}{2!6!} = 28$$

$$\text{Three Heads} \quad W = \frac{8!}{3!5!} = 56$$

$$\text{Four Heads} \quad W = \frac{8!}{4!4!} = 70$$

$$\text{Five Heads} \quad W = \frac{8!}{5!3!} = 56$$

$$\text{Six Heads} \quad W = \frac{8!}{6!2!} = 28$$

$$\text{Seven Heads} \quad W = \frac{8!}{7!1!} = 8$$

$$\text{Eight Heads} \quad W = \frac{8!}{8!0!} = 1$$

Note that the peak number of observations always occurs for the distribution that gives half heads and half tails, exactly as we would expect.

Now it will be useful to identify each of the possible outcomes of flipping the coins a "microstate" for our system. Each of these microstates is assumed to be equally likely. With the two coins there are four such microstates. With the four coins, there are 16 such microstates. When we go to eight coins, there are 256 such microstates. Now the observed results of the coin flips will be called a "macrostate" for the system. For the two coins, there are three macrostates (1, 2, or 3 heads) that can be distinguished. For the four coins, there are five macrostates that can be distinguished. And for the eight coins there are nine macrostates that can be distinguished. The probability of observing a macrostate will depend upon what fraction of the microstates possible that will produce an outcome that is observed as that macrostate.

In order to compare these distributions with distributions for larger numbers of coins it is useful to normalize them by dividing the number of observations for each number of heads by the number of observations for the most likely number of heads. When we do this, we get the following normalized values for the observation of the number of heads in each case.

Two coins		Four coins		Eight coins	
0 Heads	0.5000	0 Heads	0.1667	0 Heads	0.0143
				1 Head	0.1143
		1 Head	0.6667	2 Heads	0.4000

Two coins		Four coins		Eight coins	
				3 Heads	0.8000
1 Head	1.0000	2 Heads	1.0000	4 Heads	1.0000
				5 Heads	0.8000
		3 Heads	0.6667	6 Heads	0.4000
				7 Heads	0.1143
2 Heads	0.5000	4 Heads	0.1667	8 Heads	0.0143

A plot of the normalized data is presented in the Fig. 9.1 along with the data for 16 and 32 coins.

Note that the curve for the frequency of the number of observed heads peaks at the 50 % point and the curves get narrower as the number of coins increases. So we can say that the most likely observation is that half of the coins are heads, and as we increase the number of coins, it is more and more likely that we get closer to the 50 % heads observation. In fact by the time we have flipped 14,000 coins, a 1 % deviation from the expected observation of 7000 heads will occur less than 50 % of the time. By the time, we have flipped 46,000 coins a 1 % deviation from the expected 23,000 heads will occur less than 10 % of the time. By the time we have flipped 92,000 coins, a 1 % deviation from the expected 46,000 heads will occur less than 1 % of the time. As the number of coins increases, the probability of observing a macrostate deviation from the expected 50 % heads that is measurable continues to shrink until it is "impossible". Certainly if we flipped 10^{20} "fair" coins the probability of measuring a number of heads that deviated from the 5×10^{19} heads estimate by more than a hundredth of a percent would be negligible.

Now consider an example closer to a system of interest in thermodynamics. In a solid, the atoms vibrate about their equilibrium positions at all temperatures above absolute zero, or the zero internal energy level. Quantum mechanics tells us that the frequency of vibration is quantitized and that only certain frequencies are allowed.

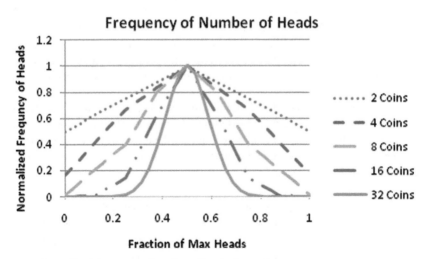

Fig. 9.1 Normalized frequency of number of heads observed

This leads to the following equation for the allowed energy levels for a vibrating atom.

$$\varepsilon_n = \left(n + \frac{1}{2}\right)hv \qquad (9.2)$$

Where v is the fundamental vibration frequency and h is Plank's constant—required to convert a frequency to energy units. The ground state for the atom corresponds to $n=0$ and each increasing integral value of **n** corresponds to the atom acquiring another quantum of energy. This model establishes a uniform energy grid with units of energy being passed from one atom to the next as a single phonon.

Now let us consider three distinguishable atoms and distribute three quanta or units of energy among them. This can be done ten possible ways.

Configuration	Atom 1	Atom 2	Atom 3
1	3	0	0
2	0	3	0
3	0	0	3
4	2	1	0
5	0	2	1
6	1	0	2
7	2	0	1
8	1	2	0
9	0	1	2
10	1	1	1

Therefore, there are ten microstates for this problem, but only three macrostates. We can observe in the first macrostate that one atom of the three has all three units of energy. In the second macrostate, one atom has two units of energy and another has one unit of energy. Finally, each of the atoms could have a single unit of energy. Three microstates contribute to the first macrostate, six microstates to the second, and one microstate to the third. If each microstate is equally likely, then the macrostates should be observed in the ratio of 3:6:1. Note that the seemingly best-ordered macrostate, one unit of energy in each atom, is the least likely.

As we increase the number of microstates, it will become important to have a generic formula to predict the number of microstates that contribute to a particular macrostate. Let us start with the second macrostate. To achieve this macrostate there are three ways that we can assign the first two quanta of energy. Once this has been done, there are two atoms left that we can assign the remaining quanta. Then we can assign zero quanta to the last atom. Therefore, there are 3 times 2 times 1 ways this macrostate can be achieved. So we will have $3 \times 2 \times 1$ or 3! Microstates. Now consider the first macrostate. There are three choices for assigning all three quanta of energy. Once that is done we can assign zero quanta to either of the remaining two atoms and then assign another zero quanta to the last atom. However, these two choices are indistinguishable. So once again, we have 3! Choices, but 2 or 2! of them are indistinguishable. Finally consider the third macrostate. We can assign the first quanta to any of the three atoms, the second quanta to either of the two remaining, and the last quanta to the last atom. There are once again 3! Ways of assigning the quanta, but 3! of them are indistinguishable. So we have

Macrostate 1: $3!/2! = 3$
Macrostate 2: $3!/1! = 6$
Macrostate 3: $3!/3! = 1$

Now let's extend this argument to N atoms. We can assign a specific number of quanta of energy to the first atom in N ways, to the second in $N-1$ ways, to the third in $N-2$ ways, etc. So the total number of ways we can assign quanta to N atoms is N!. However, if some of the numbers of quanta are identical, the resulting macrostates will be indistinguishable. So we must reduce the number of microstates used to form a macrostate by $N_a!$, where N_a is the number of times a given number of quanta is repeated. Thus the number of microstates contributing to a macrostate can be written as $N!/N_a!$. If there are two groups of quanta that contain the same number of quanta, the number of microstates contributing to this macrostate will be given by $N!/(N_a!N_b!)$. The general formula will then become, letting W equal the number of microstates contributing to a particular macrostate

$$W = \frac{N!}{\prod_{n_q} N_{n_q}!} \tag{9.3}$$

Consider now two additional examples. First, consider the observable macrostates when five quanta of energy are distributed among five atoms. The macrostates are, (Fig. 9.2).

Macrostate 1: All five quanta to one atom—$5!/(1!*4!) = 5$
Macrostate 2: Four quanta to one atom, and 1 to another—$5!/(1!*1!*3!) = 20$
Macrostate 3: Three quanta to one atom, 2 to another—$5!/(1!*1!*3!) = 20$
Macrostate 4: Three quanta to one atom, 1 to a second and 1 to a third—
              $5!/(1!*2!*2!) = 30$
Macrostate 5: Two quanta to one atom, two quanta to a second, and 1 to a third—
              $5!/(2!*1!*2!) = 30$
Macrostate 6: Two quanta to one atom, and 1 quantum to each of 3 atoms—
              $5!/(1!*3!*1!) = 20$
Macrostate 7: One quantum to each atom—$5!/(5!) = 1$

So there are a total of 126 microstates, but only 7 observable macrostates.

Now consider the case where we add five amore atoms but no more quanta. For ten atoms and five quanta, we have the same number of macrostates as above, but there are more indistinguishable microstates contributing to each because of the added atoms that do not receive a quantum of energy (Fig. 9.3). The calculation is

Macrostate 1: $10!/(1!*9!) = 10$
Macrostate 2: $10!/(1!*1!*8!) = 10*9 = 90$
Macrostate 3: $10!/(1!*1!*8!) = 10*9 = 90$
Macrostate 4: $10!/(1!*2!*7!) = 10*9*8/2 = 360$

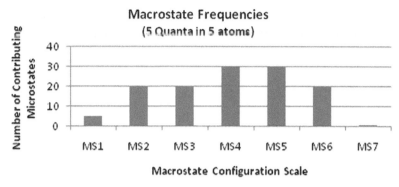

Fig. 9.2 Macrostate frequencies for five quanta in five atoms

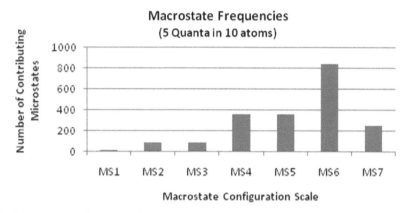

Fig. 9.3 Macrostate frequencies for five quanta in ten atoms

Macrostate 5: $10!/(2!*1!*7!) = 10*9*7/2 = 360$
Macrostate 6: $10!/(1!*3!*6!) = 10*9*8*7/3/2 = 840$
Macrostate 7: $10!/(5!*5!) = 10*9*8*7*6/5/4/3/2 = 252$

Note that we only added five atoms or doubled the original number, but the total number of microstates increased to 2002, or an increase of almost a factor of 16. Also, note that of the 2002 microstates possible, 840 or about 42 % of them contributed to the most probable macrostate. If we assign an equal probability to each microstate, Macrostate 6 is 2.33 times as likely as its nearest competitors are.

9.2   Stirling's Approximation for Large Values of N

Now since we will want to consider the number of atoms in a realistic macroscopic piece of material, which for a gas is on the order of 10^{20}–10^{27} atoms, we will need a better way to compute factorials than simply multiplying them out. Also since the magnitude of N! goes up very rapidly as N increases, it will be useful to look at

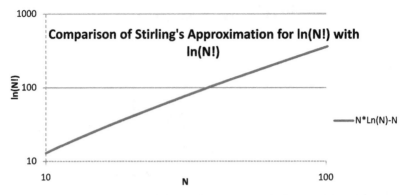

Fig. 9.4 Comparison of Stirling's approximation to ln(N!) to ln(N!) (N=10 to N=100)

ln(N!) rather than simply N!. Our formula for the natural logarithm of the number of microstates, ln(W), then becomes,

$$\ln(W) = \frac{\ln(N!)}{\sum\limits_{n_q} \ln(N_{n_q}!)} \qquad (9.4)$$

And once we have calculated ln(W) it is straightforward to calculate W, but that can often be too large of a number to store on most calculators or computers. In most cases we will simply work with ln(W). Of course we really haven't helped ourselves very much as ln(N!) is as difficult to calculate as N!. However there exists a very accurate approximation to ln(N!) developed by Stirling that gives ln(N!) as

$$\ln(N!) \sim N*\ln(N) - N \qquad (9.5)$$

A comparison of Stirling's approximation and the actual $\ln(N!)$ is given in Fig. 9.4. Since it is hard to tell the difference between the two curves in Fig. 9.4, the fractional error in Stirling's approximation is plotted in Fig. 9.5.

It can be seen that the approximation is accurate to less than a percent by the time N reaches 100. For the cases we will be interested when N approaches 10^{20} or greater, Stirling's approximation is very accurate.

9.3   The Boltzmann Distribution Law

We saw from the coin flip problem and the five quanta distributed among ten atoms that the most probable macrostate tends to dominate as the number of microstates increases. If we are going to deal with very large numbers of atoms, the most probable macrostate will tend to dominate even more. So it would be useful to be able to find this macrostate very quickly. When we do this we are asking, what is the distribution of the N_{nq} that define the most likely configuration macrostate. Now

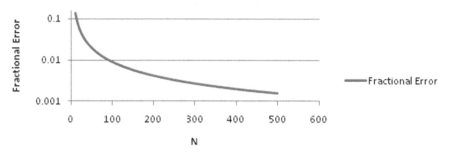

Fig. 9.5 Fractional error in Stirling's approximation ($N=10$ to $N=500$)

when we are dealing with very large numbers of microstates, we will also have very significant numbers of macrostates. These macrostates can be packed so close together and ordered in such a way that W can be thought of as a function of a configuration index for macrostates similar to the way it is presented in Fig. 9.3. We can let this variable be called C. Then if we think of W as a continuous function of C, we can find the most likely macrostate by differentiating W, or $\ln W$, as a function of C.

$$\frac{dW}{dC} \tag{9.6}$$

Our formula for W is then,

$$W = \frac{N!}{N_a!...N_l!N_m!N_n!...N_z!} \tag{9.7}$$

where the levels l, m, and n are identified, and it is not implied that only z levels are available but only there are some finite number of levels of interest. At the peak value of W we will assume that the levels l, m, and n have the occupation numbers N_l, N_m, and N_n. For a normal solid, the numbers N_l, N_m, and N_n are assumed to be quite large. To get the derivative with respect to the configuration index C, we will then perturb the numbers N_l, N_m, and N_n to see what a small change in C will have on W. Basically setting the derivative to zero means that small changes in C will have a negligible effect on W. However we must conserve energy in doing this, so let us transfer one atom from level l to level m and one atom from level n to level m. If the levels are equally spaced in energy, this will conserve energy. Now we have

$$N_l\,' = N_l - 1, \qquad N_m = N_m + 2, \quad and\ N_n\,' = N_n - 1$$

where the primes indicate the new occupation numbers for the three levels. Our new configuration then will be,

$$W = \frac{N!}{N_a!...(N_l-1)!(N_m+2)!(N_n-1)!...N_z!} \tag{9.8}$$

Since we assumed the original configuration was the max W configuration, an infinitesimal change like we have just made should not change the value of W as at the peak W, $dW/dC=0$. Therefore, these two values of W should be equal. We have,

$$\frac{N!}{N_a!...N_l!N_m!N_n!...N_z!} = \frac{N!}{N_a!...(N_l-1)!(N_m+2)!(N_n-1)!...N_z!}$$

Canceling common terms this becomes,

$$\begin{cases} N_l!*N_m!*N_n! = (N_l-1)!*(N_m+2)!*(N_n-1)! \\ \dfrac{N_l!}{(N_l-1)!}*\dfrac{N_n!}{(N_n-1)!} = \dfrac{(N_m+2)!}{N_m!} \\ \dfrac{N_l*(N_l-1)!}{(N_l-1)!}*\dfrac{N_n*(N_n-1)!}{(N_n-1)!} = \dfrac{(N_m+2)*(N_m+1)N_m!}{N_m!} \\ N_l*N_n = (N_m+2)*(N_m+1) \end{cases} \quad (9.9)$$

Now the 1 and 2 are negligible compared to N_m so this becomes

$$N_l*N_n = N_m^2 \quad (9.10a)$$

or

$$\frac{N_l}{N_m} = \frac{N_m}{N_n} \quad (9.10b)$$

and

$$\frac{N_k}{N_l} = \frac{N_l}{N_m} = \frac{N_m}{N_n} = \frac{N_n}{N_o} = \frac{N_o}{N_p} = ... \quad (9.11)$$

Basically, we have a geometric series. Now consider a system of atoms where the spacing between energy levels is not a constant. We will have,

$$\frac{\varepsilon_n - \varepsilon_m}{\varepsilon_m - \varepsilon_l} = \frac{p}{q} \quad (9.12)$$

where p and q are small integers. Now let us withdraw $p+q$ units of energy from level m and transfer q of them to level n and p of them to level l. This change will maintain the total energy constant as we can see by writing the previous equation as,

$$(\varepsilon_n - \varepsilon_m)q + p(\varepsilon_l - \varepsilon_m) = 0 \quad (9.13)$$

Now if the original configuration was the maximum W configuration, we will have

$$\frac{N!}{N_a!.....N_l!N_m!N_n!....N_z!} = \frac{N!}{N_a!.....(N_l+p)!(N_m-p-q)!(N_n+q)!....N_z!}$$

$$\frac{(N_l+p)!\,(N_n+q)!}{N_l!}\cdot\frac{}{N_n!} = \frac{N_m!}{(N_m-p-q)!}$$

$$\frac{(N_l+p)....(N_l+1)N!}{N_l!}\cdot\frac{(N_n+q).....(N_n+1)N_m!}{N_n!}$$

$$= \frac{N_m(N_m-1)....(N_m-p-q+1)(N_m-p-q)!}{(N_m-p-q)!}$$

$$(N_l+p)...(N_l+1)(N_n+q)....(N_n+1) = N_m(N_m-1)...(N_m-p-q+1) \quad (9.14)$$

Now since we chose p and q as small integer they are negligible once again when compared with N_l, N_m, and N_n. This then gives,

$$N_l^p N_n^q = N_m^{p+q} \qquad (9.15)$$

Of course the equation we derived for uniform spacing can be represented by this equation with $p=1$ and $q=1$. This equation can then be written as,

$$p\ln\left(\frac{N_l}{N_m}\right) = q\ln\left(\frac{N_m}{N_n}\right)$$

or

$$\frac{1}{\varepsilon_m - \varepsilon_l}\ln\left(\frac{N_l}{N_m}\right) = \frac{1}{\varepsilon_n - \varepsilon_m}\ln\left(\frac{N_m}{N_n}\right) = \beta \qquad (9.16)$$

where β is just a constant because we could write the same relationship for any set of adjoining levels. So we have

$$\ln\left(\frac{N_n}{N_i}\right) = \beta(\varepsilon_i - \varepsilon_n) \qquad (9.17)$$

and level i and level n are now not necessarily adjoining levels. So if we define the ground state level with the subscript o, we can write

$$\ln\left(\frac{N_n}{N_o}\right) = -\beta\varepsilon_n$$

or

$$\frac{N_n}{N_o} = e^{-\beta \varepsilon_n} \tag{9.18}$$

which is the celebrated Boltzmann distribution law for the most likely configuration energy distribution.

9.4   Estimating the Width of the Most Probable Macrostate Distribution

With the Boltzmann distribution law stated, we can now realistically estimate the width of the peak configuration in configuration space. Let us choose a set of small parameters to quantify the relative change in the population numbers for each level. We will define,

$$\alpha_n = \frac{N_n' - N_n}{N_n} \tag{9.19}$$

If the population of a level increases, α will be positive, and if the population of a level decreases, α will be negative. Since we are only shifting the configuration slightly, we would expect $|\alpha| \ll 1$. We can write,

$$N_n' - N_n = \alpha_n N_n \tag{9.20}$$

Now since we have a fixed total number of atoms we must have,

$$\Delta N = (N_o' - N_o) + (N_1' - N_1) + \ldots = \sum_n \alpha_n N_n = 0 \tag{9.21}$$

And since we are considering no change in energy, we have,

$$\Delta E = \varepsilon_o (N_o' - N_o) + \varepsilon_1 (N_1' - N_1) + \ldots = \sum_n \varepsilon_n \alpha_n N_n = 0 \tag{9.22}$$

Now we can write,

$$N_n' = N_n + \alpha_n N_n$$

and

$$\frac{W}{W'} = \frac{N! \Big/ \prod_n N_n!}{N! \Big/ \prod_n (N_n + \alpha_n N_n)!} = \frac{\prod_n (N_n + \alpha_n N_n)!}{\prod_n N_n!} \tag{9.23}$$

Taking natural logarithms of both sides gives,

$$\ln\left(\frac{W}{W'}\right) = \sum_n \ln[(N_n + \alpha_n N_n)!] - \sum_n \ln[N_n!] \qquad (9.24)$$

All of the term should be large enough to qualify for Stirling's approximation, so

$$\ln\left(\frac{W}{W'}\right) = \sum_n (N_n + \alpha_n N_n)\ln(N_n + \alpha_n N_n) - \sum_n (N_n + \alpha_n N_n) - \sum_n N_n \ln N_n + \sum_n N_n$$

This can be written,

$$\ln\left(\frac{W}{W'}\right) = \sum_n N_n \ln N_n - \sum_n N_n + \sum_n N_n \ln(1 + \alpha_n) + \sum_n \alpha_n N_n \ln N_n$$
$$+ \sum_n \alpha_n N_n \ln(1 + \alpha_n) N_n - \sum_n N_n \ln N_n + \sum_n N_n - \sum_n \alpha_n N_n$$

Cancelling terms

$$\ln\left(\frac{W}{W'}\right) = \sum_n N_n \ln(1 + \alpha_n) + \sum_n \alpha_n N_n \ln N_n + \sum_n \alpha_n N_n \ln(1 + \alpha_n) - \sum_n \alpha_n N_n$$

and for $|\alpha_n| \ll 1$, $\ln(1 + \alpha_n) \sim \alpha_n$

$$\ln\left(\frac{W}{W'}\right) = \sum_n \alpha_n N_n + \sum_n \alpha_n N_n \ln N_n + \sum_n \alpha_n^2 N_n - \sum_n \alpha_n N_n = \sum_n \alpha_n N_n \ln N_n + \sum_n \alpha_n^2 N_n$$

Now remembering the Boltzmann distribution law,

$$\ln N_n = \ln N_o = \beta \varepsilon_n$$

We have,

$$\ln\left(\frac{W}{W'}\right) = \ln N_o \sum_n \alpha_n N_n - \sum_n \beta \varepsilon_n \alpha_n N_n + \sum_n \alpha_n^2 N_n \qquad (9.25)$$

By the requirements to conserve atoms and energy, the first two summations must be identically zero. Therefore,

$$\ln\left(\frac{W}{W'}\right) = \sum_n \alpha_n^2 N_n \qquad (9.26)$$

Note that all of the terms in this summation are positive because α_n is squared. So let us define an RMS value by,

$$\bar{\alpha} = \left(\frac{\sum_n \alpha_n^2 N_n}{N} \right)^{1/2} \tag{9.27}$$

or

$$N\bar{\alpha}^2 = \sum_n \alpha_n^2 N_n \tag{9.28}$$

$$\ln\left(\frac{W}{W'} \right) = N\bar{\alpha}^2$$

$$\frac{W'}{W} = e^{-N\bar{\alpha}^2} \tag{9.29}$$

Now consider a small sample of material that only contains 6×10^{20} atoms, a milli-mole. In addition, consider a shift from the most likely configuration that involves an RMS change in occupation numbers that is one part in a billion (10^9). We have,

$$\bar{\alpha} = 1 \times 10^{-9}$$
$$\bar{\alpha}^2 = 10^{-18} \tag{9.30}$$
$$\left(\frac{W'}{W} \right) = e^{-(6 \times 10^{20} \cdot 10^{-18})} = e^{-6 \times 10^2} = e^{-600} = 10^{-260}$$

or a configuration whose occupation numbers differ from the maximum W configuration by an RMS value of one part in a billion will be 10^{-260} times less likely to be observed than the maximum W configuration. Basically, the only measurable configuration that can be observed will look like the maximum W configuration.

9.5 Estimating the Variation of W with the Total Energy

Now let us consider how W changes with changes in the total energy available to a well defined number of atoms. We have,

$$W = \frac{N!}{\prod_n N_n!} \tag{9.31}$$

$$\ln W = \ln N! - \sum_n \ln N_n!$$

Differentiating,

$$d \ln W = -\sum_n d \ln N_n !$$

Applying Stirling's approximation,

$$d \ln N_n! = d(N_n \ln N_n - N_n) = N_n \frac{dN_n}{N_n} + \ln N_n dN_n - dN_n = \ln N_n dN_n$$

$$d \ln W = -\sum_n \ln N_n dN_n \tag{9.32}$$

Moreover, for the most likely configuration, we have,

$$\ln N_n = \ln N_o - \beta \varepsilon_n \tag{9.33}$$

$$d \ln W = -\ln N_o \sum_n dN_n + \beta \sum_n \varepsilon_n dN_n \tag{9.34}$$

The first sum on the right hand side is identically zero for the fixed number of atoms under consideration. The second term on the right hand side is simply,

$$\beta \sum_n \varepsilon_n dN_n = \beta \sum_n d(\varepsilon_n N_n) = \beta d \sum_n \varepsilon_n N_n = \beta dE$$

So we have

$$d \ln W = \beta dE \tag{9.35}$$

This can be integrated for a constant β to give,

$$\ln W_2 - \ln W_1 = \beta(E_2 - E_1)$$

$$\ln \left(\frac{W_2}{W_1} \right) = \beta(E_2 - E_1) \tag{9.36}$$

$$W_2 = W_1 e^{\beta(E_2 - E_1)}$$

Thus W the number of microstates contributing to a macrostate increases exponentially with increasing energy.

It is worth pointing out at this point that we have only demonstrated that this equation holds for a system of distinguishable objects like our fixed-in-place atoms, though it does in fact hold more generally. Several other restrictions or caveats still apply at this point. The first is the assumption that by transferring energy or heat to our system, there is no shift in the quantum levels of our oscillators. This is a reasonable assumption for oscillating atoms in a solid. In a gas where it can be shown that the energy levels are a function of the volume containing the atoms or molecules, our result only applies to constant volume increases in energy. A second obvious restriction is that the addition of energy will cause no change in the number of atoms present due to chemical or nuclear reactions. Finally, we have assumed β is constant. If β changes, we have a different system.

9.6 Analyzing an Approach to Thermal Equilibrium

Now let us consider two solid bodies, X and Y, that will be brought together in thermal contact. Before they are brought together, they will be in the macrostates characterized by W_X and W_Y microstates respectively. So if there are W_X microstates in body X and W_Y microstates in body Y, when the two bodies are considered jointly, there are a total of $W_X * W_Y$ microstates possible by combining each microstate in body X with each of the microstates in body Y. Now when the two bodies are brought into thermal contact and allowed to redistribute their energy between them, the number of possible microstates for the most likely configuration will change. If they were in equilibrium before thermal contact with different values of W, what will their final values of W be.

For the approach to equilibrium and eventual equilibrium, the relevant expression is,

$$d(W_X * W_Y) \geq 0 \tag{9.37}$$

During the approach to equilibrium, the product $W_X * W_Y$ must increase or remain constant. When it remains constant, the two bodies have reached an equilibrium state. We can rewrite this as,

$$W_Y dW_X + W_X dW_Y \geq 0$$
$$\frac{dW_X}{W_X} + \frac{dW_Y}{W_Y} \geq 0$$

or

$$d \ln W_X + d \ln W_Y \geq 0 \tag{9.38}$$

and substituting for $d \ln(W)$ from above gives,

$$\beta_X dE_X + \beta_Y dE_Y \geq 0 \tag{9.39}$$

Now if X and Y constitute an isolated system such that their combined energies can neither increase nor decrease as they come into thermal equilibrium, we must have,

$$E_X + E_Y = Cons \tan t$$
$$dE_X + dE_Y = 0$$
$$dE_X = -dE_Y$$

Therefore, we have

$$\beta_X dE_X - \beta_Y dE_X \geq 0 \tag{9.40}$$

Now suppose that the approach to equilibrium starts with body Y at a higher temperature than body X so that heat or energy flows from body Y to body X. Then dE_X must be greater than zero. Or,

$$(\beta_X - \beta_Y)dE_X \geq 0$$
$$\beta_X \geq \beta_Y \tag{9.41}$$

So as long as β_X is greater that β_Y, energy will flow from body Y to body X. When β_X equals β_Y, equilibrium will be obtained.

9.7 The Physical Meaning of β

β then appears to have the behavior of an inverse temperature. So we could write,

$$\beta = \frac{1}{T} \tag{9.42}$$

but this would cause trouble dimensionally. Remember we had that

$$\frac{N_n}{N_o} = e^{-\beta \varepsilon_n} \tag{9.43}$$

Therefore β must have the dimensions of energy. This simply means that we must multiply the temperature by a constant that converts temperature to energy. We have used a constant like this before and we called it Boltzmann's constant, $k = 1.381 \times 10^{-23}$ joules/K or 8.62×10^{-5} eV/K.

If we use this constant to define β, we have $\beta = 1/kT$ and,

$$\frac{N_n}{N_o} = e^{-\frac{\varepsilon_n}{kT}} \tag{9.44}$$

We have also required that at the lowest energy level for any macroscopic system, all of the atoms must be in the ground state. We have chosen the energy of the ground state to be zero. Therefore, our formulation of β as $1/kT$ is consistent in that at absolute zero temperature, only the ground state will have a non-zero population.

9.8   The Concept of Entropy

Now for any isolated thermodynamic system we have asked that any changes that take place must have $dW \geq 0$. The direction of any spontaneous change will be that that causes W to increase and equilibrium will be attained when W attains its maximum value. Now as we have shown, any value of W for a macroscopic system will be immense, so for convenience sake we will find it easier to talk about $ln(W)$ and perform our analysis with $ln(W)$. Note that $d\ln(W) \geq 0$ whenever $dW \geq 0$ and the equilibrium arguments apply to it as well.

Now consider once again when we combine two systems, X and Y. W for the combined system will be $W_{X}{\cdot}W_{Y}$, but $ln(W)$ will be $ln(W_X) + ln(W_Y)$, so that $ln(W)$ combines like any extensive property for thermodynamic systems. So we could define an extensive parameter for a macroscopic thermodynamic system as $S = \ln(W)$. We could then write that $S_{XY} = S_X + S_Y$. However we can do somewhat better by defining $S = kS = kln(W)$. This is Boltzmann's famous equation developed even before we were certain that atoms existed.

$$S = k \cdot \ln(W) \tag{9.45}$$

Of course, S will satisfy the requirements for an **extensive** system parameter. By incorporating the constant k, we will place our statistical definition of entropy on the same scale as our classical thermodynamic entropy scale. We have,

$$d\ln W = \beta dE$$
$$kd\ln W = k\beta dE$$
$$d(k\ln W) = k\frac{1}{kT}dE$$
$$dS = \frac{dE}{T} = \frac{\delta Q}{T} = \frac{dU}{T} \tag{9.46}$$

Note also that

$$S \rightarrow 0 \ as \ T \rightarrow 0 \tag{9.47}$$

So this definition of entropy also satisfies the Third Law of Thermodynamics, or the so called Nernst Postulate.

9.9 Partition Functions

Now let us return to the Boltzmann distribution law for distinguishable atoms in a solid. We have

$$\frac{N_n}{N_o} = e^{-\beta \varepsilon_n}$$

$$N_n = N_o e^{-\beta \varepsilon_n} \tag{9.48}$$

Then the total number of atoms or molecules is given by

$$N = N_o + N_1 + N_2 + N_3 + \ldots \ldots$$

$$N = N_o e^{-\beta \varepsilon_o} + N_o e^{-\beta \varepsilon_1} + N_o e^{-\beta \varepsilon_2} + N_o e^{-\beta \varepsilon_3} \ldots \ldots$$

$$N = N_o (e^{-\beta \varepsilon_o} + e^{-\beta \varepsilon_1} + e^{-\beta \varepsilon_2} + e^{-\beta \varepsilon_3} \ldots \ldots) \tag{9.49}$$

$$N = N_o \sum_q e^{-\beta \varepsilon_q}$$

And we can write

$$N_o = \frac{N}{\sum_q e^{-\beta \varepsilon_q}}$$

$$N_n = \frac{N e^{-\beta \varepsilon_n}}{\sum_q e^{-\beta \varepsilon_q}}$$

$$Z = \sum_q e^{-\beta \varepsilon_q} = Partition\ Function$$

$$N_n = \frac{N}{Z} e^{-\beta \varepsilon_n} \tag{9.50}$$

Note that if we choose ε_0 to be equal to 0, Z will always be greater than 1.0 in magnitude and the magnitude of the summation will depend on how widely spaced the energy levels are in units of $1/\beta$.

9.10 Indistinguishable Objects

Everything we have done so far applies to distinguishable objects. That is each atom in a solid is located in one place in the crystal and it can be distinguished in space from all other atoms.

However, when we want to consider a dilute gas, all of the atoms or molecules are continuously moving about and it is impossible to assign an observational uniqueness to each of them. In addition, the energy of interest for each of the atoms or molecules is their kinetic energy. Therefore, we have,

$$\varepsilon = \frac{1}{2}mv^2 \tag{9.51}$$

The energy for a classical particle is not quantitized but rather is a continuous variable. More important is the fact that the speed and energy of a particle can have a definite value, but it can be traveling in an infinite number of directions. Thus there are many physically different locations and directions of movement that these atoms and molecules can have that are characterized by the same energy. When this happens, we say that the energy levels are *degenerate*.

To start let us constrain the atoms to move only in the x, y, or z direction. Then we will have for the occupation numbers the following,

$$N_{xi} = \frac{N}{Z}e^{-\beta\left(\frac{1}{2}mv_{xi}^2\right)}\, N_{yi} = \frac{N}{Z}e^{-\beta\left(\frac{1}{2}mv_{yi}^2\right)}\, N_{zi} = \frac{N}{Z}e^{-\beta\left(\frac{1}{2}mv_{zi}^2\right)} \tag{9.52}$$

But $v_{xi} = v_{yi} = v_{zi}$ so we can write,

$$N_{xi} = N_{yi} = N_{zi}$$

$$N_i = \frac{N}{Z}3e^{-\beta\left(\frac{1}{2}mv_i^2\right)} \tag{9.53}$$

where the three appears because the energy level is three-fold degenerate. In general, we can write this as

$$N_i = \frac{N}{Z}\omega_i e^{-\beta\varepsilon_i}$$

$$Z = \sum_i \omega_i e^{-\beta\varepsilon_i} \tag{9.54}$$

ω_i is the number of the degeneracy of the ith energy level.

Now consider the following example. Let us take a cubic meter sphere of argon at 15 °C and one atmosphere. The radius of the argon atom is approximately 0.181 nm. So it will occupy 2.5×10^{-29} m^3. Or in one cubic meter there will be room for 4.0×10^{28} cubicles that could contain an argon atom. Now consider the directions that these atoms could move. For an atom at the center of a sphere containing one cubic meter, we can compare the solid angle formed by the projection of the atom's cross sectional area on the interior surface of the sphere with the total surface are of the sphere. There are approximately 1.17×10^{19} locations that the atom moving from the center of the sphere to its inner surface that the atom could hit without overlapping with another location. That is there are 1.17×10^{19} different directions that could be observed as distinct for this atom departing from the center of the sphere towards the outer wall. Of course, not all of the atoms can be located at the center of the sphere simultaneously. But as an atom is placed away from the center of the sphere its projected area on the near wall will increase and its projected area on the far wall will decrease, providing a somewhat compensating effect.

Therefore, there are approximately 4×10^{47} locations in space and direction that an atom of argon can have in this one cubic meter sphere. At $15\,^\circ\mathrm{C}$ and one atmosphere pressure, there are about 2.5×10^{24} atoms of argon. Thus for every atom of argon there are 1.6×10^{23} locations that it can occupy with a specific energy level. The degeneracy is overwhelming.

Now if we take **n** atoms with an energy ε_i and place them in some subset of the available degenerate locations, most of the locations will be unoccupied and those that are occupied, will only contain one atom. The probability of two atoms being located in the same location in space and direction is infinitesimally small. (Obviously two atoms cannot not occupy the same physical space, so let each of the physical locations be large enough to accommodate ten atoms. Then instead of having 4×10^{47} possibilities, there will only be 4×10^{46} possibilities. The probability is still infinitesimally small of finding two atoms in one of these locations.)

The question now becomes "How many distinguishable configurations can be formed by placing **n** indistinguishable objects in **w** boxes?" The answer to this is straightforward and looks a lot like something we have seen before. The number of distinguishable configurations is given by,

$$W_i = \frac{(n_i + \omega_i)!}{n_i!\,\omega_i!} \tag{9.55}$$

And since n_i is considerably smaller than ω_i, we can express this in the following fashion

$$W_i = \frac{(n_i + \omega_i)(n_i - 1 + \omega_i)(n_i - 2 + \omega_i)\ldots(1 + \omega_i)(\omega_i)!}{n_i!\,\omega_i!}$$

$$W_i = \frac{(n_i + \omega_i)(n_i - 1 + \omega_i)(n_i - 2 + \omega_i)\ldots(1 + \omega_i)!}{n_i!} \tag{9.56}$$

ω_i is much larger than n_i so in the numerator it is reasonable to approximate every term by ω_i giving,

$$W_i = \frac{(\omega_i)^{n_i}}{n_i!} \tag{9.57}$$

This equation applies to the ith energy level, which for the moment we will leave as quantitized. Then the overall number of states possible is the product of each of the number of states for each energy level or,

$$W = \Pi W_i = \frac{\Pi(\omega_i)^{n_i}}{\Pi n_i !}$$ (9.58)

The above equation applies to an assembly of N indistinguishable units like gas molecules in free translation, whereas the previous equation

$$W = \frac{N!}{\Pi n_i !}$$ (9.59)

applies to an assembly of identical but distinguishable units, like atoms in a crystal.

Now let us select three energy levels in an assembly and require that they are spaced such that

$$\frac{\varepsilon_n - \varepsilon_m}{\varepsilon_m - \varepsilon_l} = \frac{p}{q}$$ (9.60)

where p and q are small integers and as before try to determine the most likely configuration.

We will transfer q units of energy from level m to level n and p units of energy from level m to level l.

Energy will be conserved as

$$q(\varepsilon_n - \varepsilon_m) + p(\varepsilon_l - \varepsilon_m) = 0$$ (9.61)

Then for the most likely configuration, we must have that the derivative of W with respect to small changes in the numbers of atoms with specific energies is zero. Or,

$$\frac{\omega_a^{n_a}...\omega_l^{n_l}\omega_m^{n_m}\omega_n^{n_n}...\omega_z^{n_z}}{n_a!...n_l!n_m!n_n!...n_z!} = \frac{\omega_a^{n_a}...\omega_l^{n_l+p}\omega_m^{n_m-p-q}\omega_n^{n_n+q}...\omega_z^{n_z}}{n_a!...(n_l+p)!(n_m-p-q)!(n_n+q)!...n_z!}$$

$$\frac{1}{n_l!n_m!n_n!} = \frac{\omega_l^p\omega_m^{-p-q}\omega_n^q}{(n_l+p)!(n_m-p-q)!(n_n+q)!}$$

$$\frac{1}{n_l!(n_m-1)(n_m-2)...(n_m-p-q)!n_n!} = \frac{\omega_l^p\omega_m^{-p-q}\omega_n^q}{(n_l+p)(n_l+p-1)...n_l!(n_m-p-q)!(n_n+q)(n_n+q-1)...n_n!}$$

$$\frac{1}{(n_m-1)(n_m-2)...(n_m-p-q)!} = \frac{\omega_l^p\omega_m^{-p-q}\omega_n^q}{(n_l+p)(n_l+p-1)...(n_m-p-q)!(n_n+q)(n_n+q-1)...}$$

$$\frac{1}{n_m^{p+q}} = \frac{\omega_l^p\omega_m^{-p-q}\omega_n^q}{n_l^p n_n^q} \quad \text{Or} \quad \left[\frac{n_l/\omega_l}{n_m/\omega_m}\right]^p = \left[\frac{n_m/\omega_m}{n_n/\omega_n}\right]^q$$ (9.62)

This looks like a geometric progression of the quantities in braces so the Boltzmann distribution law becomes,

$$\frac{n_i}{\omega_i} = \frac{n_0}{\omega_0} e^{-\beta \varepsilon_i} \tag{9.63}$$

This says that the average occupation of an energy level follows an exponential distribution. This is a little different from the result that we obtained for distinguishable units with localized atoms in a solid matrix. Now let us sum the number of atoms present.

$$N = n_0 + n_1 + n_2 + n_3 \ldots = n_0 + \frac{n_0}{\omega_0} \omega_1 e^{-\beta \varepsilon_1} + \frac{n_0}{\omega_0} \omega_2 e^{-\beta \varepsilon_2} \ldots$$

$$N = \frac{n_0}{\omega_0} \sum_q \omega_q e^{-\beta \varepsilon_q} = \frac{n_0}{\omega_0} Z \tag{9.64}$$

$$\frac{n_0}{\omega_0} = \frac{N}{Z} \quad Or \quad \frac{n_i}{\omega_i} = \frac{N}{Z} e^{-\beta \varepsilon_i}$$

Now note that Z is only a function of the energy levels and the degeneracies of the energy levels.

$$Z = \sum_q e^{-\beta \varepsilon_q} = \sum_i \omega_i e^{-\beta \varepsilon_i} \tag{9.65}$$

Therefore, Z clearly represents the partition function per unit. Now if we have two units, we will have $Z_T = Z_1 * Z_2$. that is the total partition function must be the product of the individual partition functions per atom. If we have N atoms than the total partition, function must be

$$Z_{total} = Z_i^N \tag{9.66}$$

However since the units are not distinct we must take into account this fact and divide the total partition function by $N!$ for the number of ways they can be rearranged without an observable difference. This gives

$$Z_{trans} = \frac{Z_i^N}{N!} \tag{9.67}$$

It should be pointed out that the $N!$ factor only applies to the condition where the occupation of any given state is very much less than 1.0 so that there are far more states available than there are atoms or molecules to occupy them, so that the highest occupancy for any state is 1 unit. If this were not the case then we would have to replace the $N!$ by $N!/\Pi n_i!$. It should be pointed out that this form of the partition function only applies to the translational states of low-density gas molecules where the number of states available far exceeds the number of atoms available to occupy them.

Now when we have multi-atom molecules it is possible for the molecules to store energy internally in terms of rotations and vibrations. Quantum mechanics states that these internal degrees of freedom are quantized and therefore the number of states that can be occupied does not greatly exceed the number of atoms available. So if we write our partition function for a single unit, we will have

$$\varepsilon_i = \varepsilon_{i,trans} + \varepsilon_{i,rot} + \varepsilon_{i,vib}$$

$$Z = \sum_i e^{-\beta \varepsilon_i} = \sum_i e^{-\beta(\varepsilon_{i,trans} + \varepsilon_{i,rot} + \varepsilon_{i,vib})}$$

$$Z = \sum_i e^{-\beta \varepsilon_{i,trans}} \sum_i e^{-\beta \varepsilon_{i,rot}} \sum_i e^{-\beta \varepsilon_{i,vib}} = Z_{trans} * Z_{rot} * Z_{vib}$$

(9.68)

Then when we consider the whole ensemble of units and take to products of the individual partition functions this becomes,

$$Z_{total} = \left[\frac{Z_{trans}^N}{N!} \right] * Z_{rot}^N * Z_{vib}^N$$

(9.69)

We do not apply the $N!$ factor to the rotation and vibration partition functions because we can distinguish which atom is vibrating or rotating. Having obtained the partition function for an Ideal gas, consider how thermodynamic parameters can be obtained from the partition function. First, consider the internal energy. We can write for the case of distinguishable units

$$U = E = n_0 \varepsilon_0 + n_1 \varepsilon_1 + n_2 \varepsilon_2 + \ldots$$

$$E = \varepsilon_0 \frac{N}{Z} e^{-\beta \varepsilon_0} + \varepsilon_1 \frac{N}{Z} e^{-\beta \varepsilon_1} + \varepsilon_2 \frac{N}{Z} e^{-\beta \varepsilon_2} + \ldots$$

(9.70)

$$E = \frac{N}{Z}(\varepsilon_0 e^{-\beta \varepsilon_0} + \varepsilon_1 e^{-\beta \varepsilon_1} + \varepsilon_2 e^{-\beta \varepsilon_2} \ldots) = N \frac{\sum_q \varepsilon_q e^{-\beta \varepsilon_q}}{\sum_q e^{-\beta \varepsilon_q}}$$

Now treating the ε_q as constants, each of the terms in the numerator can be obtained by differentiating the similar term in the denominator with respect to β.

$$-\frac{d}{d\beta} e^{-\beta \varepsilon_q} = \varepsilon_q e^{-\beta \varepsilon_q}$$

(9.71)

So we can write

$$E = -N \frac{\dfrac{d}{d\beta} \sum_q e^{-\beta \varepsilon_q}}{\sum_q e^{-\beta \varepsilon_q}} = -N \frac{1}{Z} \frac{dZ}{d\beta} = -N \frac{d \ln Z}{d\beta}$$

(9.72)

$$E = -\frac{d(N \ln Z)}{d\beta} = -\frac{d \ln Z^N}{d\beta}$$

Thus, we have a very clean expression for the internal energy in terms of the partition function. Consider now the partition function for translation for the Ideal Gas.

$$E = -\frac{d \ln \left(\dfrac{Z^N}{N!} \right)}{d\beta} = -\frac{d}{d\beta} \ln \left(\frac{1}{N!} \right) - \frac{d}{d\beta} \ln Z^N$$

(9.73)

Nevertheless, $N!$ is a constant so the first term is zero and we have,

$$E = -N \frac{d}{d\beta} \ln Z$$

(9.74)

In addition, this can be taken back a step further to obtain the original summation for the total internal energy. This works so long as the volume containing the gas is held constant. So for the translation partition function of an Ideal Gas, the constant volume restriction must be added. We have,

$$E = \left[-N \frac{d}{d\beta} \ln Z \right]_V$$

(9.75)

Now noting that $\beta = \frac{1}{kT}$

$$d\beta = -\frac{dT}{kT^2} \quad \frac{dT}{d\beta} = -kT^2$$

$$E = -\left[\frac{d \ln Z}{dT} \right]_V \left[\frac{dT}{d\beta} \right]_V$$

(9.76)

$$E = kT^2 \left[\frac{d \ln Z}{dT} \right]_V = U$$

Now consider entropy. We have

$$S = k \ln W$$

(9.77)

For distinguishable units we had

$$W = \frac{N!}{\prod_q N_q!} \tag{9.78}$$

$$\ln W = \ln N! - \sum_q \ln N_q! = N \ln N - N - \sum_q N_q \ln N_q + \sum_q N_q$$

$$\ln W = N \ln N - \sum_q N_q \ln N_q = N \ln N - \sum_q N_q \ln \frac{N}{Z} e^{-\beta \varepsilon_q}$$

$$\ln W = N \ln N - \sum_q N_q \ln \frac{N}{Z} - \sum_q N_q (-\beta \varepsilon_q)$$

$$\ln W = N \ln N - \ln \frac{N}{Z} \sum_q N_q + \beta \sum_q N_q \varepsilon_q \tag{9.79}$$

$$\ln W = N \ln N - N \ln \frac{N}{Z} + \beta E = N \ln Z + \beta E$$

$$\ln W = \ln Z^N + \beta E = \ln Z_{total} + \beta E$$

$$S = k \ln Z^N + k \beta E = Nk \ln Z + k \frac{1}{kT} E = Nk \ln Z + \frac{E}{T}$$

This gives an expression for the entropy in terms of the total energy and the partition function.

For indistinguishable units we have,

$$W = \frac{\prod_q (\omega_q)^{N_q}}{\prod_q N_q!} \tag{9.80}$$

Following a derivation similar to the above, we obtain

$$S = k \ln \left(\frac{Z^N}{N!} \right) + k \beta E$$

Or

$$S = k \ln Z_{trans} + k \beta E = k \ln Z_{trans} + \frac{E}{T} \tag{9.81}$$

Using our previous result for the internal energy gives

$$S = k \ln Z_{trans} + kT \left[\frac{d \ln Z_{trans}}{dT} \right]_V = \frac{d}{dT} [kT \ln Z_{trans}]_V \tag{9.82}$$

If we consider a gas composed of multi-atom molecules, then we must replace Z_{trans} with Z_{total}. Therefore, we have,

$$S = \frac{d}{dT}[kT \ln Z_{total}] = \frac{d}{dT}[kT \ln Z_{trans} * Z_{rot} * Z_{vib}]$$

$$S = \frac{d}{dT}[kT \ln Z_{trans} + kT \ln Z_{rot} + kT \ln Z_{vib}] \qquad (9.83)$$

$$S = S_{trans} + S_{rot} + S_{vib}$$

Therefore, the entropy is factorable and separable into terms that correspond to the different mechanisms in the molecules.

9.11 Evaluation of Partition Functions

Now let us evaluate the translational partition function for an Ideal Gas. The energy of an individual atom or molecule is given by,

$$\varepsilon = \frac{1}{2}mv^2 = \frac{m}{2}\left(v_x^2 + v_y^2 + v_z^2\right) \qquad (9.84)$$

Now quantum mechanics tells us that an atom, moving in a box can only have certain discrete energies. These energies are obtained by solving Schrodinger's wave equation in three dimensions. From a macroscopic standpoint, the discreteness of the energies is impossible to determine because the separation is infinitesimally small on compared to macroscopic dimensions. But when the wave equation is solved in a rectangular box, the allowed energies are given by the relation

$$\varepsilon = \frac{h^2}{8m}\left[\frac{n_x^2}{L_x^2} + \frac{n_y^2}{L_y^2} + \frac{n_z^2}{L_z^2}\right] \qquad (9.85)$$

h is Plank's constant and the L's are the lengths of the sides of the box. Consider a cubical box 10 cm on a side containing argon at room temperature. Then we have

$$\varepsilon = \frac{h}{8mL^2}\left[n_x^2 + n_y^2 + n_z^2\right] \qquad (9.86)$$

The average energy of an atom is 0.025 eV and the mass is 40 amu. Plugging all of this in gives

$$[n_x^2 + n_y^2 + n_z^2] \approx 5x10^{19} \qquad (9.87)$$

Therefore, the level of degeneracy is very comparable to the classical description given before for the sphere. However now we see how the volume containing the gas comes into the equation directly. The L^2 term in the denominator is essentially the volume to the two-thirds power. Much like the classical case, the energy can easily be separated into three terms, each of which will be associated with one of the three axes. Thus, we can write our partition function as a product of the partition functions for each of the three possible directions. For the x dimension we have

$$z_x = \sum_{n_x} e^{-\beta \frac{n_x^2 h^2}{8mL_x^2}} = \int_0^\infty e^{-\beta \frac{n_x^2 h^2}{8mL_x^2}} dn_x$$

$$z_x = \frac{L_x}{h}\sqrt{\frac{2\pi m}{\beta}}$$

(9.88)

The y and z dimensions are identical so,

$$z = z_x * z_y * z_z = \left[\frac{2\pi m}{\beta}\right]^{3/2} \frac{L_x L_y L_z}{h^3} = \left[\frac{2\pi m}{h^2 \beta}\right]^{3/2} V$$

(9.89)

Now applying the relation for internal energy as a function of the partition function gives,

$$E = U = -\left[\frac{d \ln Z_{trans}}{d\beta}\right] = -\frac{d}{d\beta}\left[\ln \frac{Z_{trans}^N}{N!}\right]$$

$$E = -\frac{d}{d\beta}\left[\ln \frac{1}{N!}\left(\frac{2\pi m}{h^2}\right)^{3N/2} V^N\right] - \frac{d}{d\beta}\ln\left(\frac{1}{\beta}\right)_V^{3N/2}$$

(9.90)

For a constant volume all of the terms in the square brackets are constant so we have,

$$E = U = \frac{3N}{2}\frac{d}{d\beta}\ln\beta = \frac{3N}{2\beta}$$

(9.91)

The classical definition of the constant volume specific heat is

$$C_V = \left[\frac{dE}{dT}\right]_V = \frac{d}{dT}\left[\frac{3N}{2\beta}\right]_V = \frac{3N}{2}\frac{d}{dT}\left[\frac{1}{\beta}\right]$$

(9.92)

The classical value for a monatomic gas constant volume specific heat is

$$C_V = 12.47 J/gm\text{-}mole/K = 1.247 \times 10^8 erg/gm\text{-}mole/K$$

Therefore

$$\frac{3N}{2}\frac{d}{dT}\left[\frac{1}{\beta}\right]=1.247\times10^{8}\ \frac{erg}{gm-mole*K} \tag{9.93}$$

$$\frac{d}{dT}\left[\frac{1}{\beta}\right]=\frac{d}{dT}[kT]=k=\frac{2}{3}\frac{1.247\times10^{8}\ \dfrac{erg}{gm-mole*K}}{6.022\times10^{23}\ \dfrac{atoms}{gm-mole}}=1.38\times10^{-16}\ \frac{erg}{atom*K}$$

Thus, the value of Boltzmann's constant can be derived by considering the constant volume specific heat of a monatomic gas, which is a well measured and defined quantity.

$$k=1.38\times10^{-16}\ erg/K=1.38\times10^{-23}\ Joule/K \tag{9.94}$$

Previously we had

$$E=\frac{3N}{2\beta} \tag{9.95a}$$

This now gives

$$E=\frac{3}{2}NkT \quad \text{and} \quad C_{V}=\frac{3}{2}Nk \tag{9.95b}$$

Now let us introduce the thermodynamic potential most easily expressed in terms of the partition function. Consider the Helmholtz Free Energy defined by $F=E-TS$

$$F=E-T\left(k\ln Z+\frac{E}{T}\right)=E-kT\ln Z-E \tag{9.96}$$

$$F=-kT\ln Z$$

and we have for the monatomic Ideal Gas

$$F=-kT\ln\left(\frac{z^{N}}{N!}\right)=-kT[-\ln N!+N\ln z]$$

$$F=-kT[-(N\ln N-N)+N\ln z] \tag{9.97}$$

$$F=-NkT[1+\ln(z/N)]$$

We can write the entropy as

$$S = \frac{E-F}{T} = \frac{\frac{3}{2}NkT + NkT[1+\ln(z/N)]}{T} = Nk\left(\frac{5}{2} + \ln\frac{z}{N}\right) \qquad (9.98)$$

Substituting for z gives

$$S = Nk\left\{\frac{5}{2} + \ln\left(\left[\frac{2\pi kT}{h^2}\right]^{3/2}\frac{V}{N}\right)\right\} \qquad Nk = \Re \quad Nm = M$$

$$S = \Re\left\{\frac{5}{2} + \ln\left(\left[\frac{2\pi MkT}{h^2 N}\right]^{3/2}\frac{V}{N}\right)\right\} \qquad\qquad (9.99)$$

This can be rewritten as

$$S = \Re\ln V + \frac{3}{2}\Re\ln T + \frac{3}{2}\Re\ln M + \left\{\Re\left(\frac{5}{2} + \ln\left[\frac{2\pi k}{h^2 N^{5/3}}\right]^{3/2}\right)\right\} \qquad (9.100)$$

This is a famous equation called the Sackur-Tetrode equation.

Now consider changes in S. If only the volume changes we have

$$S_2 - S_1 = \Re\ln V_2 - \Re\ln V_1 = \Re\ln\left(\frac{V_2}{V_1}\right) \qquad (9.101)$$

the exact relation we get from classical thermodynamics. If only the temperature changes, we get

$$S_2 - S_1 = \frac{3}{2}\Re\ln T_2 - \frac{3}{2}\Re\ln T_1 = \frac{3}{2}\Re\ln\left(\frac{T_2}{T_1}\right) = C_V\ln\left(\frac{T_2}{T_1}\right) \qquad (9.102)$$

Once again, this is the expected behavior from classical thermodynamics. The last two terms provide constants that allow us to calculate the entropy of a monatomic ideal gas from first principles or fundamental constants. This table taken from Nash's book and attributed to K. K. Kelley evidences the excellent agreement of this equation with the entropies determined by calorimetric measurements for the noble gases at 298 K and one atmosphere.

Entropies (cal/gm-mole/K)		
Noble gas	Calorimetric	Theoretical
Neon	35.01 +/−0.1	34.95 +/−0.01
Argon	36.95 +/−0.2	36.99 +/−0.01
Krypton	39.17 +/−0.1	39.20 +/−0.01
Xenon	40.7 +/−0.3	40.54 +/−0.01

9.12   Maxwell-Boltzmann Velocity Distribution

Our Maxwell energy distribution is of the following form,

$$\frac{N_i}{\omega_i} = \frac{N}{Z} e^{-\frac{mv_i^2}{2kT}}$$

$$N_i = \frac{N}{Z} \omega_i e^{-\frac{mv_i^2}{2kT}} \tag{9.103}$$

For the classical continuous velocity distribution, we can write this as

$$dN(v) = \frac{N}{Z} e^{-\frac{mv^2}{2kT}} d\omega(v) \tag{9.104}$$

In order to integrate this equation and determine the normalizations we need to determine a degeneracy function as a function of the velocity. The simplest answer to this question is to let the total number of degeneracies increase as the volume of velocity space. That simply says that

$$\omega(v) = \frac{4}{3} \pi v^3 \quad d\omega(v) = 4\pi v^2 dv$$

Then

$$dN(v) = \frac{N}{Z} e^{-\frac{mv^2}{2kT}} 4\pi v^2 dv$$

$$N = \int_0^\infty \frac{N}{Z} 4\pi v^2 e^{-\frac{mv^2}{2kT}} dv$$

$$1 = \frac{4\pi}{Z} \int_0^\infty v^2 e^{-\frac{mv^2}{2kT}} dv = \frac{4\pi}{Z} \left(\frac{2kT}{m}\right)^{3/2} \int_0^\infty y^2 e^{-y^2} dy \tag{9.105}$$

$$Z = 4\pi \left(\frac{2kT}{m}\right)^{3/2} \int_0^\infty y^2 e^{-y^2} dy = 4\pi \left(\frac{2kT}{m}\right)^{3/2} \frac{1}{4} \sqrt{\pi}$$

$$Z = \left(\frac{2\pi kT}{m}\right)^{3/2}$$

Then taking $dN(v) = N(v)dv$, we have

$$N(v) = N \left(\frac{m}{2\pi kT}\right)^{3/2} 4\pi v^2 e^{-\frac{mv^2}{2kT}} \tag{9.106}$$

This is the Maxwell Boltzmann velocity distribution.

The average velocity is given by

$$\bar{v} = \sqrt{\frac{8kT}{\pi m}} \tag{9.107}$$

The rms velocity is given by

$$v_{rms} = \sqrt{\frac{3kT}{m}} \tag{9.108}$$

The most probable velocity is given by

$$v_{mode} = \sqrt{\frac{2kT}{m}} \tag{9.109}$$

Problems

Problem 9.1: Calculate the number of macrostates and their frequency that will be observed in a solid with distinguishable atoms if 3 quanta are distributed among six atoms.

Problem 9.2: Consider a micro-dot of silicon that can be represented as a hemisphere 1 micron in diameter. What is the probability of observing a deviation from the most likely macrostate of one part in a trillion (10^{-12})? The density of silicon is 2.33 gm/cc and its atomic weight is 28.

Problem 9.3: What is the ratio of microstates contributing to the most likely macrostate for iron at 300 K compared to the number of microstates contributing to the most likely macrostate at 290 K? Use a constant β at 295 K.

Problem 9.4: Estimate the level of degeneracy for a hydrogen molecule moving in a 1 1 box with an average energy of 0.0253 eV. The mass of a hydrogen molecule is 2 amu.

Problem 9.5: Calculate the average velocity, rms velocity, and most probable velocity for a carbon dioxide molecule in a container at 3 MPa and 500 K.

References

1. Nash LK (2006) Elements of statistical thermodynamics, 2nd edn. Dover Publications Inc., Mineola
2. Lee JF, Francis WS, Donald LT (1973) Statistical thermodynamics. Addison-Wesley Publishing Company, Reading
3. Reif F (2009) Fundamentals of statistical and thermal physics. Waveland Press, Long Grove
4. Loeb LB (1961) The kinetic theory of gases, Dover Phoenix Edition. Dover Publications, Mineola

Chapter 10
Thermodynamic Relations

In this chapter, we will discuss mathematical relationship of thermodynamics in exact formulation using differential formation of these relations.

10.1 Thermodynamic Potentials

Classical thermodynamics has a very rich mathematical background. In a course aimed at engineering thermodynamics, it is not terribly useful to go into this analysis too deeply, but touching on some of the methods is worthwhile. Start by summarizing without proof some properties of the four common thermodynamic potentials. The first of these potentials is the internal energy, which is identified with the symbol U. The internal energy U has been used for a good portion of this study to date. Though in the past it was considered it a function of many variables, its natural variables are S and V. Normally it is written as,

$$U = U(S,V) \tag{10.1}$$

If we differentiate U with respect to S and V, holding the other variable constant we get,

$$dU = \frac{dU}{dS}\bigg|_V dS + \frac{dU}{dV}\bigg|_S dV \tag{10.2}$$

Also the combined First and Second Laws can be written

$$dU = TdS - pdV \tag{10.3}$$

Then T can be identified with $\dfrac{dU}{dS}\bigg|_V$ and $-p$ with $\dfrac{dU}{dV}\bigg|_S$.

© Springer International Publishing Switzerland 2015
B. Zohuri, P. McDaniel, *Thermodynamics In Nuclear Power Plant Systems*,
DOI 10.1007/978-3-319-13419-2_10

The enthalpy was defined as,

$$H = U(S,V) + pV = H(S,p) \tag{10.4}$$

So

$$dH = dU + pdV + Vdp = Tds - pdV + pdV + Vdp = Tds + Vdp \tag{10.5}$$

And

$$dH = \frac{dH}{dS}\bigg|_p dS + \frac{dH}{dP}\bigg|_S dp \tag{10.6}$$

Once again, T can be identified as $\dfrac{dH}{dS}\big|_p$ and V can be identified as $\dfrac{dH}{dp}\big|_S$. Remember that the reason for choosing enthalpy as the thermodynamic potential of interest is that our system will be in contact with a pressure reservoir during the process of interest.

In addition to internal energy and enthalpy, two other thermodynamic potentials are sometimes of interest. The first is the Helmholtz Potential or Helmholtz Free Energy (The term Free Energy is a poor choice of words and only causes problems.). The Helmholtz Potential is useful when our system of interest is in contact with a thermal reservoir that holds the temperature constant during the process of interest. The Helmholtz Potential is defined as,

$$F = U - TS$$
$$dF = dU - TdS - SdT = Tds - pdV - TdS - SdT = -pdV - SdT$$
$$dF = \frac{dF}{dV}\bigg|_T dV + \frac{dF}{dT}\bigg|_V dT$$
$$-S = \frac{dF}{dT}\bigg|_V$$
$$-p = \frac{dF}{dV}\bigg|_T \tag{10.7}$$

Clearly, the natural variables for the Helmholtz Potential are V and T.

The second new potential is the Gibbs Potential or Gibbs Free Energy (Bad choice of words.). This potential is useful when systems are in contact with the atmosphere and the pressure and temperature are held constant by using the atmosphere as a sink. The Gibbs potential is defined by,

$$G = H - TS$$

$$dG = dH - TdS - SdT = Tds + Vdp - TdS - SdT = Vdp - SdT$$

$$dG = \frac{dG}{dp}\bigg|_T dp + \frac{dG}{dT}\bigg|_p dT$$

$$-S = \frac{dG}{dT}\bigg|_p$$

$$V = \frac{dG}{dp}\bigg|_T \tag{10.8}$$

Now it may seem a little confusing why certain variables are chosen as the differential variables in the above equations. The simple answer is that only two variables are required to define the state of a simple system, and the two that are chosen should be the ones that make a particular process most transparent or provide the simplest solution.

The internal energy may be looked upon as the available work (or heat) from a constant volume system. The enthalpy may be looked upon as the available work from a system in contact with a pressure reservoir. The Helmholtz Potential is the available work from a system in contact with a temperature reservoir. In addition, the Gibbs Potential is the available work from a system in contact with a pressure and temperature reservoir.

The relationship for the internal energy as a function of the entropy can be inverted to give the entropy as a function of the internal energy and volume. The same can be done for the enthalpy. When either of these are accomplished, the resulting entropy function can be used to identify the direction of approach to equilibrium and the actual equilibrium state. The entropy function so defined will be maximized at equilibrium. Thus,

$$S = S(U,V)$$

$$dS = \frac{dS}{dU}\bigg|_V dU + \frac{dS}{dV}\bigg|_U dV$$

$$S = S(H,p)$$

$$dS = \frac{dS}{dH}\bigg|_p dH + \frac{dS}{dp}\bigg|_H dp \tag{10.9}$$

Both of these expressions are useful in some of the algebraic manipulations of Classical Thermodynamics.

10.2   Maxwell Relations

At this point Classical Thermodynamics drifts off into a mathematical maze that is well defined but becomes more obscure equation by equation.[1] Now having defined the four classic potentials, it is possible to apply the results of the Calculus of Several Variables to them to obtain differential relations between the various thermodynamic properties. Perhaps the most useful application of this calculus is to be able to express the derivatives, or rates of change of entropy. In terms of the rate of changes of more measurable variables, like pressure, temperature, and volume.

From the Calculus of Several Variables, remember that the order of differentiation doesn't matter when the differentiation is performed with respect to two independent variables provided the function and its derivatives are continuous. The state functions of thermodynamics satisfy these restrictions. Therefore for any $w=w(x, y)$,

$$\frac{\partial^2 w}{\partial x \partial y} = \frac{\partial^2 w}{\partial y \partial x} \tag{10.10}$$

Also, consider the situation where there are four state variables, say w, x, y, and z, any two of which can be chosen as the independent variables for a simple compressible system. Start with x as a function of y and w.

$$dx = \frac{dx}{dy}\bigg|_w dy + \frac{dx}{dw}\bigg|_y dw \tag{10.11}$$

Then choose y to be a function of z and w.

$$dy = \frac{dy}{dz}\bigg|_w dz + \frac{dy}{dw}\bigg|_z dw \tag{10.12}$$

Now substitute Eq. 10.12 into Eq. 10.11 to get

$$dx = \frac{dx}{dy}\bigg|_w \left(\frac{dy}{dz}\bigg|_w dz + \frac{dy}{dw}\bigg|_z dw \right) + \frac{dx}{dw}\bigg|_y$$

$$dw = \frac{dx}{dy}\bigg|_w \frac{dy}{dz}\bigg|_w dz + \left(\frac{dx}{dw}\bigg|_y + \frac{dx}{dy}\bigg|_w \frac{dy}{dw}\bigg|_z \right) dw \tag{10.13}$$

But

$$dx = \frac{dx}{dz}\bigg|_w dz + \frac{dx}{dw}\bigg|_z dw \tag{10.14}$$

---

[1] [1], Chap. 2.

So the following two relations must hold.

$$\frac{dx}{dw}\Big|_z = \frac{dx}{dw}\Big|_y + \frac{dx}{dy}\Big|_w \frac{dy}{dw}\Big|_z \tag{10.15}$$

$$\frac{dx}{dz}\Big|_w = \frac{dx}{dy}\Big|_w \frac{dy}{dz}\Big|_w \tag{10.16}$$

Equation 10.16 can be put in a more symmetrical form to yield.

$$\frac{dx}{dy}\Big|_w \frac{dy}{dz}\Big|_w \frac{dz}{dx}\Big|_w = 1 \tag{10.17}$$

Sometimes it is simply called the chain rule and can be extended to any number of variables.

Now consider a simple example for a real gas where its equation of state can be written as $v=v(T, p)$ or $p=p(T, v)$, before addressing the thermodynamic potentials. Differentials for these two equations can be written as

$$dv = \frac{\partial v}{\partial p}\Big|_T \, dp + \frac{\partial v}{\partial T}\Big|_p \, dT$$

$$dp = \frac{\partial p}{\partial v}\Big|_T \, dv + \frac{\partial p}{\partial T}\Big|_v \, dT \tag{10.18}$$

Eliminating dp between these two equations and collecting the coefficients of dv and dT gives,

$$\left[1 - \frac{\partial v}{\partial p}\Big|_T \frac{\partial p}{\partial v}\Big|_T\right] dv = \left[\frac{\partial v}{\partial p}\Big|_T \frac{\partial p}{\partial T}\Big|_v + \frac{\partial v}{\partial T}\Big|_p\right] dT \tag{10.19}$$

Now since changes in dv and dT are independent, both coefficients must be equal to zero. Therefore,

$$\left[1 - \frac{\partial v}{\partial p}\Big|_T \frac{\partial p}{\partial v}\Big|_T\right] = 0 \;\rightarrow\; \frac{\partial v}{\partial p}\Big|_T = \frac{1}{\frac{\partial p}{\partial v}\Big|_T}$$

$$\frac{\partial v}{\partial p}\Big|_T \frac{\partial p}{\partial T}\Big|_v + \frac{\partial v}{\partial T}\Big|_p = 0 \;\rightarrow\; \frac{\partial p}{\partial T}\Big|_v = -\frac{\frac{\partial v}{\partial T}\Big|_p}{\frac{\partial v}{\partial p}\Big|_T} = \frac{\alpha}{\kappa}$$

$$\alpha = \frac{1}{V}\frac{\partial V}{\partial T}\Big|_p \quad \textit{Coefficient of Thermal Expansion}$$

$$\kappa = -\frac{1}{V}\frac{\partial V}{\partial p}\Big|_T \quad \textit{Isothermal Compressibility} \tag{10.20}$$

Thus, the requirements for continuity of derivatives of state variables allow relationships between the derivatives themselves to be defined.

Example 10.1 Calculate the (a) Coefficient of Thermal expansion and (b) the Isothermal Compressibility for (1) an Ideal Gas and a real gas based on (2) a virial expansion, and (3) the Peng-Robinson equation of state.

Solution

a) 1) Ideal Gas $\alpha = \dfrac{1}{v}\dfrac{\partial}{\partial T}\dfrac{RT}{p}\bigg|_p = \dfrac{1}{v}\dfrac{R}{p} = \dfrac{1}{T}$

2) Virial Expansion

$$v^2 = \frac{RT}{p}(v+B)$$

$$2v\frac{\partial v}{\partial T} = \frac{RT}{p}\frac{\partial v}{\partial T} + \frac{R}{p}(v+B)$$

$$\left(2v - \frac{RT}{p}\right)\frac{\partial v}{\partial T} = \frac{R}{p}(v+B)$$

$$\frac{1}{v}\frac{\partial v}{\partial T} = \frac{R}{vp}\frac{(v+B)}{\left(2v - \dfrac{RT}{p}\right)} = \frac{R}{vp}\frac{1+B/v}{1} = \frac{1}{T}\left(1+B/v\right)$$

3) Peng-Robinson Equation

$$p = \frac{RT}{v-b} - \frac{\alpha a}{v^2+2bv-b^2}$$

$$v^3 + bv^2 - 3b^2v + b^3 = \frac{RT}{p}\left(v^2+2bv-b^2\right) - \alpha a(v-b)$$

$$(3v^2 + 2bv - 3b^2)\frac{\partial v}{\partial T} = \frac{R}{p}\left(v^2+2bv-b^2\right) + \frac{RT}{p}(2v+2b)\frac{\partial v}{\partial T} - \alpha a\frac{\partial v}{\partial T} - a(v-b)\frac{\partial \alpha}{\partial T}$$

$$(3v^2 + 2bv - 3b^2 - 2v^2 - 2bv + \alpha a)\frac{\partial v}{\partial T} = \frac{R}{p}\left(v^2+2bv-b^2\right) - a(v-b)\frac{\partial \alpha}{\partial T}$$

$$\frac{1}{v}\frac{\partial v}{\partial T} = \frac{1}{v}\frac{\dfrac{R}{p}(v^2+2bv-b^2)-a(v-b)\dfrac{\partial \alpha}{\partial T}}{(1-3b^2+\alpha a)} = \frac{1}{T}\frac{\left(v^2+2bv-b^2\right)}{(1-3b^2+\alpha a)} - \frac{a\left(1-b/v\right)}{(1-3b^2+\alpha a)}\frac{\partial \alpha}{\partial T}$$

b) 1) Ideal gas $\kappa = -\dfrac{1}{v}\dfrac{\partial v}{\partial p}\bigg|_T = -\dfrac{1}{v}\left(\dfrac{RT}{-p^2}\right) = \dfrac{1}{p}$

2) Virial Expansion

$$v^2 = \frac{RT}{p}(v+B)$$

$$2v\frac{\partial v}{\partial p} = -\frac{RT}{p^2}(v+B) + \frac{RT}{p}\frac{\partial v}{\partial p}$$

$$\left(2 - \frac{RT}{vp}\right)\frac{\partial v}{\partial p} = -\frac{RT}{vp^2}(v+B) = \frac{\partial v}{\partial p} = -\frac{1}{p}(v+B)$$

$$-\frac{1}{v}\frac{\partial v}{\partial p} = \frac{1}{p}\left(1 + \frac{B}{v}\right)$$

3) Peng Robinson

$$v^3 + v^2 b - 3b^2 v + b^3 = \frac{RT}{p}(v^2 + 2bv - b^2) - \alpha a(v-b)$$

$$(3v^2 + 2bv - 3b^2)\frac{\partial v}{\partial p} = -\frac{RT}{p^2}(v^2 + 2bv - b^2) - \alpha a\frac{\partial v}{\partial p} + \frac{RT}{p}(2v+2b)\frac{\partial v}{\partial p}$$

$$(3v^2 - 2v^2 + 2bv - 2bv - 3b^2 - \alpha a)\frac{\partial v}{\partial p} = -\frac{RT}{p^2}(v^2 + 2bv - b^2)$$

$$-\frac{1}{v}\frac{\partial v}{\partial p} = \frac{1}{p}\frac{(v^2 + 2bv - b^2)}{(1 - 3b^2 - \alpha a)}$$

Now consider the two derivative definitions that were obtained for the Helmholtz Potential.

$$-S = \frac{dF}{dT}\bigg|_V$$

$$-\frac{dS}{dV}\bigg|_T = \frac{d^2 F}{dTdV}$$

$$-p = \frac{dF}{dV}\bigg|_T$$

$$-\frac{dp}{dT}\bigg|_V = \frac{d^2 F}{dVdT}$$

$$\frac{dp}{dT}\bigg|_V = \frac{dS}{dV}\bigg|_T \qquad\qquad (10.21)$$

This is known as the Third Maxwell relation and allows estimation of the change in entropy vs. pressure at constant temperature by measuring the change in pressure vs. temperature at constant volume. The rest of the Maxwell relations are derived in a similar manner, and are given as,

$$\left.\frac{\partial T}{\partial V}\right|_S = -\left.\frac{\partial p}{\partial S}\right|_V \quad First\ Maxwell\ Relation$$

$$\left.\frac{\partial T}{\partial p}\right|_S = \left.\frac{\partial V}{\partial S}\right|_p \quad Second\ Maxwell\ Relation$$

$$\left.\frac{\partial S}{\partial p}\right|_T = -\left.\frac{\partial V}{\partial T}\right|_p \quad Fourth\ Maxwell\ Relation.$$

10.3   Clapeyron Equation

The Third Maxwell relation above can be applied to obtain h_{fg} at a point on the vapor dome identified by p_o and T_o. Start with,

$$\left.\frac{ds}{dv}\right|_{T_o} = \frac{s_g - s_f}{v_g - v_f} = \frac{s_{fg}}{v_{fg}} = \left.\frac{dp}{dT}\right|_{v_o} \tag{10.22}$$

$$dh = Tds - vdp$$
$$dp = 0$$
$$\int dh = h_{fg} = \int T_o ds = T_o \Delta s = T_o s_{fg} \tag{10.23}$$

Then

$$\left.\frac{dp}{dT}\right|_{v_o} = \frac{h_{fg}}{T_o v_{fg}}$$

$$h_{fg} = T_o v_{fg} \left.\frac{dp}{dT}\right|_{v_o} \tag{10.24}$$

Now the derivative can be evaluated from the saturated state tables using a central difference approximation.

$$\left.\frac{dp}{dT}\right|_{v_o} = \frac{h_{fg}}{T_o v_{fg}}$$

$$h_{fg} = T_o v_{fg} \left.\frac{dp}{dT}\right|_{v_o} \tag{10.25}$$

$$\left.\frac{dp}{dT}\right|_{v_o} = \frac{p_2 - p_1}{T_2 - T_1} \tag{10.26}$$

At low pressures, this becomes the Clausius-Clapeyron equation by ignoring the value of v_f and approximating v_g by RT/p (Ideal Gas).

$$\frac{dp}{dT}\bigg|_{v_o} = \frac{ph_{fg}}{RT^2} \tag{10.27}$$

This will allow extrapolating to pressures below those tabulated by rewriting this equation as,

$$\frac{dp}{p} = \frac{h_{fg}\,dT}{RT^2}$$

$$\ln\left(\frac{p_2}{p_1}\right) \cong \frac{h_{fg}}{R}\left(\frac{1}{T_1} - \frac{1}{T_2}\right) \tag{10.28}$$

10.4 Specific Heat Relations Using the Maxwell Relations

The internal energy differential can be written

$$du = \frac{du}{dT}\bigg|_v dT + \frac{du}{dv}\bigg|_T dv = C_v dT + \frac{du}{dv}\bigg|_T dv \tag{10.29}$$

And considering

$$s = s(T, v)$$

$$ds = \frac{ds}{dT}\bigg|_v dT + \frac{ds}{dv}\bigg|_T dv$$

Then

$$du = Tds - pdv$$

$$du = T\left[\frac{ds}{dT}\bigg|_v dT + \frac{ds}{dv}\bigg|_T dv\right] - pdv = T\frac{ds}{dT}\bigg|_v dT + \left[\frac{ds}{dv}\bigg|_T - p\right]dv \tag{10.30}$$

So

$$C_v = T\frac{ds}{dT}\bigg|_v$$

$$\frac{du}{dv}\bigg|_T = T\frac{ds}{dv}\bigg|_T - p = T\frac{dp}{dT}\bigg|_v - p \qquad \text{(3rd Maxwell Relation)}$$

$$du = C_v dT + \left[T\frac{dp}{dT}\bigg|_v - p\right]dv \tag{10.31}$$

This allows calculation of *du*, given an equation of state that relates **p, v,** and **T**.
Approaching enthalpy the same way and writing.

$$s = s(T, p)$$

$$ds = \frac{ds}{dT}\bigg|_p dT + \frac{ds}{dp}\bigg|_T dp$$

$$dh = Tds + vdp = T\left[\frac{ds}{dT}\bigg|_p dT + \frac{ds}{dp}\bigg|_T dp\right] + vdp$$

$$C_p = T\frac{ds}{dT}\bigg|_p$$

$$dh = C_p dT + \left[v - T\frac{dv}{dT}\bigg|_p\right]dp \qquad (4th\ Maxwell\ Relation) \qquad (10.32)$$

This can be integrated to get **dh** given an equation of state for *p, v,* and *T*.

Example 10.2 Demonstrate that the dp correction term in the dh equation for (1) an Ideal Gas is zero and (2) is non zero for the Peng Robinson equation.

Solution The *dp* term in the *dh* equation is given by $\left[v - T\frac{dv}{dT}\bigg|_p\right]dp$

1. Ideal Gas $v - T\frac{\partial v}{\partial T}\bigg|_p = v - T\frac{\partial}{\partial T}\frac{RT}{p} = v - T\frac{R}{p} = v - v = 0$

2. Peng Robinson—From Example 2.1

$$\frac{1}{v}\frac{\partial v}{\partial T} = \frac{1}{T}\frac{(v^2 + 2bv - b^2)}{(1 - 3b^2 + \alpha a)} - \frac{a\left(1 - \frac{b}{v}\right)}{(1 - 3b^2 + \alpha a)}\frac{\partial \alpha}{\partial T}$$

$$v - T\frac{\partial v}{\partial T} = v - \frac{v(v^2 + 2bv - b^2)}{(1 - 3b^2 + \alpha a)} - \frac{a\left(1 - \frac{b}{v}\right)Tv}{(1 - 3b^2 + \alpha a)}\frac{\partial \alpha}{\partial T}$$

Now revisiting the two differential entropy equations above,

$$ds = \frac{ds}{dT}\bigg|_v dT + \frac{ds}{dv}\bigg|_T dv = \frac{C_v}{T}dT + \frac{dp}{dT}\bigg|_v dv$$

And

$$ds = \frac{ds}{dT}\bigg|_p dT + \frac{ds}{dp}\bigg|_T dp = \frac{C_p}{T}dT - \frac{dv}{dT}\bigg|_p dp \qquad (10.33)$$

These two equations will allow calculation of changes in entropy for real gases requiring only the specific heats as a function of temperature and a real gas equation of state.

10.5 The Difference Between the Specific Heats for a Real Gas

Now consider the difference between the specific heats for a real gas. From above

$$C_p - C_v = T \frac{ds}{dT}\bigg|_p - T \frac{ds}{dT}\bigg|_v$$

By Eq. 10.15

$$\frac{ds}{dT}\bigg|_p = \frac{ds}{dT}\bigg|_v + \frac{ds}{dv}\bigg|_T \frac{dv}{dT}\bigg|_p$$

So

$$C_p - C_v = T \frac{ds}{dv}\bigg|_T \frac{dv}{dT}\bigg|_p$$

So

$$C_p - C_v = T \frac{ds}{dv}\bigg|_T \frac{dv}{dT}\bigg|_p$$

and applying the Third Maxwell Relation again gives,

$$C_p - C_v = T \frac{dp}{dT}\bigg|_v \frac{dv}{dT}\bigg|_p \qquad (10.34)$$

For an ideal gas

$$\frac{dp}{dT}\bigg|_v = \frac{R}{v}\frac{dv}{dT}\bigg|_p = \frac{R}{p}T\frac{dp}{dT}\bigg|_v \frac{dv}{dT}\bigg|_p = T\frac{R}{v}\frac{R}{p} = \frac{RT}{pv}R = R$$

For an Ideal gas as expected. However this is not the case for a real gas.

Example 10.3 Based on Eq. 10.33 derive an expression for the difference in specific heats for a real gas using the Peng-Robinson equation.

Solution Equation 10.33 is

$$C_p - C_v = T \frac{dp}{dT}\bigg|_v \frac{dv}{dT}\bigg|_p$$

Evaluating the two derivatives

$$p = \frac{RT}{v-b} - \frac{\alpha a}{v^2 + 2bv - b^2}$$

$$\left.\frac{\partial p}{\partial T}\right|_v = \frac{R}{v-b} - \frac{a}{v^2 + 2bv - b^2}\frac{\partial \alpha}{\partial T}$$

$$\frac{\partial \alpha}{\partial T} = \frac{\partial}{\partial T}\left[1 + \kappa\left(1 - \sqrt{T/T_C}\right)\right]^2 = 2\alpha\frac{-\kappa}{2}\left(T/T_c\right)^{-\frac{1}{2}}\frac{1}{T_c} = -\alpha\kappa\left(\frac{1}{TT_c}\right)^{\frac{1}{2}}$$

$$\left.\frac{\partial p}{\partial T}\right|_v = \frac{R}{v-b} + \frac{\alpha a\kappa}{v^2 + 2bv - b^2}\left(\frac{1}{TT_c}\right)^{\frac{1}{2}}$$

$$\left.\frac{\partial v}{\partial T}\right|_p = \frac{R\left(v^2 + 2bv - b^2\right)}{p\left(1 - 3b^2 + \alpha a\right)} - \frac{a(v-b)}{\left(1 - 3b^2 + \alpha a\right)}\left(-\alpha\kappa\left(\frac{1}{TT_c}\right)^{\frac{1}{2}}\right)$$

$$C_p - C_v = R\left[\frac{1}{v-b} + \frac{\alpha a\kappa}{R\left(v^2 + 2bv - b^2\right)}\left(\frac{1}{TT_c}\right)^{\frac{1}{2}}\right]$$

$$\left[\frac{R}{p}\frac{\left(v^2 + 2bv - b^2\right)}{\left(1 - 3b^2 + \alpha a\right)} + \frac{a\alpha\kappa(v-b)\left(\frac{1}{TT_c}\right)^{\frac{1}{2}}}{\left(1 - 3b^2 + \alpha a\right)}\right]$$

10.6   Joule-Thomson Coefficient

When a fluid passes through a throttling device, the enthalpy remains constant. Normally the temperature of the fluid will drop. However for real gases, the temperature may remain the same or it may increase. What happens depends on the value of what is called the Joule Thomson coefficient, C_{JT}

$$C_{JT} = \left.\frac{dT}{dp}\right|_h \tag{10.35}$$

If C_{JT} is positive, a temperature decrease follows a pressure decrease. If it is negative, a temperature increases results. Consider,

$$dh = C_p dT + \left[v - T \left. \frac{dv}{dT} \right|_p \right] dp$$

$$0 = C_p dT + \left[v - T \left. \frac{dv}{dT} \right|_p \right] dp \qquad (10.36)$$

$$\left. \frac{dT}{dp} \right|_h = \frac{1}{C_p} \left[T \left. \frac{dv}{dT} \right|_p - v \right]$$

With a throttling valve it is relatively easy to measure a temperature increase or decrease, so this is one way of measuring C_p. Note also that for an Ideal Gas, the Joule-Thompson coefficient is always zero.

Example 10.4 Derive an expression for the Joule Thompson Coefficient for a gas obeying the real gas equation represented by a virial expansion.

Solution The Joule Thompson Coefficient is

$$C_{JT} = \left. \frac{dT}{dp} \right|_h = \frac{1}{C_p} \left[T \left. \frac{dv}{dT} \right|_p - v \right]$$

$$C_{JT} = \frac{1}{C_p} \left[T \frac{v}{T} \left(1 + B/v \right) - v \right]$$

$$C_{JT} = \frac{v}{C_p} \left(1 + B/v \right)$$

Problems

Problem 10-1: Derive an expression for the Coefficient of Thermal Expansion based on the real gas Van der Waals equation of state.

Problem 10-2: Derive an expression for the Coefficient of Thermal Expansion based on the real gas Redlich-Kwong equation of state.

Problem 10-3: Derive an expression for the Isothermal Compressibility based on the real gas Van der Waals equation of state.

Problem 10-4: Derive an expression for the Isothermal Compressibility based on the real gas Redlich-Kwong equation of state.

Problem 10-5: Derive an expression for the dv term in the du equation based on a Van der Waals equation of state.

Problem 10-6: Derive an expression for the dv term in the du equation based on a Redlich-Kwong equation of state.

Problem 10-7: Derive an expression for the difference in specific heats for a real gas modeled by the Van der Waals equation of state.

Problem 10-8: Derive an expression for the difference in specific heats for a real gas modeled by the Redlich-Kwong equation of state.

Problem 10-9: Derive an expression for the Joule-Thompson coefficient for a real gas modeled by the Van der Waals equation of state.

Problem 10-10: Derive an expression for the Joule-Thompson coefficient for a real gas modeled by the Redlich-Kwong equation of state.

References

1. Hsieh JS (1975) Principles of thermodynamics. McGraw-Hill, New York
2. Callen HB (1985) Thermodynamics and an introduction to thermostatistics, 2nd edn. Wiley, New York
3. Potter MC, Somerton CW (2006) Thermodynamics for engineers, 2nd edn (Schaum's Outlines). McGraw-Hill, New York

Chapter 11
Combustion

Chemical combustion is the major source of energy used for transportation and the production of electricity. In this chapter, thermodynamic concepts important to the study of combustion are examined. Basic property relations for ideal gases and ideal-gas mixtures and first law of thermodynamics have been discussed in previous chapters. Some review of these concepts will be covered as they are integral to the study of combustion. Readers should refer to reference [1–5] at the end of this chapter for further information and details.

11.1 Introduction

Combustion is a rapid exothermic reaction that liberates substantial energy as heat and has the ability to propagate through a suitable medium. This propagation results from the strong coupling of the reaction with the molecular transport process. The chemistry and physics of combustion involves the destruction and rearrangement of certain molecules and a rapid energy release within a few millionths of second. Currently, the study of combustion is a mature discipline and an integral element of diverse research and development programs from fundamental studies of the physics of flames and high-temperature molecular chemistry to applied engineering projects involved with developments such as advanced coal-burning equipment and improved combustion furnaces, boilers, and engines. These developments are important in optimizing fuel use and controlling the emission of pollutants.

The study of combustion starts with the *mass* and *energy balances* that bound the combustion process. Then, the energy characteristics of various important fuel resources and their physical and chemical properties are considered. Finally the practical stoichiometry and thermochemical requirements that apply during combustion processes including chemical reactions, equilibrium compositions and temperatures are discussed.

© Springer International Publishing Switzerland 2015
B. Zohuri, P. McDaniel, *Thermodynamics In Nuclear Power Plant Systems*,
DOI 10.1007/978-3-319-13419-2_11

Combustion is the conversion of a substance called a fuel into products of combustion by combination with an oxidizer. The combustion process is an exothermic chemical reaction, i.e., a reaction that releases energy as it occurs. Thus, combustion may be represented symbolically by:

Fuel + Oxidizer $\rightarrow$ Products of combustion + Energy.

Here the fuel and the oxidizer are reactants, i.e., the substances present before the reaction takes place. This relation indicates that the reactants produce combustion products and energy. Either the chemical energy released is transferred to the surroundings as it is produced, or it remains in the combustion products in the form of elevated internal energy (temperature), or some combination thereof.

Fuels are evaluated, in part, based on the amount of energy or heat that they release per unit mass or per mole during combustion of the fuel. Such a quantity is known as the fuel's heat of reaction or heating value.

Heats of reaction may be measured in a calorimeter, a device in which chemical energy release is determined by transferring the released heat to a surrounding fluid. The amount of heat transferred to the fluid in returning the products of combustion to their initial temperature yields the heat of reaction. In combustion processes, the oxidizer is usually air but could be pure oxygen, an oxygen mixture, or a substance involving some other oxidizing element such as fluorine. Only oxygen based oxidizers will be considered in what follows. Chemical fuels exist in gaseous, liquid, or solid form. Natural gas, gasoline, and coal, are the most widely used examples of these three forms. Each is a complex mixture of reacting and inert compounds. The analysis process proceeds in three steps:

- Concepts and definitions related to element conservation,
- A definition of enthalpy that accounts for chemical bonds,
- First-law concepts defining heat of reaction, heating values, etc., and adiabatic flame temperature.

Actually, combustion is a result of dynamic, or time-dependent, events that occur on a molecular level among atoms, molecules, radicals and solid boundaries. The rapid reactions produce gradients that transport processes convert into heat and species fluxes that speed-up the reactions.

At the heart of fossil-fueled power plant operation is the combustion process. Through the combustion process, a modern power plant burns fuel to release the energy that generates steam—energy that ultimately is transformed into electricity. Yet, while the combustion process is one of a power plant's most fundamental processes, it is also one of the most complexes.

Combustion, or the conversion of fuel to useable energy, must be carefully controlled and managed. Only the heat released that is successfully captured by the steam is useful for generating power. Hence, the ability of the steam generator to successfully transfer energy from the fuel to steam is driven by the combustion process, or more precisely, the characteristics of the combustion process.

A chemical reaction may be defined as the rearrangement of atoms due to redistribution of electrons. In a chemical reaction the terms, '*reactants*' and '*products*' are frequently used. '*Reactants*' comprise the initial constituents which start the reaction while '*products*' are the final constituents which are formed by the chemical reaction. Although the basic principles, which will be discussed in this chapter, apply to any chemical reaction, the focus will be on **combustion**.

11.2 Chemical Combustion

Combustion is a chemical process in which a substance reacts rapidly with oxygen and gives off heat. The original substance is called the fuel, and the source of oxygen is called the oxidizer. The fuel can be a solid, liquid, or gas. For most forms of transportation propulsion, the fuel is usually a liquid. The oxidizer, likewise, could be a solid, liquid, or gas, but is usually a gas (air). Rockets, on the other hand, usually carry their own oxidizer in addition to their fuel.

During combustion, new chemical substances are created from the fuel and the oxidizer. These substances will be called exhaust. Most of the exhaust comes from chemical combinations of the fuel and oxygen. When a hydrogen-carbon-based fuel (like gasoline) burns, the exhaust includes water (hydrogen + oxygen) and carbon dioxide (carbon + oxygen). However, the exhaust can also include chemical combinations from the oxidizer alone. If the gasoline is burned in air, which contains 21 % oxygen and 78 % nitrogen, the exhaust can also include nitrous oxides (NOX, nitrogen + oxygen). The temperature of the exhaust is high because of the heat that is transferred to the exhaust during combustion. Because of the high temperatures, exhaust usually occurs as a gas, but there can be liquid or solid exhaust products as well. Soot, for example, is a form of solid exhaust that occurs in some combustion processes.

During the combustion process, as the fuel and oxidizer are turned into exhaust products, heat is generated. Interestingly, some source of heat is also necessary to start combustion. Gasoline and air are both present in your automobile fuel tank; but combustion does not occur because there is no source of heat. Since heat is both required to start combustion and is itself a product of combustion, we can see why combustion takes place very rapidly. Once combustion gets started, we do not have to provide the heat source because the heat of combustion will keep things going. We do not have to keep lighting a campfire, it just keep burning.

To summarize, for combustion to occur three things must be present (Fig. 11.1):

1. A fuel to be burned,
2. A source of oxygen, and
3. A source of heat.

Because of combustion, exhaust products are created and heat is released. You can control or stop the combustion process by controlling the amount of the fuel available, the amount of oxygen available, or the source of heat.

Actually, combustion is a result of dynamic, or time-dependent, events that occur on a molecular level among atoms, molecules, radicals and solid boundaries. Therefore, this chapter presents chemical kinetics that includes kinetic theory of

Combustion

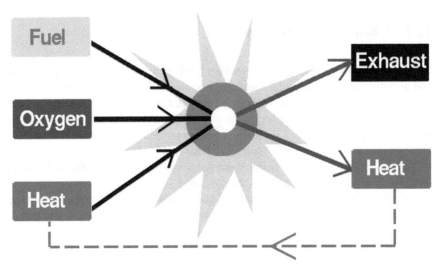

Fig. 11.1 Simple chemical combustion illustration. (Courtesy of NASA)

gases, elementary reactions and reaction rate theory. Furthermore, the rapid reactions produce gradients that transport processes convert into heat and species fluxes that speed-up the reactions.

11.3   Combustion Equations

A simple chemical-reaction equation is the combustion of propane in a pure oxygen environment. The chemical reaction is represented by;

$$C_3H_8 + 5O_2 \rightarrow 3CO_2 + 4H_2O \tag{11.1}$$

Note that the number of moles on the left-hand side may not equal the number of moles on the right-hand side. However, the number of atoms of an element must remain the same, before, after, and during a chemical reaction; this demands that the mass of each element be conserved during combustion.

In writing the equation some knowledge of the products of the reaction was assumed. *Complete combustion* was assumed. The products of *complete combustion* of a hydrocarbon fuel will be H_2O and CO_2, *Incomplete combustion* results in products that contain H_2, CO, C, and/or OH.

For a simple chemical reaction, such as Eq. 11.1, writing down a balanced chemical equation is straightforward. For reactions that are more complex the following systematic method proves useful [5]:

1. Set the number of moles of fuel equal to 1.
2. Balance CO_2 with number of C from the fuel.
3. Balance H_2O with H from the fuel.
4. Balance O_2 from CO_2 and H_2O.

For the combustion of propane, it was assumed that the process occurred in a pure oxygen environment. Actually, such a combustion process normally occurs in air. Nominally air consists of 21 % O_2 and 79 % N_2 by volume so that for each mole of O_2 in a reaction there are 3.76 mol of N_2.

$$\frac{79}{21} = 3.76 \ \frac{\text{mol } N_2}{\text{mol } O_2} \tag{11.2}$$

Thus, on the (simplistic) assumption that N_2 will not undergo any chemical reaction, Eq. 1.1 is replaced by

$$C_3H_8 + 5(O_2 + 3.76N_2) \rightarrow 3CO_2 + 4H_2O + 18.8N_2 \tag{11.3}$$

The minimum amount of air that supplies sufficient O_2 for the complete combustion of the fuel is called *Theoretical Air* or *Stoichiometric Air*. When complete combustion is achieved with theoretical air, the products contain no O_2, as in the reaction of Eq. 11.3. In practice, it is often found that if complete combustion is to occur, air must be supplied in an amount greater than theoretical air. This is due to the chemical kinetics and molecular activity of the reactants and products. The term *percent theoretical air* is used to compare the actual air provided to the combustion process compared to *stoichiometric* air (Eq. 11.4a).

$$\% \text{ theoretical air} = 100\% + \% \text{ excess air} \tag{11.4a}$$

Slightly, insufficient air results in CO being formed; some hydrocarbons may result from larger deficiencies [5]. So in summary, a mixture of air and fuel is called *Stoichiometric* if it contains just sufficient oxygen for the complete combustion of the fuel. Moreover, the percentage of excess air is given by the following (Eq. 11.4b):

$$\text{Percentage excess air} = \frac{\text{Actual (A/F) ratio} - \text{Stoichiometric (A/F) ratio}}{\text{Stoichiometric (A/F) ratio}} \tag{11.4b}$$

where A denotes Air while F denotes Fuel.

The parameter that relates the amount of air used in a combustion process is the *air-fuel ratio* (*AF*), which is the ratio of the mass of air to the mass of fuel. The reciprocal is the *fuel-air ratio* (*FA*) (Eq. 11.5). Thus

$$AF = \frac{m_{air}}{m_{fuel}} \qquad FA = \frac{m_{fuel}}{m_{air}} \tag{11.5}$$

Considering propane combustion with theoretical air as in (Eq. 11.3), the *air-fuel ratio* is (see Eq. 11.6).

$$AF = \frac{m_{air}}{m_{fuel}} = \frac{(5)(4.76)(29)}{(1)(44)} = 15.69 \frac{\text{Kg air}}{\text{Kg fuel}} \qquad (11.6)$$

where the molecular weight of air is taken as 29 kg/kmol and that of propane as 44 kg/kmol. If for the combustion of propane, $AF > 15.69$, a *Lean* or *Weak mixture* occurs; if $AF < 15.69$, a *Rich mixture* results.

For solid and liquid fuels the ratio are expressed by mass, while for gaseous fuels the ratios are normally expressed by volume. For boiler plant, the mixture is usually greater than 20% lean; for gas turbines, it can be as much as 300% lean. Petrol engines have to meet various conditions of load and speed, and operate over a wide range of mixture strengths. The following definition is then used (Eq. 11.7):

$$\text{Mixture strength} = \frac{\text{Stoichiometric (A/F) ratio}}{\text{Actual (A/F) ratio}} \qquad (11.7)$$

In this situation, the working values range between 80% (lean) and 120% (rich). Where fuels contain some oxygen (e.g. ethyl alcohol C_2H_6O) this oxygen is available for the combustion process, and so the fuel requires a smaller supply of air [6].

The combustion of hydrocarbon fuels involves H_2O in the products of combustion. The calculation of the dew point of the products is often of interest; it is the saturation temperature at the partial pressure of the water vapor. If the temperature drops below the dew point, the water vapor begins to condense. The condensate usually contains corrosive elements, and thus it is often important to ensure that the temperature of the products does not fall below the dew point.

Example 11.1 Butane is burned with dry air at an air-fuel ratio of 20. Calculate (a) the percent excess air, (b) the volume percentage of CO_2 in the products, and (c) the dew-point temperature of the products.

Solution The reaction equation for theoretical air is

$$C_4H_{10} + 6.5(O_2 + 3.76N_2) \rightarrow 4CO_2 + 5H_2O + 24.44N_2$$

a. The air-fuel ratio for theoretical air is

$$AF = \frac{m_{air}}{m_{fuel}} = \frac{(6.5)(4.76)(29)}{(1)(58)} = 15.47 \frac{\text{Kg air}}{\text{Kg fuel}}$$

This represents 100% theoretical air. The actual air-fuel ratio is 20. The excess air is then

$$\%\text{excess air} = \left(\frac{AF_{act} - AF_{th}}{AF_{th}} \right)(100\%) = \frac{20 - 15.47}{15.47}(100\%) = 29.28\%$$

b. The reaction equation with 129.28 % theoretical air is

$$C_4H_{10} + (6.5)(1.2928)(O_2 + 3.76N_2) + 4CO_2 + 5H_2O + 1.903O_2 + 31.6N_2$$

The volume percentage is obtained using the total moles in the products of combustion. For CO_2 we have

$$\%CO_2 = \left(\frac{4}{42.5}\right)(100\%) = 9.41\%$$

c. To find the dew-point temperature of the products we need the partial pressure of the water vapor. It is found using the mole fraction to be

$$p_{H_2O} = y_{H_2O} P_{atm} = \left(\frac{5}{42.5}\right) * 101.325 kPa = 11.76 kPa$$

where we have assumed an atmospheric pressure of 101.325 kPa. Using Appendix 14.2 we find the dew-point temperature to be $T_{d \cdot p} = 49\,°C$.

When a chemical reaction occurs, there may be considerable change in the chemical composition of a system. The problem this creates is that for a control volume the mixture that exits is different from the mixture that enters.

11.4   Mass and Mole Fractions

The amount of a substance present in a sample may be indicated by its mass or by the number of moles of the substance. A *mole* is defined as the mass of a substance equal to its molecular mass or molecular weight. Molecular weights for substances of interest are given in the Appendix. Compound molecular weights can be obtained by adding up the atomic weights of the constituents.

The composition of a mixture may be given as a list of the fractions of each of the substances present. Thus we define the mass fraction, of a component i, mf_i, as the ratio of the mass of the component, m_i, to the mass of the mixture (Eq. 11.8), m:

$$mf_i = \frac{m_i}{m} \tag{11.8}$$

It is evident that the sum of the mass fractions of all the components must be 1. Thus (Eq. 11.9):

$$mf_1 + mf_2 + \dots = 1 \tag{11.9}$$

The mole fraction of component i, x_i, is the ratio of the number of moles of component i, n_i, to the total number of moles in the mixture (Eq. 11.10), n:

$$x_i = \frac{n_i}{n} \tag{11.10}$$

The total number of moles, n, is the sum of the number of moles of all the components of the mixture (Eq. 11.11):

$$n = n_1 + n_2 + n_3 \dots \quad (11.11)$$

It follows that the sum of all the mole fractions of the mixture (Eq. 11.12) must equal 1.

$$x_1 + x_2 + \dots = 1 \quad (11.12)$$

The mass of component i in a mixture is the product of the number of moles of i and its molecular weight, M_i. The mass of the mixture is therefore the sum, $m = n_1 M_1 + n_2 M_2 + \cdots$ over all components of the mixture. Substituting $x_i n$ for n_i , the total mass becomes (Eq. 11.13);

$$m = (x_1 M_1 + x_2 M_2 + \dots)n \quad (11.13)$$

But the average molecular weight of the mixture is the ratio of the total mass to the total number of moles. Thus, the average molecular weight is (see Eq. 11.14).

$$M = m/n = x_1 M_1 + x_2 M_2 + \dots \quad (11.14)$$

Example 11.2 Express the mass fraction of component 1 of a mixture in terms of: (a) the number of moles of the three components of the mixture, n_1, n_2, and n_3 , and (b) the mole fractions of the three components. (c) If the mole fractions of carbon dioxide and nitrogen in a three component gas containing water vapor are 0.07 and 0.38, respectively, what are the mass fractions of the three components?

Solution

a. Because the mass of i can be written as $m_i = n_i M_i$, the mass fraction of component i can be written as:

$$mf_i = \frac{n_i M_i}{n_1 M_1 + n_2 M_2 + n_3 M_3 + \dots}$$

For the first of the three components, $i = 1$, this becomes:

$$mf_1 = \frac{n_1 M_1}{n_1 M_1 + n_2 M_2 + n_3 M_3}$$

Similarly, for $i = 2$ and $i = 3$:

$$mf_2 = \frac{n_2 M_2}{n_1 M_1 + n_2 M_2 + n_3 M_3}$$

$$mf_3 = \frac{n_3 M_3}{n_1 M_1 + n_2 M_2 + n_3 M_3}$$

b. Substituting $n_1 = x_1 n$, $n_2 = x_2 n_2$, etc. in the earlier equations and simplifying, we obtain for the mass fractions:

$$mf_1 = x_1 M_1 / (x_1 M_1 + x_2 M_2 + x_3 M_3)$$
$$mf_2 = x_2 M_2 / (x_1 M_1 + x_2 M_2 + x_3 M_3)$$
$$mf_3 = x_3 M_3 / (x_1 M_1 + x_2 M_2 + x_3 M_3)$$

c. Identifying the subscripts 1, 2, and 3 with carbon dioxide, nitrogen, and water vapor, respectively, we have $x_1 = 0.07, x_2 = 0.38$, and $x_3 = 1 - 0.07 - 0.038 = 0.55$. Then:

$$mf_1 = (0.07)(0.44) / [(0.07)(44) + (0.38)(28) + (0.55)(18)]$$
$$= (0.07)(0.44) / (23.62) = 0.1304$$
$$mf_2 = (0.38)(28) / (23.62) = 0.4505$$
$$mf_3 = (0.55)(18) / (23.62) = 0.4191$$

As a check we sum the mass fractions: $0.1304 + 0.4505 + 0.4191 = 1.0000$.

For a mixture of gases at a given temperature and pressure, the ideal gas law shows that $pV_i = n_i \Re T$ holds for any component, and $pV = n \Re T$ for the mixture as a whole. Forming the ratio of the two equations, we observe that the mole fractions have the same values as the volume fraction (Eq. 11.15):

$$x_i = V_i / V = n_i / n \tag{11.15}$$

Similarly, for a given volume of a mixture of gases at a given temperature, $pV_i = n_i \Re T$ for each component and $pV = n \Re T$ for the mixture. The ratio of the two equations shows that the *partial pressure* of any component i is the product of the mole fraction of i and the pressure of the mixture (Eq. 11.16):

$$p_i = p n_i / n = p x_i \tag{11.16}$$

Example 11.3 What is the partial pressure of water vapor in Example 11.2 if the mixture pressure is two atmospheres?

Solution The mole fraction of water vapor in the mixture of Example 11.2 is 0.55. The partial pressure of the water vapor is therefore $(0.55)(2) = 1.1$ atm.

11.5   Enthalpy of Formation

In order to deal with the heat produced in a chemical reaction, a reference point is required so that changes in enthalpy can be computed. The standard reference conditions are 25 °C (77 °F, 298 K, 537 °R) and 1 atmosphere pressure. The enthalpy of a substance at the reference state is usually identified as h^o. At these conditions many of the elements in their normal form are defined to have a 0.0 enthalpy of formation or **heat of formation,** $h^o{}_f$. Typically this includes gases like oxygen, nitrogen, and hydrogen as well as the solid form of carbon. Other gases like carbon dioxide and water vapor have a negative heat of formation at standard conditions. This means that when they are formed by burning carbon with oxygen, or hydrogen with oxygen, a certain amount of energy will be given off. The reaction is exothermic. When oxygen, nitrogen, or hydrogen is decomposed to oxygen ions, nitrogen ions, or hydrogen ions energy is required. The heat of formation for these reactions is positive because energy must be added to the molecule break it down into ions. Therefore they are endothermic reactions.

$$H_2 + \frac{1}{2}O_2 \rightarrow 2H_2O \qquad -241,820KJ/Kg\text{-mol}$$

$$C + O_2 \rightarrow CO_2 \qquad -393,520KJ/Kg\text{-mol}$$

$$O_2 \rightarrow 2O \qquad -249,170KJ/Kg\text{ - mol}$$

$$N_2 \rightarrow 2N \qquad -472,680KJ/Kg\text{ - mol}$$

$$H_2 \rightarrow 2H \qquad -218,000KJ/Kg\text{ - mol}$$

The negative sign for the heats of formation means that when the reaction occurred, energy was given up by the reactants.

The First Law for a chemical reaction can be written as Eq. 11.17;

$$Q = H_p - H_R \tag{11.17}$$

H_p is the enthalpy of the products of combustion that leave the combustion chamber and H_R is the enthalpy of the reactants that enter the combustion chamber. If the reactants are stable elements and the reaction occurs at constant temperature and pressure at the reference state (77 °F and 1 atm) then the H's represent the heats of formation for the substances involved. If the temperature deviates from the reference state, each of the enthalpies must be corrected for the temperature changes.

The general equation for a flowing system is (Eq. 11.18):

$$Q - Ws = \sum_{prod} N_p \left[h_f^o + \left(h(T) - h^o \right) \right]_p - \sum_{react} N_r \left[h_f^o + \left(h(T) - h^o \right) \right]_r \tag{11.18}$$

N_p = moles of products N_r = moles of reactants

$h(T) - h^o$ = Change in enthalpy from the reference state

The general equation for a rigid chamber is (Eq. 11.19):

$$Q - W_s = U_p - U_r$$

$$= \sum_{prod} N_p \left[h_f^o + \left(h(T) - h^o \right) - pv \right]_p - \sum_{react} N_r \left[h_f^o + \left(h(T) - h^o \right) - pv \right]_r \tag{11.19}$$

$$Q - W_s = U_p - U_r$$

$$- \sum_{prod} N_p \left[h_f^o + \left(h(T) - h^o \right) - \Re T \right]_p - \sum_{react} N_r \left[h_f^o + \left(h(T) - h^o \right) - \Re v \right]_r$$

The changes in enthalpy from the reference state can be calculated by the following techniques

For a solid or liquid $\Delta h = C\Delta T$.

For gases

1. $\Delta h = C_p \Delta T$
2. Use tabulated values for Δh
3. Use generalized charts for a real gas
4. Use tables for vapors like the steam tables

Example 11.4 Volumetric analysis of the products of combustion of an unknown hydrocarbon measured on a dry basis gives the following mole percents

$$
\begin{aligned}
CO_2 &= 10.4\% \\
CO &= 1.2\% \\
O_2 &= 2.8\% \\
N_2 &= 85.6\%
\end{aligned}
$$

Determine the composition of the hydrocarbon and the percent theoretical air.

Solution The combustion equation is

$$ C_a H_b + c(O_2 + 3.76 N_2) \rightarrow 10.4 CO_2 + 1.2 CO + 2.8 O_2 + 85.6 N_2 + d H_2 O $$

Writing equations to balance each of the species gives

C : $a = 10.4 + 1.2$ $a = 11.6$
N : $3.76c = 85.6$ $c = 22.8$
O : $2c = 10.4(2) + 1.2 + 2.8(2) + d$ $d = 2(22.8) - 20.8 - 1.2 - 5.6 = 18$
h : $b = 2d = 36$

The actual equation for 100% theoretical air must be

$$ C_{11.6} H_{36} + 20.6(O_2 + 3.76 N_2) \rightarrow 11.6 CO_2 + 77.5 N_2 + 18 H_2 O $$

22.8 mol of air were used and only 20.6 were needed, so % Theoretical Air = 110.7%

Example 11.5 Methane is burned with dry air at an air-fuel ratio of 5. Calculate the percent excess air and the percentage water vapor in the exhaust. Estimate the dew point temperature of the products.

Solution The combustion equation is

$$ CH_4 + 2O_2 + 2(3.76)N_2 \rightarrow 2H_2O + CO_2 + 2(3.76)N_2 $$

The stoichiometric air-fuel mixture is $AF_{st} = \dfrac{2(28.9669)}{16.043} = 3.611$

The percent excess air is 100*(5−3.611)/3.611=38.47%. So the balance equation is

$$CH_4 + 1.385(2O_2 + 2(3.76)N_2) \rightarrow 2H_2O + CO_2 + 0.77O_2 + 10.4N_2$$

There are 2+1+0.77+10.4=14.17 mol of products.
The mole percent of H_2O in the exhaust is 2/14.17=14.1%
The partial pressure of H_2O in the exhaust is (2/14.17)*101325=14.3 kPa
The saturation temperature (dew point) at 14.3 kPa is 326 K=53 °C.

Example 11.6 Gaseous methyl alcohol and air enter a combustion chamber at 25 °C and 1 atm and leave at 550 K and 1 atm. Assume 150% theoretical air. Estimate the heat transfer to the chamber.

Solution The reaction equation is

$$CH_3OH + 1.5\{1.5[O_2 + (3.76)N_2]\} \rightarrow 2H_2O + CO_2 + 0.75O_2 + 2.25(3.76)N_2$$

The First Law gives

$$Q = \sum_{prod} N_p (h_f^o + \Delta h)_p - \sum_{react} N_r (h_f^o + \Delta h)_r$$

Products

H_2O $h_f = -241,820$ $\Delta h = h(550) - h(298) = 17,489.1 - 8853.3 = 8635.8$
CO_2 $h_f = -393,520$ $\Delta h = h(550) - h(298) = 18,878.8 - 8378.4 = 10,500.4$
O_2 $h_f = 0$ $\Delta h = h(550) - h(298) = 15,363.4 - 7766.2 = 7597.2$
N_2 $h_f = 0$ $\Delta h = h(550) - h(298) = 15,095.3 - 7754.3 = 7431.0$

Reactants

$$CH_3OH \quad h_f = -200,890$$

The balance equation is

$$Q = 2*(-241,820 + 8635.8) + (-393,520 + 10,500.4)$$
$$+ 8.46*7431.0 + 0.75*7597.2 - (-200,890)$$
$$Q = -579,930 \frac{kJ}{kmol} \text{ of Methyl Alchol}$$

The above calculation assumes the water in the exhaust is liquid which at the temperature of the exhaust products is not very likely. So an additional enthalpy must be added to the Δh for H_2O to account for the vaporization of water.

$$\Delta h_{fg} = 2256.6 \, kJ/kg = 40,619.0 \, kJ/Kmol$$

Adding this to the negative Q gives **Q=− 539,311 kJ/kmol of Methyl Alcohol.**

Table 11.1 Selected enthalpies of combustion and enthalpies of vaporization

Substance	Formula	Higher heating value (kJ/kmol)	h_{fg}
Hydrogen	$H_2(g)$	−285,840	
Carbon	$C(s)$	−393,520	
Carbon monoxide	$CO(g)$	−282,990	
Methane	$CH_4(g)$	−890,360	
Acetylene	$C_2H_2(g)$	−1,299,600	
Ethylene	$C_2H_4(g)$	−1,410,970	
Ethane	$C_2H_6(g)$	−1,559,900	
Propylene	$C_3H_6(g)$	−2,058,500	
Propane	$C_3H_8(g)$	−2,220,000	15,060
n-Butane	$C_4H_{10}(g)$	−2,877,100	21,060
n-Pentane	$C_5H_{12}(g)$	−3,536,100	26,410
n-Hexane	$C_6H_{14}(g)$	−4,194,800	31,530
n-Heptane	$C_7H_{16}(g)$	−4,853,500	36,520
n-Octane	$C_8H_{18}(g)$	−5,512,200	41,460

Note that the heat of formation for nitrogen and oxygen is zero on both sides of the equation. If the methyl alcohol had entered as a liquid, we would have had to add its heat of vaporization to the reactant side of the equation further reducing the heat available per mole of the fuel.

11.6 Enthalpy of Combustion

With most hydrocarbons, normal combustion occurs with oxygen in the air. Therefore, the enthalpy change for the complete combustion of a substance with oxygen is called the ***heat of combustion***. Several heats of combustion are tabulated below in Table 11.1. If the products of combustion contain water in the vapor state, an allowance for the heat required to vaporize the water must be included in the change in enthalpy from the reference state. If the products of combustion include water in the liquid state, the heat of vaporization is not subtracted and this gives the ***Higher Heating Value*** for the ***heat of combustion*** for this fuel.

11.7 Adiabatic Flame Temperature

In many cases, the heat released in the combustion reaction will determine the final temperature of the products and the gases present in the combustion chamber. The temperature achieved assuming no heat transfer to the surroundings is called the Adiabatic Flame Temperature. Since the change in enthalpy of the products will depend on this temperature, the Adiabatic Flame Temperature must be found buy iteration. Note also that when a fuel is burned in air, the heat required to bring the nitrogen in the air up to the Adiabatic Flame Temperature must be included in the analysis. The Adiabatic Flame Temperature can be lowered by adding excess air above that required for complete combustion. Typically, this will be required due to peak temperature restrictions on the materials used for construction e.g. the strength of turbine blades downstream of the combustion process.

It is worth pointing out at this point that combustion in itself is an irreversible process and to achieve complete combustion typically requires a pressure loss in the combustion chamber. Since the major part of the fluid flowing through an air combustion chamber does not participate in the chemical reactions (primarily the nitrogen), the combustion process can be thought of as simply heating the working fluid. A nuclear heated heat exchanger can often accomplish the same heating with a lower pressure loss. However, in the nuclear heated system the temperature drops are in the opposite direction requiring heat exchanger walls to operate at a higher temperature than the combustion chamber walls.

Example 11.7 Calculate the enthalpy of combustion of gaseous octane and liquid octane assuming the reactants and products to be at the reference state of 25 °C and 1 atmosphere. Assume liquid water in the products exiting the steady flow combustion chamber.

Solution The reaction is

$$C_8H_{18} + 12.5(O_2 + 3.76N_2) \rightarrow 8CO_2 + 9H_2O + 47N_2$$

Products

$$
\begin{array}{ll}
H_2O & h_f = -285,830 \\
CO_2 & h_f = -393,520 \\
N_2 & h_f = 0
\end{array}
$$

Reactants

$$
\begin{array}{lll}
C_8H_{18}(l) & h_f = -208,450 & h_{fg} = 41,460 \\
C_8H_{18}(g) & h_f = -208,450 &
\end{array}
$$

Liquid Octane

$$Q = 8(-393,520) + 9(-285,830) - (208,450 - 41,460) = 5.4707 \times 10^6 \ kJ/kmol$$

Gaseous Octane

$$Q = 8(-393,520) + 9(-285,830) - (208,450) = 5.5122 \times 10^6 \ kJ/kmol$$

Example 11.8 Kerosene is burned with theoretical air in a jet engine. Estimate the adiabatic flame temperature. Kerosene can be treated as n-Dodecane—$C_{12}H_{26}$ and has a heat of formation of $-291,010$ kJ/kmol.

Solution The balance equation is

$$C_{12}H_{26} + 18.5(O_2 + 3.76N_2) \rightarrow 12CO_2 + 13H_2O + 69.5N_2$$

For the Adiabatic Flame Temperature $Q=0$, so

Products

H_2O	$h_f = -241{,}820$	$\Delta h(H2O) = ?$
CO_2	$h_f = -393{,}520$	$\Delta h(CO2) = ?$
N_2		$\Delta h(N2) = ?$

Reactants

$$C_{12}H_{26} \quad h_f = -291{,}010$$

$$Q = 0 = 12(-393{,}520 + \Delta h(CO2_1)) + 13(-241820 + \Delta h(H2O))$$
$$+ 69.5\Delta h(N2) - (-291{,}010)$$

At 25 °C we would have

$$Q = 12(-393{,}520) + 13(-241{,}820) + 291{,}010 = 7.5749 \times 10^6 \, kJ \, / \, kmol$$

Treating all of the products as nitrogen the dominant product, we would have 94.5 kg-mol of product. Thus, the change in enthalpy per mole is 80,158 kJ/kmol. At 25 °C nitrogen has an enthalpy of 7754.3 kJ/kmol. So the gas tables are entered looking for an enthalpy of 87,912 kJ/kmol. This corresponds to a temperature of 2660 K. Then evaluating the Δh's for H_2O and CO_2 at 2660 K gives

$$\Delta h(H_2O) = 116547.5 - 8853.3 = 107690.0$$
$$\Delta h(CO_2) = 140041.3 - 8378.4 = 131660.0$$
$$\Delta h(N_2) = 87846.0 - 7754.3 = 80092.0$$

At 2660 K the net heat balance is

$$Q = -7.5749 \times 10^6 + 13(107690.0) + 12(131660.0) + 69.5(80092.0) = 97{,}138 \, kJ/kmol$$

Since it is positive, the temperature does not quite reach 2660 K. Now noting that both H_2O and CO_2 had larger enthalpy changes than nitrogen, the number of moles can be adjusted to represent them with the all nitrogen model.

$$\Delta h(H_2O)/\Delta h(N_2) = 107690.0/80092 = 1.35$$
$$\Delta h(CO_2)/\Delta h(N_2) = 131660/80092 = 1.64$$

So the new number of moles will be $N = 69.5 + 1.35(13) + 1.64(12) = 106.7$.
Dividing the 25 °C enthalpy excess by 106.7 gives $7.5749 \times 10^6/106.7 = 71{,}017$ kJ/kmol.

Add this to the nitrogen enthalpy at $25\,^{\circ}C$ to get 78,771 kJ/kmol. The gas table gives a temperature of about 2410 K.

$$\Delta h(H_2O) = 103045.9 - 8853.3 = 94,193$$
$$\Delta h(CO_2) = 124671.2 - 8378.4 = 116,290.0$$
$$\Delta h(N_2) = 78,688.1 - 7754.3 = 70,934.0$$

$$Q = -7.5749 \times 10^6 + 13(94,193) + 12(116,290) + 69.5(70934) = -24,998\,kJ/kmol$$

So now the temperature is bracketed. The new mole effectiveness ratios for H_2O and CO_2 are 1.33 and 1.64. Therefore, 2410 K is very close and in fact the best answer obtained by another iteration is T=2416.3 K which is probably 2 or 3 too many digits for the round offs that have been made in the solution process.

It is obvious that this would be a very stressing temperature, and so most jet engines are running at a mixture ratio quite a bit lower than stoichiometric.

Problems

Problem 11.1: Determine:

a. The stoichmetric or theoretical air-fuel ratio for the combustion of C_5H_{12}.
b. The composition of the combustion products when C_5H_{12} is burned with 150% theoretical air.

Problem 11.2: A hydrocarbon fuel of unknown composition is burned with air. Te volumetric analysis on dry basis of the resulting flue gas is 9.27% CO_2, 2.31% CO, 4.86% O_2 and 83.56% N_2. Determine:

a. The composition of the fuel on mass basis.
b. The air-fuel ratio on mass basis, and
c. The percent excess or deficit air.

Problem 11.3: The molar composition of a particular natural gas is as follows: CH_4=67.6%; C_2H_6=31.3%, N_2=1.1%. Calculate:

(a) The theoretical air-fuel ratio on molar basis and on mass basis,
(b) Estimate the molar composition of the combustion products if the fuel is burned with 150% theoretical air

Assume complete combustion of the fuel.

Problem 11.4: Butane is burned with dry air at an air-fuel ratio of 20. Calculate

(a) The percent excess air,
(b) The volume percentage of CO_2, in the products, and
(c) The dew-point temperature of the products.

Problem 11.5: Butane is burned with dry, air and volumetric analysis of the products on a dry basis (the water vapor is not measured) gives 11.0% CO_2, 1.0% CO, 3.5% O_2 and 84.5% N_2. Determine the percent theoretical air

Problem 11.6: Volumetric analysis of the products of combustion of an unknown hydrocarbon, measured on a dry basis, gives 10.4% CO_2, 1.2% CO, 2.8% O_2, and 85.6% N_2. Determine the composition of the hydrocarbon and the percent theoretical air.

Problem 11.7: A gaseous fuel has the molar analysis of: 30% CH_4, 20% CO, and 30% N_2. Determine the theoretical air-fuel ratio on mass basis for complete combustion of the fuel gas.

Problem 11.8: A sample of dry anthracite has the following composition by mass.

$$C\ 90\%; H3\%; O2.5, N\ 1\% S0.5\%, Ash3\%$$

Calculate:

1. The stoichiometric Air/Fuel Ratio,
2. The Air/Fuel and the dry and wet analysis of the products of combustion by mass and dry volume, when 20% excess air is supplied.

Problem 11.9: The gravimetric analysis of a sample of coal is given as 80% C, 12% H, and 8% Ash. Calculate the Stoichiometric Air/Fuel (A/F) ratio and the analysis of the products by volume.

Problem 11.10: In an engine test the dry product analysis was CO_2 15.5%, O_2 2.3% and the remainder N_2. Assuming that the fuel burned was a pure hydrocarbon; calculate the ratio of carbon to hydrogen in the fuel, the A/F ratio used, and the mixture strength.

Problem 11.11: Octane vapor is burned with air. The molar analysis of the products of combustion on a dry basis is 8.45% CO_2: 1.48% CO; 7.81% O_2 and 82.8% N_2. Determine the air-fuel ratio on mass basis, used for combustion.

Problem 11.12: The ultimate analysis of a sample petrol was 85% C and 15% H. The analysis of the dry produces showed 13.5% CO_2, some CO and the remainder N_2. Calculate:

I.   The actual Air/Fuel ratio;
II. The mixture strength;
III. The mass of H_2O vapor carried by the exhaust gas per kilogram of total exhaust gas;
IV. The temperature to which the gas must be cooled before combustion of the H_2O vapor begins, if the pressure in the exhaust pipe is 1.013 bar.

References

1. El-Mahallawy F, El-Din Habik S (2002) Fundamentals technology of combustion. Elsevier, Amsterdam
2. Kirillin VA, Sychev VV, Sheindlin AE (1976) Engineering thermodynamics. Mir Publishers, Moscow

3. World Energy Outlook, WEO (1998) Chapter 3, International Energy Agency, 37. Last Edited by Jonathan Cobb at 28/01/2011
4. Potter MC, Somerton CW (2006) Thermodynamics for engineers, 2nd edn. McGraw-Hill, New York
5. Eastop TD, McConkey A (1993) Applied thermodynamics for engineering technologists, 5th edn. Pearson Education Ltd, London

Chapter 12
Heat Transfer

Thermodynamics deals with the transfer of heat to and from a working fluid and the performance of work by that fluid. Since the transfer of heat to a working fluid is central to thermodynamics, a short excursion into the technology of heat transfer is useful to tie thermodynamics to real world devices. Heat transfer processes are never ideal and a study of the technology of heat transfer will develop an understanding of the trade offs in the design of the devices that actually accomplish the heat transfer. Heat transfer technology provides the basis on which heat exchangers are designed to accomplish the actual transfer of thermal energy.

12.1 Fundamental Modes of Heat Transfer

There are three fundamental modes of heat transfer.

1. *Conduction*.
2. *Convection*.
3. *Radiation*.

The temperature distribution in any system or medium is controlled by the combined effects of these three modes of heat transfer. In most situations one mode dominates and temperature distributions and heat fluxes can be obtained very accurately by only considering that mode. In general, heat transfer can be a multidimensional time dependent phenomena. However, for heat engines, transient heat transfer effects are not usually as important as steady state heat transfer phenomena and they will be neglected in discussion that follows.

© Springer International Publishing Switzerland 2015
B. Zohuri, P. McDaniel, *Thermodynamics In Nuclear Power Plant Systems*,
DOI 10.1007/978-3-319-13419-2_12

12.2   Conduction

Conduction occurs in stationary materials as a result of the vibrations of atoms or molecules in the materials. It is governed by Fourier's law of heat conduction, which in one dimension is written as,

$$Q_x = -kA\frac{\partial T}{\partial x} \quad \text{Btu/h} \quad \text{or} \quad \text{W} \tag{12.1a}$$

or

$$q_x = \frac{Q_x}{A} = -k\frac{\partial T}{\partial x} \quad \text{Btu/h/ft}^2 \quad \text{or} \quad \text{W/m}^2 \tag{12.1b}$$

Simply stated the heat flow per unit area is proportional to the negative of the temperature gradient. The proportionality constant is called the thermal conductivity and it has units of watts/meter/K or Btus/ft/°R. The thermal conductivities of typical materials vary widely by material and it also depends on the temperature of the materials. Some typical values are given in the Appendix for solids, liquids, and gases.

12.3   Convection

Heat transfer by convection occurs as the result of a moving fluid coming in contact with a fixed surface. The moving fluid carries the heat and deposits it on the surface or draws it out of the surface. There are two types of convection. In *forced convection,* the fluid is being driven or forced along by some mechanism other than thermal gradients at the surface. In *free convection* the fluid is moved along by thermal gradients or temperature differences at the surface. Convection obeys Newton's law of cooling given by,

$$Q = hA(T_f - T_w) \tag{12.2a}$$

$$q = h(T_f - T_w) \tag{12.2b}$$

q in this case is the heat flux per unit area at the wall. The symbol h is identified as the film heat transfer coefficient. It has units of Watts/m^2/K or Btu/hr/ft^2/R. Where k in Eq. 12.1b, the thermal conductivity, is a function of only the material and its temperature, h, the film heat transfer coefficient, depends on the properties of the fluid, the temperature of the fluid, and the flow characteristics. Multiple correlations have been determined for calculating an appropriate h for most materials and flow situations.

12.4 Radiation

Radiation heat transfer takes place by means of electromagnetic waves transmitted from one body to another. It does not require a medium and so can transfer heat across a vacuum. It is governed by the Stefan-Boltzmann radiation heat transfer equation,

$$Q = \varepsilon\sigma A_{1-2}\left(T_1^4 - T_2^4\right) \qquad (12.3a)$$

$$q = \varepsilon\sigma\left(T_1^4 - T_2^4\right) \qquad (12.3b)$$

The heat transferred in this case depends on the difference in the fourth power of the temperature of the two bodies. It also depends on a universal constant, σ, call the Stefan-Boltzmann constant equal to 5.6697×10^{-8} W/m^2 K^4. The variable ε depends on the surface material of the two bodies and can depend on their temperatures as well. It is called the emissivity and varies between 0.0 and 1.0. The area factor A_{1-2}, is the area viewed by body 2 of body 1, and can become fairly difficult to calculate. Note that because the temperatures are raised to the 4th power and then differenced, radiation heat transfer must always be calculated based on absolute temperatures (K or °R). Both conduction and convection depend only on the linear differences of temperature and any consistent temperature scale will work.

Radiation heat transfer is important at very high temperatures and in a vacuum. However, for most designs involving terrestrial power plants, the heat transfer is dominated by conduction and convection, so radiation will not be treated extensively in this text. In addition, to get good quantitative results in analyzing radiation heat transfer problems, accurate calculation of the view factors, or A_{1-2}, is required. The effort involved is well beyond the level of this text. Siegel and Howell by these authors refer interested students to the text.

Before going into more detailed analysis of the modes of heat transfer, it will be useful to provide the definitions of a number of terms of importance. These are provided in Table 12.1 below.

Example 12.1 A constant temperature difference of 300 °F (166.7 °C) is maintained across the surfaces of a slab of 0.1-ft (0.0306-m) thickness. Determine the rate of heat transfer per unit area across the slab for each of the following cases. The slab material is copper ($k = 220$ Btu/h.ft.°F or 380.7 W/m °C), aluminum ($k = 130$ Btu/h.ft.°F or 225.7 W/m °C), carbon steel ($k = 10$ Btu/h.ft.°F or 17.3 W/m °C), brick ($k = 0.5$ Btu/h.ft.°F or 0.865 W/m °C), and asbestos ($k = 0.1$ Btu/h.ft.°F or 0.173 W/m °C)

Solution The Fourier law for one-dimensional heat conduction is given by Eq. 12.1b:

$$q = -k\frac{dT}{dx}$$

Table 12.1 Definitions for terms of importance for heat transfer

Blackbody	A body with a surface emissivity of 1. Such a body will emit all of the thermal radiation it can (as described by theory), and will absorb 100 % of the thermal radiation striking it. Most physical objects have surface emissivities less than 1 and hence do not have blackbody surface properties
Density, ρ	The amount of mass per unit volume. In heat transfer problems, the density works with the specific heat to determine how much energy a body can store per unit increase in temperature. Its units are kg/m^3
Emissive power	The heat per unit time (and per unit area) emitted by an object. For a blackbody, this is given by the Stefan-Boltzmann relation σT^4
Graybody	A body that emits only a fraction of the thermal energy emitted by an equivalent blackbody. By definition, a graybody has a surface emissivity less than 1, and a surface reflectivity greater than zero
Heat flux, q	The rate of heat flowing past a reference datum. Its units are W/m^2
Internal energy, e	A measure of the internal energy stored within a material per unit volume. For most heat transfer problems, this energy consists just of thermal energy. The amount of thermal energy stored in a body is manifested by its temperature
Radiation view factor, F_{12}	The fraction of thermal energy leaving the surface of object 1 and reaching the surface of object 2, determined entirely from geometrical considerations. Stated in other words, F_{12} is the fraction of object 2 visible from the surface of object 1, and ranges from zero to 1. This quantity is also known as the Radiation Shape Factor. Its units are dimensionless
Rate of heat generation, q_{gen}	A function of position that describes the rate of heat generation within a body. Typically, this new heat must be conducted to the body boundaries and removed via convection and/or radiation heat transfer. Its units are W/m^3
Specific heat, c	A material property that indicates the amount of energy a body stores for each degree increase in temperature, on a per unit mass basis. Its units are J/kg-K
Stefan-Boltzmann constant, σ	Constant of proportionality used in radiation heat transfer, whose value is 5.669×10^{-8} W/m^2-K^4. For a blackbody, the heat flux emitted is given by the product of σ and the absolute temperature to the fourth power
Surface emissivity ε	The relative emissive power of a body compared to that of an ideal blackbody. In other words, the fraction of thermal radiation emitted compared to the amount emitted if the body were a blackbody. By definition, a blackbody has a surface emissivity of 1. The emissivity is also equal to the absorption coefficient, or the fraction of any thermal energy incident on a body that is absorbed
Thermal conductivity, k	A material property that describes the rate at which heat flows within a body for a given temperature difference. Its units are W/m-k
Thermal diffusivity, α	A material property that describes the rate at which heat diffuses through a body. It is a function of the body's thermal conductivity and its specific heat. A high thermal conductivity will increase the body's thermal diffusivity, as heat will be able to conduct across the body quickly. Conversely, a high specific heat will lower the body's thermal diffusivity, since heat is preferentially stored as internal energy within the body instead of being conducted through it. Its units are m^2/s

For the problem considered here q should be constant everywhere in the medium since there are no heat sources or heat sinks in the slab. The integration of this equation across the slab for constant q and k gives;

$$qx\big|_0^L = -kT\big|_{T_1}^{T_2}$$

or

$$q = k\frac{T_1 - T_2}{L} \quad \text{Btu/h.ft}^2 (\text{or W/m}^2)$$

In the present problem $T_1 - T_2 = 300\,°F$, $L = 0.1$ ft, and k is specified for each material considered. Then, the heat fluxes, for copper, aluminum, carbon steel, brick, and asbestos, respectively, are given as 6.6×10^5, 3.9×10^4, and 3×10^2 Btu/h.ft^2 (or 20.8×10^5, 12.3×10^5, 9.5×10^4, 4.7×10^3, and 9.5×10^2 W/m^2). Note that the heat transfer rate is higher with a larger thermal conductivity.

Example 12.2 A fluid at 500 °F (260 °C) flows over a flat plate, which is kept at a uniform temperature of 100 °F (82.2 °C). If the heat transfer coefficient h for convection is 20 Btu/h.ft^2.°F (113.5 W/m^2°C), determine the heat transfer rate per unit area of the plate from the fluid into the plate.

Solution Heat transfer by convection between a fluid and a solid surface is given by Eq. 12.2b

$$q = h(T_f - T_w) \quad \text{Btu/h.ft}^2 (\text{or W/m}^2)$$

Taking $h = 20$ Btu/h.ft^2. °F (or 113.5 W/m^2°C) and $T_1 - T_2 = 500 - 100 = 400°F$ (or 222.2 °C), the heat flux at the wall becomes

$$q = 20 \times 400 = 8 \times 10^3 \text{Btu/h.ft}^2 (\text{or} 25.2 \text{kW/m}^2)$$

Example 12.3 Two identical bodies radiate heat to each other. One body is at 30 °C and the other at 250 °C. The emissivity of both is 0.7. Calculate the net heat transfer per square meter.

Solution Using Eq. 12.3a, we can write the following analysis;

$$Q = \varepsilon\sigma A\left(T_1^4 - T_2^4\right)$$
$$= 0.7 \times 56.7 \times 10^{-9} \times 1 \times (523^4 - 303^4)$$
$$= 2635 \text{ W}$$

12.5   Heat Conduction in a Slab

The most general form of the steady state heat conduction equation occurs when the material through which the heat is transported also has an internal source of heat. So consider the slab geometry defined by Fig. 12.1.

The heat balance equation for a volume given by Adx will be,

Heat conducted out of Adx—Heat conducted into Adx = Heat generated within Adx

Now assuming the overall heat flow is from left to right, the heat conducted into Adx will be given by

$$q_{in} = -kA\frac{dT}{dx}\bigg|_x$$

The heat conducted out will be

$$q_{out} = -kA\frac{dT}{dx}\bigg|_{x+dx} = -A\left[k\frac{dT}{dx}\bigg|_x + \frac{d}{dx}k\frac{dT}{dx}\bigg|_x dx + \frac{1}{2}\frac{d^2}{dx^2}k\frac{dT}{dx}\bigg|_x dx^2...\right]$$

$$q_{out} \approx -A\left[k\frac{dT}{dx}\bigg|_x + \frac{d}{dx}k\frac{dT}{dx}\bigg|_x dx\right]$$

Then identifying the heat generation rate per unit volume as Q(x) the balance equation becomes

$$q_{out} - q_{in} = Q(x)Adx = -A\left[k\frac{dT}{dx}\bigg|_x + \frac{d}{dx}k\frac{dT}{dx}\bigg|_x dx\right] + kA\frac{dT}{dx}\bigg|_x$$

$$Q(x)dx = -\frac{d}{dx}k\frac{dT}{dx}\bigg|_x dx$$

Fig. 12.1 Slab with internal heat source

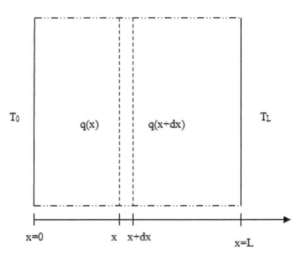

If k is not a function of position or temperature then,

$$\frac{Q(x)}{k} = -\frac{d^2T}{dx^2} = -\nabla^2 T(x) \tag{12.4}$$

This is a form of the classic Poisson's equation. The ∇^2 operator is indicated because when the equation is generalized to two or three dimensions, the two or three-dimensional version of this operator gives the correct form of the equation. Of course, if the heat source is not present then the equation reduces to Laplace's equation.

$$\nabla^2 T = 0 \tag{12.5}$$

12.6   Heat Conduction in Curve-Linear Geometries

Solving the two or three dimensional heat transfer equation is a difficult problem in the general case. However, for most heat transfer situations of interest to nuclear engineering one-dimensional solutions are usually all that is required. However, many times those solutions require the equation to be written in curvi-linear geometry, either cylindrical or spherical coordinates. Poisson's equation in cylindrical coordinates is

$$\frac{Q(r)}{k} = -\left(\frac{1}{r}\frac{d}{dr}r\frac{dT}{dr}\right) = -\left(\frac{1}{r}\frac{dT}{dr} + \frac{d^2T}{dr^2}\right) \tag{12.6}$$

And in spherical coordinates,

$$\frac{Q(\rho)}{k} = -\left(\frac{1}{\rho^2}\frac{d}{d\rho}\rho^2\frac{dT}{d\rho}\right) = -\left(\frac{2}{\rho}\frac{dT}{d\rho} + \frac{d^2T}{d\rho^2}\right) \tag{12.7}$$

Now integrating Poisson's Equation for a constant Q in cylindrical coordinates gives

$$\int_{r_1}^{r} Qr\,dr = -k\int_{r_1}^{r}\frac{1}{r}\frac{d}{dr}r\frac{dT}{dr}r\,dr = -k\left[r\frac{dT}{dr} - r_1\frac{dT}{dr}\Big|_{r_1}\right]$$

$$Q\left(\frac{r^2}{2} - \frac{r_1^2}{2}\right) = kr_1\frac{dT}{dr}\Big|_{r_1} - kr\frac{dT}{dr}$$

Note that the first term on the right hand side is simply the heat being transferred into or out of the surface at r_1. Separating terms and integrating again gives,

$$\frac{Qr^2}{2} - \frac{Qr_1^2}{2} - kr_1 \left.\frac{dT}{dr}\right|_{r_1} = -kr\left.\frac{dT}{dr}\right|$$

$$\frac{Qr}{2}dr - \left(\frac{Qr_1^2}{2} + kr_1\left.\frac{dT}{dr}\right|_{r_1}\right)\frac{dr}{r} = -kdT \qquad (12.8)$$

$$Q\left(\frac{r^2}{4} - \frac{r_1^2}{4}\right) - \left(\frac{Qr_1^2}{2} - kr_1\left.\frac{dT}{dr}\right|_{r_1}\right)\ln\left(\frac{r}{r_1}\right) = k(T_1 - T)$$

There are two unknown constants of integration in this case, namely the heat flux on the surface at $r = r_1$ and the temperature T_1 on that surface. Since the outer surface has not been specified yet, and there will be boundary conditions on it, both of the inner boundary constants cannot be specified independent of the outer surface boundary conditions. In the fuel pin case where r_1 is taken as 0, the derivative term at r_1 must be zero and the equation simplifies to,

$$\frac{Qr_2^2}{4} = k(T_1 - T_2)$$

Or   (12.9)

$$Q\pi r_2^2 = 4\pi k(T_1 - T_2)$$

The term on the left hand side is simply the linear heat rate and T_1 is a function of T_2 and this heat rate. A boundary condition at r_2 is required in order to specify T_2. Normally the boundary condition at r_2 is given in terms of some kind of a heat flux, or a derivative boundary condition. It would be unrealistic to set T_2 to a specified value because that would assume that the heat flux could be adjusted to hold it at this value. A second simplification of Eq. 12.8 occurs for the case of no heat production in the region of interest. In that case, Eq. 12.9 simplifies to

$$-\left(k\left.\frac{dT}{dr}\right|_{r_1}\right)r_1\ln\left(\frac{r}{r_1}\right) = k(T_1 - T)$$

$$T_2 - T_1 = \left.\frac{dT}{dr}\right|_{r_1} r_1\ln\left(\frac{r_2}{r_1}\right) = -\frac{q_1}{k}r_1\ln\left(\frac{r_2}{r_1}\right) \qquad (12.10)$$

Example 12.3 Consider a fuel pin with a linear heat rate of 300 W/cm and a radius of 0.41 cm and a thermal conductivity of 2.0 W/m/K. Neglect the clad for the present. It is cooled by water flowing by with a bulk temperature of 585 K and a film heat transfer coefficient 20,000 W/m²/K. Estimate the surface temperature of the fuel pin and its centerline temperature.

Solution First obtain the volumetric heat rate

$$Q = \frac{300 \text{W/cm}}{\pi (0.41)^2} = 568 \text{W/cm}^3$$

Then the heat flux into the fluid is given by

$$300 \text{W/cm} * 100 = \pi 2r * h(T_s - T_b) = \pi * 2 * 0.0041 * 20,000 * (T_s - 585)$$

$$T_s = 585 + \frac{30,000}{2 * \pi * 0.0041 * 20,000} = 643.2 \text{K}$$

Then applying Eq. 12.9 gives

$$Q\pi r^2 = 4\pi k(T_1 - T_s) = 300 = 4\pi * \frac{2.0}{100}(T_1 - 643.2)$$

$$T_1 = 643.2 + \frac{300}{4\pi * 0.02} = 1836.6 \text{K}$$

Example 12.4 Hot water is flowing through a thick tube and is being cooled by water flowing on the outside of the tube. The temperature of the hot water is 600 K. The temperature of the cold water is 400 K. The film heat transfer coefficient for the hot water is 20,000 W/m²/K and the film heat transfer coefficient for the cold water is 100,000 W/m²/K. The tube is 3 mm thick with an inner diameter of 1.5 cm and has a thermal conductivity of 16.7 W/m/K. Calculate the surface temperatures of the tube.

Solution The key is realizing that the same amount of heat is flowing through all three materials. We have

$$q = h_{cold}(T_{outer\ surface} - T_{cold\ water})$$

$$q = \frac{k}{r_1 \ln\frac{r_2}{r_1}}(T_{inner\ surface} - T_{outer\ surface})$$

$$q = h_{hot}(T_{hot\ water} - T_{inner\ surface})$$

$$20,000(600 - T_{innersurface}) = \frac{16.7}{0.00273}(T_{innersurface} - T_{outersurface})$$

$$100,000(T_{outersurface} - 400) = \frac{16.7}{0.00273}(T_{innersurface} - T_{outersurface})$$

$$3.269(600 - T_{is}) = T_{is} - T_{os} \qquad 1961.4 = 4.269T_{is} - T_{os}$$

$$T_{os} = 4.269T_{is} - 1961.4 \qquad 16.347(T_{os} - 400) = T_{is} - T_{os}$$

$$6538.8 = 17.347T_{os} - T_{is} \qquad 6538.8 = 17.347(4.269T_{is} - 1961.4) - T_{is}$$

$$6538.8 + 17.347 * 1961.4 = (17.347 * 4.269 - 1)T_{is}$$

$$T_{is} = 555.24 \text{ K} \qquad T_{os} = 4.269 * 555.24 - 1961.4 = 408.9 \text{ K}$$

Now consider the integral of the Poisson Equation in spherical geometry.

$$\int_{r_1}^{r} Qr^2\,dr = -k \int_{r_1}^{r} \frac{1}{r^2}\frac{d}{dr} r^2 \frac{dT}{dr} r^2\,dr$$

$$Q\left(\frac{r^3}{3} - \frac{r_1^3}{3}\right) = -k\left(r^2 \frac{dT}{dr} - r_1^2 \frac{dT}{dr}\Big|_{r_1} \right)$$

$$\frac{Qr}{3}dr - \left(\frac{Qr_1^3}{3} + kr_1^2 \frac{dT}{dr}\Big|_{r_1}\right)\frac{dr}{r^2} = -kdT$$

$$Q\left(\frac{r^2}{6} - \frac{r_1^2}{6}\right) + \left(\frac{Qr_1^3}{3} + kr_1^2 \frac{dT}{dr}\Big|_{r_1}\right)\left(\frac{1}{r} - \frac{1}{r_1}\right) = k(T_1 - T)$$

Then for the case where $r_1 = 0$, this reduces to

$$\frac{Qr_2^2}{6} = k(T_1 - T_2) \qquad\qquad (12.11)$$

Note that a heat rate per sphere could be defined as

$$Q_{sphere} = \frac{4}{3}\pi r^3 Q$$

In addition, Eq. 12.11 can be written as

$$Q_{sphere} = 8\pi r_2 k(T_1 - T_2)$$

And for the case of no heat generation within the shell,

$$k\frac{dT}{dr}\Big|_{r_1}\, r_1^2\left(\frac{1}{r_2} - \frac{1}{r_1}\right) = k(T_1 - T_2) \qquad\qquad (12.12)$$

Example 12.5 Estimate the maximum amount of heat that can be generated per spherical particle with a radius of 0.5 mm contained in a shell of pyrolytic carbon 0.1 mm thick. The outer surface of the particle will be maintained at 700 °K by a very efficient heat removal system. The peak temperature allowed for the uranium dioxide in the fuel particle is 1800 °K. The average thermal conductivity for uranium dioxide is 3.0 W/m/K. The average thermal conductivity for pyrolytic carbon is 2.0 W/m/K.

Solution For the sphere

$$Q_{sphere,max} = 8\pi r_2 k(T_1 - T_2) = 8\pi * 0.0005 * 3.0 * (1800 - T_s) = 0.0377(1800 - T_s)$$

For the shell

$$k\frac{dT}{dr}\bigg|_{r_1} r_1^2\left(\frac{1}{r_2}-\frac{1}{r_1}\right) = k(T_1-T_2)$$

$$4\pi r_1^2 k\frac{dT}{dr}\bigg|_{r_1} = 4\pi r_1^2 \frac{k}{r_1^2\left(\dfrac{1}{r_2}-\dfrac{1}{r_1}\right)}(T_s-700)$$

$$Q_{sphere,\max} = -4\pi\frac{2.0}{\left(\dfrac{1}{0.0006}-\dfrac{1}{0.0005}\right)}(T_s-700)$$

$$Q_{sphere,\max} = 0.0754(T_s-700)$$

Then

$$T_s = 700+\frac{Q_{sphere,\max}}{0.0754}$$

$$Q_{sphere,\max} = 0.0377\left(1800-700-\frac{Q_{sphere,\max}}{0.0754}\right)$$

$$1.5Q_{sphere,\max} = 41.47$$

$$Q_{sphere,\max} = 27.6\mathrm{W}$$

12.7 Convection

Convection is more complicated than conduction and depends much more on the configuration of the heat transfer media than conduction. However, it has been analyzed in detail for many different configurations and as a result, efficient prediction of convection results depends on finding the correct analysis for the problem under consideration.

There are several ways of categorizing convection results. The most basic is dividing convection into forced convection and free convection. In forced convection, a pressure drop drives the heat transport media or fluid and the velocity of the fluid does not depend on the heat transfer or thermal gradients. In free convection, there is no pressure drop driving the fluid and its velocity develops because of the buoyancy of the fluid and thermal gradients. Forced convection is by far the most common in heat exchangers and nuclear reactors.

The second way of dividing convection is into internal flows and external flows. Internal flows are typically flows with in a pipe or pipe-like structure. The flow is confined and there is no free surface. External flows deal with flows on a surface and are generally bounded on only one side. However, pipes or other confining media can be so large that what happens on one surface is not influenced by other confining surfaces and therefore the flow can be treated as an external flow.

12.8   Boundary Layer Concept

In most convection situations, the transition from the bulk temperature of the fluid to the temperature of the confining surface occurs over a very short distance very close to the surface. The velocity of the fluid and its temperature must match the velocity of the wall, usually zero, and the temperature of the wall. The transition from the bulk fluid velocity and the bulk fluid temperature occurs over a very thin region in the fluid called the boundary layer. A physical description of the boundary layer is provided in Fig. 12.2.

Figure 12.2 describes the development of the velocity and thermal boundary layers over a flat plate. The outer boundaries of each are defined by where the velocity reaches 99 % of its free stream value and the temperature difference between the fluid and the wall reaches 99 % of its ultimate value. There is no reason that the two boundary layers should be the same thickness, but sometimes they are taken to be so to simplify the analysis. In the case of the external flow over a flat plate, both boundary layers start at zero thickness at the start of the plate and grow continuously along the plate. In order to predict the heat transfer on the plate, it is important to track the growth of the velocity boundary layer as well as the thermal boundary layer.

At the start of the flat plate, the velocity profile varies in a very smooth manner almost parabolic in the distance from the plate. The fluid flows in smooth layers over the plate with each layer moving in a stable manner. This is called laminar flow. The heat transfer is basically by a conduction mechanism between layers. As the fluid travels down the plate, eventually the layers become unstable and packets of fluid move transverse to the plate in random motions. The flow becomes turbulent. In this case, the packets of fluid transport heat as they move within the boundary layer. The transition from laminar flow to turbulent flow is controlled by the ration of the inertia forces in the boundary layer to the viscous forces in the boundary layer. This ratio is given by

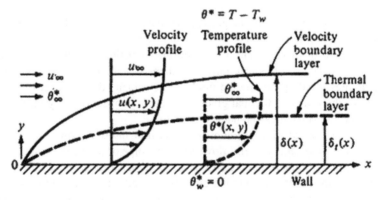

Fig. 12.2 Velocity and thermal boundary layers for laminar flow over a flat plane [5]

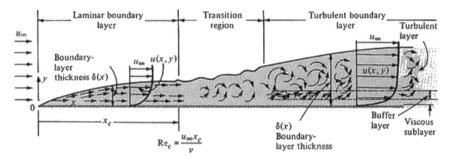

Fig. 12.3 Boundary layer concept for flow along a flat plate [5]

$$\text{Re} = \frac{InertiaForce}{ViscousForce} = \frac{\rho V^2/L}{\mu V/L^2} = \frac{\rho VL}{\mu} = \frac{VL}{\nu} \qquad (12.13)$$

Moreover, it is called the Reynolds number after Osborne Reynolds who first recognized its importance in characterizing the flow through pipes. Note that it is dimensionless and is simply a number. Typically around $\text{Re} = 2300$ flow over a smooth plate will transition from laminar to turbulent flow. Of course, this can vary depending on the roughness of the plate and the smoothness of the flow upstream of the plate. A better description of the phenomena in a transitioning boundary layer is presented in Fig. 12.3.

Since heat transfer in the turbulent boundary layer is greater, it is often desirable to "trip" the boundary layer near the front of the plate by adding roughness or irregularities to cause an immediate transition to turbulent flow. However, turbulent flow does cause a greater drag on the flat plate and will cause a greater friction force attempting to slow the flow over the plate. The trade between increased heat transfer and increased drag will be important to the design of heat exchangers. The drop in pressure required to obtain the desired heat transfer represents the mechanical cost of the heat transfer.

The flow over a curved surface complicates the boundary layer flow even more. Consider Fig. 12.4. The fluid approaches the curved body from the left. As the fluid

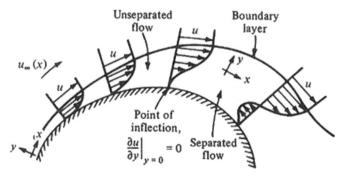

Fig. 12.4 Boundary layer separation for flow over a curved body [5]

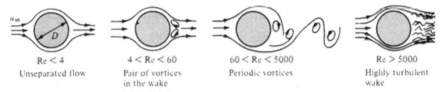

Re < 4	4 < Re < 60	60 < Re < 5000	Re > 5000
Unseparated flow	Pair of vortices in the wake	Periodic vortices	Highly turbulent wake

Fig. 12.5 Flow around a circular cylinder as Reynolds number varies [5]

approaches, the pressure builds up in the fluid as the body slows it down. Then as the fluid flows over the body, the pressure decreases due to flow speeding up. The boundary layer on the front surface is compressed because of the higher pressure but expands as the pressure decreases. Eventually as the flow turns the corner and moves down the backside of the body, the flow separates due to a loss of energy in the boundary layer. This produces a low pressure on the backside of the curved body increasing the drag that the fluid sees significantly.

A quantitative description of the flows across a cylinder as a function of the Reynolds number is given by Fig. 12.5. Note that in this case the Reynolds number used to categorize the flows is based on the diameter of the cylinder.

$$Re = \frac{\rho V D}{\mu} \qquad (12.14)$$

Flow over a flat plate, or over a curved surface (circular cylinder), provide the basis for most film heat transfer correlations for forced convection in external flows. Free convection produces significantly different boundary layer phenomena and additional variables come into play. The free convection scenario of most interest is on a vertical flat plate, or tube, and gravitational forces and the buoyancy of the fluid become important. Two typical scenarios are described in Fig. 12.6.

The flow is significantly more complicated than that in forced convection.

Moving on to internal flow heat transfer, the development of the flow pattern for laminar flow is described in Fig. 12.7.

At the entrance to the tube, the flow is uniform across the tube surface area. However, the boundary layer builds up on all internal surfaces until there is no flow outside of the boundary layer. When this occurs, the velocity profile within the tube takes on a parabolic shape and maintains that shape for the rest of the tube length. Typically, the region near the entrance while the flow is developing varies from 10 to 60 diameters for very smooth tubes. If the flow is turbulent, the velocity profile is much flatter. Once again, the transition from laminar to turbulent flow occurs at a Reynolds number of approximately 2300. Since turbulent flows generally give the better heat transfer, roughness at the entrance may be included to transition to turbulent flow at a significantly lower Reynolds number.

Heat transfer to or from the fluid flowing in the tube depends on the surface conditions of the tube. Two surface conditions are usually used to characterize the heat transfer to the fluid. The first condition is that of a uniform surface temperature

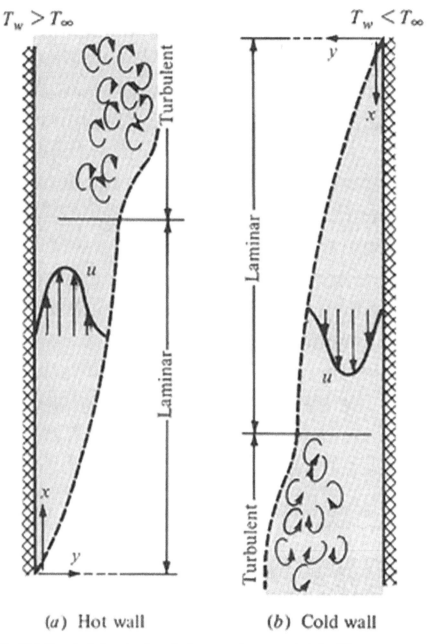

Fig. 12.6 The velocity boundary layer for free convection on a vertical plate [5]

for the tube. In this case, since the tube surface temperature does not change the difference between the fluid mean temperature and the surface temperature decays

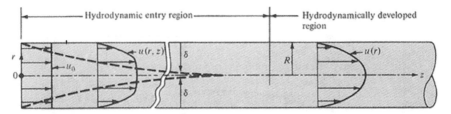

Fig. 12.7 Development of the flow pattern at the entrance of a circular tube [5]

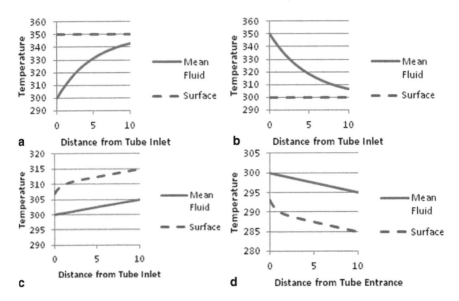

Fig. 12.8 a Mean fluid and surface temperatures—constant surface temperature. **b** Mean fluid and surface temperatures—constant heat flux

exponentially from the difference at the inlet. The second condition is that of a uniform heat flux into or out of the tube along its length. For this case, the mean fluid temperature increases or decreases linearly from the inlet, and the surface temperature of the tube moves to a constant difference from the mean fluid temperature after entrance effects have died out. Typical results for both conditions are shown in Fig. 12.8a and b.

As should be obvious, heat transfer by convection can be very different depending on the geometry of the situation and the flow pattern. Fortunately, a great deal of work has been competed in the past and heat transfer correlations are available for most problems of interest. These correlations or solutions for various flow patterns and conditions are normally expressed in terms of dimensionless groups of parameters relevant to the convection problem of interest.

12.9 Dimensionless Numbers or Groups

Much like the Reynolds number that represents the ratio of inertial forces to viscous forces there are a number of dimensionless groups that allow analysis of many convective heat transfer problems based on general solutions. The second of these dimensionless numbers that is of interest is the Prandtl number. It is the ratio of the kinematic viscosity to the thermal diffusivity or the ratio of the momentum and thermal diffusivities.

$$Pr = \frac{\nu}{\alpha} = \frac{\mu/\rho}{k/\rho C_p} = \frac{\mu C_p}{k} \tag{12.15}$$

For a Prandtl number of 1.0, the thermal boundary layer and velocity boundary layer are the same thickness.

Typical values for some fluids are,

Air	~0.71
Water	~0.8–2.0
Liquid metals	~0.003–0.03
Oils	~100–50,000

Flow patterns and heat transfers are generally so different for the three ranges of Prandtl numbers above that different heat transfer correlations will be required for each of the three ranges. Corrections based on Prandtl numbers will be useful within the three different ranges.

Perhaps the most important dimensionless group for convection heat transfer is the Nusselt number. The Nusselt number represents the ratio of convection to pure conduction heat transfer and is given by,

$$Nu = \frac{hL}{k} \tag{12.16}$$

Most heat transfer correlations would relate the Nusselt number to other dimensionless numbers characterizing the flow pattern of interest. Then the film heat transfer coefficient h, can be calculated knowing the thermal conductivity of the fluid and the appropriate length of interest. Often the length of interest will be the diameter of the tube through which the flow is passing or an equivalent diameter for non-circular flow paths. This introduces a term that allows conversion of non-circular flow paths to equivalent ones called the *hydraulic diameter*. The *hydraulic diameter* is defined as four times the flow area divided by the wetted perimeter of the flow channel.

$$D_h = \frac{4A}{P_{wetted}} \tag{12.17}$$

The Nusselt number is the direct connection to the calculation of the film heat transfer coefficient.

A modified form of the Nusselt number that is often used instead is the Stanton number. The Stanton number is defined as the Nusselt number divided by the Reynolds number and the Prandtl number.

$$St = \frac{Nu}{Re * Pr} \quad Nu = St * Re * Pr \tag{12.18}$$

$$St = \frac{h}{\rho V C_p} \tag{12.19}$$

The Stanton number is particularly useful for fluids that have Prandtl numbers close to 1.0.

An additional number that is often used to correlate heat transfer is the Peclet number. It is simply the product of the Reynolds number and the Prandtl number.

$$Pe = Re * Pr = \frac{\rho V L}{\mu} * \frac{C_p \mu}{k} = \frac{VL}{C_p / \rho k} = \frac{VL}{\alpha} \tag{12.20}$$

Since the pressure drop through a channel is the mechanical energy that must be paid in order to accomplish heat transfer, it is essential to have correlations that predict the pressure drops. This introduces another set of dimensionless numbers. The first of these is the friction factor given by

$$f = \frac{\Delta p}{\left(\frac{L}{D}\right)\left(\frac{\rho V^2}{2}\right)} \tag{12.21}$$

Note that L/D is also a dimensionless number and in using any correlation, it is important to know if this ratio was included when the friction factor correlation was developed. Obviously the pressure drop will be greater the longer the flow channel.

The second parameter of interest for predicting pressure drops is the coefficient of friction. It is defined by

$$C_f = \frac{\tau_s}{\rho V^2 / 2} \tag{12.22}$$

τ_s is the wall shear stress. The coefficient of friction is more fundamental than the friction factor and can be applied to geometries other that internal flow in tubes.

These dimensionless parameters are enough to address most forced convection heat transfer situations. Additional parameters will be required for high speed flows or for free convection. For details on these parameters and how they are used refer to a standard heat transfer text such as Incropera et al. in the References section.

Example 12.6 Estimate the friction pressure drop in a PWR flow channel if the flow rate of the water is 1.25 m/s and the effective temperature is 600 K. The hydraulic diameter for the channel is 1.18 cm and its length is 363 cm. The friction factor is given by

$$f = 0.184 \, Re_D^{-\frac{1}{5}}$$

Solution Calculating the fluid properties

$$\rho = 648.9 \text{ kg} / \text{m}^3$$
$$C_p = 7000.0$$
$$\mu = 8.10E - 5$$
$$k = 0.497$$
$$Re = \frac{648.9 * 1.25 * 0.0118}{8.10E - 5} = 118,164$$

$$f = 0.184 \, Re_D^{-0.2} = \frac{0.184}{118,163^{0.2}} = 0.0178$$

$$\Delta p = \frac{L}{D} \frac{\rho V^2}{2} f = \frac{363}{1.18} \frac{648.9 * 1.25^2}{2} 0.0178 = 5.55 \, kPa$$

12.10 Correlations for Common Geometries

Over a Flat Plate The two correlations of interest for a flat plate will be those that give the average coefficient of friction and the average Nusselt number for a plate of length L [1]. For laminar flow and an isothermal plate they are

$$C_{f,L} = 1.328 \, Re_L^{-\frac{1}{2}} \qquad\qquad (12.23)$$

$$Nu_L = 0.664 \, Re_L^{\frac{1}{2}} Pr^{\frac{1}{3}} \qquad\qquad (12.24)$$

These equations apply to gases and liquids with Prandtl numbers near 1.0. The properties should be evaluated at the average temperature between that of the fluid and that of the plate. For fluids with very small Prandtl numbers, a better Nusselt number correlations is,

$$Nu_L = 1.13 Pe_L^{\frac{1}{3}} \qquad\qquad (12.25)$$

which is good for Prandtl numbers less than 0.05 and Peclet numbers greater than 100.

For turbulent flow over a flat isothermal plate the friction factor is given by,

$$C_{f,L} = 0.1184\,\mathrm{Re}^{-\frac{1}{5}} \tag{12.26}$$

The Nusselt number is given by

$$Nu_L = 0.0592\,\mathrm{Re}_L^{\frac{4}{5}}\,\mathrm{Pr}^{\frac{1}{3}} \tag{12.27}$$

Of course, for smooth flow at the front of the flat plate the first part of the heat transfer will have to be calculated with the laminar equations up until the transition point (nominally Re~2300) and then the turbulent equations used thereafter. If the front edge of the flat plate is rough, enough no laminar regions will exist and the turbulent equations can be used for the whole length of the plate.

For the case of a flat plate with a constant heat flux, the equation for the Nusselt number in laminar flow is,

$$Nu_L = 0.906\,\mathrm{Re}_L^{\frac{1}{2}}\,\mathrm{Pr}^{\frac{1}{3}} \tag{12.28}$$

And the equation for turbulent flow with a constant heat flux is,

$$Nu_L = 0.616\,\mathrm{Re}_L^{\frac{4}{5}}\,\mathrm{Pr}^{\frac{1}{3}} \tag{12.29}$$

Example 12.7 Many early reactors were built with flat plate aluminum fuel elements. Calculate the film heat transfer coefficient and coefficient of friction for flow across a flat plate 0.5 m in length at a water velocity of 1.5 m/s. The bulk water temperature is 350 K and the fuel plate surface temperature is 365 K. Assume a constant heat flux from the plate.

Solution Evaluating the water properties at the mid film temperature 357.5 K

$$\rho = 973.7 - \frac{357.5 - 350}{400 - 350}(973.7 - 937.2) = 973.7 - 0.15*36.5 = 968.2\ \mathrm{kg/m^3}$$

$$C_p = 4195.0 + 0.15*(4256 - 4195) = 4204.2\ \mathrm{J/kg/K}$$

$$\mu = 3.43E - 04 - 0.15*(3.43 - 2.17)E - 04 = 3.24E - 04\ \mathrm{N*s/m^2}$$

$$k = 0.668 + 0.15*(0.688 - 0.668) = 0.672\ \mathrm{W/m/K}$$

$$\mathrm{Re} = \frac{968.2*1.5*0.5}{3.24E - 4} = 2,241,204$$

$$\mathrm{Pr} = \frac{3.24E - 4*4204.2}{0.672} = 2.027$$

Clearly the flow is turbulent and for a constant heat flux,

$$Nu_L = 0.616\,\mathrm{Re}_L^{\frac{4}{5}}\,\mathrm{Pr}^{\frac{1}{3}} = 0.616*2241204^{0.8}2.027^{0.3333} = 93,810.3$$

$$h = \frac{0.672}{0.5}93,810.3 = 126,081\mathrm{W/m^2/K}$$

The coefficient of friction is

$$C_{f,L} = 0.1184 \, \text{Re}^{-\frac{1}{5}} = \frac{0.1184}{2241204^{0.2}} = 0.00636$$

Along Circular Tubes The same two correlations are of interest for flow in and around circular tubes, though the friction factor usually replaces the coefficient of friction. The friction factor for fully developed laminar flow is given by [1],

$$f = \frac{64}{\text{Re}_D} \tag{12.30}$$

For fully developed turbulent flow there are two correlations.

$$f = 0.316 \, \text{Re}_D^{-\frac{1}{4}} \quad below \, \text{Re} = 20,000 \tag{12.31}$$

$$f = 0.184 \, \text{Re}_D^{-\frac{1}{5}} \quad above \, \text{Re} = 20,000$$

For an isothermal boundary condition on the inside of the tube, the laminar flow correlation for fully developed flow is,

$$Nu_D = 3.66 \tag{12.32}$$

For a constant heat flux, the laminar flow correlation is,

$$Nu_D = 4.36 \tag{12.33}$$

For turbulent flow with small to moderate temperature differences between the surface and the fluid mean temperature the *Dittus-Boelter* correlation is,

$$Nu_D = 0.023 \, \text{Re}_D^{\frac{4}{5}} \, \text{Pr}^n \tag{12.34}$$

Where $n = 0.4$ if the fluid is being heated and $n = 0.3$ if the fluid is being cooled. This correlation is satisfactory if $0.6 < Pr < 160$, $Re_D > 10,000$, and $L/D > 10$.

For large temperature differences, a slightly better correlation is given by

$$Nu_D = 0.027 \, \text{Re}_D^{\frac{4}{5}} \, \text{Pr}^{\frac{1}{3}} \left(\frac{\mu}{\mu_s} \right) \tag{12.35}$$

which, is good for $0.7 < Pr < 16,700$, $Re_D > 10,000$, and $L/D > 10$. All properties except μ_s are evaluated at the mean fluid temperature.

In this case these two correlations may be applied to *both a constant surface temperature condition and a constant heat flux condition*.

Fig. 12.9 Flow channel
between four tubes

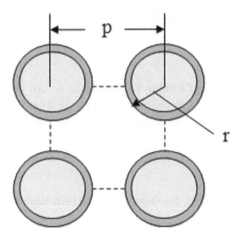

All of the above correlations were developed for flow within tubes. For the flow along the outside of tubes in an array, a simple trick is used to apply these correlations to this case. Consider the array of four tubes in Fig. 12.9 below.

The flow channel between the tubes can be treated as flow within a cylindrical tube if the hydraulic diameter is adjusted properly. The hydraulic diameter is calculated as,

$$D_h = \frac{4 * Area}{Wetted\ Perimeter} = \frac{4 * (p^2 - \pi r^2)}{2\pi r} \qquad (12.36)$$

Then all of the above correlations apply. Note that this model assumes no cross flow, all of the flow is exactly along the tubes.

Example 12.8 At a point in a Boiling Water Reactor (BWR) the water temperature is 270 °C and the clad surface temperature is 300 °C. The diameter of the fuel rods is 1.25 cm and they are spaced on a square matrix with a pitch of 1.62 cm. The water is flowing at the rate of 2 m/s. Calculate the film heat transfer coefficient using the correlation for a large temperature drop across the thermal boundary layer.

Solution For a large temperature drop, the properties are evaluated at the mean fluid temperature, in this case 285 °C.

$$\rho = 831.3 - \frac{558 - 500}{600 - 500}(831.3 - 648.9) = 831.3 - 0.58 * 182.4 = 725.5 \ kg/m^3$$

$$C_p = 4660.0 + 0.58 * (7000 - 4660) = 6017.2 \ J/kg/K$$

$$\mu = 1.18E - 04 - 0.58 * (1.18 - 0.810)E - 4 = 9.65E - 5 \ N * s/m^2$$

$$k = 0.642 - 0.58 * (0.642 - 0.497) = 0.558 \ W/m/K$$

$$Pr = \frac{\mu C_p}{k} = \frac{9.65E - 5 * 6017.2}{0.558} = 1.0406$$

$$\mu_s = 1.18E - 04 - 0.73 * (1.18 - 0.810)E - 4 = 9.099E - 5 \ N * s/m^2$$

The hydraulic diameter is calculated as

$$D_h = \frac{4*(p^2 - \pi r^2)}{2\pi r} = \frac{4\left(1.62^2 - \pi\left(\dfrac{1.25}{2}\right)^2\right)}{\pi 1.25} = 1.423 \text{ cm}$$

$$\text{Re} = \frac{725.5*2*0.01423}{9.65E-5} = 213,966$$

$$Nu_D = 0.027\,\text{Re}_D^{4/5}\,\text{Pr}^{1/3}\left(\frac{\mu}{\mu_s}\right) = 0.027*213,966^{0.8}1.0406^{0.333}\frac{9.65}{9.099} = 533.25$$

$$h = \frac{k}{D_h}\,Nu_D = \frac{0.558}{0.01423}533.25 = 20,910 \text{ W/m}^2/\text{K}$$

Across a Circular Tube For flow across a circular tube the concept of separately describing laminar and turbulent coefficients does not make sense. The flow rapidly transitions and separates at all but the lowest Reynolds numbers. The data for the drag coefficient defined as

$$C_D = \frac{F_D}{A_f \rho \dfrac{V^2}{2}} \quad A_f = tube\ frontal\ area, F_D = drag\ force \qquad (12.37)$$

A plot of the drag coefficient vs, Reynolds number is given in Fig. 12.10.
    An engineering correlation for the heat transfer for a single tube in cross flow is given by,

$$Nu_D = C\,\text{Re}_D^m\,\text{Pr}^{1/3} \qquad (12.38)$$

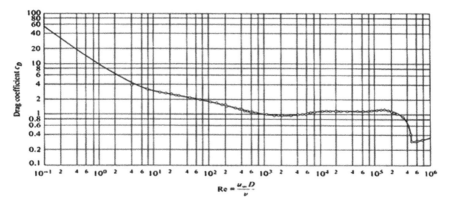

Fig. 12.10 Drag coefficient for flow across a single circular [5]

Table 12.2 Coefficients for
Eq. 12.37

Re_D	C	m
0.4–4	0.989	0.330
4–40	0.911	0.385
40–4000	0.683	0.466
4000–40,000	0.193	0.618
40,000–400,000	0.027	0.805

where C and m are given in the following table for ranges of Reynolds numbers (Table 12.2).

This correlation is good for both a constant surface temperature and a constant heat flux.

Across Tube Bundles In addition to flow through tubes and across one tube, a common geometry is flow across banks of tubes. In this case, there are two common configurations as described in Fig. 12.11. Either the tubes are lined up directly behind each other, or they are staggered. The relevant parameters to describe the geometry are the tube's diameter, D, the spacing between the tubes in a plane perpendicular to the flow, S_T, and the spacing between rows, S_L. The Nusselt number is given by

$$Nu_D = 1.13 C_1 \, Re_{D,max}^m \, Pr^{1/3} \tag{12.39}$$

The Reynolds number is based on the maximum fluid velocity computed from

$$V_{max} = \frac{S_T}{S_T - D} V_\infty \tag{12.40}$$

for the inline configuration and for the staggered configuration if A1 is less than A2. It will occur at A2 if

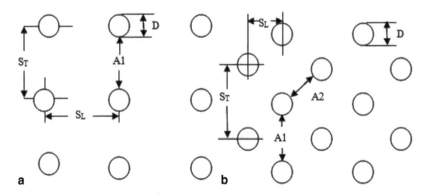

Fig. 12.11 Tube banks. **a** Inline. **b** Staggered

Table 12.3 Constants for Eq. 12.39 [1]

S_L/D	S_T/D							
	1.25		1.5		2.0		3.0	
	C_1	m	C_1	m	C_1	m	C_1	m
Aligned								
1.25	0.348	0.592	0.275	0.608	0.100	0.704	0.0633	0.752
1.50	0.367	0.586	0.250	0.620	0.101	0.702	0.0678	0.744
2.00	0.418	0.570	0.299	0.602	0.229	0.632	0.198	0.648
3.00	0.290	0.601	0.357	0.584	0.374	0.581	0.286	0.608
Staggered								
0.600	–	–	–	–	–	–	0.213	0.636
0.900	–	–	–	–	0.446	0.571	0.401	0.581
1.000	–	–	0.497	0.558	–	–	–	–
1.125	–	–	–	–	0.478	0.565	0.518	0.560
1.250	0.518	0.556	0.505	0.554	0.519	0.556	0.522	0.562
1.500	0.451	0.568	0.460	0.562	0.452	0.568	0.488	0.568
2.000	0.404	0.572	0.416	0.568	0.482	0.556	0.449	0.570
3.000	0.310	0.592	0.356	0.580	0.440	0.562	0.428	0.574

Table 12.4 Correction factors for Eq. 12.42 for less than ten rows [1]

NL	1	2	3	4	5	6	7	8	9
Aligned	0.64	0.80	0.87	0.90	0.92	0.94	0.96	0.98	0.99
Staggered	0.68	0.75	0.83	0.89	0.92	0.95	0.97	0.98	0.99

$$2(S_D - D) < (S_T - D) \quad S_D = \sqrt{S_T^2 + S_L^2}$$

In which case

$$V_{max} = \frac{S_T}{2(S_D - D)} V \tag{12.41}$$

The constants for Eq. 12.39 are given in Table 12.3 above. This equation and table were developed for banks of tubes with more than ten rows. For less than ten rows, the Nusselt number must be corrected according to Eq. 12.42 below and Table 12.4.

$$Nu_{D,<10} = C_2 Nu_{D,\geq10} \tag{12.42}$$

The table of values for C_2 is given below.

The pressure drop across a bank of tubes is given by

$$\Delta p = N_L \chi \left(\frac{\rho V_{max}^2}{2} \right) f \tag{12.43}$$

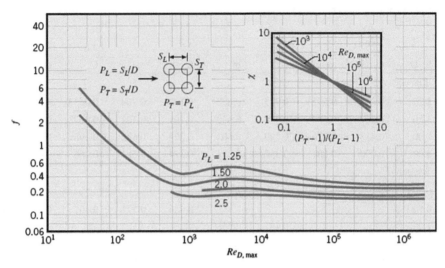

Fig. 12.12 Friction factor and χ for aligned tube banks [1]

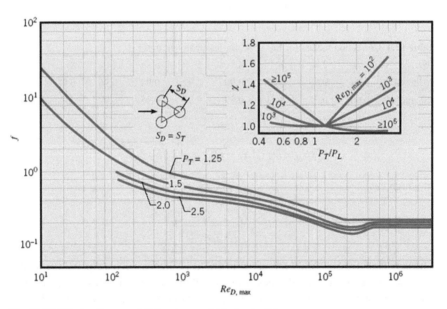

Fig. 12.13 Friction factor and χ for staggered tube banks [1]

The friction factor for aligned tube banks is given in Fig. 12.12 and for staggered tube banks in Fig. 12.13.

The parameter χ is also given in the inserted sub-plots in Figs. 12.12 and 12.13.

Example 12.9 Calculate the effective film heat transfer coefficient and the friction factor for a tube bank of staggered tubes with the following dimensions.

Tube Diameter=3 cm
Spacing Transverse to the Flow=6 cm
Spacing Parallel to the Flow=6 cm
Entering Flow Velocity=50 m/s
Fluid is Water (steam) at 500 K and 15 kPa
Eight rows of tubes

Solution Begin by calculating $V_{max} = 6/(6-3) = 2*V_{enter} = 100$ m/s

Then $Re_{d,max} = \dfrac{100*0.03}{5.75965*1.70E-5} = 30,639$

$S_T/D = 2.0$ $S_L/D = 2.0$ For staggered tubes $C_1 = 0.482$ m$=0.556$

Equation 12.39 gives

$$Nu = 1.13 C_1\, Re^m_{D,max}\, Pr^{\frac{1}{3}} = 1.13*0.482*30,639^{0.556}*1$$

$$Nu = 170$$

$$\text{Correcting for 8 tubes gives}\quad Nu = 0.98*170 = 166.6$$

Then $h = \dfrac{k}{D} Nu = \dfrac{0.0339}{0.03} 166.6 = 188.2 \text{W/m}^2/\text{K}$

The pressure drop is given by Eq. 12.43.

$$\Delta p = N_L \chi \left(\rho V_{max}^2 \Big/ 2 \right) f$$

From Fig. 12.13 at a Re of $30,639\, f = 0.21$ and $\chi = 1.05$

$$\Delta p = 8*1.05*\frac{100^2}{5.75965*2}*0.21 = 1.531 kPa$$

12.11   Enhanced Heat Transfer

Newton's law of cooling is given by

$$Q = Ah(T_f - T_w)$$

When no more improvements in the film heat transfer coefficient, h, can be made, the heat transfer rate can still be increased by increasing the heat transfer area. This is usually accomplished by adding fins to the heat transfer surfaces. This is very common when one fluid is a liquid and the other fluid is a gas. The automobile

Fig. 12.14 Fin on a vertical
surface

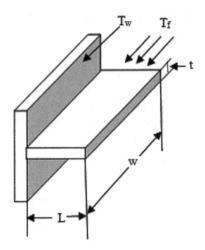

radiator is a common example. A typical geometry for a fin attached to a vertical
surface is shown in Fig. 12.14.

The cross sectional area of the fin is given by $A_c = w*t$, and the perimeter of the
fin is given by $P = 2w + 2t$. Detailed analysis of the heat transfer from the fin to the
fluid is given in Incropera et al. for many different boundary conditions, but a good
approximation for many finned situations is given by the solution,

$$q_f = M \tanh mL_c$$

$$M = \sqrt{hPkA_c}\,(T_f - T_w) \quad m = \sqrt{\frac{hP}{kA_c}} \qquad (12.44)$$

$$L_c = L + \frac{t}{2}$$

It is also a reasonable approximation for fins around a tube that extend out an
amount equal to the radius of the tube.

The utility of adding fins to a surface has to be evaluated in terms their heat
transfer effectiveness. The *fin heat transfer effectiveness* is defined by

$$\varepsilon_f = \frac{Heat\ transferred\ through\ the\ fin}{Heat\ transferred\ through\ the\ base\ area\ without\ the\ fin}$$

$$\varepsilon_f = \frac{q_f}{hA_c(T_f - T_w)} = \frac{\sqrt{hPkA_c}\,(T_f - T_w)}{hA_c(T_f - T_w)} = \sqrt{\frac{Pk}{hA_c}} \qquad (12.45)$$

Another measure of fin performance is given by the *fin efficiency*. In this case the
heat transferred by the fin is compared with the heat that could be transferred if the
entire fin were at the wall temperature.

$$\eta_f = \frac{M \tanh mL}{hPL(T_f - T_w)} = \frac{\sqrt{hPkA_c}\,(T_f - T_w)\tanh mL}{hPL(T_f - T_w)} = \frac{\tanh mL}{mL} \qquad (12.46)$$

In the end, the important parameter of interest in transferring heat is the overall performance of the heat transfer surface with the fins attached. Since adding one fin is not the optimal solution usually, an array of fins will be added with the spacing between the fins identified as S. Then the overall surface efficiency is given by

$$\eta_o = \frac{q_{total}}{q_{max}} = \frac{q_{wall\ surface} + q_{fins}}{h(A_{wall\ surface} + A_{fins})(T_f - T_w)} = \frac{[N\eta_f\, hA_f + Nh(S*w)](T_f - T_w)}{NhA_f + Nh(S*w)(T_f - T_w)}$$

$$\eta_o = \frac{\eta_f\, A_f + S*w}{A_f + S*w} = \frac{N\,\eta_f\, A_f + S*w}{N\ A_f + S*w} = \frac{N\eta_f\, A_f + (A_t - NA_f)}{A_t} = 1 - \frac{NA_f}{A_t}(1 - \eta_f) \qquad (12.47)$$

Note that A_t is the total surface area of the fins and the remaining wall surface that is exposed to the fluid. A_f is simply the surface are of one fin. Then Newton's law of cooling can be written as

$$Q = \eta_o hA_t (T_f - T_w) \qquad (12.48)$$

and the increased heat transfer effectiveness is represented by the increased surface are and the overall surface efficiency.

Example 12.10 Estimate the overall surface effectiveness for heat transfer from a 1 m by 1 m copper radiation that has 3 cm fins extending into the airflow. The 3 cm fins are 1 mm thick and run the length of the 1 m dimension. They are spaced 9 mm apart. The wall temperature is 600 K and the air temperature is 350 K. Assume turbulent flow at a Reynolds number of 5000.

Solution The Nusselt number**Error! Bookmark not defined.** for turbulent airflow over a flat plate is

$$Nu_L = 0.0592\, Re^{4/5}\, Pr^{1/3} = 0.0592 * 5000^{4/5}\, 0.7^{1/3} = 47.85$$

$$h = \frac{k}{L} Nu = \frac{0.03}{1} 47.85 = 1.44 \quad m = \sqrt{\frac{hP}{kA_c}}$$

P=2*1 m=2 m A_c=1 m*0.001 m=0.001 m²   k=379 W/m/K

$$m = \sqrt{\frac{1.44 * 2}{379 * 0.001}} = 2.757 \quad \eta_f = \frac{\tanh mL}{mL} = \frac{\tanh 2.757 * 0.03}{2.757 * 0.03} = 0.9977$$

For a 1 m high plate there will be 100 fins spaced 9 mm apart with a thickness of 1 mm. The total area will be

$A_t = 100*(0.009 + 2*0.03)1.0 = 6.9 \; sqm$
Then the overall effectiveness is given by

$$\eta_o = 1.0 - \frac{100*2*0.03*1}{6.9}(1 - 0.9977) = 99.8$$

12.12   Pool Boiling and Forced Convection Boiling

A good starting point to analyze mechanism of heat transfer in a boiling system is given by the subject of *pool boiling*. The most common situation and the simplest form of boiling is pool boiling in which boiling occurs is where a solid surface in contact with a liquid is brought to a temperature above the saturation temperature of the liquid and immersed below the free surface of the liquid causes boiling. This process is also called *saturated* (or *bulk*) *boiling* because liquid is maintained at saturation temperature. On the other hand, when the main body of the liquid in the immediate vicinity of the heated surface is at, slightly above, or below the saturation temperature the situation is called *sub-cooled* (or *local*) *boiling* because the vapor bubbles that are formed at the hot metal surface either collapse without leaving the surface or collapse immediately upon leaving the surface.

When the liquid is significantly below saturation temperature and the heater temperature is low, that is the only time that *free convection* occurs.

Although the heat transfer mechanism of pool boiling is reasonably understood by engineers and investigators of this field, and their findings were extensively discussed in different literatures, it is still not an easy task to theoretically analyze and predict characteristic of an even simple boiling system. For example, one of the pioneers of this subject Nukiyama experimentally managed to establish the characteristic of pool boiling phenomena. He deduced both the heat flux and the temperature from measurements of current and voltage of a submersed electrical element inside a body of saturated water and initiated boiling on the surface of the wire. Since then other scientists have done further investigation of this phenomenon and their fact findings for pool boiling of water at atmospheric pressure is illustrating in Fig. 12.15. This figure demonstrate the variation of the heat transfer coefficient as a function of the temperature difference between the wall surface T_w and the liquid saturation T_s temperatures in the pool boiling of a liquid at saturation temperature

The Fig. 12.15 shows six different regions where the plot going through each division forms slope which is different from those in other region. Events that taking in each region is described as below;

1. In this region, no vapor bubbles are formed because the energy transfer from the heated surface to the saturated liquid is by free convection from the free surface.
2. In this region, bubbles begin to form at the hot surface of the wire, but as soon as they are separated from the surface, they will dissipate in the liquid.
3. In this region, bubbles are detached from the electrical wire surface and they rise to the surface of the liquid where they will dissipate.

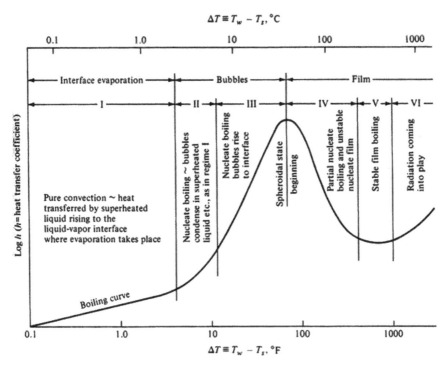

Fig. 12.15 Principal boiling regimes in pool boiling of water at atmospheric pressure and saturation temperature T_s from an electrically heated platinum wire. (From Faber and Scorah [5])

4. In this region, the bubble formation is taking place rapidly, that bubbles begin to coalesce before they detached, consequently a large fraction of the heating surface is blanketed by an unstable film of vapor, which causes an abrupt loss in the heat flux and in the boiling heat transfer coefficient. This is an *unstable region* and it signifies transition from *nucleate boiling* to *film boiling*. (Fig. 12.17).

5. In this region, the heat flux decreases to a minimum level and the wire surface is blanketed with a stable film of vapor and it is called the *stable film boiling* region.

6. In this region, both heat flux and the heat transfer coefficient increases with $T_w - T_s$ because the wire surface temperature in this region is sufficiently high for thermal radiation effects to augment heat transfer through the vapor film. It is significant to mention that in this region, the boiling also takes place as *stable film boiling* but radiation effects are dominant.

If the heat flux q is also plotted as a function of the temperature difference $T_w - T_s$, the general shape of the heat-flux curve (Fig. 12.16) would be similar to that of the heat transfer coefficient curve that is shown in Fig. 12.15.

Free convection is the process that the heat transfer from the heater (i.e. electrical element or wire) surface to the saturated liquid takes place. In this region, the heater surface is only a few degrees above the saturation temperature of the liquid, but the

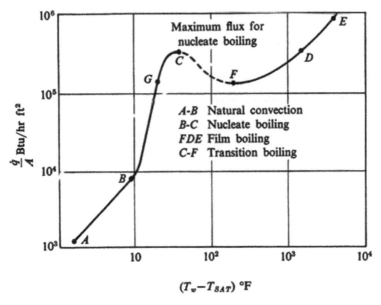

Fig. 12.16 The poll boiling curve [5]

flow produced by free convection in the liquid is sufficient to remove the heat from the surface. The heat transfer correlation, then are in form of

$$Nu = f(Gr, Pr) \qquad (12.49)$$

And as soon as we can determine the heat transfer coefficient h, then the heat flux for the free convection regime can be obtained from the following relationship as;

$$q = h(T_w - T_{sat}) \qquad (12.50)$$

Free convection, transition to boiling, and changes in the boiling mechanism can best be seen in Fig. 12.16.

The figure shows a plot of heat flux versus the temperature difference $(T_w - T_{sat})$ between a horizontal cylindrical heater surface and the saturation temperature of the surrounding liquid. This curve presents the results of common pool boiling experiment of electrically conduction wire, which is submerged in a stationary pool of water at saturation temperature and atmospheric pressure. As we can see in Fig. 12.16 for values of $(T_w - T_{sat}) < 10\,°F$ (point A to B on curve) the heat flux is the value that is predicated for free convection with no phase change. Evaporation takes place at the pool free surface and if we consider the water as liquid then this event occurs at the surface of electric heater less than approximately $10\,°F$ above the saturation temperature of water.

The heater surface in most cases exceeds the saturation temperature of the liquid (i.e. water) by some perceptible amount before noticeable boiling commences. There requires and excessive temperature that is explained by the Eq. 12.51 below, depends on the type of fluid and the pressure.

Fig. 12.17 Nucleate pool boiling from a horizontal cylindrical heater at low heat flux [5]

$$R = \frac{2\sigma T_{sat}}{\rho_v h_{fg}(T_v - T_{sat})} \qquad (12.51)$$

Where

R : Bubble radius
σ : Surface tension
T_{sat} : Saturated temperature
T_v : Vapor temperature
ρ_v : Vapor density
h_{fg} : Specific enthalpy of vaporization

Additionally, the type and condition of filament surface is significant since these factors determine the number and the size of the small vapor and gas bubbles from which bubble growth should begin.

Between point B and C as it can be seen on Fig. 12.16 where the temperature difference $(T_w - T_{sat})$ is approximately equal to $10\,^\circ$F, the curve shows a sharp increase in the slope, which is indication of large increase in the heat transfer coefficient, which is due to the inception of *nucleate boiling* at the heater surface. Figure 12.17 is showing such nucleate boiling regime from a horizontal cylindrical heater where the liquid near the heater has gained sufficient superheat to cause significant numbers of bubble to come to live and grow.

When a liquid changes phase and converts to vapor at the free surface of a liquid, the process is referred to as *evaporation*. However, when the phase changes in a vigorous manner within the bulk of a liquid, the process is called *boiling*.

A boiling process may take place in certain designated purposes such as;

1. The generation of large quantities of vapor, as in a steam power plant or in chemical processing, or
2. The production of large heat fluxes with moderate temperature differences.

From what we know these two objectives may not be related, since heat fluxes are required to provide the latent heat of vaporization designated as h_{fg} and necessary when generating large quantities of vapor within instrument of reasonable size. When the phase change in a liquid is taking place without significant rising of liquid temperature due to energy being put into it, then the large heat fluxes in boiling are possible.

The second general situation under which boiling occurs called *forced convection boiling*. In this case, there is the added complexity of the forced flow of liquid through the tube or past the surface on which boiling occurs.

12.13   Nucleate Boiling Regime

The process of originating a bubble is referred to as *nucleation*. The term nucleate boiling is used since the vapor bubbles are formed at discrete nucleation sites and Fig. 12.17 has picture such process from a horizontal electric heater.

The nucleate boiling regime can be separated into two distinct regions in which bubbles are formed on the surface of the electrical heater. Utilizing Fig. 12.15 where it shows a region designated as II, bubbles start to form at the favored sites on the heater surface, but as soon as the bubbles are detached from the surface, they dissipated in the liquid. In region III, the nucleation sites are numerous and the bubbles are generating at very high rate that causes continuous vapor columns to appear. As result, very high heat fluxes can be obtained in this region. In the nucleate boiling regime, the heat flux increases rapidly with increasing temperature difference $(T_w - T_{sat})$ until the peak heat flux is reached and the burnout point can be defined at that peak, which is departure from nucleate boiling (DNB), or critical heat flux (CHF). As soon as the peak heat flux is exceeded, an extremely large temperature difference is needed to realize the resulting heat flux. Such extreme temperature difference may cause the burning up, or melting down of conducting heating filament. If the number of active nucleation sites increases, the interaction between the bubbles may become important and will affect the heat transfer within the region where this process takes place. Although one should consider other variables such as state of the fluid and properties of filament materials in a heat transfer analysis among other factors in the nucleate boiling regime. One empirical equation that is correlating the heat flux in the entire nucleate boiling regime with various parameters in relation to forced convection presented by Rohsenow as follows;

$$\frac{c_{pl}\,\Delta T}{h_{gf}\,\mathrm{Pr}_l^n} = C_{sf}\left[\frac{q}{\mu_l h_{fg}}\sqrt{\frac{\sigma^*}{g(\rho_l - \rho_v)}}\right]^{0.33} \tag{12.52}$$

Where

c_{pl} h   Specific heat of saturation liquid, J/(kg.°C)

C_{sf} g   Constant, to be determined from experimental data that depends on heating surface-fluid combination.

h_{gf} o   Latent heat of vaporization, J/kg.

Table 12.5 Values of the coefficient C_{sf} of Eq. 12.52 for water surface combination. (Source: Evaluation of constants for the Rohsenow pool-boiling correlation. J Heat Transf 90:239–247)

Liquid-surface combination	C_{sf}
Water copper	0.0130
Water—scored copper	0.0068
Water—emery-polished copper	0.0128
Water—emery-polished, paraffin-treated copper	0.0147
Water—chemically etched stainless steel	0.0133
Water—mechanically polished stainless steel	0.0132
Water—ground and polished stainless steel	0.0080
Water—Teflon pitted stainless steel	0.0058
Water—platinum	0.0130
Water—brass	0.0060

Table 12.6 Values of liquid-vapor surface tension σ^* for water

Saturation temperature °C	Surface tension $\sigma^* \times 10^3$ N/m
0.00	75.6
15.56	73.2
37.78	60.7
93.34	60.1
100.00	58.8
160.00	46.1
226.7	31.9
293.3	16.2
360.00	1.46
374.11	0.00

g a	Gravitational acceleration, m/s^2.
$Pr_l = c_{pl}\mu_l k_l =$	Prandtl number of saturated liquid.
q o	Boiling heat flux, W/m^2.
$\Delta T = T_w - T_{sat}$,	Temperature difference between wall and saturation temperature, °C.
α_q	Viscosity of saturated liquid, kg/(m s).
$\rho_l, \rho_v =$	Density of liquid and saturated vapor, respectively, kg/m^3.
σ^*	Surface tension of liquid-vapor interface, N/m.

In Eq. 12.52 the exponent n and the coefficient C_{sf} are the two provisions for adjusting the correlation for the liquid-surface combination. Table 12.5 lists the experimentally determined values C_{sf} for water boiling on a variety of surfaces. The value of n for water should be taken as 1.

Table 12.6 gives the value of vapor-liquid surface tension for water at different saturation temperatures.

Example 12.11 During the boiling of saturated water at $T_s = 100\degree$ C with an electric heating element, a heat flux of $q = 7 \times 10^5$ W/m^2 is achieved with a temperature difference of $\Delta T = T_w - T_{sat} = 10.4\degree$ C. What is the value of the constant $\mathbf{C_{sf}}$ in Eq. 12.52?

Solution The physical properties of saturated water and vapor are taken as

$$c_{pl} = 4216, J/(kg\,^{\circ}C) \qquad h_{gf} = 2257 \times 10^3\,J/kg.$$

$$\rho_l = 960.6 kg/m^3 \qquad\qquad \rho_v = 0.6 kg/m$$

$$Pr_l = 1.74 \qquad\qquad \mu_l = 0.282 \times 10^{-3}\,kg/(ms)$$

$$\sigma^* = 58.8 \times 10^{-3}\,N/m \qquad g = 9.81 m/s^2$$

These numerical values are substitute into Eq. 12.52 with $n = 1$, we get

$$\frac{c_{pl}\,\Delta T}{h_{gf}\,Pr_l^n} = C_{sf}\left[\frac{q}{\mu_l h_{fg}}\sqrt{\frac{\sigma^*}{g(\rho_l - \rho_v)}}\right]^{0.33}$$

$$\frac{(4216)(10.4)}{(2257 \times 10^3)(1.74)} = C_{sf}\left[\frac{(7 \times 10^5)}{(0.282 \times 10^{-3})(2257 \times 10^3)}\sqrt{\frac{58.8 \times 10^{-3}}{9.81(960.6 - 0.6)}}\right]^{0.33}$$

$$C_{sf} = 0.008$$

Example 12.12 A brass heating element of surface area $A = 0.04$ m², maintained at a uniform temperature $T_w = 112\,^{\circ}C$, is immersed in a saturated water at atmospheric pressure at temperature $T_s = 100\,^{\circ}C$. Calculate the rate of evaporation.

Solution The physical properties of saturated water and vapor at $100\,^{\circ}C$ were given in Example 12.11. Introducing these properties into Eq. 12.52 with $n = 1$ and $\Delta T = T_w - T_{sat} = 112 - 100 = 12\,^{\circ}C$, obtaining the coefficient C_{sf} for water brass from Table 12.5 as $C_{sf} = 0.006$, the heat flux becomes;

$$q = \left(\frac{c_{pl}\,\Delta T}{h_{gf}\,Pr_l^n\,C_{sf}}\right)^3 \mu_l h_{fg}\sqrt{\frac{g(\rho_l - \rho_v)}{\sigma^*}}$$

$$= \left[\frac{(4216)(12)}{(2257 \times 10^3)}\right]^3 (0.282 \times 10^{-3})(2257 \times 10^3)\sqrt{\frac{(9.81)(060.6 - 0.6)}{(58.8 \times 10^{-3})}}$$

$$= 2521.23\,kW/m^2$$

The total rate of heat transfer is

$$Q = area \times q$$
$$= (0.04)(2521.23) = 100.85\,kW$$

The rate of evaporation is

$$\dot{m} = \frac{Q}{h_{fg}}$$

$$= \frac{100.85 \times 10^3}{2257 \times 10^3} = 0.0447\,\text{kg/s} = 160.9\,\text{kg/h}$$

12.14 Peak Heat Flux

The correlation that is found in Eq. 12.52 is enough to provide information about the heat flux in nucleate boiling, but it cannot to predicate the peak heat flux. To be able to have the knowledge of such information in a nucleate boiling regime is of interest because of burnout consideration. As we described before, if the heat flux exceeds the peak heat flux, the burnout situation takes place. This is the circumstances that the transition from the nucleate to stable film boiling regime takes place, in which, depending on the kind of fluid, boiling may occur at temperature difference well above the melting point of the heating surface. The following correlation by expert in the field for determination of the peak flux heat was proposed:

$$q_{max} = F(L') \times 0.131 \rho_v^{1/2} h_{fg} [\sigma^* g (\rho_l - \rho_v)]^{1/4} \tag{12.53}$$

where

$h_{gf} =$ Latent heat of vaporization, J/kg.
$g =$ Gravitational acceleration, m/s².
$q_{max} =$ Peak heat flux, W/m².
$\rho_l, \rho_v =$ density of liquid and saturated vapor, respectively, kg/m³.
$\sigma^* =$ Surface tension of liquid-vapor interface, N/m.

and $F(L')$ is a correction factor that depends on heat geometry and is provided in Table 12.7.

The dimensionless characteristic length of L' of the heater is defined as;

$$L' = L \sqrt{\frac{g(\rho_l - \rho_v)}{\sigma^*}} \tag{12.54}$$

Table 12.7 Correction factor $F(L')$ for use in Eq. 12.53. (Based on Lienhard and coworkers)

Heater geometry	F(L')	Remarks
Infinite flat plate facing up	1.14	$L' \geq 2.7$; L is the heat width or diameter
Horizontal cylinder	$0.89 + 2.27 e^{-3.44\sqrt{L'}}$	$L' \geq 0.15$; L is the cylinder radius
Large sphere	0.84	$L' \geq 4.26$; L is the sphere radius
Small sphere	$1.734/(L')^{1/2}$	$0.15 \leq L' \leq 4.26$: L is the sphere radius
Large finite body	~ 0.90	$L' \geq 4$; $L =$ (volume)/(surface area)

where L is the characteristic dimension of the heater and other quantities are defined as before. In Eq. 12.53 the physical properties of the vapor are evaluated at $T_f = 1/2(T_w + T_{sat})$. The enthalpy of evaporation h_{fg} and the liquid properties must be evaluated at the saturated temperature of the liquid.

Example 12.13 Water at saturation temperature and atmospheric pressure is boiled in the nucleate boiling regime with a large plate heating element facing up. Calculate the peak heat flux.

Solution The physical properties of saturated water and vapor at 100 °C were given in Example 12.11. Introducing these properties into Eq. 12.53, with the correction factor $F(L') = 1.14$ obtained from Table 12.7 for a large plate heating element facing up, the peak heat flux q_{max} is determined as;

$$q_{max} = F(L') \times 0.131 \rho_v^{\frac{1}{2}} h_{fg} [\sigma^* g (\rho_l - \rho_v)]^{\frac{1}{4}}$$

$$= (1.14)(0.131)(0.6)^{\frac{1}{2}} (2257 \times 10^3)$$

$$\times \{(58.8 \times 10^{-3})(9.81)(960.6 - 0.6)\}^{\frac{1}{4}}$$

$$= 1.27 \times 10^6 \ W/m^2 = 1.27 \ MW/m^2$$

Example 12.14 Water at atmospheric pressure and saturation temperature is boiled in a 25-cm diameter, eclectically heated, mechanically polished, stainless-steel pan. The heated surface of the pan is maintained at a uniform temperature $T_w = 116\,°C$.

a. Calculate the surface heat flux.
b. Calculate the rate of evaporation from the pan.
c. Calculate the peak heat flux.

Solution The physical properties of saturated water and vapor for $\Delta T = T_w - T_v = 16°C$ are taken as;

$$c_{pl} = 4216, \ J/(kg\,°C) \qquad h_{gf} = 2257 \times 10^3 \ J/kg.$$

$$\rho_l = 960.6 kg/m^3 \qquad \rho_v = 0.6 kg/m^3$$

$$Pr_l = 1.74 \qquad \mu_l = 0.282 \times 10^{-3} kg/(ms)$$

$$\sigma^* = 58.8 \times 10^{-3} N/m \qquad g = 9.81 m/s^2$$

a. Equating 12.52 is used to compute the surface heat flux for water $n = 1$ and using Table 12.5, we have $C_{sf} = 0.0132$ for water-mechanically polished stainless steel. Introducing all the numerical values into Eq. 12.53, we obtain;

$$\frac{(4216)(16)}{(2257 \times 10^3)(1.74)} = 0.0132 \left[\frac{q}{(0.282 \times 10^{-3})(2257 \times 10^3)} \sqrt{\frac{58.8 \times 10^{-3}}{(9.81)(960.6 - 0.6}} \right]^{0.33}$$

Then the surface heat flux becomes

$$q = 5.61 \times 10^5 \text{ W/m}^2$$

b. The total rate of heat transfer becomes

$$Q = \text{Area} \times q = \left(\frac{\pi}{4} \times (0.25)^2 \right)(5.61 \times 10^5)$$

$$= 0.275 \times 10^5 \text{ W or J/s}$$

The rate of evaporation becomes

$$\dot{m} = \frac{Q}{h_{fg}}$$

$$= \frac{0.275 \times 10^5}{22.57 \times 10^3} \times 3600 = 43.9 \text{kg/h}$$

c. To calculate the peak heat flux, we use Eq. 12.53 with the factor $F(L')$ taken from Table 12.7 for infinite flat plate facing up being equal to 1.14, which is valid for

$$L' = L \sqrt{\frac{g(\rho_l - \rho_v)}{\sigma^*}} \geq 2.7$$

For $L = 0.25$ m and other quantities as given above, we have $L' = 100$, which is larger than the specified lower bound 2.7.

Hence for q_{max} we have;

$$q_{max} = F(L') \times 0.131 \rho_v^{1/2} h_{fg} [\sigma^* g(\rho_l - \rho_v)]^{1/4}$$

$$= (1.14)(0.13)(0.6)^{1/2} (2257 \times 10^3)$$

$$\times \{ (58.8 \times 10^{-3})(9.81)(960.6 - 0.6) \}^{1/4}$$

$$= 1.27 \times 10^6 \text{ W/m}^2 = 1.27 \text{ MW/m}^2$$

Note that $q = 5.61 \times 10^5 \text{ W/m}^2$ is well below the peak heat flux $1.27 \times 10^6 \text{ W/m}^2$ or J/(m².s).

12.15   Film Boiling Regime

As soon as the nucleate boiling region ends and the unstable film boiling region begins after the peak heat flux is reached. Using Fig. 12.15, knowing that there are no correlations are available for the prediction of heat flux in this unstable region until the minimum point in the boiling curve is reached and the stable film boiling region starts. In the region V and VI where we are within the stable film-boiling region, the heating surface is separated from the liquid by a vapor layer across which heat must be transferred. Due to low thermal conductivity of vapor in this region, large temperature differences are needed in order to have proper heat transfer, therefore, heat transfer in this region is generally avoided when high temperatures are involved. However, stable film boiling has numerous applications in the boiling of cryogenic fluids.

Bromley developed a theory to predict the heat transfer coefficient for stable film boiling on the outside of a horizontal cylinder and the basic analysis is very similar to Nusselt's theory for film-wise consideration on a horizontal tube. The result of his study for the average heat transfer coefficient h_0 for stable film boiling, in absence of radiation, is given by;

$$h_0 = 0.62 \left[\frac{k_v^3(\rho_l - \rho_v)gh_{fg}}{\mu_v D_0 \Delta T} \left(1 + \frac{0.4c_{pv}\Delta T}{h_{fg}} \right) \right]^{1/4} \tag{12.55}$$

where

$h_0 =$ Average boiling heat transfer coefficient in absence of radiation, W/(m².°C)

$h_{gf} =$ Latent heat of vaporization, J/kg.

$g =$ Gravitational acceleration, m/s².

$k_v =$ Thermal conductivity of saturated vapor, W/(m.°C).

$\rho_l, \rho_v =$ density of liquid and saturated vapor, respectively, kg/m³.

$D_0 =$ Outside diameter of tube, m.

$c_{pv} =$ Specific heat of saturated vapor, J/(kg.°C).

$\mu_v =$ Viscosity of saturated vapor, kg/(m s).

$\Delta T = T_w - T_{sat},$ temperature difference between wall and saturation temperatures, °C.

and the physical properties of vapor must be evaluated at $T_f = 1/2(T_w + T_{sat})$. In addition, the enthalpy of evaporation h_{fg} and the liquid density ρ_l should be evaluated at the saturation temperature T_{sat} of the liquid.

Example 12.15 Water at saturation temperature $T_s = 100\,°C$ and atmospheric pressure is boiled with an electrically heated horizontal platinum wire of diameter $D = 0.2$ cm. Boiling takes place with a temperature difference of $T_w - T_s = 454\,°C$ in the stable film boiling range. Calculate the film boiling heat transfer coefficient and the heat flux, in the absence of radiation.

Solution The physical properties of vapor are evaluated at $T_f = 1/2(T_w + T_{sat}) = (554 + 100)/2 = 327\,°C = 600\,°K$

$$c_{pv} = 20262026 \text{ j/kg}\,°C \qquad k_v = 0.0422 \text{W/(m}°C)$$

$$\mu_w = 2067 \times 10^{-5} \text{kg/(ms)} \qquad \rho_v = 0.365 \text{ kg/m}^3$$

and the liquid density and h_{gf} are evaluated at the saturation temperature $T_s = 100\,^\circ\text{C}$ as;

$$\rho_l = 960.6\,\text{kg/m}^3 \qquad h_{gf} = 2257 \times 10^3\,\text{J/kg}$$

The heat transfer coefficient h_0 for stable film boiling without the radiation effects is computed from Eq. 12.55 as:

$$h_0 = 0.62 \left[\frac{k_v^3 (\rho_l - \rho_v) g h_{fg}}{\mu_v D_0 \Delta T} \left(1 + \frac{0.4 c_{pv} \Delta T}{h_{fg}} \right) \right]^{1/4}$$

$$= 0.62 \left[\frac{(0.0422)^3 (0.365)(960.6 - 0.365)(98.1)(2257 \times 10^3)}{(2.067 \times 10^{-5})(0.002)(454)} \right.$$

$$\left. \times \left(1 + \frac{(0.4)(2026)(454)}{2257 \times 10^3} \right) \right]^{1/4}$$

$$= 270.3\ \text{W/(m}^2.\,^\circ\text{C)}$$

Problems

Problem 12.1: A brick wall 20 cm thick with thermal conductivity 1.4 W/(m. °C) is maintained at 40 °C at one face 240 °C at the other face. Calculate the heat transfer rate across 6-m² surface area of the wall.

Problem 12.2: The heat flow rate across an insulating material of thickness 4 cm with thermal conductivity 1.0 W/(m. °C) is.250 W/m². If the hot surface temperature is 180 °C, what is the temperature of the cold surface?

Problem 12.3: The heat flow rate through a 4-cm thick wood board for a temperature difference of 30 °C between the inner and outer surface is 80 W/m². What is the thermal conductivity of the wood?

Problem 12.4: A temperature of 550 °C is applied across a block of 10 cm thick with thermal conductivity 1.0 W/(m. °C). Calculate the heat transfer rate per square meter area.

Problem 12.5: By conduction 1000 W is transferred through a 0.5-m² section of a 5-cm thick insulating material. Determine the temperature difference across the insulating layer if the thermal conductivity is 0.1 W/(m. °C).

Problem 12.6: Using Fig. 12.18, consider an insulated rod of length L and constant cross section area A in which a steady state has been reached between two fixed ends of temperature T_1 and T_2 that are separated by this rod. Compute the rate of heat transfer through the rod.

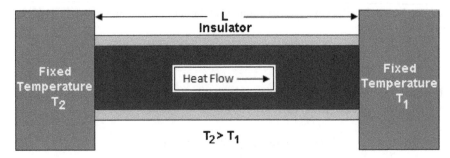

Fig. 12.18 Conduction of heat through an insulated conducting bar

Problem 12.7: A compound slab is shown in figure below which is consisting of two different materials having two separate thicknesses as L_1 and L_2 respectively with different conductivity k_1 and k_2 accordingly. If the temperatures of the outer surfaces are kept at T_2 and T_1, calculate the heat transfer through the compound slab in a steady state situation (Fig. 12.19).

Problem 12.8: A thin metal plate 0.2 by 0.2 m is placed in an evacuated container whose walls are kept at 400 °K. The bottom surface of the plate is insulated, and the top surface is maintained at 600 °K because of electric heating. If the emissivity of the surface of the plate is $\varepsilon = 0.1$, what is the rate of heat exchange between the plate and the walls of the container? Take Boltzmann constant to be $\sigma = 5.67 \times 10^{-8}\,\mathrm{W}/(\mathrm{m}^2 \cdot \mathrm{K}^4)$.

Problem 12.9: Two large parallel plates, one at a uniform temperature 600 °K and the other at 1000 °K, are separated by a nonparticipating gas. Assume that the surfaces of the plates are perfect emitters and that the convection is negligible;

Figure 12.19 Conduction of heat through two layers of matter with different thermal conductivities

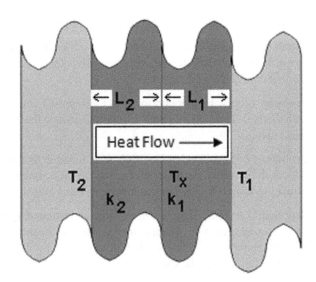

determine the rate of heat exchange between the surfaces per square meter. Take Boltzmann constant to be $\sigma = 5.67 \times 10^{-8}\, W / (m^2 \cdot K^4)$

Problem 12.10: A thin metal sheet separates two large parallel plates, one at a uniform temperature of $1200\,°K$ and the other $600\,°K$. Blackbody conditions can be assumed for all surfaces, and heat transfer can be assumed to be by radiation only. Calculate the temperature of the separating sheet.

Problem 12.11: Two very large, perfectly black parallel plates, one maintained at temperature $1000\,°K$ and the other at $500\,°K$, exchanging heat by radiation (i.e. convection is negligible). Determine the heat transfer rate per 1-m^2 surface. Take Boltzmann constant to be $\sigma = 5.67 \times 10^{-8}\, W / (m^2 \cdot K^4)$.

Problem 12.12: One surface of a thin plate is exposed to a uniform heat flux of $500\ W/m^2$, and the other side dissipates heat by radiation to an environment at $T_\infty = -10\,°C$. Determine the temperature of the plate T_p. Assume a blackbody conditions for radiation and take Boltzmann constant to be $\sigma = 5.67 \times 10^{-8}\, W / (m^2 \cdot K^4)$ and that the convection is negligible.

Problem 12.13: In many practical situations, a surface loses or receives heat by convection and radiation simultaneously where the two methods act in parallel to determine the total heat transfer. Assume a horizontal steel pipe having an outer diameter of 80 mm is maintained at a temperature of $60\,°C$ in a large room where the air and wall temperature are at $20\,°C$. The average free convection heat transfer coefficient between the outer surface of the pipe and the surrounding air is 6.5 W/ $(m^2\ K)$, and the surface emissivity of steel is 0.8. Calculate the total heat loss by the pipe per length. Use the following figure to have some concept of the idea of combined convection and radiation heat transfer (Fig. 12.20).

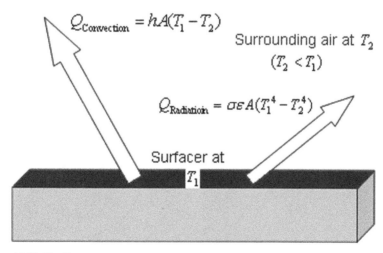

Figure 12.20 The System

Figure 12.21 One-Dimensional Layout in Cartesian

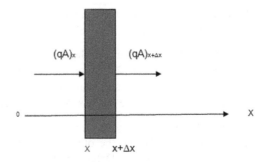

$(qA)_x$ $(qA)_{x+\Delta x}$

0 ──────   ──────────→ X

X   x+Δx

Problem 12.14: Using the figure below and writing an energy balance for a differential volume element, derive the one-dimensional time-dependent heat conduction equation with internal energy generation g and variable thermal conductivity in the rectangular coordinate system for the x variable (Fig. 12.21).

Problem 12.15: Using the figure below and writing an energy balance for a differential cylindrical volume element r variable, derive the one-dimensional time-dependent heat conduction equation with internal heat generation g and variable thermal conductivity in the cylindrical coordinate system for the r variable (Fig. 12.22).

Problem 12.16: The local drag coefficient c_x can be determined by the following relationship;

$$c_x = \frac{2v}{u_\infty^2} \frac{\partial u(x,y)}{\partial y}\bigg|_{y=0} \tag{a}$$

If the velocity profile $u(x,y)$ for boundary layer flow over a flat plate is given by;

Figure 12.22 One-Dimensional Layout in Cylindrical Coordinate

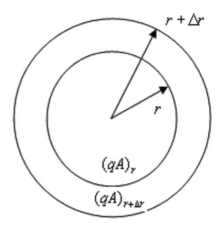

$r + \Delta r$

r

$(qA)_r$

$(qA)_{r+\Delta r}$

$$\frac{u(x,y)}{u_\infty} = \frac{3}{2}\left[\frac{y}{\delta(x)}\right] - \frac{1}{2}\left[\frac{y}{\delta(x)}\right]^3 \qquad (b)$$

Where the boundary-layer thickness $\delta(x)$ is

$$\delta(x) = \sqrt{\frac{280}{13}\frac{vx}{u_\infty}} \qquad (c)$$

And assume that average drag coefficient c_m over a distance $0 \le x \le L$ is also given by the following equation;

$$c_m = \frac{1}{L}\int_{x=0}^{L} c_x dx \qquad (d)$$

Develop an expression for the local drag coefficient c_x.

Develop an expression for the average drag coefficient c_m over a distance $x = L$ from the leading edge of the plate.

Problem 12.17: The exact expression for the local drag coefficient c_x for laminar flow over a flat plate is given by the following relation and $Re_x^{1/2}$ is square root of Reynolds number

$$c_x = \frac{0.664}{Re_x^{1/2}}$$

knowing that the mean value of the drag coefficient c_m over $x = 0$ to $x = L$ is defined as

$$c_m = \frac{1}{L}\int_{x=0}^{L} c_x dx$$

And drag force acting on the same plate from $x = 0$ to $x = L$ for width can be found by

$$F = wLc_m \frac{\rho u_\infty^2}{2}$$

Air at atmospheric pressure and $T_\infty = 300\,\text{K}$ flows with a velocity of $u_\infty = 1.5\,\text{m/s}$ along the plate. Determine the distance from the leading edge of the plate where transition begins from laminar to turbulent flow. Calculate the drag force F acting per 1-m width of the plate over the distance from $x = 0$ to where the transition starts.

Problem 12.18: Air at atmospheric pressure and 100 °F (37.8 °C) temperature flows with a velocity of $u_\infty=3$ ft/s (0.915 m/s) along a flat plate. Determine the boundary-layer thickness $\delta(x)$ and the local-drag coefficient c_x at a distance $x = 2$ ft (0.61 m) from the leading edge of the plate. What is the mean drag coefficient over the length $x = 0$ to 2 ft, and the drag force acting on the plate over the length $x = 0$ to 2 ft per foot width of the plate? Use the exact solution for boundary layer Thickness and the local drag coefficient for laminar flow along a flat plate as $\delta(x) = (4.96x)/\sqrt{Re_x}$ and $c_x = (0.664)/\sqrt{Re_x}$ respectively. Assume the mean the mean value of the drag coefficient $c_{m,L} = 2c_x$ in this case and drag force F acting on the plate over given length is $x = 0$ to $x = L$ and width w described as;

$$F = wLc_{m,L}\frac{\rho u_\infty^2}{2g_c}lb_f \text{ or } (N)$$

Problem 12.19: Air at atmospheric pressure and at a temperature 150 °F (65.6 °C) flows with a velocity of $u_\infty=3$ ft/s (0.915 m/s) along a flat plate which is kept at a uniform temperature 250 °F (121.1 °C). Determine the local heat transfer coefficient $h(x)$ at a distance $x = 2$ ft (0.61 m) from the leading edge of the plate and the average heat transfer coefficient h_m over the length $x = 0$ to 2 ft (0.61 m). Calculate the total heat transfer rate from the plate to the air over the region $x = 0$ to 2 ft per foot width of the plate. Use solution that is also provided by Pohlhausen as

$$Nu_x = \frac{h(x)x}{k} = 0.332\,Pr^{1/3}\,Re_x^{1/2}.$$

Problem 12.20: Air at atmospheric pressure and at a temperature 24.6 °C flows with a velocity of $u_\infty=10$ m/s along a flat plate $L=4$ m which is kept at a uniform temperature 130 °C. Assume $Re_c = 2.0\times10^5$. Using Figure below and show that the flow is Turbulent and use experimental correlation for turbulent boundary layer along a flat plate as (Fig. 12.23)

$$Nu_x = \frac{h(x)x}{k} = 0.029\,Pr^{0.43}\,Re_x^{0.8}$$

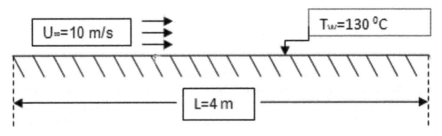

Figure 12.23 Flow over a Flat Plate

a. Calculate the local heat transfer coefficient at $x = 2, 3$, and 4 m from the leading edge of the plate. Assume Reynolds number is $\mathrm{Re}_x = \dfrac{u_\infty L}{\nu}$

b. Find the average heat transfer coefficient over $L = 4$ m. Assume that,

$$\frac{h_m L}{k} = \mathrm{Nu}_m = 0.036\,\mathrm{Pr}^{0.43}\,(\mathrm{Re}_L^{0.8} - 9200)\left(\frac{\mu_\infty}{\mu_W}\right)^{0.25}$$ and neglect the viscosity correction and it is equal to unity.

c. Determine the heat transfer rate from the plate to the air per meter width of the plate.

Problem 12.21: Helium at 1 atm, $u_\infty = 30$ m/s, and $300\,^\circ$K flows over a flat plate $L = 5$ m long and $W = 1$ m wide which is maintained at a uniform temperature of $600\,^\circ$K. Calculate the average heat transfer coefficient and the total heat rate. Use the following figure and assume $\mathrm{Re}_c = 2 \times 10^5$ further assume that,

$$\frac{h_m L}{k} = \mathrm{Nu}_m = 0.036\,\mathrm{Pr}^{0.43}\,(\mathrm{Re}_L^{0.8} - 9200)\left(\frac{\mu_\infty}{\mu_W}\right)^{0.25}$$ and neglect the viscosity correction and it is equal to unity (Fig. 12.24).

Problem 12.22: A fluid at $27\,^\circ$C flows with a velocity of 10 m/s across a 5-cm OD tube whose surface is kept a uniform temperature of $120\,^\circ$C. Determine the average heat transfer coefficients and the heat transfer rates per meter length of the tube for;

a. Air at atmospheric pressure. Use given correlation for part (c) and ignore viscosity correction part.

b. Water. Use general correlation for the average heat transfer coefficient h_m for flow across a single cylinder as;

$$\mathrm{Nu}_m = 0.3 + \frac{0.62\,\mathrm{Re}^{1/2}\,\mathrm{Pr}^{1/3}}{[1 + (0.4/\mathrm{Pr})^{2/3}]^{1/4}}\left[1 + \left(\frac{\mathrm{Re}}{282000}\right)^{5/8}\right]^{4/5}$$

c. Ethylene glycol. Use general correlation for the average heat transfer coefficient h_m for flow across a single cylinder as below with viscosity correction from Appendix Table for given conditions (Fig. 12.25);

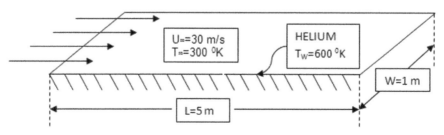

Figure 12.24 Flow over a Flat Plate

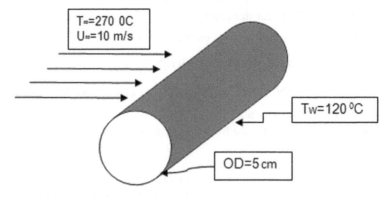

Figure 12.25 Flow across a Single Cylinder

$$\mathrm{Nu}_m = \frac{h_m D}{k} = (0.4\,\mathrm{Re}^{0.5} + 0.06\,\mathrm{Re}^{2/3})\,\mathrm{Pr}^{0.4}\left(\frac{\mu_\infty}{\mu_W}\right)^{9.25}$$

Problem 12.23: A very long, 10 mm diameter copper rod (k=370 W/(m. °K) is exposed to an environment at 20 °C. The base temperature of the rod is maintained at 120 °C. The heat transfer coefficient between the rod and the surrounding air is 10 W/(m2. °K).

a. Determine the heat loss at the end, and us the following relationship for the rate of loss from the fin as

$$Q = \sqrt{hPkA}\,\theta_b\,\frac{\sinh(mL) - (h/mk)\cosh(mL)}{\cosh(mL) + (h/mk)\sinh(mL)}$$

b. Compare the results with that of an infinitely long fin whose tip temperature equals the environment temperature of 20 °C. For an inifite long rod use the following heat transfer equation

$$Q = -kA\left(\frac{dT}{dx}\right)_{x=0}$$
$$= kAm(T_b - T_\infty)$$
$$= kAm\theta_b$$
$$= \sqrt{hPkA}(T_b - T_\infty)$$

where $\theta_b = T_b - T_\infty$

Problem 12.24: In a specific application, a stack (see figure below) that is 300 mm wide and 200 mm deep contains 60 fins each of length $L=12$ mm. The entire stack is made of aluminum which is everywhere 1.0 mm thick. The temperature limitations associated with electrical components joined to opposite plates dictate the maximum allowable plate temperature of $T_b=400\,°$K and $T_L=350\,°$K. Determine the rate of heat loss from the plate at $400\,°$K, give $h=150$ W/(m². °K) and $T_\infty=300\,°$K. Take $k_{Aluminum}=230$ W/(m². °K). Use the rate of heat loss from the fin can be determined by making use of the following equation (Fig. 12.26).

$$Q = \sqrt{hPkA}\theta_b \, \frac{\cosh(mL)-(\theta_L/\theta_b)}{\sinh(mL)}$$

Problem 12.25: Saturated water at $T_{sat}=100\,°$C is boiled inside a copper pan having a heating surface $A=5\times10^{-2}$ m² which is maintained at a uniform $T_w=100\,°$C. Calculate

a. The surface heat flux and
b. The rate of evaporation,

Problem 12.26: Repeat Problem 12.25 assuming for a pan made of brass

Problem 12.27: Water at atmospheric pressure and saturation temperature is boiled by using an electrically heated, circular disk of diameter $D=20$ cm with the heated

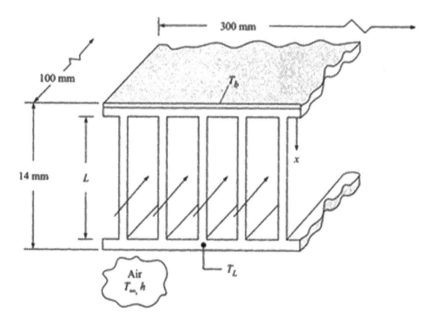

Figure 12.26 A Stack Containing Fins as Explained in the Problem

surface facing up. The surface of the element is maintained at a uniform temperature $T_w = 110 \,°C$. Calculate

a. The surface heat flux
b. The rate of evaporation, and
c. The peak heat flux

Problem 12.27: Saturated water at $T_v = 100 \,°C$ is boiled with a copper heating element having a heating surface $A = 4 \times 10^{-2}$ m^2 which is maintained at a uniform temperature $T_w = 115 \,°C$. Calculate the surface heat flux and the rate of evaporation.

Problem 12.28: In problem 12.25, if the heating element were made of brass instead of copper, what would be the heat flux at the surface of the heater?

Solution 12.28: The problem is exactly the same as that in Problem 12.25, except that $C_{sf} = 0.013$ should be replaced by $C_{sf} = 0.006$ according to Table 12.5. Then from Eq. 12.51 we write

$$\frac{q_{water - brass}}{q_{water - copper}} = \left(\frac{C_{sf, water - brass}}{C_{sf, water - copper}} \right)^3$$

Substituting the numerical values and taking result of Problem 12.25 for q as 4.84×10^5 W/m^2, we have;

$$q_{water - brass} = (4.84 \times 10^5)(\text{W/m}^2) \left(\frac{0.013}{0.006} \right)^3 = 4.93 \times 10^5 \text{ W/m}^2$$

Problem 12.29: Water at saturation temperature and atmospheric pressure is boiled with an electrically heated, horizontal platinum wire of $D_0 = 0.127$ cm diameter. Compute the boiling heat transfer coefficient h_m and the heat flux for a temperature difference $T_w - T_{sat} = 650°\text{C}$. Assume no radiation and then consider radiation effects using the following formulas

$$h_r = \frac{1}{1/\varepsilon + 1/\alpha - 1} \frac{\sigma(T_w^4 - T_{sat}^4)}{T_w - T_{sat}}$$

You may also assume that h_m is given by empirical equation as follow;

$$h_m = h_0 + \frac{3}{4} h_r$$

Hint: You may also assume that $\varepsilon = \alpha \approx 1$.

Problem 12.30: Water at saturation temperature and atmospheric pressure is boiled with an electrically heated, horizontal platinum wire of $D_0 = 0.2$ cm diameter, $\varepsilon = 1$ in stable film boiling regime with a temperature difference $T_w - T_{sat} = 654°C$.

Compute the film boiling heat transfer coefficient h_m and the heat flux. Assume no radiation and then consider radiation effects using the following formulas

$$h_r = \frac{1}{1/\varepsilon + 1/\alpha - 1} \frac{\sigma(T_w^4 - T_{sat}^4)}{T_w - T_{sat}}$$

You may also assume that h_m is given by empirical equation as follow;

$$h_m = h_0 + \frac{3}{4} h_r$$

Hint: You may also assume that $\varepsilon = \alpha \approx 1$.

References

1. Incropera FP, Dewitt DP, Bergman TL, Lavine AS (2007) Introduction to heat transfer. Wiley, Danvers
2. Necati Ozisik M (1977) Basic heat transfer. McGraw-Hill Book Company, New York

Chapter 13
Heat Exchangers

A heat exchanger is a heat transfer device that exchanges heat between two or more process fluids. Heat exchangers have widespread industrial and domestic applications. Many types of heat exchangers have been developed for use in steam power plants, chemical processing plants, building heat and air conditioning systems, transportation power systems and refrigeration units.

The actual design of heat exchangers is a complicated problem. It involves more than heat-transfer analysis alone. Cost of fabrication and installation, weight, and size play important roles in the selection of the final design from a total cost of ownership point of view. In many cases, although cost is an important consideration, size and footprint often tend to be the dominant factors in choosing a design.

13.1 Heat Exchangers Types

Most heat exchangers may be classified as one of several basic types. The four most common types, based on flow path configuration, are illustrated in Fig. 13.1 below [1].

1. In *concurrent,* or *parallel-flow*, units the two fluid streams enter together at one end, flow through in the same direction, and leave together at the other end;
2. In *countercurrent*, or *counter-flow,* units the two streams move in opposite directions.
3. In *single-pass crossflow* units one fluid moves through the heat transfer matrix at right angles to the flow path of other fluid.
4. In *multipass crossflow* units one fluid stream shuttles back and forth across the flow path of the other fluid stream, usually giving a crossflow approximation to counterflow.

© Springer International Publishing Switzerland 2015
B. Zohuri, P. McDaniel, *Thermodynamics In Nuclear Power Plant Systems,*
DOI 10.1007/978-3-319-13419-2_13

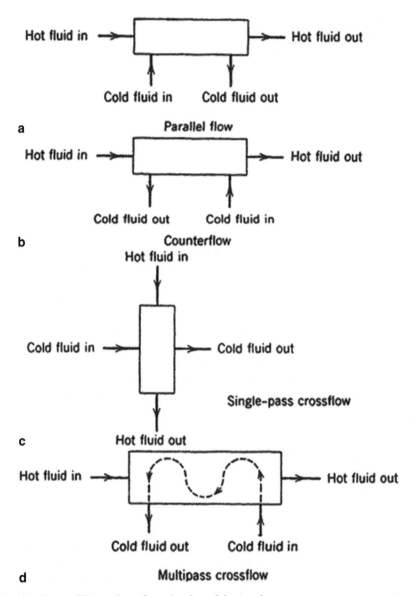

Fig. 13.1 Types of flow path configuration through heat exchanger

The most important difference between these four basic types lies in the relative amounts of heat transfer surface area required to transfer the desired amount of heat between the two fluids.

Figure 13.2 below shows the relative area required for each type as a function of the change in temperature of the fluid with the largest temperature change requirement for a typical set of conditions. In the region in which the fluid temperature

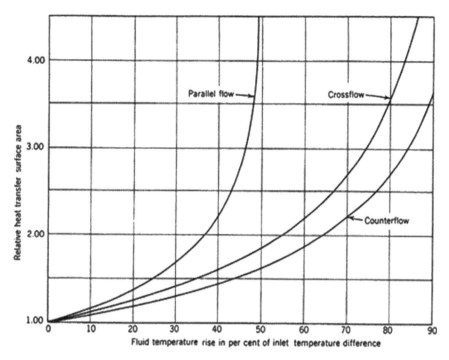

Fig. 13.2 The required relative heat transfer surface area as a function of the ratio of the temperature rise (or drop) in the fluid stream having the greater change in temperature to the difference in temperature between the inlet streams

change across the heat exchanger is a small percentage of the difference in temperature between the two entering fluid streams, all the units require roughly the same area. The parallel-flow heat exchanger is of interest primarily for applications in this region. Cross-flow units have a somewhat broader range of application, and are peculiarly suited to some types of heat exchanger construction that have special advantages. The counter-flow heat exchanger requires the least area. Furthermore, it is the only type that can be employed in the region in which the temperature change in one or both of the fluid streams closely approaches the temperature difference between the entering fluids streams.

In addition, heat exchangers may be classified as direct contact or indirect contact. In the direct-contact type, heat transfer takes place between two immiscible fluids, such as a gas and a liquid, coming into direct contact. For example, cooling towers, jet condensers for water vapor, and other vapors utilizing water spray are typical examples of direct-contact exchangers.

An *Immiscible Fluids* are incapable of is being mixed or blended together. Immiscible liquids that are shaken together eventually separate into layers. Oil and Water are typical immiscible fluids.

In the indirect-contact type of heat exchangers, such as automobile radiators, the hot and cold fluids are separated by an impervious surface, and they are referred to as *surface heat exchangers*. There is no mixing of the two fluids.

13.2   Classification of Heat Exchanger by Construction Type

Heat exchangers also can be classified according to their construction features. For example, there are tubular, plate, plate-fin, tube-fin, and regenerative exchangers. An important performance factor for all heat exchangers is the amount of heat transfer surface area within the volume of the heat exchanger. This is called its *compactness factor* and is measured in square meters per cubic meter.

13.2.1   Tubular Heat Exchangers

Tubular exchangers are widely used, and they are manufactured in many sizes, flow arrangements, and types. They can accommodate a wide range of operating pressures and temperatures. The ease of manufacturing and their relatively low cost have been the principal reason for their widespread use in engineering applications. A commonly used design, called the *shell-and-tube* exchanger, consists of round tubes mounted on a cylindrical shell with their axes parallel to that of the shell.

Figure 13.3 illustrates the main features of a shell-and tube exchanger having one fluid flowing inside the tubes and the other flowing outside the tubes. The principle components of this type of heat exchanger are the tube bundle, shell, front and rear end headers, and baffles. The baffles are used to support the tubes, to direct the fluid flow approximately normal to the tubes, and to increase the turbulence of the shell fluid. There are various types of baffles, and the choice of baffle type, spacing, and geometry depends on the flow rate allowable shell-side pressure drop, tube support requirement, and the flow-induced vibrations. Many variations of shell-and-tube exchanger are available; the differences lie in the arrangement of flow configurations and in the details of construction.

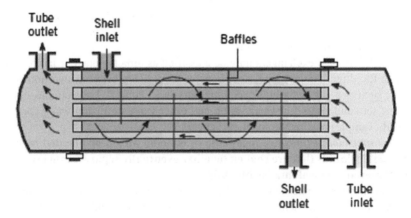

Fig. 13.3 A shell-and-tube heat exchanger; one shell pass and one tube pass [2]

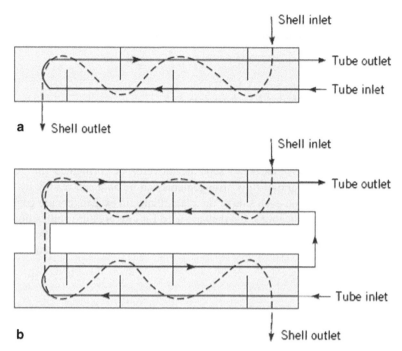

Fig. 13.4 Shell-and-tube heat exchangers. **a** One shell pass and two tube passes. **b** Two shell passes and four tube passes [3]

Baffled heat exchangers with one shell pass and two tubes passes and with two shell passes and four tube passes are shown in Figs. 13.4a [3] and b [3], respectively [3].

The character of the fluids may be *liquid-to-liquid*, *liquid-to-gas*, or *gas-to-gas*. Liquid-to-liquid exchangers have the most common applications. Both fluids are pumped through the exchangers; hence, the heat transfer on both the tube side and the shell side is by forced convection. Since the heat transfer coefficient is high with the liquid flow, generally there is no need to use fins [2].

The liquid-to-gas arrangement is also commonly used; in such cases, the fins usually are added on the gas side of the tubes, where the heat transfer coefficient is low.

Gas-to-gas exchangers are used in the exhaust-gas and air preheating recuperators for gas gas-turbine systems, cryogenic gas-liquefaction systems, and steel furnaces. Internal and external fins generally are used in the tubes to enhance heat transfer.

13.2.2   Plate Heat Exchangers

As the name implies, plate heat exchangers usually are constructed of thin plates. The plates may be smooth or may have some form of corrugation. Since the plate geometry cannot accommodate as high pressure and/or temperature differentials as a circular tube, it is generally designated for moderate temperature and/or pressure differentials. The compactness factor for plate exchangers ranges from about 120 to 230 m^2/m^3.

13.2.3   Plate Fin Heat Exchangers

The compactness factor can be significantly improved (i.e., up to about 6000 m^2/m^3) by using the plate-fin type of heat exchanger. Figure 13.5 illustrates typical plate-fin configurations. Flat plates separate louvered or corrugated fins. Cross-flow, counterflow, or parallel-flow arrangements can be obtained readily by properly arranging the fins on each side of the plate. Plate-fin exchangers are generally used for gas-to-gas applications, but they are used for low-pressure applications not exceeding about 10 atm (that is, 1000 kPa). The maximum operating temperatures are limited to about 800 °C. Plate-fin heat exchangers have also been used for cryogenic applications.

13.2.4   Tube Fin Heat Exchangers

When a high operating pressure or an extended surface is needed on one side, tube-fin exchangers are used. Figure 13.6 illustrates two typical configurations, one with round tubes and the other with flat tubes. Tube-fin exchangers can be used for a wide range of tube fluid operating pressures not exceeding about 30 atm and operating temperatures from low cryogenic applications to about 870 °C. The maximum compactness ratio is somewhat less than that obtainable with plate-fin exchangers.

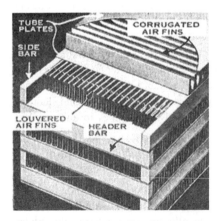 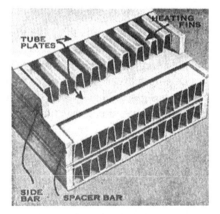

Fig. 13.5 Plate-fin heat exchangers. (Courtesy of Harrison Radiator Division of General Motors Corporation)

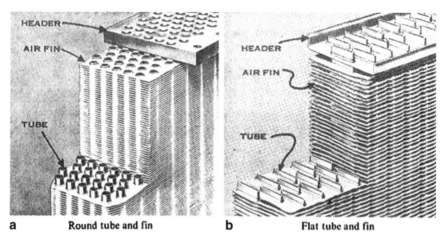

a   Round tube and fin   **b**   Flat tube and fin

Fig. 13.6 Tube fin heat exchangers. (Courtesy of Harrison Radiator Division of General Motors Corporation)

The tube-fin heat exchangers are used in gas turbine, nuclear, fuel cell, automobile, airplane, heat pump, refrigeration, electronics, cryogenics, air conditioning, and many other applications.

13.2.5 Regenerative Heat Exchangers

Regenerative heat exchangers use a heat transfer matrix that is heated by one fluid and then cooled by the second fluid. The flow over the matrix is switched as a function of time with both fluids flowing over the same surfaces of the matrix. They have seen little use in fixed station power plants and will not be emphasized here.

13.3 Condensers

Condensers are used for such varied applications as steam power plants, chemical processing plants, and nuclear electric plants for space vehicles. The major types include the *surface condensers, jet condensers,* and *evaporative condensers.* The most common type is the surface condenser, which has the feed-water system [4]. Figure 13.7 shows a section through a typical two-pass surface condenser for a large steam turbine in a power plant. Since the steam pressure at the turbine exit is only 1.0–2.0 in Hg absolute, the steam density is very low and the volume rate of flow is extremely large. To minimize the pressure loss in transferring steam from the turbine to the condenser, the condenser is normally mounted beneath and attached to the turbine. Cooling water flows horizontally inside the tubes, while the steam

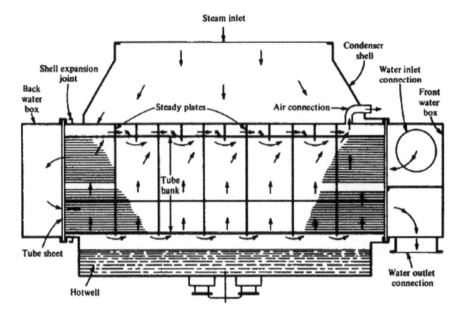

Fig. 13.7 Section through a typical two-pass surface condenser for a large plant. (Courtesy of Allis-Chalmers Manufacturing Company)

flows vertically downward from the large opening at the top and passes transversely over the tubes. Note that provision is made to aspirate cool air from the regions just above the center of the hot well. This is important because the presence of noncondensable gas in the steam reduces the heat transfer coefficient for condensation.

13.4   Boilers

Steam boilers are one of the earliest applications of heat exchangers. The term *steam generator* is often applied to boiler in which the heat source is a hot fluid stream rather than the products of combustion.

An enormous variety of boiler types exist, ranging from small units for house heating applications to huge, complex, expensive units for modern power stations.

13.5   Classification According to Compactness

The ratio of the heat transfer surface area on one side of the heat exchanger to the volume can be used as a measure of the compactness of heat exchangers. A heat exchanger having a surface area density on any one side greater than about $700 \text{ m}^2/\text{m}^3$ quite arbitrarily is referred to as a compact heat exchanger regardless of its structural design. For example, automobile radiators having an area density approximately $1100\ \mathrm{m^2/m^3}$ and the glass ceramic heat exchangers for some vehicular gas-turbine engines having an area density approximately $6600\ \mathrm{m^2/m^3}$ are compact heat exchangers. The human lungs, with an area density of about $20,000\ \mathrm{m^2/m^3}$, are the most compact heat-and-mass exchanger. The very fine matrix regenerator for the Stirling engine has an area density approaching that of the human lung.

On the other hand extreme of the compactness scale, plane tubular and shell-and-tube type exchangers, having an area density in the range of $70–500\ \mathrm{m^2/m^3}$, are not considered compact [2].

The incentive for using compact heat exchangers lies in the fact that a high value of compactness reduces the volume for a specified heat exchanger performance.

When heat exchangers are to be employed for automobiles, marine uses, aircraft, aerospace vehicles, cryogenic systems, and refrigeration and air conditioning, the weight and size—hence the compactness—become important. To increase the effectiveness or the compactness of heat exchangers, fins are used. In a gas-to-liquid heat exchanger, for example, the heat transfer coefficient on the gas side is an order of magnitude lower than for the liquid side. Therefore, fins are used on the gas side to obtain a balanced design; the heat transfer surface on the gas side becomes much more compact.

13.6   Types of Applications

Heat exchangers are often classified based on the application for which they are intended, and special terms are employed for major types. These terms include *boiler, steam generator, condenser, radiator, evaporator, cooling tower, regenerator, recuperator, heater, and cooler.* The specialized requirements of the various applications have led to the development of many types of construction, some of which are unique to particular applications [5].

13.7   Cooling Towers

In locations where the supply of water is limited, heat may be rejected to the atmosphere very effectively by means of cooling towers such as that Fig. 13.8 here. A fraction of the water sprayed into these towers evaporates, thus cooling the balance. Because of the high heat of vaporization of water, the water consumption is only about 1 % as much as would be the case if water were taken from a lake or a stream and heated 10 or 20 °F.

Cooling towers may be designed so that the air moves through them by thermal convection, or fans may be employed to provide forced air circulation. To avoid contamination of the process water, shell-and-tube heat exchangers are sometimes employed to transmit heat from the process water to the water recirculated through the cooling tower.

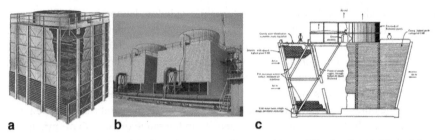

a   **b**   **c**

Fig. 13.8 a Vertical induced draft-cooling tower. (Courtesy Foster Wheeler Corp.) [1]. **b** Schematic of cooling tower. **c** Forced convection cooling tower with draft induced by a fan [1]

13.8   Regenerators and Recuperators

The thermal efficiency of both gas-turbine power plants can be greatly increased if heat can be extracted from the hot gases that are leaving the gas turbine and added to the air being supplied to the combustion chamber. For a major gain in thermal efficiency, it is necessary to employ a very large amount of heat transfer surface area. This is particularly noticeable in gas-turbine plants, where even with counterflow the size of the heat exchanger required for good performance is inclined to be large compared to the size of the turbine and compressor. This characteristic can be observed even in the small, portable gas turbine (about 3 ft in diameter) shown in Fig. 13.9. Note that in this device the hot combination gases leave the radial in-flow turbine wheel at the right end of the shaft and enter a set of heat exchanger cores arranged in parallel around the central axis.

Figure 13.10 shows a close-up view of one of these cores. In each core, the hot gases from the turbine flow roughly radially outward through one set of gas passages. Air from the centrifugal compressor wheel at the center of the shaft flows to the right through the space just inside of the outer casing and axially into the other set of gas passages through the core. The air being heated makes two passes, flowing first to the right in the outer portion of the core and then back to the left through the inner portion, thus giving a two-pass crossflow approximation to counterflow (The flow passages through the combustion chamber are not shown in this view).

As can be seen in Fig. 13.10, the heat exchanger core is constructed of alternate layers of flat and corrugated sheets. The flat sheets separate the hot and cold fluid streams, while the corrugated sheets act as fins that roughly triple the heat transfer

Fig. 13.9 A small gas-turbine power plant fitted with a recuperator to improve the fuel economy. (Courtesy of AiResearch Manufacturing Co.) [5]

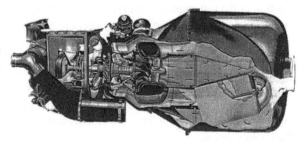

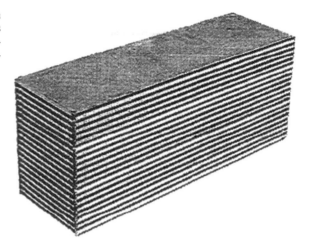

Fig. 13.10 A brazed plate-fin recuperator core for the gas turbine of Fig. 13.9. (Courtesy of AiResearch Manufacturing) [5]

surface area per unit of volume. Note also that the axis of the corrugations is at right angles in alternate layers to provide a crossflow pattern for the two fluid streams.

One of several recuperator units to be mounted in parallel in a much larger gas turbine plants is shown in Fig. 13.11. The hot exhaust gas from the turbine enters vertically at the bottom, flows upward through the heat transfer matrix, and discharges vertically from the top. The air from the compressor enters a large circular port at the top at the right end, flows vertically downward in pure counterflow, and leaves a second circular port at the bottom to flow to the combustion chamber. The hot exhaust gas passages are formed by corrugated sheets sandwiched between flat plates that extend all the way from the bottom to the top of the unit. The air to be

Fig. 13.11 A welded steel recuperator for a large gas-turbine power plant. (Courtesy Harrison Radiator Division, General Motors Corp) [5]

heated flows horizontally from the long plenum at the top into the spaces between the walls of the exhaust gas passages. Curved space strips guide the air through a 90° bend and then downward between the heated walls. A similar header arrangement is used at the bottom. Note that both the flow passage area and the heat transfer surface area for the hot exhaust gas are about three times as great as the corresponding values for the air being heated. This comes about because the two fluid streams differ in density by a factor of about four.

The air pre-heaters in steam power plants are usually quite different from the units just described for gas turbines. Rotary regenerators are often used. These consist of a cylindrical drum filled with a heat transfer matrix made of alternately flat and corrugated sheets. The drum is mounted so that the hot gas heats a portion of the matrix as it passes from the furnace to the stack. The balance of the matrix gives up its stored heat to the fresh air in route from the forced draft fans to the furnace. The ducts are arranged so that the two gas streams move through the drum in counterflow fashion while it is rotated, so that the temperature of any given element of the metal matrix fluctuates relatively little as it is cycled from the hot to the cold gas streams.

In the steam- and gas-turbine power plant fields a distinction is sometimes made between air pre-heaters that involve a conventional heat transfer matrix with continuous flow on both sides of a stationary heat transfer surface and those through which the fluids flow periodically, the hot fluid heating one section of the matrix while the cold fluid is removing heat from another section. Where this distinction is made, the term regenerator is applied to the periodic-flow type of heat exchanger, since this term has long been applied to units of this type employed for blast furnaces and steel furnaces, whereas the term recuperator is applied to units through which the flow is continuous.

Recuperators are used for gas turbine, but the gas turbines installed until the mid-1970's suffered from low efficiency and poor reliability. In the past, large coal and nuclear power plants dominated the base-load electric power generation (Point 1 in Fig. 13.12). Base load units are on line at full capacity or near full capacity almost all of the time. They are not easily nor quickly adjusted for varying large amounts of load because of their characteristics of operation [6]. However, there has been a historic shift toward natural gas-fired turbines because of their higher efficiencies, lower capital costs, shorter installation times, better emission characteristics, the abundance of natural gas supplies, and shorter start up times (Point 1 in Fig. 13.12). Now electric utilities are using gas turbines for base-load power production as well as for peaking, making capacity at maximum load periods and in emergencies, situations because they are easily brought on line or off line (Point 2 in Fig. 13.12). The construction costs for gas-turbine power plants are roughly half that of comparable conventional fossil fuel steam power plants, which were the primary base-load power plants until the early 1980's, but peaking units are much higher in energy output costs. A recent gas turbine manufactured by General Electric uses a turbine inlet temperature of 1425 °C (2600 °F) and produces up to 282 MW while achieving a thermal efficiency of 39.5 % in the simple-cycle mode. Over half of all power plants to be installed in the near future are forecast to be gas turbine or combined gas-steam turbine types (Fig. 13.12).

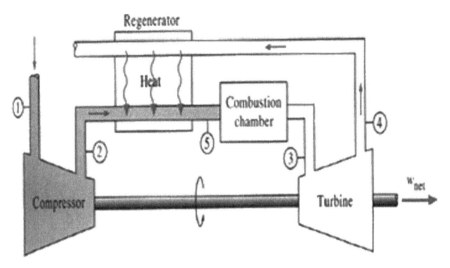

Fig. 13.12 A gas-turbine engine with recuperator

In gas turbine engines with the Brayton Cycle that includes recuperator, the temperature of the exhaust gas leaving the turbine is often considerably higher than the temperature of the air leaving the compressor. Therefore, the high-pressure air leaving the compressor can be heated by transferring heat to it from the hot exhaust gases in a counter-flow heat exchanger, which is also known as a regenerator or recuperator [See Fig. 13.13 on point-1]. Gas turbine regenerators are usually con-

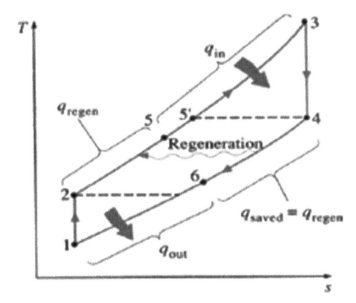

Fig. 13.13 T-s diagram of a Brayton cycle with regeneration

structed as shell-and-tube type heat exchangers using very small diameter tubes, with the high-pressure air inside the tubes and low-pressure exhaust gas in multiple passes outside the tubes [7]. The thermal efficiency of the Brayton cycle increases as a result of regeneration since the portion of energy of the exhaust gases that is normally rejected to the surroundings is now used to preheat the air entering the combustion chamber. This, in turn, decreases the heat input (thus fuel) requirements for the same net work output. Note, however, that the use of a regenerator is recommended only when the turbine exhaust temperature is higher than the compressor exit temperature. Otherwise, heat will flow in the reverse direction (to the exhaust gases), decreasing the efficiency. This situation is encountered in gas turbines operating at very high-pressure ratios [Point 1 on Fig. 5].

The highest temperature occurring within the regenerator is the temperature of the exhaust gases leaving the turbine and entering the regenerator (Point 1 on Fig. 13.13). The gas turbine recuperator receives air from the turbine compressor at pressures ranging from 73.5 to 117 psia and temperatures from 350 to 450 °F (Point 3 on Fig. 13.13). Under no conditions can the air be preheated in the regenerator to a temperature above this value. In the limiting (ideal) case, the air will exit the regenerator at the inlet temperature of the exhaust gases. Air normally leaves the regenerator at a lower temperature (Point 1 on Fig. 13.13). Gas turbine exhaust gas passes over the other side of the recuperator at exhaust temperatures ranging from 750 to 1000 °F. Compressor air temperatures are now raised to higher temperatures up to about 750–900 °F as it enters the combustor. Turbine exhaust gases are then reduced to between 500 and 650 °F from the original 750 to 1000 °F. This heat recovery contributes appreciably to the turbine fuel rate reduction and increase in efficiency (Point 3 on Fig. 13.13). The regenerator is well insulated and any changes in kinetic and potential energies are neglected.

A regenerator with a higher effectiveness will save a greater amount of fuel since it will preheat the air to a higher temperature prior to combustion [Point 1 on Fig. 4]. However, achieving a higher effectiveness requires the use of a larger regenerator, which carries a higher price tag and causes a larger pressure drop because shaft horsepower is reduced. Pressure drop through the regenerator or recuperator is important and should be kept as low as practical on both sides. Generally, the air pressure drop on the high-pressure side should be held below 2 % of the compressor total discharge pressure. The gas pressure drop on the exhaust side (hot side) should be held below 4 in. of water. Therefore, the use of a regenerator with a very high effectiveness cannot be justified economically unless the savings from the fuel costs exceed the additional expenses involved. The effectiveness of most regenerators used in practice is below 0.85. The thermal efficiency of an ideal Brayton cycle with regeneration depends on the ratio of the minimum to maximum temperatures as well as the pressure ratio. Regeneration is most effective at lower pressure ratios and low minimum-to-maximum temperature ratios.

Gas-to-air recuperators (or regenerators) are also used on marine type industrial, and utility open-cycle gas turbine applications. In this application, recuperator receives air from the turbine compressor at pressure and temperature ranging as above, where gas turbine exhaust gas passes over the other side of the recuperator at exhaust temperature, depending on the turbine. The air side (high pressure side) of the recuperator is in the system between the compressor and the combustor and compressor air is raised to a higher temperature up what is mentioned in above as it enters the combustor. Obviously, pressure drop through the regenerator or recuperator is important and should be kept as low as practical on both sides.

13.9   Heat Exchanger Analysis: Use of the LMTD

Utilizing the Log Mean Temperature Difference (LMTD) method is one way to design or to predict the performance of a heat exchanger, it is essential to relate the total heat transfer rate to measurable quantities such as the inlet and outlet fluid temperatures, the overall heat transfer coefficient, and the total surface area for heat transfer. Two such relations may readily be obtained by applying overall energy balances to the hot and cold fluids, as shown in Fig. 13.14. In particular, if q is the total rate of heat transfer between the hot and cold fluids and there is negligible heat transfer between the exchanger and its surroundings, as well as negligible potential and kinetic energy changes, application of the steady flow energy equation, gives;

$$q_{total} = \dot{m}_h (h_{h,i} - h_{h,o}) \qquad (13.1a)$$

and

$$q_{total} = \dot{m}_c (h_{c,o} - h_{c,i}) \qquad (13.1b)$$

Where h is the fluid enthalpy. The subscripts h and c refer to the hot and cold fluids, whereas the subscripts i and o designate the fluid inlet and outlet conditions. If the fluids are not undergoing a phase change and constant specific heats are assumed, these expressions reduce to

$$q_{total} = \dot{m}_h c_{p,h} (T_{h,i} - T_{h,o}) \qquad (13.2a)$$

and

$$q_{total} = \dot{m}_c c_{p,c} (T_{c,o} - T_{c,i}) \qquad (13.2b)$$

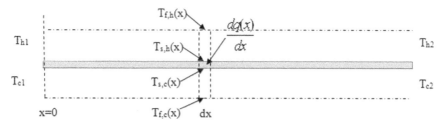

Fig. 13.14 Heat transfer between two moving fluids separated by a solid boundary

where the temperatures appearing in the expressions refer to the *mean* fluid temperatures at the designated locations. Note that Eqs. 13.1 and 13.2 are independent of the flow arrangement and heat exchanger type.

Now consider the heat transfer at a particular point, x, on the heat transfer surface. At x there will be a bulk hot fluid temperature given by $T_{f,h}(x)$, a wall surface temperature on the hot fluid side given by $T_{s,h}(x)$, a wall surface temperature on the cold fluid side given by $T_{s,c}(x)$, and a cold fluid bulk temperature given by $T_{f,c}(x)$. The total temperature drop from the hot fluid at x to the cold fluid at x is given by

$$\Delta T = T_{f,h}(x) - T_{f,c}(x) = T_{f,h}(x) - T_{s,h}(x) + T_{s,h}(x) - T_{s,c}(x) + T_{s,c}(x) - T_{f,c}(x)$$
$$= \Delta T_{f,h} + \Delta T_s + \Delta T_{f,c}$$

Then the heat flux leaving the hot fluid is given by

$$\frac{dq(x)}{dx} = h_{f,h} A_{f,h}(T_{f,h}(x) - T_{s,h}(x)) = h_{f,h} dA_{f,h} \Delta T_{f,h}(x) \quad \Delta T_{f,h}(x) = \frac{\frac{dq(x)}{dx}}{h_{f,h} dA_{f,h}}$$

The heat flux crossing the wall between the two fluids is given by,

$$\frac{dq(x)}{dx} = \frac{k_s}{\delta_s} dA_s (T_{s,h}(x) - T_{s,c}(x)) = \frac{k_s}{\delta_s} dA_s \Delta T_s(x) \quad \Delta T_s(x) = \frac{\frac{dq(x)}{dx}}{\frac{k_s}{\delta_s} dA_s}$$

And the heat flux into the cold fluid is given by,

$$\frac{dq(x)}{dx} = h_{f,c} dA_{f,c}(T_{s,c}(x) - T_{f,c}(x)) = h_{f,c} dA_{f,c} \Delta T_{f,c}(x) \quad \Delta T_{f,c}(x) = \frac{\frac{dq(x)}{dx}}{h_{f,c} dA_{f,c}}$$

Then the difference in the bulk temperatures of the two fluids can be written as

$$T_{f,h}(x) - T_{f,c}(x) = \frac{\frac{dq(x)}{dx}}{h_{f,h} dA_{f,h}} + \frac{\frac{dq(x)}{dx}}{\frac{k_s}{\delta_s} dA_s} + \frac{\frac{dq(x)}{dx}}{h_{f,c} dA_{f,c}}$$

$$= \frac{dq(x)}{dx} \left[\frac{1}{h_{f,h} dA_{f,h}} + \frac{\delta_s}{k_s dA_s} + \frac{1}{h_{f,c} dA_{f,c}} \right]$$

(13.3)

Note that δ_s will depend on the geometry. For slab or plate geometry

$$\delta_s = \Delta t \quad \text{the wall thickness}$$

For cylindrical geometry typical of tubes,

$$\delta_s = r_{in} \ln\left(\frac{r_{out}}{r_{in}}\right) \quad r_{out} - r_{in} = \text{the tube wall thickness}$$

Also note that the differential areas do not all have to be equal. There will be a slight difference if the bounding surface is a tube, but the addition of fins to either the hot or the cold side could change the effective area significantly and that is the area that must be used in Eq. 13.3. Also, note that the areas are areas per unit length. That is why they have been written as dA.

$$T_{f,h}(x) - T_{f,c}(x) = \frac{dq(x)}{dx}\left(\frac{1}{U\frac{dA}{dx}}\right) \quad \frac{1}{U\frac{dA}{dx}} = \frac{1}{h_{f,h}dA_{f,h}} + \frac{\delta_s}{k_s dA_s} + \frac{1}{h_{f,c}dA_{f,c}}$$

$$\frac{dq(x)}{dx} = U\frac{dA}{dx}(T_{f,h} - T_{f,c})$$

Then the heat lost by the hot fluid is given by

$$\frac{dq(x)}{dx} = -\dot{m}_{f,h}C_{p,h}\frac{dT_{f,h}(x)}{dx} \tag{13.4}$$

And the heat gained by the cold fluid is given by

$$\frac{dq(x)}{dx} = \dot{m}_{f,c}C_{p,c}\frac{dT_{f,c}(x)}{dx} \tag{13.5}$$

Combining these two equations gives

$$\frac{dT_{f,h}(x)}{dx} - \frac{dT_{f,c}(x)}{dx} = -\frac{dq(x)}{dx}\left(\frac{1}{\dot{m}_h C_{p,h}} + \frac{1}{\dot{m}_c C_{p,c}}\right) = -U\frac{dA}{dx}(T_{f,h} - T_{f,c})$$

$$\Delta T(x) = T_{f,h}(x) - T_{f,c}(x)$$

$$\frac{d\Delta T(x)}{dx} = -U\frac{dA}{dx}\Delta T\left(\frac{1}{\dot{m}_h C_{p,h}} + \frac{1}{\dot{m}_c C_{p,c}}\right)$$

$$\frac{d\Delta T(x)}{\Delta T(x)} = -U\left(\frac{1}{\dot{m}_h C_{p,h}} + \frac{1}{\dot{m}_c C_{p,c}}\right)dA \quad \frac{dA}{dx}dx = dA$$

Integrating gives

$$\ln\left(\frac{\Delta T_2}{\Delta T_1}\right) = -UA\left(\frac{1}{\dot{m}_h C_{p,h}} + \frac{1}{\dot{m}_c C_{p,c}}\right) \tag{13.6}$$

Now for the hot fluid flowing from left to right Eq. 13.2a becomes

$$q_{total} = \dot{m}_h C_{p,h}(T_{f,h,1} - T_{f,h,2}) \quad \frac{1}{\dot{m}_h C_{p,h}} = \frac{(T_{f,h,1} - T_{f,h,2})}{q_{total}}$$

In addition, for the cold fluid also flowing from left to right (parallel flow) Eq. 13.2b becomes

$$q_{total} = \dot{m}_c C_{p,c}(T_{f,c,2} - T_{f,c,1}) \quad \frac{1}{\dot{m}_c C_{p,c}} = \frac{(T_{f,c,2} - T_{f,c,1})}{q_{total}}$$

Plugging these into Eq. 13.3 gives

$$\ln\left(\frac{\Delta T_2}{\Delta T_1}\right) = -UA\left(\frac{T_{f,h,1} - T_{f,h,2}}{q_{total}} + \frac{T_{f,c,2} - T_{f,c,1}}{q_{total}}\right)$$

$$= \frac{UA}{q_{total}}(T_{f,h,2} - T_{f,c,2} - T_{f,h,1} + T_{f,c,1})$$

$$\ln\left(\frac{\Delta T_2}{\Delta T_1}\right) = \frac{UA}{q_{total}}(\Delta T_2 - \Delta T_1) \quad q_{total} = UA\frac{(\Delta T_2 - \Delta T_1)}{\ln\left(\frac{\Delta T_2}{\Delta T_1}\right)}$$

$$q_{total} = UA\Delta T_{lm} \quad \Delta T_{lm} = \frac{(\Delta T_2 - \Delta T_1)}{\ln\left(\frac{\Delta T_2}{\Delta T_1}\right)} \tag{13.7}$$

This looks a lot like Newton's law of cooling with ΔT_{lm} playing the role of the standard ΔT. ΔT_{lm} is called the log-mean temperature difference.

Now consider the counter flow arrangement. In this case Eq. 13.5 becomes,

$$\frac{dq(x)}{dx} = -\dot{m}_{f,c} C_{p,c} \frac{dT_{f,c}(x)}{dx}$$

Moreover, Eq. 13.6 becomes,

$$\ln\left(\frac{\Delta T_2}{\Delta T_1}\right) = -UA\left(\frac{1}{\dot{m}_h C_{p,h}} - \frac{1}{\dot{m}_c C_{p,c}}\right)$$

Then Eq. 13.2b becomes

$$q_{total} = \dot{m}_c c_{p,c} (T_{f,c,1} - T_{f,c,2})$$

This gives

$$\ln\left(\frac{\Delta T_2}{\Delta T_1}\right) = -UA\left(\frac{T_{f,h,1} - T_{f,h,2}}{q_{total}} - \frac{T_{f,c,1} - T_{f,c,2}}{q_{total}}\right)$$

$$= \frac{UA}{q_{total}}(T_{f,h,2} - T_{f,c,2} - T_{f,h,1} + T_{f,c,1})$$

$$\ln\left(\frac{\Delta T_2}{\Delta T_1}\right) = \frac{UA}{q_{total}}(\Delta T_2 - \Delta T_1) \qquad q_{total} = UA\frac{(\Delta T_2 - \Delta T_1)}{\ln\left(\frac{\Delta T_2}{\Delta T_1}\right)}$$

which is the identical equation for the parallel flow heat exchanger. It is important to remember how the ΔT's are defined.

$$\text{Parallel Flow } \Delta T_1 = T_{f,h.in} - T_{f,c,in} \quad \Delta T_2 = T_{f,h,out} - T_{f,c,out}$$

$$\text{Counter Flow } \Delta T_1 = T_{f,h.in} - T_{f,c,out} \quad \Delta T_2 = T_{f,h,out} - T_{f,c,in}$$

Example 13.1 A counterflow, concentric tube heat exchanger is used to cool the lubricating oil for a large industrial gas turbine engine. The flow rate of cooling water through the inner tube ($D_i = 25$ mm) is 0.2 kg/s, while the flow rate of oil through the outer annulus ($D_0 = 45$ mm) is 0.1 kg/s. The oil and water enter at temperatures of 100 and 30 °C, respectively. How long must the tube be made if the outlet temperature of the oil is to be 60 °C? (The steel tube that separates the two flows is so thin that the temperature drop across it may be neglected.). See Fig. 13.15.

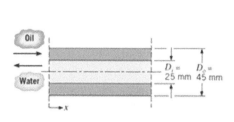
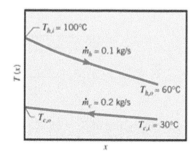

Fig. 13.15 Schematic of Example 13.1 [5]

Solution

Known

Fluid flow rates and inlet temperature for a counter-flow, concentric tube heat exchanger of prescribed inner and outer diameter

Find

Tube length to achieve a desired hot fluid outlet temperature.

Assumptions

1.0	Negligible heat loss to the surroundings
2.0	Negligible kinetic and potential energy changes
3.0	Constant properties
4.0	Negligible tube wall thermal resistance and fouling factors
5.0	Fully developed conditions for the water and oil (U independent of x)

Properties Table A-5, unused engine oil ($\overline{T}_h = 80\ ^{0}C = 353K$):

$$c_p = 2131\ j/kg \cdot K,\ \mu = 3.25 \times 10^{-2}\ N \cdot s/m^2,\ k = 0.138\ W/m \cdot K.$$

Table A-6, water ($\overline{T}_c \approx 35\ ^{0}C = 308K$):

$$c_p = 4178\ j/kg \cdot K,\ \mu = 725 \times 10^{-6}\ N \cdot s/m^2,\ k = 0.625\ W/m \cdot K,\ Pr = 4.85.$$

Analysis The required heat transfer rate may be obtained from the overall energy balance for the hot fluid, Eq. 13.1a

$$q = \dot{m}_h c_{p,h}(T_{h,i} - T_{h,o})$$

$$q = 0.1\ kg/s \times 2131\ J/kg \cdot K(100 - 60)\ ^{0}C = 8524\ W$$

Applying Eq. 13.2b, the water outlet temperature is

$$T_{c,o} = \frac{q}{\dot{m}_c c_{p,c}} + T_{c,i}$$

$$T_{c,o} = \frac{8524\ W}{0.2\ kg/s \times 2131\ J/kg \cdot K} + 30\ ^{0}C = 40.2\ ^{0}C$$

Accordingly, use of $\overline{T}_c = 35\ ^{0}C$ to evaluate the water properties was a good choice. The required heat exchanger length may now be obtained from Eq. 13.7,

$$q = UA\Delta T_{lm}$$

where $A = \pi D_i L$

$$\Delta T_{lm} = \frac{(T_{h,i} - T_{c,o}) - (T_{h,o} - T_{c,i})}{\ln[(T_{h,i} - T_{c,o})/(T_{h,o} - T_{c,i})]} = \frac{59.8 - 30}{\ln(59.8/30)} = 43.2\ ^{\circ}C$$

The overall heat transfer coefficient is

$$U = \frac{1}{(1/h_i) + (1/h_o)}$$

For water flow through the tube,

$$\dot{m} = \rho A V \quad \frac{\dot{m}}{A} = \rho V \quad Re_D = \frac{\dot{m}_c D_i}{A\mu} = \frac{4\dot{m}}{\pi D_i \mu}$$

$$= \frac{4 \times 0.2\ \text{kg/s}}{\pi(0.025\ \text{m})725 \times 10^{-6}\ \text{N} \cdot \text{s/m}^2} = 14,050$$

Accordingly, the flow is turbulent and the convection coefficient may be computed from Eq. 12.34.

$$Nu_D = 0.023\,Re_D^{4/5}\,Pr^{0.4}$$
$$Nu_D = 0.023(14,050)^{4/5}(4.85)^{0.4} = 90$$

Hence

$$h_i = Nu_D \frac{k}{D_i} = \frac{90 \times 0.625\ \text{W/m} \cdot \text{K}}{0.025\ \text{m}} = 2250\ \text{W/m}^2 \cdot \text{K}$$

For the flow of oil through the annulus, the hydraulic diameter is,

$$D_h = \frac{4A}{P} = \frac{4(\pi r_o^2 - \pi r_i^2)}{2\pi(r_o - r_i)} = 2(r_o - r_i) = D_o - D_i$$

and the Reynolds number is

$$Re_D = \frac{\rho u_m D_h}{\mu} = \frac{\dot{m}(D_o - D_i)}{A\mu} = \frac{\dot{m}(D_o - D_i)}{\mu\pi(D_o^2 - D_i^2)/4}$$

$$Re_D = \frac{4\dot{m}}{\pi(D_o + D_i)\mu} = \frac{4(0.1\ \text{kg/s})}{\pi(0.045 + 0.025)3.25 \times 10^{-2}\ \text{kg/s·m}} = 56.0$$

The annular flow is therefore laminar. For a constant heat flux the laminar correlation is Eq. 12.33.

$$Nu_i = \frac{h_o D_h}{k} = 4.36$$

and

$$h_o = 4.36 \frac{0.138 \text{ W/m} \cdot \text{K}}{0.020 \text{ m}} = 30.1 \text{ W/m}^2 \cdot \text{K}$$

The overall convection coefficient is then

$$U = \frac{1}{(1/2250 \text{ W/m}^2 \cdot \text{K}) + (1/30.1 \text{ W/m}^2 \cdot \text{K})} = 29.7 \text{ W/m}^2 \cdot \text{K}$$

and from the rate equation it follows that

$$L = \frac{q}{U \pi D_i \Delta T_{lm}} = \frac{8524 \text{ W}}{29.7 \text{ W/m}^2 \cdot \text{K} \pi (0.025 \text{ m})(43.2 \,^0\text{C})} = 84.6 \text{ m}$$

Comments
1. The hot side convection coefficient controls the rate of heat transfer between the two fluids, and the low value of h_o is responsible for the large value of L.
2. Because $h_i \gg h_o$, the tube wall temperature will follow closely that of the coolant water.

Many heat exchangers have been designed based on a log-mean temperature difference. However, there are some problems with proceeding based on the log-mean temperature difference equation. First, it says nothing about cross flow heat exchangers, which are very common due to the ease of construction of this type of exchanger. Secondly, it often requires iterative calculations for a design if all of the inlet and exit temperatures are not known *a priori*. In the example above, had the outlet temperature of the oil not been specified, an iterative solution would have been required. Iterative solutions can certainly be accurate, but they often require more work. Thirdly, the log-mean temperature difference method does not provide a feel for the maximum heat transfer possible given the entering conditions of the fluids. Sometimes this is an important parameter to understand, if the design is to be optimized.

13.10   Effectiveness-NTU Method for Heat Exchanger Design

A better method has been developed for heat exchanger design that uses some of the preceding analysis. This method is called the effectiveness-NTU (Number of Transfer Unit) method.[1] It starts by considering the fluid heat transfer capacity rates defined as

$$Cold\ Fluid\ Capacity\ Rate \quad C_c = \dot{m}_c C_{p,c} \quad W/K$$
$$Hot\ Fluid\ Capacity\ Rate \quad C_h = \dot{m}_h C_{p,h} \quad W/K \tag{13.8}$$

Then the maximum amount of heat that can be transferred between the two fluids is the Minimum Fluid Capacity Rate times the difference in temperature of the hot fluid entering the exchanger and the cold fluid entering the exchanger. Or

$$C_{min} = min(C_c, C_h)$$
$$q_{max} = C_{min}(T_{h,in} - T_{c,in}) \tag{13.9}$$

Then the heat exchanger effectiveness is defined as

$$\varepsilon = \frac{q_{act}}{q_{max}} = \frac{C_h(T_{h,in} - T_{h,out})}{C_{min}(T_{h,in} - T_{c,in})} = \frac{C_c(T_{c,in} - T_{c,out})}{C_{min}(T_{h,in} - T_{c,in})} \tag{13.10}$$

The number of heat exchanger transfer units is then defined as

$$NTU = \frac{UA}{C_{min}} \tag{13.11}$$

where U and A are defined as above. The heat Capacity-Rate Ratio is defined as,

$$C_r = \frac{C_{min}}{C_{max}} \tag{13.12}$$

Then in general it is possible to express the effectiveness as

$$\varepsilon = \varepsilon(NTU, C_r, Flow\ Arrangement)$$

[1] Kays, W. M. and A. L. London, Compact Heat Exchangers, 3rd Ed. Krieger Publishing Company, Malabar FL, 1998.

Different functions for ε have been developed for many flow arrangements.[2] The three main flow arrangements of interest are parallel flow, counter flow, and cross flow.

Parallel Flow   For parallel flow the expression is

$$\varepsilon = \frac{1-e^{-NTU(1+C_r)}}{1+C_r} \tag{13.13}$$

There are two interesting limits,

$$C_r \to 0 \quad \varepsilon = 1-e^{-NTU}$$
$$C_r \to 1 \quad \varepsilon = \frac{1}{2} \tag{13.14}$$

The $C_r = 0$ limit corresponds to one fluid vaporizing or condensing and the heat capacity rate for this fluid becomes immense. The other limit $C_r = 1.0$ corresponds to both fluids having the same Heat Capacity Rate.

Counter Flow

$$\varepsilon = \frac{1-e^{-NTU(1-C_r)}}{1-C_r e^{-NTU(1-C_r)}} \tag{13.15}$$

Addressing the same two limits

$$C_r \to 0 \quad \varepsilon = 1-e^{-NTU}$$
$$C_r \to 1 \quad \varepsilon = \frac{NTU}{1+NTU} \tag{13.16}$$

The $C_r = 0$ limit is the same but the $C_r = 1$ limit is twice as effective for large values of NTU. This essentially is the known performance advantage of counter flow heat exchangers.

Cross Flow   Cross flow has to be broken down into three different types. The performance is different depending on whether or not the fluids are allowed to mix with themselves as they move through the exchanger. A typical tube and shell exchanger would have the fluid moving through the tubes described as unmixed and the fluid moving through the shell would be mixed.

---

[2] Kays, and London, Op. Cit.

Cross Flow—Both Fluids Unmixed This case requires a series numerical solution and the curves for values of various C_r values are given in Kays and London. For the case of $C_r = 0$, the solution is the same as for counter flow and parallel flow.

$$C_r = 0 \quad \varepsilon = 1 - e^{-NTU}$$

However all of the curves for any value of C_r asymptotically approach 1.0 like the counter flow exchanger. For all $C_r > 0$ the effectiveness is less than for a counter flow exchanger with the same C_r.

Cross Flow—One Fluid Mixed For the case of

$$C_{max} = C_{unmixed} \quad C_{min} = C_{mixed}$$

$$\varepsilon = 1 - e^{-\Gamma/C_r} \quad \Gamma = 1 - e^{-NTUC_r} \tag{13.17}$$

And for the case of

$$C_{max} = C_{mixed} \quad C_{min} = C_{unmixed}$$

$$\varepsilon = \frac{1 - e^{-\Gamma'C_r}}{C_r} \quad \Gamma' = 1 - e^{-NTU} \tag{13.18}$$

Once again for $C_r = 0$, this gives the same behavior as the counter flow heat exchanger. For $C_r = 1.0$, it gets complicated but it is important to note that if a choice is possible it is better to have the fluid with the smaller heat capacity rate to be the mixed fluid (Eq. 13.17).

Cross Flow—Both Fluids Mixed The closed form solution is

$$\varepsilon = \frac{NTU}{\dfrac{NTU}{(1 - e^{-NTU})} + \dfrac{C_r NTU}{1 - e^{-NTUC_r}} - 1} \tag{13.19}$$

As always, for the case of $C_r = 0$, the results are the same as the counter flow exchanger.

For $C_r = 1.0$ as NTU becomes large, the effectiveness goes to *1/2*. However this is the only case that a better effectiveness can be obtained at a lower NTU. The effectiveness actually decreases after an NTU of about 3–5.

There are many other configurations reported by Kays and London, but these three are the most important. The availability of solutions for the common cross flow case of one fluid mixed makes this technique very useful.

Example 13.2 Consider a gas-to-gas recuperator of the shell and tube design.

The tubes are 2 cm diameter tubes spaced on 4 cm centers with a 2 mm thickness made of aluminum. The flow cross section is a 2 m by 2 m². The pressure ratio for the compressor is 20. Both fluids are air and the cold fluid is in the tubes. The hot fluid enters at 783 °K and exits at 670 °K and is atmospheric pressure. The cold fluid enters at 655 °K and exits at 768 °K and is at 20 atmospheres. The flow rate is 2.5E+5 kg/h, for a 10 MW power plant.

Solution Start with the hot fluid—Calculate Re No.

$$N\ tubes = 2401,\ A_{flow} = 2^2 - 2401 * \pi (0.012)^2 = 2.9138\ m^2$$

$$\dot{m} = \rho AV \quad \rho V = \frac{\dot{m}}{A} = \frac{69.44 kg/s}{2.9138} = 23.8\ kg/s/m^2$$

$$\mu = 3.65 E - 5 \quad D = \frac{4(0.04^2 - \pi 0.012^2)}{2\pi 0.012} = 0.0609\ m$$

$$Re = \frac{23.8 * 0.0609}{3.7E - 05} = 39,789.2$$

This is clearly in the turbulent range.
   Using the same equations, the cold fluid $Re = 56,731.1$

$$Pr_{hot} = C_p m / k = 1078.8 * 3.65 E - 5 / 0.0564 = 0.697$$

$$Pr_{cold} = C_p m / k = 1076.7 * 3.25 E - 5 / 0.050 = 0.700$$

$$Nu_{hot} = 0.023 * 39,789.2^{0.8} * 0.697^{0.3} = 98.7$$

$$h_{hot} = 98.7 * 0.0564 / 0.0609 = 91.5\ w/m^2/K$$

$$Nu_{cold} = 0.023 * 56,731.1^{0.8} * 0.700^{0.4} = 127.0$$

$$h_{cold} = 127.0 * 0.0500 / 0.02 = 316.0\ w/m^2/K$$

This allows us to calculate UA as a function of L

$$A_{hot} = 2\pi (0.012) * 2401 * L = 724.8 * L\ m^2$$

$$A_{cold} = 2\pi(0.01)*2401*L = 603.4*L \ m^2$$

$$A_{tube} = 2\pi(0.011)*2401*L = 663.74*L \ m^2 \ \otimes$$

$$\frac{1}{UA} = \left(\frac{1}{h_{hot}A} + \frac{t}{kA} + \frac{1}{h_{cold}A} \right)$$

$$= \frac{1}{91.5(724.8)L} + \frac{0.002}{218(663.74)L} + \frac{1}{316(603.4)L}$$

$$= (1.508E-5 + 1.38E-8 + 5.25E-6)/L = \frac{2.03E-5}{L} \ w/K$$

$$\dot{Q} = 8.46E+6 \ w \quad \Delta T_{in} = 15K \quad \Delta T_{out} = 15K$$

$$\Delta T_{lmn} = 15K \quad \dot{Q} = \left(\frac{UA}{L} \right) L \Delta T_{lmn} \quad L = \frac{\dot{Q}}{(UA/L)\Delta T_{lmn}}$$

$$L = \frac{8.46E+6}{4.92E+4*15} = 11.5 \ m$$

Now try the NTU-Effectiveness method

$$C_{hot} = 2.5E+5/3600*1078.8 = 7.49E+04$$

$$C_{cold} = 2.5E+5/3600*1076.7 = 7.48E+04 = C_{min}$$

$$C_r = 0.998 \sim 1.0 \quad \varepsilon = NTU/(NTU+1) \quad NTU = \varepsilon/(1-\varepsilon)$$

$$\varepsilon = \frac{C_{hot}(T_{h,in}-T_{h,out})}{C_{min}(T_{h,in}-T_{c,in})} = \frac{7.49E+4*(783-670)}{7.48E+4*(783-655)} = 0.8846$$

$$NTU = 7.67 \quad \frac{NTU}{L} = \frac{UA/L}{C_{min}} = 0.6576 \quad L = \frac{7.67}{0.6576} = 11.66 \ m$$

Note that the largest resistance to heat transfer was in the hot side convection and the resistance of the tube wall was negligible.

So add fins to the hot side channel by putting a 2 mm thick web between the tubes. See Fig. 13.16.

First, recalculate the hydraulic diameter for the hot side

Treating the webs as wetted perimeter gives a new $D_h = 0.0329$ m

Fig. 13.16 Web channel

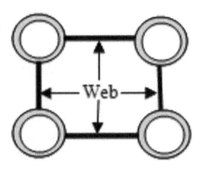

This gives a new

$$Re = 21,521.3$$

$$Nu = 60.4$$

$$h = 104.0.$$

Dividing the web in half for a fin for each tube,

$$w = 0.01, \ t = 0.002, \ P = 0.02, \ A = 0.00002, \ m = 21.79 \ L = 0.008$$

$$mL = 0.1743 \ h_{fin} = 0.9899 \ h_o = 0.9915 \ A_{tot} = 0.1394$$

$$h_o * A_{tot} = 0.1382 \ A_{hot} \ / \ A_{cold} = 2.1997 \ A_{hot} \ / \ L = 1327.3$$

The cold side and tube resistances don't change, so

$$\frac{1}{UA} = \frac{1}{1.25E-5*L} L = \frac{\dot{Q}}{UA_{q \ L} \Delta T_{lmn}} = \frac{8.47E+6}{7.98E+4*15} = 7.1 \ m$$

Now to get the pressure drops

$$C_{f,hot} = 0.046 \, \text{Re}^{-0.2} = 0.046 * 21521.3^{-0.2} = 6.25E-3$$
$$C_{f,cold} = 0.046 \, \text{Re}^{-0.2} = 0.046 * 56731.1^{-0.2} = 5.15E-3$$

This gives

$$\tau_{hot} = 6.25E - 3 * \frac{1}{2} \rho V^2 = \frac{3.13E - 3}{\rho_{hot}} \left(\frac{\dot{m}}{A}\right)^2_{hot}$$

$$\tau_{cold} = 5.15E - 3 * \frac{1}{2} \rho V^2 = \frac{2.58E - 3}{\rho_{cold}} \left(\frac{\dot{m}}{A}\right)^2_{cold}$$

$$\rho_{hot} = \frac{101325 * 28.9669}{8314.4 * 783} = 0.451 kg / m^3$$

$$\rho_{cold} = \frac{20 * 101325 * 28.9669}{768 * 8314.4} = 9.193 kg / m^3$$

$$\tau_{hot} = \frac{3.13E - 3}{0.451} (23.83)^2_{hot} = 3.941 Pascals$$

$$\tau_{cold} = \frac{2.58E - 3}{9.193} (92.1)^2_{cold} = 2.38 Pascals$$

Clearly these pressure drops are negligible compared to atmospheric pressure. This says that the heat exchanger could be made a lot more compact by adding more surface area per unit volume - m^2/m^3.

13.11 Special Operating Conditions

It is useful to note certain special conditions under which heat exchangers may be operated. Figure 13.14a shows temperature distributions for a heat exchanger in which the hot fluid has a heat capacity rate, $C_h \equiv \dot{m}_h C_{p,h}$, which is much larger than that of the cold fluid, $C_c \equiv \dot{m}_c C_{p,c}$.

For this case the temperature of the hot fluid remains approximately constant throughout the heat exchanger, while the temperature of the cold fluid increases. The same condition is achieved if the hot fluid is a condensing vapor. Condensation occurs at constant temperature, and, for all practical purposes, $C_h \to \infty$. Conversely, in an evaporator or a boiler (Fig. 13.14b), it is the cold fluid that experiences a change in phase and remains at a nearly uniform temperature ($C_c \to \infty$). The same effect is achieved without phase change if $C_h \ll C_c$. Note that, with condensation or evaporation, the heat rate is given by Eqs. 13.1a or 13.1b, respectively. Conditions illustrated in Fig. 13.17a or b also characterize an internal tube flow (or *single stream heat exchanger*) exchanging heat with a surface at constant temperature or an external fluid at constant temperature.

The third special case (Fig. 13.17c) involves a counterflow heat exchanger for which the heat capacity rates are equal ($C_h = C_c$). The temperature difference ΔT must then be constant throughout the exchanger, in which case $\Delta T_1 = \Delta T_2 = \Delta T_{lm}$.

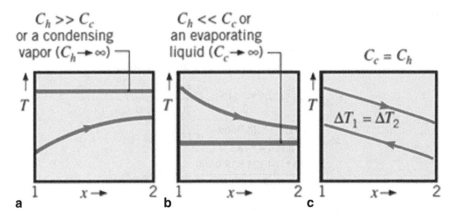

Fig. 13.17 Special heat exchanger conditions. **a** $C_h \gg C_c$ or a condensing vapor. **b** An evaporating liquid or $C_h \ll C_c$. **c** A counterflow heat exchanger with equivalent fluid heat capacities $C_h = C_c$ [5]

13.12   Compact Heat Exchangers

A heat exchanger is quite arbitrarily referred to as a *compact heat exchanger*, providing that it has a surface area density greater than about 700 m²/m³.

A special and important class of heat exchangers is used to achieve a very large (≥ 400 m²/m³ for liquids and ≥ 700 m²/m³ for gases) heat transfer surface area per unit volume. Termed compact heat exchangers, these devices have dense arrays of finned tubes or plates and are typically used when at least one of the fluids is a gas, and hence is characterized by a small convection coefficient. The tubes may be flat or circular, as in Figs. 13.18a, b, and c, respectively, and the fins may be plate or circular, as in Figs. 13.15a, b and c, respectively. Parallel-plate heat exchangers may be finned or corrugated and may be used in single-pass (Fig. 13.15d) or multipass (Fig. 13.15e) modes of operation. Flow passages associated with compact heat exchangers are typically small ($D_h \leq 5$ mm), D_h is the magnitude of the *hydraulic diameter* and the flow is often laminar. Many of the geometries are far too complicated to apply deterministic methods to predict their performance. So many for these compact heat exchangers have had their performance determined experimentally.

Kays and London [7] have studied a wide variety of configurations for heat transfer matrices, and catalogued their heat transfer and pressure drop characteristics. Figure 13.19 shows typical heat transfer materials for compact heat exchangers [5]. Figure 13.19a shows a *circular finned-tube array* with fins on individual tubes; Fig. 13.19b shows a *plain plate-fin matrix* formed by corrugation, and Fig. 13.19c shows a *finned flat-tube* matrix [5].

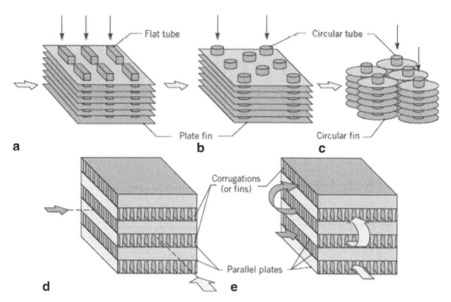

Fig. 13.18 Compact heat exchanger cores. **a** Fin–tube (flat tubes, continuous plate fins). **b** Fin–tube (circular tubes, continuous plate fins). **c** Fin–tube (circular tubes, circular fins). **d** Plate–fin (single pass). **e** Plate–fin (multipass) [5]

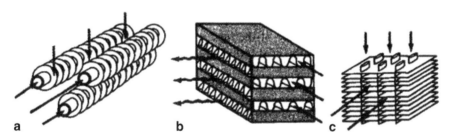

Fig. 13.19 Typical heat transfer matrices for compact heat exchangers. **a** Circular finned-tube matrix. **b** Plain plate-fin matrix. **c** Finned flat-tube matrix [5]

The heat transfer and pressure drop characteristics of such configurations for use as compact heat exchangers have been determined experimentally as explained above. Figs. 13.20, 13.21 and 13.22 show typical heat transfer and friction factor data for three different configurations.

Note that the principal dimensionless groups governing these correlations are the Stanton, Prandtl, and Reynolds numbers [5].

$$St = \frac{h}{Gc_p} \quad Pr = \frac{c_p \mu}{k} \quad Re = \frac{GD_h}{\mu} \tag{13.20}$$

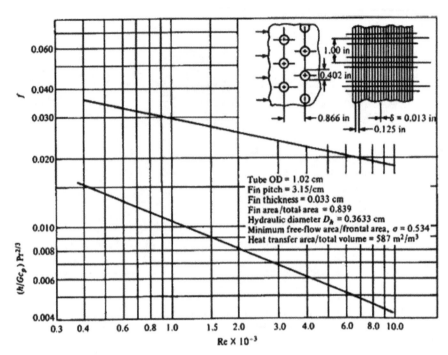

Fig. 13.20 Heat transfer and friction factor for flow across plate-finned circular tube matrix. (Courtsy of Kays and London)

Here G is the *mass velocity* defined as

$$G = \frac{m}{A_{min}} \; kg / (m^2 .s) \tag{13.21}$$

where m = total mass flow rate of fluid (kg/s) and A_{min} = minimum free-flow cross-sectional area (m²) regardless of where this minimum occurs.

The magnitude of the *hydraulic diameter* D_h for each configuration is specified on Figs. 13.17–13.19. The hydraulic D_h is defined as;

$$D_h = 4\frac{LA_{min}}{A} \tag{13.22}$$

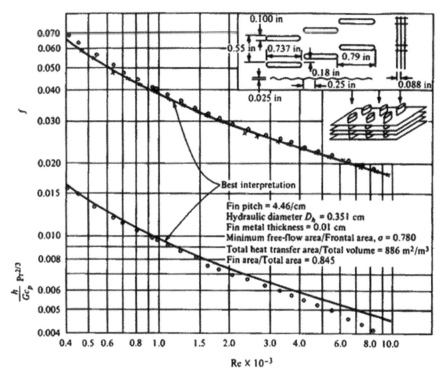

Fig. 13.21 Heat transfer and friction factor for flow across finned flat-tube matrix. (Courtsy of Kays and London)

where A is the total heat transfer area and the quantity LA_{min} can be regarded as the minimum free-flow passage volume, since L is the flow length of the heat exchanger matrix.

Thus, once the heat transfer and the friction factor charts such as those shown in Fig. 13.20 are available for a specified matrix and the Reynolds number Re for the flow is given, the heat transfer coefficient h and the friction f for flow across the matrix can be evaluated. Then the rating and sizing problem associated with the heat exchanger matrix can be performed by utilizing either the LMTD or the effectiveness-NTU method of analysis.

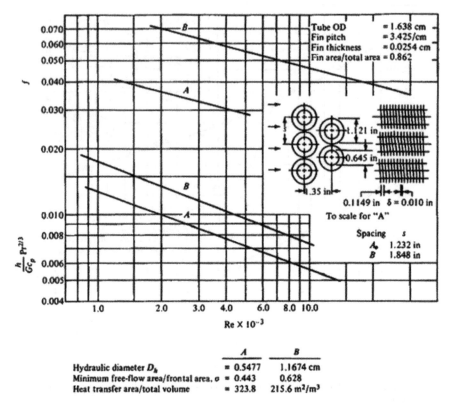

Fig. 13.22 Heat transfer and friction factor for flow across circular finned-tube matrix. (Courtsey of Kays and London)

Problems

Problem 13.1: Rework Example 13.1 using the effectiveness-NTU method of analysis.

Problem 13.2: Instead of a single tube in Example 13.1, assume the water is in a tube bank that is 15 tubes on a side with the same dimensions as the tube in Example 13.1 and spaced on 35 mm centers. The oil now flows between the tubes and an effective hydraulic diameter must be calculated. All other specifications of the problem remain the same. Estimate the length of the tube bank required.

Problem 13.3: A tube and shell recuperator for a gas turbine engine has been designed with 1.5 cm diameter tubes on 2.5 cm centers (pitch=2.5 cm) for the cold air flow. The tube material is copper and is 2 mm thick. The hot air flows through the shell in a counter flow arrangement. The overall length of the active section is 5 m. It has a cross section of 2 m by 2m. The cold flow enters at 15 atm pressure and 600 K. The hot flow enters at 1 atm and 800 K. Both flows are 2.0 kg/sec. Estimate the hot and cold flow exit temperatures and the total heat transferred based on the log-mean-temperature method. (Note: This will require iteration)

Problem 13.4: A tube and shell recuperator for a gas turbine engine has been designed with 1.5 cm diameter tubes on 2.5 cm centers (pitch = 2.5 cm) for the cold air flow. The tube material is copper and is 2 mm thick. The hot air flows through the shell in a counter flow arrangement. The overall length of the active section is 5 m. It has a cross section of 2 m by 2 m. The cold flow enters at 15 atm pressure and 600 K. The hot flow enters at 1 atm and 800 K. Both flows are 2.0 kg/sec. Estimate the hot and cold flow exit temperatures and the total heat transferred using the effectiveness-NTU method.

Problem 13.5: Add 1 mm copper webs between the tubes, like in Example 13.2, to the recuperator in problem 4 to improve its performance and recalculate the exit temperatures and total heat transferred.

Problem 13.6: A cross flow tube and shell condenser is installed below a 10 MW steam turbine that exhausts saturated steam at 20 kPa and 5.5 kg/s. The cooling water flows through copper tubes 1.5 cm in diameter on a 2.5 cm pitch in a square array. The tubes are 2 mm thick. The tube length is 3 m and the shell cross sectional area is 2 m by 3 m. The cooling water enters at 17 °C and flows at 10 kg/s. Estimate the cooling water exit temperature and the quality of the steam exiting the condenser.

References

1. Fraas AP (1989) Heat exchangers design, 2nd edn. Wiley, New York
2. Necati Ozisik M (1985) Heat transfer a basic approach. McGraw-Hill, New York
3. Incropera F, Dewitt D, Bergman T, Lavine A (2007) Introduction to heat transfer, 5th edn. Wiley, New York
4. Kays WM, London AL (1964) Compact heat exchangers, 2nd edn. McGraw-Hill, New York
5. Incropera F, Dewitt D, Bergman T, Lavine A (2011) Fundamentals of heat and mass transfer, 7th edn. Wiley, New York
6. Pansini AJ, Smalling KD, (1991) Guide to electric power generation. The Fairmont Press, Liburn
7. Boyen JL (1975) Practical heat recovery. Wiley, New York

Chapter 14
Gas Power Cycles

An important application of thermodynamics is the analysis of power cycles through which the energy absorbed as heat can be continuously converted into mechanical work. A thermodynamic analysis of the heat engine cycles provides valuable information regarding the design of new cycles or for improving the existing cycles. In this chapter, various gas power cycles are analyzed under some simplifying assumptions.

14.1 Introduction

Two most important areas application of thermodynamics are power generation and refrigeration and they both are usually accomplished by system that operate on thermodynamic a cycle.

Thermodynamically, the word of "cycle" is used in a procedure or arrangement in which some material goes through a cyclic process and one form of energy, such as heat at an elevated temperature from combustion of a fuel, is in part converted to another form, such as mechanical energy of a shaft, the remainder being rejected to a lower temperature sink that is also known as heat cycle.

Thermodynamic cycle is defined as a process in which a working fluid undergoes a series of state changes and finally returns to its initial state. A cycle plotted on any diagram of properties forms a closed curve (Fig. 14.1).

Note that a reversible cycle consists only of reversible processes. The area enclosed by the curve plotted for a reversible cycle on a p-v diagram represents the net work of the cycle as we explained in preceding chapters as follow:

- The work is done on the system, if the state changes happen in an anticlockwise manner.
- The work is done by the system, if the state changes happen in a clockwise manner.

© Springer International Publishing Switzerland 2015
B. Zohuri, P. McDaniel, *Thermodynamics In Nuclear Power Plant Systems*,
DOI 10.1007/978-3-319-13419-2_14

Fig. 14.1 Schematic of a
closed cycle

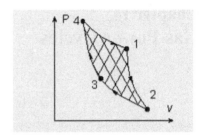

The purpose of a thermodynamic cycle is either to produce power, or to produce re-frigeration/pumping of heat. Therefore, the cycles are broadly classified as follows:

A. Heat engine or power cycles.
B. Refrigeration/heat pump cycles.

A thermodynamic cycle requires, in addition to the supply of incoming energy;

1. A working substance, usually a gas or vapor;
2. A mechanism in which the processes or phases can be carried through sequentially; and
3. A thermodynamic sink to which the *residual heat* can be rejected.

The cycle itself is a repetitive series of operations.

Any thermodynamic cycle is essentially a *closed cycle* in which, the working substance undergoes a series of processes and is always brought back to the initial state.

However, some of the power cycles operate on *open cycle*. It means that the working substance is taken into the unit from the atmosphere at one end and is discharged into the atmosphere after undergoing a series of processes at the other end. The following are illustrations of heat engines operating on open cycle:

- Petrol and diesel engines in which the air and fuel are taken into the engine from a fuel tank and products of combustion are exhausted into the atmosphere.
- Steam locomotives in which the water is taken in the boiler from a tank and steam is exhausted into the atmosphere.

The basic processes of the cycle, either is open or closed, are heat addition, heat rejection, expansion, and compression. These processes are always present in a cycle even though there may be differences in working substance, the individual processes, pressure ranges, temperature ranges, mechanisms, and heat transfer arrangements.

Many cyclic arrangements, using various combinations of phases but all seeking to convert heat into work, were proposed by many investigators whose names are attached to their proposals. For example, the *Diesel, Otto, Rankine, Brayton, Stirling, Ericsson,* and *Atkinson* cycles. Not all proposals are equally efficient in the conversion of heat into work. However, they may offer other advantages, which have led to their practical development for various applications. See also Brayton cycle; Carnot cycle; Diesel cycle; Otto cycle; Stirling engine; Thermodynamic processes.

Fig. 14.2 Graphic illustration
of thermal efficiency

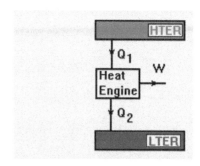

Essentially, such devices do not form a cycle. However, they can be analyzed by adding imaginary processes to bring the state of the working substance, thus completing a cyclic. Note that the terms *closed* and *open* cycles that are used here do not mean closed system cycle and open system cycle. In fact, the processes both in closed and open cycles could either be closed or open system processes.

There is a basic pattern of processes common to power-producing cycles. There is a compression process wherein the working substance undergoes an increase in pressure and therefore density. There is an addition of thermal energy from a source such as a *fossil fuel*, a *fissile fuel* (fissile material is one that is capable of sustaining a *chain reaction* of nuclear fission), or solar radiation. Work is done by the system on the surroundings during an expansion process. There is a rejection process where thermal energy is transferred to the surroundings. The algebraic sum of the energy additions and abstractions is such that some of the thermal energy is converted into mechanical work.

Different types of working fluids are employed in the power plants. The nature of the working fluids can be classified into two groups:

a. Vapors.
b. Gases.

The power cycles are accordingly classified into two groups as:

1. Vapor power cycles in which the working fluid undergoes a phase change during the cyclic process.
2. Gas power cycles in which the working fluid does not undergo any phase change.

In the thermodynamic analysis of power cycles, our main interest lies in estimating the energy conversion efficiency or the thermal efficiency. The thermal efficiency of a heat engine is defined as the ratio of the network output W delivered to the energy absorbed as heat Q and mathematically is presented by symbol η and can be written as:

$$\eta = \frac{W}{Q} \tag{14.1}$$

and it can be illustrated as Fig. 14.2 below;
In this depiction, we identify the followings;

Fig. 14.3 Illustration of
actual vs. ideal cycle in a
$P - \upsilon$ diagram

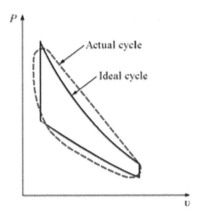

LTER = Low Temperature Energy Reservoir
HTER = High Temperature Energy Reservoir

Using these definitions and refer to Fig. 14.2, the Equation of 14.1 can be written
more precisely as follow;

$$\eta = \frac{W}{Q_1} \tag{14.2}$$

where Q_1 is the heat supplied at high temperature.

A procedure or arrangement in which one form of energy, such as heat at an
elevated temperature from combustion of a fuel, is in part converted to another
form, such as mechanical energy on a shaft, and the remainder is rejected to a lower-
temperature sink as low-grade heat.

Heat engines, depending on how the heat is supplied to the working fluid, are
categorized in two types as;

A. *External combustion.*
B. *Internal combustion.*

In external combustion engines, such as steam power plants, heat is supplied to
the working fluid from an external source such as a furnace, a geothermal well, a
nuclear reactor, or even the sun [1].

In internal combustion engines, such as automobile engines, this is done by burning the fuel within the system boundaries [1].

Our study of gas power cycles will involve the study of those heat engines in
which the working fluid remains in the gaseous state throughout the cycle. We often
study the ideal cycle in which internal irreversibilities and complexities (the actual
intake of air and fuel, the actual combustion process, and the exhaust of products of
combustion among others) are removed. We will be concerned with how the major
parameters of the cycle affect the performance of heat engines. The performance is
often measured in terms of the cycle efficiency of η_{th} as ratio of network W_{net} and
energy as heat of Q_{in}. See Fig. 14.3 where one can observe *Actual Cycle* versus and

Fig. 14.4 Basic thermodynamic cycle

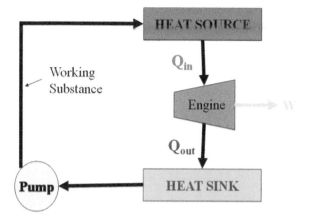

Ideal Cycle in a $P-v$ diagram and using the Eq. 14.1, and referring to Fig. 14.4 below, mathematically we can show that;

$$\eta_{th} = \frac{W_{net}}{Q_{in}} \qquad (14.3)$$

Several cycles utilize a gas as the working substance, the most common being the Otto cycle and the diesel cycle used in internal combustion engines. We are touching upon some of these cycles in this chapter such as Otto and others such as Brayton, Carnot, etc. and we will expand on them among other cycles for further evaluation as well.

14.1.1   Open Cycle

When internal combustion engine operation is examined, it is seen to differ in the process of heat supply for a typical heat engine cycle because there is a permanent change in the working fluid during combustion. Therefore, the fluid does not pass through a cycle so the internal combustion engine is often referred to as an "open cycle" device, not a cyclic thermodynamic heat engine.

The term "open cycle", while meaningless from a thermodynamic perspective, refers to the fact that energy is supplied to the engine from outside in the form of petroleum fuel and the unconverted portion of energy remaining in the spent combustion mixture is exhausted to the environment. "Closing the cycle", i.e. returning the rejected products to the starting point where they can be reused, is left for nature to accomplish—hence the term "open cycle" is coming to play.

An internal combustion engine is therefore a device for releasing mechanical energy from petroleum fuel using air as the working medium rather than a heat engine for processing air in a thermodynamic cycle. Heat, as such, is not supplied to the

Fig. 14.5 Illustration of a
thermodynamic closed cycle

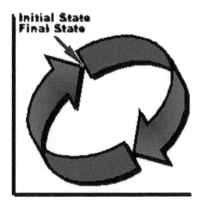

internal combustion engine, so it cannot be a heat engine in the sense described in
most thermodynamic references.

A simulated heat engine cycle can be constructed to correspond approximately
to the operation of an internal combustion engine by substitution of analogous heat
transfer processes for some of the actual engine processes. The specific mechanism
of such heat transfer is neglected because the simulation is only a theoretical model
of the engine, not an actual device. Such cycles, called *air standard cycles*, which
are subject of study in Sect. 13.3 of this chapter, are useful in the elementary study
of internal combustion engines.

14.1.2   Closed Cycle

Thermodynamic cycles can be categorized yet another way closed and open cycles
and open cycle was described in Sect. 13.1.1. In closed cycles, the working fluid is
returning to the initial state at the end of the cycle and is recirculated. By the same
talking, in open cycles, the working fluid is renewed at the end of each cycle instead
of being recirculated. For example, in automobile engines, the combustion gases
are exhausted and replaced by fresh air-fuel mixture at the end of each cycle. The
engine operates on a mechanical cycle, but the working fluid does not go through a
complete thermodynamic cycle [1] (Fig. 14.5).

As we said before, any thermodynamic cycle is essentially a *closed cycle* in
which, the working substance undergoes a series of processes and is always brought
back to the initial state.

14.2   Gas Compressors and Brayton Cycle

The work in a gas compressor is calculated by,

$$\dot{W}_{comp} = \dot{m}(h_e - h_i)$$ (14.4)

If we assume that the gas in the compressor is calorically perfect, then we have,

$$\dot{W}_{comp} = \dot{m}C_p(T_e - T_i) \tag{14.5}$$

In many cases, this is a reasonable approximation. For noble gases, it is very accurate because they are calorically perfect. For air and similar working fluids, it is reasonable because the temperature rise is not that great and an average value of C_p is usually adequate. However the average value of C_p should be chosen based on a temperature between T_e and T, not one at 300 K.

If then we assume that a compressor operates isentropically (adiabatic and reversible), the exit temperature can be related to the pressure rise in the compressor as shown below in Eq. 14.6;

$$T_e = T_i \left(\frac{p_e}{p_i}\right)^{\frac{\gamma-1}{\gamma}}$$

$$\dot{W} = \dot{m}C_p[T_e - T_i] = \dot{m}C_pT_i\left[\frac{T_e}{T_i}-1\right] = \dot{m}\frac{\gamma R}{\gamma-1}T_i\left[\left(\frac{p_e}{p_i}\right)^{\frac{\gamma-1}{\gamma}}-1\right] \tag{14.6}$$

There are basically three types of compressors, reciprocating, centrifugal flow, and axial flow. In a reciprocating or positive displacement compressor, a piston slides in a cylinder and valves open and close to admit low-pressure fluid and exhaust high-pressure fluid. In centrifugal flow and axial flow compressors, the fluid enters at one end and is compressed by rotating blades and exits at the opposite end of the compressor. In the centrifugal flow compressor, the flow is in a radially outward direction and the compression is achieved by forcing the flow against the outer annulus of the compressor. In an axial flow compressor, a set of rotating blades move the flow through the compressor, acting as airfoils. They force the flow through an increasingly narrower channel, thus increasing the density and pressure. Gasoline and diesel engines are examples of reciprocating compressors, as are positive displacement pumps. Water pumps are examples of centrifugal flow compressors, similar to the rotor in a washing machine. Jet engine compressors are typically axial flow compressors. Reciprocating compressors require no priming and can reach very high pressures, but only moderate flow rates. Centrifugal flow and axial flow compressors usually require priming and can reach very high flow rates, but moderate pressures.

In this section we discuss the **Brayton Thermodynamic Cycle** which is used in all gas turbine engines. The Fig. 14.6 shows a T-s diagram of the Brayton Cycle. Using the turbine engine station numbering system, we begin with free stream conditions at station **0**. In cruising flight, the inlet slows the air stream as it is brought to the compressor face at **station 2**. As the flow slows, some of the energy associated with the aircraft velocity increases the static pressure of the air and the flow is compressed. Ideally, the compression is isentropic and the static temperature is also

Ideal Brayton Cycle
T–s diagram

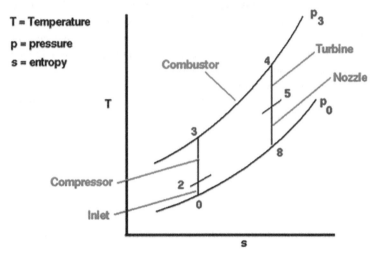

Fig. 14.6 Illustration of Brayton cycle. (Courtesy of NASA)

increased as shown on the plot. The compressor does work on the gas and increases the pressure and temperature isentropically to **station 3** the compressor exit. Since the compression is ideally isentropic, a vertical line on the T-s diagram describes the process. *In reality, the compression is not isentropic and the compression process line leans to the right because of the increase in entropy of the flow.* The combustion process in the burner occurs at constant pressure from **station 3** to **station 4**. The temperature increase depends on the type of fuel used and the fuel-air ratio. The hot exhaust is then passed through the power turbine in which work is done by the flow from **station 4** to **station 5**. Because the turbine and compressor are on the same shaft, the work done on the turbine is exactly equal to the work done by the compressor and, ideally, the temperature change is the same. The nozzle then brings the flow isentropically (adiabatic and reversible) back to free stream pressure from **station 5** to **station 8**. Externally, the flow conditions return to free stream conditions, which complete the cycle. The area under the T-s diagram is proportional to the useful work and thrust generated by the engine. The T-s diagram for the ideal Brayton Cycle is shown here:

The Brayton cycle analysis is used to predict the thermodynamic performance of gas turbine engines.

As we know the gas turbine is another mechanical system that produces power and it may operate on a cycle when used as an automobile or truck engine, or on a closed cycle when used in a nuclear power plant [2].

Usage of Brayton process in a simple gas turbine cycle can be describe by first in an open cycle operation where air enters the compressor, and passes through a

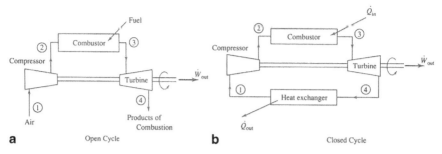

Fig. 14.7 Illustration of Brayton components for open and closed cycles

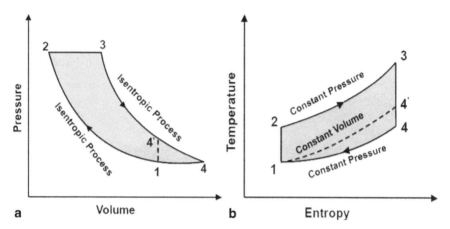

Fig. 14.8 Illustration of the Brayton cycle on P-V and T-s diagram

constant-pressure combustion chamber, then goes through turbine, and then exits as product of combustion to the atmosphere, as shown in Fig. 14.7a. Similar situation can be studied when the combustion chamber a heat exchanger gets add on to the loop of Fig. 14.7a, in order to organize a closed cycle as it can be seen in Fig. 14.7b. Energy from some external source enters the cycle and additional heat exchanger that has been added on into the loop, transfers heat from the cycle so that the air can be returned to its initial state, as clearly can be seen in Fig. 14.7b.

The Brayton cycle is a theoretical cycle for simple gas turbine. This cycle consists of two isentropic and two constant pressure processes. Figure 14.8 shows the Brayton cycle on *P-V* and *T-s* coordinates. The cycle is similar to the Diesel cycle in compression and heat addition. The isentropic expansion of the Diesel cycle is further extended followed by constant pressure heat rejection.

The following notation gives the thermal efficiency in mathematical format for ideal cycle used to model the gas turbine, which utilizing isentropic compression and expansion in Brayton process as:

$$\eta_{th} = \frac{\text{Heat added - Heat rejected}}{\text{Heat added}} = \frac{\dot{Q}_{out}}{\dot{Q}_{in}} \tag{14.7a}$$

$$\eta_{th} = \frac{mC_p(T_3 - T_1) - mC_p(T_4 - T_1)}{mC_p(T_3 - T_2)}$$

$$= 1 - \frac{T_4 - T_1}{T_3 - T_2} \tag{14.7b}$$

$$= 1 - \frac{T_1}{T_2} \frac{(T_4/T_1) - 1}{(T_3/T_2) - 1}$$

Using the following isentropic process and relations that we have;

$$\frac{T_2}{T_1} = \left(\frac{P_2}{P_1}\right)^{\frac{\gamma-1}{\gamma}} \text{ and } \frac{T_3}{T_4} = \left(\frac{P_3}{P_4}\right)^{\frac{\gamma-1}{\gamma}} \tag{14.7c}$$

For ideal gas and observation P-V diagram of Fig. 14.8a obviously shows that we can state $P_2 = P_3$ and $P_1 = P_4$ as result, using the Eq. 14.7c, will induced the following;

$$\frac{T_2}{T_1} = \frac{T_3}{T_4} \text{ or } \frac{T_4}{T_1} = \frac{T_3}{T_2} \tag{14.7d}$$

Then the Thermal efficiency η_{th} from Eq. 14.7a can be reduced to the following form as;

$$\eta_{th} = 1 - \frac{T_4}{T_3} = 1 - \frac{T_1}{T_2} \tag{14.7e}$$

Now if we introduce a term of the pressure ration $r_2 = P_2/P_1$ the thermal efficiency from Eq. 14.7e will take a form of the following;

$$\frac{T_4}{T_3} = \frac{T_4}{T_3} = \frac{V_2}{V_1} = \frac{1}{r_p^{\gamma-1}} \tag{14.7f}$$

$$\frac{1}{r_p^{\gamma-1}} = \left(\frac{V_2}{V_1}\right)^{\gamma-1} \left\{\left(\frac{P_1}{P_2}\right)\right\}^{(\gamma-1)/\gamma} = (r_p)^{\frac{\gamma-1}{\gamma}} \tag{14.7g}$$

$$\eta_{th} = 1 - \frac{T_1}{T_2} = 1 - \left(\frac{P_1}{P_2}\right)^{(\gamma-1)/\gamma} \tag{14.7h}$$

or

$$\eta_{th} = 1 - r_p^{(1-\gamma)/\gamma} \tag{14.7i}$$

Note that the above final expression for thermal efficiency η_{th} in both form of Eqs. 14.7h and 14.7i were obtained based on assumption of using constant specific heats. For more accurate calculations the gas tables should be utilized.

In actual gas turbine the compressor and the turbine are not isentropic and some losses are taking place. These losses, usually in the neighborhood of 85%, significantly reduce the efficiency of the gas turbine engine [3].

Considering all the above we can see that the back work ratio is defined for a Brayton system as W_{comp} / W_{turb}. This an important feature of the gas turbine that limits thermal efficiency, that is required for compressor to have high work and is measured by this ration. This can actually be fairly large approaching 1.0. If the compressor is too inefficient, the Brayton Cycle will not work. Only when efficient air compressors were developed was the jet engine feasible.

Example 14.1   Air enters the compressor of a gas turbine at 100 kPa and 25 °C. For a pressure ratio of 5 and a maximum temperature of 850 °C determine the back work ratio and the thermal efficiency using the Brayton cycle.

Solution   To find the back work ration we can see that;

$$\frac{W_{comp}}{W_{turb}} = \frac{C_p(T_2 - T_1)}{C_p(T_3 - T_4)} = \frac{T_2 - T_1}{T_3 - T_4}$$

The temperatures are $T_1 = 273 + 25 = 298$ K, $T_2 = 273 + 850 = 1123$ K, and

$$T_2 = T_1 \left(\frac{P_2}{P_1}\right)^{(\gamma-1)/\gamma} = (298)(5)^{0.2857} = 472.0K$$

and

$$T_4 = T_3 \left(\frac{P_4}{P_5}\right)^{(\gamma-1)/\gamma} = (1123)\left(\frac{1}{5}\right)^{0.2857} = 709.1K$$

The back work ratio is then given by;

$$\frac{W_{comp}}{W_{turb}} = \frac{472.0 - 298}{1123 - 709} = 0.420 \text{ or } 42.0\%$$

The thermal efficiency is

$$\eta_{th} = 1 - r^{(1-\gamma)/\gamma} = 1 - (5)^{-0.2857} = 0.369 \text{ or } 36.9\%$$

Example 14.2   In an air-standard Brayton cycle with a pressure ratio of 8, the pressure and temperature at the start of the compression are 100 kPa and 300 K. The maximum allowed temperature in the cycle is 1200 K. Determine the energy added per kg air, the work done per kg air and the thermal efficiency of the cycle. Use figure below (Fig. 14.9).

Solution   The air-standard Brayton cycle is shown on a P-V diagram in Figure above. The pressure ratio r_p of the Brayton cycle is given by;

Fig. 14.9 Sketch of Example
14.2

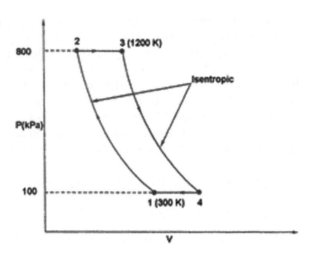

$$r_p = \frac{P_2}{P_1} = 8$$

In addition to pressure ratio, the other data given is $P_1 = 100\,\text{kPa}$, $T_1 = 300\,\text{K}$, $T_3 = 1200\,\text{K}$.
The temperature T_2 of air after isentropic compression process $1 \rightarrow 2$ is given by

$$T_2 = T_1 \left(\frac{P_2}{P_1} \right)^{\frac{\gamma-1}{\gamma}} = 300(8)^{\frac{0.4}{1.4}} = 543.4\text{K}$$

The energy

$$q_1 = C_p(T_3 - T_2) = 1.0047(1200 - 543.4) = 659.69\text{kJ}/\text{kg}$$

The thermal efficiency η_{th} of the Brayton cycle is given by;

$$\eta_{th} = 1 - \left(\frac{1}{r_p} \right)^{(1-\gamma)/\gamma} = 1 - \left(\frac{1}{8} \right)^{\frac{0.4}{1.4}} = 0.448$$

The network done W_{net} per kg air is given by;

$$W_{net} = q_1\eta_{th} = (6.59.69) \times (0.448) = 295.54\text{kJ/kg}$$

14.3   The Non-Ideal Brayton Cycle

The Ideal Air Standard Brayton Cycle assumes isentropic compression and expansion processes. So far this has not been achieved in any real device. The isentropic efficiency for these processes is defined as

$$Isentropic\ efficiency\ (compression) = \frac{\Delta h_{isentropic}}{\Delta h_{actual}}$$

$$Isentropic\ efficiency\ (expansion) = \frac{\Delta h_{actual}}{\Delta h_{isentropic}}$$

Unfortunately the isentropic efficiency of a compressor or turbine will depend on the pressure ratio for the device. In doing parametric or design studies it is more useful to define an efficiency that does not depend on the pressure ratio, but only on the manufacturing tolerances and efficiencies of individual stages. This small stage, or infinitesimal stage, efficiency is called the polytropic efficiency.

Consider the combined First and Second Law for an infinitesimal process.

$$dh = vdp + Tds \tag{14.8}$$

The term Tds represents a heat flow for the process. During a compression, the inefficiency of the process represents a heat flow into the system. For an expansion the inefficiency represents a heat flow out of the system. So on an infinitesimal basis we can write,

$$dh = vdp + (Tds) = \frac{vdp}{e_{c,poly}} \qquad for\ a\ compressor$$
$$dh = vdp + (-Tds) = e_{t,poly} * vdp \quad for\ a\ turbine \tag{14.9}$$

then these two equations can be integrated similar to the way the isentropic relation was integrated. For an isentropic expansion of a calorically perfect ideal gas we have,

$$vdp = \frac{RT}{p}dp$$

$$dh = c_p dT = \frac{RT}{p}dp$$

$$\frac{dT}{T} = \frac{R}{C_p}\frac{dp}{p} = \frac{\gamma-1}{\gamma}\frac{dp}{p}$$

$$\left(\frac{T_2}{T_1}\right) = \left(\frac{p_2}{p_1}\right)^{\frac{\gamma-1}{\gamma}}$$

For a polytropic compression we have,

$$vdp = \frac{RT}{e_{c,poly}\,p}dp$$

$$dh = c_p dT = \frac{RT}{e_{c,poly}\,p}dp$$

$$\frac{dT}{T} = \frac{R}{C_p}\frac{dp}{p} = \frac{\gamma-1}{e_{c,poly}\gamma}\frac{dp}{p}$$

$$\left(\frac{T_2}{T_1}\right) = \left(\frac{p_2}{p_1}\right)^{\frac{\gamma-1}{e_{c,poly}\gamma}}$$

And for a polytropic expansion we have,

$$vdp = \frac{e_{t,poly}RT}{p}dp$$

$$dh = c_p dT = \frac{e_{t,poly}RT}{p}dp$$

$$\frac{dT}{T} = \frac{R}{C_p}\frac{dp}{p} = \frac{e_{t,poly}(\gamma-1)}{\gamma}\frac{dp}{p}$$

$$\left(\frac{T_2}{T_1}\right) = \left(\frac{p_2}{p_1}\right)^{\frac{e_{t,poly}(\gamma-1)}{\gamma}}$$

Now for a calorically perfect gas, the isentropic efficiency of a compressor is given by,

$$\eta_{c,isen} = \frac{C_p(T_{out,isen}-T_{in})}{C_p(T_{out,actual}-T_{in})} = \frac{\dfrac{T_{out,isen}}{T_{in}}-1}{\dfrac{T_{out,actual}}{T_{in}}-1} = \frac{\left(\dfrac{p_{out}}{p_{in}}\right)^{\frac{\gamma-1}{\gamma}}-1}{\left(\dfrac{p_{out}}{p_{in}}\right)^{\frac{\gamma-1}{\gamma e_{c,poly}}}}$$

And the isentropic efficiency of a turbine is given by,

$$\eta_{t,isen} = \frac{C_p(T_{out,actual}-T_{in})}{C_p(T_{out,isen}-T_{in})} = \frac{\dfrac{T_{out,actual}}{T_{in}}-1}{\dfrac{T_{out,isen}}{T_{in}}-1} = \frac{\left(\dfrac{p_{out}}{p_{in}}\right)^{\frac{e_{t,poly}(\gamma-1)}{\gamma}}-1}{\left(\dfrac{p_{out}}{p_{in}}\right)^{\frac{\gamma-1}{\gamma e_{c,poly}}}}$$

Fig. 14.10 Gas turbine cycle on a *T-s* diagram for Example 14.3

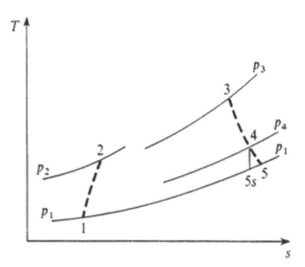

Example 14.3 A gas turbine operates in ambient conditions of 14.677 psi, 17 °C, and the maximum cycle temperature is limited to 1000 K. The compressor, which has a polytropic efficiency of 88 %, is driven by the gas generator turbine, and a separate power turbine is geared to the power output on a separate shaft; both turbines have polytropic efficiencies of 90 %. There is a pressure loss of 2.9 psi bar between the compressor and the gas generator turbine inlet. Neglecting all other losses, and assuming negligible kinetic energy changes, calculates:

a. The compressor pressure ratio which will give maximum specific power output;

b. The isentropic efficiency of the power turbine.

For the gases in both turbines, take $c_p = 1.15$ kJ/kg K and $\gamma = 1.33$.
For air take $c_p = 1.005$ kJ/kg K and $\gamma = 1.4$.

Solution

a. The cycle is shown on a *T-s* diagram in Fig. 14.10.

Let $\dfrac{p_2}{p_1} = r$. From Eq. 14.17, we can write;

$$T_2 = T_1 r^{(\gamma-1)/\gamma\eta_{\infty c}} = (17+273) \times r^{0.4/1.4 \times 0.88} = 290 r^{0.325}$$

Now

$$p_3 = p_2 - 0.2 = (p_1 \times r) - 0.2 = 1.012 \times r - 0.2$$

$$p_5 = p_1 = 14.677$$

$$\text{or} \frac{p_3}{p_5} = \frac{14.677r - 2.9}{14.677} = r - 0.198$$

Since the polytropic efficiency of both turbines is the same then, using Eq. 14.14, provides;

$$\frac{T_3}{T_5} = \frac{(p_3)^{(\gamma-1)\eta_{\infty e}/\gamma}}{(p_1)} = (r - 0.198)^{0.9/4}$$

$$T_5 = \frac{1000}{(r - 0.198)^{0.225}}$$

Turbine specific power output $= c_p(T_3 - T_5)$

$$= 1.15\left(1000 - \frac{1000}{(r - 0.198)^{0.225}}\right)$$

$$= 1150[1 - (r - 0.198)^{-0.225}]$$

Compressor specific power output $= c_p(T_2 - T_1)$

$$= 1.005(290r^{0.325} - 290)$$

$$= 291.5(r^{0.325} - 1)$$

Net specific power output

$$\dot{W} = 1150[1 - (r - 0.198)^{-0.225}] - 291.5(r^{0.325} - 1)$$

To find the maximum of the above relation, we set $d\dot{W}/dr = 0$ so we get

$$0.225 \times 1150 \times (r - 0.198)^{-1.225} = 0.325 \times 291.5 \times r^{-0.675}$$

Trial and error, or graphical, solution gives $r = 6.65$

The compressor pressure ratio for maximum specific power output $= 6.65$

b. $T_2 = 290r^{0.325} = 290(6.65)^{0.325} = 536.8\text{K}$

Now we can write the following
Gas turbine power output = Compressor power input

Therefore

$$1.15(1000 - T_4) = 1.005(536.8 - 290)$$

i.e. $T_4 = 784.3\text{K}$

Then

$$\frac{p_3}{p_4} = \frac{(T_3)^{\gamma/(\gamma-1)\eta_{\infty e}}}{(T_4)} = \frac{(1000)^{4/0.9}}{(7843.3)} = 2.944$$

Also

$$p_3 = 6.65 p_1 - 0.290 = (6.65 \times 1.012) - 0.290 = 4.73 \text{bar}$$

Therefore

$$p_4 = \frac{4.73}{2.944} = 1.607 \text{bar}$$

Then

$$\frac{T_4}{T_5} = \frac{(p_4)^{(\gamma-1)/\eta_{\infty e} q}{\gamma}}{(p_5)} = \frac{(1.607)^{0.9/4}}{(14.677)} = 1.110$$

$$\frac{T_4}{T_{5s}} = \frac{(p_4)^{(\gamma-1)/\gamma}}{(p_5)} = \frac{(1.607)^{0.25}}{(14.677)} = 1.123$$

Using the relationship for turbine isentropic efficiency, η_T as below;

Power turbine isentropic efficiency $\eta_{PT} = \dfrac{c_p(T_4 - T_5)}{c_p(T_4 - T_{5s})} = \dfrac{(T_4 - T_5)}{(T_4 - T_{5s})}$

Then we can write

$$\eta_{PT} = \frac{(T_4 - T_5)}{(T_4 - T_{5s})} = \frac{1 - (T_5/T_4)}{1 - (T_{5s}/T_4)} = \frac{1 - (1/1.11)}{1 - (1/1.123)} = 0.905 \text{ or } 90.5\%$$

14.4   The Air Standard Cycle

It is important to note that air standard cycle apply to the performance of an internal combustion engine because once the fuel ignites; it releases its energy as heat. If the process of combustion is ignored and the heat released is considered as heat applied during the appropriate portion of an air standard cycle, the heat conversion process in the internal combustion engine can be examined with standard thermodynamic methods.

It is equally important to remember, however, that the air standard cycle is not an internal combustion engine, so one must be careful not to carry the analogy too far. Some individuals attempt to apply limitations and requirements for closed cycles to processes that are not closed. This can easily lead to an incorrect analysis of the open process because an open process by definition can gain or lose heat in the system by means that a closed cycle cannot.

To further analyze an *air cycle* in studying internal combustion engine performance characteristics through use of air standard cycles involves making a number of simplifying assumptions. It involves simulating engine operation with the help of thermodynamics to formulate mathematical expressions which can then be solved in order to obtain the relevant information.

The method of solution will depend upon the complexity of the formulation of the mathematical expressions, which in turn will depend upon the assumptions that have been introduced in order to analyze the processes in the engine. The more the assumptions, the simpler will be the mathematical expressions and the easier the calculations, but the lesser will be the accuracy of the final results.

Any device that operated in a thermodynamic cycle, absorbs thermal energy from a source, rejects a part of it to a sink and presents the difference between the energy absorbed and energy rejected as work to the surroundings is called a heat engine.

A heat engine is, thus, a device that produces work. In order to achieve this purpose, the air cycle heat engine-working medium undergoes the following processes:

1. A compression process where the working medium absorbs energy as work
2. A heat addition process where the working medium absorbs energy as heat from a source.
3. An expansion process where the working medium transfers energy as work to the surroundings.
4. A heat rejection process where the working medium rejects energy as heat to a sink.

If the working medium does not undergo any change of phase during its passage through the cycle, the heat engine is said to operate in a non-phase change cycle. A phase change cycle is one in which the working medium undergoes changes of phase. Air standard cycles, using air as the working medium are examples of non-phase change cycles while the steam and vapor compression refrigeration cycles are examples of phase change cycles.

We will consider several engines that operate in a cycle using primarily air as a working fluid. To do this we will make several approximations:

1. Air is the working fluid throughout the entire cycle. The mass of the small amount of injected fuel is negligible.
2. There is no inlet or exhaust process.
3. Combustion is replaced by a heat transfer process.
4. Heat transfer to the surroundings replaces the exhaust process used to restore air to its original state. Since we will start by considering open cycles, the return to initial conditions relies on the atmosphere to complete this part of the cycle. When we analyze a nuclear Brayton system, we will close the loop and completely contain the main loop working fluid. We will use a heat exchanger to bring the working fluid back to the compressor inlet conditions.

5. All processes will be assumed to be quasi equilibrium.
6. Air will be assumed to be a calorically perfect gas, so that we can use constant specific heats. For nuclear systems that use helium or a combination of noble gases this is not an approximation.

When we talk about reciprocating engines, one of the parameters that will be of most interest is the compression ratio. It is the ratio of the volume at bottom dead center of the piston movement divided by the volume at top dead center of the piston movement.

$$r = \frac{V_{BDC}}{V_{TDC}} \tag{14.10}$$

Another characteristic of reciprocating engines is the Mean Effective Pressure. It is defined in terms of the work output of a cycle and the difference in volumes for the piston.

$$W_{cycle} = MEP(V_{BDC} - V_{TDC}) \tag{14.11}$$

Since the air standard analysis is the simplest and most idealistic way of modeling an internal combustion engine, such cycles are also called ideal cycles and the engine running on such cycles are called ideal engines.

In order that the analysis such as above to be made as simple as possible, certain assumptions have to be made, which are listed below. These assumptions result in an analysis that is far from correct for most actual combustion engine processes, but the analysis is of considerable value for indicating the upper limit of performance. The analysis is also a simple means for indicating the relative effects of principal variables of the cycle and the relative size of the apparatus.

Assumptions:

1. The working medium is a perfect gas with constant specific heats and molecular weight corresponding to values at room temperature.
2. No chemical reactions occur during the cycle. The heat addition and heat rejection processes are merely heat transfer processes.
3. The processes are reversible.
4. Losses by heat transfer from the apparatus to the atmosphere are assumed to be zero in this analysis.
5. The working medium at the end of the process (cycle) is unchanged and is at the same condition as at the beginning of the process (cycle).

When selecting an idealized process one is always faced with the fact that the simpler the assumptions, the easier the analysis, but the farther the result from reality. The air cycle has the advantage of being based on a few simple assumptions and of lending itself to rapid and easy mathematical handling without recourse to thermodynamic charts or tables or complicated calculations. On the other hand, there is

Fig. 14.11 Sketch of
Example 14.4

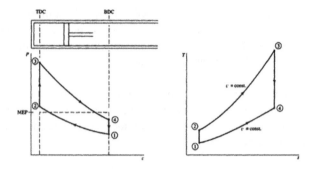

always the danger of losing sight of its limitations and of trying to employ it beyond its real usefulness.

Example 14.4 An engine operates with air on the cycle in figure below with isentropic $1 \rightarrow 2$ and $3 \rightarrow 4$. If the compression ratio is 12, the minimum pressure is 200 kPa, and the maximum pressure is 10 MPa determine (a) the percent clearance and (b) the MEP (Fig. 14.11).

Solution

a. The percent clearance is given by;

$$c = \frac{V_2}{V_1 - V_2}(100)$$

But the compression ratio is $r = V_1 / V_2$. Thus we have,

$$c = \frac{V_2}{12 V_2 - V_2}(100) = \frac{100}{11} = 9.09\%$$

b. To determine the MEP we must calculate the area under the $P\text{-}V$ diagram. This is equivalent to calculating the work. The work from $3 \rightarrow 4$ is, using $PV^{\gamma} = C$;

$$W_{3 \rightarrow 4} = \int P dV = C \int \frac{dV}{V^{\gamma}} = \frac{C}{1 - \gamma}\left(V_4^{1-\gamma} - V_3^{1-\gamma}\right) = \frac{P_4 V_4 - P_3 V_3}{1 - \gamma}$$

where $C = P_4 V_4^{\gamma} = P_3 V_3^{\gamma}$. But we know that $V_4 / V_3 = 12$, so

$$W_{3 \rightarrow 4} = \frac{V_3}{1 - \gamma}(12 P_4 - P_3)$$

Likewise, the work from $1 \rightarrow 2$ is

$$W_{1 \rightarrow 2} = \frac{V_2}{1 - \gamma}(P_2 - 12 P_1)$$

Since no work occurs in the two constant-volume processes, we find, using $V_2 = V_3$

$$W_{cycle} = \frac{V_2}{1-\gamma}(12P_4 - P_3 + P_2 - 12P_1)$$

The pressures P_2 and P_4 are found as follows;

$$P_2 = P_1\left(\frac{V_1}{V_2}\right)^{\gamma} = (200)(12)^{1.4} = 1665\text{kPa}$$

and

$$P_4 = P_3\left(\frac{V_3}{V_4}\right)^{\gamma} = (1000)\left(\frac{1}{12}\right)^{1.4} = 308$$

hence

$$W_{cycle} = \frac{V_2}{-0.4}[(12)(308) - 1000 + 1665 - (12)(200)] = 20070V_2$$

But $W_{cycle} = (\text{MEP})(V_1 - V_2) = (\text{MEP})(12V_1 - V_2)$; equating the two expressions yields

$$\text{MEP} = \frac{20070}{11} = 1824\text{kPa}$$

14.5   Equivalent Air Cycle

A particular air cycle is usually taken to represent an approximation of some real set of processes which the user has in mind. Generally speaking, the air cycle representing a given real cycle is called *an Equivalent Air Cycle*. The equivalent cycle has, in general, the following characteristics in common with the real cycle, which it approximates:

1. A similar sequence of processes.
2. Same ratio of maximum to minimum volume for reciprocating engines or maximum to minimum pressure for gas turbine engines.
3. The same pressure and temperature at a given reference point.
4. An appropriate value of heat addition per unit mass of air.

Now under above circumstances that that both Air Standard Cycles and Equivalent Air Cycle are mentioned and taken under consideration, we now can study and describe other known gas power engine cycles, which are subject of next few sections.

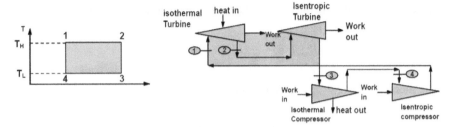

Fig. 14.12 Steady flow Carnot engine

Fig. 14.13 Reciprocating
Carnot engine

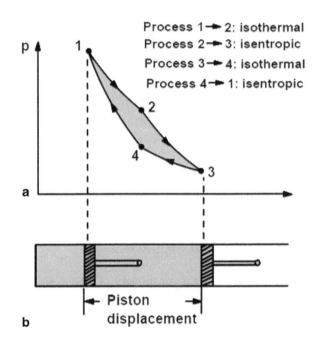

14.6   Carnot Cycle

A Carnot gas cycle operating in a given temperature range is shown in the *T-s* diagram in Fig. 14.12a. One way to carry out the processes of this cycle is through the
use of steady-state, steady-flow devices as shown in Fig. 14.12b. The isentropic
expansion process 2-3 and the isentropic compression process 4-1 can be simulated
quite well by a well-designed turbine and compressor respectively, but the isothermal expansion process 1-2 and the isothermal compression process 3-4 are most
difficult to achieve. Because of these difficulties, a steady-flow Carnot gas cycle is
not practical.

The Carnot gas cycle could also be achieved in a cylinder-piston apparatus (a
reciprocating engine) as shown in Fig. 14.13b. The Carnot cycle on the p-v diagram

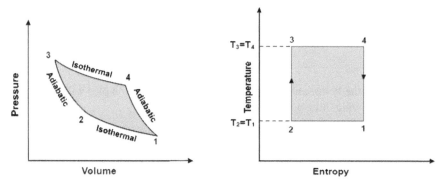

Fig. 14.14 Carnot cycle on *P-V* and *T-s* diagrams

is as shown in Fig. 14.13a, in which processes 1-2 and 3-4 are isothermal while processes 2-3 and 4-1 are isentropic. We know that the Carnot cycle efficiency is given by the expression.

$$\eta_{th} = 1 - \frac{T_L}{T_H} = 1 - \frac{T_4}{T_1} = 1 - \frac{T_3}{T_2} \tag{14.12}$$

The cycle associated with the Carnot engine is shown in Fig. 14.13 in above, using an ideal gas as the working substance. It is composed of the following four reversible processes [1]:

Process $1 \rightarrow 2$ *An Isothermal Expansion*. Heat is transferred reversibly from the high-temperature reservoir at the constant temperature T_H . The piston in the cylinder is withdrawn and the volume increases

Process $2 \rightarrow 3$ *An Isentropic (Adiabatic Reversible Expansion)*. The cylinder is completely insulated so that no heat transfer occurs during this reversible process. The piston continues to be withdrawn, with the volume increasing

Process $3 \rightarrow 4$ An isothermal compression. Heat is transferred reversibly to the low-temperature reservoir at the constant temperature T_L . The piston compresses the working substance, with the volume decreasing

Process $4 \rightarrow 1$ *An adiabatic reversible compression*. The completely insulated cylinder allows no heat transfer during this reversible process. The piston continues to compress the working substance until the original volume, temperature, and pressures are reached, thereby completing the cycle

Applying the first law of thermodynamic to the Carnot cycle presented in Fig. 14.14, we see that;

$$W_{net} = Q_H - Q_L \tag{14.13}$$

where Q_L is assumed to be a positive value for the heat transfer to the low-temperature reservoir. This allows us to write the thermal efficiency for the Carnot cycle as;

$$\eta_{th} = \frac{Q_H - Q_L}{Q_H} = 1 - \frac{Q_L}{Q_H} \tag{14.14}$$

Under the above condition Eq. 14.10 sometimes can be expressed as the following form;

$$\eta_{Carnot} = 1 - \frac{T_L}{T_H} \tag{14.15}$$

Also based on the Potter and Somerton [1], the following examples will be used to prove the following three postulates:

1. It is impossible to construct an engine, operating between two given temperature reservoirs, that is more efficient than the Carnot engine.
2. The efficiency of a Carnot engine is not dependent on the working substance used or any particular design feature of the engine.
3. All reversible engines, operation between two given temperature reservoir, have the same efficiency as a Carnot engine operating between the same two temperature reservoirs.

Since the working fluid is an ideal gas with constant specific heats, we have, for the isentropic process; we have the following relationship as;

$$\frac{T_1}{T_4} = \left(\frac{V_4}{V_1}\right)^{\gamma-1} \quad \text{and} \quad \frac{T_2}{T_3} = \left(\frac{V_3}{V_2}\right)^{\gamma-1} \tag{14.16}$$

Now, $T_1 = T_2$ and $T_4 = T_3$, therefore we can conclude that;

$$\frac{V_4}{V_1} = \frac{V_3}{V_2} = r_v = \text{Compression or expansion volume ratio} \tag{14.17}$$

Then, Carnot cycle efficiency using the ratio in above, may be written as;

$$\eta_{th} = 1 - \frac{1}{r_v^{\gamma-1}} \tag{14.18}$$

From the Eq. 14.18 in above, it can be observed that the Carnot cycle efficiency increases as r increases. This implies that the high thermal efficiency of a Carnot cycle is obtained at the expense of large piston displacement. In addition, for isentropic processes we have,

$$\frac{T_1}{T_4} = \left(\frac{P_1}{P_4}\right)^{\frac{\gamma-1}{\gamma}} \quad \text{and} \quad \frac{T_2}{T_3} = \left(\frac{P_2}{P_3}\right)^{\frac{\gamma-1}{\gamma}} \tag{14.19}$$

Since, $T_1 = T_2$ and $T_4 = T_3$, we have

$$\frac{P_1}{P_4} = \frac{P_2}{P_3} = r_p = \text{Pressure ratio} \tag{14.20}$$

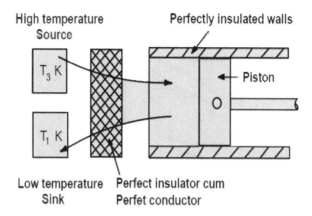

Fig. 14.15 Working of Carnot engine

Therefore, Carnot cycle efficiency may be written as;

$$\eta_{th} = 1 - \frac{1}{r_p^{\frac{\gamma-1}{\gamma}}}$$ (14.21)

From the Eq. 14.21 in above, it can be observed that, the Carnot cycle efficiency can be increased by increasing the pressure ratio. This means that Carnot cycle should be operated at high peak pressure to obtain large efficiency. See Fig. 14.15.

Example 14.5 A Carnot engine delivers 10 kW of power while absorbing energy as heat from a source at 1000 °C. It rejects energy as heat to a sink at 27 °C. Determine the energy absorbed and the energy rejected per second by the engine.

Solution We know that the thermal efficiency η_{th} of a Carnot engine is given by;

$$\eta_{th} = 1 - \frac{T_2}{T_1} = \frac{\dot{W}}{\dot{Q}_1} = 1 - \frac{(273+27)^\circ \text{K}}{(273+1000)^\circ \text{K}} = 0.7643$$

$$= 0.7643 = \frac{\dot{W}}{\dot{Q}_1} = \frac{10 \times 10^3}{\dot{Q}_1}$$

$$\dot{Q}_1 = 13.084 \text{kW}$$

But $\dot{W} = \dot{Q}_1 - \dot{Q}_2$ or $\dot{Q}_2 = \dot{Q}_1 - \dot{W} = 13,084 - 10 = 3.084 \text{kW}$.

Therefore the energy absorbed per second by the engine is equal to 13.084 kJ/s. The energy rejected per second by the engine is equal to 3.084 kJ/s.

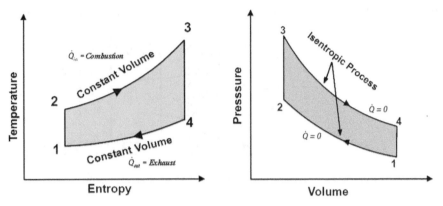

Fig. 14.16 Otto cycle on *P-V* and *T-s* diagrams

14.7   Otto Cycle

The air-standard-Otto cycle is the idealized cycle for the spark-ignition internal combustion engines. This cycle is shown above on *P-V* and *T-s* diagrams. The *spark-ignition engine* is modeled with this Otto cycle.

The Otto cycle $1 \rightarrow 2 \rightarrow 3 \rightarrow 4$ consists of following four processes:

Process $1 \rightarrow 2$ Reversible adiabatic compression of air

Process $2 \rightarrow 3$ Heat addition at constant volume

Process $3 \rightarrow 4$ Reversible adiabatic expansion of air

Process $4 \rightarrow 1$ Heat rejection at constant volume

Since processes $1 \rightarrow 2$ and $3 \rightarrow 4$ are adiabatic processes, the heat transfer during the cycle takes place only during processes $2 \rightarrow 3$ and $4 \rightarrow 1$ respectively. Mathematically speaking the processes for the Otto cycle are described on the two charts that are shown in Fig. 14.16. Therefore, thermal efficiency η_{th} can be written as;

$$\eta_{th} = \frac{\text{Net Workdone}}{\text{Net Heat Added}} = \frac{\dot{W}_{net}}{\dot{Q}_{in}} \qquad (14.22)$$

Note that the two heat transfer processes is taking place during the constant-volume processes, for which the work is zero and this results that Eq. 14.22 to take the following form of;

$$\eta_{th} = \frac{\text{Heat Added} - \text{Heat Rejected}}{\text{Heat Added}} = \frac{\dot{Q}_{in} - \dot{Q}_{out}}{\dot{Q}_{in}} 1 - \frac{\dot{Q}_{out}}{\dot{Q}_{in}} \qquad (14.23)$$

Consider '*m*' kg of working fluid, and assume each quantity in Eq. 14.23 to be positive process, then we have;

$$\begin{cases} \text{Heat Added} = \dot{Q}_{in} = mC_v(T_3 - T_2) \\ \text{Heat Rejected} = \dot{Q}_{out} = mC_v(T_4 - T_1) \end{cases} \quad (14.24)$$

Substitution of the set of Eq. 14.24 into 14.23, we can write;

$$\eta_{th} = 1 - \frac{\dot{Q}_{out}}{\dot{Q}_{in}} = 1 - \frac{mC_v(T_3 - T_2)}{mC_v(T_4 - T_1)} = 1 - \frac{T_3 - T_2}{T_4 - T_1} \quad (14.25)$$

Equation 14.25 can be rearranged and be presented as follow;

$$\eta_{th} = 1 - \frac{T_1}{T_2}\left[\frac{T_4/T_1 - 1}{T_3/T_2 - 1}\right] \quad (14.26)$$

For an isentropic (reversible adiabatic) processes $3 \to 4$ and $1 \to 2$, we can write the following form for Eq. 14.26;

$$\frac{T_2}{T_1} = \left(\frac{V_1}{V_2}\right)^{\gamma-1} \text{ and } \frac{T_3}{T_4} = \left(\frac{V_4}{V_3}\right)^{\gamma-1} \quad (14.27)$$

But using the fact that $V_1 = V_4$ and $V_3 = V_2$, we see that

$$\frac{T_2}{T_1} = \frac{T_3}{T_4} \text{ or } \frac{T_1}{T_2} = \frac{T_4}{T_3} \quad (14.28)$$

Utilizing the relationships of Eq. 14.28 allows writing the Eq. 14.24 to obtain the final form of thermal efficiency η_{th} as;

$$\eta_{th} = 1 - \frac{T_1}{T_2} = 1 - \left(\frac{V_1}{V_2}\right)^{\gamma-1} \quad (14.29)$$

Defining the parameter $r = V_1/V_2$ that is called as *Compression Ratio* (CR), then Eq. 14.28 can be simplifies as;

$$\eta_{th} = 1 - \frac{1}{r^{\gamma-1}} \quad (14.30)$$

From the Eq. 14.15, it can be observed that the efficiency of the Otto cycle is mainly the function of compression ratio for the given ratio of C_p and C_v. If we plot the variations of the thermal efficiency with increase in compression ratio for different gases, the curves are obtained as shown in Fig. 14.17. Beyond certain values of compression ratios, the increase in the thermal efficiency is very small, because the curve tends to be asymptotic. However, practically the compression ratio of petrol

Fig. 14.17 Effect of CR and
γ on efficiency for Otto
cycle

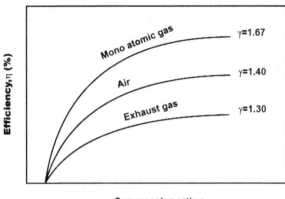

engines is restricted to maximum of nine or ten due to the phenomenon of knocking
at high compression ratios. Same equation also indicates that, the higher the compression ratio, the higher the thermal efficiency.

Example 14.6 A spark-ignition engine is proposed to have a compression ratio of
10 while operating with a low temperature of 200 °C and a low pressure of 200 kPa.
If the work output is to be 1000 kJ/kg, calculate the maximum possible thermal
efficiency and compare with that of a Carnot cycle. Also calculate the MEP (See
next Section for MEP).

Solution The Otto cycle provides the model for this engine. The maximum possible thermal efficiency for the engine would be;

$$\eta_{th} - 1 - \left(\frac{1}{r}\right)^{\gamma-1} = 1 - \left(\frac{1}{10}\right)^{0.4} = 0.602 \text{ or } 60.2\%$$

Since processes $1 \rightarrow 2$ is isentropic, we find that

$$T_2 = T_1 \left(\frac{V_1}{V_2}\right)^{\gamma-1} = (473)(8)^{0.4} = 1188 \text{K}$$

The net work for the cycle is given by;

$$W_{net} = W_{1\rightarrow2} + W_{2\rightarrow3}^0 + W_{3\rightarrow4} + W_{4\rightarrow1}^0 = C_v(T_1 - T_2) + C_v(T_1 - T_2)$$

or

$$1000 = (0.717)(473 - 1188 + T_3 - T_4)$$

But, for the isentropic process $3 \rightarrow 4$;

$$T_3 = T_4 \left(\frac{V_4}{V_3}\right)^{\gamma-1} = (T_4)(10)^{0.4} = 2.512 T_4$$

Solving the last two equations simultaneously (a practice left for the readers), we find $T_3 = 3508\text{K}$ and $T_4 = 1397\text{K}$, so that (applying Eq. 14.15);

$$\eta_{carnot} = 1 - \frac{T_L}{T_H} = 1 - \frac{473}{3508} = 0.865 \text{ or } 86.5\%$$

The Otto cycle efficiency is less than that of a Carnot cycle operating between the limiting temperatures because the heat transfer processes in the Otto cycle are not isothermal.

$$W_{net} = (\text{MEP})(V_1 - V_2)$$

We have

$$V_1 = \frac{RT_1}{P_1} = \frac{(0.287)(473)}{200} = 0.6788 \text{ m}^3/\text{kg} \quad \text{or} \quad V_2 = \frac{V_1}{10}$$

Thus;

$$\text{MEP} = \frac{W_{net}}{V_1 - V_2} = \frac{1000}{(0.9)(0.6788)} = 1640\text{kPa}$$

14.7.1   Mean Effective Pressure (Otto Cycle)

Generally, it is defined as the ratio of the net work done to the displacement volume of the piston.
  Let us consider 'm m' kg of working substance.

$$\text{Net Work Done} = mC_v\{(T_3 - T_2) - (T_4 - T_1)\} \tag{14.31}$$

$$\text{Displacement Volume} = (V_1 - V_2)$$

$$= V_1\left(1 - \frac{1}{r}\right) = \frac{mRT_1}{P_1}\left(\frac{r-1}{r}\right) \tag{14.32}$$

$$= \frac{mC_v(\gamma-1)T_1}{P_1}\left\{\frac{r-1}{r}\right\}$$

Knowing that by now R is Universal Ideal Gas Constant and $\gamma = C_p/C_v$ is the Specific Heat Index, their relation between each other is defined as;

$$R = C_p - C_v = C_v(\gamma-1) \tag{14.33}$$

Substituting for Eq. 14.33 into 14.32 will result in an equation and relation for *Mean Efficiency Pressure* (**MEP**) defined as;

$$MEP = \frac{m C_v[(T_3 - T_2) - (T_4 - T_1)]}{\frac{m C_v(\gamma - 1)}{P_1}\left\{\left(\frac{r-1}{r}\right)\right\}}$$ (14.34)

$$= \left(\frac{1}{\gamma - 1}\right)\left(\frac{P_1}{T_1}\right)\left(\frac{r}{r-1}\right)\{(T_3 - T_2) - (T_4 - T_1)\}$$

Now,

$$T_2 = T_1(r)^{\gamma - 1}$$ (14.35)

Let,

$$r_p = \frac{P_3}{P_2} = \frac{T_3}{T_2} = \text{Pressure Ratio}$$ (14.36)

Or,

$$T_3 = \frac{P_3}{P_2}T_2 = r_p T_2 = r_p r^{\gamma - 1}T_1 \text{ for } V = C$$ (14.37)

So,

$$T_4 = T_3\left(\frac{1}{r}\right)^{\gamma - 1} = r_p r^{\gamma - 1}T_1\left(\frac{1}{r}\right)^{\gamma - 1} = r_p T_1$$ (14.38)

Then,

$$MEP = \frac{P_1 r}{(r-1)(\gamma - 1)}\{(r_p r^{\gamma - 1} - r^{\gamma - 1}) - (r_p - 1)\}$$

$$= P_1 r\left\{\left(\frac{r^{\gamma - 1}(r_p - 1) - (r_p - 1)}{(\gamma - 1)(r - 1)}\right)\right\}$$ (14.39)

$$MEP = P_1 r\left\{\frac{(r^{\gamma - 1} - 1)(r_p - 1)}{(r - 1)(\gamma - 1)}\right\}$$ (14.40)

Example 14.7 Consider an air-standard Otto cycle with a compression ratio of 8. At the beginning of the compression stroke, the temperature is 300 K and pressure is 100 kPa. Thermal energy q is supplied at 1840 kJ/kg. Given that C_v for air is 0.7176 kJ/kg K, determine:

a. The temperature and pressure at the terminal points of all the processes.

b. The thermal efficiency.

c. The work done per kilogram air, and

d. The mean effective pressure.

Solution It is given that $T_1 = 300\text{K}$; $P_1 = 100\text{kPa}$ and $r_0 = \dfrac{V_1}{V_2} = 8$

a. Process $1 \rightarrow 2$ is isentropic. Therefore,

$$PV^\gamma = \text{constant}$$

$$T_2 = T_1 \left(\frac{V_1}{V_2}\right)^{\gamma-1} = 300(8)^{0.4} = 689.2\text{K}$$

$$P_2 = P_1 \left(\frac{V_1}{V_2}\right)^{\gamma} = 100(8)^{1.4} = 1837.9\text{kPa}$$

Process $2 \rightarrow 3$ is a constant volume process. Hence,

$$q = u_3 - u_2 = C_v(T_3 - T_2)$$

Therefore,

$$1840 = 0.7176(T_3 - 689.2) \text{or} T_3 = 3253.5\text{K}$$

For a constant volume process;

$$\frac{P_3}{T_3} = \frac{P_2}{T_2}$$

$$P_3 = \frac{T_3 P_2}{T_2} = \frac{3253.5 \times 1837.9}{689.2} = 8676.1\text{kPa}$$

$3 \rightarrow 4$ is an isentropic process. Hence

$$T_4 = T_3 \left(\frac{V_3}{V_4}\right)^{\gamma-1} = 3253.5 \left(\frac{1}{8}\right)^{0.4} = 1416.2\text{K}$$

$$P_4 = P_3 \left(\frac{V_3}{V_4}\right)^{\gamma-1} = 8676.1 \left(\frac{1}{8}\right)^{1.4} = 472.1\text{kPa}$$

b. The thermal efficiency η_{th} is given by Eq. 14.29, so we have;

$$\eta_{th} = 1 - \left(\frac{1}{r}\right)^{\gamma-1} = 1 - \left(\frac{1}{8}\right)^{0.4} = 0.565$$

c. The work done per kilogram air is;

$$\text{Workdone} = q\eta_{th} = 1840(0.565) = 1039.6$$

d. The mean effective pressure can be calculated as follow;

$$V_1 = \frac{NRT_1}{P_1} = \frac{8.314 \times 10^3 \times 300}{28.97 \times 10^5} = 0.861 \text{m}^3/\text{kg}$$

$$V_2 = {}^{V_1}/_r = 0.861/8 = 0.1076 \text{m}^3/\text{kg}$$

$$\text{MEP} = {}^{1039.6}/_{(0.861-0.1076)} = 1379.88 \text{kPa}$$

14.8 Diesel Cycle

The engines in use today, which are known as diesel engines, are far diverged from the original design by Diesel in 1892 who worked on the idea of spontaneous ignition of powered coal, which was blasted into the cylinder by compressed air. Air standard diesel cycle is an idealized cycle for diesel engines and if air is compressed to a high enough pressure ratios, typically above 14 or greater, no spark will be required to ignite the fuel. This process is as shown on P-V and T-s diagrams in Fig. 14.15 below.

The Diesel cycle $1 \rightarrow 2 \rightarrow 3 \rightarrow 4$ consists of following four processes:

Process $1 \rightarrow 2$ Reversible adiabatic compression

Process $2 \rightarrow 3$ Constant pressure heat addition

Process $3 \rightarrow 4$ Reversible adiabatic compression

Process $4 \rightarrow 1$ Constant volume heat rejection

Consider 'm' kilogram of fluid. Since the compression and expansion processes are reversible adiabatic processes, we can write,

As before the, using Fig. 14.18, we can write, the thermal efficient η_{th} as the following form;

$$\eta_{th} = \frac{\dot{W}_{net}}{\dot{Q}_{in}} = \frac{\dot{Q}_{in} - \dot{Q}_{out}}{\dot{Q}_{in}} = 1 + \frac{\dot{Q}_{out}}{\dot{Q}_{in}} \tag{14.41}$$

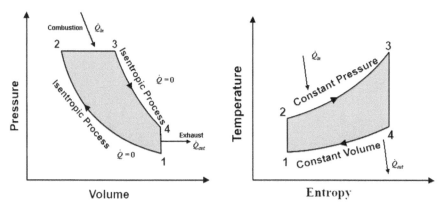

Fig. 14.18 Illustration of a diesel cycle

For constant-volume and the constant-pressure process, we have;

$$\dot{Q}_{out} = \dot{m}C_v(T_4 - T_1) \text{and} \dot{Q}_{in} = \dot{m}C_p(T_3 - T_2) \qquad (14.42)$$

Substituting for Eq. 14.41 the quantities that are define in Eq. 14.42, we have;

$$\eta_{th} = 1 - \frac{C_v(T_4 - T_1)}{Cp(T_3 - T_2)} = 1 - \frac{(T_4 - T_1)}{\gamma(T_3 - T_2)} = 1 - \frac{1}{\lambda}\left(\frac{(T_4 - T_1)}{(T_3 - T_2)}\right) \qquad (14.43)$$

Equation 14.43 can be written as;

$$\eta_{th} = 1 - \frac{T_1}{\gamma T_2}\left[\frac{T_4/_{T_1} - 1}{T_3/_{T_2} - 1}\right] = 1 - \frac{1}{r^{\gamma-1}}\frac{r-1}{\lambda(r-1)} \qquad (14.44)$$

From Eq. 14.44 we can conclude the followings;

$$T_2 = T_1 r^{\gamma-1} \text{or} r = \frac{V_1}{V_2} = \frac{V_4}{V_2} \qquad (14.45)$$

Defining the cutoff ratio as;

$$\text{cutoffratio} = r_c = \frac{T_3}{T_2} = \frac{V_3}{V_2} \qquad (14.46)$$

Then we have;

$$T_3 = r_c T_2 = r_c T_1 r^{\gamma-1} \qquad (14.47)$$

14   Gas Power Cycles

$$T_4 = T_3 \left(\frac{V_3}{V_4} \right)^{\gamma-1} = T_3 \left(\frac{V_4}{V_3} \right)^{\gamma-1}$$

$$= T_3 \left(\frac{V_4}{V_2} \frac{V_2}{V_3} \right)^{\gamma-1} = T_3 \left(\frac{r}{r_c} \right)^{1-\gamma}$$

(14.48)

$$= r_c T_1 r^{\gamma-1} \left(\frac{r}{r_c} \right)^{1-\gamma}$$

Simplified version of the above equation will be;

$$T_4 = r_c^{\gamma} T_1$$

(14.49)

Hence

$$\eta_{th} = 1 - \frac{1}{\gamma} \left\{ \frac{r_c^{\gamma} T_1 - T_1}{r_c r^{\gamma-1} T_1 - r^{\gamma-1} T_1} \right\}$$

$$= 1 - r^{1-\gamma} \left\{ \frac{r_c^{\gamma} - 1}{\gamma(r_c - 1)} \right\}$$

(14.50)

From the above equation, it is observed that, the thermal efficiency of the diesel engine can be increased by increasing the compression ratio, r, by decreasing the cut-off ratio, a_2, or by using a gas with large value of γ. Since the quantity $(r^{\gamma}-1)/\gamma(r_p-1)$ in above equation is always greater than unity, the efficiency of a Diesel cycle is always lower than that of an Otto cycle having the same compression ratio. However, practical Diesel engines uses higher compression ratios compared to petrol engines.

Example 14.8 A diesel engine has an inlet temperature and pressure of 15 °C and 1 bar respectively. The compression ratio is 12/1 and the maximum cycle temperature is 1100 °C. Calculate the air standard thermal efficiency based on the diesel cycle.

Solution Referring to Figure below, $T_1 = 15 + 273 = 288$K and $T_3 = 1100 + 273 = 1373$K. From Eq. 14.48, we that (Fig. 14.19);

$$\frac{T_2}{T_1} = \left(\frac{V_1}{V_2} \right)^{\gamma-1} = r_c^{\gamma-1} = 12^{0.4} = 2.7$$

Then we have;

$$T_2 = 2.7 \times 288 = 778 \text{K}$$

At constant pressure from 2 to 3, since $PV = RT$ for a perfect gas, then

Fig. 14.19 Illustration of
Example 14.8

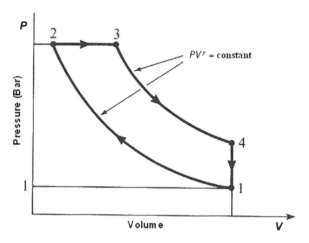

$$\frac{T_3}{T_2} = \frac{V_3}{V_2}$$

i.e.

$$\frac{V_3}{V_2} = \frac{1373}{778} = 1.765$$

Therefore

$$\frac{V_4}{V_3} = \frac{V_4 V_2}{V_2 V_4} = \frac{V_1 V_2}{V_2 V_3} = 12 \times \frac{1}{1.765} = 6.8$$

Then using Eq. 14.47 again;

$$\frac{T_3}{T_4} = \left(\frac{V_4}{V_3}\right)^{\gamma-1} = r_c^{\gamma-1} = 6.8^{0.4} = 2.153$$

i.e.

$$T_4 = \frac{1373}{2.153} = 638K$$

Then from Eq. 14.42, per kilogram of air, we can write;

$$Q_1 = C_p(T_3 - T_2) = 1.005(1373 - 778) = 598kJ/kg$$

Also from same equation, per kilogram of air, the heat rejected is

$$Q_2 = C_v(T_4 - T_1) = 0.718(638 - 288) = 251$$

Therefore from Eq. 14.42, we can write;

$$\eta_{th} = \frac{598-251}{598} = 0.58 \, \text{or} \, 58\%$$

14.8.1 Mean Effective Pressure (Diesel Cycle)

We just show the steps without any description of each step since the process is straight forward;

$$MEP = \frac{Network\,done}{Displacement\,volume}$$ (14.51)

$$= \frac{mC_p(T_3-T_2)-mC_v(T_4-T_1)}{V_1-V_2}$$

$$V_1-V_2 = V_1\left(1-\frac{V_2}{V_1}\right) = V_1\left(1-\frac{1}{r}\right)$$

$$= mRT_1\left(\frac{r-1}{r}\right)$$ (14.52)

$$= \frac{mC_v(\gamma-1)}{P_1}\left(\frac{r-1}{r}\right)$$

$$MEP = \frac{mC_p(T_3-T_2)-mC_v(T_4-T_1)}{mC_vT_1\left(\frac{\gamma-1}{P_1}\right)\left(\frac{r-1}{r}\right)}$$ (14.53)

$$= \left(\frac{P_1 r}{r-1}\right)\left(\frac{1}{\gamma-1}\right)\left\{\gamma\left(\frac{T_3-T_2}{T_1}\right)-\left(\frac{T_4-T_1}{T_1}\right)\right\}$$

$$= P_1 r\left\{\frac{\gamma r^{\gamma-1}(r_c-1)-(r_c^\gamma-1)}{(r-1)(\gamma-1)}\right\}$$

Example 14.9 A Diesel cycle operates with a compression ratio of 16 and a cut-off ratio of 2. At the beginning of the compression stoke, the air is at 1 bar and 300 K. Determine;

a. The maximum temperature and pressure of the cycle.

b. The energy added per kg air, and

c. The mean effective pressure of the (MEP) cycle.

Solution The given data for the air-standard Diesel cycle is;

$$r_0 = 16 r_c = 2 P_1 = 1 T_1 = 300 \text{K}$$

a. The temperature of air after isentropic compression, using Eq. 14.45 is given by

$$T_2 = T_1 r_0^{\gamma-1} = 300(16)^{0.4} = 909.43$$

We know that the cut-off ratio r_c is given by Eq. 14.45 as $r_c = \dfrac{V_3}{V_2} = \dfrac{T_3}{T_2}$.

Therefore,

$$T_3 = T_2 r_c = 909 \times 2 = 1819.86 \text{K}$$

That is, the maximum temperature T_3 of the cycle is 1818.06 K.

The pressure of air after isentropic compression is given by $P_2 = P_1 r_0^{\gamma} = 1(16)^{1.4} = 48.5$ bar.

The energy addition, in the Diesel cycle takes place at constant pressure. Therefore, $P_3 = P_2$. Hence, the maximum pressure P_3 of the cycle is 48.5 bar.

b. The heat energy added is given;

$$q_1 = C_p(T_3 - T_2) = 1.0047(1818.86 - 909.43) = 913.7 \text{kJ/kgair}$$

c. The mean efficient effective pressure (MEP), p_m is given by Eq. 14.52;

$$p_m = \frac{1(1.4)(16)^{1.4}(2-1) - 16(2^{1.4} - 1)}{0.4 \times 15} = 6.947 \text{bar}$$

14.9   Comparison of Otto and Diesel Cycles

The Otto and Diesel cycles can be compared on the basis of either the compression ratio or the maximum temperature and pressure of the cycle. Figure 14.20 shows the Otto and Diesel Cycles having the same compression ratio, both of them reject the same quantity of energy Q_2 to the surroundings. We know that;

$$\int dQ = \int T ds \qquad (14.54)$$

That is, the area under the curve on a T-s diagram represents the heat interaction. In Fig. 14.20, the Otto cycle is representing by 1234 and the Diesel cycle is shown by 123'4. Either cycles have the same compression ratio V_2/V_1 and reject the same

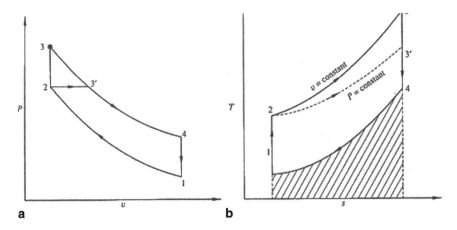

Fig. 14.20 a $P\text{-}V$ diagram for Otto $(1 \to 2 \to 3 \to 4)$ and diesel cycle $(1 \to 2 \to 3' \to 4)$. **b** $T\text{-}s$ diagram for Otto cycle $(1 \to 2 \to 3 \to 4)$ and diesel cycle $(1 \to 2 \to 3' \to 4)$

Table 14.1 Difference between actual diesel and the Otto engines

Otto engine	Diesel engine
Homogenous mixture of fuel and air formed in the carburetor is supplied to engine cylinder	No carburetor is used. Air alone is supplied to the engine cylinder. Fuel is injected directly into the engine cylinder at the end of compression stroke by means of a fuel injector. Fuel-air mixture is heterogeneous
Ignition is initiated by means of an electric spark plug	No spark plug is used. Compression ratio is high and the high temperature of air ignites fuel
Power output is controlled by varying the mass of fuel-air mixture by means of a throttle valve in the carburetor	No throttle value is used. Power output is controlled only by means of the mass of fuel injected by the fuel injector

quantity of energy Q_2 during the constant volume process $4 \to 1$ (shown by the hatched area under the curve $4 \to 1$ on the $T\text{-}s$ diagram) [4].

The Table 14.1 below summarizes these comparisons.

The efficiency of the cycle then is given by;

$$\eta = 1 - \frac{Q_2}{Q_1} \tag{14.55}$$

From this equation we see that, for the same value of Q_2 , the efficiency will be higher for higher values of Q_1 . On the $T\text{-}s$ diagram, the area under the curve $2 \to 3$ represents Q_1 for the Otto cycle and the area under the curve $2 \to 3'$ represents Q_1 , for the Diesel cycle. Since Q_1 for the Otto cycle is greater than for Diesel cycle, we find that;

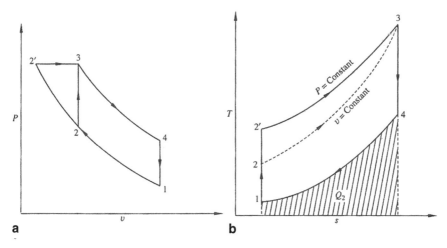

Fig. 14.21 a P-V diagram for Otto cycle $(1 \rightarrow 2 \rightarrow 3 \rightarrow 4)$ and diesel cycle $(1 \rightarrow 2' \rightarrow 3 \rightarrow 4)$
. **b** T-s diagram for Otto cycle $(1 \rightarrow 2 \rightarrow 3 \rightarrow 4)$ and diesel cycle $(1 \rightarrow 2' \rightarrow 3 \rightarrow 4)$

$$\eta_{Otto} > \eta_{Diesel} \qquad (14.56)$$

The Otto and Diesel cycles for the same maximum temperature and pressure (state 3) and for the same quantity of energy rejection Q_2 are shown in Fig. 14.21. The Otto cycle is represented by 1234 and the Diesel cycle is illustrated as 1234. From the T-s diagram it can be observed that Q_1, the area under the curve $2' \rightarrow 3$ for the Diesel cycle, is greater than Q_1, the area under the curve $2 \rightarrow 3$ for the Otto cycle when Q_2 is the same for both the cycles. Therefore;

$$\eta_{Diesel} > \eta_{Otto} \qquad (14.57)$$

14.10 Dual Cycle

With increasing crude oil prices, concern about energy security and calls for reduction in greenhouse gases, energy use in the future will be a large priority. One method for reducing oil consumption is through increased efficiency of the internal combustion engine. A significant loss in this engine is the exhaust energy which is approximately 20–40 % of the fuel potential. Research has indicated that there exists a capability to increase the brake thermal efficiency of the engine system by using this exhaust in a separate thermodynamic cycle. By simulating an engine and dual cycle system in a novel architecture, this project intends to demonstrate a proof of concept that can reduce engine emissions while increasing thermal efficiency.

Combustion in the Otto cycle is based on a constant volume process; in Diesel cycle, it is based on a constant pressure process. But combustion in actual spark-ignition engine requires a finite amount of time if the process is to be completed.

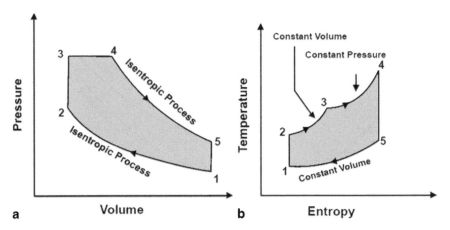

Fig. 14.22 Illustration of dual cycle on *P-V* and *T-s* diagrams

For this reason, combustion in Otto cycle does not actually occur under the constant volume condition. Similarly, in compression-ignition engines, combustion in Diesel cycle does not actually occur under the constant pressure condition, because of the rapid and uncontrolled combustion process.

The operation of the reciprocating internal combustion engines represent a compromise between the Otto and the Diesel cycle, and can be described as a Dual combustion cycle. Heat transfer to the system may be considered to occur first at constant volume and then at constant pressure. Such a cycle is called Dual cycle. Sometime another name for Dual Cycle is known as *Limited Pressure Cycle*, which is shown in Fig. 14.22 based on *P-V* and *T-s* diagrams. Here the heat addition occurs partly at constant volume and partly at constant pressure. This cycle is a closer approximation to the behavior of the actual Otto and Diesel engines because in the actual engines, the combustion process does not occur exactly at constant volume or at constant pressure but rather as in the dual cycle.

The Dual cycle is composed of the following five processes and they are as follows:

Process $1 \rightarrow 2$ Reversible adiabatic compression (isentropic compression)

Process $2 \rightarrow 3$ Constant volume heat addition

Process $3 \rightarrow 4$ Constant pressure heat addition

Process $4 \rightarrow 5$ Reversible adiabatic expansion (isentropic expansion)

Process $5 \rightarrow 1$ Constant volume heat rejection (removing)

$$\text{Heat Energy Supplied} = mC_v(T_3 - T_2) + mC_p(T_4 - T_3) \qquad (14.58a)$$

$$\text{Heat Energy Rejected} = mC_v(T_5 - T_1) \qquad (14.58b)$$

$$\text{Net Work Done} = mC_v(T_3 - T_2) + mC_p(T_4 - T_3) - mC_v(T_5 - T_1) \quad (14.59)$$

Then Thermal efficiency η_{th} is given by;

$$\eta_{th} = \frac{mC_v(T_3 - T_2) + mC_p(T_4 - T_3) - mC_v(T_5 - T_1)}{mC_v(T_3 - T_2) + mC_p(T_4 - T_3)} \quad (14.60a)$$

$$\eta_{th} = 1 - \frac{(T_5 - T_1)}{(T_3 - T_2) + \gamma(T_4 - T_3)} \quad (14.60b)$$

Let $r_p = \dfrac{P_3}{P_2}, r_c = \dfrac{V_4}{V_3}$ and $r = \dfrac{V_1}{V_2}$, then we have

$$T_2 = T_1 r^{\gamma-1} \quad (14.61a)$$

$$T_3 = T_2 r_p = T_1 r^{\gamma-1} r_p \quad (14.61b)$$

$$T_4 = T_3 r_c = T_1 r^{\gamma-1} r_p r_c \quad (14.61c)$$

$$\frac{T_5}{T_4} = \left(\frac{V_4}{V_5}\right)^{\gamma-1} = \left(\frac{V_4}{V_2} \cdot \frac{V_2}{V_5}\right)^{\gamma-1} = \left(\frac{r_c}{r}\right)^{\gamma-1} \quad (14.61d)$$

$$T_5 = T_4 \left(\frac{r_c}{r}\right)^{\gamma-1} \quad (14.61e)$$

Form all the above analysis we can write the following steps to calculate the final form of the thermal efficiency η_{th} as;

$$
\begin{aligned}
\eta_{th} &= 1 - \frac{\left(T_1 r^{\gamma} r_p r_c - T_1\right)}{\left\{\left(T_1 r^{\gamma-1} r_p - T_1 r^{\gamma-1}\right) + \gamma\left(T_1 r^{\gamma-1} r_p r_c - T_1 r^{\gamma-1} r_p\right)\right\}} \\[2mm]
&= 1 - \frac{\left(r^{\gamma} r_p r_c - 1\right)}{\left\{\left(r^{\gamma-1} r_p - r^{\gamma-1}\right) + \gamma\left(r^{\gamma-1} r_p r_c - r^{\gamma-1} r_p\right)\right\}} \quad (14.62) \\[2mm]
&= 1 - \frac{1}{r^{\gamma-1}}\left\{\frac{\left(r^{\gamma} r_p r_c - 1\right)}{\left(r_p - 1\right) \gamma r_p \left(r_c - 1\right)}\right\}
\end{aligned}
$$

From the above equation, it is observed that, a value of $r_p > 1$ results in an increased efficiency for a given value of r_c and γ. Thus the efficiency of the dual cycle lies between that of the Otto cycle and the Diesel cycle having the same compression ratio.

14.10.1   Mean Effective Pressure for Dual Cycle

We just going to show analytical process here for **MEP**;

$$\text{MEP} = \frac{\text{WorkDone}}{\text{DisplacementVolume}} \qquad (14.63a)$$

$$\text{MEP} = \frac{mC_v(T_3 - T_2) + mC_p(T_4 - T_3) - mC_v(T_5 - T_1)}{(V_1 - V_2)} \qquad (14.63b)$$

But

$$V_1 - V_2 = \frac{mC_v(\gamma - 1)T_1}{P_1}\left(\frac{r-1}{r}\right) \qquad (14.63c)$$

Then we have;

$$\begin{aligned}
\text{MEP} &= \frac{P_1 r}{(r-1)(\gamma-1)}\left\{\frac{(T_3-T_2)}{T_1} + \frac{\gamma(T_4-T_3)}{T_1} - \frac{(T_2-T_1)}{T_1}\right\} \\
&= \frac{P_1 r}{(r-1)(\gamma-1)}\left\{r^{\gamma-1}(r_p-1)+\gamma r^{\gamma-1}r_p(r_c-1)-\left(r_p r_c^{\gamma}-1\right)\right\} \quad (14.63d) \\
&= \frac{P_1 r}{(r-1)(\gamma-1)}\left\{r^{\gamma-1}(r_p-1)+\gamma r_p(r_c-1)-\left(r_p r_c^{\gamma}-1\right)\right\}
\end{aligned}$$

Example 14.10 A Dual cycle, which operates on air with a compression ratio of 16, has a low pressure of 200 kPa and a low temperature of 200 °C. If the cutoff ratio is 2 and the pressure ratio is 1.3, calculate;

a. The thermal efficiency,

b. The heat input

c. The work output, and

d. The MEP

Solution

a. Using Eq. 14.60, we have;

$$\eta_{th} = 1 - \frac{1}{(16)^{0.4}}\frac{(1.3)(2)^{1.4}-1}{(1.4)(1.3)(2-1)+1.3-1} = 0.622 \text{ or } 62.2\%$$

b. The heat input is found from Eq. 14.56, where we also use Eq. 14.59, then we have;

$$T_2 = T_1 \left(\frac{V_1}{V_2} \right)^{\gamma-1} = (473)(16)^{0.4} = 1434\text{K}$$

$$T_3 = T_2 \left(\frac{P_3}{P_2} \right) = (1434)(1.3) = 1864\text{K}$$

$$T_4 = T_3 \left(\frac{V_4}{V_3} \right) = (1864)(2) = 3728\text{K}$$

Therefore

$$q_{in} = (0.717)(1864-1434) + (1.00)(3728-1864) = 2172\text{kJ/kg}$$

c. The work output is found from Eq. 14.57 as;

$$w_{out} = \eta_{th} q_{in} = (0.622)(2172) = 3728\text{kJ/kg}$$

d. Finally to find the MEP, since we have;

$$V_1 = \frac{RT_1}{P_1} = \frac{(0.287)(473)}{200} = 0.6788\text{m}^3/\text{kg}$$

Then

$$\text{MEP} = \frac{w_{out}}{V_1(1-V_2/V_1)} = \frac{1350}{(0.6788)(15/16)} = 2120\text{kPa}$$

14.11   Stirling Cycle

The Stirling and Ericsson cycles, although not extensively used to model actual engines, are presented here and next section respectively, to illustrate the effective use of *regenerate* and that is why, sometime are called *Regenerative Cycle*, a heat exchanger which utilizes waste heat.

The Stirling and Ericsson cycles are not used to model real engines as they are difficult to achieve in practice. The advantage of both is that they can achieve efficiencies approaching the true Carnot efficiency. They do this by extracting the heat produced in the compressor and transferring it to the turbine. This is done through a device called a regenerator. But the compressor and turbine are treated as isothermal, which is very difficult to achieve in practice.

The Carnot cycle has a low mean effective pressure because of its very low work output. Hence, one of the modified forms of the cycle to produce higher mean

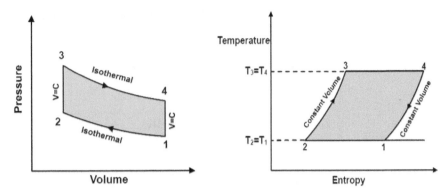

Fig. 14.23 Illustration of stirling cycle processes on *P-V* and *T-s* diagrams

effective pressure whilst theoretically achieving full Carnot cycle efficiency is the Stirling cycle. It consists of two isothermal and two constant volume processes. The heat rejection and addition take place at constant temperature. The *P-V* and *T-s* diagrams for the Stirling cycle are shown in Fig. 14.23.

The Stirling Cycle Processes are describes as follows;

a. The air is compressed isothermally from state 1 to 2 (T_L to T_H).
b. The air at state-2 is passed into the regenerator from the top at a temperature T_1. The air passing through the regenerator matrix gets heated from T_L to T_H.
c. The air at state-3 expands isothermally in the cylinder until it reaches state-4.
d. The air coming out of the engine at temperature T_H (condition 4) enters into regenerator from the bottom and gets cooled while passing through the regenerator matrix at constant volume and it comes out at a temperature T_L, at condition 1 and the cycle is repeated.
e. It can be shown that the heat absorbed by the air from the regenerator matrix during the process $2 \rightarrow 3$ is equal to the heat given by the air to the regenerator matrix during the process $4 \rightarrow 3$, and then the exchange of heat with external source will be only during the isothermal processes.

Now we can write, Net work done to be;

$$W_{net} = Q_S - Q_R \qquad (14.64)$$

Where Q_s is the heat supplied during the isothermal process $3 \rightarrow 4$ and Q_R is the Heat rejected during the isothermal compression process $1 \rightarrow 2$.

$$Q_s = P_3 V_3 \ln\left(\frac{V_4}{V_3}\right) r = \frac{V_4}{V_3} = CR \qquad (14.65)$$
$$= mRT_H \ln(r)$$

and

$$Q_R = P_1 V_1 \ln\left(\frac{V_1}{V_2}\right) r = \frac{V_1}{V_2} = CR \qquad (14.66)$$
$$= mRT_L \ln(r)$$

Then equation by substituting Eqs. 14.63 and 14.64 into Eq. 14.64, we have;

$$W_{net} = mR \ln(r)[T_H - T_L] \qquad (14.67)$$

Now, the thermal efficiency η_{th} is given by;

$$\eta_{th} = \frac{W_{net}}{Q_S} = \frac{mR \ln(r)(T_H - T_L)}{mR \ln(r)(T_H)} \frac{(T_H - T_L)}{(T_H)} \qquad (14.68a)$$

or

$$\eta_{th} = 1 - \frac{T_L}{T_H} \qquad (14.68b)$$

It is interesting to note that the thermal efficiency of the String cycle is the same as the efficiency of a Carnot cycle when both are working with the same temperature limits. It is not possible to obtain 100% efficient regenerator and hence there will be always 10–20% loss of heat in the regenerator, which decreases the cycle efficiency. Considering regenerator efficiency, the efficiency of the cycle can be written as;

$$\eta_{th} = \frac{R \ln(r)(T_H - T_L)}{RT_H \ln(r) + (1 - \eta_R)C_V(T_H - T_L)} \qquad (14.69)$$

Where, η_R is the *Regenerator Efficiency*.

Example 14.11 A Stirling cycle, which operates on air with a compression ratio of 10. If the low pressure is 30 psia, the low temperature is 200 °F, and the high temperature is 1000 °F, calculate the work done and the heat input.

Solution For the Stirling cycle using *P-V* diagram of Fig. 14.19 the work output is;

$$W_{out} = W_{3\to4} + W_{1\to2} = RT_3 \ln\frac{V_4}{V_3} + RT_1 \ln\frac{V_2}{V_1}$$
$$= (53.3)[1460 \ln(10) + 660 \ln(0.1)]$$

For the isothermal process and using Eq. 14.66 will provide us;

$$\eta_{th} = 1 - \frac{T_L}{T_H} = 1 - \frac{660}{1460} = 0.548$$

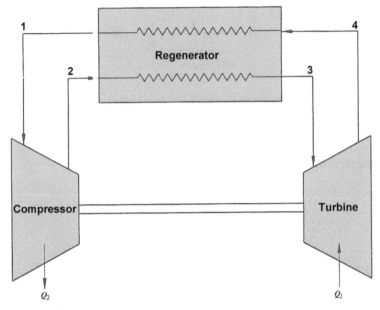

Fig. 14.24 Schematic diagram of air-standard Ericsson cycle

14.12   Ericsson Cycle

To overcome the difficulty associated with the constant volume regeneration, Ericsson developed a constant pressure regeneration cycle, which is named after him. The Air-Standard schematic layout and Ericsson cycle are shown in Figs. 14.24 and 14.25 respectively.

The thermal efficiency of the Ericsson cycle can be calculated easily and steps are shown as follows.

The Ericsson cycle consists of two isothermal and two constant pressure processes.

The processes are:

Process $1 \to 2$	Reversible isothermal compression
Process $2 \to 3$	Constant pressure heat addition
Process $3 \to 4$	Reversible isothermal expansion
Process $4 \to 1$	Constant pressure heat rejection

The heat addition and rejection take place at constant pressure as well as isothermal processes. Since the process 2-3 and 3-4 are parallel to each other on the T-s diagram, the net effect is that the heat need to be added only at constant temperature $T_3 = T_4$ and rejected at the constant temperature $T_1 = T_2$. The cycle is shown on P-V and T-s diagrams in Fig. 14.22. The advantage of the Ericsson cycle over the

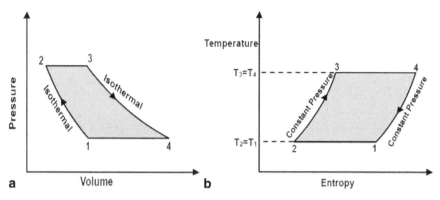

Fig. 14.25 Illustration of Ericsson **a** *P-V* diagram and **b** *T-S* diagram

Carnot and Stirling cycles is its smaller pressure ratio for a given ratio of maximum to minimum specific volume with higher mean effective pressure.

The thermal efficiency of Ericsson cycle is given by, (derivation is same as that of Stirling cycle),

$$\eta_{th} = \frac{T_H - T_L}{T_H} = \left[1 - \frac{T_L}{T_H}\right] \tag{14.70}$$

The Ericsson cycle does not find practical application in piston engines but is approached by a gas turbine employing a large number of stages with heat exchangers, insulators and reheaters.

Example 14.12 An Ericsson cycle operates on air with a compression ratio of 10. For a low pressure 200 kPa, low temperatures of 100 °C, and a high temperature of 600 °C, calculate the work output and the heat input.

Solution For the Ericsson cycle the work output is

$$W_{out} = W_{1\rightarrow 2} + W_{2\rightarrow 3} + W_{3\rightarrow 4} + W_{4\rightarrow 1} = RT_1 \ln\frac{V_2}{V_1} + P_2(V_3 - V_2) + RT_3 \ln\frac{V_4}{V_3} + P_1(V_1 - V_4)$$

We must calculate P_2, V_1, V_2, V_3, and V_4. Then we have;

$$V_1 = \frac{RT_1}{P_1} = \frac{(0.287)(373)}{200} = 0.5353 \text{m}^3/\text{kg}$$

For the constant pressure process $4 \rightarrow 1$,

$$\frac{T_4}{V_4} = \frac{T_1}{V_1} \text{ or } \frac{873}{V_4} = \frac{373_1}{0.5353_1} \Rightarrow V_4 = 1.253 \text{m}^3/\text{kg}$$

From the definition of the compression ratio, $V_4/V_2 = 10$, giving $V_2 = 0.1253\text{m}^3/\text{kg}$. Using the ideal-gas law, we have

$$P_3 = P_2 = \frac{RT_2}{V_2} = \frac{(0.25)(373)}{0.1253} = 854.4\text{kPa}$$

The final necessary property is $V_3 = RT_3/P_3 = (0.287)(873)/854.4 = 0.2932\text{m}^3/\text{kg}$. The expression for work output gives;

$$W_{out} = (0.287)(373)\ln\frac{0.1253}{0.5353} + (854.4)(0.2932 - 0.1253)$$

$$+0.287 \times 873\ln\frac{1.253}{0.2932} + (200)(0.5353 - 1253) = 208\text{kJ/kg}$$

Finally

$$\eta_{th} = 1 - \frac{T_L}{T_H} = 1 - \frac{378}{873} = 0.573 \text{ and } q_{in} = \frac{Wout}{\eta_{th}} = \frac{208}{0.573}\text{kJ/kg}$$

14.13 Atkinson Cycle

Atkinson cycle is an ideal cycle for Otto engine exhausting to a gas turbine. In this cycle the isentropic expansion $(3 \rightarrow 4)$ of an Otto cycle $(1 \rightarrow 2 \rightarrow 3 \rightarrow 4)$ is further allowed to proceed to the lowest cycle pressure to increase the work output. With this modification the cycle is known as Atkinson cycle. The cycle is shown on P-V and T-s diagrams in Fig. 14.26.
  Processes involved are:

Process $1 \rightarrow 2$ Reversible adiabatic compression (v1 to v2)

Process $2 \rightarrow 3$ Constant volume heat addition

Process $3 \rightarrow 4$ Reversible adiabatic expansion (v3 to v4)

Process $4 \rightarrow 1$ Constant pressure heat rejection

The thermal efficiency is calculated as follows;

$$\text{Heat supplied} = C_v(T_3 - T_2) \qquad (14.71a)$$

$$\text{Heat rejected} = C_p(T_4 - T_1) \qquad (14.71b)$$

$$\text{Net work done} = C_v(T_3 - T_2) - C_p(T_4 - T_1) \qquad (14.72)$$

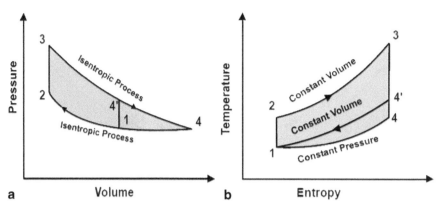

Fig. 14.26 Atkinson cycle on *P-V* and *T-s* diagrams

$$\eta_{th} = \frac{C_v(T_3 - T_2) - C_p(T_4 - T_1)}{C_v(T_3 - T_2)} \qquad (14.73a)$$

$$\eta_{th} = 1 - \frac{\gamma(T_4 - T_1)}{(T_3 - T_2)} \qquad (14.73b)$$

Let, $r = \dfrac{V_1}{V_2} = CR$ then we can write

$$T_2 = T_1 r^{\gamma-1} \qquad (14.74)$$

$$\frac{T_3}{T_2} = \frac{P_3}{P_2} = r_p = \text{Pressure ration} \qquad (14.75)$$

$$T_3 = T_2 r_p = T_1 r^{\gamma-1} r_p \qquad (14.76)$$

$$\frac{T_3}{T_4} = \left(\frac{P_3}{P_4}\right)^{\frac{\gamma-1}{\gamma}} = \left(\frac{P_3}{P_1}\right)^{\frac{\gamma-1}{\gamma}} = \left(\frac{P_3}{P_2}\cdot\frac{P_2}{P_1}\right)^{\frac{\gamma-1}{\gamma}} = r_p^{\frac{\gamma-1}{\gamma}} \cdot r^{\gamma-1} \qquad (14.77)$$

Since,

$$\frac{P_2}{P_1} = \left(\frac{V_1}{V_2}\right)^{\gamma} = r^{\gamma} \qquad (14.78)$$

$$T_4 = \frac{T_3}{r_p^{\frac{\gamma-1}{\gamma}} r^{\gamma-1}} = \frac{T_1 r_p r^{\gamma-1}}{r_p^{\frac{\gamma-1}{\gamma}} r^{\gamma-1}} = T_1 r_p^{\frac{1}{\gamma}} \qquad (14.79)$$

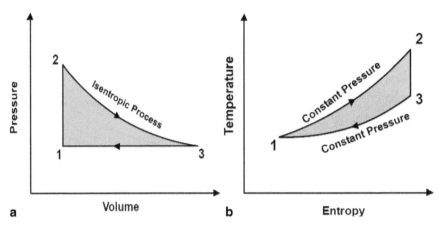

Fig. 14.27 Lenoir cycle on P-V and T-s diagrams

$$\eta_{th} = 1 - \lambda \left[\frac{r_p^{\frac{1}{\gamma}} - 1}{(r_p - 1)r^{\gamma - 1}} \right] \tag{14.80}$$

14.14 Lenoir Cycle

The Lenoir cycle is an idealized thermodynamic cycle often utilized to model a pulse jet engine. The Lenoir cycle is of interest because combustion (or heat addition) occurs without compression of the charge. Fig. 14.27a and b show the P-V and T-s diagrams.

The Lenoir cycle consists of the following processes:

Process $1 \to 2$ Constant volume heat addition

Process $2 \to 3$ Reversible adiabatic expansion

Process $3 \to 4$ Constant pressure heat rejection

No compression process

According to the cycle, the piston is at the top dead center, point 1, when the charge is ignited (or heat is added). The process is at constant volume so the pressure rises to point 2. From 2 to 3, expansion takes place and from 3 to 1 heat is rejected at constant pressure.

The thermal efficiency can be derived as follows:

$$\text{Heat supplied} = Q_s = C_v(T_2 - T_1) \tag{14.81a}$$

$$\text{Heat rejected} = Q_r = C_p(T_3 - T_1) \tag{14.81b}$$

Since Net work done $= W_{net} = Q_s - Q_r = C_v(T_2 - T_1) - C_p(T_3 - T_1)$ (14.82)

Thus

$$\eta_{th} = \frac{C_v(T_2 - T_1) - C_p(T_3 - T_1)}{C_v(T_2 - T_1)} \qquad (14.83a)$$

$$\eta_{th} = 1 - \frac{c_p(T_3 - T_1)}{c_v(T_2 - T_1)} \qquad (14.83b)$$

$$\eta_{th} = 1 - \frac{\gamma\left(\dfrac{T_3}{T_1} - 1\right)}{\left(\dfrac{T_2}{T_1} - 1\right)} \qquad (14.83c)$$

Since

$$\frac{T_2}{T_1} = \frac{P_2}{P_1} = r_p, \frac{T_3}{T_1} = \frac{V_3}{V_1} \text{ and } P_2 V_2{}^\gamma = P_3 V_3{}^\gamma \qquad (14.84)$$

So

$$\frac{P_2}{P_3} = \left(\frac{V_3}{V_2}\right)^\gamma = \left(\frac{V_3}{V_1}\right)^\gamma = \frac{P_2}{P_1} = \frac{T_2}{T_1} \qquad (14.85)$$

Therefore

$$\eta_{th} = 1 - \frac{\gamma\left(\dfrac{V_3}{V_1} - 1\right)}{\left[\left(\dfrac{V_3}{V_1}\right)^\gamma - 1\right]} \qquad (14.86a)$$

$$\eta_{th} = 1 - \frac{\gamma\left(r_p{}^{\frac{1}{\gamma}} - 1\right)}{(r_p - 1)} \qquad (14.86b)$$

$$\eta_{th} = 1 - \frac{\gamma(r_e - 1)}{(r_e{}^\gamma - 1)} \qquad (14.86c)$$

Here, $r_e = V_3 / V_1$, the volumetric expansion ratio. Eq. 3.18 indicates that the thermal efficiency of the Lenoir cycle depends primarily on the expansion ratio and the ratio of specific heats.

The intermittent-flow engine which powered the German V-1 buzz-bomb in 1942 during World War II operated on a modified Lenoir cycle. A few engines running on the Lenoir cycle were built in the late nineteenth century until the early twentieth century.

14.15   Deviation of Actual Cycles from Air Standard Cycles

The actual Otto and Diesel engines show marked deviations from the air-standard cycles described above. Above figure shows p-v-diagram for a high-speed diesel engine would be very similar in appearance. The main differences between the actual and theoretical cycles are as follows.

a. Compression and expansion are not friction less adiabatic processes. A Certain amount of friction is always present and there is considerable heat transfer between the gases and cylinder wall.
b. Combustion does not occur either at constant volume or at constant pressure.
c. The thermodynamics properties of the gases after combustion are different than those of the fuel-air mixture before combustion.
d. The combustion may be incomplete.
e. The specific heats of the working fluid are not constant but increases with temperature.
f. The cylinder pressure during exhaust process is higher than the atmosphere. As a result, more work has to be done by the piston on the gases to expel them out of the cylinder, than work done by the gases on the piston during the intake stroke. This difference in work, called pumping work, is represented by the pumping loop shown by hatched area. Note that this work is negative and represents loss of work called pumping loss.

14.16   Recuperated Cycle

A recuperated turbine is similar to a simple-cycle gas turbine, except for the inclusion of a special heat exchanger—called a recuperator—that captures exhaust thermal energy to preheat compressed air before the burner. Capturing exhaust energy helps increase electrical efficiency compared to a simple-cycle gas turbine. Figure 14.28 is illustration of a typical Recuperated Gas Turbine.

Typical turbines can accommodate only a limited additional mass flow—from 5 to 15%, depending on the design of the original gas turbine. Steam-injected gas

Fig. 14.28 Recuperated gas
turbine

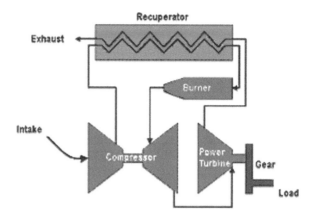

turbines enable the host to use the steam for industrial purposes, space heating, or
for the generation of additional power.

A group of cycles under consideration for development involves the use of
adiabatic saturators to provide steam at compressor discharge pressure to augment
greatly the mass flow through the turbine, and consequently increase cycle power
and efficiency. In the adiabatic saturator, water flows in a countercurrent path to the
compressor discharge air in a mass transfer tower. Such equipment is often used
in the chemical processing industries. The saturated air is preheated in a turbine
exhaust heat recuperator. This cycle is called the *Humid Air Turbine*, or HAT, cycle.
The HAT cycle is particularly useful in using the low-temperature heat generated in
coal-gasification-fueled gas turbine power plants. As the mass flow through the turbine is significantly augmented, engineers can no longer use the expansion turbine,
which was matched to the compressor in a conventional simple cycle gas turbine.

The critical cycle modification in an engine is the presence of heat exchanger
modules in the hot exhausts, with the purpose of bringing back thermal energy from
engine exit to combustion chamber. Given a fixed *High Pressure Turbine Entry
Temperature* (HPT TET), this concept allows fuel savings, because the exhausts,
which are discharged at a lower temperature, provide part of the required burner
temperature rise. In order for the heat exchange process to be effective, there must
be a sufficient temperature difference between compressor exit and turbine exit
temperature at all operating points. A low *High Pressure Compressor* (HPC) exit
temperature can be obtained with the use of inter-cooling, as can be clearly seen
in Fig. 14.25. A high turbine exit temperature at different operating points can be
obtained with a variable geometry turbine system (Sect. 5.4). The combined effect
of the two heat-exchange processes maximizes the heat transfer in the recuperator,
therefore improving the efficiency benefit. Figure 14.29 shows the thermodynamic
cycle for a typical Intercooled Recuperated Aro (IRA) engine cycle in the *h-S* plane
for *Overall Pressure Ratio* (OPR = 30) and T41 = 1800 K. The chart illustrates that
the amount of recuperated heat can be substantially increased by using the intercooler: the dotted line shows the compression process without inter-cooling.

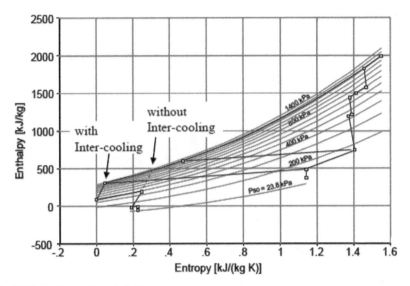

Fig. 14.29 Engine cycle in *h-S* diagram

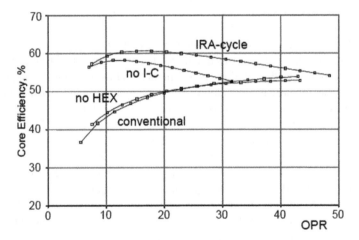

Fig. 14.30 Core efficiency illustrations

The effect of the different cycle innovations is shown in Fig. 14.30, where the core efficiency is calculated for different but consistent cycles (consistent technology level and same simulation tool). A considerable improvement in the performance of the recuperated engine can be achieved by adding the inter-cooling module, despite the detrimental effect of additional pressure losses.

Problems

Problem 14.1: In a power plant operating on a Rankin cycle, steam at 100 bar enters a turbine and expands to 0.1 bar. Calculate the minimum temperature of the steam, which enters the turbine to ensure a quality of 0.9 at the exit of the turbine.

Problem 14.2: In a thermal power plant steam enters a turbine at 500 °C and the condenser is maintained at 0.1 bar. It is required that the quality of steam at the turbine exit should be at least 0.9. Determine the pressure at which steam should be supplied to the turbine.

Problem 14.3: In a thermal power plant operating on a Rankine cycle, superheated steam is produced at 3 MPa and 300 °C and feed to a turbine where it expands to the condenser pressure of 5 kPa. The saturated liquid coming out of the condenser is fed to a pump, which the isentropic efficiency of 0.80 has been achieved. Calculate the thermal efficiency of the power plant if the isentropic efficiency of the turbine is 0.85. Determine the rate of steam production if the power output of the plant is 1 MW. Also calculate the efficiency of the corresponding ideal Rankine cycle.

Problem 14.4: In a power plant, employing a Rankin cycle with reheat modification, the steam enters the turbine at 3 MPa and 500 °C. After expansion to 0.6 MPa, the steam is reheated to 500 °C and expanded in a second turbine to a condenser pressure of 5 kPa. The steam leaves the condenser as saturated liquid. Calculate the thermal efficiency of the plant if the isentropic efficiency of the pump is 0.6 and the isentropic efficiency of the turbine is 0.8. Use Fig. 14.31, below.

Problem 14.5: Determine the efficiency of a steam power plant operating on a Carnot cycle with isothermal energy addition as heat at 3 MPa and an isothermal energy rejection as heat at 5 kPa. At the beginning of the isothermal energy addition, the fluid is a saturated liquid and at the end of the isothermal energy addition, it is in a saturated vapor state. Use Fig. 14.32 below.

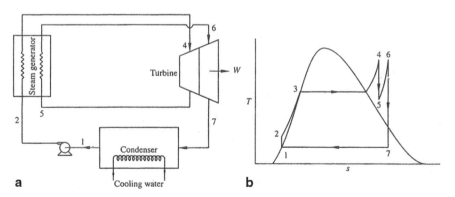

Fig. 14.31 Sketch of problem 2

Fig. 14.32 Sketch of problem 14.5. Carnot Cycle on a
T-s Diagram

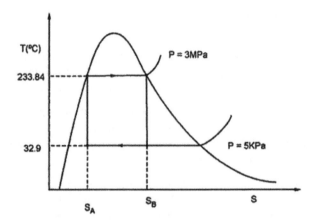

Fig. 14.32 Sketch of problem 14.5. Carnot Cycle on a
T-s Diagram

Problem 14.6: In the test of a turbine in a thermal power plant, the following data
has been recorded. Superheated steam at 3 MPa and 300 °C enters the turbine. The
steam leaves the turbine at 5 kPa with a moisture content of 0.15. Determine the
isentropic efficiency of the turbine. Use Fig. 14.33 below.

Problem 14.7: A gas turbine unit has a pressure ratio of 10/1 and a maximum cycle
temperature of 700 °C. The isentropic efficiencies of the compressor and turbine
are 0.82 and 0.85 respectively. Calculate the power output of an electric generator geared to the turbine when the air enters the compressor at 15 °C at the rate of
15 kg/s. Take $C_p = 1.005$ kJ/kg K and $\gamma = 1.4$ for the compression process, and take
$C_p = 1.11$ kJ/kg K and $\gamma = 1.333$ for the expansion process.

Problem 14.8: Calculate cycle efficiency and the work ratio of the plant in Problem
14.7, assuming that C_p for the combustion process is 1.11 kJ/kg K.

Fig. 14.33 Sketch of problem 14.6. Rankin cycle on a
T-s Diagram

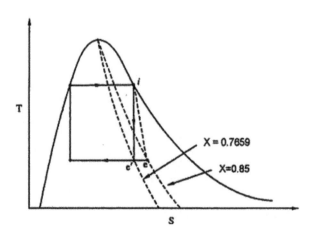

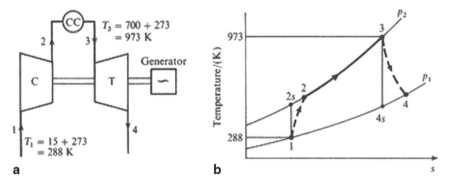

Fig. 14.34 Sketch of problem 14.7

Problem 14.9: A gas turbine has an overall pressure ration of 5 and a maximum cycle temperature of 550 °C. The turbine drives the compressor and an electric generator, the mechanical efficiency of the drive being 97 %. The ambient temperature is 20 °C and air enters the compressor at a rate of 15 kg/s; the isentropic efficiencies of the compressor and turbine are 80 and 83 %. Neglecting changes in kinetic energy, the mass flow rate of fuel, and all pressure losses, using Fig. 14.35 below, calculate;

i.   The power output,
ii. The cycle efficiency,
iii. The work ratio.

Problem 14.10: In a marine gas turbine unit a High Pressure (HP) stage turbine drives the compressor, and a Low Pressure (LP) stage turbine drives the propeller through suitable gearing. The overall pressure ratio is 4/1, the mass flow rate is 60 kg/s, the maximum temperature is 650 °C, and the air intake turbine, and LP

Fig. 14.35 Sketch of problem 14.9

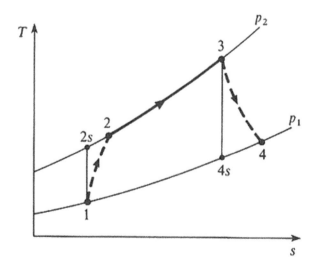

Fig. 14.36 Sketch of problem 14.10

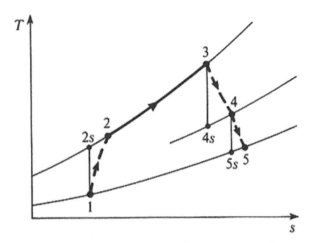

turbine, is 0.8, 0.83, and 0.85 respectively, and the mechanical efficiency of both
shafts is 98%. Neglecting kinetic energy changes, and the pressure loss in combustion, using Fig. 14.36, calculate:

i. The pressure between turbine stages,
ii. The cycle efficiency,
iii. The shaft power.

Problem 14.11: For the unit of Problem 14.11, calculate the cycle efficiency obtainable when a heat exchange is fitted. Assume a thermal ratio of 0.75

Problem 14.12: In a gas turbine generating set two stages of compression are used
with an intercooler between stages. The High Pressure (HP) turbine drives the HP
compressor, and the Low Pressure (LP) turbine drives the LP compressor and the
generator. The exhaust from the LP turbine passes through a heat exchanger which
transfers heat to the air leaving the HP compressor. There is a reheat combustion
chamber between turbine stages which raises the gas temperature to 600 °C, which
is also the gas temperature at entry to the HP turbine. The overall pressure ratio is
10/1, each compressor having the same pressure ratio, and the air temperature at
entry to the unit is 20 °C. The heat exchanger thermal ratio may be taken as 0.7, and
inter-cooling is complete between compressor stages. Assume isentropic efficiencies of 0.8 for both compressor stages and 0.85 for both turbine stages and the 2 %
of the work of each turbine is used in overcoming friction. Neglecting all losses in
pressure, and assuming that velocity changes are negligibly small, using Fig. 14.37,
calculate:

i. The power output in kilowatts for a mass flow of 115 kg/s.
ii. The overall cycle efficiency of the plant

Problem 14.13: A motor car gas turbine unit has two centrifugal compressor in
series giving an overall pressure ratio of 6/1. The air leaving the High Pressure
(HP) compressor passes through a heat exchanger before entering the combustion

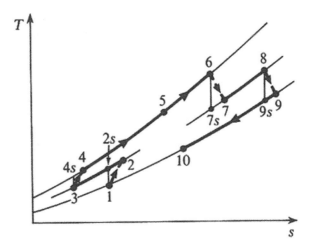

Fig. 14.37 Sketch of problem 14.11

chamber. The expansion is in two turbine stages, the first stage driving the compressors and the second stage driving the car through gearing. The gases leaving the Low Pressure (LP) turbine pass through the heat exchanger before exhausting to atmosphere. The HP turbine inlet temperature is 800 °C and the air inlet temperature to the unit is 15 °C. The isentropic efficiency of the compression is 0.8, and that of each turbine is 0.85; the mechanical efficiency of each shaft is 98 %. The heat exchanger thermal ratio may be assumed to be 0.65. Neglecting pressure losses and changes in kinetic energy, using Fig. 14.38, calculate:

i.   The overall cycle efficiency,
ii. The power developed when the air mass flow is 0.7 kg/s,
iii. The specific fuel consumption when the calorific value of the fuel used is 42.600 kJ/kg, and the combustion efficiency is 97 %.

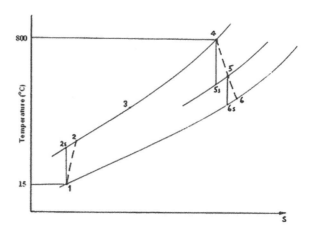

Fig. 14.38 Sketch of problem 14.12

Problem 14.14: In a gas turbine, generating station the overall compression ratio is 12/1, performed in three stag with pressure ratios of 2.5/1, 24/1, and 2/1 respectively. The air inlet temperature to the plant is 25 °C and inter-cooling between stages reduces the temperature to 40 °C. The High Pressure (HP).drives the HP and intermediate-pressure compressor stages; the LP turbine drives the LP compressor and the generator. The gases leaving the LP turbine are passed through a heat exchanger, which heats the air leaving the HP compressor. The temperature at inlet to the HP turbine is 650 °C, and reheating between turbine stages raises the temperature to 650 °C. The gases leave the heat exchanger at a temperature of 200 °C. The isentropic efficiency of each compressor stage is 0.83, and the isentropic efficiencies of the HP and LP turbines are 0.85 and 0.88 respectively. Take the mechanical efficiency of each shaft as 98 %. The air mass flow is 140 kg/s. Neglecting pressure losses and changes in kinetic energy, and taking the specific heat of water as 4.19 kJ/kg K, using Fig. 14.39 below, calculates:

i.   The power output in kilowatts,
ii.   The cycle efficiency,
iii. The flow of cooling water required for the intercoolers when the rise in water temperature must not exceed 30 °K,
iv.   The heat exchanger thermal ratio

Problem 14.15: A simple ideal Brayton cycle with air as the working fluids has a pressure ratio of 11. The air enters the compressor at 300 °K and the turbine at 1200 °K. Accounting for the variation of the specific heats with temperature determine (a) the air temperature at the compressor and turbine exits, (b) the back work ratio, and (c) the thermal efficiency. Assume that;

i.   Steady operating conditions exit.
ii.   The air-standard assumptions are applicable.
iii. Kinetic and potential energy changes are negligible.
iv.   The variation of specific heat with temperature is to be considered.

Fig. 14.39 Sketch of problem 14.14

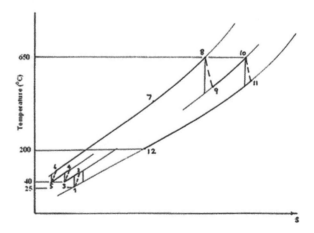

Problem 14.16: Consider and ideal gas-turbine cycle with two stages of compression and two stages of expansion. The pressure ratio across each stage of the com pressor and turbine is 3. The air enters each stage of the compressor at 300 °K and each stage of the turbine at 1200 °K. Determine the back work ratio and the thermal efficiency of the cycle, assuming (a) no regenerator is used and (b) a regenerator with 75 % effectiveness is used. Use constant specific heats at room temperature. Assume that;

i.   Steady operating conditions exit.
ii. The air-standard assumptions are applicable.
iii. Kinetic and potential energy changes are negligible.
iv. The variation of specific heat with temperature is to be considered.

References

1. Cengel YA, Boles MA (2011) Thermodynamics an engineering approach, 7th edn. McGraw Hill, New York
2. Potter MC, Somerton CW (2006) Thermodynamics for engineers, 2nd edn. McGraw-Hill, New York
3. Zemansky MW, Dittman RH (1997) Heat and thermodynamic, 7th edn. McGraw-Hill, New York
4. Rao YVC (2001) An introduction to thermodynamics, 2nd edn. Universities Press, Delhi

Chapter 15
Vapor Power Cycles

Vapor (or Rankine) power cycles are by far the most common basis for the generation of electricity in large fixed plant operations. They were one of the first developed for steam engines and have been adapted to many applications. They have also been modified in a number of ways to improve their thermal efficiency and better utilize combustible fuels.

15.1 The Basic Rankine Cycle

Although the Carnot cycle is the most efficient cycle, but its work ratio is low and there are practical difficulties to follow it. Hence the ideal cycle, which is more suitable as a criterion for actual steam cycles than the Carnot cycle, is called *Rankine cycle*. The basic Rankine (vapor) Cycle consists of 4 processes. The state points are identified in Fig. 15.1.

> Process 1 to 2—Isentropic compression of a liquid in a pump.
> Process 2 to 3—Constant pressure heat addition in a boiler or heat exchanger— creates vapor.
> Process 3 to 4—Isentropic expansion of the vapor in a turbine.
> Process 4 to 1—Constant pressure heat extraction in a condenser—returns vapor to liquid.

The general lay out of the components is pictured in Fig. 15.1. The steam generator, or boiler, consists of two generally separate heat exchangers—the economizer to heat the high pressure water to its boiling temperature, and the evaporator to convert the saturated liquid to steam. The heat transfer processes are different for the two different states of the liquid/vapor so this usually necessitates a different design for each region.

The temperature entropy diagram for the basic cycle is presented in Fig. 15.2. The work output of the system is represented by the area enclosed in the cycle on the T-s diagram by the curve **1-2-3-4-1.** The heat input to the system is represented

© Springer International Publishing Switzerland 2015
B. Zohuri, P. McDaniel, *Thermodynamics In Nuclear Power Plant Systems,*
DOI 10.1007/978-3-319-13419-2_15

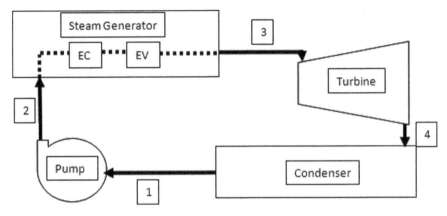

Fig. 15.1 Major components of a basic Rankine cycle

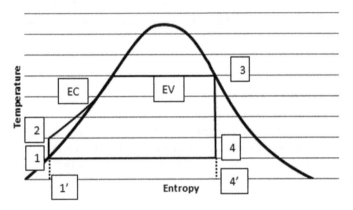

Fig. 15.2 T-s diagram for a basic Rankine cycle

by the curve **1'-1-2-3-4-4'-1'**. Therefore, the net thermal efficiency of the cycle is represented by the ratio of the two areas.

$$\eta_{th} = \frac{area\ 1-2-3-4-1}{area\ 1'-2-3-4'-1'} \tag{15.1}$$

The efficiency of a Rankine cycle must be less than that of a Carnot cycle because the heat addition does not all occur at the peak temperature of the cycle.

One of the major advantages of the Rankine cycle is that the compression process in the pump takes place on a liquid. The energy consumption of this process is almost negligible compared to the work output of the turbine, so a less than isentropic efficiency for the liquid pump is not a major energy loss for the cycle.

Consider the two ways of improving the Rankine cycle efficiency by adding all of the heat at the peak temperature as indicated in Figs. 15.3a and b. In Fig. 15.3a,

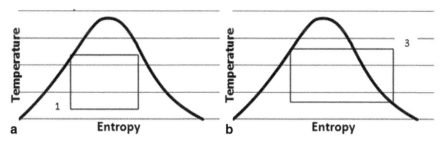

Fig. 15.3 Alternative cycles

the cycle will require pumping on a vapor at state point 1, which is much less efficient than pumping on a liquid. In Fig. 15.3b, to reach state point 3 at a constant temperature from the saturation line, the pressure will have to be continuously lowered, which is very difficult to accomplish. In the vapor region, a constant pressure curve departs the saturation line with a very significant upward slope as will soon be shown. Also by lowering the pressure to maintain a constant temperature with heat addition, some of the mechanical energy that was used in the pump would be sacrificed. There are several ways of increasing the work output of a Rankine cycle, but they do not improve on the efficiency relative to a Carnot cycle.

Now consider how the cycle efficiency is calculated. The heat input is given by

$$q_{in} = h_3 - h_2 \tag{15.2}$$

The turbine work is given by

$$w_t = h_3 - h_4 \tag{15.3}$$

The enthalpy at state 4 is found by assuming an isentropic expansion process across the turbine so the entropy at state point 4 must be the same as the entropy at state point 3. Once the pressure and entropy at state point 4 are known, the rest of the thermodynamic variables can be found including the enthalpy. Typically, a constant entropy expansion down to the pressure at state point 4 will put the fluid under the vapor dome and in the two-phase region. The entropy is then used to find the quality of the steam in the two-phase region. Once the quality is known, the enthalpy at state 4 can be easily obtained by the standard property expression for variables in the two phase regime.

The pump work is calculated by

$$w_p = h_2 - h_1 = \Delta pv = (p_2 - p_1)v_1 \tag{15.4}$$

Therefore, the cycle thermal efficiency is given by

$$\eta_{th} = \frac{h_3 - h_4 - (p_2 - p_1)v_1}{h_3 - h_2} = \frac{h_3 - h_4 - (p_2 - p_1)v_1}{h_3 - [h_1 + (p_2 - p_1)v_1]} \tag{15.5}$$

Note that for a closed system, the atmospheric pressure exiting the turbine and in the condenser will be significantly below atmospheric pressure as the liquid is being condensed to a temperature slightly above the cooling water temperature, which is near the ambient atmospheric temperature.

This can cause atmospheric gases to leak into the condenser and decrease the efficiency of the other components in the system, notably the pump. Therefore, a leak tight condenser is a very desirable component.

Example 15.1 Consider a basic Rankine power cycle with the saturated steam delivered to the turbine at 1000 psi and the condensate water delivered to the pump at 2.5 psi. Assume an isentropic turbine and pump. Calculate the cycle thermal efficiency.

Solution At 1000 psi $T_{sat} = 1004.3$ R, $h_3 = 1195.1$ Btu/lbm, $s_3 = 1.39548$ Btu/lbm/R at 2.5 psi, $T_{sat} = 594.1$ R, $v_f = 0.0163$ m³/kg, $h_f = 102.6$ Btu/lbm, $h_{fg} = 1019.1$ Btu/lbm, $s_f = 0.18954$ Btu/lbm, $s_{fg} = 1.71547$ Btu/lbm/R

Then

$$Carnot\ Efficiency \quad \eta_C = 1 - \frac{594.1}{1004.3} = 0.4084$$

$$x_4 = \frac{1.39548 - 0.18954}{1.71547} = 0.7030 \quad h_4 = 102.6 + 0.7030 * 1019.1 = 819.0$$

$$\Delta h_{3-4} = 1195.1 - 819.0 = 376.1\ Btu/lbm$$

$$\Delta h_{pump} = (1000 - 2.5) * 144 * 0.0163/778 = 3.0\ Btu/lbm$$

$$h_2 = 102.6 + 3.0 = 105.6\ Btu/lbm \quad \Delta h_{2-3} = 1195.1 - 105.6 = 1089.5\ Btu/lbm$$

$$\eta_{th} = \frac{376.1 - 3.0}{1089.5} = 0.3425 \quad \eta_{th} = 34.25\% \quad \eta_{II} = \frac{0.3425}{0.4084} = 0.8386$$

In addition to lossess required to drive the pump, inefficiencies in the turbine and pump can also affect the thermodynamic cycle efficiency. The change in enthalpy across a non-ideal turbine is given by

$$\Delta h_{actual} = \eta_t \Delta h_{isentropic} \tag{15.6}$$

And for a non-ideal pump the actual change in enthalpy is given by

$$\Delta h_{actual} = \frac{\Delta h_{isentropic}}{\eta_p} \tag{15.7}$$

Example 15.2 Re-evaluate the thermal efficiency of the Rankine cycle in Example 15.1 assuming a turbine with an adiabatic efficiency of 90% and a pump with an adiabatic efficiency of 80%.

Solution The only changes are the Δh's across the pump and the turbine.

$$\Delta h_{pump} = \frac{3.0}{0.8} = 3.75 Btu/lbm \quad \Delta h_{3-4} = 0.9*376.1 = 338.5 \ Btu/lbm$$

$$\eta_{th} = \frac{338.5 - 3.75}{1089.5} = 0.3073 \quad \eta_{th} = 30.73\%$$

Now consider possible ways of increasing the efficiency of the basic Rankine Cycle. There are really three possibilities—(1) increase the boiler pressure, (2) decrease the condensing temperature, and therefore the pressure, for the condensate entering the pump, and (3) increase the peak temperature of the cycle.

Example 15.3 Change the turbine inlet pressure for Example 15.1 to 1200 psi and re-evaluate the cycle efficiency.

Solution At 1200 psi, $T_{sat} = 1026.9$, $h_3 = 1186.8$ Btu/lbm, $s_3 = 1.37253$ Btu/lbm/R

$$Carnot \ Efficiency \quad \eta_C = 1 - \frac{594.1}{1026.9} = 0.4215$$

$$x_4 = \frac{1.37253 - 0.18954}{1.71547} = 0.6896 \quad h_4 = 102.6 + 0.6896*1019.7 = 805.8 Btu/lbm$$

$$\Delta h_{3-4} = 1186.8 - 805.8 = 381.0 \ Btu/lbm \quad \Delta h_{2-3} = 1186.8 - 105.6 = 1081.2 \ Btu/lbm$$

$$\eta_{th} = \frac{381.0 - 3.0}{1081.2} = 0.3496 \quad \eta_{th} = 34.96\% \quad \eta_{II} = \frac{0.3496}{0.4215} = 0.8294$$

Example 15.4 Change the condensate pump input pressure in Example 15.1 to 1.5 psi and re-evaluate the thermal efficiency of the cycle.

Solution At 1.5 psi, $T_{sat} = 575.3$ R, $v_f = 0.0162$ m^3/kg, $h_f = 83.8$ Btu/lbm, $h_{fg} = 1029.9$ Btu/lbm, $s_f = 0.15746$ Btu/lbm, $s_{fg} = 1.79023$ Btu/lbm/R

$$Carnot \ Efficiency \quad \eta_C = 1 - \frac{575.3}{1004.3} = 0.4272$$

$$x_4 = \frac{1.39548 - 0.15746}{1.79023} = 0.6915$$

$$h_4 = 83.8 + 0.6915*1029.9 = 796.0 \ Btu/lbm$$

$$\Delta h_{3-4} = 1195.1 - 796.0 = 399.1 \ Btu/lbm$$

$$\Delta h_{pump} = (1000 - 1.5)*144*0.0162/778 = 2.99 \ Btu/lbm$$

$$\Delta h_{2-3} = 1195.1 - (83.8 + 2.99) = 1108.3 \ Btu/lbm$$

$$\eta_{th} = \frac{399.1 - 2.99}{1108.3} = 0.3574 \quad \eta_{th} = 35.74\% \quad \eta_{II} = \frac{0.3574}{0.4272} = 0.8366$$

15.2   Process Efficiency

Actual expansion process is irreversible, as shown by line 1–2 in Fig. 15.4 [1]. Similarly the actual compression of the water is irreversible, as indicated by line 3–4. The *isentropic efficiency* of a process is defined by;

$$\text{Isentropic efficiency} = \frac{\text{actual work}}{\text{isentropic work}} \quad \text{for an expansion process}$$

and

$$\text{Isentropic efficiency} = \frac{\text{isentropic work}}{\text{actual work}} \text{ for an expansion process}$$

Hence

$$\text{Turbine isentropic efficiency} = \frac{-W_{12}}{-W_{12s}} = \frac{h_1 - h_2}{h_1 - h_{2s}} \tag{15.8}$$

In analyzing the performance of steam plant another criterion also known as *Specific Steam Consumption* (SSC). This relates the power output to the steam flow necessary to produce it. The steam flow indicates the size of plant and its component parts, and the SSC is a means whereby the relative sizes of different plants can be compared.

The ssc is the steam flow required to develop unit power output. The power output is given by $-m\sum W$, therefore we can write;

$$SSC = \frac{m}{-m\sum W} = \frac{1}{-\sum W} \tag{15.9}$$

Fig. 15.4 Rankine cycle showing real processes on a *T-s* diagram

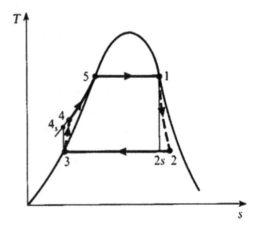

Neglecting the feed pump work we have

$$-\sum W = (h_1 - h_2) - (h_4 - h_3)$$

Therefore

$$SSC = \frac{1}{(h_1 - h_2) - (h_4 - h_3)}$$

Note that when h_1 and h_2 are expressed in kilojoules per kilogram then the units of SSC are kg/kJ or kg/kW h.

Example 15.5 A steam power plant operates between a boiler pressure of 609.126 psi and a condenser pressure of 0.507605 psi. Calculate for these limits the cycle efficiency, the work ratio, and the Specific Steam Consumption (SSC):

a. For a Carnot cycle using wet steam;
b. For a Rankine cycle with dry saturated steam at entry to the turbine;
c. For the Rankine cycle of (b), when the expansion process has an isentropic efficiency of 80%.

Solution

a. A Carnot cycle is shown in Fig. 15.5a

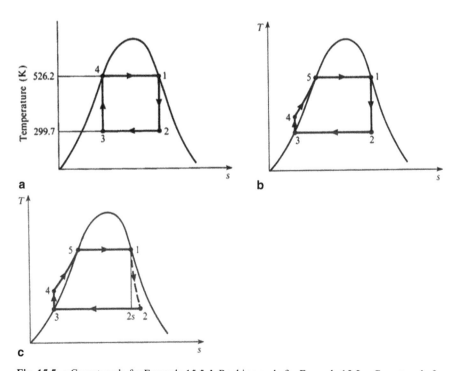

Fig. 15.5   a Carnot cycle for Example 15.5. **b** Rankine cycle for Example 15.5. **c** Carnot cycle for Example 15.5

Consider the following;

$$T_1 = \text{ saturation temperature at } 609.126 \, \text{psi}$$
$$= 253.2 + 273 = 526.2 \, \text{K}$$

$$T_2 = \text{ saturation temperature at } 0.507605 \, \text{psi}$$
$$= 26.7 + 273 = 299.7 \, \text{K}$$

Using Eq. 7.10 in the following form, we can write

$$\eta_{Carnot} = 1 - \frac{T_2}{T_1} = \frac{T_1 - T_2}{T_1} = \frac{526.2 - 299.7}{526.2} = 0.432 \text{ or } 43.2\%$$

Also heat supplied $= h_1 - h_4 = h_{fg}$ at 42 bar $= 1698 \, \text{kJ/kg}$

Then

$$\eta_{Carnot} = \frac{\text{Net work output} \left(-\sum W \right)}{\text{Turbine heat supplied}} = 0.432$$

Therefore

$$\left(-\sum W \right) = 0.432 \times 1689$$

i.e. Net work output,

$$\left(-\sum W \right) = 734 \, \text{kJ/kg}$$

To find the turbine work of the expansion process it is necessary to calculate h_2, using the fact that $s_1 = s_2$. From steam tables we can see that $h_1 = 2800 \, \text{kJ/kg}$ and $s_1 = s_2 = 6.049 \, \text{kJ/kg K}$. Using the entropy of wet steam that is given by the entropy of the water in the mixture plus the entropy of the dry steam in the mixture. For wet steam with quality x, we have

$$s = s_f + x s_{fg}$$

The quality is given by

$$x = \frac{s - s_f}{s_{fg}}$$

Applying all these relationship to part a of this example we can continue to write that;

$$s_2 = 6.049 = s_{f2} + x_2 s_{fg} = 0.391 + x_2 8.13$$

Therefore

$$x_2 = \frac{6.049 - 0.391}{8.13} = 0.696$$

Therefore

$$h = h_f + xh_{fg}$$

Having the quality for an isotropic expansion the enthalpy is simply given by:

$$h_2 = h_{f2} + x_2 h_{fg2} = 112 + (0.696 \times 2436) = 1808 \text{kJ/kg}$$

So the turbine work for an isentropic expansion is given by:

$$W_{12} = h_2 - h_1$$

Or work output is going to be $-W_{12} = h_1 - h_2$, thus we can write

$$-W_{12} = h_1 - h_2 = 2800 - 1808 = 992 \text{kJ/kg}$$

Using Eq. 15.6 we can write that

$$SSC = \frac{1}{\text{Net work output}(-\sum W)}$$

$$= \frac{1}{734 \text{kJ/kg}} = 0.00136 \, \text{kg}\Big/ \text{kWsec}$$

$$= 4.91 \, \text{kg}\Big/ \text{kWh}$$

b. The Rankine cycle is shown in Fig. 15.5b

As in part (a)

$$h_1 = 2800 \text{kJ/kg} \qquad \text{and} \qquad h_2 = 1808 \text{ kJ/kg}$$

Also, $h_3 = h_f$ at 0.507605 psi $= 112$ kJ/kg

But for a reversible adiabatic process and using second law of thermodynamics, we can write that $dQ = dh - \upsilon dp = 0$, therefore $dh = \upsilon dp$, and;

$$\int_3^4 dh = \int_3^4 \upsilon dp \qquad \text{and} \qquad h_4 - h_3 = \upsilon \int_3^4 dp = \upsilon(p_4 - p_3)$$

i.e. $h_4 - h_3 = \upsilon(p_4 - p_3)$

Therefore, the pump work input is given by;

$$\text{Pump input work} = h_4 - h_3 = \upsilon(p_4 - p_3)$$

With assumption of $\upsilon = \upsilon_f$ at 0.507605 psi and using the above relation, we can write

$$\text{Pump input work} = \upsilon_f(p_4 - p_3) = 0.001 \times (609.126 - 0.507605) \times \frac{10^5}{10^3} = 4.2 \text{kJ/kg}$$

Using the relationship for turbine work output that we found in part (a) of this example, namely $-W_{12} = h_1 - h_2$, we can write that

$$-W_{12} = h_1 - h_2 = 2800 - 1808 = 992 \text{kJ/kg}$$

If we define the Rankin efficiency is defined as η_R, then we can write

$$\text{Rankine efficiency} = \eta_R = \frac{\text{Net work output}}{\text{Heat supplied in the boiler}}$$

or

$$\eta_R = \frac{(h_1 - h_2) - (h_4 - h_3)}{(h_1 - h_3) - (h_4 - h3)} = \frac{99.2 - 4.2}{(2800 - 112) - 4.2} = 0.368$$

or $\eta_R = 36.8\%$

And using Eq. 15.6 will give us the following result;

$$SSC = \frac{1}{\text{Net work output}(-\sum W)}$$

$$= \frac{1}{(992 - 4.2)\text{kJ/kg}} = 0.00101 \; {}^{kg}\!/_{kW\,sec}$$

$$= 3.64 \; {}^{kg}\!/_{kWh}$$

c. The cycle with an irreversible expansion process is shown in Fig. 15.5c. Using Eq. 15.6 we can write

$$\text{Isentropic efficiency} = \frac{h_1 - h_2}{h_1 - h_{2s}} = \frac{-W_{12}}{-W_{12s}}$$

Therefore

$$0.8 = \frac{-W_{12}}{992} \text{ or } -W_{12} = 0.8 \times 992 = 793.6 \text{kJ/kg}$$

Then the cycle efficiency is given by

$$\text{Cycle efficiency} = \frac{(h_1 - h_2) - (h_4 - h_3)}{\text{gross heat supplied}}$$

$$= \frac{793.6 - 4.2}{(2800 - 112) - 4.2} = 0.294$$

$$= 29.4\%$$

Also, we can write for SSC as before

$$SSC = \frac{1}{793.6 - 4.2} = 0.001267\,^{kg}\!\!\Big/_{\!kW\,sec} = 4.56\,^{kg}\!\!\Big/_{\!kW\,h}$$

The feed-pump term has been included in the above calculation, but an inspection of the comparative values shows that it could have been neglected without having a noticeable effect on the results.

15.3   The Rankine Cycle with a Superheater

The only way to increase the peak temperature of the Rankine cycle without in-creasing the boiler pressure is to heat the steam itself to a higher temperature. This requires the addition of another type of heat exchanger called a super heater. This third heat exchanger is added to the steam generator as described in Fig. 15.6. The T-s diagram is provided in Fig. 15.7.

The only difference that the superheated Rankine cycle presents when evaluat-ing the cycle thermal efficiency is that the enthalpy at state point 3, the inlet to the turbine, must be obtained from the superheated steam tables.

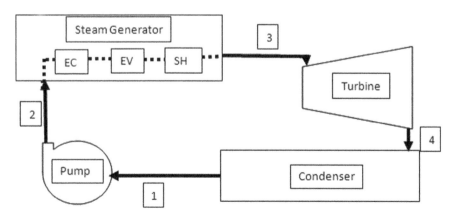

Fig. 15.6 Components for a superheated Rankine cycle

428 15 Vapor Power Cycles

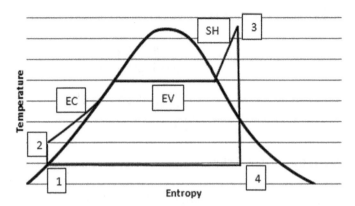

Fig. 15.7 T-s diagram for a Rankine cycle with a superheater

Example 15.6 Add a superheater to the cycle described in Example 15.1 that takes the turbine inlet temperature up to 1500 °R.

Solution At 1000 psi and 1500 °R, $h_3 = 1532.5$ Btu/lbm, $s_3 = 1.67285$ Btu/lbm/R

At 2.5 psi, $T_{sat} = 594.1$ R, $v_f = 0.0163$ m³/kg, $h_f = 102.6$ Btu/lbm, $h_{fg} = 1019.1$ Btu/lbm, $s_f = 0.18954$ Btu/lbm, $s_{fg} = 1.71547$ Btu/lbm/R

$$Carnot\ Efficiency \quad \eta_C = 1 - \frac{594.1}{1500} = 0.6039$$

$$x_4 = \frac{1.67258 - 0.18954}{1.71547} = 0.8645$$

$$h_4 = 102.6 + 0.8645 * 10191 = 983.6 Btu\ /\ lbm$$

$$\Delta h_{3-4} = 1535.5 - 983.6 = 548.9 Btu/lbm$$

$$\Delta h_{2-3} = 1532.5 - 105.6 = 1426.9 Btu/lbm$$

$$\eta_{th} = \frac{548.9 - 3.0}{1426.9} = 0.3826 \quad \eta_{th} = 38.26\% \quad \eta_{II} = \frac{0.3826}{0.6039} = 0.6335$$

Note that increasing the superheater outlet temperature further will increase the cycle efficiency further, but will decrease the second law efficiency relative to a Carnot cycle.

Example 15.7 Increase the superheater exit temperature from Example 15.6 to 1600 °R and estimate the increase in cycle efficiency.

Solution At 1000 psi and 1600 °R, $h_3 = 1589.3$ Btu/lbm, $s_3 = 1.70929$ Btu/lbm/R

$$Carnot\ Efficiency \quad \eta = 1 - \frac{594.1}{1600} = 0.6287$$

$$x_4 = \frac{1.70929 - 0.18954}{1.71547} = 0.8859$$

$$h_3 = 102.6 + 08859 * 1019.1 = 1005.4\,Btu/lbm$$

$$\Delta h_{3-4} = 1589.3 - 1005.4 = 583.9\,Btu/lbm$$

$$\Delta h_{2-3} = 1589.3 - 105.6 = 1492.7\,Btu/lbm$$

$$\eta_{th} = \frac{583.9 - 3.0}{1492.7} = 0.3892 \quad \eta_{th} = 38.92\% \quad \eta_{II} = \frac{0.3892}{0.6287} = 0.6190$$

Obviously, the reason the second law efficiency goes down is because the peak temperature of the cycle has been increased, but an even smaller fraction of the total heat input to the cycle has occurred near the peak temperature of the cycle.

15.4   External Reversibilities

No science or design can completely start at the beginning and proceed to develop everything in a uniform sequence. So backup and consider the two main heat exchange processes that go on in a typical Rankine cycle power plant. In the boiler, the feed water is heated to the turbine inlet conditions. In the condenser, the water is cooled from turbine exit conditions to a saturated liquid for delivery to the pump. In both processes, the heat transfer is driven by temperature differences. In the boiler, it is the difference between the hot gases in a combustion plant or primary coolant water in a nuclear reactor. In the condenser, it is the difference in temperature of the condensate and the external cooling water. These temperature differences are over plotted on a vapor dome representing a typical simple Rankine cycle in Fig. 15.8 below.

The hot side of the boiler is represented by the line *b–a* and the external cooling flow is represented by the line *c–d*. There is very little that can be done about the irreversibilities in the external coolant flow other than to try to bring the differences c-3 and d-2 closer together. This can be accomplished by building a bigger more effective condenser, but at some point the cost of the equipment takes over and further reductions in irreversibilities are not cost effective.

However, on the hot side there are some things that can be done. Note that the way the diagram is drawn, the flow of the hot gases or water in the boiler is from *a* to *b*. It is essentially a counter flow heat exchanger. It would be possible to flow the hot fluids in the opposite direction as in a parallel flow heat exchanger, but that would only increase the temperature differences. This can be seen by comparing the two schematics below in Fig. 15.9, where the first (a) represents the temperature differences for a parallel flow exchanger and the second (b) represents a counter flow exchanger. The horizontal axis in both is the distance along the boiler.

Fig. 15.8 T-s diagram
for basic Rankine cycle
with overplot of external
temperatures

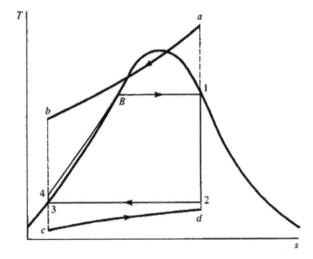

Fig. 15.9 Temperature
differences for parallel (**a**)
and Counter (**b**) flow heat
exchangers

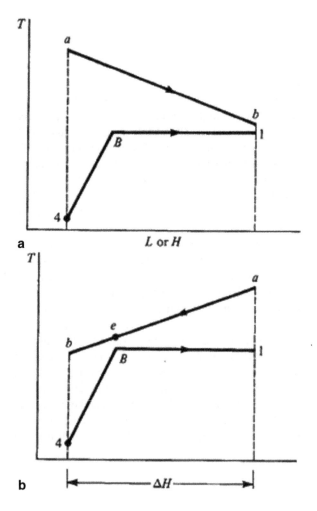

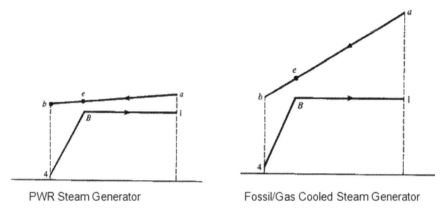

PWR Steam Generator   Fossil/Gas Cooled Steam Generator

Fig. 15.10 Temperature gradients for nuclear and fossil steam generators

It should be clear why only the counter flow approach is ever considered. Now consider the difference between a nuclear steam generator and a fossil or gas steam generator. In the nuclear steam generator, the heat carried and transferred per unit volume is greater so the temperature slope on the hot side of the exchanger is significantly lower. The temperature drop on the hot side comes much closer to paralleling the constant temperature line across the vapor dome as can be seen in Fig. 15.10 below.

For a combustion or gas, cooled system there is a much stronger advantage to coming out of the vapor dome and superheating the working fluid.

Note that the minimum temperature drop between the two curves occurs at the point where the working fluid reaches the liquid saturation point identified on the graphs as the separation *e-B*. This point is called the *"pinch point"* and the fluid providing the heat must always maintain a positive separation here in order to stably transfer heat to the working fluid.

15.5   Superheated Rankine Cycle with Reheaters

In order to deliver more of the heat at a temperature close to the peak of the cycle, the steam is usually extracted from the turbine after a less than maximum expansion and reheated to a higher temperature. This involves splitting the turbine into a high pressure and a low pressure turbine or into high, intermediate, and low pressure turbines. Three turbines are not uncommon in many plants. They may all be on the same shaft to drive a common generator, but they will have separate cases. With a reheater, the flow is extracted after a partial expansion, run back through the boiler to heat it back up to the peak temperature, and then passed to the lower pressure turbine. A component diagram for a single reheater is provided in Fig. 15.11. The fluid is extracted after a partial expansion in the high pressure turbine at state 5.

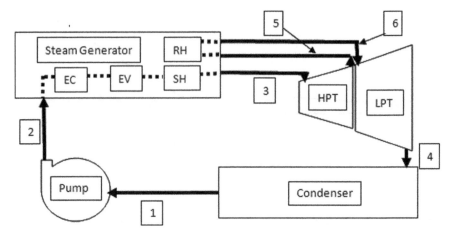

Fig. 15.11 Superheated Rankine cycle with single reheater

Typically, the extraction pressure is around one-fourth of the peak pressure. It is rerouted to the boiler and reheated to the peak temperature at state 6. It may even be heated to a higher temperature as the components may be able to stand a higher temperature at the reduced pressure. It is then routed back to the intermediate or low-pressure turbine to be expanded to condenser pressure.

The temperature-entropy diagram is provided in Fig. 15.12. Note that state point 5 may or may not be under the vapor dome. If it is not under the vapor dome, the superheated steam tables must be interpolated based on the known pressure and entropy to find the enthalpy at state point 5.

Most large steam plants today typically use three turbines and two reheaters. The analysis follows that for a single reheater but is repeated second time. In each case, a rule of thumb is that the pressure out of the turbine should be approximately one-fourth of the inlet pressure.

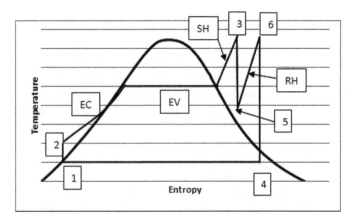

Fig. 15.12 T-s diagram for superheated Rankine cycle with single reheater

Example 15.8 Add a reheater to the cycle of Example 15.6 after the superheated steam has been expanded to 250 psi.

Solution At 250 psi and $s_3 = 1.67258$ Btu/lbm interpolate to find h_5

$$s(200\,psi, 1000R) = 1.65013, h(200\,psi, 1000R) = 1293.6$$

$$s(300\,psi, 1000R) = 1.59813, h(300\,psi, 1000R) = 1284.2$$

$$s(250\,psi, 1000R) = 1.64213, h(250\,psi, 1000R) = 1288.9$$

$$s(200\,psi, 1100R) = 1.7002, h(200\,psi, 1100R) = 1346.2$$

$$s(300\,psi, 1100R) = 1.65091, h(300\,psi, 1000R) = 1339.6$$

$$s(250\,psi, 1100R) = 1.67556, h(250\,psi, 1100R) = 1342.9$$

$$y = \frac{1.67258 - 1.64213}{1.67556 - 1.64213} = 0.911 \quad h_5 = 1288.9 = 0.911*(1342.9 - 1288.9)$$

$$= 1388.1\,Btu/lbm$$

$$\Delta h_{3-5} = 1532.5 - 1388.1 = 144.4\,Btu/lbm \quad h_6(250\,psi,1500R)$$

$$= 1552.8\,Btu/lbm \quad s_6 = 1.83603\,Btu/lbm/R$$

$$\Delta h_{5-6} = 1552.8 - 1388.1 = 164.7\,Btu/lbm$$

$$x_4 = \frac{1.83603 - 0.18954}{1.71547} = 0.9598 \quad h_4 = 102.6 + 0.9598*1019.1$$

$$= 1080.7\,Btu/lbm$$

$$\Delta h_{6-4} = 1552.8 - 1080.7 = 472.1\,Btu/lbm$$

$$\Delta h_{2-3} = 1532.5 - 105.6$$

$$= 1426.9\,Btu/lbm$$

$$\eta_{th} = \frac{144.4 + 472.1 - 3.0}{1426.9 + 164.7} = 0.3855 \quad \eta_{th} = 38.55\%$$

$$\eta_{II} = \frac{0.3855}{0.6039} = 0.6384$$

Note that the current generation of LWRs only produces saturated steam and cannot use nuclear driven superheaters or reheaters. Advanced reactors including liquid metal reactors, gas cooled reactors and molten salt reactors may be able to take advantage of superheating and reheating to improve their cycle efficiencies. From a cost of operation perspective, improvements in efficiency are not as important for nuclear plants as they are for fossil plants because the cost of fuel is a much smaller fraction of the total operating cost for a nuclear plant. However, it still pays to improve thermal efficiency in all cases.

15.6   Feed Water Heaters

With the addition of the reheaters, the external irreversibilities above the saturated water point break point, or pinch point, in the Rankine cycle have been dealt with as completely as possible. Next, the irreversibilities below the saturated water break point, or pinch point, can be addressed. This is normally taken care of with devices called Feed Water Heaters. There are three types of Feed Water Heaters and some plants use at least one of each. The three types are (1) Open or Direct Contact Feedwater Heaters, (2) Closed Feedwater Heaters with Drains Cascaded Backwards, and (3) Closed Feedwater Heaters with Drains Pumped Forward.

15.6.1   Open or Direct Contact Feedwater Heaters

A typical system with an Open Feedwater Heater is shown in Fig. 15.13 below.

The main turbine exit flow is drawn off at state point 4 as before. It passes through the condenser and enters the pump at state point 1 where its pressure is raised to the pressure of the bleed flow from the turbine at state point 2. The small bleed flow for the heater is bled off at state point 5. It does not have to be at the exit of a turbine however. Since the flow is only being bled off and not to be returned, it can be bled off at any desired pressure. These two flows are then mixed in the open feed water heater. The resulting liquid then enters the feed water pump at state 7.

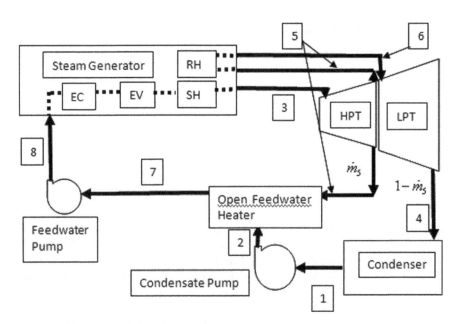

Fig. 15.13 Open feed water heater cycle

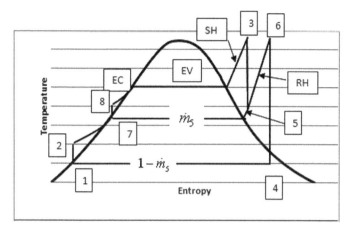

Fig. 15.14 T-s diagram for a cycle with an open feed water heater

The feed water pump then raises the pressure of the hot water to the required system pressure to enter the boiler at state point 8. Note that the boiler must heat the working fluid from the temperature at the exit of the feed water heater to the turbine throttle conditions at state point 3. But the heating from the condensate temperature to the temperature at state point 8 has been accomplished with lower temperature water (steam) so some of the irreversibility associated with the heat transfer has been eliminated (Fig. 15.14).

The main liability of the open feed water heater system is that it requires a pump after every heater that must pass most of the turbine flow. One advantage other than the decrease in heat transfer is that mixing of the two streams serves to release trapped gases in the feedwater and serves as an aerator. This is an advantage for fossil fueled plants but is normally not taken advantage of by nuclear plants due to concerns about releases of radioactivity.

Example 15.9 Add an open feed water heater to the cycle of Example 15.8 with the steam being extracted between the two turbines. Estimate the new cycle efficiency.

Solution

$$h_7 = \dot{m}_5 h_5 + (1 - \dot{m}_5)h_2 \quad \dot{m}_5 = \frac{h_7 - h_2}{h_5 - h_2}$$

$$h_2 = 102.6 + (250 - 2.5)*144*0.0162/778 = 102.6 + 0.742 = 103.3 Btu/lbm$$

$$h_5 = 1388.1 Btu/lbm(Ex.6) \quad h_7 = 3.77.0 Btu/lbm \,(Sat\ Liquid)$$

$$\Delta h_{feedwater\ pump} = (1000 - 250)*144*0.0187/778 = 2.596 Btu/lbm$$

$$h_8 = h_7 + h_{feedwater\ pump} = 3.77.0 + 2.596 = 379.6 Btu/lbm$$

$$\dot{m}_5 = \frac{377.0 - 103.3}{1388.1 - 103.3} = 0.213$$

$$\eta_{th} = \frac{144.4 + (1.0 - 0.213) * 472.1 - 2.596 - (1.0 - 0.213) * 0.742}{[1532.5 - (379.6)] + (1.0 - 0.213) * 164.7}$$

$$= \frac{512.76}{1282.52} = 0.3998$$

$$\eta_{th} = 39.98\% \quad \eta_{II} = \frac{0.3998}{0.6039} = 0.6620$$

15.6.2   Closed Feed Water Heaters with Drain Pumped Forward Second Type

The second type of feed water heater is a closed feed water heater with drains cascaded backward. The closed feed water heater is simply a two fluid heat exchanger (Figs. 15.15 and 15.16).

Steam is extracted from the turbine and passed through the feed water heat exchanger. It then must pass through an expansion valve to get it back down to the condenser pressure or the next lower feed water heater in the chain. Typically the steam that is extracted is very close to saturation and so the dominant heat transfer activity is the condensation of this steam to heat the high pressure water out of the condensate pump. It is possible that the closed feed water heater could contain three separate types of heat exchanger. The first component would be a "de-superheater" that transfers heat from the superheated steam to the high-pressure water. The second component would be a condenser, which condenses the saturated steam to a liquid to heat the high-pressure water. The third component would be a "drain cooler"

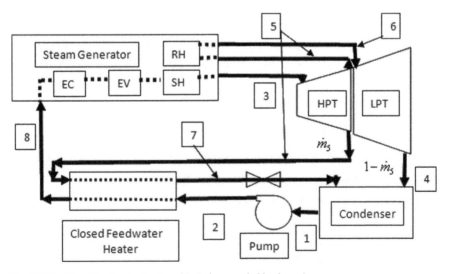

Fig. 15.15 Closed feed water heater with drain cascaded backwards

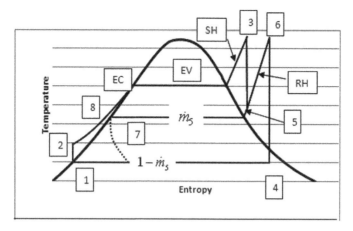

Fig. 15.16 T-s diagram for closed feed water heater with drain cascaded backward

that heats to cold high-pressure water with the hot liquid condensate. The extracted steam would pass through each of these components in sequence. The cold high-pressure water passes through them in the opposite sequence in the typical counter flow heat exchanger.

Because the dominant heat transfer mechanism is the condensation of the steam, the Terminal Temperature Difference (TTD) is defined as the difference in temperature between the high-pressure water exiting the heater and the condensation temperature of the high pressure steam. If there is no de-superheater component or a small one, this temperature difference will be positive. If there is a significant de-superheater section this temperature, difference could be negative which means that the high pressure water would exit the feed water heater at a higher temperature than the saturation temperature of the high pressure steam entering it.

Example 15.10 Replace the open feedwater heater of Example 15.9 with a closed feedwater heater with drain cascaded backward, extracting the steam between the two turbines. Assume a Terminal Temperature Difference (TTD) of 15 °R. Estimate the new cycle efficiency.

Solution

$$m_5(h_5 - h_7) = (h_8 - h_2)$$

$T_8 = 800R \ Interpolating \ in \ the \ sub - cooled \ liquid \ table$

$h_8(1000 \, psi, 800R) = 313.7 \, Btu/lbm$

$h_7 \ comes \ from \ the \ sub - cooled \ liquid \ table \ for \ 594.1 + 15$

$\quad = 609R \ and \ 250 \, psi \ \ h_7 = 119.2 \, Btu/lbm$

$h_5 = 1388.1 \, Btu/lbm (Ex.6) \ h_2 = 105.6 \, Btu/lbm$

$$m_5 = \frac{(h_8 = h_2)}{h_5 - h_7)} = \frac{313.7 - 105.6}{1388.1 - 119.2} = 0.164$$

$$\Delta h_{3-8} = 1532.5 - 313.7$$

$$= 1218.8 Btu / lbm$$

$$\eta_{th} = \frac{\Delta h_{3-5} + (1 - m_5)\Delta h_{6-4} - \Delta h_{pump}}{\Delta h_{3-8} + (1 - m_5)\Delta h_{5-6}} = \frac{144.4 + 0.836 * 472.1 - 3.0}{1218.8 + 0.836 * 164.7} = 0.3952$$

$$\eta_{th} = 39.52\%   \quad \eta_{II} = \frac{0.3952}{0.6039} = 0.6544$$

15.6.3   Closed Feed Water Heaters with Drain Pumped Forward Third Type

The third type of feed water heater is a closed feed water heater with the drain pumped forward. A layout of components is described in Fig. 15.17 and the T-s diagram in Fig. 15.18. For this configuration, the closed feed water heater is once again a two fluid counter flow heat exchanger. On the other hand, instead of the condensed steam being expanded back to condensate pressure, it is pumped up to the full system pressure with an added pump. This pump is smaller than the main condensate

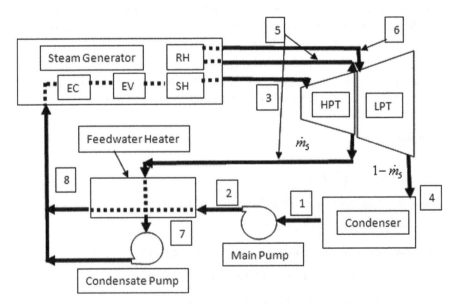

Fig. 15.17 System layout for a closed feed water heater with drain pumped forward

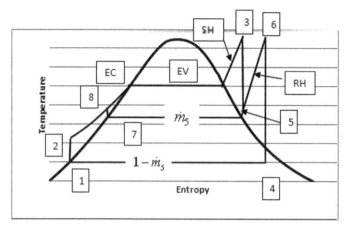

Fig. 15.18 T-s diagram for closed feed water heater with drain pumped forward

pump because it processes a smaller amount of water and it starts at a much higher pressure. But it is another active component that could fail or require maintenance.

The cycle thermodynamic efficiency is improved very slightly over the closed feed water heater with the drain cascaded backward.

Example 15.11 Replace the closed feed water heater in Example 15.10 with a closed feed water heater with drain pumped forward extracting the steam between the two turbines. Estimate the new cycle efficiency.

Solution

Balancing the heat transfer across the FWH

$$(1 - \dot{m}_5)(h_8 = h_2) = \dot{m}_5(h_5 - h_7) \ h_8 = h_7 + h_{cp}$$

$$h_{cp} = (1000 - 250)*0.0179*144/778 = 2.5 \, Btu/lbm$$

$$\dot{m}_5 = \frac{h_8 - h_2}{h_5 - h_2 + h_{cp}} = \frac{313.7 - 105.6}{1388.1 - 105.6 + 2.5} = 0.1619$$

$$\eta_{th} = \frac{144.4 + 0.8381*472.1 - 0.8381*3 - 0.1619*2.5}{1421.5 + 0.8381*164.7} = 0.3959$$

$$\eta_{th} = 39.59\% \quad \eta_{II} = \frac{0.3959}{0.6039} = 0.6556$$

Typically large steam power plants have multiple feed water heaters cascaded in increasing temperature ranges to heat the feed water from the temperature exiting the condensate pump up to near the system saturation pressure. As many as eight might be used in a typical 1000 MW(t) plant. In a fossil plant one will be an open feed water heater to release dissolved gases, but nuclear plants have to contain the radioactivity so typically they are all closed feed water heaters.

The pressures at which the steam is extracted are usually arranged so that the temperature range between the saturated temperature of the condensate-exiting condenser and the saturated temperature at the full system pressure is divided into n+1 equal intervals, where n is the number of feed water heaters. Then the extraction pressures are chosen to match these temperatures. For instance if the steam generator pressure is 1000 psi with a saturation temperature of 1004.3 °R and the condenser pressure is 2.5 psi with a saturation pressure of 594.1 °R and five feed water heaters will be used, the recommended saturation temperatures are,

$$\Delta T = (1004.3 - 594.1)/(5+1) = 68.3°R$$
$$T_1 = 662.5°R \quad p_1 = 12.5\,psi$$
$$T_2 = 730.8°R \quad p_2 = 42.5\,psi$$
$$T_3 = 799.2°R \quad p_3 = 118.5\,psi$$
$$T_4 = 867.6°R \quad p_4 = 268\,psi$$
$$T_5 = 935.9°R \quad p_5 = 546.5\,psi$$

Example 15.12 Reconsider the LWR circuit of Example 15.1. Add two feedwater heaters with drains cascaded backwards to this cycle and compute the new thermal efficiency. Assume the TTD for both heaters is 15 °R. Assume the hot fluid exiting both heaters is saturated liquid (Fig. 15.19).

Solution At 1000 psi $T_{sat} = 1004.3$ R, $h_3 = 1195.1$ Btu/lbm, $s_3 = 1.39548$ Btu/lbm/R at 2.5 psi, $T_{sat} = 594.1$ R, $v_f = 0.0163$ m³/kg, $h_f = 102.6$ Btu/lbm, $h_{fg} = 1019.1$ Btu/lbm, $s_f = 0.18954$ Btu/lbm, $s_{fg} = 1.71547$ Btu/lbm/R

The extraction pressures will be given by dividingup the range from 594.1 to 1004.3 °R into three intervals. The saturation temperatures would be 730.8 and 867.6 °R. These correspond to saturation pressures of 42 and 270 psi. Therefore, at 270 psi and 1.39548 Btu/lbm/R the steam is in the saturated two-phase region.

$$s_f = 0.57675 \text{ Btu/lbm/R}, \ s_{fg} = 0.94760 \text{ Btu/lbm/R},$$
$$h_f = 373.7 \text{ Btu/lbm}, \ h_{fg} = 821.3 \text{ Btu/lbm}$$

$$x_5 = \frac{1.39548 - 0.57675}{0.94760} = 0.864$$
$$h_5 = 383.7 + 0.864 * 821.3 = 1093.3 Btu/lbm$$

And at 42 psi,

$$s_f = 0.39734 \text{ Btu/lbm/R}, \ s_{fg} = 1.27880 \text{ Btu/lbm/R},$$
$$h_f = 239.8 \text{ Btu/lbm}, \ h_{fg} = 933.5 \text{ Btu/lbm}$$

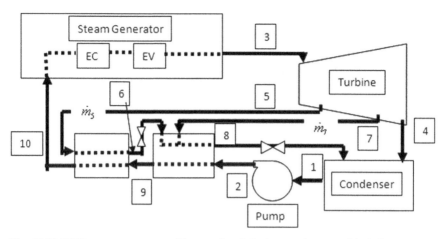

Fig. 15.19 LWR steam generator with two closed feed water heaters with drains cascaded backwards

$$x_7 = \frac{1.39548 - 0.39734}{1.27880} = 0.781$$

$$h_7 = 239.8 + 0.781 * 933.5 = 968.4 Btu/lbm$$

$$T_9 = 730.8 - 15 = 715\,^{\circ}R, \; h_9 = 226.6 \text{ Btu/lbm } T_{10} = 867.5 - 15 = 852.5\,^{\circ}R,$$

$$h_{10} = 369.1 \text{ Btu/lbm}$$

$$h_6 = 383.7 \text{ Btu/lbm } h_8 = 239.8 \text{ Btu/lbm}$$

Then

$$m_5(h_5 - h_6) = h_{10} - h_9 \; m_5 = \frac{h_{10} - h_9}{h_5 - h_6} = \frac{369.1 - 226.6}{1093.3 - 383.7} = 0.2008$$

$$m_5 h_6 + m_7 h_7 - (m_5 + m_7)h_8 = h_9 - h_2 \; m_7 = \frac{h_9 - h_2 - m_5(h_6 - h_8)}{h_7 - h_8}$$

$$m_7 = \frac{226.6 - 105.6 - 0.2008 * (383.7 - 239.8)}{986.4 - 239.8} = 0.1234$$

$$n_{th} = \frac{1195.1 - 1093.3 + (1.0 - 0.2008)(1093.3 - 968.4)}{1195.1 - 369.1} +$$

$$\frac{(1.0 - 0.2008 - 0.1234) * (968.4 - 819.0) - 3.0}{1195.1 - 369.1}$$

$$\eta_{th} = \frac{302.52}{826} = 0.3662 \quad \eta_{th} = 36.62\%$$

15.7   The Supercritical Rankine Cycle

The major advantage of the Rankine cycle is that the compression process works on a liquid. It is certainly possible to compress the liquid to a pressure greater than its critical pressure and heat it above the critical temperature. The expansion process could completely take place above the vapor dome, only dropping into the two-phase region near the exit from the turbine. The cycle efficiency would be improved slightly, but the stresses in the components would necessitate much more expensive structures. In addition, the safety consequences of a pipe rupture would increase significantly. So no supercritical cycle devices have been attempted with water as the working fluid, but other fluids, such as carbon dioxide, have been considered.

Problems

Problem 15.1: Rework Example 15.3 assuming a turbine efficiency of 90 %.

Problem 15.2: Rework Example 15.6 with two Reheaters. The pressure of the first Reheater should be 250 ψ and the second Reheater pressure should be 62.5 ψ. Assume both reheat to 1500 °R.

Problem 15.3: Rework Example 15.11 with an open feed water heater and a closed feed water heater with the drain cascaded backwards. Assume the open reheater operates at the lower pressure.

Problem 15.4: Rework Example 15.11 with two closed feed water heaters with the drains pumped forward.

Problem 15.5: Rework Example 15.11 with a 90 % efficiency turbine.

Problem 15.6: A power plant is to be operated on an ideal Rankine cycle with the superheated steam exiting the boiler at 4 MPa and 500 C. (one super-heater and no feed water heaters) Calculate the thermal efficiency and the quality at the turbine outlet if the condenser pressure is (a) 20 kPa, (b) 10 kPa, and (c) 8 kPa.

Problem 15.7: Coal with a heating value of 2500 Btu/lbm, is used to provide energy to the working fluid in a boiler which is 85 % efficient. (The working fluid acquires 85 % of the energy available from the coal.) Determine the minimum mass flux of coal in lbm/hr, that would be necessary for the turbine output to be 100 MW. The pump receives water at 2 psia, in the simple Rankine cycle, and delivers it to the boiler at 2000 psia. Superheated steam leaves the boiler at 1000 F. (one super-heater and no feed water heaters):

Reference

1. Eastop T, McConkey A (1993) Applied thermodynamics for engineering technologiests, 5th edn. Pearson Prentice Hall, New Jersey

Chapter 16
Circulating Water Systems

Nuclear power plants are usually built next to lakes, rivers, and oceans. Not for the scenic views that such locales provide, but because water can absorb the waste heat produced by the plants. Nuclear power plants consume vast amounts of water during normal operation to absorb the waste heat left over after making electricity and to cool the equipment and buildings used in generating that electricity. In event of an accident, nuclear power plants need water to remove the decay heat produced by the reactor core and to cool the equipment and buildings used to provide the core's heat removal. This chapter describes the reliance of nuclear power plants on nearby bodies of water during normal operation and under accident conditions.

16.1 Introduction

A power plant generates electricity by boiling water to steam, which turns a turbine that then turns a generator. Cooling systems are required to cool the steam back to water so the cycle can continue.

Most power plants use one of two types of cooling water systems. The two modes of cooling are used to remove the waste heat from electrical generation:

1. Once through cooling and,
2. Closed cycle cooling.

In the *once-through cooling* system, water from the nearby lake, river or ocean flows through thousands of metal tubes inside the condenser. Steam flowing through the condenser outside the tubes gets cooled down and converted back into water. The condensed water is re-used by the plant to make more steam. The water exits the condenser tubes warmed up to 30 °F higher than the lake, river, or ocean temperature and returns to that water body (Fig. 16.1).

As we said in once-through cooling withdraws water from a water body and circulates it within the plant to condense the steam from the turbine into water through heat absorption. In a wet cooling tower system, circulating water from the

© Springer International Publishing Switzerland 2015   443
B. Zohuri, P. McDaniel, *Thermodynamics In Nuclear Power Plant Systems,*
DOI 10.1007/978-3-319-13419-2_16

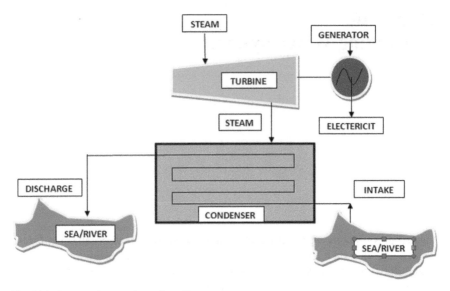

Fig. 16.1 Open cycle once through cooling system

plant moves through the tower and is cooled by evaporation, which is a closed-cycle cooling system.

In the *closed-cycle cooling* system, water flows through thousands of metal tubes inside the condenser. Steam flowing through the condenser outside the tubes gets cooled down and converted back into water. The condensed water is re-used by the plant to make more steam.

Water exits the condenser tubes warmed nearly 30°F higher than upon entry to the condenser tubes. The water leaving the condenser tubes flows to a cooling tower. Air moving upward past the water spraying downward inside the cooling tower cools the water. The water collected in the cooling tower basin is pumped back to the condenser for re-use. Water from the nearby lake, river, or ocean is needed to make-up for the water vapor carried away by the air leaving the cooling tower (Fig. 16.2).

For energy sources, water consumption is calculated in terms of the amount of water consumed to produce a standard measure of electricity generated—a mega-watt-hour (MWh).

Nuclear energy consumes 400 gallons/MWh with once-through cooling and 720 gallons/MWh with wet cooling towers. Coal consumes less, ranging from about 300 gallons/MWh for plants with minimal pollution controls and once-through cool-ing to 714 gallons/MWh for plants with advanced pollution control system and wet cooling towers. Natural gas-fueled power plants consume even less, at 100 gallons/MWh for once-through, 370 gallons/MWh for combined-cycle plants with cooling towers and none for dry cooling.

Hydropower's typical water consumption is the most significant at 4500 gallons/MWh, due in large part to evaporation from reservoirs. Renewable energy sources

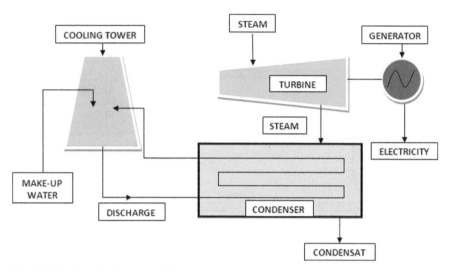

Fig. 16.2 Typical closed cycle cooling system

such as geothermal and solar thermal consume two to four times more water than nuclear power plants.

All of the 104 nuclear reactors currently licensed to operate in the United States are light2 water reactors. Sixty-nine (69) are pressurized water reactors (PWRs) and 35 are boiling water reactors (BWRs).

In a BWR, water flowing through the reactor core is heated by its thermal energy and boils. The steam flows from the reactor vessel to the turbine. The steam spins the turbine, which is connected to a generator that produces electricity. The steam exits the turbine into the condenser.

Water from the nearby lake, river, or ocean flows through thousands of metal tubes in the condenser. Steam flowing past outside these tubes cools and changes back into water. The condensed water flows back to the reactor vessel for another cycle. The water leaves the condenser tubes nearly 30 °F warmer and returns to the nearby lake, river, or ocean (Fig. 16.3).

In a PWR, water flowing through the reactor core is heated by its thermal energy. Because this water is maintained under high pressure (over 2000 pounds/sq in.), it does not boil even when heated to over 500 °F. The hot water flows from the reactor vessel and enters thousands of metal tubes within the steam generator. Heat passes through the thin tube walls to boil water at lower pressure that surrounds the tubes.

The water leaves the tubes about 10 °F cooler and returns to the reactor vessel for another cycle. Steam leaves the steam generator and enters the turbine. The steam spins the turbine, which is connected to a generator that produces electricity. The steam exits the turbine into the condenser. Water from the nearby lake, river, or ocean flows through thousands of metal tubes in the condenser. Steam flowing past outside these tubes cools and changes back into water.

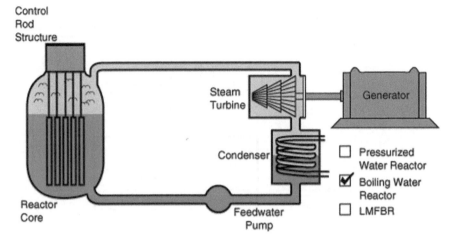

Fig. 16.3 Schematic of BWR. (In the boiling water reactor the same water loop serves as moderator, coolant for the core, and steam source for the turbine. Courtesy of DOE)

The condensed water flows back to the steam generator for another cycle. The water leaves the condenser tubes up to 30 °F (Fig. 16.4).

The water needs of and usage by PWRs and BWRs are essentially identical. Both types of nuclear power reactors are about 33 % efficient, meaning that for every three units of thermal energy generated by the reactor core, one unit of electrical energy goes out to the grid and two units of waste heat go out to the environment.

Note that An exception to the rule is the Palo Verde nuclear plant. Built in the arid region west of Phoenix, Arizona, cooling water is brought to the facility.

Also Note that Another type of reactor uses "heavy" water—water enriched in deuterium, an isotope of hydrogen that makes it heavier than regular or "light" water.

16.2   Cooling Power Plants

The most common types of nuclear power plants use water for cooling in two ways:

1. To convey heat from the reactor core to the steam turbines.
2. To remove and dump surplus heat from this steam circuit. (In any steam/ Rankine cycle plant such as present-day coal and nuclear plants there is a loss of about two-thirds of the energy due to the intrinsic limitations of turning heat into mechanical energy.)

The bigger the temperature difference between the internal heat source and the external environment (pinch point) where the surplus heat is dumped, the more

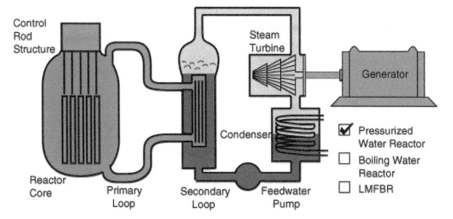

Fig. 16.4 Schematic of PWR. (In the pressurized water reactor, the water which flows through the reactor core is isolated from the turbine. Courtesy of DOE)

efficient is the process in achieving mechanical work—in this case, turning a generator [Carnot efficiency]. Hence the desirability of having a high temperature internally and a low temperature in the external environment. This consideration gives rise to desirably sitting power plants alongside very cold water.

Note that Many power plants, fossil and nuclear, have higher net output in winter than summer due to differences in cooling water temperature.

Also it is sufficed to say that;

- The amount of cooling required by any steam-cycle power plant (of a given size) is determined by its thermal efficiency. It has essentially nothing to do with whether it is fuelled by coal, gas or uranium.
- However, currently operating nuclear plants often do have slightly lower thermal efficiency than coal counterparts of similar age, and coal plants discharge some waste heat with combustion gases, whereas nuclear plants rely on water.
- Nuclear power plants have greater flexibility in location than coal-fired plants due to fuel logistics, giving them more potential for their sitting to be determined by cooling considerations.

16.2.1   Steam Cycle Heat Transfer

For the purpose of heat transfer from the core, the water is circulated continuously in a closed loop steam cycle and hardly any is lost. It is turned to steam by the primary heat source in order to drive the turbine to do work making electricity [Rankine Cycle], and it is then condensed and retuned under pressure to the heat source in a closed system [The function of the condenser is to condense exhaust steam from the steam turbine by losing the latent heat of vaporization to the cooling water (or possibly air) passing through the condenser] A very small amount of make-up water is required in any such system. The water needs to be clean and fairly pure. [Within a nuclear reactor water or heavy water must be maintained at very high pressure (1000–2200 psi, 7–15 MPa) to enable it to remain liquid above 100 °C, as in present reactors].

This function is much the same whether the power plant is nuclear, coal-fired, or conventionally gas-fired. Any steam cycle power plant functions in this way. At least 90 % of the non-hydro electricity in every country is produced thus.

In a nuclear plant there is an additional requirement. When a fossil fuel plant is shut down, the source of heat is removed. When a nuclear plant is shut down some heat continues to be generated from radioactive decay, though the fission has ceased. This needs to be removed reliably, and the plant is designed to enable and assure this, both with routine cooling and also Emergency Core Cooling Systems (ECCS) provided in case of major problem with primary cooling. The routine cooling is initially with the main stream supply circuit bypassing the turbine and dumping heat into the condenser. After pressure drops, a residual heat removal system is relied upon with its own heat exchanger. The intensity of this decay heat diminishes with time, rapidly at first, and after a day or two ceases to be a problem if circulation is maintained.

Note that When Kashiwazaki-Kariwa seven nuclear reactor automatically shut down because of a severe earthquake in 2007, it took 16 h for the coolant temperature to diminish from 287 to 100 °C so that it would no longer boil. "Cold shutdown" is when the primary circuit is at atmospheric pressure and not boiling (Fig. 16.5).

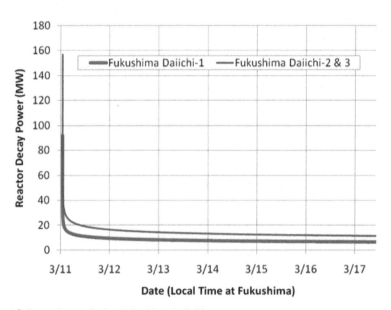

Fig. 16.5 Decay heat in fuel at Fukushima Daiichi reactors

16.2.2   Cooling to Condense the Steam and Discharge Surplus Heat

The second function for water in such a power plant is to cool the system so as to condense the low-pressure steam and recycle it. As the steam in the internal circuit condenses back to water, the surplus (waste) heat which is removed from it needs to be discharged by transfer to the air or to a body of water. This is a major consideration in sitting power plants, and in the UK sitting study in 2009 for nuclear plants all recommendations were for sites within 2 km of abundant water—sea or estuary.

This cooling function to condense the steam may be done in one of three ways:

Direct or "once-through" cooling. If the power plant is next to the sea, a big river, or large inland water body it may be done simply by running a large amount of water through the condensers in a single pass and discharging it back into the sea, lake or river a few degrees warmer and without much loss from the amount withdrawn. That is the simplest method. The water may be salt or fresh. Some small amount of evaporation will occur off site due to the water being a few degrees warmer.

- **Recirculating or indirect cooling**. If the power plant does not have access to abundant water, cooling may be done by passing the steam through the condenser and then using a cooling tower, where an up draught of air through water droplets cools the water. Sometimes an on-site pond or canal may be sufficient for cooling the water. Normally the cooling is chiefly through evaporation, with simple heat transfer to the air being of less significance. The cooling tower evaporates up to 5 % of the flow and the cooled water is then returned to the power plant's condenser. The 3–5 % or so is effectively consumed, and must be continually replaced. This is the main type of recirculating or indirect cooling.
- **Dry cooling**. A few power plants are cooled simply by air, without relying on the physics of evaporation. This may involve cooling towers with a closed circuit, or high forced draft air flow through a finned assembly like a car radiator.

With a fossil-fuel power plant some of the heat discharged is in the flue gases. With a large coal-fired plant some 15 % of the waste heat is through the stack, whereas in a nuclear power plant virtually all the waste heat has to be dumped into the condenser cooling water. This gives rise to some difference in water consumption or use between a nuclear and a coal plant. (A gas turbine plant will discharge most of its waste heat in the exhaust.)

Beyond this, and apart from size, any differences between plants is due to thermal efficiency, i.e. how much heat has to be discharged into the environment, which in turn largely depends on the operating temperature in the steam generators. In a coal-fired or conventionally gas-fired plant it is possible to run the internal boilers at higher temperatures than those with finely-engineered nuclear fuel assemblies which must avoid damage. This means that the efficiency of modern coal-fired plants is typically higher than that of nuclear plants, though this intrinsic advantage may be offset by emission controls such as Flue Gas Desulfurization (FGD) and in the future, Carbon Capture and Storage (CCS) [2].

A nuclear or coal plant running at 33 % thermal efficiency will need to dump about 14 % more heat than one at 36 % efficiency. Nuclear plants currently being built have about 34–36 % thermal efficiency, depending on site (especially water temperature). Older ones are often only 32–33 % efficient. The relatively new Stanwell coal-fired plant in Queensland runs at 36 %, but some new coal-fired plants approach 40 % and one of the new nuclear reactors claims 39 %.

16.3   Circulating Water Systems

Brief description of circulating water systems for any power plant is presented here [3].

1. Functions

Circulating Water Systems at any power plant have two important functions:

- Filter water before it is pumped to and through the condenser
- Cool the condenser

2. Major Components

Intake (Supply) Basin Water is supplied from an abundant source—river, lake, sea, or ocean—to a storage basin which, in turn, supplies the large pumps.

Trash Racks These racks remove large debris such as seaweed, grass, or logs. These racks must be periodically cleaned to remove debris that builds up.

Traveling Screens These are large steel screens up to 30 ft long that rotate and filter out larger debris that passes through the trash racks.

Condenser Cleaning System A system which usually uses small abrasive balls pumped through the condenser tubes for cleaning. Usually these systems are very effective at cleaning the tubes. Several times per year the plant power level must be reduced so that the tubes can be manually cleaned. This operation is performed when plant efficiency decreases to the point where the plant's operations and engineering staffs determine the tube cleaning is needed to improve plant performance (Fig. 16.6).

These large pumps supply water at over 100,000 gallons/min to the condenser. The pump is usually over 15 ft deep. The motor assembly may be 8–10 ft high.

3. Simplified Diagram

The diagram below illustrates the arrangement of components within the system and the major flow paths (Fig. 16.7).

The green flow paths show how the water is taken from a river (yellow) to an intake supply basin (green) that the Circ Water Pumps take suction from. The water is then pumped to the Condenser where the water is heated. The water is then sent to an exit distribution basin where the water then can be returned to the river and/or pumped by the Cooling Tower Pumps to the Cooling Towers then the water returned

Fig. 16.6 Circulating water pumps [3]

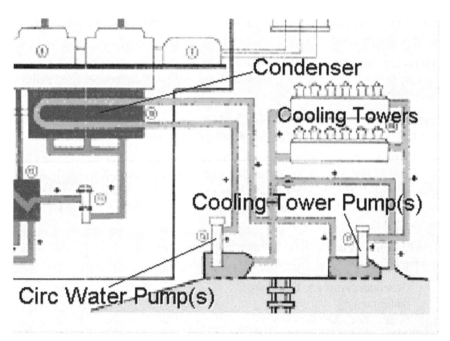

Fig. 16.7 The arrangement of components within the system and the major flow paths [3]

to the intake supply basin where the water can be reused. The Cooling Tower System is discussed elsewhere in site materials.

16.4   Service or Cooling Water Systems

Brief description of service or cooling water systems for any power plant is presented here [3].

1. **Functions**

Service Water Systems at any power plant has one major function

- Cool the multitude of heat exchangers or coolers in the power house-other than the condenser.

This system is referred to by a variety of names—Service Water, Cooling Water, and Salt Water. Sometimes the system is broken into separate building systems as Turbine Building, Auxiliary Building, and Reactor Building. Often the system is broken into safety and non-safety portions. For a number of plants the safety portion is referred to as the Essential Service Water System. The equipment in the safety portion of the system is powered by independent sources, e.g. diesel-driven pumps and diesel generators to supply electrical power.

2. **Major Components**

Normal and Emergency Intake (Supply) Basins Water is supplied from an abundant source—river, lake, sea, or ocean—to a storage basin, which, in turn, supplies the large pumps.

Emergency Intake Pipe An emergency intake pipe may be used to provide a direct alternate path to the emergency portion of the system if the normal flow path is unavailable.

Emergency or Safeguards Traveling Screens These are large steel screens up to 30 ft long that rotate and filter out larger debris that passes through the trash racks.

Service Water Pumps For many plants there are an average of three pumps for single unit plants and five pumps for dual unit facilities. These pumps each supply water at about 10,000–15,000 gallons/min to a two (2) header system. Often the pumps are broken into safety and non-safety classes. For safety-related components this method assures redundancy.

Heat Exchangers or Coolers Components cooled or served by the Service Water System include:

- Generator Hydrogen Gas Cooler.
- Stator cooling.
- Generator exciter cooler.
- Air compressor after cooler.

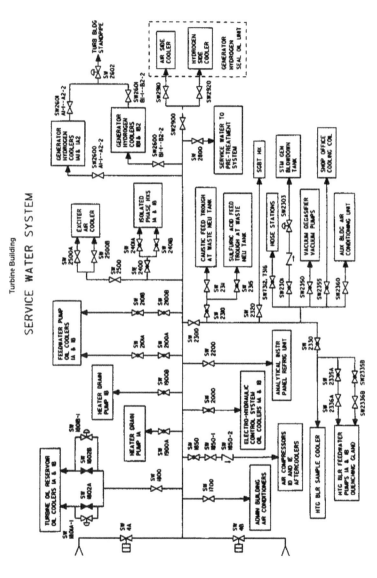

Fig. 16.8 Service water system

- Component Cooling Heat Exchanger.
- Containment Fan Cooler Units.
- Feed water Pump Oil and Motor Coolers.
- Condensate Pump Oil and motor Coolers.
- Turbine Oil Cooler
- Heater Drain Pump Oil and Motor Cooler.
- Diesel Generators
- Circulating Pump Seal Water

In addition, the Service Water system may provide the backup water source to the emergency feed water pump.

3. Simplified Diagrams

The diagram below illustrates the major flow paths for a typical Service Water System in the Turbine Building where major nonessential heat loads are cooled (Fig. 16.8).

Chapter 17
Electrical System

17.1 Introduction

A power reactor generates heat that is converted into steam. The steam can be used directly for power, as in a nuclear submarine. It can also be used to generate electric power—for example, in a commercial nuclear power plant. In the early twenty-first century more than 400 nuclear plants were generating

The interface between nuclear power plants (NPPs) and electric grids deserves careful attention especially in small electric power systems. Electric grids have an important impact on the safe and reliable operation as well as startup/shutdown of NPPs. Conversely, tripping a nuclear reactor may give rise to sizable disturbances in grid parameter (frequency and voltage), which require automated action on the part of grid to maintain the quality of electricity [1].

17.2 Balancing the Circuit to Maximize the Energy Delivered to the Load

The electrical generator driven by the steam, or gas, turbine is generally connected to a large electrical network with many components and branches. However for the sake of discussion this network can be reduced to a simple circuit as described in Fig. 17.1.

Vo represents the electrical generator. L represents the lumped inductive load in the network. R represents the resistive load in the network. C represents the lumped capacitance load in the network. Though there is a voltage drop across all three lumped parameters, the actual work is accomplished by the resistive load. The voltage drops across the lumped parameters must add up to the voltage introduced by the electrical generator.

© Springer International Publishing Switzerland 2015   455
B. Zohuri, P. McDaniel, *Thermodynamics In Nuclear Power Plant Systems*,
DOI 10.1007/978-3-319-13419-2_17

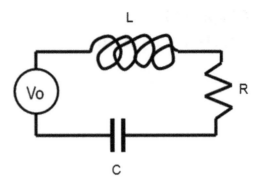

$$V_o(t) = V_L(t) + V_R(t) + V_C(t) \tag{17.1}$$

Since the generator is trying to maintain a constant frequency, either 50 Hz or 60 Hz,
its voltage can be represented by

$$V_o(t) = V_o \sin \omega t \quad \omega = 2\pi f \tag{17.2}$$

The voltage drop across an inductive load is proportional to the rate of change of the
current passing through the load.

$$V_L(t) = L\frac{di(t)}{dt} \tag{17.3}$$

The voltage drop across a resistive load is proportional to the current passing
through the load.

$$V_R(t) = i(t)R \tag{17.4}$$

The voltage generated across a capacitive load is proportional to the integral of the
current into the capacitor.

$$V_C(t) = \frac{1}{C}\int i(t)dt \tag{17.5}$$

Inserting each of these into the voltage drop equation gives,

$$V_o \sin \omega t = L\frac{di(t)}{dt} + i(t)R + \frac{1}{C}\int i(t)dt \tag{17.6}$$

Now the current must oscillate at the same frequency as the driving voltage at steady
state, but it can be out of phase with the driving voltage so it can be represented by

$$i(t)I \cos \omega t + J \sin \omega t \tag{17.7}$$

Then

$$V_o \sin \omega t = L\omega\left(-\sin \omega t + J \cos \omega t\right) + R\left(I \cos \omega t + J \sin \omega t\right) + \frac{1}{\omega C}\left(I \sin \omega t - J \cos \omega t\right) \quad (17.8)$$

Now since the $\sin \omega t$ and $\cos \omega t$ are independent functions, this equation can be split into two equations in their coefficients.

$$V_o = -IL\omega + RJ + \frac{I}{\omega C} \quad 0 = JL\omega + RI - \frac{J}{\omega C} \quad (17.9)$$

Now let

$$\Delta = \frac{1}{\omega C} - L\omega \quad (17.10)$$

This gives

$$V_o = I\Delta + RJ \quad 0 = J\Delta - RI \quad (17.11)$$

and

$$I = \frac{J\Delta}{R}, \quad V_o = \frac{\Delta^2 J}{R} + RJ \quad V_o = \left(\frac{\Delta^2 J}{R} + R\right)J \quad (17.12)$$

$$J = \frac{V_o R}{\Delta^2 + R^2} \quad I = \frac{V_o \Delta}{\Delta^2 + R^2}$$

$$i(t) = V_o\left(\frac{\Delta}{\Delta^2 + R^2}\cos \omega t + \frac{R}{\Delta^2 + R^2}\sin \omega t\right)$$

Now the power delivered to the load is

$$P_{Load} = i^2(t)R$$

$$P_{Load} = RV_o^2\left[\frac{\Delta^2 \cos^2 \omega t + 2\Delta R \cos \omega t \sin \omega t + R^2 \sin^2 \omega t}{\left(\Delta^2 + R^2\right)^2}\right] \quad (17.13)$$

Now to get the work accomplished in one cycle,

$$W_{Load} = \int_0^{2\pi} P_{Load}\, dt = RV_o^2\left[\frac{\Delta^2 \pi + R^2 \pi}{\left(\Delta^2 + R^2\right)^2}\right] = \frac{\pi RV_o^2}{\left(\Delta^2 + R^2\right)} \quad (17.14)$$

This can be maximized by setting

$$\Delta = 0 \quad or \quad \omega L = \frac{1}{\omega C} \quad LC = \frac{1}{\omega^2} = \frac{1}{4\pi^2 f^2}$$

The units on f are $1/sec$, so the units on LC should be sec^2. Typically, C is measured in *farads*, and L is measured in *henrys*.

$$1\,farad = \frac{1\,coulomb}{1\,volt} = \frac{1\,coulomb}{\dfrac{1\,joule}{1\,coulomb}} = \frac{1\,coulomb^2}{1\,joule}$$

$$1\,henry = \frac{1\,joule}{1\,amp^2} = \frac{1\,joule}{\left(\dfrac{1\,coulomb}{1\,sec}\right)^2}$$

$$1\,henry * 1\,farad = \frac{1\,joule * sec^2}{1\,coulomb^2} * \frac{1\,coulomb^2}{1\,joule} = sec^2$$

Then

$$W_{Load} = \frac{\pi V_o^2}{R} \tag{17.15}$$

Typically the dominant industrial electric load for a utility consists of large motors that add inductance to the network. To compensate for this and meet the matching condition

$$\omega L = \frac{1}{\omega C} \tag{17.16}$$

a utility will have on hand large capacitance banks that can be switched into and out of the circuit. This enables them to deliver the maximum work to the total load and efficiently use the electrical energy that has been produced

17.3   Optimizing the Transmission of Energy to the Load

The lumped resistance identified in the previous section includes all of the resistances in the network. The resistances in the consumer's devices are where the utility makes money. But the resistances in the network getting the electrical energy to the consumer also consume energy. So it is important to minimize transmission line losses. Transmission line losses are minimized by transmitting the electrical energy at the highest voltage possible consistent with safety and reliability. Note,

$$P = IV = I^2 R \ for\ a\ balanced\ network \tag{17.17}$$

Since R is fixed for a particular transmission line material, it is important to have the minimum current. The power dissipated in the transmission line material is proportional to the square of the current. Since the power is also the product of the current times the voltage, upping the voltage decreases the current. So long distance transmission lines carry the power at the highest voltage possible to minimize resistance losses in the lines.

The voltage can be increased or decreased by passage through a transformer. Typically the voltage coming out of the driven electrical generator is increased significantly to the transmission line voltage and then transmitted most of the distance to consumer centers. Then it passes through one or more transformers to drop it down to the levels required by individual consumers.

17.4   Overview of an Electrical Grid System

Since all electrical power plants cannot operate 100 % of the time, they are hooked to a network grid that has multiple power plants supplying power to multiple consumers. Consumer requirements also fluctuate, so some power plants operate continuously and some only operate when the demand is high. Obviously utilities would like the power plants that operate the cheapest to be the ones that operate continuously and the ones that are cycled to meet peak demands can be more expensive to operate. Typically hydro, coal and nuclear are operated continuously and gas and oil plants are operated intermittently to satisfy peak demands. With increasing penetration by renewable energy systems into the market and the advent of consumer power plants (solar cells and small turbines) the management of a stable electricity supply will become more complicated.

Typically, the electricity supply system comprises a single interconnected alternating current network that connects together all the power plants and other sources of electrical power with electricity to consumers. The electricity supply system uses a range of voltage levels: extra high voltage lines (e.g. 400 kV and higher three-phase) are used to transmit large amounts of power long distances; medium voltage lines are used to transmit power shorter distances and to connect large industrial consumers; and low voltages (such as 230 V single phase or 120 V single phase) are used to provide supplies to domestic consumers.

The main components of the electricity supply system include:

• Power plants of various kinds and sizes;
• Overhead lines operated at various voltages;
• Underground cables operated at various voltages;
• Substations (switchyards) with switching facilities where overhead lines and underground cables are interconnected, and where power plants may be connected;
• Transformers (often located in substations) to connect parts of the network operating at different voltages;
• HVDC connections (comprising a converter station at each end, and an overhead line or underground cable connection between them);
• Electrical protection, monitoring and metering equipment;
• Communication and control systems;
• One or more control centers.

The electricity system can be considered to comprise two significantly different kinds of networks: the transmission system(s), and the distribution systems(s). Because of their different purposes, the transmission systems and distribution systems are designed, operated and controlled in different ways.

The transmission system (also commonly called the 'grid system') comprises those parts of the system (overhead lines, underground cables, substations, transformers) operating at very high voltage (generally greater than 100 kV), which are used to transmit large amounts of power long distances between large power plants and load centers. The large majority of circuits on the transmission system are overhead lines, because of the ease and speed of installation and the much lower cost compared to underground cables. Overhead lines on the transmission system are normally carried on tall steel lattice towers.

The distribution systems comprise those parts of the system operated at lower voltages (less than 100 kV), and used to transmit smaller amounts of power shorter distances from the high voltage transmission network to individual customers at the voltage level the customer needs. The overhead lines on distribution systems may be carried on smaller steel lattice towers, or wooden or concrete poles. In urban areas underground cables are often used instead of overhead lines. Typically the transmission and distribution systems are owned and operated by different companies or organizations.

17.5   How Power Grids System Work

Power travels from the power plant to your house through a complex system called the power distribution grid. The grid system is exposed in public—if you live in a rural, urban and suburban area, chances are it is right out in the open for all to see. It is so public, in fact, that you probably do not even notice it anymore. Your brain likely ignores all of the power lines because it has seen them so often. In this section, we will look at all of the equipment that brings electrical power to your home.

Electrical power starts at the power plant. In almost all cases, the power plant consists of a spinning electrical generator. Something has to spin that generator—it might be a water wheel in a hydroelectric dam, a large diesel engine or a gas turbine. But in most cases, the element spinning the generator is a steam turbine. The steam might be created by burning coal, oil or natural gas. Or the steam may come from a nuclear reactor.

Extra high voltage electricity is delivered through the power grid from where it is generated and along the way converted into manageable voltage levels to be used by the consumer. No matter what it is that spins the generator, commercial electrical generators of any size generate what is called 3-phase AC power. To understand 3-phase AC power, it is helpful to understand single-phase power first (Fig. 17.2 and 17.3).

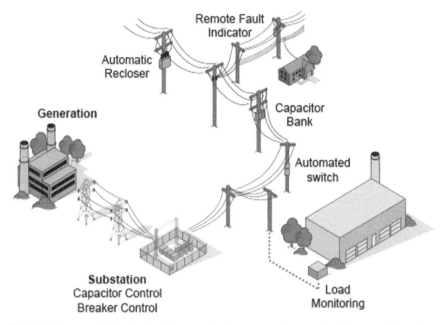

Fig. 17.2 Power grid distribution lines can be above or under ground. (Courtesy of American Electric Power)

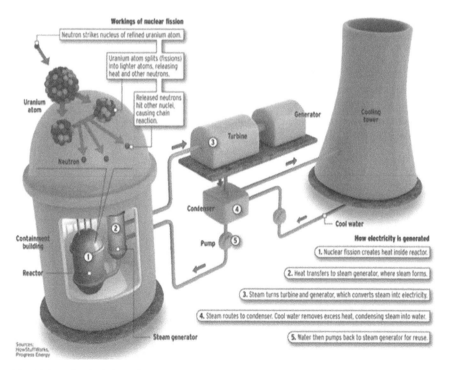

Fig. 17.3 Artist depiction of nuclear power components. (Courtesy of Micheal Barts/The News & Observer)

17.5.1   Electrical Alternating (AC)

Most household electrical services are as single-phase, 120-V AC service. If you use an oscilloscope and look at the power found at a normal wall-plate outlet in your house, what you will find is that the power at the wall plate looks like a sine wave, and that wave oscillates between -170 V and 170 V (the peaks are indeed at 170 V; it is the effective (rms) voltage that is 120 V). The rate of oscillation for the sine wave is 60 cycles per second.

Oscillating power like this is generally referred to as Alternating Current (AC), or alternating current. The alternative to AC is Direct Current (DC), or direct current. Batteries produce DC: A steady stream of electrons flows in one direction only, from the negative to the positive terminal of the battery. AC has at least three advantages over DC in a power distribution grid:

1. Large electrical generators happen to generate AC naturally, so conversion to DC would involve an extra step.
2. Transformers must have alternating current to operate, and we will see that the power distribution grid depends on transformers.
3. It is easy to convert AC to DC but expensive to convert DC to AC, so if you were going to pick one or the other AC would be the better choice.

The power plant, therefore, produces AC. On the next page, you will learn about the AC power produced at the power plant. Most notably, it is produced in three phases.

17.5.2   Three-Phase Power

The power plant produces three different phases of AC power simultaneously, and the three phases are offset $120°$ from each other. There are four wires coming out of every power plant: the three phases plus a neutral or ground common to all three. If you were to look at the three phases on a graph, they would look like this relative to ground (Fig. 17.4):

Fig. 17.4 Graph of 3-Phase Power. (Courtesy of Group Builder, Inc.)

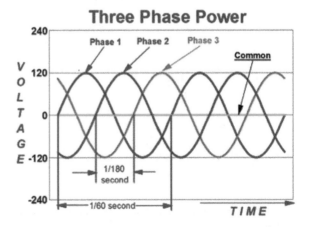

There is nothing magical about three-phase power. It is simply three single phases synchronized and offset by 120°.

Why three phases? Why not one or two or four? In 1-phase and 2-phase power, there are 120 moments per second when a sine wave is crossing zero volts. In 3-phase power, at any given moment one of the three phases is nearing a peak. High-power 3-phase motors (used in industrial applications) and things like 3-phase welding equipment therefore have even power output. Four phases would not significantly improve things but would add a fourth wire, so 3-phase is the natural settling point.

In addition, what about this "ground," as mentioned above? The power company essentially uses the earth as one of the wires in the power system. The earth is a pretty good conductor and it is huge, so it makes a good return path for electrons. (Car manufacturers do something similar; they use the metal body of the car as one of the wires in the car's electrical system and attach the negative pole of the battery to the car's body.) "Ground" in the power distribution grid is literally "the ground" that's all around you when you are walking outside. It is the dirt, rocks, groundwater, etc., of the earth.

17.5.3 Transmission System

Perhaps the most recognizable public part of the power grid is the high voltage network of transmission lines, supported by large metal pylons, that threads its way across the country. Other components of the grid include terminal stations, zone substations and sub-transmission lines, carrying low voltage electricity to the common consumers who may live in the city or the country, by using the power grid.

Extra high voltage transmission lines are needed to carry large amount of electricity over long distances. Most of the transmission lines operate at voltage of 500 kV. These extra high voltage lines have higher energy efficiency than a large number of lower voltage lines. They are more economic to construct, operate and maintain (Fig. 17.5).

Most transmission lines are overhead lines with conductors supported on steel lattice towers. The conductors are insulated from the towers by porcelain, glass or synthetic insulators. Transmission lines can also be constructed with underground cable. Extra high voltage underground cables are considerably more expensive than overhead lines of equivalent capacity; therefore they are not very cost effective form ownership point of view by utility companies.

17.5.4 Substation (Terminal Station) System

The power carried over the extra high voltage lines must be changed to a lower voltage before it can be used in the home or industry. This takes place in several stages. Firstly, the electricity is delivered to substations either to change the voltage level

Fig. 17.5 High voltage
transmission lines. (Courtesy
of STAV Publishing)

of the power or to provide a switching point for a number of transmission lines. The voltage of the electricity is lowered at the terminal station to 66,000 V (66 kV) by a transformer.

17.5.5 Zone Substation System

After the terminal station is where electricity distribution businesses such as city power and power core elements come into play. The electricity is distributed on the sub-transmission system, which is made up of very tall concrete or wooden power poles and power lines, or sometimes-underground power lines.

The sub-transmission system transports the electricity to one of zone substations where it is converted from 66,000 V (66 kV) to 22,000 V (22 kV) or 11,000 V (11 kV). Electricity at this voltage can be then be distributed on smaller, lighter power poles.

From this point, high voltage distribution lines transfer the electricity from the zone substations to distribution substation, where it is transformed to 230–240 V for supply to customers (Fig. 17.6).

Fig. 17.6 A typical zone substation. (Courtesy of STAV Publishing)

Fig. 17.7 A typical regulator
bank. (Courtesy of Langly
Engineering)

17.5.6   Regulator Bank System

You will also find regulator banks located either along the line, underground or in
the air. They regulate the voltage on the line to prevent undervoltage and overvolt-
age conditions (Fig. 17.7).

Up toward the top are three switches that allow this regulator bank to be discon-
nected for maintenance when necessary (Fig. 17.8):

Fig. 17.8 Regulator bank
switches. (Courtesy of S&C
Electric Company)

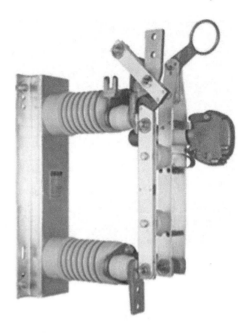

Fig. 17.9 A typical line
voltage. (Courtesy of Scott
Engineering)

At this point, we have typical line voltage at something like 7200 V running
through the neighborhood on three wires (with a fourth ground wire lower on the
pole) (Fig. 17.9):

17.5.7 Taps System

A house needs only one of the three phases, so typically you will see three wires
running down a main road, and taps for one or two of the phases running off on side
streets. Pictured below is a 3-phase to 2-phase tap, with the two phases running off
to the right (Fig. 17.10):

Here is a 2-phase to 1-phase tap, with the single phase running out to the right
(Fig. 17.11):

Fig. 17.10 A 3-Phase
or 2-Phase tap picture.
(Courtesy of Cooper Power
System)

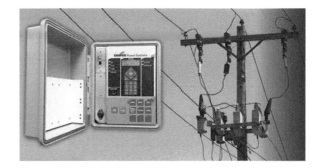

Fig. 17.11 Pole mounted
single-phase with three-wire
center tapped picture. (Courtesy of Wikipedia)

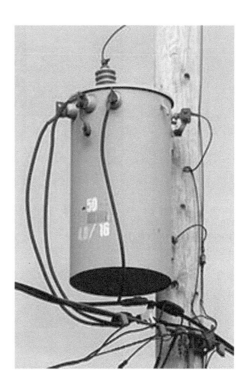

17.5.8   At the House Level

Finally, we are down to the wire that brings power to your house! Past a typical
house runs a set of poles with one phase of power (at 7200 V) and a ground wire
(although sometimes there will be two or three phases on the pole, depending on
where the house is located in the distribution grid). At each house, there is a transformer drum attached to the pole, like this (Fig. 17.12):

In many suburban neighborhoods, the distribution lines are underground and
there are green transformer boxes at every house or two. Here is some detail on
what is going on at the pole (Fig. 17.13):

Fig. 17.12 Typical poles
with 1-Phase power (7200 V)
at the house street picture.
(Courtesy of epb)

Fig. 17.13 Distribution lines.
(Courtesy of Wikipedia)

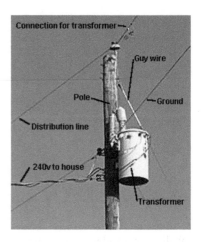

The transformer's job is to reduce the 7200 V down to the 240/230 V that makes up normal household electrical service. Let's look at this pole one more time, from the bottom, to see what is going on (Fig. 17.14):

There are two things to notice in this picture:

Fig. 17.14 Transformer pole.
(Courtesy of Wikipedia)

1. There is a bare wire running down the pole. This is a grounding wire. Every utility pole on the planet has one. If you ever watch the power company install a new pole, you will see that the end of that bare wire is stapled in a coil to the base of the pole and therefore is in direct contact with the earth, running 6–10 ft (1.8–3 m) underground. It is a good, solid ground connection. If you examine a pole carefully, you will see that the ground wire running between poles (and often the guy wires) are attached to this direct connection to ground.

2. There are two wires running out of the transformer and three wires running to the house. The two from the transformer are insulated, and the third one is bare. The bare wire is the ground wire. The two insulated wires each carry 120 V, but they are 180° out of phase so the difference between them is 240 V. This arrangement allows a homeowner to use both 120-V and 240-V appliances. The transformer is wired in this sort of configuration (Fig. 17.15):

The 240 V enters your house through a typical watt-hour meter like this one below (Fig. 17.16):

The meter lets the power company charge you for putting up all of these wires.

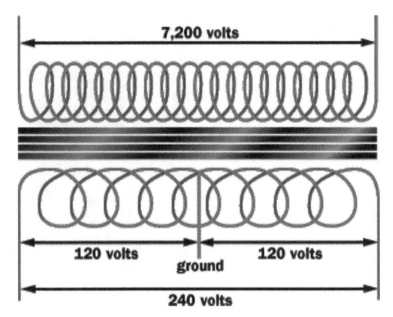

Fig. 17.15 Transformer wiring system

Fig. 17.16 Typical power
meter. (Courtesy of PG&E)

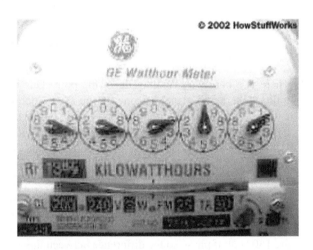

© 2002 HowStuffWorks

17.5.9   Safety Devices: Fuses, Circuit Breakers, Plugs and Outlets

Fuses and circuit breakers are safety devices. Let us say that you did not have fuses or circuit breakers in your house and something "went wrong." What could possibly go wrong? Here are some examples:

Fig. 17.17 Typical circuit
breaker panel

© 2002 HowStuffWorks

- A fan motor burn out a bearing, seizes, overheats and melts, causing a direct connection between power and ground.
- A wire comes loose in a lamp and directly connects power to ground.
- A mouse chews through the insulation in a wire and directly connects power to
  ground.
- Someone accidentally vacuums up a lamp wire with the vacuum cleaner, cutting
  it in the process and directly connecting power to ground.
- A person is hanging a picture in the living room and the nail used for said picture happens to puncture a power line in the wall, directly connecting power to
  ground.

When a 120-V power line connects directly to ground, its goal in life is to pump as
much electricity as possible through the connection. Either the device or the wire
in the wall will burst into flames in such a situation. (The wire in the wall will get
hot as the element in an electric oven gets hot, which is to say very hot!). A fuse is
a simple device designed to overheat and burn out extremely rapidly in such a situation. In a fuse, a thin piece of foil or wire quickly vaporizes when an overload of
current runs through it. This kills the power to the wire immediately, protecting it
from overheating. Fuses must be replaced each time they burn out. A circuit breaker
uses the heat from an overload to trip a switch, and circuit breakers are therefore
resettable (Fig. 17.17).

The power then enters the home through a typical circuit breaker panel like the
one above.

Fig. 17.18 Typical circuit
breaker or fuses. (Courtesy of
Mission Electric Service)

Fig. 17.19 Typical circuit outlet
in USA. (Courtesy of ACE Hard-
ware Store)

Inside the circuit breaker panel, you can see the two primary wires from the
transformer entering the main circuit breaker at the top. The main breaker lets you
cut power to the entire panel when necessary. Within this overall setup, all of the
wires for the different outlets and lights in the house each have a separate circuit
breaker or fuse (Fig. 17.18):

If the circuit breaker is on, then power flows through the wire in the wall and
makes its way eventually to its final destination, the outlet (Fig. 17.19).

Fig. 17.20 Control center operators monitor the flow of electricity

17.5.10 Control Centers

The flow of electricity through the power grid is controlled from several control centers throughout the State. These control centers use computers and communication systems to enable the operators to monitor both the flow of electricity and the condition of the power grid (Fig. 17.20).

17.5.11 Interstate Power Grids

Electricity power delivery does not recognize state boundaries. Power grid is joined to similar grids in entire net of power stations. This allows the buying and selling of electricity between states. The interconnection grid enables the more efficient use of resources and a reduced need for generating power plants because generation capacity can be shared between the stakeholders and states (Fig. 17.21).

17.6 United States Power Grid

Initially, power is delivered from one of about 10,000 power plants in the U.S. where the electricity itself is created, to a portion of the power distribution grid. Though we may have the idea that there is just one "grid" in the United States, in fact, there are many. What we refer to as the power grid is simply the interconnected lines of these transmission stations throughout the country.

Currently, the U.S. electrical grid is undergoing changes in the same way that cell phones have become "smart phones." Utility companies are looking at ways to use technology to increase the reliability and effectiveness of electricity supply. At

How electricity is delivered

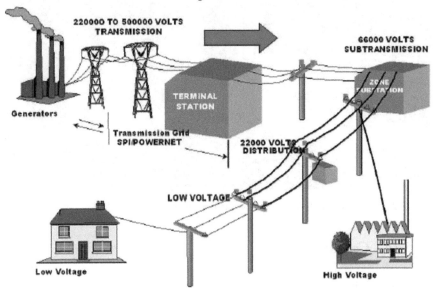

Fig. 17.21 How electricity is delivered (Courtesy of STAV Publishing)

this time, sections of the electrical grid are divided into Eastern, Western and Texas (ERCOT) areas, with electrical supply in all of those areas independently owned by many different companies.

One of the biggest challenges with the U.S. power grid and how it works is it's age and changes in regulation. With a base of development during the 1960's, power needs in America have changed radically. Originally, power was created and distributed locally. Now, large, multi-state organizations create and sell their power across vast regions of America. Adding to the confusion is the requirement that many states have created to deregulate power. Now, deregulation has caused confusion as to who is to maintain these transmission lines across the grid.

Ultimately, the electrical grid that we think of is nothing but a series of transmission lines that are all linked together into three separate regional areas in the US. While there is a government-led movement to update the way electricity is managed, currently, the grid operates quite simply. First, power is developed at a power plant, by burning fossil fuels, using solar power or even by positioning turbines under the falling water of a hydroelectric dam. Once the power is created, it is sent to a transmission sub-station near the power plant where it is broken down into smaller units, where it is then transmitted over high-voltage transmission lines. Once the electricity is moved over high-voltage transmission lines, it it deposited at a power substation. Because the power at the power substation is so potent, it is then transformed, using a "transformer," to be sent over local power lines. Finally, the power is transformed one more time, at either a transformer drum or ground-based transformer to be delivered into your home.

Fig. 17.22 Electrical grid distribution in United States Department of Energy Graphic. (Courtesy of Department of Energy)

Knowing that the electricity you use is part of the United States power grid simply gives you the information as to what region your electric service resides in. With upgrades into a "smart" power grid in the future, electric users can count on a more reliable and efficient electric supply throughout the United States.

As we said, the electric grid delivers electricity from points of generation to consumers, and the electricity delivery network functions via two primary systems: the transmission system and the distribution system. The transmission system delivers electricity from power plants to distribution substations, while the distribution system delivers electricity from distribution substations to consumers. The grid also encompasses myriads of local area networks that use distributed energy resources to serve local loads and/or to meet specific application requirements for remote power, village or district power, premium power, and critical loads protection.

It is important to note that there is no "national power grid" in the United States. In fact, the continental United States is divided into three main power grids (Fig. 17.22):

1. The Eastern Interconnected System, or the Eastern Interconnect.
2. The Western Interconnected System, or the Western Interconnect.
3. The Texas Interconnected System, or the Texas Interconnect.

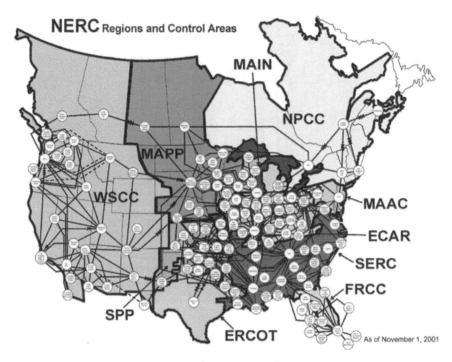

Fig. 17.23 Dynamically controlled generation. (Courtesy of Department of Energy)

ECAR	East Central Area Reliability Coordination Agreement
ERCOT	Electric Reliability Council of Texas
FRCC	Florida Reliability Coordinating Council
MAAC	Mid-Atlantic Area Council
MAIN	Mid-America Interconnected Network
MAPP	Mid-Continent Area Power Pool
NPCC	Northeast Power Coordinating Council
SERC	Southeastern Electric Reliability Council
SPP	Southwest Power Pool
WSCC	Western Systems Coordinating Council

The Eastern and Western Interconnects have limited interconnections to each other, and the Texas Interconnect is only linked to the others via direct current lines. Both the Western and Texas Interconnects are linked with Mexico, and the Eastern and Western Interconnects are strongly interconnected with Canada. All electric utilities in the mainland United States are connected to at least one other utility via these power grids.

The grid systems in Hawaii and Alaska are much different from on the U.S. mainland. Alaska has an interconnected grid system, but it connects only Anchorage, Fairbanks, and the Kenai Peninsula. Much of the rest of the state depends on small diesel generators, although there are a few minigrids in the state as well. Hawaii also depends on minigrids to serve each island's inhabitants (Fig. 17.23).

Power grids are inherently prone to big blackouts. The United States electric power grid is growing increasingly complex and interconnected, with a greater number of power buyers and sellers making a burgeoning number of transactions.

The U.S. power grid is outmoded, with too few transmission lines to handle Americans' ever-growing appetite for electricity.

17.7   Smart Power Grid (SG)

"Smart Grid" generally refers to a class of technology people are using to bring utility electricity delivery systems into the twenty-first century, using computer-based remote control and automation. These systems are made possible by two-way communication technology and computer processing that has been used for decades in other industries. They are beginning to be used on electricity networks, from the power plants and wind farms all the way to the consumers of electricity in homes and businesses. They offer many benefits to utilities and consumers—mostly seen in big improvements in energy efficiency on the electricity grid and in the energy users' homes and offices.

For a century, utility companies have had to send workers out to gather much of the data needed to provide electricity. The workers read meters, look for broken equipment and measure voltage, for example. Most of the devices utilities use to deliver electricity have yet to be automated and computerized. Now, many options and products are being made available to the electricity industry to modernize it.

The "grid" amounts to the networks that carry electricity from the plants where it is generated to consumers. The grid includes wires, substations, transformers, switches and much more.

Much in the way that a "smart" phone these days means a phone with a computer in it, smart grid means "computerizing" the electric utility grid. It includes adding two-way digital communication technology to devices associated with the grid. Each device on the network can be given sensors to gather data (power meters, voltage sensors, fault detectors, etc.), plus two-way digital communication between the device in the field and the utility's network operations center. A key feature of the smart grid is automation technology that lets the utility adjust and control each individual device or millions of devices from a central location (Fig. 17.24).

Smart Grid is about the big picture of improved energy delivery, informed consumption and reduced environmental impact. The new Smart Grid (SG) by various companies is offering a complete spectrum of products, solutions and services for the protection, automation, planning, monitoring and diagnosis of grid infrastructure, including products, complete solutions and services for rail electrification.

Companies in this technology are bringing to market Smart Grid solutions that can evolve over time, are scalable and are compliant with industry standards for interoperability and security. They are making the Smart Grid a reality by partnering with utilities and cities to implement their Smart Grid visions.

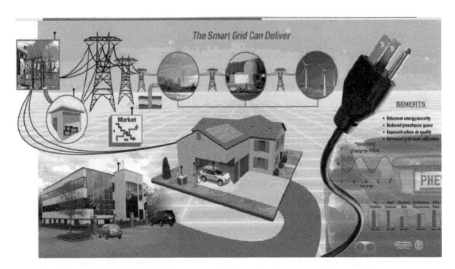

Fig. 17.24 Smart grid illustration

The number of applications that can be used on the smart grid once the data communications technology is deployed is growing as fast as inventive companies can create and produce them. Benefits include enhanced cyber-security, handling sources of electricity like wind and solar power and even integrating electric vehicles onto the grid. The companies making smart grid technology or offering such services include technology giants, established communication firms and even brand new technology firms.

Reference

1. International Atomic Energy Agency (2012a) Electric grid reliability and interface with nuclear power plants, IAEA Nuclear Energy Series No. NG-T-3.8, IAEA, Vienna

Chapter 18
Nuclear Power Plants

Currently, about half of all nuclear power plants are located in the US. There are many different kinds of nuclear power plants, and we will discuss a few important designs in this text. A nuclear power plant harnesses the energy inside atoms themselves and converts this to electricity. All of us use this electricity. In Sect. 18.1 of this chapter we show you should the idea of the fission process and how it works. A nuclear power plant uses controlled nuclear fission. In this chapter, we will explore how a nuclear power plant operates and the manner in which nuclear reactions are controlled. There are several different designs for nuclear reactors. Most of them have the same basic function, but one's implementation of this function separates it from another. There is several classification systems used to distinguish between reactor types. Below is a list of common reactor types and classification systems found throughout the world and they are briefly explained down below according to three types of classification either; (1) Classified by Moderator Material or (2) Classified by Coolant Material and (3) Classified by Reaction Type.

18.1 Fission Energy Generation

There is strategic as well as economic necessity for nuclear power in the United States and indeed most of the world. The strategic importance lies primarily in the fact that one large nuclear power plant saves more than 50,000 barrels of oil per day. At $ 30–40 per barrel (1982), such a power plant would pay for its capital cost in a few short years. For those countries that now rely on but do not have oil, or must reduce the importation of foreign oil, these strategic and economic advantages are obvious. For those countries that are oil exporters, nuclear power represents an insurance against the day when oil is depleted. A modest start now will assure that they would not be left behind when the time comes to have to use nuclear technology.

The unit costs per kilowatt-hour for nuclear energy are now comparable to or lower than the unit costs for coal in most parts of the world. Other advantages are the lack of environmental problems that are associated with coal or oil-fired power

© Springer International Publishing Switzerland 2015
B. Zohuri, P. McDaniel, *Thermodynamics In Nuclear Power Plant Systems*,
DOI 10.1007/978-3-319-13419-2_18

plants and the near absence of issues of mine safety, labor problems, and transportation bottlenecks. Natural gas is a good, relatively clean-burning fuel, but it has some availability problems in many countries and should, in any case, be conserved for small-scale industrial and domestic uses. Thus, nuclear power is bound to become the social choice relative to other societal risks and overall health and safety risks.

Nuclear fission is the process of splitting atoms, or fissioning them. This page will explain

18.2   The First Chain Reaction

Early in World War II the scientific community in the United States, including those Europeans now calling the US their safe home, pursued the idea that uranium fission and the production of excess neutrons could be the source of extraordinary new weapons. They knew Lisa Meitner's interpretation, in Sweden, of Hahn's experiments would likely be known in Germany. Clearly there might now be a race commencing for the development and production of a new, super weapon based on the fission of $^{235}U_{92}$ or $^{239}Pu_{94}$.

By early 1942, it was known that the two naturally occurring isotopes of uranium reacted with neutrons as follows:

$$^{235}U_{92} + {}^1n_0 \rightarrow \text{fission products} + (2.5)^1 n_0 + 200 \text{ MeV Energy}$$

$$^{238}U_{92} + {}^1n_0 \rightarrow {}^{239}U_{92}$$

$$^{239}U_{92} \rightarrow {}^{239}Np_{93} + \beta^{-1} \quad t_{1/2} = 23.5 \text{ min.}$$

$$^{239}Np_{93} \rightarrow {}^{239}Pu_{94} + \beta^{-1} \quad t_{1/2} = 2.33 \text{ days}$$

Each U-235 that undergoes fission produces an average of 2.5 neutrons. In contrast, some U-238 nuclei capture neutrons, become U-239, and subsequently emit two beta particles to produce Pu-239. The plutonium was fissile also and would produce energy by the same mechanism as the uranium. A flow sheet for uranium fission is shown in Fig. 18.1 below.

The answers to two questions were critical to the production of plutonium for atomic bombs:

1. Is it possible, using natural uranium (99.3 % U-238 and 0.7 % U-235), to achieve a controlled chain reaction on a large scale? If so, some of the excess neutrons produced by the fission of U-235 would be absorbed by U-238 and produce fissile Pu-239.

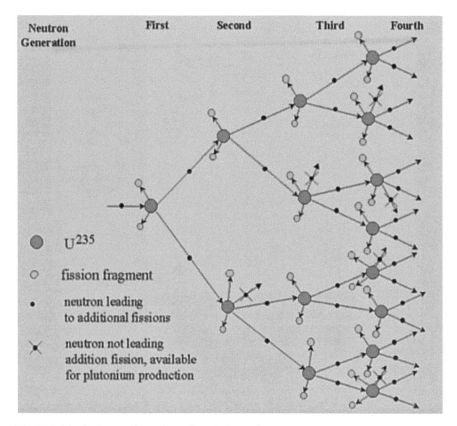

Neutron Generation

First   Second   Third   Fourth

🔴 U²³⁵

⭕ fission fragment

• neutron leading
to additional fissions

✕ neutron not leading
addition fission, available
for plutonium production

Fig. 18.1 The first generations of a nuclear chain reaction

2. How can we separate (in a reasonable period of time) the relatively small quantities of Pu-239 from the unreacted uranium and the highly radioactive fission-product elements?

Although fission had been observed on a small scale in many laboratories, no one had carried out a controlled chain reaction that would provide continuous production of plutonium for isolation.

Enrico Fermi thought that he could achieve a controlled chain reaction using natural uranium. He had started this work with Leo Szilard at Columbia University, but moved to the University of Chicago in early 1942.

The first nuclear reactor, called a pile, was a daring and sophisticated experiment that required nearly 50 t of machined and shaped uranium and uranium oxide pellets along with 385 t—the equivalent of four railroad coal hoppers—of graphite blocks, machined on site.

The pile itself was assembled in a squash court under the football field at the University of Chicago from the layered graphite blocks and uranium and uranium oxide lumps (Fermi's term) arranged roughly in a sphere with an anticipated

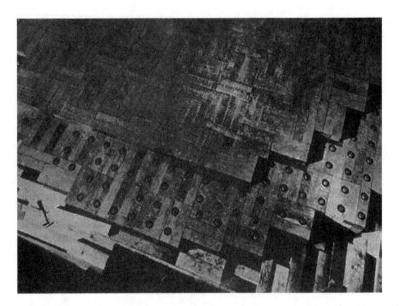

Fig. 18.2 CP-1-Graphite blocks with 3 inch diameter uranium cylinders inserted—part of a layer of CP-1, the first nuclear reactor. A layer of graphite blocks without inserted uranium is seen covering the active layer

13 foot radius. Neutron absorbing, cadmium coated control rods were inserted in the pile. By slowly withdrawing the rods, neutron activity within the pile was expected to increase and at some point, Fermi predicted, there would be one neutron produced for each neutron absorbed in either producing fission or by the control rods (Fig. 18.2).

On December 2, 1942, with 57 of the anticipated 75 layers in place, Fermi began the first controlled nuclear chain reaction occurred. At around 3:20 p.m. the reactor went critical; that is, it produced one neutron for every neutron absorbed by the uranium nuclei. Fermi allowed the reaction to continue for the next 27 min before inserting the neutron-absorbing control rods. The energy releasing nuclear chain reaction stopped as Fermi predicted it would.

In addition to excess neutrons and energy, the pile also produced a small amount of Pu-239, the other known fissile material (Fig. 18.3).

The achievement of the first sustained nuclear reaction was the beginning of a new age in nuclear physics and the study of the atom. Humankind could now use the tremendous potential energy contained in the nucleus of the atom. However, while a controlled chain reaction was achieved with natural uranium, and could produce plutonium, it would be necessary to separate U-235 from U-238 to build a uranium bomb.

On December 28, 1942, upon reviewing a report from his advisors, President Franklin Roosevelt recommended building full-scale plants to produce both U-235 and Pu-239.

Fig. 18.3 The first controlled chain reaction, Stagg Field, Chicago, December 2, 1942. (Courtesy of the Argonne National Laboratory)

This changed the effort to develop nuclear weapons from experimental work in academic laboratories administered by the U.S. Office of Scientific Research and Development to a huge effort by private industry. This work, supervised by the U.S. Army Corps of Engineers, was codenamed the Manhattan Project. It spread throughout the entire United States, with the facilities for uranium and plutonium production being located at Oak Ridge, Tennessee, and Hanford, Washington, respectively. Work on plutonium production continued at the University of Chicago, at what became known as the Metallurgical Laboratory or Met Lab. A new laboratory at Los Alamos, New Mexico, became the focal point for development of the uranium and plutonium bombs.

18.3 Concepts in Nuclear Criticality

A nuclear reactor works on the principle of a chain reaction. An initial neutron is absorbed by a fissile nuclide and during the process of fission; additional neutrons are released to replace the neutron that was consumed. If more neutrons are produced than are consumed, then the neutron population grows. If fewer neutrons are produced than are consumed, the neutron population shrinks. The number of fissions caused by the neutron population determines the energy released.

In order to quantify this concept let us define a multiplication factor k. We will define k as the ratio of the production to consumption of neutrons.

$$k = \frac{Production}{Consumption} \tag{18.1}$$

18.4 Fundamental of Fission Nuclear Reactors

Today many nations are considering an expanded role for nuclear power in their energy portfolios. This expansion is driven by concerns about global warming, growth in energy demand, and relative costs of alternative energy sources. In 2008, 435 nuclear reactors in 30 countries provided 16% of the world's electricity. In January 2009, 43 reactors were under construction in 11 countries, with several hundred more projected to come on line globally by 2030.

Concerns over energy resource availability, climate change, air quality, and energy security suggest an important role for nuclear power in future energy supplies. While the current Generation II and III nuclear power plant designs provide a secure and low-cost electricity supply in many markets, further advances in nuclear energy system design can broaden the opportunities for the use of nuclear energy. To explore these opportunities, the U.S. Department of Energy's Office of Nuclear Energy has engaged governments, industry, and the research community worldwide in a wide ranging discussion on the development of next generation nuclear energy systems known as "Generation IV." See Sect. 18.4 of this Chapter for more information on New Generation of Power Plant know as Gen IV (Fig. 18.4).

Nuclear reactors produce energy through a controlled fission chain reaction (See Sect. 1.1 in above: The First Chain Reaction). While most reactors generate electric power, some can also produce plutonium for weapons and reactor fuel. Power reactors use the heat from fission to produce steam, which turns turbines to generate electricity. In this respect, they are similar to plants fueled by coal and natural gas. The components common to all nuclear reactors include a fuel assembly, control rods, a coolant, a pressure vessel, a containment structure, and an external cooling facility.

In a nuclear reactor neutrons interact with the nuclei of the surrounding atoms. For some nuclei (e.g. U-235) an interaction with a neutron can lead to fission: the nucleus is split into two parts, giving rise to two new nuclei (the so-called fission products), energy and a number of new highly energetic neutrons. Other possible interactions are absorption (the neutron is removed from the system), and simple collisions, where the incident neutron transfers energy to the nucleus, either elastically (hard sphere collision) or inelastically.

The speed of the neutrons in the chain reaction determines the reactor type (See Fig. 16.5). Thermal reactors use slow neutrons to maintain the reaction. These reactors require a moderator to reduce the speed of neutrons produced by fission. Fast

Fig. 18.4 A nuclear power plant. (Courtesy of R2 Controls)

Fig. 18.5 Types of nuclear reactors. (Courtesy of Chem Cases)

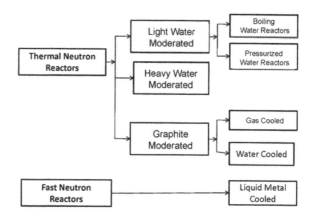

neutron reactors, also known as Fast Breeder Reactors (FBR), use high speed, un-moderated neutrons to sustain the chain reaction (Fig. 18.5).

Thermal reactors operate on the principle that uranium-235 undergoes fission more readily with slow neutrons than with fast ones. Light water (H^2O), Heavy water (D_2O), and carbon in the form of graphite are the most common moderators. Since slow neutron reactors are highly efficient in producing fission in uranium-235, they use fuel assemblies containing either natural uranium (0.7 % U-235) or slightly enriched uranium (0.9–2.0 % U-235) fuel. Rods composed of neutron-absorbing material such as cadmium or boron are inserted into the fuel assembly. The position of these control rods in the reactor core determines the rate of the fission chain reaction. The coolant is a liquid or gas that removes the heat from the core and produces steam to drive the turbines. In reactors using either light water or heavy water, the coolant also serves as the moderator. Reactors employing gaseous coolants (CO_2 or He) use graphite as the moderator. The pressure vessel, made of heavy-duty steel, holds the reactor core containing the fuel assembly, control rods, moderator, and coolant. The containment structure, composed of thick concrete and steel, inhibits the release of radiation in case of an accident and also secures components of the reactor from potential intruders. Finally, the most obvious components of many nuclear power plants are the cooling towers, the external components, which provide cool water for condensing the steam to water for recycling into the containment structure. Cooling towers are also employed with coal and natural gas plants.

18.5   Reactor Fundamentals

It is important to realize that while the U-235 in the fuel assembly of a thermal reactor is undergoing fission, some of the fertile U-238 present in the assembly is also absorbing neutrons to produce fissile Pu-239. Approximately one third of the energy produced by a thermal power reactor comes from fission of this plutonium. Power reactors and those used to produce plutonium for weapons operate in dif-

Fig. 18.6 The fate of plutonium in a thermal reactor.
(Courtesy of Chem Cases)

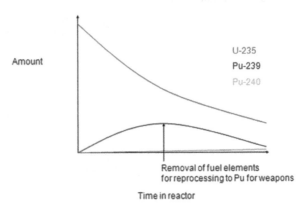

Production of plutonium in a nuclear reactor

ferent ways to achieve their goals. Production reactors produce less energy and
thus consume less fuel than power reactors. The removal of fuel assemblies from a
production reactor is timed to maximize the amount of plutonium in the spent fuel
(See Fig. 15.6). Fuel rods are removed from production reactors after only several
months in order to recover the maximum amount of plutonium-239. The fuel assemblies remain in the core of a power reactors for up to 3 years to maximize the
energy produced. However, it is possible to recover some plutonium from the spent
fuel assemblies of a power reactor (Fig. 18.6).

The power output or capacity of a reactor used to generate electricity is measured
in megawatts of electricity, MW(e). However, due to the inefficiency of converting
heat into electricity, this represents only about one third of the total thermal energy,
MW(t), produced by the reactor. Plutonium production is related to MW(t). A production reactor operating at 100 MW(t) can produce 100 g of plutonium per day or
enough for one weapon every 2 months.

Another important property of a reactor is its capacity factor. This is the ratio
of its actual output of electricity for a period of time to its output if it had been
operated at its full capacity. The capacity factor is affected by the time required for
maintenance and repair and for the removal and replacement of fuel assemblies.
The average capacity factor for U.S. reactors has increased from 50% in the early
1970s to over 90% today. This increase in production from existing reactors has
kept electricity affordable.

18.6   Thermal Reactors

Currently the majority of nuclear power plants in the world are water-moderated,
thermal reactors. They are categorized as either light water or heavy water reactors. Light water reactors use purified natural water (H_2O) as the coolant/modera-
tor, while heavy water reactors employ heavy water, deuterium oxide (D_2O). In

light water reactors, the water is either pressured to keep it in superheated form (in a pressurized water reactors, PWR) or allowed to vaporize, forming a mixture of water and steam (in a boiling water reactors, BWR). In a PWR (Fig. 16.10), superheated water flowing through tubes in the reactor core transfers the heat generated by fission to a heat exchanger, which produces steam in a secondary loop to generate electricity. None of the water flowing through the reactor core leaves the containment structure. In a BWR (Fig. 16.12), the water flowing through the core is converted to directly to steam and leaves the containment structure to drive the turbines. Light water reactors use low enriched Uranium as fuel. Enriched fuel is required because natural water absorbs some of the neutrons, reducing the number of nuclear fissions. All of the 103 nuclear power plants in the United States are light water reactors; 69 are PWRs and 34 are BWRs.

18.7   Nuclear Power Plants and Their Classifications

A nuclear power plant uses controlled nuclear fission. In this section, we will explore how a nuclear power plant operates and the manner in which, nuclear reactions are controlled. There are several different designs for nuclear reactors. Most of them have the same basic function, but one's implementation of this function separates it from another. There is several classification systems used to distinguish between reactor types. Below is a list of common reactor types and classification systems found throughout the world and they are briefly explained down below according to three types of classification either;

1. Classified by Moderator Material [i.e. Light Water Reactor, or Graphite Moderated Reactor, and Heavy Water Reactor].
2. Classified by Coolant Material [i.e. Pressurized Water Reactor, or Boiling Water Reactor, and Gas Cooled Reactor].
3. Classified by Reaction Type [i.e. Fast Neutron Reactor, or Thermal Neutron Reactor, and Liquid Metal Fast Breeder Reactor].

18.8   Classified by Moderator Material

These types of reactors and their general description are presented below;

18.8.1   Light Water Reactors (LWR)

A light water reactor is a type of thermal reactor that uses "light water" (plain water) as a neutron moderator or coolant instead of using deuterium oxide (^2H_2O); light water reactors are the most commonly used among thermal reactors. Light water reactors are contained in highly pressurized steel vessels called reactor vessels. Heat

Fig. 18.7 A pumpless light
water reactor

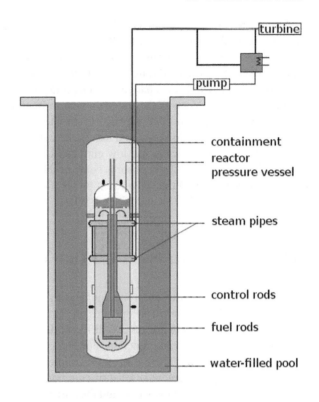

is generated by means of nuclear fission within the core of the reactor. The hundreds
into a "fuel assembly" about 12 ft in length and about as thin as a pencil, group the
nuclear fuel rods, each. Each fuel rod contains pellets of an oxidized form of Uranium (UO_2). A light water fuel reactor uses ordinary water to keep the system cool.
The water is circulated past the core of the reactor to absorb the generated heat. The
heated water then travels away from the reactor where it leaves the system as nothing more than water vapor. This is the method used in all LWRs except the BWR
for in that specific system water is boiled directly by the reactor core (Fig. 18.7).

18.8.2   Graphite Moderated Reactors (GMR)

A Graphite Moderated Reactor (GMR) is a type of reactor that is moderated with
graphite. The first ongoing nuclear reaction carried out by Enrico Fermi at The
University of Chicago was of this type, as well as the reactor associated with the
Chernobyl accident. GMRs share a valuable property with heavy water reactors,
in that natural un-enriched Uranium may be used. Another highlight for the GMR
is a low power density, which is ideal if power were to suddenly stop; this would
not waste as much power/fuel. The common criticisms for this design are a lack

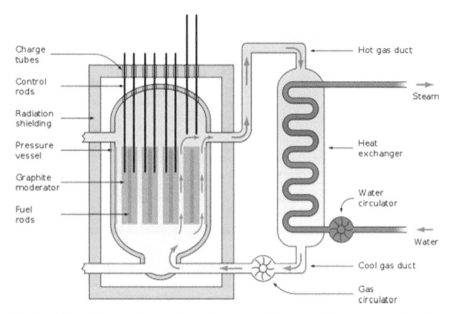

Fig. 18.8 A typical core layout of graphite moderated Reactor. (Courtesy Osterreichisches Ökologie-Institut)

of room for steam suppression and the limited safety precautions available to the design (Fig. 18.8).

18.8.3   Heavy Water Reactors (HWR)

Heavy water reactors (HWR) are a class of fission reactor that uses heavy water as a *neutron moderator*. Heavy water is deuterium oxide, D_2O. Neutrons in a nuclear reactor that uses uranium must be slowed down so that they are more likely to split other atoms and get more neutrons released to split other atoms. Light water can be used, as in a light water reactor (LWR), but since it absorbs neutrons, the uranium must be enriched for criticality to be possible. The most common pressurized heavy water reactor (PHWR) is the CANDU reactor.

Usually the heavy water is also used as the coolant but as example, the Lucens reactor was gas cooled. Advantages of this type reactor are that they can operate with unenriched uranium fuel. Although the opponents of heavy water reactors suggest that such reactors pose a much greater risk of nuclear proliferation because of two characteristics:

1. They use unenriched uranium as fuel, the acquisition of which is free from supervision of international institutions on uranium enrichment.
2. They produce more plutonium and tritium as by-products than light water reactors, these are hazardous radioactive substances that can be used in the produc-

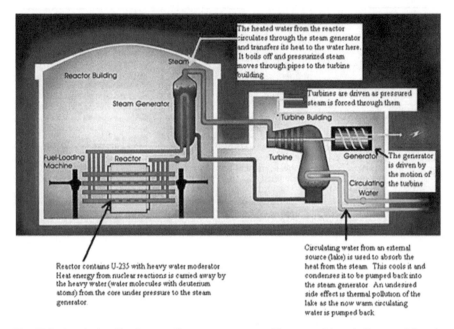

The heated water from the reactor circulates through the steam generator and transfers its heat to the water here. It boils off and pressurized steam moves through pipes to the turbine building

Reactor Building

Steam

Steam Generator

Turbines are driven as pressured steam is forced through them

Turbine Building

Fuel-Loading Machine

Reactor

Turbine

Generator

Generator

The generator is driven by the motion of the turbine

Circulating Water

Reactor contains U-235 with heavy water moderator Heat energy from nuclear reactions is carried away by the heavy water (water molecules with deuterium atoms) from the core under pressure to the steam generator.

Circulating water from an external source (lake) is used to absorb the heat from the steam. This cools it and condenses it to be pumped back into the steam generator. An undesired side effect is thermal pollution of the lake as the now warm circulating water is pumped back.

Fig. 18.9 A typical outline layout of heavy water reactor (Courtesy of Atomic Energy of Canada Limited)

tion of modern nuclear weapons such as fission, boosted fission, and neutron bombs as well as the primary stages of thermonuclear weapons. For instance, India produced its plutonium for Operation Smiling Buddha, its first nuclear weapon test, by extraction from the spent fuel of a heavy water research reactor known as "CIRUS (Canada India Research Utility Services)". It is advocated that safeguards need to be established to prevent exploitation of heavy water reactors in such a fashion.

In heavy water reactors, both the moderator and coolant are heavy water (D_2O). A great disadvantage of this type comes from this fact: heavy water is one of the most expensive liquids. However, it is worth its price: this is the best moderator. Therefore, the fuel of HWRs can be slightly (1–2 %) enriched or even natural uranium. Heavy water is not allowed to boil, so in the primary circuit very high pressure, similar to that of PWRs, exists (Fig. 18.9).

The main representative of the heavy water type is the Canadian CANDU reactor. In these reactors, the moderator and coolant are spatially separated: the moderator is in a large tank (calandria), in which there are pressure tubes surrounding the fuel assemblies. The coolant flows in these tubes only.

The advantage of this construction is that the whole tank need not be kept under high pressure; it is sufficient to pressurize the coolant flowing in the tubes. This arrangement is called pressurized tube reactor. Warming up of the moderator is much less than that of the coolant; it is simply lost for heat generation or steam production. The high temperature and high-pressure coolant, similarly to PWRs, goes to the steam generator where it boils the secondary side light water. Another advantage of this type is that fuel can be replaced during operation and thus there is no need for outages.

The other type of heavy water reactor is the Pressurized Heavy Water Reactor (PHWR). In this type, the moderator and coolant are the same and the reactor pressure vessel has to stand the high pressure.

Heavy water reactors produce cca. 6 % of the total NPP power of the world; however 13.2 % of the under construction nuclear power plant capacity is accounted for by this type. One reason for this is the safety of the type; another is the high conversion factor, which means that during operation a large amount of fissile material is produced from U-238 by neutron capture.

18.9   Classified by Coolant Material

The descriptions of these types of reactors are as follow;

18.9.1   Pressurized Water Reactors (PWR)

A Pressurized Water Reactor (PWR) is Westinghouse Bettis Atomic Power Laboratory has used a type of light water reactor for decades in designs for military ship applications, now the primary manufacturers are Framatome-ANP and Westinghouse for present day power plant reactors. The pressurized water reactor is unique in that although water passes through the reactor core to act as moderator and coolant it does not flow in to the turbine. Instead of the conventional flow cycle, the water passes into a pressurized primary loop. This step in the PWR cycle produces steam in a secondary loop that drives the turbine. Advantages of the PWR include zero fuel leaks of radioactive material into the turbine or environment, and the ability to with stand higher pressures and temperatures to higher the Carnot efficiency. Disadvantages include complex reactor designs and costs. This reactor type accounts for the majority of reactors located in the U.S (Fig. 18.10).

Pressurized Water Reactor (PWR) is a type of a nuclear power reactor that uses enriched Uranium as a fuel which in turn heats the light water used for producing steam. The main feature which differentiates it from a BWR nuclear reactor is that a PWR has a separate arrangement to make steam in the form of a heat exchanger

18.9.1.1   The Arrangement of PWR

A Pressurized Water Reactor (PWR) is a type of power plant reactor consisting of two basic circuits having light water as the working fluid. In one of the circuits water is heated to a high temperature and kept at high pressure as well, so that it does

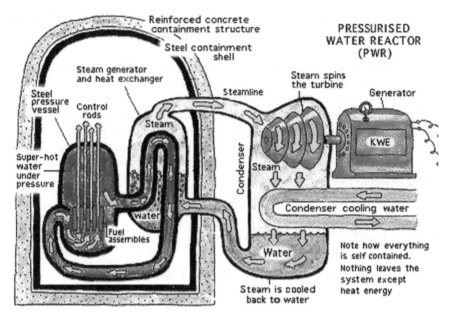

Fig. 18.10 A typical pressurized water reactor. (Courtesy of the Uranium Information Centre)

not get converted into a gaseous state. This superheated water is used as a coolant and a moderator for the nuclear reactor core hence the name PWR or pressurized water reactor.

The secondary circuit consists of water at high pressure in the gaseous state i.e. steam which is used to run the turbine-alternator arrangement. The point of interaction between these two circuits is the heat exchanger or the boiler wherein heat from the superheated high-pressure water converts the water in the secondary circuit to steam.

18.9.1.2   Advantages of PWR

- Much fewer control rods are required in a PWR. In fact, for a typical 1000 MW plant just around five dozen control rods are sufficient.
- Since the two circuits are independent of each other, it makes it very easy for the maintenance staff to inspect the components of the secondary circuit without having to shut down the power plant entirely.
- A PWR has got a high power density and this, combined with the fact that enriched Uranium is used as fuel instead of normal Uranium, leads to the construction of very compact core size for a given power output.
- One feature, which makes a PWR reactor very suitable for practical applications, is its positive demand coefficient, which serves to increase the output as a direct proportion to demand of power.
- The water used in the primary circuit is different from that used in the secondary circuit and there is no intermixing between the two, except for heat transfer, which takes place in the boiler or heat exchanger. This means that the water used

in the turbine side is free from radioactive steam hence the piping on that side is not required to be clad with special shielding materials.

18.9.1.3   Drawbacks of PWR

- The primary circuit consists of high temperature, high pressure water which accelerates corrosion. This means that the vessel should be constructed of very strong material such as stainless steel, which adds to construction costs of PWR.
- PWR fuel charging requires the plant to be shut down and this certainly requires a long time period of the order of at least a couple of months.
- The pressure in the secondary circuit is relatively quite low as compared to the primary circuit hence the thermodynamic efficiency of PWR reactors is quite low of the order of 20

18.9.1.4   Pressuriser

One important point to note here is that despite the changing loads the pressure in the primary circuit needs to be maintained at a constant value. This is achieved by installing a device known as pressuriser in the primary circuit. It basically, consists of a dome shaped structure which has heating coils which, are used to increase or decrease pressure as and when required depending on varied load conditions.

Note that in the Pressurized Water Reactor (PWR), the water, which passes over the reactor core to act as moderator and coolant, does not flow to the turbine, but is contained in a pressurized primary loop. The primary loop water produces steam in the secondary loop, which drives the turbine. The obvious advantage to this is that a fuel leak in the core would not pass any radioactive contaminants to the turbine and condenser (Fig. 18.11).

Another advantage is that the PWR can operate at higher pressure and temperature, about 160 atmospheres and about 315 C. This provides a higher *Carnot efficiency* than the BWR, but the reactor is more complicated and more costly to construct. Most of the U.S. reactors are pressurized water reactors.

18.9.2   Boiling Water Reactor (BWR)

The Boiling Water Reactor (BWR) date back to their General Electric introduction in the 1950s. The distinguishing feature in the BWR is the boiling method for steam. In this type of reactor, water passes over the core as a coolant to expand and become steam source for a turbine placed directly above. Advantages of this design type include a simpler reactor design, a smaller reactor system, and lower costs. Disadvantages found are the increase of radioactive materials in the turbine and a greater chance for fuel to burn out as water quickly evaporates to expose fuel rods to an atmosphere absent of a coolant. BWRs have found fame all over the world due to the cheap simple design.

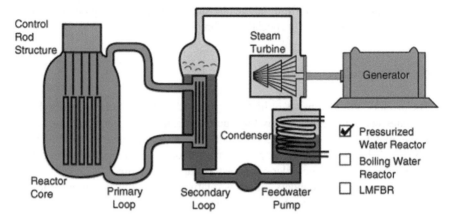

Fig. 18.11 A typical outline of pressurized water reactor

In Fig. 18.12 (1) the core inside the reactor vessel creates heat, (2) a steam-water mixture is produced when very pure water (reactor coolant) moves upward through the core, absorbing heat, (3) the steam-water mixture leaves the top of the core and enters the two stages of moisture separation where water droplets are removed before the steam is allowed to enter the steam line, and (4) the steam line directs the steam to the main turbine, causing it to turn the turbine generator, which produces electricity.

Note that in the Boiling Water reactor (BWR), the water, which, passes over the reactor core to act as moderator and coolant, is also the steam source for the turbine. The disadvantage of this is that any fuel leak might make the water radioactive and that radioactivity would reach the turbine and the rest of the loop (Fig. 18.13).

A typical operating pressure for such reactors is about 70 atmospheres at which pressure the water boils at about 285 °C. This operating temperature gives a *Carnot efficiency* of only 42 % with a practical operating efficiency of around 32 %, somewhat less than the PWR.

18.9.3   Gas Cooled Reactors (GCR)

The Gas Cooled Reactor (GCR) or the gas-graphite reactors operate using graphite as moderator and some gas (mostly CO_2, lately helium) as coolant. This belongs to the oldest reactor types. The first GGR was the Calder Hall power plant reactor, which was built in 1955 in England. This type is called MAGNOX after the special magnesium alloy (Magnox), of which the fuel cladding was made. The fuel is natural uranium. These reactors account for 1.1 % of the total NPP power of the world and are not built any more (Fig. 18.14).

The Advanced Gas cooled Reactor (AGR) is a development from MAGNOX: the cladding is not Magnox and the fuel is slightly enriched. The moderator is also graphite and the coolant is CO_2. Contribution to total world capacity is 2.5 %. This type is not manufactured any longer (Fig. 18.15).

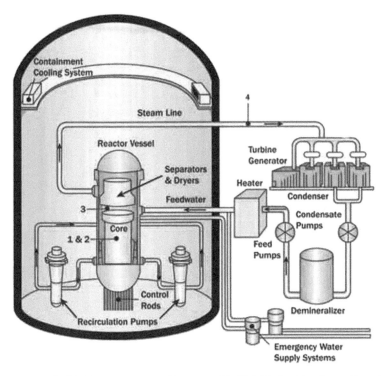

Fig. 18.12 A typical boiling water reactor. (Courtesy of the U.S. Nuclear Regulatory Commission)

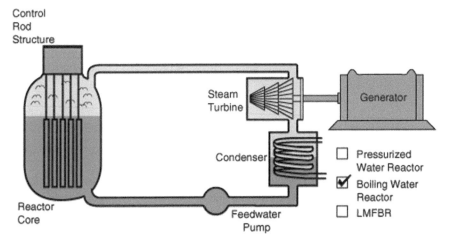

Fig. 18.13 A typical layout of boiling water reactor

Fig. 18.14 A typical core
layout of gas cooled reactor

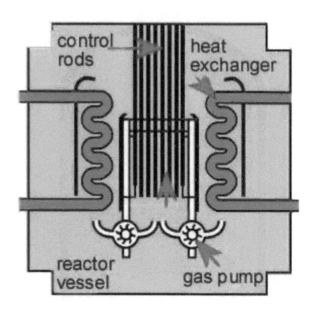

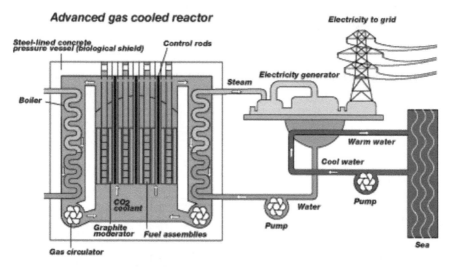

Fig. 18.15 A typical outline layout of gas cooled reactor

The newest gas cooled reactor type is the High Temperature Gas cooled Reactor (HTGR), which is cooled by helium and moderated by graphite. In this reactor as high as 950 °C coolant temperature can be achieved. The efficiency of a newly developed type, the Gas Turbine Modular Helium Reactor (GT-MHR) might be as high as almost 50 %.

Gas Cooled Reactors (GCR) and Advanced Gas Cooled Reactors (AGR) use carbon dioxide as the coolant to carry the heat to the turbine, and graphite as the moderator. Like heavy water, a graphite moderator allows natural uranium (GCR) or slightly enriched uranium (AGR) to be used as fuel.

18.10   Classified by Reaction Type

The descriptions of each of these reactors are given as follows;

18.10.1   Fast Neutron Reactor (FNR)

Fast Neutron Reactors (FNR), also known as Fast Breeder Reactors (FBR), use depleted nuclear waste as a form of energy. Uranium, which is composed of 0.7 % Uranium-235 and 99.3 % Uranium-238, is processed in the fast neutron reactors into isotopes of usable plutonium of plutonium 239 and 241. Fast neutron reactors are 60 % more efficient than normal reactors; a fast neutron reactor uses liquid metal as its coolant as opposed to water, which makes the reactor safer to use and its fuel is metallic, which keeps the reactors under control more easily. Some cons of fast neutron reactors though are that they are very unpredictable, making them more tedious to use. Bubbles are more present in processes, so fast neutron reactors tend to heat up more rather than cool down and the coolant that it requires is much more exotic, such liquid sodium and bismuth eutectic.

Several countries have research and development programs for improved Fast Breeder Reactors (FBR), which are a type of Fast Neutron Reactors. These use the uranium-238 in reactor fuel as well as the fissile U-235 isotope used in most reactors.

Natural uranium contains about 0.7 % U-235 and 99.3 % U-238. In any reactor, the U-238 component is turned into several isotopes of plutonium during its operation. Two of these, Pu 239 and Pu 241, then undergo fission in the same way as U 235 to produce heat. In a fast neutron reactor this process is optimized so that it can 'breed' fuel, often using a depleted uranium blanket around the core. FBRs can utilize uranium at least 60 times more efficiently than a normal reactor.

Fast-neutron reactors could extract much more energy from recycled nuclear fuel, minimize the risks of weapons proliferation and markedly reduce the time nuclear waste must be isolated.

If developed sensibly, nuclear power could be truly sustainable and essentially inexhaustible and could operate without contributing to climate change. In particular, a relatively new form of nuclear technology could overcome the principal drawbacks of current methods—namely, worries about reactor accidents, the potential for diversion of nuclear fuel into highly destructive weapons, the management of dangerous, long-lived radioactive waste, and the depletion of global reserves of economically available uranium. This nuclear fuel cycle would combine two innovations: pyrometallurgical processing (a high-temperature method of recycling reactor waste into fuel) and advanced fast-neutron reactors capable of burning that fuel. With this approach, the radioactivity from the generated waste could drop to safe levels in a few hundred years, thereby eliminating the need to segregate waste for tens of thousands of years.

Fast Reactor Technology: A Path to Long-Term Energy Sustainability
Position Statement November 2005

"The American Nuclear Society believes that the development and deployment of advanced nuclear reactors based on fast-neutron fission technology is important to the sustainability, reliability, and security of the world's long-term energy supply. Of the known and proven energy technologies, only nuclear fission can provide the large quantities of energy required by industrial societies in a sustainable and environmentally acceptable manner".

"Natural uranium mined from the earth's crust is composed primarily of two isotopes: 99.3% is U-238, and 0.7% is the fissile U-235. Nearly all current power reactors are of the "thermal neutron" design, and their capability to extract the potential energy in the uranium fuel is limited to less than 1% of that available. The remainder of the potential energy is left unused in the spent fuel and in the uranium, depleted in U-235 that remains from the process of enriching the natural uranium in the isotope U-235 for use in thermal reactors. With known fast reactor technology, this unutilized energy can be harvested, thereby extending by a hundred-fold the amount of energy extracted from the same amount of mined uranium".

"Fast reactors can convert U-238 into fissile material at rates faster than it is consumed making it economically feasible to utilize ores with very low uranium concentrations and potentially even uranium found in the oceans [1–3]. A suitable technology has already been proven on a small scale [4]. Used fuel from thermal reactors and the depleted uranium from the enrichment process can be utilized in fast reactors, and that energy alone would be sufficient to supply the nation's needs for several hundred years".

"Fast reactors in conjunction with fuel recycling can diminish the cost and duration of storing and managing reactor waste with an offsetting increase in the fuel cycle cost due to reprocessing and fuel prefabrications. Virtually all long-lived heavy elements are eliminated during fast reactor operation, leaving a small amount of fission product waste that requires assured isolation from the environment for less than 500 years [5].

"Although fast reactors do not eliminate the need for international proliferation safeguards, they make the task easier by segregating and consuming the plutonium as it is created. The use of onsite reprocessing makes illicit diversion from within the process highly impractical. The combination of fast reactors and reprocessing is a promising option for reasons of safety, resource utilization, and proliferation resistance [5].

"Reaping the full benefits of fast reactor technology will take a decade or more for a demonstration reactor, followed by buildup of a fleet of operating power stations. For now and in the intermediate-term future, the looming short-term energy shortage must be met by building improved, proven thermal-reactor power plants. To assure longer-term energy sustainability and security, the American Nuclear Society sees a need for cooperative international efforts with the goal of building a fast reactor demonstration unit with onsite reprocessing of spent fuel."

18.10.2   Thermal Neutron Reactor

Thermal reactors go through the same process as fast neutron reactors, but in a thermal reactor the process of obtaining plutonium is slower. These types of reactors use a neutron moderator to slow neutrons until they approach the average kinetic energy of the surrounding particles, that is, to reduce the speed of the neutrons to low velocity thermal neutrons. The nuclear cross section of uranium-235 for slow thermal neutrons is about 1000 barns. For fast neutrons, it is in the order of 1 barn. In a thermal reactor, the neutrons that undergo the reaction process have significantly lower electron-volt energy, so the neutrons are considered to be slower. A neutron's speed will determine its chances to interact with the nucleus of an atom; the slower its speed the bigger its fission cross section becomes and thus the higher its chance of interacting with the nucleus becomes (Fig. 18.16).

This figure gives the value of the fission cross section for some fissile isotopes. Note that both axes are logarithmic. The thermal and fast energy regions are indicated. For thermal energies, the fission cross section equals several thousand barn, at high energies the fission cross section is of the order of 1–10 barn.

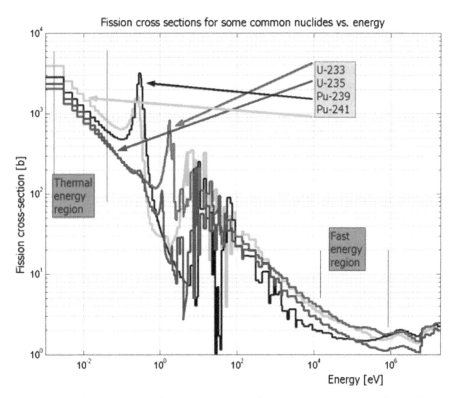

Fig. 18.16 Fission cross section for some common nuclides vs. energy. (Courtesy of TUDelft)

The fact that the fission cross section is rather large for low-energy neutrons has an important effect on the design of a nuclear reactor: in a reactor where the neutrons have a low energy, not much fissile material is required, because the probability of an interaction is very large. The lowest energy a neutron can have in a nuclear reactor is the energy at which it is in equilibrium with its environment. The movement of the neutron is then identical to the thermal movement of the atoms that constitute the reactor. The neutrons have slowed down from the high energy (2 MeV) where they are born to this equilibrium energy are called 'thermal neutrons'. The average energy of a neutron in thermal equilibrium is 0.025 eV—the neutron is slowed down over nine decades, more than a billion times. Reactors in which, most fissions are induced by thermal neutrons are called thermal reactors. Thermal reactors are by far the most widely used reactors in the world today. Most reactors use water, heavy water or graphite as moderator. The reason for the choice of thermal reactors is a simple one: a thermal reactor requires a small amount of fuel to become critical, and thus the fuel is cheap.

18.10.3   *Liquid Metal Fast Breeder Reactors (LMFBR)*

The plutonium-239 breeder reactor is commonly called a fast breeder reactor, and the cooling and a liquid metal does heat transfer. The metals, which can accomplish this, are sodium and lithium, with sodium being the most abundant and most commonly used. The construction of the fast breeder requires a higher enrichment of U-235 than a light-water reactor, typically 15–30 %. The reactor fuel is surrounded by a "blanket" of non-fissionable U-238. No moderator is used in the breeder reactor since fast neutrons are more efficient in transmuting U-238 to Pu-239. At this concentration of U-235, the cross-section for fission with fast neutrons is sufficient to sustain the chain-reaction. Using water as coolant would slow down the neutrons, but the use of liquid sodium avoids that moderation and provides a very efficient heat transfer medium (Fig. 18.17).

The Super-Phenix was the first large-scale breeder reactor. It was put into service in France in 1984. It ceased operation as a commercial power plant in 1997. Such a reactor can produce about 20 % more fuel than it consumes by the breeding reaction. Enough excess fuel is produced over about 20 years to fuel another such reactor. Optimum breeding allows about 75 % of the energy of the natural uranium to be used compared to 1 % in the standard light water reactor (Fig. 18.18).

Under appropriate operating conditions, the neutrons given off by fission reactions can "breed" more fuel from otherwise non-fissionable isotopes. The most common *breeding reaction* is that of plutonium-239 from non-fissile uranium-238. The term "fast breeder" refers to the types of configurations, which can actually produce more fissionable fuel than they use, such as the LMFBR. This scenario is possible because the non-fissile uranium-238 is 140 times more abundant than the fissionable U-235 and can be efficiently converted into Pu-239 by the neutrons from a fission chain reaction.

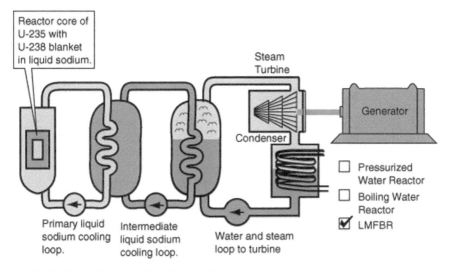

Fig. 18.17 A typical layout of Liquid metal fast breeder reactor

Fig. 18.18 This is a photo of a model of the containment vessel of the Super-Phenix. It is displayed at the National Museum of Nuclear Science and Technology in Albuquerque, NM

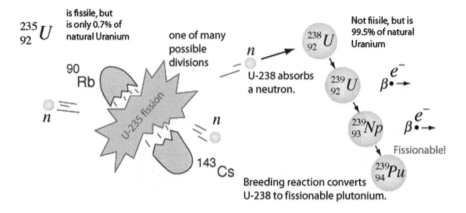

France has made the largest implementation of breeder reactors with its large Super-Phenix reactor (today is not in production line) and an intermediate Russian scale reactor (BN-600) on the Caspian Sea for electric power and desalinization.

Breeding Plutonium-239 can be accomplished from non-fissile uranium-238 by the reaction illustrated.

$n \longrightarrow {}^{238}_{92}U$

$T_{1/2} = 23.5\,min$ ${}^{239}_{92}U$ $\beta \bullet \overset{e^-}{\longrightarrow}$

$T_{1/2} = 2.35\,days$ ${}^{239}_{93}Np$ $\beta \bullet \overset{e^-}{\longrightarrow}$

$T_{1/2} = 2.44x10^4\,yrs$ ${}^{239}_{94}Pu$

The bombardment of Uranium-238 with neutrons triggers two successive *beta decays* with the production of plutonium. The amount of plutonium produced depends on the *breeding ratio*.

The concept of breading ratio of Plutonium-239 can be defined in following. In the breeding of plutonium fuel in breeder reactors, an important concept is the breeding ratio, the amount of fissile plutonium-239 produced compared to the amount of fissile fuel (like U-235) used to produce it. In the liquid-metal, fast-breeder reactor (LMFBR), the target-breeding ratio is 1.4 but the results achieved have been about 1.2. This is based on 2.4 neutrons produced per U-235 fission, with one neutron used to sustain the reaction.

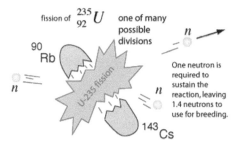

This particular fission path yields three neutrons,
but the average neutron yield is 2.4 neutrons.

The time required for a breeder reactor to produce enough material to fuel a second reactor is called its doubling time, and present design plans target about 10 years as a doubling time. A reactor could use the heat of the reaction to produce energy for 10 years, and at the end of that time have enough fuel to fuel another reactor for 10 years.

Liquid sodium is used as the coolant and heat-transfer medium in the LMFBR reactor. That immediately raised the question of safety since sodium metal is an extremely reactive chemical and burns on contact with air or water (sometimes explosively on contact with water). It is true that the liquid sodium must be protected from contact with air or water at all times, kept in a sealed system. However, it has been found that the safety issues are not significantly greater than those with high-pressure water and steam in the light-water reactors.

Sodium is a solid at room temperature but liquefies at 98 °C. It has a wide working temperature since it does not boil until 892 °C. That brackets the range of operating temperatures for the reactor so that it does not need to be pressurized as does a water-steam coolant system. It has a large *specific heat* so that it is an efficient heat-transfer fluid.

In practice, those reactors, which have used liquid metal coolants, have been fast-neutron reactors. The liquid metal coolant has a major advantage there because water as a coolant also moderates or slows down the neutrons. Such fast-neutron reactors require a higher degree of enrichment of the uranium fuel than do the water-moderated reactors.

18.11   Nuclear Fission Power Generation

Nuclear fission energy is today a competitive and mature low-carbon technology, operating at very high levels of safety. The installed nuclear electricity capacity in the Europe (EU) for example, is 132 GWe, which provides one third of the EU's generated electricity. Most of the current designs are Light Water Reactors (LWR) of the second generation, capable of providing base-load electricity often with availability factors of over 90%. There have been only a few new nuclear power plants connected to the grid in the last two decades, and as a result of decommissioning of old plants the total number of reactors in Europe has decreased. Nevertheless, electricity supply from nuclear has remained constant and the levelised cost has decreased owing to improved efficiency, power upgrade and improved availability factor.

More recently, there has been a renewed interest in nuclear energy, referred to as "nuclear renaissance", mainly driven by concerns over climate change, security and independence of supply and energy costs.

18.12   Generation IV Nuclear Energy Systems

Concerns over energy resource availability, climate change, air quality, and energy security suggest an important role for nuclear power in future energy supplies. While the current Generation II and III nuclear power plant designs provide a secure and low-cost electricity supply in many markets, further advances in nuclear energy system design can broaden the opportunities for the use of nuclear energy. To explore these opportunities, the U.S. Department of Energy's Office of Nuclear Energy has engaged governments, industry, and the research community worldwide in a wide ranging discussion on the development of next generation nuclear energy systems known as "Generation IV".

The goal of the Gen IV Nuclear Energy Systems is to address the fundamental research and development (R&D) issues necessary to establish the viability of next-generation nuclear energy system concepts to meet tomorrow's needs for clean and reliable electricity, and non-traditional applications of nuclear energy. Successfully addressing the fundamental Research and Development (R&D) issues will allow Gen IV concepts that excel in safety, sustainability, cost-effectiveness, and pro-liferation risk reduction to be considered for future commercial development and deployment by the private sector (Fig. 18.19).

Gen IV reactor concepts are being developed to use advanced fuels, fashioned from recycled reactor fuel and capable of high-burnups. The corresponding fuel cycle strategies allow for efficient utilization of domestic uranium resources while minimizing waste. Reduction of proliferation risk and improvements in physical protection are being designed into Gen IV concepts to help thwart those who would target nuclear power plants for terrorist acts or use them improperly to develop materials for nuclear weapons. Gen IV concepts will feature advances in safety and reliability to improve public confidence in nuclear energy while providing enhanced investment protection for plant owners. Competitive life-cycle costs and acceptable financial risk are being factored into Gen IV concepts with high-efficiency electricity generation systems, modular construction, and shortened development schedules before plant startup.

Gen IV is also an active participant in the International Project on Innovative Nuclear Reactors and Fuel Cycles (INPRO). INPRO was established in 2001 in response to a resolution by the IAEA General Conference to help to ensure that

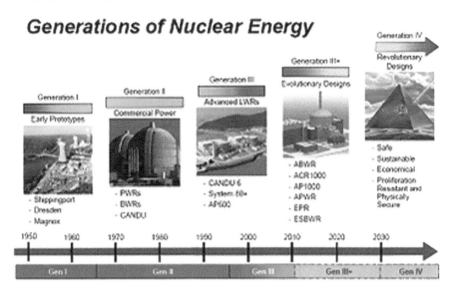

Fig. 18.19 The evolution of nuclear power

nuclear energy is available to contribute, in a sustainable manner, to meeting the energy needs of the twenty-first century and to bring together technology holders and users so that they can consider jointly the international and national actions required for achieving desired innovations in nuclear reactors and fuel cycles. INPRO provides a forum for discussion for experts and policy makers from industrialized and developing countries on all aspects of nuclear energy planning as well as on the development and deployment of innovative nuclear energy systems in the twenty-first century.

The Generation IV International Forum (GIF) was chartered in May 2001, to lead the collaborative efforts of the world's leading nuclear technology nations to develop the next generation of nuclear energy systems. The initial efforts of GIF resulted in the identification of the six most promising reactor concepts to be investigated by this international research community and are documented in the Generation IV Technology Roadmap. Thirteen members have signed the GIF Charter: Argentina, Brazil, Canada, People's Republic of China, Euratom, France, Japan, Republic of Korea, the Russian Federation, Republic of South Africa, Switzerland, the United Kingdom and the United States. This unique international effort reached a major milestone on February 28, 2005, as five of the Forum's member countries (Canada, France, Japan, United Kingdom, and United States) signed the world's first multi-lateral agreement aimed at the international development of advanced nuclear energy systems—the Framework Agreement for International Collaboration on Research and Development of Generation IV Nuclear Energy Systems. Subsequent signatories to the Framework Agreement included People's Republic of China, Euratom, Republic of Korea, Republic of South Africa, and Switzerland.

Fig. 18.20 Map of member countries

The United Kingdom is a signatory of the Framework but is currently a non-active member. Argentina and Brazil have not ratified the Framework Agreement and are therefore considered non-active. The Russian Federation is working on the necessary approvals for its accession to the Framework (Fig. 18.20).

As detailed in its Charter and subsequent GIF Policy Statements, GIF is led by the Policy Group (PG), which is responsible for the overall coordination of GIF's Research and Development (R&D) collaboration, policy formation and for interactions with other organizations. France with currently chairs the Policy Group vice chairs from the U.S. and Japan. An Experts Group and the Senior Industry Advisory Panel advises the Policy Group on (R&D) strategy, priorities, and methodology and on evaluating research plans for each Generation IV System. The Framework Agreement establishes two levels of implementing arrangements in order to conduct the joint (R&D). The first level consists of a System Arrangement for each Generation IV reactor concept directed by a System Steering Committee (SSC). Under each SSC, Project Arrangements are established with Project Management Boards to manage and implement the joint (R&D).

18.13   Technological State of the Art and Anticipated Developments

It has been demonstrated that Generation-II plants can be safely and economically operated for up to 60 years through the development of improved harmonized Plant-Life Management technologies and Plant License Extension practices (PLIM/ PLEX) and that developments in fuel technologies can still lead to improvements in reactor performance. The first Generation-III reactors, which are an evolution of thermal reactors with even further improved safety characteristics and economy, are now being built. In the coming decades, nuclear electricity generation should increase or at least maintain its current level by a combination of lifetime extension and power upgrades of Generation-II reactors and new build of Generation-III reactors. Two 1.6 GWe Generation-III reactors are presently under construction in Finland and France, targeted for connection to the grid in 2012.

The Finnish reactor was a First-of-a-Kind (FOAK) and the construction has suffered delays with the Overnight Cost increasing from 2000 to 3100 €/kWe, whereas the Overnight Cost for the second reactor in France is now 2400 €/kWe. In series production, the industry expects the cost to be 2000 ± 500 €/GWe, which is in line with recent international studies. An additional capacity of 100 GWe of Generation-III reactors over the next 25 years is a reasonable estimate, which would require an investment in the range of € 200–280 billion. The capital costs represent typically 60–70 % of the levelised cost for nuclear electricity, operation and maintenance 20–25 % and fuel 10–15 %. The front-loaded cost profile means that the levelised cost is very sensitive to construction time and the financial schemes for the investment. Estimates in 2007 for UK resulted in range of 31–44 £/MWh (37–53 €/MWh).

Though uranium is relatively abundant in the Earth's crust and oceans, estimates of natural reserves are always related to the cost of mineral extraction. As the price of uranium increases on world markets, the number of economically exploitable deposits also increases. The most recent estimates [6] identified 5.5 Million t of Uranium (MtU) that could be exploited below 130 $/kg. The total amount of undiscovered resources (reasonably assured and speculative) available at an extraction cost below 130 $/kgU is estimated at 10.5 MtU. Unconventional resources, from which uranium is extracted as a by-product only, e.g. in phosphate production, lie between 7 and 22 MtU, and reserves in sea water are estimated to be 4000 MtU. Japanese studies suggest that uranium from sea water can be extracted at 300 €/kg. At a conservative estimate, 25,000 t of the uranium are required to produce the fuel to generate 1000 TWhe in an open fuel cycle. The global electricity supplied by nuclear is 2600 TWhe, which means that the conventional resources below 130 $/kgU at the current rate of consumption would last for at least 85 years with the already identified resources (5.5 MtU) and 246 years, if the undiscovered are also included (5.5 + 10.5 MtU). In addition to uranium, it is also possible to use thorium, which is three times more abundant in the Earth's crust, though would require different reactors and fuel cycles. Nonetheless, natural resources are plentiful and do not pose an immediate limiting factor for the development of nuclear energy.

However, in a scenario with a large expansion of nuclear energy, resources will become an issue much earlier, especially since new plants have at least a 60-year lifetime and utilities will need assurances when ordering new build that uranium supply can be maintained for the full period of operation. Eventually, known conventional reserves will all be earmarked for current plants or those under construction, and this could happen by the middle of this century. This underlines the need to develop the technology for a new generation, the so-called Generation-IV, of reactors and fuel cycles that are more sustainable. In particular, fast neutron breeder reactors could produce up to 100 times more energy from the same quantity of uranium than current designs and may significantly reduce the amount of ultimate radioactive waste for disposal.

Fast reactors convert non-fissile material (U-238) in the fuel into fissile material (Pu-239) during reactor operation so that the net amount of fissile material increases (breeding). After re-processing of the spent fuel, the extracted fissile materials are then re-cycled as new fuels. Reduction of the radio-toxicity and heat load of the waste is achieved by separating some long-lived radionuclides, the minor actinides, which could then be "burned" in fast reactors or alternatively in Accelerator Driven Systems (ADS), through transmutation. The fast reactor concept has been demonstrated in research programs and national prototypes in the past, but further R&D is needed to make it commercially viable and to develop the designs in compliance with true Generation-IV criteria. Major issues involve new materials that can withstand higher temperatures, higher burn-ups and neutron doses, corrosive coolants; reactor designs that eliminate severe accidents; and development of fuel cycles for waste minimization and elimination of proliferation risks. Fast Reactors are expected to be commercially available from 2040.

So far nuclear power has primarily been used to produce electricity, but it can also be used for process heat applications [7]. Currently, LWRs are already being used to a limited extent for some lower temperature applications (200 °C), such as district heating and desalination of seawater. Existing designs of High-Temperature Reactors (HTR) that can reach 800 °C can be deployed in the coming decades and Very-High Temperature Reactors (VHTR) that can reach gas coolant temperatures beyond 1000 °C are being studied as a Generation-IV concept for later deployment. Process heat applications include petroleum refinery applications (400 °C), recovery of oil from tar sands (600–700 °C), synthetic fuel from CO_2 and hydrogen (600–1000 °C), hydrogen production (600–1000 °C) and coal gasification (900–1200 °C). Small reactors that can be inherently safe and used to support specific high energy applications and often in remote areas are another very interesting application that is receiving more attention, in particular in the IAEA INPRO Initiative.

The management of radioactive waste, as either spent fuel or ultimate waste, depending on the national strategy, is a key issue for public acceptance of nuclear energy. There is scientific consensus that geological disposal is the only safe long-term solution for the management of ultimate waste. After a long period of intensive research and development coupled with in-depth political and social engagement, the world's first deep geological repositories for nuclear waste will be in operation in Sweden and Finland by 2020, with France following a few years later, demonstrating that practical solutions exist for the safe long-term management of hazardous waste from the operation of nuclear power plants. Though there will also be ultimate waste from Generation-IV fast reactor fuel cycles after reprocessing, the volumes and heat loads will be greatly reduced thereby facilitating disposal operations and optimizing use of space in available geological repositories.

18.14   Next Generation Nuclear Plant (NGNP)

The Next Generation Nuclear Plant (NGNP) demonstration project forms the basis for an entirely new generation of advanced nuclear plants capable of meeting the Nation's emerging need for greenhouse-gas-free process heat and electricity. The NGNP is based on the Very-High-Temperature gas-cooled Reactor (VHTR) technology, which was determined to be the most promising for the U.S. in the medium term. The determination is documented as part of the Generation IV implementation strategy in a report submitted to Congress in 20031 following an extensive international technical evaluation effort. The VHTR technology incorporates substantive safety and operational enhancements over existing nuclear technologies. As required by the Energy Policy Act of 2005 (EPAct), the NGNP will be a prototype nuclear power plant, built at the Idaho National Laboratory (INL). Future commercial versions of the NGNP will meet or exceed the reliability, safety, proliferation-resistance, and economy of existing commercial nuclear plants.

It is envisioned that these advanced nuclear plants would be able to supply cost-competitive process heat that can be used to power a variety of energy intensive industries, such as the generation of electricity, hydrogen, enhanced oil recovery, refineries, coal-to-liquids and coal-to-gas plants, chemical plants, and fertilizer plants.

The U.S. Nuclear Regulatory Commission (NRC) is responsible for licensing and regulating the construction and operation of the NGNP. The EPAct authorizes the U.S. Department of Energy (DOE) to build the NGNP at the Idaho National Laboratory and charges INL with responsibility for leading the project development. The project's completion depends on the collaborative efforts of DOE and its national laboratories, commercial industry participants, U.S. universities, and international government agencies as well as successful licensing by the NRC. At present, and pending further evaluation as the NGNP proceeds through Phase 1 in cost-shared collaboration with industry as required by the EPAct, DOE has not made a final determination on whether the license applicant will be DOE or one or more entities that reflect a partnership between DOE and private sector firms.

Under the provisions of Section 644 of the EPAct, the Secretary of Energy and the Chairman of the Nuclear Regulatory Commission are to jointly submit to Congress a licensing strategy for the NGNP within 3 years of the enactment of the Act on August 8, 2005. This report addresses the requirement by outlining a NGNP licensing strategy jointly developed by the NRC and DOE. The scope of the document includes all four elements of the NGNP licensing strategy described in Section 644 (b) of the EPAct:

1. A description of the ways in which current NRC light-water reactor (LWR) licensing requirements need to be adapted for the types of reactors considered for the project.
2. A description of the analytical tools that the NRC will need to develop in order to independently verify the NGNP design and its safety performance.
3. A description of other research or development activities that the NRC will need to conduct for the review of an NGNP license application.
4. A budget estimate associated with the licensing strategy.

DOE has determined that the NGNP nuclear reactor will be a very-high-temperature gas-cooled reactor (VHTR) for the production of electricity, process heat, and hydrogen. The VHTR can provide high-temperature process heat (up to 950 °C) that can be used as a substitute for the burning of fossil fuels for a wide range of commercial applications. Since the VHTR is a new and unproven reactor design, the NRC will need to adapt its licensing requirements and process, which have historically evolved around light-water reactor (LWR) designs, for licensing the NGNP nuclear reactor. Thus, Section 644 of the EPAct recognized the need for an alternative licensing strategy. This report provides the recommended NGNP licensing strategy, jointly developed by the NRC and DOE. As the technology matures, the government/industry partnership evolves, and input is provided by the general public, revisions to the strategy may be necessary and appropriate.

The report addresses the four elements of the licensing strategy set forth in Section 644(b) of the EPAct. These elements are summarized above.

18.15   Generation IV Systems

The world's population is expected to expand from 6.7 billion people today to over 9 billion people by the year 2050, all striving for a better quality of life. As the earth's population grows, so does the demand for energy and the benefits that it brings: improved standards of living, better health and longer life expectancy, improved literacy and opportunity, and many others. Simply expanding the use of energy along the same mix of today's production options, however, does not satisfactorily address concerns over climate change and depletion of fossil resources. For the earth to support its population while ensuring the sustainability of humanity's development, we must increase the use of energy supplies that are clean, safe, cost effective, and which could serve for both basic electricity production and other primary energy needs. Prominent among these supplies is nuclear energy.

There is currently 370 GWe of nuclear power capacity in operation around the world, producing 3000 TWh each year—15 % of the world's electricity—the largest share provided by any non-greenhouse gas- emitting source. This reduces significantly the environmental impact of today's electricity generation and affords a greater diversity of electricity generation that enhances energy security.

For more than a decade, Generation IV International Forum (GIF) has led international collaborative efforts to develop next-generation nuclear energy systems that can help meet the world's future energy needs. Generation-IV designs will use fuel more efficiently, reduce waste production, be economically competitive, and meet stringent standards of safety and proliferation resistance.

As, we said the Generation IV International Forum (GIF) was initiated in May 2001 and formally chartered in mid 2001. It is an international collective representing government of 13 countries where nuclear energy is significant now and also seen as vital for the future. Most are committed to joint development of the next generation of nuclear technology. Led by the USA, Argentina, Brazil, Canada,

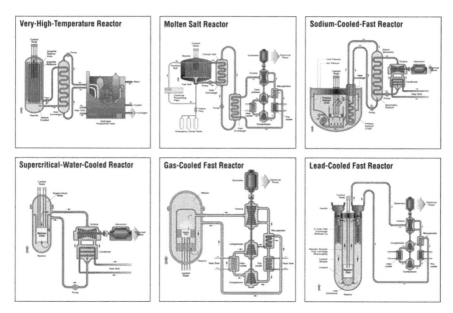

Fig. 18.21 Six reactor technologies of generation IV. (Courtesy of the Generation IV International Forum)

China, France, Japan, Russia, South Korea, South Africa, Switzerland, and the UK are charter members of the GIF, along with the EU (Euratom). Most of these are party to the Framework Agreement (FA), which formally commits them to participate in the development of one or more Generation IV systems selected by GIF for further R&D. Argentina, and Brazil did not sign the FA, and the UK withdrew from it; accordingly, within the GIF, these three are designated as "inactive Members." Russia formalized its accession to the FA in August 2009 as its tenth member, with Rosatom as implementing agent. In 2011 the 13 members decided to modify and extend the GIF charter indefinitely

With these goals in mind, some 100 experts evaluated 130 reactor concepts before GIF selected six reactor technologies for further research and development.

These include:

1. Very High Temperature Reactor (VHTR)
2. the Molten Salt Reactor (MSR),
3. the Sodium-cooled Fast Reactor (SFR),
4. the Super Critical Water-cooled Reactor (SCWR),
5. the Gas-cooled Fast Reactor (GFR), and
6. the Lead-cooled Fast Reactor (LFR)

Figure 18.21 below, is illustration of the six types of reactors that are considered as part of Generation IV power plant. More details of each of these reactors are provided in later sections.

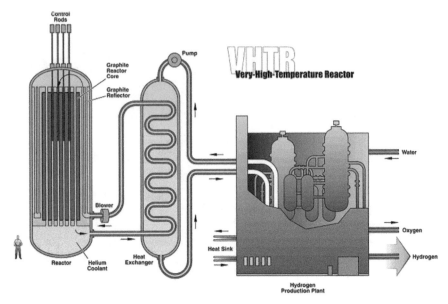

Fig. 18.22 Very high temperature reactor. (Courtesy of the Generation IV International Forum)

18.15.1   Very High Temperature Reactor (VHTR)

Among the six candidates of the Gen IV nuclear systems in the technical roadmap
of Gen IV International Forum (GIF), the Very High Temperature Reactor (VHTR)
is primarily dedicated to the cogeneration of electricity and hydrogen, the latter
being extracted from water by using thermo-chemical, electro-chemical or hybrid
processes. Its high outlet temperature makes it attractive also for the chemical, oil
and iron industries. Original target of outlet temperature of 1000 °C from VHTR
can support the efficient production of hydrogen by thermo-chemical processes.
The technical basis for VHTR is the TRISO coated particle fuel, the graphite as the
core structure, helium coolant, as well as the dedicated core layout and lower power
density to removal decay heat in a natural way. The VHTR has potential for inherent
safety, high thermal efficiency, process heat application capability, low operation
and maintenance costs, and modular construction (Fig. 18.22).

The VHTR is a next step in the evolutionary development of high-temperature
gas-cooled reactors. It is a graphite-moderated, helium-cooled reactor with thermal
neutron spectrum. It can supply nuclear heat and electricity over a range of core
outlet temperatures between 700 and 950 °C, or more than 1000 °C in future. The
reactor core type of the VHTR can be a prismatic block core such as the Japanese
HTTR, or a pebble-bed core such as the Chinese HTR-10. For electricity generation, a helium gas turbine system can be directly set in the primary coolant loop,
which is called a direct cycle or at the lower end of the outlet temperature range,
a steam generator can be used with a conventional rankine cycle. For nuclear heat
applications such as process heat for refineries, petrochemistry, metallurgy, and hy-

drogen production, the heat application process is generally coupled with the reactor through an intermediate heat exchanger (IHX), the so-called indirect cycle. The VHTR can produce hydrogen from only heat and water by using thermochemical processes (such as the sulfur-iodine (S-I) process or the hybrid sulfur process), high temperature steam electrolysis (HTSE), or from heat, water, and natural gas by applying the steam reformer technology.

While, the original approach for VHTR at the start of the Generation IV program focused on very high outlet temperatures and hydrogen production, current market assessments have indicated that electricity production and industrial processes based on high temperature steam that require modest outlet temperatures (700–850 °C) have the greatest potential for application in the next decade. This also reduces technical risk associated with higher outlet temperatures. As a result, over the past decade, the focus has moved from higher outlet temperature designs such as GT-MHR and PBMR to lower outlet temperature designs such as HTR-PM in China and the NGNP in the US.

The VHTR has two typical reactor configurations, namely:

1. the pebble bed type and
2. the prismatic block type

Although the shape of the fuel element for two configurations are different, the technical basis for both configuration is same, such as the TRISO coated particle fuel in the graphite matrix, full ceramic (graphite) core structure, helium coolant, and low power density.

This will allow achieving high outlet temperature and the retention of fission production inside the coated particle under normal operation condition and accident condition. The VHTR can support alternative fuel cycles such as U-Pu, Pu, MOX, U-Th.

18.15.2 Molten Salt Reactor (MSR)

The MSR is distinguished by its core in, which the fuel is dissolved in molten fluoride salt. The technology was first studied more than 50 years ago. Modern interest is on fast reactor concepts as a long-term alternative to solid-fuelled fast neutrons reactors. The onsite fuel-reprocessing unit using pyrochemistry allows breeding plutonium or uranium-233 from thorium. R&D progresses toward resolving feasibility issues and assessing safety and performance of the design concepts. Key feasibility issues focus on a dedicated safety approach and the development of salt redox potential measurement and control tools in order to limit corrosion rate of structural materials. Further work on the batch wise online salt processing is required. Much work is needed on molten salt technology and related equipments.

Molten Salt Reactor (MSR) technology was partly developed, including two demonstration reactors, in the 1950s and 1960s in the USA (Oak Ridge National Laboratory). The demonstrations MSRs were thermal-neutron-spectrum graphite-moderated concepts. Since 2005, R&D has focused on the development of fast-spectrum MSR concepts (MSFR) combining the generic assets of fast neutron reactors (extended resource utilization, waste minimization) with those relating to molten salt fluorides as fluid fuel and coolant (low pressure and high boiling temperature, optical transparency).

In contrast to most other molten salt reactors previously studied, the MSFR does not include any solid moderator (usually graphite) in the core. This design choice is motivated by the study of parameters such as feedback coefficient, breeding ratio, graphite lifespan and 233U initial inventory. MSFR exhibit large negative temperature and void reactivity coefficients, a unique safety characteristic not found in solid-fuel fast reactors.

Compared with solid-fuel reactors, MSFR systems have lower fissile inventories, no radiation damage constraint on attainable fuel burn-up, no requirement to fabricate and handle solid fuel, and a homogeneous isotopic composition of fuel in the reactor. These and other characteristics give MSFRs potentially unique capabilities for actinide burning and extending fuel resources.

MSR developments in Russia on the Molten Salt Actinide Recycler and Transmuter (MOSART) aim to be used as efficient burners of transuranic (TRU) waste from spent UOX and MOX light water reactor (LWR) fuel without any uranium and thorium support and also with it. Other advanced reactor concepts are being studied, which use the liquid salt technology, as a primary coolant for Fluoride salt-cooled High-temperature Reactors (FHRs), and coated particle fuels similar to high temperature gas-cooled reactors (Fig. 18.23).

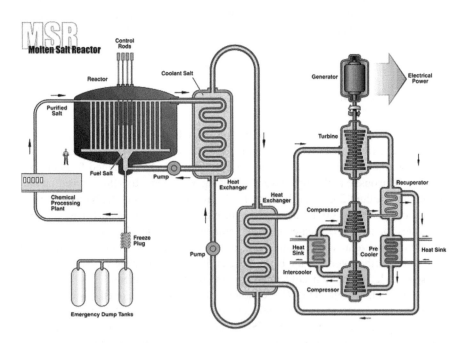

Fig. 18.23 Molten salt reactor. (Courtesy of the Generation IV International Forum)

More generally, there has been a significant renewal of interest in the use of liquid salt as a coolant for nuclear and non-nuclear applications. These salts could facilitate heat transfer for nuclear hydrogen production concepts, concentrated solar electricity generation, oil refineries, and shale oil processing facilities amongst other applications.

18.15.3   Sodium Cooled Fast Reactor (SFR)

The Sodium Cooled Fast Reactor (SFR) uses liquid sodium as the reactor coolant, allowing high power density with low coolant volume fraction and operation at low pressure. While the oxygen-free environment prevents corrosion, sodium reacts chemically with air and water and requires a sealed coolant system.

Plant size options under considerations are ranging from small, 50 to 300 MWe, modular reactors to larger plants up to 1500 MWe. The outlet temperature is 500–550 °C for the options, which allows the use of the materials developed and proven in prior fast reactor programs.

The SFR closed fuel cycle enables regeneration of fissile fuel and facilitates management of minor actinides. However, this requires that recycle fuels be developed and qualified for use. Important safety features of the Generation IV system include a long thermal response time, a reasonable margin to coolant boiling, a primary system that operates near atmospheric pressure, and an intermediate sodium system between the radioactive sodium in the primary system and the power conversion system. Water/steam, supercritical carbon-dioxide or nitrogen can be considered as working fluids for the power conversion system to achieve high performance in terms of thermal efficiency, safety and reliability. With innovations to reduce capital cost, the SFR is aimed to be economically competitive in future electricity markets. In addition, the fast neutron spectrum greatly extends the uranium resources compared to thermal reactors. The SFR is considered to be the nearest-term deployable system for actinide management (Fig. 18.24).

Much of the basic technology for the SFR has been established in former fast reactor programmes, and is being confirmed by the Phenix end-of-life tests in France, the restart of Monju in Japan and the lifetime extension of BN-600 in Russia. New programs involving SFR technology include the Chinese experimental fast reactor (CEFR) which was connected to the grid in July 2011, and India's prototype fast breeder reactor (PFBR) which is currently planned to go critical in 2013.

The SFR is an attractive energy source for nations that desire to make the best use of limited nuclear fuel resources and manage nuclear waste by closing the fuel cycle.

Fast reactors hold a unique role in the actinide management mission because they operate with high energy neutrons that are more effective at fissioning actinides. The main characteristics of the SFR for actinide management mission are:

- Consumption of transuranics in a closed fuel cycle, thus reducing the radiotoxicity and heat load, which facilitates waste disposal and geologic isolation.

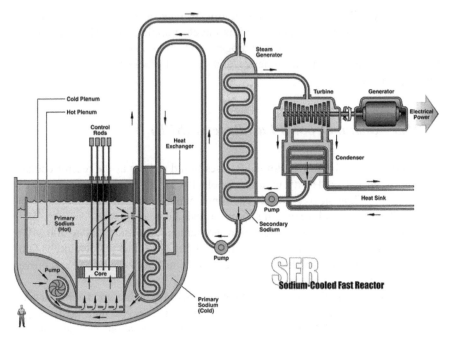

Fig. 18.24 Sodium cooled fast reactor. (Courtesy of the Generation IV International Forum)

- Enhanced utilization of uranium resources through efficient management of fissile materials and multi-recycle.

High level of safety achieved through inherent and passive means also allows accommodation of transients and bounding events with significant safety margins.

The reactor unit can be arranged in a pool layout or a compact loop layout. Three options are considered:

- A large size (600–1500 MWe) loop-type reactor with mixed uranium-plutonium oxide fuel and potentially minor actinides, supported by a fuel cycle based upon advanced aqueous processing at a central location serving a number of reactors.
- An intermediate-to-large size (300–1500 MWe) pool-type reactor with oxide or metal fuel.
- A small size (50–150 MWe) modular-type reactor with uranium-plutonium-minor-actinide-zirconium metal alloy fuel, supported by a fuel cycle based on pyro-metallurgical processing in facilities integrated with the reactor.

18.15.4 Super Critical Water Cooled Reactor (SCWR)

The Super Critical Water Cooled (SCWRs) are high temperature, high-pressure, light-water-cooled reactors that operate above the thermodynamic critical point of water (374 °C, 22.1 MPa).

The reactor core may have a thermal or a fast-neutron spectrum, depending on the core design. The concept may be based on current pressure vessel or on pressure tube reactors, and thus use light water or heavy water as moderator. Unlike current water-cooled reactors, the coolant will experience a significantly higher enthalpy rise in the core, which reduces the core mass flow for a given thermal power and increases the core outlet enthalpy to superheated conditions. For both pressure vessel and pressure-tube designs, a once through steam cycle has been envisaged, omitting any coolant recirculation inside the reactor. As in a boiling water reactor, the superheated steam will be supplied directly to the high-pressure steam turbine and the feed water from the steam cycle will be supplied back to the core. Thus, the SCWR concepts combine the design and operation experiences gained from hundreds of water-cooled reactors with those experiences from hundreds of fossil-fired power plants operated with supercritical water (SCW). In contrast to some of the other Generation IV nuclear systems, the SCWR can be developed incrementally step-by-step from current water-cooled reactors.

a. **Advantage and Challenges**

Such SCWR designs have unique features that offer many advantages compared to state-of the-art water-cooled reactors:

- SCWRs offer increases in thermal efficiency relative to current-generation water-cooled reactors. The efficiency of a SCWR can approach 44 % or more, compared to 34–36 % for current reactors.
- Reactor coolant pumps are not required. The only pumps driving the coolant under normal operating conditions are the feed water pumps and the condensate extraction pumps.
- The steam generators used in pressurized water reactors and the steam separators and dryers used in boiling water reactors can be omitted since the coolant is superheated in the core.
- Containment, designed with pressure suppression pools and with emergency cooling and residual heat removal systems, can be significantly smaller than those of current water-cooled reactors can.
- The higher steam enthalpy allows to decrease the size of the turbine system and thus to lower the capital costs of the conventional island.

These general features offer the potential of lower capital costs for a given electric power of the plant and of better fuel utilization, and thus a clear economic advantage compared with current light water reactors.

However, there are several technological challenges associated with the development of the SCWR, and particularly the need to validate transient heat transfer models (for describing the depressurization from supercritical to sub-critical conditions), qualification of materials (namely advanced steels for cladding), and demonstration of the passive safety systems.

b. **GIF Progress up to 2012**

Pre-conceptual core design studies for a core outlet temperature of more than 500 °C have been performed in Japan, assuming either a thermal neutron spectrum or a fast neutron spectrum. Both options are based on a coolant heat-up in two steps with intermediate mixing underneath the core. Additional moderator for a thermal neutron spectrum is provided by feed water inside water rods. The fast-spectrum option uses zirconium-hydride (ZrH2) layers to minimize hardening of the neutron spectrum in case of core voiding. A pre-conceptual design of safety systems for both options has been studied with transient analyses.

A pre-conceptual plant design with 1700 MW net electric power based on a pressure-vessel-type reactor has been studied by Yamada et al. and has been assessed with respect to efficiency, safety and cost. The study confirms the target net efficiency of 44 % and estimates a cost reduction potential of 30 % compared with current pressurized water reactors. Safety features are expected to be similar to advanced boiling water reactors.

A pre-conceptual design of a pressure-vessel-type reactor with a 500 °C core outlet temperature and 1000 MW electric power has been developed in Europe, as summarized by Schulenberg and Starflinger. The core design is based on coolant heat-up in three steps. Additional moderator for the thermal neutron spectrum is provided in water rods and in gaps between assembly boxes. The design of the nuclear island and of the balance of the plant confirms results obtained in Japan, namely an efficiency improvement up to 43.5 % and a cost reduction potential of 20–30 % compared with latest boiling water reactors. Safety features as defined by the stringent European Utility Requirements are expected to be met.

Canada is developing a pressure-tube-type SCWR concept with a 625 °C core outlet temperature at the pressure of 25 MPa. The concept is designed to generate 1200 MW electric power (a 300 MW concept is also being considered). It has a modular fuel channel configuration with separate coolant and moderator. A high-efficiency fuel channel is incorporated to house the fuel assembly. The heavy-water moderator directly contacts the pressure tube and is contained inside a low-pressure calandria vessel. In addition to providing moderation during normal operation, it is designed to remove decay heat from the high-efficiency fuel channel during long-term cooling using a passive moderator cooling system. A mixture of thorium oxide and plutonium is introduced as the reference fuel, which aligns with the GIF position paper on thorium fuel. The safety system design of the Canadian SCWR is similar to that of the ESBWR. However, the introduction of the passive moderator cooling system coupled with the high-efficiency channel could reduce significantly the core damage frequency during postulated severe accidents such as large-break loss-of-coolant or station blackout events.

Pre-conceptual designs of three variants of pressure vessel supercritical reactors with thermal, mixed and fast neutron spectrum have been developed in Russia, which joined the SCWR System Arrangement in 2011.

Outside of the GIF framework, two conceptual SCWR designs with thermal and mixed neutron spectrum cores have been established by some research institutes in China. This is done, under framework of the Chinese national R&D projects from 2007 to 2012, covering some basic research projects on materials and thermo hydraulics, the core/fuel design, the main system design (including the conventional part), safety systems design, reactor structure design and fuel assembly structure design. The related feasibility studies have also been completed, and show that the design concept has promising prospects in terms of the overall performance, integration of design, component structure feasibility and manufacturability.

Prediction of heat transfer in SCW can be based on data from fossil fired power plants as discussed by Pioro et al. Computational tools for more complex geometries like fuel assemblies are available but still need to be validated with bundle experiments. System codes for transient safety analyses have been upgraded to include SCW, including depressurization transients to subcritical conditions. Flow stability in the core has been studied numerically. As in boiling water reactors, flow stability can be ensured using suitable inlet orifices in fuel assemblies. A number of candidate cladding materials have been tested in capsules, autoclaves and recirculating loops up to 700 °C at a pressure of 25 MPa. Stainless steels with more than 20 % chromium (Cr) are expected to have the required corrosion resistance up to a peak cladding temperature of 650 °C. More work is needed to develop alloys suitable for use at the design peak cladding temperatures of 850 °C for the Canadian SCWR concept. Further work is also needed to better identify the coolant conditions that lead to stress corrosion cracking. It has been shown that the creep resistance of existing alloys can be improved by adding small amounts of elements, such as zirconium (Zr), as reported by Kaneda et al. In the longer term, the steel experimental oxide dispersion strengthened (ODS) alloys offer an even higher potential, whereas nickel-base alloys are being considered for use in ultra supercritical fossil fired plants are less favorable for use in SCWRs due to their high neutron absorption and associated swelling and embrittlement.

Key water chemistry issues have been identified by Guzonas et al.; predicting and controlling water radiolysis and corrosion product transport (including fission products) remain the major R&D areas. In this regard, the operating experience using nuclear steam reheat at the Beloyarsk nuclear power plant in Russia is extremely valuable (Fig. 18.25).

18.15.5 Gas Cooled Fast Reactor (GFR)

The Gas Cooled Reactor (GFR) system is a high-temperature helium-cooled fast-spectrum reactor with a closed fuel cycle. It combines the advantages of fast-spectrum systems for long-term sustainability of uranium resources and waste minimization (through fuel multiple reprocessing and fission of long-lived actinides), with those of high-temperature systems (high thermal cycle efficiency and industrial use of the generated heat, for hydrogen production for example).

The GFR uses the same fuel recycling processes as the SFR and the same reactor technology as the VHTR. Therefore, its development approach is to rely, in so far as feasible, on technologies developed for the VHTR for structures, materials, components and power conversion system. Nevertheless, it calls for specific R&D

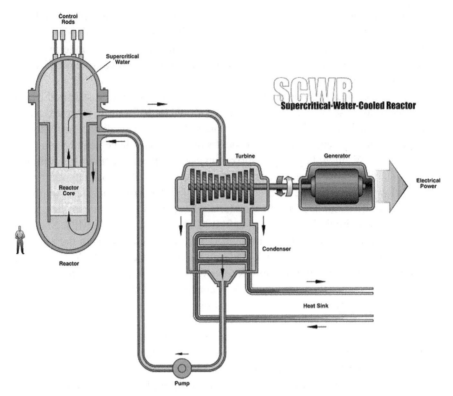

Fig. 18.25 Super critical water cooled reactor. (Courtesy of the Generation IV International Forum)

beyond the current and foreseen work on the VHTR system, mainly on core design and safety approach.

The reference design for GFR is based around a 2400 MWth reactor core contained within a steel pressure vessel. The core consists of an assembly of hexagonal fuel elements, each consisting of ceramic-clad, mixed-carbide-fuelled pins contained within a ceramic hex-tube. The favoured material at the moment for the pin clad and hex-tubes is silicon carbide fibre reinforced silicon carbide. The figure below shows the reactor core located within its fabricated steel pressure vessel surrounded by main heat exchangers and decay heat removal loops. The whole of the primary circuit is contained within a secondary pressure boundary, the guard containment (Figs. 18.26 and 18.27).

The coolant is helium and the core outlet temperature will be of the order of 850 °C. A heat exchanger transfers the heat from the primary helium coolant to a secondary gas cycle containing a helium-nitrogen mixture, which, in turn drives a closed cycle gas turbine. The waste heat from the gas turbine exhaust is used to raise steam in a steam generator, which is then used to drive a steam turbine. Such a combined cycle is common practice in natural gas-fired power plant so represents an established technology, with the only difference in the GFR case being the use of a closed cycle gas turbine.

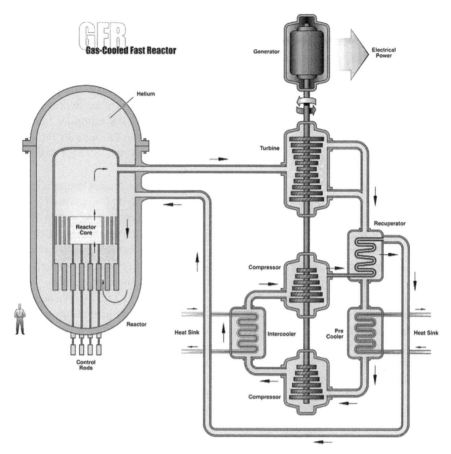

Fig. 18.26 Gas cooled fast reactor. (Courtesy of the Generation IV International Forum)

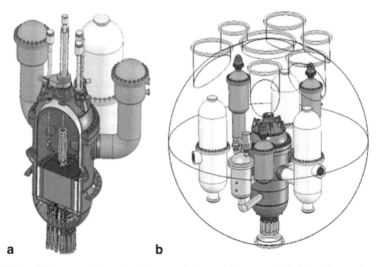

Fig. 18.27 **a** GFR-reactor, decay heat loops, main heat exchangers and fuel handling equipment. **b** GFR spherical guard vessel. (Courtesy of the Generation IV International Forum)

18.15.6   Lead Cooled Fast Reactor (LFR)

The Lead-cooled Fast Reactor (LFR) features a fast neutron spectrum, high temperature operation, and cooling by molten lead or Lead-Bismuth Eutectic (LBE),
low-pressure, chemically inert liquids with very good thermodynamic properties. It
would have multiple applications including production of electricity, hydrogen and
process heat. System concepts represented in plans of the Generation-IV International Forum (GIF) System Research Plan (SRP) are based on Europe's ELFR lead-
cooled system, Russia's BREST-OD-300 and the SSTAR system concept designed
in the US.

The LFR has excellent materials management capabilities since it operates in the
fast-neutron spectrum and uses a closed fuel cycle for efficient conversion of fertile
uranium. It can also be used as a burner to consume actinides from spent LWR
fuel and as a burner/breeder with thorium matrices. An important feature of the
LFR is the enhanced safety that results from the choice of molten lead as a chemically inert and low-pressure coolant. In terms of sustainability, lead is abundant and
hence available, even in case of deployment of a large number of reactors. More
importantly, as with other fast systems, fuel sustainability is greatly enhanced by
the conversion capabilities of the LFR fuel cycle. Because they incorporate a liquid
coolant with a very high margin to boiling and benign interaction with air or water,
LFR concepts offer substantial potential in terms of safety, design simplification,
proliferation resistance and the resulting economic performance. An important factor is the potential for benign end state to severe accidents (Fig. 18.28).

The LFR has development needs in the areas of fuels, materials performance,
and corrosion control. During the next 5 years progress is expected on materials,
system design, and operating parameters. Significant test and demonstration activities are underway and planned during this period.

18.16   Next Generation of Nuclear Power Reactors
for Power Production

Experts are projecting worldwide electricity consumption will increase substantially in the coming decades, especially in the development world, accompanying
economic growth and social progress that has direct impact on rising electricity
prices have focused fresh attention on nuclear power plants. New, safer and more
economical nuclear reactors could not only satisfy many of our future energy needs
but could combat global warming as well. Today's existing nuclear power plants on
line in the United States provide fifth of the nation's total electrical output.

Taking into account the expected increase in energy demand worldwide and the
growing awareness about global warming, climate change issues and sustainable
development, nuclear energy will be needed to meet future global energy demand.

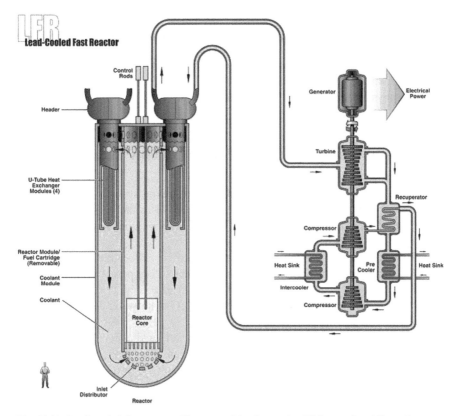

Fig. 18.28 Lead cooled fast reactor. (Courtesy of the Generation IV International Forum)

Nuclear power plant technology has evolved as distinct design generations as we mentioned in previous section and briefly summarized here again as follows:

- First Generation: prototypes, and first realizations (~ 1950–1970)
- Second Generation: current operating plants (~ 197–2030)
- Third generation: deployable improvements to current reactors (~ 2000 and on)
- Fourth generation: advanced and new reactor systems (2030 and beyond)

The Generation IV International Forum, or GIF, was chartered in July 2001 to lead the collaborative efforts of the world's leading nuclear technology nations to develop next generation nuclear energy systems to meet the world's future energy needs.

Eight technology goals have been defined for Generation IV systems in four broad areas:

1. Sustainability,
2. Economics,
3. Safety and Reliability, and finally,
4. Proliferation resistance and Physical protection.

A large number of countries share these ambitious goals as they aim at responding to economic, environmental and social requirements of the twenty-first century. They establish a framework and identify concrete targets for focusing GIF R&D efforts

Eight technology goals have been defined for Generation IV systems in four broad areas: sustainability, economics, safety and reliability, and proliferation resistance and physical protection.

18.17   Goals for Generation IV Nuclear Energy Systems

The next generation ("Generation IV") of nuclear energy systems is intended to meet the below goals (while being at least as effective as the "third" generation in terms of economic competitiveness, safety and reliability) in order to provide a sustainable development of nuclear energy.

In principle, the Generation IV Systems should be marketable or deployable from 2030 onwards. The systems should also offer a true potential for new applications compatible with an expanded use of nuclear energy, in particular in the fields of hydrogen or synthetic hydrocarbon production, seawater desalination and process heat production.

It has been recognized that these objectives, widely and officially shared by a large number of countries, should be at the basis of an internationally shared R&D program, which allows keeping open and consolidating the technical options, and avoiding any early or premature down selection.

Sustainability—1	Generation IV nuclear energy systems will provide sustainable energy generation that meets clean air objectives and provides long term availability of systems and effective fuel utilization for worldwide energy production.
Sustainability—2	Generation IV nuclear energy systems will minimize and manage their nuclear waste and notably reduce the long term stewardship burden, thereby improving protection for the public health and the environment.
Economics—1	Generation IV nuclear energy systems will have a clear life cycle cost advantage over other energy sources.
Economics—2	Generation IV nuclear energy systems will have a level of financial risk comparable to other energy projects.
Safety and reliability—1	Generation IV nuclear energy systems operations will excel in safety and reliability.
Safety and reliability—2	Generation IV nuclear systems will have a very low likelihood and degree of reactor core damage.
Safety and reliability—3	Generation IV nuclear energy systems will eliminate the need for offsite emergency response.
Proliferation resistance and physical protection	Generation IV nuclear energy systems will increase the assurance that they are very unattractive and the least desirable route for diversion or theft of weapons usable materials, and provide increased physical protection against acts of terrorism.

Evolution of Nuclear Power

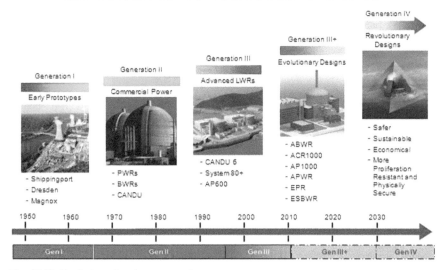

Fig. 18.29 Evolution of nuclear power plants

In fact, because the next generation nuclear energy systems will address needed areas of improvement and offer great potential, many countries share a common interest in advanced R&D that will support their development. The international research community should explore such development benefits from the identification of promising research areas and collaborative efforts that. The collaboration on R&D by many nations on the development of advanced next generation nuclear energy systems will in principle aid the progress toward the realization of such systems, by leveraging resources, providing synergistic opportunities, avoiding unnecessary duplication and enhancing collaboration (Fig. 18.29).

In 2009, the Experts Group published an outlook on Generation IV R&D, to provide a view of what GIF members hope to achieve collectively in the period 2010–2014. All Generation IV systems have features aiming at performance improvement, new applications of nuclear energy, and/or more sustainable approaches to the management of nuclear materials. High-temperature systems offer the possibility of efficient process heat applications and eventually hydrogen production. Enhanced sustainability is achieved primarily through adoption of a closed fuel cycle with reprocessing and recycling of plutonium, uranium and minor actinides using fast reactors; this approach provides significant reduction in waste generation and uranium resource requirements. The following Table summarizes the main characteristics of the six Generation IV systems.

System	Neutron spectrum	Coolant	Temp. °C	Fuel cycle	Size (MWe)
VHTR (Very high temperature gas reactor)	Thermal	Helium	900–1000	Open	250–300
SFR (Sodium-cooled fast reactor)	Fast	Sodium	550	Closed	30–150, 300–1500 1000–2000
SCWR (Supercritical water—cooled reactor)	Thermal/fast	Water	510–625	Open/Closed	300–700 1000–2000
GFR (Gas—cooled fast reactor)	Fast	Helium	850	Closed	1200
LFR (Lead—cooled fast reactor)	Fast	Lead	480–800	Closed	20–180 300–1200 600–1000
MSR (Molten salt reactor)	Epithermal	Fluoride salt	700–800	Closed	1000

18.18   Why We Need to Consider the Future Role of Nuclear Power Now

The following reasoning's are some arguments that show why we need to consider the future role in design of new nuclear power plant;

1. Nuclear power has been part of the global energy need mix for the past five decades. Currently it provides about 18 % of the electricity we use in our homes and workplaces. For example in the UK, about one third of our emissions of carbon dioxide come from electricity generation. The vast majority of those emissions come from coal and gas power plants.
2. Energy companies will need to invest in around 30–35 GW of new electricity generating capacity—as coal and nuclear plants retire—over the next two decades, with around two-thirds needed by 2020. This is equivalent to about one-third of our existing capacity. The world needs a clear and stable regulatory framework to reduce uncertainty for business to help ensure sufficient and timely investment in technologies that contribute to our energy goals.
3. Of the capacity that is likely to close over the two decades, two thirds is from carbon intensive fossil fuel generation and about 10 GW is nuclear and therefore low carbon. So companies' decisions on the type of power stations they invest in to replace this capacity will have significant implications for the level of carbon emissions. As an illustration, if our existing nuclear power stations were all replaced with fossil fuel fired power stations, our emissions would be between 8 and 16 MtC (million tons of carbon) a year higher as a result (depending on the mix of gas and coal-fired power stations). This would be equivalent to about 30–60 % of the total carbon savings we project to achieve under our central

scenario from all the measures we are bringing forward in the Energy White Paper. Our gas demand would also be higher, at a time when we are becoming more dependent on imported sources of fossil fuels.

4. Electricity demand in the United States is expected to grow significantly in the future. Over the past decade, Americans used 17 % more electricity, but domestic capacity rose only 2.3 % (National Energy Policy, May 2001). Unless the United States significantly increases its generating capacity, the country will face an energy shortage that is projected to adversely affect our economy, our standard of living, and our national security. Coupled with this challenge is the need to improve our environment.

5. New nuclear power stations have long lead times. This time is necessary to secure the relevant regulatory and development consents, which must be obtained before construction can begin, and there is also a long construction period compared to other generating technologies. Our conservative assumption is that for the first new nuclear plant the pre-construction period would last around 8 years (to secure the necessary consents) and the construction period would last around 5 years. For subsequent plants, this is assumed to be 5 and 5 years; respectively. New nuclear power stations are therefore unlikely to make a significant contribution to the need for new capacity before 2020.

6. Even with our expectations that the share of renewable will grow, it is likely that fossil fuel generation will meet some of this need. However, beyond that date there are still significant amounts of new capacity needed; for example, in 2023 one third or 3 GW of our nuclear capacity will still be operational, based on published lifetimes. Given the likely increase in fossil fuel generation before this date, it is important that much of this capacity is replaced with low carbon technologies. New nuclear power stations could make an important contribution, as outlined in this consultation document, to meeting our needs for low carbon electricity generation and energy security in this period and beyond to 2050. Because of the lead-times, without clarity now we will foreclose the opportunity for nuclear power.

7. The existing approach on new nuclear build was set out in 200311:"Nuclear power is currently an important source of carbon-free electricity. However, its current economics make it an unattractive option for new, carbon-free generating capacity and there are also important issues of nuclear waste to be resolved. These issues include our legacy waste and continued waste arising from other sources. This white paper does not contain specific proposals for building new nuclear power stations. However, we do not rule out the possibility that at some point in the future new nuclear build might be necessary if we are to meet our carbon targets. Before any decision to proceed with the building of new nuclear power stations, there will need to be the fullest public consultation and the publication of a further white paper setting out our proposals."

8. Since 2003 there have been a number of developments, which have led the Government to consider afresh the potential contribution of new nuclear power stations. Firstly, there has been significant progress in tackling the legacy waste issue:

- we have technical solutions for waste disposal that scientific consensus and experience from abroad suggest could accommodate all types of wastes from existing and new nuclear power stations;
- there is now an implementing body (the Nuclear Decommissioning Authority), with expertise in this area, and Government is reconstituting the Committee on Radioactive Waste Management (CoRWM) in order to provide continued independent scrutiny and advice; and
- a framework for implementing long-term waste disposal in a geological repository will be consulted on in the coming months.

9. The Government has also made progress in considering the issue of waste management in relation to potential new nuclear power stations:

   - This consultation provides the opportunity to discuss the ethical, intergenerational and public acceptability issues associated with a decision to allow the private sector to invest in new nuclear power stations and generate new nuclear waste;
   - The Government is developing specific proposals to protect the taxpayer. Under these proposals, private sector developers would meet the full decommissioning costs and full share of waste management costs. The proposals would be implemented in the event that we conclude that energy companies should be allowed to invest in new nuclear power stations. They would need to be in place before proposals for new power stations could go ahead.

10. Secondly, the high-level economic analysis of nuclear power, prepared for the Energy Review, concluded that under likely scenarios for gas and carbon prices and taking prudent estimates of nuclear costs, nuclear power would offer general economic benefit to the UK in terms of reduced carbon emissions and security of supply benefits. Therefore, the Government believes that it has a potential contribution to make, alongside other low-carbon generating technologies.

11. Thirdly, some energy companies have expressed a strong interest in investing in new nuclear power stations. They assess that new nuclear power stations could be an economically attractive low-carbon investment, which could help diversify their generation portfolios. Their renewed interest reflects assessments that with carbon being priced to reflect its impacts and gas prices likely to be higher than previously expected, the economics of new nuclear power stations are becoming more favorable.

12. Nuclear power stations have long lead times. If they are to be an option to replace the capacity closing over the next two decades, and in particular after 2020, a decision on whether allowing energy companies the option of investing in new nuclear power stations would be in the public interest, needs to be taken now. Energy companies would need to begin their initial preparations in the near future in order to have a reasonable prospect of building new generation in this period. Not taking the public interest decision now would foreclose the option of new nuclear being one of our options for tackling climate change and achieving energy security.

18.19 The Generation IV Roadmap Project

As the Generation IV goals were being finalized, preparations were made to develop the Generation IV technology roadmap. The organization of the roadmap is shown in the Fig. 15.21 below. The Roadmap Integration Team (RIT) is the executive group. Groups of international experts were organized to undertake identification and evaluation of candidate systems, and to define R&D to support them (Fig. 18.30).

In a first step, an Evaluation Methodology Group was formed to develop a process to systematically evaluate the potential of proposed Generation IV nuclear energy systems to meet the Generation IV goals. A discussion of the Evaluation Methodology Group's evaluation methodology is included in this report. At the same time, a solicitation was issued worldwide, requesting that concept proponents submit information on nuclear energy systems that they believe could meet some or all of the Generation IV goals. Nearly 100 concepts and ideas were received from researchers in a dozen countries [8].

Technical Working Groups (TWGs) were formed—covering nuclear energy systems employing water-cooled, gas-cooled, liquid-metal-cooled, and non-classical reactor concepts—to review the proposed systems and evaluate their potential using the tools developed by the Evaluation Methodology Group. Because of the large number of system concepts submitted, the TWGs collected their concepts into sets of concepts with similar attributes. The TWGs conducted an initial screening, *termed screening for potential*, to eliminate those concepts or concept sets that did not have reasonable potential for advancing the goals, or were too distant or technically infeasible [9].

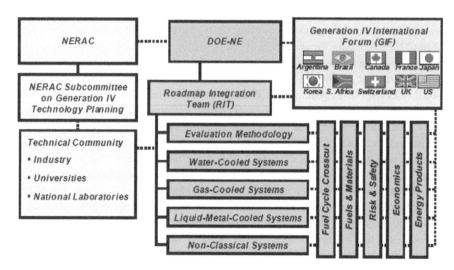

Fig. 18.30 The roadmap organization

A Fuel Cycle Crosscut Group (FCCG) was also formed at a very early stage to explore the impact of the choice of fuel cycle on major elements of sustainability—especially waste management and fuel utilization. Their members were equally drawn from the working groups, allowing them to compare their insights and findings directly. Later, other Crosscut Groups were formed covering economics, risk and safety, fuels and materials, and energy products. The Crosscut Groups reviewed the TWG reports for consistency in the technical evaluations and subject treatment, and continued to make recommendations regarding the scope and priority for crosscutting R&D in their subject areas. Finally, the TWGs and Crosscut Groups worked together to report on the R&D needs and priorities of the most promising concepts.

The international experts that contributed to this roadmap represented all ten GIF countries, the Organization for Economic Cooperation and Development Nuclear Energy Agency, the European Commission, and the International Atomic Energy Agency.

18.20   Licensing Strategy Components

A DOE and NRC working group was formed to develop the licensing strategy. This group conducted an in-depth analysis of LWR licensing process and technical requirements options, which was performed by the experienced senior staff of the two agencies. The methodology used in formulating the NGNP licensing strategy alternatives also included development of a phenomena identification and ranking table (PIRT) for a prototypical NGNP by subject matter experts in the nuclear field. The PIRT results assisted in the identification of key R&D needs. Based on the detailed analysis of these alternatives and balancing schedule considerations with licensing risk and other pertinent factors, the Secretary of Energy and the Commission concluded that the following NGNP licensing strategy provides the best opportunity for meeting the 2021 date for initial operation of a prototype NGNP, which details of such analysis can be found in NGNP report to Congress.

NGNP reactor technology will differ from that of commercial LWRs currently used for electric power generation. LWRs have a well-established framework of regulatory requirements, a technical basis for these requirements, and supporting regulatory guidance on acceptable approaches an applicant can take to show that NRC requirements are met. The NRC uses a Standard Review Plan to review licensing applications for these reactor designs. Additionally, the NRC has a well-established set of validated analytical codes and methods and a well-established infrastructure for conducting safety research needed to support its independent safety review of an LWR plant design and the technical adequacy of a licensing application.

New nuclear power plants can be licensed under either of two existing regulatory approaches. The first approach is the traditional "two-step" process described in Title 10, Part 50, "Domestic Licensing of Production and Utilization Facilities," of the *Code of Federal Regulations* (10 CFR Part 50), which requires both a Construction Permit (CP) and a separate Operating License (OL). The second approach is the new "one-step" licensing process described in 10 CFR Part 52, "Licenses, Certifications, and Approvals for Nuclear Power Plants," which incorporates a combined Construction and Operating License (COL). Both of these processes allow a deterministic or risk-informed performance-based approach to technical requirements.

Many of the regulatory requirements and supporting review guidance for LWRs are technology-neutral; that is, they are applicable to non-LWR designs as well as LWR designs. However, certain LWR requirements may not apply to the unique aspects of a VHTR design. Accordingly, in developing the NGNP licensing strategy, the NRC and DOE considered the various options available to the NRC staff for adapting current NRC LWR licensing requirements for the NGNP VHTR. These options related to legal, process, technical, research, and regulatory infrastructure matters and included an examination of historical licensing activities. These considerations led to selection of a licensing strategy that would comply best with the considerations identified in the EPAct.

The licensing strategy outlined in this report is composed of two distinct aspects. The first aspect is a recommended approach for how the NRC will adapt the current LWR technical requirements to apply to a VHTR. The second aspect is a recommended licensing process alternative that identifies which of the procedural alternatives in the NRC regulations would be best for licensing the NGNP. To arrive at these recommendations, NRC and DOE evaluated a number of options and alternatives.

18.21   Market and Industry Status and Potentials

Europe plays a leading role in the development of nuclear energy and has 35 % of the globally installed capacity. The reactors in Europe have been in operation for 27 years on average. Current plans in most EU member countries are to extend their lifetime on a case by case basis beyond 40 years, and even beyond 60 years in some cases, in combination with power upgrades. The first two Generation-III reactors, European Pressurized-water Reactor (EPR) are currently being constructed.

The global growth of the nuclear energy can be measured by the increasing number of reactors (three more in 2005 and 2006; seven in 2007 and ten in 2008), but with a strong concentration in Asia. Nevertheless a number of these reactors are of European design. There are presently four reactors under construction in Europe: the EPRs in Finland and France and two smaller reactors of Generation-II type (VVER 440) in Slovakia and with plans to build new reactors in France, Romania, Bulgaria and Lithuania. Perhaps more importantly the UK has taken concrete steps towards new build with bidding beginning in 2009 from leading utilities, and Italy has declared that it intends to start a nuclear program with a target to produce 25 % of the electricity by 2030. The estimated maximum potential installed capacities of nuclear fission power for the EU-27, (150 GWe by 2020 and 200 GWe by 2030) appear more realistic than the baseline (115 GWe in 2020 and 100 GWe in 2030).

Programs to build fast reactor and high-temperature reactor demonstrators are being implemented in Russia and several Asian countries. Although these are not Generation-IV designs, transfer of knowledge and experience from operation will contribute significantly to future Generation-IV development. In Europe, a concerted effort is proposed in the form of a European Industrial Initiative in sustainable nuclear fission as part of the Community's SET-Plan. The EII has singled out the Sodium Fast Reactor (SFR) as its primary system with the basic design selected by 2012 and construction of a prototype of 250–600 MWe that is connected to the grid and operational by 2020.

In parallel, a gas- or lead-cooled fast reactor (GFR/LFR) will also be investigated. The selection of the alternative fast reactor technology is scheduled for 2012 on the basis of a current program of pre-conceptual design research. The reactor will be a 50–100 MWth demonstrator reactor that should also be in operation by 2020. The SFR prototype and LFR/GFR demonstrator will be complemented by a fuel fabrication workshop that should serve both systems, and by a range of new or refurbished supporting experimental facilities for qualification of safety systems, components, materials and codes. A commercial deployment for a SFR reactor is expected from 2040 and for the alternative design a decade later.

High temperature reactors dedicated to cogeneration of process heat for the production of synthetic fuels or industrial energy products could be available to meet market needs by 2025, which would trigger requirements to construct "first of a kind" demonstrators in the next few years. Indeed, such programs are currently being set up in some countries (USA, Japan, South Africa and China). The key aspect is the demonstration of the coupling with the conventional industrial plant. Supercritical water reactors and molten salt reactors, as well as accelerator driven sub-critical systems dedicated to transmutation of nuclear waste, are currently being assessed in terms of feasibility and performance, though possible industrial applications have yet to be clearly identified.

18.22   Barriers

The high capital cost of nuclear energy in combination with uncertain long-term conditions constitutes a financial risk for utilities and investors. The lack of widespread support in the EU Member States may undermine the strength of EU industry for the development of new technologies. Harmonized regulations, codes and standards at the EU-level would strengthen the competitiveness of Europe's nuclear sector and promote deployment of Generation-III technology in the near term. The industry, infrastructures and services that support nuclear power has shrunk significantly during the last decades. This situation in Europe is not unique but it may pose a bottleneck for the deployment of reactors in the relatively near future. One example is large forgings needed for pressure vessel heads. World capacity is limited and even at the present new build construction rate there is a waiting list for delivery of these components.

Public acceptance remains an important issue, but even though opinion is not very favorable in a number of Member States, there are signs that the mood is changing. Nevertheless, concerted efforts are still required, based on objective and open dialogue amongst all stakeholders. International cooperation currently exists at the level of research, and this is being facilitated in the area of Generation-IV systems by the Generation-IV International Forum (GIF). However, EU industry is facing stiff competition, especially in Asia where strong corporate support for R&D is putting industry in a better position to gain leadership in the near future. Another significant potential barrier for nuclear fission is the shortage of qualified engineers and scientists as a result of the lack of interest in nuclear careers during the 1990s and the reduced availability of specialist courses at universities. Preservation of nuclear knowledge remains a major issue, especially since most of the current generations of nuclear experts are nearing retirement.

18.23   Needs

The high initial capital investments and sensitive nature of the technology involved means that renewed deployment of currently available nuclear technology can only take place in a stable (or, at least, predictable) regulatory, economic and political environment. In June 2009, the EU established a common binding framework on nuclear safety with the adoption of the Council Directive establishing a Community framework for the safety of nuclear installations [8, 9]. This is the first binding EU legislation in this field.

In order to retain its leading position and to overcome bottlenecks in the supply chain, Europe also needs to re-invigorate the industrial supply chains supporting the nuclear sector. Apart from this overriding requirement for a clear European strategy on nuclear energy, a new research and innovation system is needed that can assure additional funding, especially for the development of Generation-IV technology. In this context the Sustainable Nuclear Energy Technology Platform plays a key role. The timescales involved, and the fact that key political and strategic decisions are yet to be taken regarding this technology, mean that a significant part of this additional funding must be public. The launch of the European Sustainable Nuclear Industrial Initiative under the Community's SET-Plan, bringing together key industrial and R&D organizations would be a very significant step towards the construction and operation of the necessary demonstrators and prototypes.

High temperature reactors based on existing technology can also be deployed in the near future with the aim of demonstrating the co-generation of process heat and the coupling with industrial processes. This would need to be built and funded through a European or International consortium, which should also include the process heat end-user industries. In the meantime, an enhanced research effort is needed to ensure Europe's leadership in sustainable nuclear energy technologies that include continuous innovation in LWRs, qualification and development of materials, closed fuel cycle with U-Pu multi-recycling and (very) high temperature reactors and related fuel technology.

Breakthroughs are especially sought in the fields of materials to enhance safety, nuclear fuels and fuel cycle processes. Additionally, there is a need for harmonization of European standards and a strategic planning of national and European research infrastructures for use by the European research community. The implementation of geological disposal of high-level waste is also being pursued as part of national waste management programs, though some countries are not as advanced as others. The new Implementing Geological Disposal Technology Platform, launched in November 2009, is coordinating the remaining necessary applied research in Europe leading up to the start of operation of the first geological repositories for high-level and long-lived waste around 2020, and will facilitate progress in and technology transfer with other national programs.

More effort is needed to inform and interact with the public and other stakeholders, and the education and training of a new generation of nuclear scientists and engineers and transfer of knowledge from the generation that designed and built reactors in the seventies and eighties needs urgent attention. The European Nuclear Energy Forum (ENEF) provides a unique platform for a broad open discussion on the role nuclear power plays today and could play in the low carbon economy of the future. ENEF analyses and discusses the opportunities (competitiveness, financing, grid, etc) and risks (safety, waste) and need for education and training associated with the use of nuclear power and proposes effective ways to foster communication with and participation of the public.

18.24   Synergies with Other Sectors

Nuclear energy provides a very stable base-load electricity supply and can therefore work in synergy with renewable energies that are more intermittent. Nuclear energy should also contribute significantly to a low-carbon transport sector as high temperature applications can provide synthetic fuel and hydrogen, while generated electricity could provide a large share of the energy for electrical cars. Interactions are anticipated with activities in "Hydrogen Energy and Fuel Cells" through the potential of nuclear hydrogen production and with "grids" from the characteristics of nuclear electricity generation. With respect to basic materials research, there should be synergies with other applications, such as "Biofuels" and "Clean Coal", where materials are subjected to extreme environments. In addition, the opportunities for important common research with the fusion program, especially in the area of materials, need to be fully exploited. The European Energy Research Alliance under the SET-Plan is also expected to provide opportunities for synergies and collaborative work in the area of nuclear materials. In general, cross cutting research would benefit from more clearly defined channels of interaction, responsibilities and increased flexibility regarding funding and programming.

Fig. 18.31 Sketch for Problem 18.1

Problems

Problem 18.1: A 1000 MW power plant is powered by nuclear fuel. Determined the amount of nuclear fuel consumed per year. See Fig. 18.31 below.

Problem 18.2: A 1000 MW power plant is powered by burning coal. Calculate the amount of coal consumed per year.

Problem 18.3: A power plant that burns coal produces 1.1 kg of carbon dioxide (CO_2) per kWh. Determine the amount of CO_2 production that is due to the refrigerators in a city. Assume that the city uses electricity produced by a coal power plant.

Problem 18.4: A person trades in his Ford Taurus for a Ford Explorer. Calculate the amount of CO_2 emitted by the Explorer within 5 years. Assume the Explorer is assumed to use 940 gallons a year compared to 715 gallons for Taurus.

Problem 18.5: A power plant that burns natural gas produces 0.59 kg of carbon dioxide (CO_2) per kWh. Calculate the amount of CO_2 production that is due to the refrigerators in the city. Assume the city uses electricity produced by a natural gas power plant. Give the fact that 0.59 kg of CO_2 is produced per kWh of electricity generated. Noting that there are 200,000 households in the city and each household consumes 700 kWh of electricity for refrigeration.

References

1. Cohen BL (1983, Jan) Breeder reactors: a renewable energy source. Am J Phys 51(1)
2. Weinberg A (1989) The second fifty years of nuclear fission. Proceedings of the special symposium: 50 years of nuclear fission in review, Canadian Nuclear Society, Ontario, Canada, June 5, 1989
3. Seko N (2003, Nov) Aquaculture of uranium in seawater by a fabric-adsorbent submerged system. Nucl Technol 144:274
4. Some Physics of Uranium. http://www.worldnuclear.org/education/phys.html. Accessed Dec 2005

5. Hannum WH (1997) The technology of the integral fast reactor and its associated fuel cycle. Prog Nucl Energy 31:1

6. Uranium 2007: Resources, Production and Demand. OECD Nuclear Energy Agency and the International Atomic Energy Agency, OECD 2008 NEA N 6345

7. Nuclear Energy Outlook 2008, OECD/NEA Report No. 6348, 2008, Nuclear Energy Agency, Paris

8. Proposal for a COUNCIL DIRECTIVE (EURATOM) setting up a Community framework for nuclear safety COM (2008) 790/3, November 2008

9. COUNCIL OF THE EUROPEAN UNION Legislative Acts and Other Instruments 10667/09, June 2009

10. Projected Costs of Generating Electricity, 2005 Update, NEA/OECD, 2005

11. Strategic and Policy Issues Raised by the Transition from Thermal to Fast Nuclear Systems, 2009, OECD/NEA Report No. 6352

12. Raja AK, Srivastava AP, Dwivedi M (2006) Power plant engineering. New Age International (P) Limited, New Delhi

13. Mathieu L, Heuer D, Merle-Lucotte E et al (2009) Possible configurations for the thorium molten salt reactor and advantages of the fast non-moderated version. Nucl Sci Eng 161:78–89

14. Forsberg CW et al (2007) Liquid salt applications and molten salt reactors. Rev Gén Nucl 4(2007):63

15. Merle-Lucotte E, Heuer D et al (2008) Introduction of the physics of molten salt reactor, materials issues for generation IV systems, NATO science for peace and security series'B. Springer, Berlin, pp 501–521

16. Merle-Lucotte E, Heuer D et al (2009) Minimizing the fissile inventory of the molten salt fast reactor. Proceedings of the advances in nuclear fuel management IV (ANFM 2009), Hilton Head Island, USA

17. Renault C, Hron M, Konings R, Holcomb DE (2009) The molten salt reactor (MSR) in Generation IV: overview and perspectives, GIF Symposium proceeding, Paris, France

18. Ignatiev V et al (2005) Characteristics of molten salt actinide recycler and transmuter system. In: Proceedings of international conference on emerging nuclear energy systems, Brussels, Belgium, 21–26 Aug: paper ICQ064

19. ISTC# 1606 final report, International scientific technical centre, Moscow, July, 2007

20. Forsberg CW, Peterson PF, Kochendarfer RA (2008) Design options for the advanced high-temperature reactor. Proceedings of the 2008 international congress on advances in nuclear power plants (ICAPP'08), Anaheim, USA, June 8–12, 2008

21. Bardet P et al (2008) Design, analysis and development of the modular PB-AHTR. Proceedings of the 2008 international congress on advances in nuclear power plants (ICAPP'08), Anaheim, USA, June 8–12, 2008

22. Ignatiev V, Surenkov A (2012) Material performance in molten salts. In: Comprehensive nuclear materials, vol 5. pp 221–250

23. Bene O, Konings RJM (2012) Molten salt reactor fuel and coolant. In: Comprehensive Nuclear Materials, vol 3. pp 359–389

24. Merle-Lucotte E, Heuer D, Allibert M, Brovchenko M, Capellan N, Ghetta V (2011) Launching the thorium fuel cycle with the molten salt fast reactor, Contribution 11190, international congress on advances in nuclear power plants (ICAPP), Nice, France

25. Merle-Lucotte E, Heuer D, Allibert M, Doligez X, Ghetta V (2009) Optimizing the burning efficiency and the deployment capacities of the molten salt fast reactor, contribution 9149, Global 2009, The nuclear fuel cycle: sustainable options & industrial perspectives, Paris, France

26. Doligez et al (2009) Numerical tools for molten salt reactors simulations. Proceedings of the international conference global 2009—The nuclear fuel cycle: sustainable options & industrial perspectives, Paris, France

27. Merle-Lucotte E, Heuer D, Allibert M, Doligez X, Ghetta V (2010) Simulation tools and new developments of the molten salt fast reactor, Contribution A0115, European nuclear conference ENC2010, Barcelone, Espagne

28. Brovchenko M, Heuer D, Merle-Lucotte E, Allibert M, Capellan N, Ghetta V, Laureau A (2012) Preliminary safety calculations to improve the design of Molten Salt Fast Reactor, PHYSOR 2012 Advances in Reactor Physics Linking Research, Industry, and Education, Knoxville, Tennessee, USA, April 15–20, 2012, on CD–ROM

29. Delpech S, Merle-Lucotte E, Augé T, Heuer D (2009) MSFR: material issued and the effect of chemistry control. Generation IV international forum symposium, Paris, France

30. Beilmann M, Bene O, Konings RJM, Fanghänel Th (Feb 2013) Thermodynamic assessment of the (LiF + UF3) and (NaF + UF3) systems. J Chem Thermodyn 57:22–31

31. Bene O, Beilmann M, Konings RJM (15 Oct 2010) Thermodynamic assessment of the LiF-NaF-ThF4-UF4. J Nucl Mater 405(2):186–198

32. Delpech S, Merle-Lucotte E, Heuer D, Allibert M, Ghetta V, Le-Brun C, Mathieu L, Picard G (2009) Reactor physics and reprocessing scheme for innovative molten salt reactor system. J Fluorine Chem 130(1):11–17

33. Jaskierowicz S, Delpech S, Fichet P, Colin C, Slim C, Picard G (2011) Pyrochemical reprocessing of thorium-based fuel. Proceeding of ICAPP2011, Nice, France

34. Ignatiev V et al (2012) Molten salt reactor: new possibilities, problems and solutions. At Energ 112(3):135

35. Ignatiev V et al (2012) Progress in development of MOSART concept with Th support, ICAPP'12, Chicago, USA, June 24–28, 2012, Paper No. 12394

36. Afonichkin V, Bovet A, Shishkin V (2011) Salts purification and voltammetric study of the electro reduction of U(IV) to U(III) in molten LiF?ThF4?. J Nucl Mater 419(1–3):347

37. Baque F, Paumel K, Cornloup G, Ploix MA, Augem JM (2011) Non-destructive examination of immersed structures within liquid sodium, ANIMMA 2011, Ghent, June 6–9, 2011

38. Joo YS, Park CG, Kim JB, Lim SH (2011) Development of ultrasonic waveguide sensor for under-sodium inspection in a sodium-cooled fast reactor, NDT & E International 44, pp 239–246

39. Floyd J, Alpy N, Haubensack D, Avakian G, Rodriguez G (2011) On-design efficiency reference charts for the supercritical CO2 Brayton cycle coupled to a SFR. Proc. ICAPP2011, Nice, France, 2–5 May, 2011, Paper 11054

40. Moisseytsev A, Sienicki JJ (2012) Dynamic simulation and control of the S-CO2 cycle: from full power to decay heat removal. Proc. ATH'12, embedded topical meeting of ANS 2012 Winter Meeting, San Diego, CA, USA, 11–15 Nov, 2012, Paper 6461

41. Sienicki JJ et al (2013) Synthesis of results obtained on sodium components and technology through the generation IV international forum SFR component design and balance-of-plant project. Proc. FR13, Paris, France, 4–7, March, 2013

42. Delage F et al (2013) Status of advanced fuel candidates for sodium fast reactor within the Generation IV international forum. J Nucl Mater NUMA46668 (to be published)

43. Oka Y, Koshizuka S, Ishiwatari Y, Yamaji A (2010) Super light water reactors and super fast reactors. Springer, Berlin

44. Yamada K, Sakurai S, Asanuma Y, Hamazaki R, Ishiwatari Y, Kitoh K (2011) Overview of the Japanese SCWR concept developed under the GIF collaboration. Proc. ISSCWR-5, Vancouver, Canada, March 13–16

45. Schulenberg T, Starflinger J (2012) High performance light water reactor design and analyses. KIT Scientific Publishing,

46. Yetisir M, Diamond W, Leung LKH, Martin D, Duffey R (2011) Conceptual mechanical design for a pressure-tube type supercritical water-cooled reactor. Proc. 5th international symposium on supercritical water-cooled reactors, Vancouver, Canada, March 13–17

47. Ryzhov SB, Mokhov VA, Nikitenko MP, Podshibyakin AK, Schekin IG, Churkin AN Advanced designs of VVER reactor plant, the 8th international topical meeting on nuclear thermal-hydraulics, operation and safety (NUTHOS-8), Oct 10–14, 2010, Shanghai, China, Paper N8P0184

48. Ryzhov SB, Kirillov PL et al (2011) Concept of a single-circuit RP with vessel type supercritical water-cooled reactor. Proc. ISSCWR-5, Vancouver, Canada, March 13–16

49. Pioro IL, Duffey RB (2007) Heat transfer and hydraulic resistance at supercritical pressures in power engineering applications. ASME Press,

50. Kaneda J, Kasahara S, Kano F, Saito N, Shikama T, Matsui H (2011) Material development for supercritical water-cooled reactor. Proc. ISSCWR-5, Vancouver, Canada, March 13–16

51. Guzonas D, Brosseau F, Tremaine P, Meesungnoen J, Jay-Gerin J-P (2012) Water chemistry in a supercritical water-cooled pressure tube reactor. Nucl Technol 179

52. Stainsby R, Garnier JC, Guedeney P, Mikityuk K, Mizuno T, Poette C, Pouchon M, Rini M, Somers J, Touron E (2011) The Generation IV gas-cooled fast reactor. Paper 11321, Proc. ICAPP 2011 Nice, France, 2–5 May 2011

53. Perkó Z, Kloosterman JL, Fehér S (2012, Jan) Minor actinide transmutation in GFR600. Nucl Technol 177

54. Stainsby R, Peers K, Mitchell C, Poette C, Mikityuk K, Somers J (2011) Gas cooled fast reactor research in Europe. Nucl Eng Des 241:3481–3489

55. Epiney A, Alpy N, Mikityuk K, Chawla R (Jan 2012) A standalone decay heat removal device for the gas-cooled fast reactor for intermediate to atmospheric pressure conditions. Nucl Eng Des 242:267–284

56. Smith RR, Cissei DW (1978) Fast reactor operation in the United States. International symposium on design, construction, and operating experience of demonstration LMFBRs, Bologna, Italy, April 10–14, 1978

57. U.S. Nuclear Energy Research Advisory Committee (NERAC) and the Generation IV International Forum (GIF), Generation IV Technology Roadmap, Report GIF-002–00, Dec 2002

58. Alan EW, Todd DR, Tsvetkov PV (2012) Fast spectrum reactors. Springer, New York

59. De Bruyn D, Maes D, Mansani L, Giraud B (2007) From MYRRHA to XT-ADS: the design evolution of an experimental ADS system, AccApp'07, Pocatello, Idaho, July 29–Aug 2, 2007

60. Cinotti L, Locatelli G, Aït Abderrahim H, Monti S, Benamati G, Tucek K, Struwe D, Orden A, Corsini G, Le Carpentier D (2008) The ELSY Project, Paper 377. Proceedings of the international conference on the physics of reactors (PHYSOR), Interlaken, Switzerland, 14–19 Sep 2008

61. Alemberti et al (2012) The European lead fast reactor: design, safety approach and safety characteristics, IAEA Technical meeting on impact of Fukushima event on current and future FR designs, Dresden, Germany

62. Alemberti et al (2013) The lead fast reactor demonstrator (ALFRED) and ELFR design, international conference on fast reactor and nuclear fuel cycle (FR13), Paris, France

63. Takahashi M (2012) LFR development in Japan, 11th LFR Prov. SSC Meeting, Pisa, Italy, 16 April 2012

64. Takahashi M et al (2005) Pb-Bi-cooled direct contact boiling water small reactor. Prog Nucl Energy. 47:190–201

65. Sekimoto H, Nagata A (2008) Fuel cycle for CANDLE reactors. Proc. of workshop on advanced reactors with innovative fuels ARWIF-2008, Tsuruga/Fukui, 20–22 Feb 2008

66. Kim WJ et al (2006) Supercritical carbon dioxide Brayton power conversion cycle design for optimized battery-type integral reactor system. Paper 6142, ICAPP-06, Reno, NV, USA, June 4–8, 2006

67. Hwang IS (2006) A sustainable regional waste transmutation system: PEACER, plenary invited paper, ICAPP-06, Reno, NV, U.S.A., June 4–6, 2006

68. Smith C, Halsey W, Brown N, Sienicki J, Moisseytsev A, Wade D (15 June 2008) SSTAR: the US lead-cooled fast reactor (LFR). J Nucl Mater 376(3):255–259

69. Short MP, Ballinger RG (2010) Design of a functionally graded composite for service in high temperature lead and lead-bismuth cooled nuclear reactors, MIT-ANP-TR-131, 2010

70. GIF-LFR Provisional System Steering Committee (PSSC), Draft System Research Plan for the Lead-cooled Fast Reactor (LFR), 2008

71. Toshinsky GI, Komlev OG, Tormyshev IV et al (2011) Effect of potential energy stored in reactor facility coolant on NPP safety and economic parameters. Proceedings of ICAPP, Nice, France, May 2–5, 2011, Paper 11465

Chapter 19
Nuclear Fuel Cycle

Nuclear power has unresolved challenges in long-term management of radioactive wastes. A critical factor for the future of an expanded nuclear power industry is the choice of the fuel cycle—what type of fuel is used, what types of reactors 'burn' the fuel, and the method of disposal of spent fuel.

19.1 The Nuclear Fuel Cycle

The *nuclear fuel cycle* is dealing with all steps in the life history of the reactor fuel, starting with the mining of the uranium ore. The nuclear fuel cycle, which is an industrial process involving various activities to produce electricity from uranium in nuclear power reactors. The nuclear fuel cycle starts with the mining of uranium and ends with the disposal of nuclear waste. With the reprocessing of used fuel as an option for nuclear energy, the stages form a true cycle. The complete fuel cycle loop can be divided into two fold steps and they are classified as (Fig. 19.1);

1. **Front End:** The steps that are making up this cycle are starting with to prepare uranium for use in a nuclear reactor, and it undergoes the steps of mining and milling, conversion, enrichment and fuel fabrication. These steps make up the *'front end'* of the nuclear fuel cycle.
2. **Back End:** The steps that are making up this part of cycle are ending up with uranium that has been spent about 3 years in a reactor to produce electricity, and then used fuel may undergo a further series of steps including temporary storage, reprocessing, and recycling before wastes are disposed. Collectively these steps are known as the *'back end'* of the fuel cycle.

As it mentioned the cycle starts with the mining of uranium and ends with the disposal of nuclear waste.

The uranium is extracted from the ore and is converted in a series of steps into actual fuel material, usually pure uranium dioxide and the raw material for today's most commercial nuclear reactors fuel is uranium. The conclusion of these series of

© Springer International Publishing Switzerland 2015   539
B. Zohuri, P. McDaniel, *Thermodynamics In Nuclear Power Plant Systems,*
DOI 10.1007/978-3-319-13419-2_19

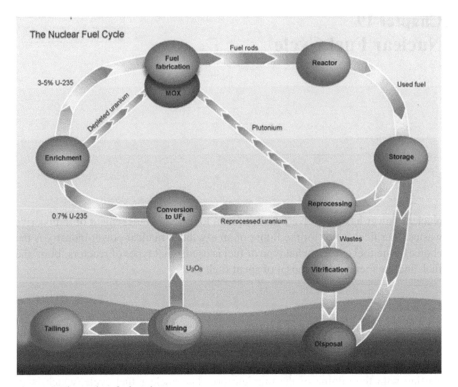

Fig. 19.1 The nuclear fuel cycle

steps will produce an efficient fuel for generating electricity. Therefore, a lengthily isotope separation process is part of the fuel preparation effort. Natural uranium consists primarily of two isotopes, 99.3 % is U_{238} and 0.7 % is U_{235}.

Uranium is a common metal that can be found throughout the world. It is present in most rocks and soils, in many rivers and in sea water. Uranium is about 500 times more abundant than gold and about as common as tin and it is a common, slightly radioactive material that occurs naturally in the Earth's crust.

There are three ways to mine uranium:

1. Open pit mines,
2. Underground mines and
3. In situ leaching where the uranium is leached directly from the ore.

The largest producers of uranium ore are Kazakhstan, Canada and Australia. The concentration of uranium in the ore could range from 0.03 up to 20 %.

Uranium is enriched in U_{235} by introducing the gas in fast-spinning cylinders ('centrifuges'), where heavier isotopes are pushed out to the cylinder walls. Uranium can also be enriched using older technology by pumping UF_6 gas through porous membranes that allow U_{235} to pass through more easily than heavier isotopes, such as U_{238}. Enrichment increases the proportion of the U_{235} isotope.

Fig. 19.2 Enrichment increases the proportion of the U_{235} isotope

The fission process which heat releases energy in a nuclear reactor, takes place mainly in U_{235}. Most nuclear power plants require fuel with U_{235} enriched to a level of 3–5 %. To increase the concentration of U_{235}, uranium must be enriched. Since enrichment happens in gaseous form, yellow cake is converted to uranium hexafluoride gas (UF_6) at a conversion facility. UF_6 gas is filled into large cylinders where it solidifies. The cylinders are loaded into strong metal containers and shipped to an enrichment plant (Fig. 19.2).

Enriched uranium (UF_6) cannot be directly used in reactors, as it does not withstand high temperatures or pressures. It is therefore, converted into uranium oxide (UO_2). Fuel pellets are formed by pressing UO_2, which is sintered (baked) at temperatures of over 1400 °C to achieve high density and stability.

The pellets are cylindrical and are typically 8–15 mm in diameter and 10–15 mm long. They are packed in long metal tubes to form fuel rods, which are grouped in 'fuel assemblies' for introduction into a reactor (Fig. 19.3).

Once the fuel is loaded inside a nuclear reactor, controlled fission can occur. Fission means that the U_{235} atoms are split. The splitting releases heat energy that is used

Fig. 19.3 Reactor fuel is in form of ceramic pellets

to heat water and produce high-pressure steam. The steam turns a turbine connected to a generator, which generates electricity. The fuel is used in the reactor for 3–6 years. About once a year, 25–30% of the fuel is unloaded and replaced with fresh fuel.

A nuclear power plant produces electricity by:

1. Heating water to
2. Generate steam that
3. Makes the turbine rotate to
4. Enable the generator to produce electricity.

The useful lifetime of the fuel in a reactor may be limited by dimensional changes in solid fuel elements, by the accumulation of neutron absorbing fission product poisons (especially in thermal reactors), and by deletion of the fissile material. As a rule of thumb, the fuel will require replacement when only a few percent of the total fissile and fertile species have been consumed.

If the unused materials were to be recovered and recycled, the spent fuel would be subjected to a complex, chemical reprocessing procedure. The spent fuel assemblies removed from the reactor are very hot and radioactive. Therefore the spent fuel is stored under water, which provides both cooling and radiation shielding. After a few years, spent fuel can be transferred to an interim storage facility. This facility can involve either wet storage, where spent fuel is kept in water pools, or dry storage, where spent fuel is kept in casks (Fig. 19.4).

Both the heat and radioactivity decrease over time and after 40 years in storage, the fuel's radioactivity will be about a thousand times lower than when it was removed from the reactor. The intense radioactivity of the fission products present introduces a special problem and a *"cooling period"* of at least 120 days must generally be allowed for the radioactivity to decay to a sufficient extent to permit the reprocessing operations.

Used fuel also needs to be taken care of for reuse and disposal. The nuclear fuel cycle includes the 'front end', i.e. preparation of the fuel, the 'service period' in which fuel is used during reactor operation to generate electricity, and the 'back end', i.e. the safe management of spent nuclear fuel including reprocessing and reuse and disposal. If spent fuel is not reprocessed, the fuel cycle is referred to as an *'open'* or *'once-through'* fuel cycle; if spent fuel is reprocessed, and partly reused, it is referred to as a *'closed'* nuclear fuel cycle.

Fig. 19.4 During temporary storage, the radioactivity and heat of spent nuclear fuel decreases significantly

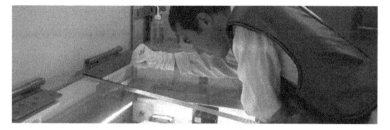

Fig. 19.5 Spent fuel can be recycled to produce more power

The spent fuel contains uranium (96%), plutonium (1%) and high-level waste products (3%).

The uranium, with less than 1% fissile U_{235} and the plutonium can be reused. Some countries chemically reprocess usable uranium and plutonium to separate them from unusable waste. Recovered uranium from reprocessing can be returned to the conversion plant, converted to UF_6 and subsequently re-enriched. Recovered plutonium, mixed with uranium, can be used to fabricate mixed oxide fuel (MOX) (Fig. 19.5).

19.2 Fuel Cycle Choices

In the world today, nuclear energy comprises ~ 15% of the total electricity mix. This power source produces no CO^2 during the operation of the plants and can provide bulk power to industry and households 24/7. Concerns about CO^2 emissions from burning fossil fuels (coal, oil and natural gas) have caused the world demand of nuclear power to rise. With increasing interest in nuclear energy, we are facing a problem of sharing the readily available uranium resources. To increase the energy utilization and decrease the radio toxicity of the waste advanced concepts for fuel reprocessing are being investigated on a global scale. This is a multi disciplinary research field spanning from organic synthesis to particle physics and no single research group can hope to cover all aspects.

Collaboration is crucial to the success of these projects. For a successful advanced nuclear fuel cycle, one or several chemical separations steps are required to fractionate the different elements in spent fuel. These separations processes are of varying complexity and use a range of different chemicals making it challenging to implement them on industrial scale. Furthermore, some of the processes under development are still in the experimental stage and require supporting fundamental studies to succeed. In this presentation, similarities and differences between the different separations schemes are discussed and emphasizing the challenges that will be faced before an advanced nuclear fuel cycle can be implemented in industry.

Nuclear fuel cycle policies have been rather inconsistent throughout history, arguably due to the complex interdependencies of technical, social, economic and

environmental issues involved. Sophisticated fuel cycle simulation tools are being developed to aid the decision makers with defining future fuel cycle strategies.
However, the value of these tools for policy making has yet to be demonstrated.
This limited success may be partially attributed to the fact that the large amount of
technical data generated by these tools obscures the actual policy tradeoffs. From
the recent simplified analyses, it was discovered, not surprisingly, that in order to
address nuclear waste and resource sustainability issues, the construction of fuel
reprocessing infrastructure along with fast spectrum reactors has to be pursued as
soon as possible. This policy, however, has a significant economic penalty due to
the high cost of fast reactors and reprocessing plants. Decoupling these two parts
of the infrastructure would thus allow their gradual introduction, reducing the economic burden and increasing the chances of success.

Collaboration is crucial to the success of these projects. For a successful advanced nuclear fuel cycle, one or several chemical separations steps are required
to fractionate the different elements in spent fuel. These separations processes are
of varying complexity and use a range of different chemicals making it challenging
to implement them on industrial scale. Furthermore, some of the processes under
development are still in the experimental stage and require supporting fundamental
studies to succeed. In this presentation, similarities and differences between the
different separations schemes are discussed and emphasizing the challenges that
will be faced before an advanced nuclear fuel cycle can be implemented in industry.

Nuclear fuel cycle policies have been rather inconsistent throughout history, arguably due to the complex interdependencies of technical, social, economic and
environmental issues involved. Sophisticated fuel cycle simulation tools are being developed to aid the decision makers with defining future fuel cycle strategies.
However, the value of these tools for policy making has yet to be demonstrated.
This limited success may be partially attributed to the fact that the large amount of
technical data generated by these tools obscures the actual policy tradeoffs. From
the recent simplified analyses, it was discovered, not surprisingly, that in order to
address nuclear waste and resource sustainability issues, the construction of fuel
reprocessing infrastructure along with fast spectrum reactors has to be pursued as
soon as possible. This policy, however, has a significant economic penalty due to
the high cost of fast reactors and reprocessing plants. Decoupling these two parts
of the infrastructure would thus allow their gradual introduction, reducing the economic burden and increasing the chances of success.

One study and analysis by team of experts at MIT separates fuel cycle into two
classes of choices and they are as;

1. Open Cycle, and
2. Closed Cycle.

In the open or once-through fuel cycle, the spent fuel discharged from the nuclear
reactor is treated as waste. Figure 19.6 is depiction of an open or one-through fuel
that is projected to year 2050.

In the closed fuel cycle today, the spent fuel discharged from the reactor is reprocessed, and the products are partitioned into Uranium (U) and Plutonium (Pu)

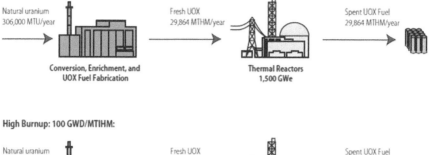

Fig. 19.6 Open fuel cycle or once through fuel—projected to 2050

suitable for fabrication into oxide fuel or Mixed Oxide fuel (MOX) for recycle back into a reactor (See Fig. 19.7). The rest of the spent fuel is treated as High-Level Waste (HLW). In the future, closed fuel cycles could include use of a dedicated reactor that would be used to transmute selected isotopes that have been separated from spent fuel (See Fig. 19.8). The dedicated reactor also may be used as a breeder to produce new fissile fuel by neutron absorption at a rate that exceeds the consumption of fissile fuel by the neutron chain reaction. In such fuel cycles the waste stream will contain less actinides which, will significantly reduce the long-term radioactivity of the nuclear waste.

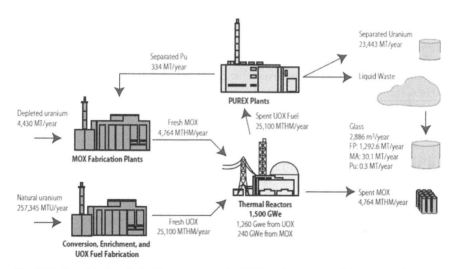

Fig. 19.7 Closed fuel cycle for Plutonium recycle (MOX option—one cycle)—projected 2050

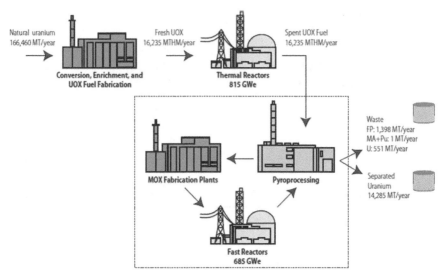

Fig. 19.8 Closed fuel cycle a full actinide recycle—projected to 2050

In general these MIT experts expect that the once-through fuel cycle to have an advantage in terms of cost and proliferation resistance (since there is no reprocessing and separation of actinides), compared to the closed cycle. Closed cycles have an advantage over the once-through cycle in terms of resource utilization (since the recycled actinides reduce the requirement for enriched uranium), which in the limit of very high ore prices would be more economical. Some argue that closed cycles also have an advantage for long-term waste disposal, since long-lived actinides can be separated from the fission products and transmuted in a reactor.

Both once-through and closed cycles can operate on U (Uranium) or Th (Thorium) fuel and can involve different reactor types, e.g., Light Water Reactors (LWRs), Heavy Water Reactors (HWRs), Supercritical water reactors (SCWRs), High Temperature and Very High Temperature Gas Cooled Reactors (HTGRs), Liquid Metal and Gas Fast Reactors (LMFRs and GFRs), or Molten Salt Reactors (MSR) of various sizes. Today, almost all deployed reactors are of the LWR type. The introduction of new reactors or fuel cycles will require considerable development resources and some period of operating experience before initial deployment.

19.3   In Core Fuel Management

The primary aim for the in-core fuel management in a nuclear reactor is to achieve higher fuel utilization without compromising the safety during operation. Loading pattern optimization during each refueling stage is the main challenge in light water reactors like pressurized water reactors (PWRs) and boiling water reactors (BWRs). Whereas, in case of heavy water reactors like pressurized heavy water reactors (PHWRs), the development of refueling scheme is relatively simple mainly due to on-line refueling, use of small length bundle and natural uranium as fuel. The use of small length bundle and flexibility of multi-bundle shift scheme helps in controlling the ripples in power due to refueling.

As part of nuclear fuel cycle process, considerations of fuel management elements are in order and they are indicating as follow:

- Deliver the required power level at lowest costs.
- Meet the safety requirements (SAR) for the core.
- Mechanical and thermal design of fuel assemblies.
- Amount and attributes of fresh assemblies to purchase.
- Reuse of partially burnt fuel.
- Core loading pattern.
- Achieve the desired burn—up.
- Consider impact on subsequent reload cycle.
- With one exception all elements need to be considered for each reload cycle.

There are computer codes written to validate all these points in order to obtain operation license commercially as part of the process. This includes, fuel cycle considerations, reactor core parameters, average power per assembly, fuel management and its objectives, and complying with nuclear waste policies in place.

19.4   Nuclear Fuel and Waste Management

All the steps of the nuclear fuel cycle generate radioactive waste. Nuclear waste is classified according to the level of radioactivity into four broad categories: Very Low Level Waste (**VLLW**), Low Level Waste (**LLW**), Intermediate Level Waste (**ILW**) and High Level Waste (**HLW**). Each of these categories at very high level can be describes as follows;

1. **Very Low Level Waste:** (VLLW) is an exempt waste (VLLW) contains radioactive materials at a level which is not considered harmful to people or the surrounding environment. It consists mainly of demolished material (such as concrete, plaster, bricks, metal, valves, piping etc.) produced during rehabilitation or dismantling operations on nuclear industrial sites. Other industries, such as food processing, chemical, steel etc also produce VLLW because of the concentration of natural radioactivity present in certain minerals used in their manufacturing processes. The waste is therefore disposed of with domestic refuse, although countries such as France are currently developing facilities to store VLLW in specifically designed VLLW disposal facilities. Radioactive materials which, occur naturally and where human activities increase the exposure of people to ionizing radiation are known by the acronym 'NORM'. NORM is an acronym for Naturally Occurring Radioactive Material, which potentially includes all radioactive elements found in the environment. Long-lived radioactive elements such as uranium, thorium and potassium and any of their decay products,

such as radium and radon are examples of NORM. These elements have always been present in the Earth's crust and atmosphere. The term NORM exists also to distinguish 'natural radioactive material' from anthropogenic sources of radio-active material, such as those produced by nuclear power and used in nuclear medicine, where incidentally the radioactive properties of a material maybe what make it useful. However, from the perspective of radiation doses to people, such a distinction is completely arbitrary.

2. **Low Level Waste:** (LLW) is generated from hospitals and industry, as well as the nuclear fuel cycle. It comprises paper, rags, tools, clothing, and filters, etc., which contain small amounts of mostly short-lived radioactivity. It does not require shielding during handling and transport and is suitable for shallow land burial. To reduce its volume, it is often compacted or incinerated before disposal. It comprises some 90% of the volume but only 1% of the radioactivity of all radioactive waste.

3. **Intermediate Level Waste:** (ILW) contains higher amounts of radioactivity and some requires shielding. It typically comprises resins, chemical sludges and metal fuel cladding, as well as contaminated materials from reactor decom-missioning. Smaller items and any non-solids may be solidified in concrete or bitumen for disposal. It makes up some 7% of the volume and has 4% of the radioactivity of all radwaste.

4. **High Level Waste:** (HLW) arises from the 'burning' of uranium fuel in a nuclear reactor. HLW contains the fission products and transuranic elements generated in the reactor core. It is highly radioactive and hot, so requires cooling and shield-ing. It can be considered as the 'ash' from 'burning' uranium. HLW accounts for over 95% of the total radioactivity produced in the process of electricity generation.

There are two distinct kinds of HLW:

- Used fuel itself.
- Separated waste from reprocessing the used fuel (as described in section on Managing HLW from used fuel below).

HLW has both long-lived and short-lived components, depending on the length of time it will take for the radioactivity of particular radionuclide's to decrease to lev-els that are considered no longer hazardous for people and the surrounding environ-ment. If generally short-lived fission products can be separated from long-lived actinides, this distinction becomes important in management and disposal of HLW.

19.4.1   Managing HLW from Used Fuel

Most countries operating nuclear power plants have developed or continue to de-velop strategies to deal with waste. In many countries, disposal facilities are already available for LLW and, in some, for ILW.

More than 95 % of the total radioactivity in radioactive wastes is contained in HLW (spent nuclear fuel or the most radioactive residues of reprocessing), even though HLW accounts for less than 5 % of the total volume of waste. A typical 1000-MW nuclear power plant produces 10 m^3 of spent fuel per year, when packaged for disposal. If this spent fuel is reprocessed, about 2.5 m^3 of vitrified waste is produced. Today, spent fuel and HLW are stored in special purpose interim storage facilities.

Large-scale reprocessing facilities are currently operational in France, Russia and the United Kingdom. The main Japanese reprocessing plant is still being commissioned; although a small plant is in operation, (most of Japanese reprocessing to date has taken place in France and the UK). Utilities in a few European countries (including Belgium, Germany, the Netherlands, Sweden and Switzerland) have had a significant amount of spent fuel reprocessed in France and the UK. In most cases, these contracts have now ended, following changes in policy in these countries, but the power companies or countries concerned have a contractual obligation to take back the HLW produced for eventual disposal (as well as the separated plutonium and uranium). India has plans for commercial reprocessing as part of a thorium-uranium fuel cycle, but this is at the development stage. Other countries may reconsider the reprocessing option in future if alternative reprocessing technologies are developed or if reprocessing appears to be more economically attractive than direct disposal.

New reactor designs and fuel cycles are being developed with this consideration in mind. There are relevant international cooperation programmers, with the United States taking a major role, as well as those countries, which today reprocess.

Used fuel gives rise to HLW, which may be either the used fuel itself in fuel rods, or the separated waste arising from reprocessing this (see next section on Recycling used fuel). In either case, the amount is modest—as noted above, a typical reactor generates about 27 t of spent fuel or 3 m^3/year of vitrified waste. Both can be effectively and economically isolated, and have been handled and stored safely since nuclear power began.

Storage is mostly in ponds at reactor sites, or occasionally at a central site. See later section below.

If the used fuel is reprocessed, as is that from UK, French, Japanese and German reactors, HLW comprises highly radioactive fission products and some transuranic elements with long-lived radioactivity. These are separated from the used fuel, enabling the uranium and plutonium to be recycled. Liquid HLW from reprocessing must be solidified. The HLW also generates a considerable amount of heat and requires cooling. It is vitrified into borosilicate (Pyrex) glass, encapsulated into heavy stainless steel cylinders about 1.3 m high and stored for eventual disposal deep underground. This material has no conceivable future use and is unequivocally waste. The hulls and end-fittings of the reprocessed fuel assemblies are compacted, to reduce volume, and usually incorporated into cement prior to disposal as ILW. France has two commercial plants to vitrify HLW left over from reprocessing oxide fuel, and there are plants in the UK and Belgium as well. The capacity of these Western European plants is 2500 canisters (1000 t) a year and some have been operating for three decades.

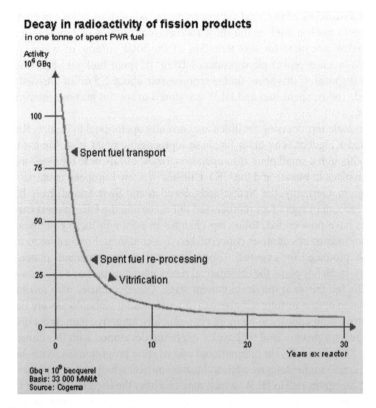

Fig. 19.9 Decay in radioactivity of fission products

If used reactor fuel is not reprocessed, it will still contain all the highly radioactive isotopes, and then the entire fuel assembly is treated as HLW for direct disposal. It too generates a lot of heat and requires cooling. However, since it largely consists of uranium (with a little plutonium), it represents a potentially valuable resource and there is an increasing reluctance to dispose of it irretrievably.

Either way, after 40–50 years the heat and radioactivity have fallen to one thousandth of the level at removal. This provides a technical incentive to delay further action with HLW until the radioactivity has reduced to about 0.1 % of its original level.

After storage for about 40 years the used fuel assemblies are ready for encapsulation or loading into casks ready for indefinite storage or permanent disposal underground.

Direct disposal of used fuel has been chosen by the USA and Sweden among others, although evolving concepts lean towards making it recoverable if future generations see it as a resource. This means allowing for a period of management and oversight before a repository is closed (Fig. 19.9).

HLW disposal is more contentious than disposal of lower-level wastes and no country today has an operating disposal site for high-level waste. Though wide technical consensus exists on the adequacy of geological disposal of HLW, it has not yet won general public consent. In some countries, however, there are volunteer

Table 19.1 Examples of high-level waste disposal strategies

	Facilities and progress towards final repositories
Belgium	Underground laboratory in Boom Clay at Mol since 1984. Repository has not been selected yet
Canada	Owners of used fuel required by law to develop strategy. Ultimate disposal in geological formation proposed but no sites have been selected
Czech Republic	Decision for final HLW repository after 2010
Finland	Construction of underground research laboratory. Resulting HLW repository expected to start operation in 2020
France	HLW from spent fuel reprocessing vitrified and stored at La Hague and Marcoule (new waste stored at La Hague). Three research directions: partitioning/transmutation, reversible deep repository and storage. Studies under way for site selection and storage conception. Storage operational by 2025
Germany	Used fuel storage at Ahaus and Gorleben. Expects to have a final HLW repository in operation around 2030
Hungary	Site in Boda Claystone Formation selected. Surface exploration commenced in 2004. Underground research laboratory in 2010
India	Research on deep geological disposal for HLW
Japan	Vitrified HLW stored at Mutsu-Ogawara since 1995. Ongoing research for deep geological repository site. Operation expected in mid-2030s
Netherlands	Temporarily surface storage is only allowed for existing plant. Study announced for final disposal of waste of existing plant and of any new plant. Decision expected in 2016
Slovak Republic	Research for deep geological disposal started in 1996. Four areas have been proposed for detailed exploration
Republic of Korea	Central interim HLW storage planned for 2016. Ongoing development of a repository concept

communities to host repositories. Table 19.1 provides examples of strategies to deal with HLW. The search for politically acceptable solutions continues.

19.4.2   Recycling Used Fuel

Any used fuel will still contain some of the original U_{235} as well as various plutonium isotopes, which have been formed inside the reactor core, and the U_{238}. In total, these account for some 96 % of the original uranium and over half of the original energy content (ignoring U_{238}). Reprocessing, undertaken in Europe and Russia, separates this uranium and plutonium from the wastes so that they can be recycled for re-use in a nuclear reactor. Plutonium arising from reprocessing is recycled through a MOX fuel fabrication plant where it is mixed with depleted uranium oxide to make fresh fuel. European reactors currently use over 5 t of plutonium a year in fresh MOX fuel.

Major commercial reprocessing plants operate in France, UK, and Russia with a capacity of some 5000 t/year and cumulative civilian experience of 80,000 t over

Fig. 19.10 Storage pond for used fuel at the thermal oxide reprocessing plant at the UK's Sellafield site (Sellafield Ltd)

50 years. A new reprocessing plant with an 800 t/year capacity at Rokkasho in Japan is undergoing commissioning.

France and UK also undertake reprocessing for utilities in other countries, notably Japan, which has made over 140 shipments of used fuel to Europe since 1979. Until now most Japanese used fuel has been reprocessed in Europe, with the vitrified waste and the recovered uranium and plutonium (as MOX fuel) being returned to Japan to be used in fresh fuel. Russia also reprocesses some spent fuel from Soviet-designed reactors in other countries.

There are several proposed developments of reprocessing technologies (Described in the section of this chapter on Processing of Used Nuclear Fuel). One technology under development would separate plutonium along with the minor actinides as one product. This however cannot be simply put into MOX fuel and recycled in conventional reactors; it requires fast neutron reactors, which are yet few and far in between. On the other hand, it would make disposal of high-level wastes easier (Fig. 19.10).

19.4.3   Storage and Disposal of Used Fuel and Other HLW

Spent nuclear fuel or high-level waste can be safely disposed of deep underground, in stable rock formations such as granite, thus eliminating the health risk to people and the environment. The first disposal facilities will be in operation around 2020.

HLW is the liquid waste that results when spent fuel is reprocessed to recover unfissioned uranium and plutonium. During this process, strong chemicals, and these results in liquid HLW dissolve the fuel. Plans are to solidify these liquids into a

Fig. 19.11 Waste management disposal facilities

form that is suitable for disposal. Solidification is still in the planning stages. While currently there are no commercial facilities in this country that reprocess spent fuel, spent fuel from defense program reactors has been routinely reprocessed for use in producing nuclear weapons or for reuse in new fuel (Fig. 19.11).

Compared to the total inventory of HLW, the volume of commercial HLW from the reprocessing of commercial spent fuel is almost insignificant, less than one percent. Defense-related HLW comprises greater than ninety-nine percent of the volume of HLW. Figure 19.12 shows the historical and projected volume of de-

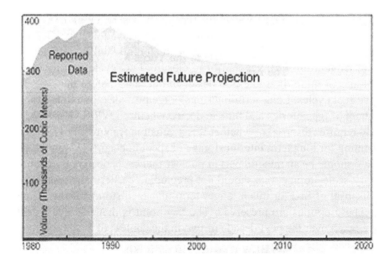

Fig. 19.12 Historical and projected inventories of defense high-level radioactive waste. (Reference: DOE/RW-0006, Rev. 7)

Fig. 19.13 Nuclear waste can be buried deep underground

fense-related HLW through the year 2020. The effect of the end of the "Cold War" on these projections is uncertain.

HLW is now stored in underground tanks or stainless steel silos on federal reservations in South Carolina, Idaho, and Washington and at the Nuclear Fuel Services Plant in West Valley, NY. These facilities have begun programs to solidify and structurally stabilize the waste in preparation for disposal at a national repository.

Waste will be packed in long-lasting containers and buried deep in the geological formations chosen for their favorable stability and geochemistry, including limited water movement. These geological formations have stability over hundreds of millions of years, far longer than the waste is dangerous (Fig. 19.13).

Historically the United States has not considered Spent Nuclear Fuel (SNF) storage as a major component of fuel cycle policy. However, repository programs worldwide have adopted a policy of storing SNF (or the HLW from reprocessing) for 40–60 years before disposal in a geological repository to reduce the radioactivity and decay heat. This reduces repository costs and performance uncertainties. Countries such as France with its partly closed fuel cycle and Sweden with its open fuel cycle built storage facilities several decades ago for this purpose. The failure to include long-term storage as part of the spent fuel management has had major impacts on the design of the proposed Yucca Mountain Repository (YMR). Due to the heat load of SNF, the repository was required to be ventilated to remove decay heat while the SNF cooled. The YMR would have, after 30 years of filling, become functionally an underground storage facility with active ventilation for an additional 50 years prior to closure. Fuel cycle transitions require a half-century or more. It is likely to be several decades before the U.S. deploys alternative fuel cycles. Long-term interim storage provides time to assure proper development of repositories and time to decide whether LWR SNF is a waste that ultimately requires disposal or whether it is a valuable resource. For multiple reasons, planning for long term interim storage of spent nuclear fuel—on the scale of a century—should be an integral part of nuclear fuel cycle design. In recommending century-scale storage, one should not precluding earlier reprocessing or geological disposal of SNF or much longer term managed storage if the technology permits. These options are preserved. The key point is that fuel cycle decisions should be taken over the next decade or two in the context of a century time scale for managed storage.

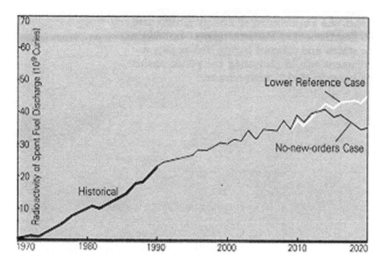

Fig. 19.14 Projected accumulated radioactivity of commercial spent fuel discharges for the DOE/EIA No-New-Orders and lower reference cases. Integrated data base for 1991: U.S. spent fuel and radioactive waste inventory projections and characteristics, DOE/ORNL, Oct 1991. (DOE/RW-0006, Rev. 7)

Until a disposal or long-term storage facility is operational, most spent fuel is stored in water pools at the reactor site where it was produced. The water removes leftover heat generated by the spent fuel and serves as a radiation shield to protect workers at the site.

The operation of nuclear reactors over the last 20 years has substantially added to the amount of radioactive waste in this country. As shown in Fig. 19.14, by the year 2020, the total amount of spent fuel is expected to increase significantly.

The process of selecting appropriate deep geological repositories is now underway in several countries. Finland and Sweden are well advanced with plans for direct disposal of used fuel, since their parliaments decided to proceed on the basis that it was safe, using existing technology. Both countries have selected sites, in Sweden, after competition between two municipalities. The USA has opted for a final repository at Yucca Mountain in Nevada, though this is now stalled due to political decision. There have also been proposals for international HLW repositories in optimum geology (Fig. 19.15).

A pending question is whether wastes should be emplaced so that they are readily retrievable from repositories. There are sound reasons for keeping such options open—in particular, it is possible that future generations might consider the buried waste to be a valuable resource. On the other hand, permanent closure might increase long-term security of the facility. After being buried for about 1000 years most of the radioactivity will have decayed. The amount of radioactivity then remaining would be similar to that of the naturally occurring uranium ore from which it originated, though it would be more concentrated.

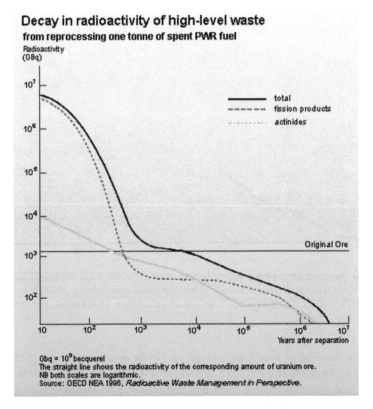

Fig. 19.15 Decay in radioactivity of high level waste. (Courtesy of DOE)

19.4.4   Regulation of Disposal

The fuel for most nuclear reactors consists of pellets of ceramic uranium dioxide that are sealed in hundreds of metal rods. These rods are bundled together to form what is known as a "fuel assembly." Depending upon the type and size of the reactor, a fuel assembly can weigh up to 1500 pounds. As the nuclear reactor operates, uranium atoms fission (split apart) and release energy.

When most of the usable uranium has fissioned, the "spent" fuel assembly is removed from the reactor. Some elements, such as plutonium, in HLW and spent fuel are highly radioactive and remain so for thousands of years. Therefore, the safe disposal of this waste is one of the most controversial environmental subjects facing the federal government and affected states.

The federal government (the EPA, the DOE, and the NRC) has overall responsibility for the safe disposal of HLW and spent fuel. The EPA is responsible for developing environmental standards that apply to both DOE-operated and NRC-licensed facilities. Currently, the NRC is responsible for licensing such facilities and ensuring their compliance with the EPA standards. DOE is responsible for developing the deep geologic repository, which has been authorized by Congress for disposing of spent fuel and high level waste. Both the NRC and the Department of Transportation are responsible for regulating the transportation of these wastes to storage and disposal sites.

19.5 Processing of Used Nuclear Fuel

Over the last 50 years, the principal reason for reprocessing used fuel has been to recover unused uranium and plutonium in the used fuel elements and thereby close the fuel cycle, gaining some 25% more energy from the original uranium in the process and thus contributing to energy security. A secondary reason is to reduce the volume of material to be disposed of as high-level waste to about one fifth. In addition, the level of radioactivity in the waste from reprocessing is much smaller and after about 100 years falls much more rapidly than in used fuel itself.

A key, nearly unique, characteristic of nuclear energy is that used fuel may be reprocessed to recover fissile and fertile materials in order to provide fresh fuel for existing and future nuclear power plants. Several European countries, Russia and Japan have had a policy to reprocess used nuclear fuel, although government policies in many other countries have not yet addressed the various aspects of reprocessing.

In the last decade interest has grown in recovering all long-lived actinides together (*i.e.* with plutonium) to recycle them in fast reactors so that they end up as short-lived fission products. This policy is driven by two factors: reducing the long-term radioactivity in high-level wastes, and reducing the possibility of plutonium being diverted from civil use—thereby increasing proliferation resistance of the fuel cycle. If used fuel is not reprocessed, then in a century or two the built-in radiological protection will have diminished, allowing the plutonium to be recovered for illicit use (though it is unsuitable for weapons due to the non-fissile isotopes present).

Reprocessing used fuel to recover uranium (as reprocessed uranium, or RepU) and plutonium (Pu) avoids the wastage of a valuable resource. Most of it—about 96%—is uranium, of which less than 1% is the fissile U_{235} (often 0.4–0.8%); and up to 1% is plutonium. Both can be recycled as fresh fuel, saving up to 30% of the natural uranium otherwise required. The materials potentially available for recycling (but locked up in stored used fuel) could conceivably run the US reactor fleet of about 100 GWe for almost 30 years with no new uranium input.

So far, almost 90,000 t (of 290,000 t discharged) of used fuel from commercial power reactors has been reprocessed. Annual reprocessing capacity is now some 4000 t/year for normal oxide fuels, but not all of it is operational.

Between now and 2030 some 400,000 t of used fuel is expected to be generated worldwide, including 60,000 t in North America and 69,000 t in Europe (Table 19.2).

19.5.1 Reprocessing Policies

Conceptually reprocessing can take several courses, separating certain elements from the remainder, which becomes high-level waste. Reprocessing options include:

Table 19.2 World commercial reprocessing capacity (Tons per year)

LWR fuel	France, La Hague	1700	
	UK, Sellafield (THORP)	900	
	Russia, Ozersl (Mayak)	400	
	Japan (Rokkasho)	800	
Total LWR (approximation)			3800
Other nuclear fuels	UK, Sellafield (Magnox)	1500	
	India, (PHWR, 4 plants)	330	
Total others (approximation)			1830
Total civil capacity			5630

- Separate U, Pu, (as today).
- Separate U, Pu + U (small amount of U).
- Separate U, Pu, minor actinides.
- Separate U, Pu + Np, Am + Cm.
- Separate U + Pu all together.
- Separate U, Pu + actinides, certain fission products.

In today's reactors, reprocessed uranium (RepU) needs to be enriched, whereas plutonium goes straight to mixed oxide (MOX) fuel fabrication. This situation has two perceived problems: the separated plutonium is a potential proliferation risk, and the minor actinides remain in the separated waste, which means that its radioactivity is longer-lived than if it comprised fission products only.

As there is no destruction of minor actinides, recycling through light water reactors delivers only part of the potential waste management benefit. For the future, the focus is on removing the actinides from the final waste and burning them with the recycled uranium and plutonium in fast neutron reactors. (The longer-lived fission products may also be separated from the waste and transmuted in some other way.) Hence, the combination of reprocessing followed by recycling in today's reactors should be seen as an interim phase of nuclear power development, pending widespread use of fast neutron reactors.

All but one of the six Generation IV reactors being developed has closed fuel cycles, which recycle all the actinides. Although US policy has been to avoid reprocessing, the US budget process for 2006 included $ 50 million to develop a plan for "integrated spent fuel recycling facilities", and a program to achieve this with fast reactors has become more explicit since.

19.6   Back End of Fuel Cycle

As part of back-end of fuel cycle is concerned the study that is done by World Nuclear Association talks about nuclear fuel and waste management that was discussed in previous section of this chapter and they can be summarized as;

- Nuclear power is the only large-scale energy-producing technology which, takes full responsibility for all its wastes and fully costs this into the product.
- The amount of radioactive wastes is very small relative to wastes produced by fossil fuel electricity generation.
- Used nuclear fuel may be treated as a resource or simply as a waste.
- Nuclear wastes are neither particularly hazardous nor hard to manage relative to other toxic industrial wastes.
- Safe methods for the final disposal of high-level radioactive waste are technically proven; the international consensus is that this should be geological disposal.

All parts of the nuclear fuel cycle produce some radioactive waste (radwaste) and the relatively modest cost of managing and disposing of this is part of the electricity cost, i.e. it is internalised and paid for by the electricity consumers.

At each stage of the fuel cycle there are proven technologies to dispose of the radioactive wastes safely. For low- and intermediate-level wastes these are mostly being implemented. For high-level wastes some countries await the accumulation of enough of it to warrant building geological repositories; others, such as the USA, have encountered political delays.

Unlike other industrial wastes, the level of hazard of all nuclear waste—its radioactivity—diminishes with time. Each radionuclide contained in the waste has a half-life—the time taken for half of its atoms to decay and thus for it to lose half of its radioactivity. Radionuclides with long half-lives tend to be alpha and beta emitters—making their handling easier—while those with short half-lives tend to emit the more penetrating gamma rays. Eventually all radioactive wastes decay into non-radioactive elements. The more radioactive an isotope is, the faster it decays.

The main objective in managing and disposing of radioactive (or other) waste is to protect people and the environment. This means isolating or diluting the waste so that the rate or concentration of any radionuclides returned to the biosphere is harmless. To achieve this, practically all wastes are contained and managed—some clearly need deep and permanent burial. From nuclear power generation, none is allowed to cause harmful pollution.

All toxic wastes need to be dealt with safely, not just radioactive wastes. In countries with nuclear power, radioactive wastes comprise less than 1 % of total industrial toxic wastes (the balance of which remains hazardous indefinitely).

Chapter 20
The Economic Future of Nuclear Power

From the global viewpoint and urgent need to support rising demand for electricity, many countries recognize the substantial role which nuclear power has played in satisfying various policy objectives, including energy security of supply, reducing import dependence and reducing greenhouse gas or polluting emissions. Nevertheless, as such considerations are far from being fully accounted for in liberalized power markets, nuclear plants must demonstrate their viability on normal commercial criteria as well as their life cycle advantages.

20.1 Introduction

Of all factors affecting prospects for the substantial growth of nuclear power in the twenty-first century, cost is the most fundamental. What are the essential economics associated with the construction and operation of advanced state-of-the-art nuclear power plants?.

Certainly, other factors will affect the pace of the global nuclear renaissance now under way. The nuclear debate continues to feature expressions of concern about nuclear arms, terrorism, operational and transport safety, and effective waste management and disposal. In addressing all of these concerns, however, the combined efforts of science, diplomacy and industry have achieved substantial advance in ensuring that civil nuclear power can be used without substantial human or environmental risk.

Studies by organizations such as Nuclear Energy Institute (NEI) and World Nuclear Association (WNA) are showing an excellent demonstration and analysis of the economic future of nuclear power plants both from total cost of ownership and return on investment of such plants and some of these analyses are showing here. For further detailed and more up to date, we urge readers to refer to the sites of these two organizations.

Today, given the urgent environmental imperative of achieving a global clean-energy revolution, public policy has sound and urgent justification for placing a

© Springer International Publishing Switzerland 2015
B. Zohuri, P. McDaniel, *Thermodynamics In Nuclear Power Plant Systems,*
DOI 10.1007/978-3-319-13419-2_20

sizeable premium on clean technologies. Such environmentally driven incentives can come through carbon taxes, emissions trading, or subsidies for non-emitting generators of power.

This achievement in meeting legitimate public concerns has provided the foundation for the nuclear renaissance by prompting governments in countries representing the preponderance of world population and economic activity to consider a wider exploitation of the benefits of nuclear energy.

These benefits fall into two categories:

- **National:** price stability and security of energy supply
- **Global environmental:** near-zero greenhouse gas emissions tolerances.

Nuclear energy is, in many places, competitive with fossil fuels for electricity generation, despite relatively high capital costs and the need to internalize all waste disposal and decommissioning costs. If the social, health and environmental costs of fossil fuels are also taken into account, the economics of nuclear power are outstanding.

The research and development work that was undertaken in the early stages of nuclear power development was a challenging project for government research organizations as well as the industrial sector. The optimum technical solutions were progressively uncovered through multiple and various demonstration programs developed in the 1950s and 1960s under government funding and, at the same time, by increasingly scaling up the reactor ratings to compete more easily with fossil fuels. Designs were mainly motivated by the search for higher thermal efficiency, lower system pressure, the ability to stay on line continuously and better utilization of uranium resources. The breakthrough in the commercialization of nuclear power was reached when unit ratings exceeded several hundreds of MWe in the mid-1960s.

Since the late 1980s, on the electricity power supply side, governments have steadily moved away from direct regulation in energy markets to concentrate on establishing the framework for a competitive supply system. There are significant differences in regulatory regimes with some countries retaining a substantial regulated element. Electricity market liberalization itself comes in many guises, but the industry today recognizes that all plants must demonstrate that they are cost-effective and that this must be achieved while still maintaining very high safety and environmental standards. Safety and the best economic operation tend, in any case, to go hand in hand.

20.2   Overall Costs: Fuel, Operation and Waste Disposal

With nuclear energy's higher capital cost and longer development and construction period, investors will focus on how risks can be managed, mitigated and risk allocations optimized. The business case for nuclear will ultimately depend on the structure of risk allocation between operators, investors, suppliers and customers.

Although new nuclear power plants require large capital investment, they are hardly unique by the standards of the overall energy industry, where oil platforms and Liquefied Natural Gas (LNG) liquefaction facilities cost many billions of dollars. Main aspects of these costs are due to imposed federal and state government's regulation for standard operating procedures and safety factors that are required be implemented and followed. Projects of similar magnitude can be found in the building of new roads, bridges and other elements of infrastructure. Many of the risk-control and project management techniques developed for these projects are equally applicable to building nuclear power stations.

Risks that are specific to nuclear plants are those surrounding the management of radioactive waste and used fuel and the liability for nuclear accidents. As with many other industrial risks, public authorities must be involved in setting the regulatory framework. The combined goal must be public safety and the stable policy environment necessary for investment.

Assessing the relative costs of new generating plants utilizing different technologies is a complex matter and the results depend crucially on location. Coal is, and will probably remain, economically attractive in countries such as China, the USA and Australia with abundant and accessible domestic coal resources as long as carbon emissions are cost-free. Gas is also competitive for base-load power in many places, particularly using combined-cycle plants, though rising gas prices have removed much of the advantage.

As part cost study analyzing WNA and NEI organizations, the following points were considered and taken into account as;

- Nuclear power is cost competitive with other forms of electricity generation, except where there is direct access to low-cost fossil fuels.
- Fuel costs for nuclear plants are a minor proportion of total generating costs, though capital costs are greater than those for coal-fired plants are and much greater than those for gas-fired plants.
- In assessing the economics of nuclear power, decommissioning and waste disposal costs are fully taken into account.

To support new-build projects, projects must be structured to share risks amongst key stakeholders in a way that is both equitable and that encourages each project participant to fulfill its responsibilities.

20.2.1 Fuel Costs

This is the total annual cost associated with the "burn-up" of nuclear fuel resulting from the operation of the unit. This cost is based upon the amortized costs associated with the purchasing of uranium, conversion, enrichment, and fabrication services along with storage and shipment costs, and inventory (including interest) charges less any expected salvage value.

For a typical 1000 MWe BWR or PWR, the approximate cost of fuel for one reload (replacing one third of the core) is about $ 40 million, based on an 18-month refueling cycle. The average fuel cost at a nuclear power plant in 2011 was 0.68 cents/kWh.

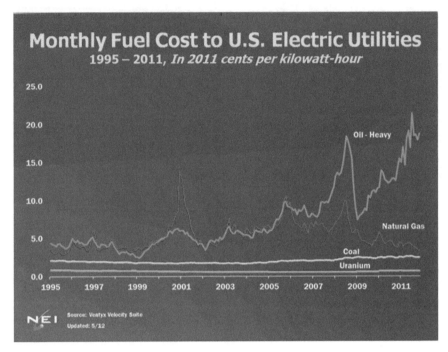

Fig. 20.1 Monthly fuel cost to U.S. electric utilities

Because nuclear plants refuel every 18–24 months, they are not subject to fuel price volatility like natural gas and oil power plants. The following chart is presentation of 'Monthly Fuel Cost to U.S. Electric Utilities'. This line graph shows U.S. electric utilities' monthly fuel costs for nuclear, coal, gas, and oil between 1995 and 2011 (Fig. 20.1).

From the outset, the basic attraction of nuclear energy has been its low fuel costs compared with coal, oil and gas-fired plants. Uranium, however, has to be processed, enriched and fabricated into fuel elements, and about half of the cost is due to enrichment and fabrication. In the assessment of the economics of nuclear power allowances must also be made for the management of radioactive used fuel and the ultimate disposal of this used fuel or the wastes separated from it. But even with these included, the total fuel costs of a nuclear power plant in the Convention on the Organization for Economic Co-operation and Development (OECD) are typically about a third of those for a coal-fired plant and between a quarter and a fifth of those for a gas combined-cycle plant. The US Nuclear Energy Institute suggests that for a coal-fired plant 78 % of the cost is the fuel, for a gas-fired plant the figure is 89 %, and for nuclear the uranium is about 14 %, or double that to include all front end costs.

In March 2011, the approx. US $ cost to get 1 kg **of uranium as UO₂ reactor fuel** (at current spot uranium price) (Table 20.1):

At 45,000 MWd/t burn-up this gives 360,000 kWh electrical per kg, hence fuel cost: 0.77 cents/kWh.

Table 20.1 Source World Nuclear Association

Uranium	8.9 kg U$_3$O$_8$ × $ 146	US$ 1300
Conversion	7.5 kg U × $ 13	US$ 98
Enrichment	7.3 SWU × $ 155	US$ 1132
Fuel fabrication	per kg	US$ 240
Total, approx		US$ 2770

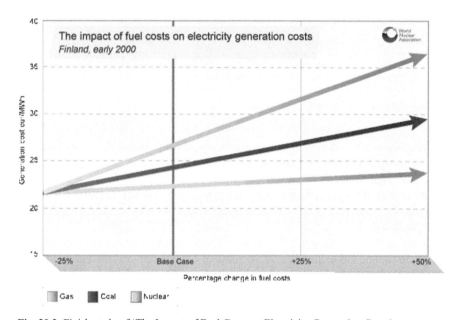

Fig. 20.2 Finish study of 'The Impact of Fuel Costs on Electricity Generation Costs'

Fuel costs are one area of steadily increasing efficiency and cost reduction. For instance, in Spain the nuclear electricity cost was reduced by 29 % over 1995–2001. This involved boosting enrichment levels and burn-up to achieve 40 % fuel cost reduction. Prospectively, a further 8 % increase in burn-up will give another 5 % reduction in fuel cost.

Uranium has the advantage of being a highly concentrated source of energy, which is easily and cheaply transportable. The quantities needed are very much less than for coal or oil. One kilogram of natural uranium will yield about 20,000 times as much energy as the same amount of coal. It is therefore intrinsically a very portable and tradable commodity.

The fuel's contribution to the overall cost of the electricity produced is relatively small, so even a large fuel price escalation will have relatively little effect (see below).

A Finnish study in 2000 also quantified fuel price sensitivity to electricity costs (Fig. 20.2):

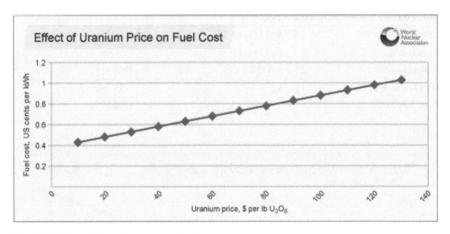

Fig. 20.3 Effect of Uranium price on fuel cost

These show that a doubling of fuel prices would result in the electricity cost for nuclear rising about 9%, for coal rising 31% and for gas 66%. Gas prices have since risen significantly.

The impact of varying the uranium price in isolation is shown below in a worked example of a typical US plant, assuming no alteration in the tails assay at the enrichment plant (Fig. 20.3).

Doubling the uranium price (say from $ 25 to 50/lb U_{308}) takes the fuel cost up from 0.50 to 0.62 US cents/kWh, an increase of one quarter, and the expected cost of generation of the best US plants from 1.3 to 1.42 US cents/kWh (an increase of almost 10%). So while there is some impact, it is comparatively minor, especially by comparison with the impact of gas prices on the economics of gas generating plants. In these, 90% of the marginal costs can be fuel. Only if uranium prices rise to above $ 100/lb U_{308} ($ 260/kgU) and stay there for a prolonged period (which seems very unlikely) will the impact on nuclear generating costs be considerable.

Nevertheless, for nuclear power plants operating in competitive power markets where it is impossible to pass on any fuel price increases (i.e. the utility is a price-taker), higher uranium prices will cut corporate profitability. Yet fuel costs have been relatively stable over time—the rise in the world uranium price between 2003 and 2007 added to generation costs, but conversion, enrichment and fuel fabrication costs did not followed the same trend.

For prospective new nuclear plants, the fuel element is even less significant (see below). The typical front-end nuclear fuel cost is typically only 15–20% of the total, as opposed to 30–40% for operating nuclear plants.

There are other possible savings. For example, if used fuel is reprocessed and the recovered plutonium and uranium is used in mixed oxide (MOX) fuel, more energy can be extracted. The costs of achieving this are large, but are offset by MOX fuel not needing enrichment and particularly by the smaller amount of high-level wastes produced at the end. Seven UO_2 fuel assemblies give rise to one MOX assembly plus some vitrified high-level waste, resulting in only about 35% of the volume, mass and cost of disposal.

20.2.2   Future Cost Competitiveness

Understanding the cost of new generating capacity and its output requires careful analysis of what is in any set of figures. There are three broad components: capital, finance and operating costs. Capital and financing costs make up the project cost.

Capital costs comprise several things: the bare plant cost (usually identified as Engineering-Procurement-Construction—EPC—cost), the owner's costs (land, cooling infrastructure, administration and associated buildings, site works, switch-yards, project management, licenses, etc), cost escalation and inflation. Owner's costs may include transmission infrastructure. The term "overnight capital cost" is often used, meaning EPC plus owners' costs and excluding financing, escalation due to increased material and labor costs, and inflation. Construction cost—some-times called "all-in cost", adds to overnight cost any escalation and interest during construction and up to the start of construction. It is expressed in the same units as overnight cost and is useful for identifying the total cost of construction and for determining the effects of construction delays. In general the construction costs of nuclear power plants are significantly higher than for coal- or gas-fired plants because of the need to use special materials, and to incorporate sophisticated safety features and back-up control equipment. These contribute much of the nuclear generation cost, but once the plant is built the cost variables are minor.

Long construction periods will push up financing costs, and in the past they have done so spectacularly. In Asia construction times have tended to be shorter, for instance the new-generation 1300 MWe Japanese reactors which began operating in 1996 and 1997 were built in a little over 4 years, and 48–54 months is typical projection for plants today.

Decommissioning costs are about 9–15% of the initial capital cost of a nuclear power plant. But when discounted, they contribute only a few percent to the invest-ment cost and even less to the generation cost. In the USA, they account for 0.1–0.2 cents/kWh, which is no more than 5% of the cost of the electricity produced.

Financing costs will depend on the rate of interest on debt, the debt-equity ratio, and if it is regulated, how the capital costs are recovered. There must also be an al-lowance for a rate of return on equity, which is risk capital.

Operating costs include operating and maintenance (O&M) plus fuel. Fuel cost figures include used fuel management and final waste disposal. These costs, while usually external for other technologies, are internal for nuclear power (i.e. they have to be paid or set aside securely by the utility generating the power, and the cost passed on to the customer in the actual tariff).

This "back-end" of the fuel cycle, including used fuel storage or disposal in a waste repository, contributes up to 10% of the overall costs per kWh,—rather less if there is direct disposal of used fuel rather than reprocessing. The $ 26 billion US used fuel program is funded by a 0.1 cents/kWh levy.

Calculations of relative generating costs are made using levelised costs, meaning average costs of producing electricity including capital, finance, and owner's costs on site, fuel and operation over a plant's lifetime, with provision for decommission-ing and waste disposal.

It is important to note that capital cost figures quoted by reactor vendors, or which are general and not site-specific, will usually just be for EPC costs. This is because owner's costs will vary hugely, most of all according to whether a plant is Greenfield or at an established site, perhaps replacing an old plant.

There are several possible sources of variation, which preclude confident comparison of overnight, or EPC (Engineering, Procurement and Construction) capital costs—e.g. whether initial core load of fuel is included. Much more obvious is whether the price is for the nuclear island alone (Nuclear Steam Supply System) or the whole plant including turbines and generators—all the above figures include these. Further differences relate to site works such as cooling towers as well as land and permitting—usually they are all owners' costs as outlined earlier in this section. Financing costs are additional, adding typically around 30%, and finally there is the question of whether cost figures are in current (or specified year) dollar values or in those of the year in which spending occurs.

20.2.3   Major Studies on Future Cost Competitiveness

There have been many studies carried out examining the economics of various future generation options, and the following are merely the most important and also focus on the nuclear element.

The **2010 OECD** study Projected Costs of generating Electricity compared 2009 data for generating base-load electricity by 2015 as well as costs of power from renewables, and showed that nuclear power was very competitive at $ 30/t CO_2 cost and low discount rate. The study comprised data for 190 power plants from 17 OECD countries as well as some data from Brazil, China, Russia and South Africa. It used levelised lifetime costs with carbon price internalized (OECD only) and discounted cash flow at 5 and 10%, as previously. The precise competitiveness of different base-load technologies depended very much on local circumstances and the costs of financing and fuels.

Nuclear overnight capital costs in OECD ranged from US$ 1556/kW for APR-1400 in South Korea through $ 3009 for ABWR in Japan, $ 3382/kW for Gen III + in USA, $ 3860 for EPR at Flamanville in France to $ 5863/kW for EPR in Switzerland, with world median $ 4100/kW. Belgium, Netherlands, Czech Rep and Hungary were all over $ 5000/kW. In China, overnight costs were $ 1748/kW for CPR-1000 and $ 2302/kW for AP1000, and in Russia $ 2933/kW for VVER-1150. EPRI (USA) gave $ 2970/kW for APWR or ABWR Eurelectric gave $ 4724/kW for EPR. OECD black coal plants were costed at $ 807–2719/kW, those with carbon capture and compression (tabulated as CCS, but the cost not including storage) at $ 3223–5811/kW, brown coal $ 1802–3485, gas plants $ 635–1747/kW and onshore wind capacity $ 1821–3716/kW. (Overnight costs were defined here as EPC, owner's costs and contingency, but excluding interest during construction) (Table 20.2).

Table 20.2 OECD electricity generating cost projections for year 2010 on—5% discount rate, cents/kWh. (Source: [1])

Country	Nuclear	Coal	Coal with CCS	Gas CCGT	Onshore wind
Belgium	6.1	8.2	–	9.0	9.6
Czech R	7.0	8.5–9.4	8.8–9.3	9.2	14.6
France	5.6	–	–	–	9.0
Germany	5.0	7.0–7.9	6.8–8.5	8.5	10.6
Hungary	8.2	–	–	–	–
Japan	5.0	8.8	–	10.5	–
Korea	2.9–3.3	6.6–6.8	–	9.1	–
Netherlands	6.3	8.2	–	7.8	8.6
Slovakia	6.3	12.0	–	–	–
Switzerland	5.5–7.8	–	–	9.4	16.3
USA	4.9	7.2–7.5	6.8	7.7	4.8
China[a]	3.0–3.6	5.5	–	4.9	5.1–8.9
Russia[a]	4.3	7.5	8.7	7.1	6.3
EPRI (USA)	4.8	7.2	–	7.9	6.2
Eurelectric	6.0	6.3–7.4	7.5	8.6	11.3

[a] For China and Russia: 2.5c is added to coal and 1.3c to gas as carbon emission cost to enable sensible comparison with other data in those fuel/technology categories, though within those countries coal and gas will in fact be cheaper than the Table above suggests

At 5% discount rate comparative costs are as shown above. Nuclear is comfortably cheaper than coal and gas in all countries. At 10% discount rate (below) nuclear is still cheaper than coal in all but the Eurelectric estimate and three EU countries, but in these three gas becomes cheaper still. Coal with carbon capture is mostly more expensive than either nuclear or paying the $ 30/t for CO_2 emissions, though the report points out "great uncertainties" in the cost of projected CCS. Also, investment cost becomes a much greater proportion of power cost than with 5% discount rate (Table 20.3).

A 2004 report from the University of Chicago, funded by the US Department of Energy, compared the levelised power costs of future nuclear, coal, and gas-fired power generation in the USA. Various nuclear options were covered, and for an initial ABWR or AP1000 they range from 4.3 to 5.0 cents/kWh on the basis of overnight capital costs of $ 1200 to 1500/kW, 60 year plant life, 5 year construction and 90% capacity. Coal gives 3.5–4.1 cents/kWh and gas (CCGT) 3.5–4.5 cents/kWh, depending greatly on fuel price.

The levelised nuclear power cost figures include up to 29% of the overnight capital cost as interest, and the report notes that up to another 24% of the overnight capital cost needs to be added for the initial unit of a first-of-a-kind advanced design such as the AP1000, defining the high end of the range above. For more advanced plants such as the EPR or SWR1000, overnight capital cost of $ 1800/kW is as-

Table 20.3 OECD electricity generating cost projections for year 2010 on—10% discount rate, cents/kWh. (Source: [1])

Country	Nuclear	Coal	Coal with CCS	Gas CCGT	Onshore wind
Belgium	10.9	10.0	–	9.3–9.9	13.6
Czech R	11.5	11.4–13.3	13.6–14.1	10.4	21.9
France	9.2	–	–	–	12.2
Germany	8.3	8.7–9.4	9.5–11.0	9.3	14.3
Hungary	12.2	–	–	–	–
Japan	7.6	10.7	–	12.0	–
Korea	4.2–4.8	7.1–7.4	–	9.5	–
Netherlands	10.5	10.0	–	8.2	12.2
Slovakia	9.8	14.2	–	–	–
Switzerland	9.0–13.6	–	–	10.5	23.4
USA	7.7	8.8–9.3	9.4	8.3	7.0
China[a]	4.4–5.5	5.8	–	5.2	7.2–12.6
Russia[a]	6.8	9.0	11.8	7.8	9.0
EPRI (USA)	7.3	8.8	–	8.3	9.1
Eurelectric	10.6	8.0–9.0	10.2	9.4	15.5

[a] For China and Russia: 2.5c is added to coal and 1.3c to gas as carbon emission cost to enable sensible comparison with other data in those fuel/technology categories, though within those countries coal and gas will in fact be cheaper than the Table above suggests

sumed and power costs are projected beyond the range above. However, considering a series of eight units of the same kind and assuming increased efficiency due to experience which lowers overnight capital cost, the levelised power costs drop 20% from those quoted above and where first-of-a-kind engineering costs are amortized (e.g. the $ 1500/kW case above), they drop 32%, making them competitive at about 3.4 cents/kWh (Table 20.4).

The study also shows that with a minimal carbon control cost impact of 1.5 cents/kWh for coal and 1.0 cents/kWh for gas superimposed on the above figures, nuclear is even more competitive. But more importantly it goes on to explore other policy options which would offset investment risks and compensate for first-of-a-kind engineering costs to encourage new nuclear investment, including investment tax breaks, and production tax credits phasing out after 8 years. (US wind energy gets a production tax credit which, has risen to 2.1 cents/kWh.)

In May 2009, an update of a heavily referenced 2003 MIT study was published. This said that "since 2003 construction costs for all types of large-scale engineered projects have escalated dramatically. The estimated cost of constructing a nuclear power plant has increased at a rate of 15% per year heading into the current economic downturn. This is based both on the cost of actual builds in Japan and Korea and on the projected cost of new plants planned for in the United States. Capital costs for both coal and natural gas have increased as well, although not by as much.

Table 20.4 Nuclear plant: projected electricity costs (cents/kWh). (Source: [1])

Overnight capital cost $/kW		1200	1500	1800
First unit	7 year build, 40 year life	5.3	6.2	7.1
	5 year build, 60 year life	4.3	5.0	5.8
Fourth unit	7 year build, 40 year life	4.5	4.5	5.3
	5 year build, 60 year life[a]	3.7	3.7	4.3
Eighth unit	7 year build, 40 year life	4.2	4.2	4.9
	5 year build, 60 year life[a]	3.4	3.4	4.0

[a] calculated from above data

The cost of natural gas and coal that peaked sharply is now receding. Taken together, these escalating costs leave the situation [of relative costs] close to where it was in 2003." The overnight capital cost was given as $ 4000/kW, in 2007 dollars. Applying the same cost of capital to nuclear as to coal and gas, nuclear came out at 6.6 cents/kWh, coal at 8.3 cents and gas at 7.4 cents, assuming a charge of $ 25/t CO2 on the latter.

Escalating capital costs were also highlighted in the US Energy Information Administration (EIA) 2010 report "Updated Capital Cost Estimates for Electricity Generation Plants". The US cost estimate for new nuclear was revised upwards from $ 3902/kW by 37% to a value of $ 5339/kW for 2011 by the EIA. This is in contrast to coal, which increases by only 25%, and gas, which actually shows a 3% decrease in cost. Renewables estimates show solar dropping by 25% while onshore wind increases by about 21%. The only option to increase faster than nuclear is offshore wind at 49%, while the increase in coal with CCS is about the same as nuclear. In the previous year's estimate, EIA assumed that the cost of nuclear would drop with time and experience, and that by 2030 the cost of nuclear would drop by almost 30% in constant dollars.

By way of contrast, China is stating that it expects its costs for plants under construction to come in at less than $ 2000/kW and that subsequent units should be in the range of $ 1600/kW. These estimates are for the AP1000 design, the same as used by EIA for the USA. This would mean that an AP1000 in the USA would cost about three times as much as the same plant built in China. Different labor rates in the two countries are only part of the explanation. Standardized design, numerous units being built, and increased localization are all significant factors in China.

The **French Energy and Climate Directorate** published in November 2008 an update of its earlier regular studies on relative electricity generating costs. This shied away from cash figures to a large extent due to rapid changes in both fuel and capital, but showed that at anything over 6000 h production per year (68% capacity factor), nuclear was cheaper than coal or gas combined cycle (CCG). At 100% capacity CCG was 25% more expensive than nuclear. At less than 4700 h/year CCG was cheapest, all without taking CO_2 cost into account.

With the nuclear plant, fixed costs were almost 75 % of the total, with CCG they were less than 25 % including allowance for CO_2 at $ 20/t. Other assumptions were 8 % discount rate, gas at 6.85 $/GJ, coal at EUR 60/t. The reference nuclear unit is the EPR of 1630 MWe net, sited on the coast, assuming all development costs being borne by Flamanville 3, coming on line in 2020 and operating only 40 of its planned 60 years. Capital cost apparently EUR 2000/kW. Capacity factor 91 %, fuel enrichment is 5 %, burnup 60 GWd/t and used fuel is reprocessed with MOX recycle. In looking at overall fuel cost, uranium at $ 52/lb made up about 45 % of it, and even though 3 % discount rate was used for back-end, the study confirmed the very low cost of waste in the total—about 13 % of fuel cost, mostly for reprocessing.

At the end of 2008 EdF updated the overnight cost estimate for Flamanville 3 EPR (the first French EPR, but with some supply contracts locked in before escalation) to EUR 4 billion in 2008 Euros (EUR 2434/kW), and electricity cost 5.4 cents/kWh (compared with 6.8 cents/kWh for CCGT and 7.0 cents/kWh for coal, "with lowest assumptions" for CO_2 cost). These costs were confirmed in mid 2009, when EdF had spent nearly EUR 2 Billion. In July 2010, EdF revised the overnight cost to about EUR 5 Billion.

In May 2008, **South Carolina Electric and Gas Co**. and Santee Cooper locked in the price and schedule of new reactors for their summer plant in South Carolina at $ 9.8 billion. (The budgeted cost earlier in the process was $ 10.8 billion, but some construction and material costs ended up less than projected.) The EPC contract for completing two 1117-MW AP1000s is with Westinghouse and the Shaw Group. Beyond the cost of the actual plants, the figure includes forecast inflation and owners' costs for site preparation, contingencies and project financing. The units are expected to be in commercial operation in 2016 and 2019.

In November 2008, **Duke Energy** Carolinas raised the cost estimate for its Lee plant (2 × 1117 MWe AP1000) to $ 11 billion, excluding finance and inflation, but apparently including other owners costs.

In November 2008, **TVA** updated its estimates for Bellefonte units 3 and 4 for which it had submitted a COL application for twin AP1000 reactors, total 2234 MWe. It said that overnight capital cost estimates ranged from $ 2516 to 4649/kW for a combined construction cost of $ 5.6–10.4 billion. Total cost to the owners would be $ 9.9–17.5 billion.

In 2013, the **Nuclear Energy Institute** announced the results of its financial modeling of comparative costs in the USA, based on figures from the US Energy Information Administration's 2013 Annual Energy Outlook. NEI assumed 5 % cost of debt, 15 % return on equity and a 70/30 debt equity capital structure. The figures are tabulated below. The report went on to show that with nuclear plant license renewal beyond 60 years, power costs would be $ 53–60/MWh (Table 20.5).

Table 20.5 NEI 2013 financial modeling

	EPC cost (per kW)	Capacity (%)	Electricity cost (per MWh)
Gas combined cycle, gas @ $ 3.70/GJ	$ 1000	90	$ 44.00
Gas combined cycle, gas @ $ 5.28/GJ	$ 1000	90	$ 54.70
Gas combined cycle, gas @ $ 6.70/GJ	$ 1000	90	$ 61.70
Gas combined cycle, gas @ $ 6.70/GJ, 50–50 debt-equity	$ 1000	90	c $ 70
Supercritical pulverised coal, 1300 MWe	$ 3000	85	$ 75.70
Integrated gasification combined cycle coal, 1200 MWe	$ 3800	85	$ 94.30
Nuclear, 1400 MWe (EIA's EPC figure)	$ 5500	90	$ 121.90
Nuclear, 1400 MWe (NEI suggested EPC figure)	$ 4500–5000	90	$ 85–90
Wind farm, 100 MWe	$ 1000	30	112.90

5 % cost of debt, 15 % return on equity and a 70-30 debt equity capital structure

Regarding bare plant costs, some recent figures apparently for overnight capital cost (or Engineering, Procurement and Construction—EPC—cost) quoted from reputable sources but not necessarily comparable are:

- EdF Flamanville EPR: EUR 4 billion/$ 5.6 billion, so EUR 2434/kW or $ 3400/kW
- Bruce Power Alberta 2 × 1100 MWe ACR, $ 6.2 billion, so $ 2800/kW
- CGNPC Hongyanhe 4 × 1080 CPR-1000 $ 6.6 billion, so $ 1530/kW
- AEO Novovronezh 6 and 7 2136 MWe net for $ 5 billion, so $ 2340/kW
- AEP Volgodonsk 3 and 4, 2 × 1200 MWe VVER $ 4.8 billion, so $ 2000/kW
- KHNP Shin Kori 3 and 4 1350 MWe APR-1400 for $ 5 billion, so $ 1850/kW
- FPL Turkey Point 2 × 1100 MWe AP1000 $ 2444–3582/kW
- Progress Energy Levy county 2 × 1105 MWe AP1000 $ 3462/kW
- NRG South Texas 2 × 1350 MWe ABWR $ 8 billion, so $ 2900/kW
- ENEC for UAE from Kepco, 4 × 1400 MWe APR-1400 $ 20.4 billion, so $ 3643/kW

A striking indication of the impact of **financing costs** is given by Georgia Power, which said in mid 2008 that twin 1100 MWe AP1000 reactors would cost $ 9.6 billion if they could be financed progressively by ratepayers, or $ 14 billion if not. This gives $ 4363 or 6360/kW including all other owners' costs.

Finally, in the USA the question of whether a project is subject to regulated cost recovery or is a merchant plant is relevant, since it introduces political, financial and tactical factors. If the new build cost escalates (or is inflated), some cost recovery may be possible through higher rates can be charged by the utility if those costs are deemed prudent by the relevant regulator. By way of contrast, a merchant plant has to sell all its power competitively, so must convince its shareholders that it has a good economic case for moving forward with a new nuclear unit.

20.2.4   Operations and Maintenance (O&M) Costs

This is the annual cost associated with the operation, maintenance, administration, and support of a nuclear power plant. Included are costs related to labor, material and supplies, contractor services, licensing fees, and miscellaneous costs such as employee expenses and regulatory fees. The average non-fuel O&M cost for a nuclear power plant in 2011 was 1.51 cents/kWh.

These costs are much easier to quantify and are independently verified as they relate directly to the profitability of the Utilities, which operate them. Any discrepancies are soon discovered through accounting audits. Companies that operate the USA's nuclear power reactors have made excellent profits over the last 5 years. The US Nuclear Power industry has at last lived up to its promise made in 1970's to produce electricity reliably and cheaply. Since 1987 the cost of producing electricity has decreased from 3.63 to 1.68 cents/kWh in 2004 and plant availability has increased from 67 to over 90%. The operating cost includes a charge of 0.2 cents/kWh to fund the eventual disposal of waste from the reactor and for decommissioning the reactor. The price of Uranium Ore contributes approximately 0.05 cents/kWh.

As part of management of nuclear power plant operations, it is clear from both the French and US experience that pro-active Industry organizations are vital in obtaining efficient plant utilization and in minimizing running costs. In the US in the late 1980s and early 1990s there was little pooling of knowledge and experience amongst Nuclear Power Operators. A combination of industry inexperience, the lack of standardized designs and the fragmentation of the industry caused this.

Once again, this was in contrast to the French experience where the uniform design and the single state-owned organization allowed knowledge to be more easily shared.

The US industry has since gone through several cycles of consolidation and the operation of the USA's fleet of Nuclear Reactors has mostly been taken over by specialist companies that specialize in this activity. In addition, the industry has learned the benefits of pooling knowledge. This combination has demonstrably improved the performance of the US reactor fleet and is reflected in the share price of the nuclear operation companies.

20.3   Production Costs

Operation and Maintenance (O&M) and Fuel Costs are part of production cost at a power plant. Since 2001, nuclear power plants have achieved the lowest production costs between coal, natural gas and oil.

Fuel costs make up 30% of the overall production costs of nuclear power plants. Fuel costs for coal, natural gas and oil, however, make up about 80% of the production costs. The following charts and plots are summary of all costs for United States

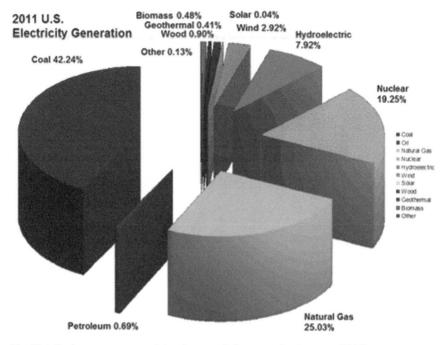

Fig. 20.4 Fuel as a percentage of electric power industry production costs (2011)

electric companies that, one need to take under consideration in order to own and operate a power plant to generate electricity.

• Fuel as a Percentage of Electric Power Industry Production Costs
• U.S. Nuclear Industry Production Costs by Quartile
• U.S. Electricity Production Costs
• U.S. Electricity Production Costs and Components (Table)

Note The following bar chart shows the percentage of electric industry production costs attributable to fuel for nuclear, coal, gas and oil in 2011 (Fig. 20.4).

Note The following bar chart depicts a 3-year rolling average of production costs by quartile in 2011 cents/kWh between 2007 and 2011 (Fig. 20.5).

Note The following line graph shows average U.S. electricity production costs between 1995 and 2011 for nuclear, coal, gas, and oil (Fig. 20.6).

US figures for 2013 published by NEI show the general picture, with nuclear generating power at 1.87 cents/kW (See Fig. 20.7).

Note The data given in Fig. 20.7 refer to fuel plus the operation and maintenance costs only; they exclude capital, since this varies greatly among utilities and states, as well as with the age of the plant.

Note The following table shows annual average U.S. electricity production, operations and maintenance (O&M), and fuel costs—from 1995 to 2011 for nuclear, coal, gas and oil (Table 20.6).

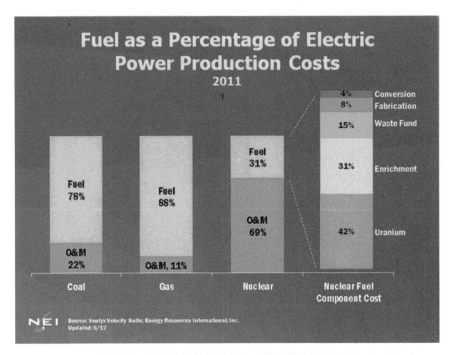

Fig. 20.5 U.S. nuclear industry production costs by quartile (2007–2011)

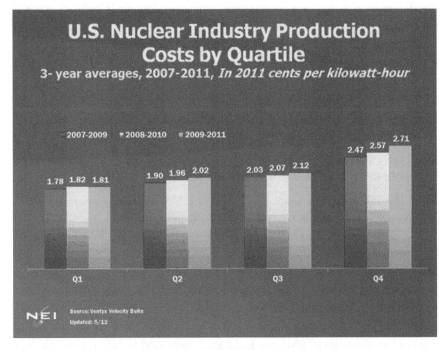

Fig. 20.6 U.S. electricity production costs (1995–2011)

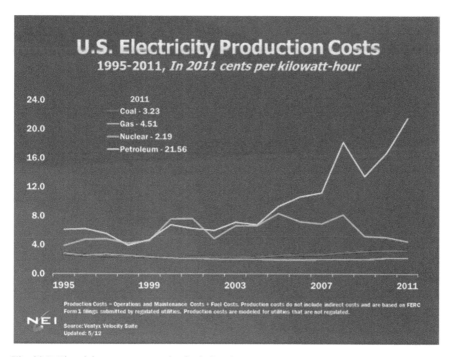

Fig. 20.7 Electricity power generation by industries

20.3.1   Costs Related to Waste Management

In the USA, Nuclear Power operators are charged 0.1 cents/kWh for the disposal of Nuclear Waste. In Sweden, this cost is 0.13 US cents/kWh. These Countries have utilized these funds to pursue research into Geologic disposal of waste and both now have mature proposals for the task.

In France, the cost of waste disposal and decommissioning is estimated to be 10 % of the construction cost. So far, provisions of 71 billion Euros have been acquired for this from the sale of electricity. As part of the cost related to waste management, two following important key issues should be take under consideration and they are;

1. **Committed Funds for the Nuclear Waste Management**
   $ 35.8 billion (1/10th of a cent per kWh of electricity generated at nuclear power plants plus interest since 1983). Of the $ 35.8 billion, $ 10.8 billion has been spent. Payments to the Nuclear Waste Fund are included in the fuel costs. See the table below (Table 20.7).

Table 20.6 U.S. electricity production costs and components (1995–2011). (Source: Ventyx velocity suite)

Year	Total production costs				Operations and maintenance costs				Fuel costs			
	Coal	Gas	Nuclear	Petroleum	Coal	Gas	Nuclear	Petroleum	Coal	Gas	Nuclear	Petroleum
1995	2.69	3.91	2.80	6.10	0.64	0.74	1.96	1.74	2.05	3.16	0.84	4.36
1996	2.53	4.76	2.52	6.23	0.56	0.73	1.76	1.43	1.97	4.03	0.76	4.80
1997	2.44	4.84	2.68	5.54	0.54	0.70	1.94	1.17	1.89	4.13	0.74	4.37
1998	2.39	4.21	2.52	3.93	0.58	0.63	1.79	0.76	1.81	3.58	0.73	3.17
1999	2.30	4.57	2.32	4.70	0.55	0.54	1.65	1.06	1.75	4.03	0.67	3.64
2000	2.25	7.60	2.22	6.80	0.54	0.60	1.59	0.84	1.71	7.01	0.63	5.96
2001	2.31	7.64	2.14	6.26	0.57	0.67	1.55	0.85	1.74	6.98	0.59	5.41
2002	2.28	4.86	2.10	6.01	0.58	0.65	1.55	0.97	1.71	4.21	0.55	5.04
2003	2.25	6.68	2.07	7.16	0.57	0.69	1.51	1.12	1.68	6.00	0.56	6.03
2004	2.34	6.69	2.02	6.79	0.60	0.57	1.47	1.00	1.74	6.12	0.55	5.79
2005	2.54	8.37	1.95	9.33	0.60	0.55	1.44	1.00	1.94	7.83	0.51	8.33
2006	2.64	7.23	1.98	10.69	0.62	0.56	1.47	1.41	2.02	6.66	0.51	9.28
2007	2.69	6.97	2.00	11.24	0.63	0.55	1.48	1.50	2.06	6.42	0.52	9.74
2008	2.94	8.20	2.04	18.25	0.63	0.58	1.52	1.97	2.30	7.62	0.52	16.29
2009	3.11	5.23	2.02	13.48	0.71	0.63	1.44	2.76	2.40	4.60	0.57	10.72
2010	3.18	5.05	2.20	16.75	0.72	0.58	1.53	2.21	2.45	4.48	0.67	14.54
2011	3.23	4.51	2.19	21.56	0.71	0.53	1.51	2.23	2.52	3.98	0.68	19.33

Production Costs = Operations and Maintenance Costs + Fuel Costs, Production costs do not include indirect costs and are based on FERC Form 1 filings submitted by regulated utilities. Production costs are modeled for utilities that are not regulated
Updated: 5/12

Table 20.7 Nuclear waste fund payment information by state

State	Total NWF contributions	One-time fee outstanding	Allocation of interest on NWF	Total liabilities
Alabama	811.1	0.0	655.7	1466.8
Arizona	579.8	0.0	468.6	1048.4
Arkansas	319.9	180.9	258.6	759.5
California	873.0	0.7	705.7	1579.4
Colorado	0.2	0.0	0.2	0.4
Connecticut	391.7	480.3	316.6	1188.6
Florida	810.1	0.0	654.3	1464.9
Georgia	732.8	0.0	592.2	1324.8
Illinois	1936.4	1018.2	1565.2	4519.8
Lowa	120.5	0.0	97.4	217.9
Kansas	201.1	0.0	162.6	363.7
Louisiana	350.2	0.0	283.1	633.3
Maine	65.5	185.5	52.9	303.9
Maryland	377.8	0.0	305.4	683.3
Massachusetts	170.3	0.0	137.7	308.9
Michigan	567.6	427.9	458.8	1454.3
Minnesota	406.0	0.0	328.2	734.3
Mississippi	217.9	0.0	176.1	394.0
Missouri	209.5	0.0	169.4	378.9
Nebraska	275.5	0.0	222.7	498.2
New Hampshire	169.2	0.0	136.8	306.0
New Jersey	655.0	174.5	529.5	1359.0
New York	864.9	505.6	699.1	2069.5
North Carolina	897.3	0.0	725.3	1622.6
Ohio	325.7	32.7	263.3	621.7
Oregon	75.5	0.0	61.0	136.5
Pennsylvania	1687.8	88.9	1364.3	3141.0
South Carolina	1319.3	0.0	1066.4	2385.8
Tennessee	505.2	0.0	408.4	913.6
Texas	677.8	0.0	547.8	1225.6
Vermont	101.6	144.6	82.2	328.4
Virginia	739.4	0.0	597.6	1337.0
Washington	172.3	0.0	139.3	311.5
Wisconsin	374.3	0.0	302.5	676.8
Other	7.6	0.0	6.1	13.7

2. **Estimated Cost of Decommissioning**

The US industry average cost for decommissioning a power plant is USD $ 300 million. The funds for this activity are accumulated in the operating cost of the plant. The French and Swedish Nuclear Industries expect decommissioning costs to be 10–15 % of the construction costs and budget this into the price charged for electricity. On the other hand, the British decommissioning costs have been projected to be around 1 billion pounds/reactor. Cleaning up the Hanford Nuclear Weapons reactor is budgeted at $ 5.6 billion but may cost two to three times this much.

Per plant $ 300–500 million—includes estimated radiological, used fuel and site restoration costs—about $ 300 million, $ 100–150 million and $ 50 million, respectively.

Industry $ 31.9 billion—about $ 300 million/reactor. Decommissioning costs are not included in production costs.

Of the total $ 31.9 billion estimated to decommission all eligible nuclear plants at an average cost of $ 300 million, $ 22.5 billion or about two-thirds have already been funded. The remaining $ 9.4 billion will be funded over the next 20 years (the average nuclear plant is licensed for 40 years).

20.3.2   Life-Cycle Costs (U.S. Figures)

These are the estimated total costs versus the electricity output over the lifetime of a new power plant. Costs include construction, operations and maintenance, fuel and decommissioning. For life cycle costs of nuclear, coal, and gas, NEI's white paper, The Cost of New Generating Capacity in Perspective is a good source to refer to it.

Although nuclear project costs are undeniably large, total project cost does not measure a project's economic viability. The relevant metric is the cost of the electricity produced by the nuclear project relative to alternative sources of electricity and relative to the market price of electricity at the time the nuclear plant comes into service. As illustrated by the detailed financial modeling cited above, new nuclear power plants can be competitive, even with total project costs exceeding $ 6000/kWe, including EPC and owners' costs and financing costs.

20.3.3   Construction Costs

Construction costs are very difficult to quantify but dominate the cost of Nuclear Power. The main difficulty is that third generation power plants now proposed are claimed to be both substantially cheaper and faster to construct than the second-generation power plants now in operation throughout the world. The Nuclear Industry says its learned the lessons of economy-of-volume demonstrated by the French Nuclear Program, and that these will be employed for the new power plants. In 2005, Westinghouse claimed its Advanced PWR reactor, the AP1000, will cost USD $ 1400/kW for the first reactor and fall in price for subsequent reactors. A more technical description is here. Proponents of the CANDU ACR and Gas Cooled pebble bed reactors made similar or stronger claims. However, the first wave of new generation plants in the USA is expected to cost over $ 3500/kW of capacity. Additional costs increase the price even more.

The General Electric ABWR was the first third generation power plant approved. The first two ABWR's were commissioned in Japan in 1996 and 1997. These took just over 3 years to construct and were completed on budget. Their construction costs were around $ 2000/kW. Two additional ABWR's are being constructed in Taiwan. However, these have faced unexpected delays and are now at least 2 years behind schedule.

Meanwhile the Chinese Nuclear Power Industry has won contracts to build new plants of their own design at capital costs reported to be $ 1500 and 1300/kW at sites in Southeast and North-East China. If completed on budget these facilities will be formidable competitors to the Western Nuclear Power Industry.

Given the history of Nuclear Plant construction in the U.S.A., the financial industry sees the construction of the new generation of reactors as a risky investment and demands a premium on capital lent for the purpose. The Energy Bill recently passed by the US Congress assumes this risk and provides production credits of 1.8 cents/kWh for the first 3 years of operation. This subsidy is equivalent to what is paid to Wind Power companies and is designed to encourage new nuclear reactor construction in the USA.

If the AP1000 lives up to, its promises of $ 1000/kW construction cost and 3 year construction time, it will provide cheaper electricity than any other Fossil Fuel based generating facility, including Australian Coal power, even with no sequestration charges. This promise appears to have been unfulfilled. The cost of the first AP1000 is expected to be over $ 3500/kW.

20.4   Comparing the Economics of Different Forms of Electricity Generation

A 2010 OECD study Projected Costs of generating Electricity set out some actual costs of electricity generation, from which the following figures are taken (Table 20.8):

It is important to distinguish between the economics of nuclear plants already in operation and those at the planning stage. Once capital investment costs are effectively "sunk", existing plants operate at very low costs and are effectively "cash machines". Their operations and maintenance (O&M) and fuel costs (including used fuel management) are, along with hydropower plants, at the low end of the spectrum and make them very suitable as base-load power suppliers. This is irrespective of whether the investment costs are amortized or depreciated in corporate financial accounts. This cost is based on, assuming the forward or marginal costs of operation are below the power price and the plant will operate.

Table 20.8 Actual costs of electricity (US cents/kWh). (Source: [1], projected costs of generating electricity, Tables 3.7 this shows the levelized cost, which is the average cost of producing electricity including capital, finance, owner's costs on site, fuel and operation over a plant's lifetime)

Technology	Region or country	At 10 % discount rate	At 5 % discount rate
Nuclear	OECD Europe	8.3–13.7	5.0–8.2
	China	4.4–5.5	3.0–3.6
Black coal with CCS	OECD Europe	11.0	8.5
Brown coal with CCS	OECD Europe	9.5–14.3	6.8–9.3
CCGT with CCS	OECD Europe	11.8	9.8
Large hydro-electric	OECD Europe	14.0–45.9	7.4–23.1
	China: Three gorges	5.2	2.9
	China: other	2.3–3.3	1.2–1.7
Onshore wind	OECD Europe	12.2–23.0	9.0–14.6
	China	7.2–12.6	5.1–8.9
Offshore wind	OECD Europe	18.7–26.1	13.8–18.8
Solar photovoltaic	OECD Europe	38.8–61.6	28.7–41.0
	China	18.7–28.3	12.3–18.6

20.5   System Cost

System Costs are external to the building and operation of any power plant, but must be paid by the electricity consumer, usually as part of the transmission and distribution cost. From a government policy point of view they are just as significant as the actual generation cost, but are seldom factored in to comparisons among different supply options, especially comparing base-load with dispersed renewables. In fact that the total system cost should be analysed when introducing new power generating capacity on the grid. Any new power plant likely requires changes to the grid, and hence incurs a significant cost for power supply that must be accounted for. But this cost for large base-load plants is small compared with integrating renewables to the grid.

See also paper on Electricity Transmission Grids.

20.6   External costs

External costs are not included in the building and operation of any power plant, and are not paid by the electricity consumer, but by the community generally. The external costs are defined as those actually incurred in relation to health and the environment, and which are quantifiable but not built into the cost of the electricity.

The report of a major European study of the external costs of various fuel cycles, focusing on coal and nuclear, was released in mid 2001—ExternE. It shows that in clear cash terms nuclear energy incurs about one tenth of the costs of coal. If these costs were in fact included, the EU price of electricity from coal would double and that from gas would increase 30%. These are without attempting to include the external costs of global warming.

The European Commission launched the project in 1991 in collaboration with the US Department of Energy, and it was the first research project of its kind "to put plausible financial figures against damage resulting from different forms of electricity production for the entire EU". The methodology considers emissions, dispersion and ultimate impact. With nuclear energy the risk of accidents is factored in along with high estimates of radiological impacts from mine tailings (waste management and decommissioning being already within the cost to the consumer). Nuclear energy averages 0.4-euro cents/kWh, much the same as hydro, coal is over 4.0 cents (4.1–7.3), gas ranges 1.3–2.3 cents and only wind shows up better than nuclear, at 0.1–0.2 cents/kWh average. NB these are the external costs only.

As part of Energy Subsidies and External Costs, the following points are in process

- Substantial amounts have been invested in energy R&D over the last 30 years. Much of this has been directed at developing nuclear energy—which now supplies 14% of world electricity.
- Today, apart from Japan and France, there is about twice as much R&D investment in renewables than nuclear, but with rather less to show for it and with less potential for electricity supply.
- Nowhere in the world is nuclear power subsidized per unit of production. In some countries however it is taxed because production costs are so low.
- Renewables receive heavy direct subsidies in the market, fossil fuels receive indirect subsidies in their waste disposal as well as some direct subsidies.
- Nuclear energy fully accounts for its waste disposal and decommissioning costs in financial evaluations.

There are three main areas where, broadly speaking, subsidies or other support for energy may apply:

1. Government R&D for particular technologies,
2. Subsidies for power generation per unit of production (or conceivably per unit of capacity), including costs imposed on disincentivised alternatives, and the
3. Allowance of external costs, which are either paid by the community at large or picked up later by governments.

In recent years, some controversy has surrounded the question of the relative levels of R&D expenditure on nuclear energy and on new renewables (essentially technologies to harness wind and solar energy). Unfortunately, IEA data available for the first edition of this paper is no longer available, hence some of the following is dated. There are some plots and data resources presented here, for more detailed, and update information readers should refer themselves to web site of world nuclear organization at http://www.world-nuclear.org

Serbia published new FITs early in 2013, which will be valid for 12 years from project commissioning and will be corrected annually every February in line

Table 20.9 Expenditure by IEA countries on energy R&D

Year	1975	1980	1985	1990	1995	2000	2005
Conservation	333	955	725	510	1240	1497	1075
Fossil fuels	587	2564	1510	1793	1050	612	1007
Renewables	208	1914	843	563	809	773	1113
Nuclear fission	4808	6794	6575	4199	3616	3406	3168
Nuclear fusion	597	1221	1470	1055	1120	893	715
Other	893	1160	787	916			
Total energy R&D	7563	15034	12186	9394	9483	9070	9586
Total: Japan	1508	3438	3738	3452	3672	3721	3905
Total: excluding Japan	6055	11596	8448	5842	5811	5349	5681

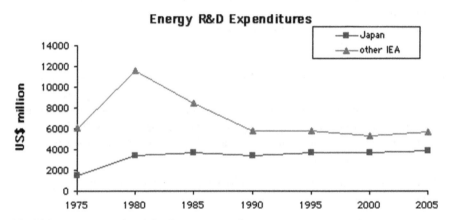

Fig. 20.8 Energy research and development expenditures

with the level of inflation in the Eurozone. They include EUR 9.2 cents/kWh for wind and 16.25 cents/kWh for solar, but with low caps on the capacity covered (Table 20.9) (Fig. 20.8).

The above table and graph are from the OECD International Energy Agency's database (IEA, 2001 and 2006) regarding government expenditure in the 26 IEA member countries. The database does not include information about private companies' expenditure, nor funds spent by non-IEA countries, such as China, Russia or India.

The total amount of energy R&D expenditure by governments of IEA countries rose in response to the oil price shocks of the early 1970s and then fell away as associated concerns abated, with the conspicuous exception of Japan. Private R&D investment has apparently followed the same pattern outside Japan.

Table 20.10 Expenditure by IEA countries on fission R&D. (2005 US$ millions)

	UK	France	Japan	USA	Other IEA countries	All IEA countries
1975	929	0	763	2164	952	4808
1980	741	0	2098	2410	1160	6794
1985	638	895	2259	1241	1542	6575
1990	253	555	2298	737	356	4199
1995	17	599	2455	103	442	3616
2000	0	666	2393	39	308	3406
2005	4	?	2398	171	?	3168

Throughout the period, the expenditure on nuclear fission dominated the overall figures, though falling from 64 % of the total in 1975 to 33 % in 2005. However, Table 20.10 shows that in most IEA countries (apart from Japan), government R&D expenditure on nuclear fission fell significantly through the 1990s, to trivial levels—in fact below that spent on renewables, which has averaged about US$ 700 million per year for the last two decades but is now rising (Table 20.10).

IEA data shows R&D on nuclear fission peaking around 1980 and after 1985 declining steadily to less than half that level. Since 1990, Japan alone has been responsible for some two thirds of IEA R&D expenditure on nuclear fission, with France accounting for most of the remainder. If the French and Japanese figures are excluded, fission R&D expenditure in the rest of the IEA countries totaled US$ 308 million in 2000 (Fig. 20.9).

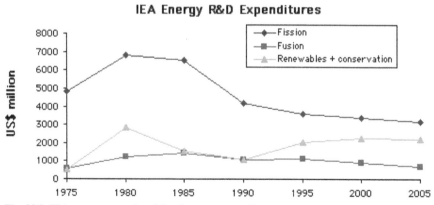

Fig. 20.9 IEA energy research and development expenditures

Table 20.11 Other US DOE R&D data is as follows ($ millions). (EIA 2008 and 2011)

Years	Currency	Renewables	Coal	Nuclear	End use
1988–2007	$ 2007	6271	7593	13,500	5737
FY 2007	$ 2010	717	582	1017	509
FY 2010	$ 2010	1409	663	1169	832

US Department of Energy figures show the renewables total in the US R&D budget as $ 505 million in FY2007 and energy efficiency $ 676 million, compared with nuclear power at $ 300 million (double the 2003 level) and fossil fuels at $ 397 million. Nuclear fusion is additional at $ 319 million (Table 20.11).

In FY 2007, relating the support to actual energy produced, the figures are: wind 2.34 cents/kWh, "clean coal" 2.98 cents/kWh, gas, coal 0.044 cents/kWh, nuclear 0.16 cents/kWh.

Outside the IEA, Russia, India and China have substantial nuclear fission programs and as the European Union also funds an amount of fission R&D, the worldwide totals for fission will be rather higher than the figure above. Nonetheless, given that the bulk of government-sponsored R&D into nuclear fission focuses on waste management and other fuel cycle back-end processes, it is clear that little is being spent at present by governments on new reactor designs.

Reference

1. OECD/IEA NEA (2010) Projected Costs of Generating Electricity

Chapter 21
Safety, Waste Disposal, Containment, and Accidents

The public acceptance of nuclear energy is still greatly dependent on the risk of radiological consequences in case of severe accidents. Such consequences were recently emphasized with the Fkushima-Daiichi accident in 2011. The nation's nuclear power plants are among the safest and most secure industrial facilities in the United States. Multiple layers of physical security, together with high levels of operational performance, protect plant workers, the public and the environment.

21.1 Safety

As a matter of fact, despite the highly-efficient prevention measures adopted for the current plants, some accident scenarios may, with a low probability, result in a severe accident, potentially leading to core melting, plant damage and dispersal of radioactive materials out of the plant containment.

Even if the Japanese power station was not equipped with the newest devices for the prevention or mitigation of severe accidents, Fukushima, as well as the Three Mile Island accident in 1979, confirmed the key role of the containment barrier in the significant mitigation of radioactive releases.

The improvement of the nuclear designs and the set-up of adequate accident management trategies require confinement structures and emergency system that must be properly dimensioned (configuration, choice of materials and cooling circuits), to guarantee the integrity of the safety barriers and avoid the release of radioactive gasses and aerosols to the outside environment.

In this chapter, schematics of typical nuclear power plants are preliminarily shown, together with the main concepts of nuclear reactions and the formation of radioactive isotopes.

The risk of radioactive release to the external environment because of accidents is then pointed out, in order to explain the reasons of safety criteria that characterize these kinds of installations.

© Springer International Publishing Switzerland 2015

B. Zohuri, P. McDaniel, *Thermodynamics In Nuclear Power Plant Systems,*
DOI 10.1007/978-3-319-13419-2_21

Attention is paid to the needs of maintaining the structural integrity of both cooling circuits and confinement structures. Seismic analyses are cited, as well as the risk of containment building damaging due to over-pressurization in case of severe accident.

U.S. nuclear plants are well designed, operated by trained personnel, defended against attack and prepared in the event of an emergency and following measurements are taken under consideration;

1. **Emergency Preparedness:** Every nuclear power plant in the country has a detailed plan for responding in the event of an emergency. Operators test that plan regularly, with the participation of local and state emergency response organizations.
2. **Operational Safety:** Stringent federal regulation, automated, redundant safety systems and the industry's commitment to comprehensive safety procedures keep nuclear power plants and their communities safe.
3. **Personnel Training and Screening:** Operators receive rigorous training and must hold valid federal licenses. All nuclear power plant staff are subject to background and criminal history checks before they are granted access to the plant.
4. **Plant Security:** Each nuclear power plant has extensive security measures in place to protect the facility from intruders. Since Sept. 11, 2001, the nuclear energy industry has substantially enhanced security at nuclear plants.

An illustration of the area of a nuclear power plant protected by armed guards, physical barriers and surveillance equipment from top-level point of view is depicted below (Fig. 21.1).

21.2   Nuclear Waste Disposal

Most used fuel from nuclear power plants is stored in steel-lined concrete pools filled with water, or in airtight steel or concrete-and-steel containers as pictured below (Fig. 21.2).

Used nuclear fuel is a solid material safely stored at nuclear plant sites. This storage is only temporary—one component of an integrated used fuel management system that addresses all facets of storing, recycling and disposal. In summary all the measurements that are being looked at are as follows;

1. **Integrated Used Fuel Management:** Under an integrated management approach, used nuclear fuel will remain stored at nuclear power plants in the near term. Eventually, the government will recycle it and place the unusable end product in a deep geologic repository.
2. **Low-Level Radioactive Waste:** Low-level waste is a byproduct of the beneficial uses of a wide range of radioactive materials. These include electricity generation, medical diagnosis and treatment, and various other medical processes.

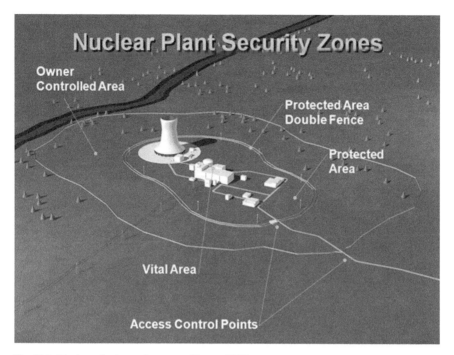

Fig. 21.1 Nuclear plant security zones. (Source NEI)

Fig. 21.2 Nuclear waste disposal container. (Source NEI)

3. **Recycling Used Nuclear Fuel:** The federal government plans to develop advanced recycling technologies to take full advantage of the unused energy in the used fuel and reduce the amount and toxicity of byproducts requiring disposal.
4. **Storage of Used Nuclear Fuel:** Currently, used nuclear fuel is stored at the nation's nuclear power plants in steel-lined, concrete pools or basins filled with water or in massive, airtight steel or concrete-and-steel canisters.
5. **Transportation:** The U.S. Department of Energy will transport used nuclear fuel to the repository by rail and road, inside massive, sealed containers that have undergone safety and durability testing.

Fig. 21.3 Aerial view of Yucca mountain. (Source NEI)

6. **Repository Development:** Under any used fuel management scenario, disposal of high-level radioactive byproducts in a permanent geologic repository is necessary.

The following image is the aerial view of Yucca Mountain, Nevada, site of national repository (Fig. 21.3).

21.3   Contamination

Environmental effects and impacts in recent years has been a main concerns of any major construction including new planning for deployment of nuclear power plant within United Sates or throughout the world. Therefore, before taking a nuclear power plant effects and impacts with environment, under consideration, it would be fruitful to have some understanding of contamination and what we mean, by environmental contamination both form nuclear and fossil fueled energy sources.

There is a synonymous common denominator between *contamination* and *pollution* yet from environmental point of view there is distinction between them. Introduction of a foreign substance within environment can be known as contamination, while pollution is a level of contamination, which is harmful to human and its environment.

Every human being is continuously exposed to different forms of radiation every moment of their life. In fact, the use of radiation in medicine, electricity generation and many other common applications has improved, extended and saved the lives of millions of Americans. Studies by the United Nations Scientific Committee on the Effects of Atomic Radiation, the National Research Council's BEIR VII study group and the National Council on Radiation Protection and Measurements all show that the risk associated with low-dose radiation from natural and manmade sources, including nuclear power plants, is extremely small. Researchers with the U.S. Department of Energy's Lawrence Berkeley National Laboratory, through a combination of state-of-the-art time-lapse live imaging and mathematical modeling of a special line of human breast cells, found evidence that for low-dose levels of ionizing radiation, cancer risks may not be directly proportional to dose. The data show that at lower doses of ionizing radiation, DNA repair mechanisms work much better than at higher doses. This contradicts the standard model for predicting biological damage from ionizing radiation—the linear-no-threshold hypothesis or LNT—that holds that risk is directly proportional to dose at all levels of irradiation.

Nuclear power plants have controlled and monitored emissions of radiation, but the amount is extremely small and poses no threat to the public or the environment. The Nuclear Regulatory Commission reports that people living close to a nuclear power plant receive, at most, an additional 1 millirem of radiation exposure a year. To put this in perspective, 1 millirem is one thousandth of the radiation exposure from a single whole-body CAT scan. The average American is exposed to 620 millirem of radiation every year. Three hundred millirem comes from natural sources, such as cosmic rays, uranium in the Earth's crust and radon gas in the atmosphere. Most of the rest comes from medical procedures such as CAT scans and consumer products. The radiation exposure from living near a nuclear power plant is insignificant and is no threat to the health of the public. After more than 3600 reactor years of operation, there is no scientific or medical evidence that shows anyone has been harmed by the radiation from any of America's commercial nuclear energy facilities, including the accident at Three Mile Island 32 years ago.

After more than a half-century of radiological monitoring and medical research, there is no evidence linking U.S. nuclear energy plants to negative effects on the health of the public or workers. Claims that radioactivity from nuclear plants has caused negative health effects on public/workers, have been refuted by the United Nations Scientific Committee of the Effects of Atomic Radiation, National Research Council's BEIR VII study group, the National Cancer Institute, the American Cancer Society, the American Academy of Pediatrics, numerous state departments of health and other independent studies.

After 32 years, there is no evidence that the nation's worst nuclear power plant accident harmed a single person or had any negative effect on the environment. More than a dozen health studies and continuous environmental monitoring have found no effect on the health of the people or the environment around the Three Mile Island nuclear plant in Pennsylvania.

Radioactive materials from nuclear plants, including used nuclear fuel, are highly regulated, strictly controlled and monitored and can be safely stored indefinitely. No member of the public has ever been harmed by the handling, transportation, storage or disposal of any of the radioactive material from the nation's nuclear power plants.

Note At the same concentration, industrial waste products such as hydrogen cyanide and arsenic are more toxic to humans than any of the materials used or produced at a nuclear plant.

There is no evidence that any person has died because of radiation exposure associated with the accident at the Fukushima Daiichi nuclear facility. In Japan, a small group of nuclear workers received radiation doses that may increase the risk of cancer over their lifetimes, but none of the workers' exposure is considered life threatening. Protective actions being taken by the Japanese government, including long-term evacuation of nearby residents, area decontamination and extensive radiation monitoring, are expected to avert significant radiological health consequences among the citizens of Japan. In the United States, extensive radiation monitoring by the U.S. Environmental Protection Agency's national radiation monitoring network at American commercial nuclear power plants and U.S. Department of Energy facilities detected extremely low levels of radiation thousands of times below government limits that pose no threat to human health.

21.4 Accidents

Nuclear power reactors are repository of enormous fission products and possible potential for a major accidents whether natural disaster or manmade. Fortunately, there have been no such accidents to date.

In addition to engineering and procedures, which reduce the risk and severity of accidents, all plants have guidelines for Severe Accident Management or Mitigation (SAM). These conspicuously came into play after the Fukushima accident, where staff had immense challenges in the absence of power and with disabled cooling systems following damage done by the tsunami. The experience following that accident is being applied not only in design but also in such guidelines, and peer reviews on nuclear plants will focus more on these than previously.

In mid 2011, the IAEA Incident and Emergency Centre launched a new secure web-based communications platform to unify and simplify information exchange during nuclear or radiological emergencies. The Unified System for Information Exchange on Incidents and Emergencies (USIE) has been under development since 2009 but was actually launched during the emergency response to the accident at Fukushima.

At Fukushima Daiichi in March 2011 the three operating reactors shut down automatically, and were being cooled as designed by the normal residual heat removal system using power from the back-up generators, until the tsunami swamped them an hour later. The emergency core cooling systems then failed. Days later, a separate problem emerged as spent fuel ponds lost water. Detailed analysis of the accident continues, but the main results include more attention being given to siting criteria and the design of back-up power and post-shutdown cooling, as well as provision for venting the containment of that kind of reactor and other emergency management procedures.

Nuclear plants have Severe Accident Mitigation Guidelines (SAMG, or in Japan: SAG), and most of these, including all those in the USA, address what should be done for accidents beyond design basis, and where several systems may be disabled. See section below.

In 2007, the US NRC launched a research program to assess the possible consequences of a serious reactor accident. Its draft report was released nearly a year after the Fukushima accident had partly confirmed its findings. The State-of-the-Art Reactor Consequences Analysis (SOARCA) showed that a severe accident at a US nuclear power plant (PWR or BWR) would not be likely to cause any immediate deaths, and the risks of fatal cancers would be vastly less than the general risks of cancer. SOARCA's main conclusions fall into three areas: how a reactor accident progresses; how existing systems and emergency measures can affect an accident's outcome; and how an accident would affect the public's health. The principal conclusion is that existing resources and procedures can stop an accident, slow it down or reduce its impact before it can affect the public, but even if accidents proceed without such mitigation they take much longer to happen and release much less radioactive material than earlier analyses suggested.

This was borne out at Fukushima, where there was ample time for evacuation—3 days—before any significant radioactive releases.

The April 1986 disaster at the Chernobyl nuclear power plant in the Ukraine was the result of major design deficiencies in the RBMK type of reactor, the violation of operating procedures and the absence of a safety culture. One peculiar feature of the RBMK design was that coolant failure could lead to a strong increase in power output from the fission process (positive void coefficient). However, this was not the prime cause of the Chernobyl accident.

The accident destroyed the reactor and killed 56 people, 28 of whom died within weeks from radiation exposure. It also caused radiation sickness in a further 200–300 staff and firefighters, and contaminated large areas of Belarus, Ukraine, Russia and beyond. It is estimated that at least 5% of the total radioactive material in the Chernobyl-4 reactor core was released from the plant, due to the lack of any containment structure. Most of this was deposited as dust close by. Some was carried by wind over a wide area.

About 130,000 people received significant radiation doses (i.e. above internationally accepted ICRP limits) and continue to be monitored. About 4000 cases of thyroid cancer in children have been linked to the accident. Most of these were curable, though about nine were fatal. No increase in leukemia or other cancers have yet shown up, but some is expected. The World Health Organization is closely monitoring most of those affected.

The Chernobyl accident was a unique event and the only time in the history of commercial nuclear power that radiation-related fatalities occurred.

The destroyed unit 4 was enclosed in a concrete shelter, which now requires remedial work.

An OECD expert report on it concluded that "the Chernobyl accident has not brought to light any new, previously unknown phenomena or safety issues that are not resolved or otherwise covered by current reactor safety programs for commercial power reactors in OECD Member countries". In other words, the concept of 'defense in depth' was conspicuous by its absence, and tragically shown to be vitally important.

Apart from the RBMK reactor design, an early Russian PWR design, the VVER-440/V-230, gave rise to concerns in Europe, and a program was initiated to close these down as a condition of EU accession, along with Lithuania's two RBMK units. See related papers on Early Soviet Reactors and EU Accession, and RBMK Reactors.

References

1. http://www.world-nuclear.org
2. http://www.nei.org

Erratum to: Thermodynamics In Nuclear Power Plant Systems

Bahman Zohuri and Patrick McDaniel

Erratum to:
B. Zohuri and P. McDaniel, Thermodynamics In Nuclear Power Plant Systems,
DOI 10.1007/978-3-319-13419-2

The publisher regrets to inform the omission of the problem file from the volume in the first printing, as well as the below mentioned errors which were spotted by the author.

1. Page v, Line 4: "Ben Pollared" should read "Colonel Ben Pollard"
2. Page xxii, Line 23: "adjunct and research" should read "research"
3. Page xxii, Line 29: "Phillips Laboratory has" should read "Phillips Laboratory's"
4. Page xxii, Line 37: "McDanile hold" should read "McDaniel holds"
5. Page xxii, Line 39: "Purde" should read "Purdue"
6. Page 151, Line 27: "volume it occupies at" should be "volume they occupy at"
7. Page 157, Equation 6.23: "(v - - b)" should be "(v - b)"
8. Page 160, Equation 6.32: "k = 5/3" should be "$\gamma = 5/3$"
9. Page 160, Equation 6.33: "K = 9/7" should be "$\gamma = 9/7$"

The online version of the original book can be found under
DOI 10.1007/978-3-319-13419-2

Bahman Zohuri
Department of Nuclear Engineering
University of New Mexico
Albuquerque, USA
e-mail: BahmanZ@aol.com

Patrick McDaniel
Department of Chemical and Nuclear Engineering
University of New Mexico
Albuquerque, USA
e-mail: McDanielPK@aol.com

© Springer International Publishing Switzerland 2015
B. Zohuri, P. McDaniel, *Thermodynamics In Nuclear Power Plant Systems,*
DOI 10.1007/978-3-319-13419-2_22

10. Page 160, 2nd line from bottom: "for k of 7/5" should be "for γ of 7/5"
11. Page 162, 2nd line above 6.7: "This end the" should be "This ends the"
12. Page 166, Line 20: "than the other does." should be "than the other."
13. Page 170, Line 6: "k$_T$T" should be "k$_B$T"
14. Page 173, End of Paragraph 7.1: "More statements that are elegant" should be "More rigorous statements"
15. Page 173, 2nd line from bottom: "it is virtually to transfer" should be "it is virtually impossible to transfer"
16. Page 175, Line 2: "and produces heat" should be "and transfers heat"
17. Page 176, Line 7: "No process is reversible, but a process can reversed if the energy losses are so small as to be negligible" should be "No real process is reversible, but a process can be treated as reversible if the energy losses are so small as to be negligible."
18. Page 177, Line 4: "form" should be "from"
19. Page 185, Line 19: "are listed in Appendix B and Appendix 9(Bahman)" should be "are listed in Appendices A.8 and A.9"
20. Page 186, Line 170: "QH" should be "Q_H".
21. Page 187, Line 11: "Does this cycle the" should be "Does this cycle satisfy the"
22. Page 189, Example 7.9: both times "below" should be "above"
23. Page 203, Line 10: "Prof. Leonard K. Nash, Dover, 2006, bases this discussion on the excellent text, "Elements of Statistical Thermodynamics [1–4]." should be "This discussion closely follows the excellent text "Elements of Statistical Thermodynamics by L. K. Nash, Dover 2006 [1–4]."
24. Page 203, Line 21, "~1020" should be "~10 <superscript> 20"
25. Page 206, Line 4: "and the curve get" should be "and the curves get"
26. Page 210, Equation 9.5: *"In(N!)~N*In(N)-N"* should be *"ln(N!)~N*ln(N)-N"*
27. Page 211, Line 20: "one atom from level n to level l" should be "one atom from level n to level m"
28. Page 218, Line 8: "WX and WY" should be "W$_X$ and W$_Y$"
29. Page 219, Line 15: "Therefore, b must have the" should be "Therefore β must have the"

Appendix A: Table and Graphs Compilations

The following tables provide samples of engineering data for materials of interest to Thermodynamics. For any detailed design work, more extensive handbooks should be consulted. The thermophysical property data for water, carbon dioxide and sodium were generated for this book. They represent the best fit to the latest physical measurements. More extensive tables are available from the National Institutes of Standards and Technology (NIST).

A.1 Physical Constants

Acceleration of gravity	$g = 9.80665 \text{m/s}^2$ or 32.174 ft/s^2
Sea level atmospheric pressure	$p_{SL} = 101.325$ kPa
	$= 1.01325$ bar
	$= 14.696$ psia
	$= 760$ mm Hg $(0°C)$
	$= 29.9213$ in Hg $(32°F)$
	$= 10.3323$ m H2O $(4°C)$
Boltzman's constant	$k = 1.3806503 \times 10^{-23}$ J/K
Avogadro's number	$N_a = 6.02214199 \times 10^{23}$ 1/g mole
Electronic charge	$q_e = 1.60217646 \times 10^{-19} °C$
Electron volt	$\varepsilon = 1.60217646 \times 10^{-19}$ J
Atomic mass unit	$amu = 1.6605402 \times 10^{-27}$ kg
	$= 931.49432$ MeV/c^2
Plank's constant	$h = 6.62606876 \times 10$
	$= h/2\pi = 1.054571596 \times 10$ J-s

© Springer International Publishing Switzerland 2015
B. Zohuri, P. McDaniel, *Thermodynamics In Nuclear Power Plant Systems,*
DOI 10.1007/978-3-319-13419-2

Stefan Boltzman constant	$\sigma = 0.17123 \times 10^{-8}$ Btu/(h-ft^2-R^4)
	$\sigma = 5.670400 \times 10^{-8}$ w/(m^2-K^4)
Speed of Light in Vacuum	$c = 2.99792458 \times 10^8$ m/s
Solar constant	$q_{solar} = 429$ Btu/(h-ft^2)
	$q^{solar} = 1{,}353$ W/m^2
Universal gas constant	$R = 8.31447$ kj/(kgmol-K)
	$= 8.31447$ kPa-m^3(kgmol-K)
	$= 0.0831447$ bar-m^3(kgmol-K)
	$= 82.05$ L-atm(kgmol-K)
	$= 1.98583$ Btu/(lbmol-R)
	$= 1545.37$ ft-1bf/(lbmol-R)
	$= 10.73$ psia-ft^3/(lbmol-R)

A.2 Conversion Factors

Distance			
	To convert	*To*	*Multiply by*
	Millimeters	Meters	10^{-3}
	Centimeters	Meters	10^{-2}
	Kilometers	Meters	10^3
	Inches	Feet	1/12
	Feet	Yards	1/3
	Feet	Miles	1/5,280
	Yards	Miles	1/1,760
	Inches	Centimeters	2.54
	Feet	Meters	0.3048
Area			
	Millimeters^2	Meters^2	10^{-6}
	Centimeters^2	Meters^2	10^{-4}
	Kilometers^2	Meters^2	10^6
	Inches^2	Feet^2	1/144
	Inches^2	Centimeters^2	6.4516
	Centimeters^2	Inches^2	0.1550
	Feet^2	Meters^2	0.092903
	Meters^2	Feet^2	10.7639
Volume			
	Milimeters^3	Meters^3	10^{-9}
	Centimeters^3	Meters^3	10^{-6}

	Liters	Meters^3	10^{-3}
	Kilometers^3	Meters^3	10^9
	Inches^3	Centimeters^3	16.3871
	Centimeters^3	Inches^3	0.061024
	Feet^3	Meters^3	0.0283168
	Meters^3	Feet^3	35.3147
Force			
	Kg-m/sec^2	Newtons	1.0
	Pound-mass-ft/sec^2	Pounds	1/32.1745
	Newtons	Pounds	0.4482
	Pounds-force	Newtons	0.2248
Mass and density			
	Grams	Kilograms	10^{-3}
	Metric tonnes	Kilograms	10^3
	Ounces	Pounds-mass	1/16
	Tons	Pounds-mass	2000
	Grams	Ounces	1/28.35
	Ounces	Grams	28.35
	Pounds-mass	Kilograms	0.4536
	Kilograms	Pounds-mass	2.2046
	Tons	Metric tonnes	0.9072
	Metric tonnes	Tons	1.1023
	Grams/cc	Lbm/ft^3	62.427
	Grams/cc	Kg/m^3	10^3
	Lb/ft^3	Kg/m^3	16.0187
Pressure			
	psi	in Hg	2.036
	psi	in H2O	27.7
	atm	in Hg	29.92
	atm	ft H2O	33.93
	atm	Pa	101,320
	atm	bar	1.0133
	atm	psi	14.69
	kPa	psi	0.145
	psi	kPa	6.895
Energy			
	joules	ergs	107
	lbf-ft	joules	1.356
	Btu	joules	1055
	Btu	ft-lbf	778
	Cal	Btu	0.003968

Cal	ft-lbf	3.088
Btu	W-hr	0.2930
kW-hr	Btu	3412
Cal	joules	4.1868
kj/kg	Btu/lbm	0.4992
Btu/lbm	kj/kg	2.3260
Cal/gm	Btu/lbm	1.8000
Power		
hp	ft-lbf/s	550
hp	Btu/hr	2545
hp	kW	0.7455
W	Btu/hr	3.412
kW	hp	1.341
ton	Btu/hr	12,000
ton	kW	3.517
Heat Transfer		
W/m/K	Btu/hr/ft/R	0.57779
kcal/h/m/K	W/m/K	1.1630
kcal/h/m/K	Btu/h/ft/R	0.67197
Btu/h/ft/R	W/m/K	1.7307
$W/m^2/K$	$Btu/h/ft^2/R$	0.17611
$Btu/hr/ft^2/R$	$W/m^2/K$	5.6783
Fluid Flow		
m^3/s	ft^3/s	35.3147
g/cm/sec (poise)	$lbf\text{-}sec/ft^2$	0.002088
$lbf\text{-}sec/ft^2$	g/cm/sec	478.96
$lbf\text{-}sec/ft^2$	kg/m/hr	172,400
lbm/ft/sec	gm/cm/sec	14.882
m^2/sec	ft^2/sec	10.7639
cm^2/sec	ft^2/hr	3.875
ft^2/sec	m^2/hr	334.45

A.3 Standard Atmosphere

Table A.3.1 SI Units

Altitude (km)	P/P_{SL}	T/T_{SL}	$\rho\rho_{SL}$
0	1	1	1
1	0.887	0.9774	0.9075
2	0.7846	0.9549	0.8217
3	0.692	0.9324	0.7423

Altitude (km)	P/P_{SL}	T/T_{SL}	$\rho\rho_{SL}$
4	0.6085	0.9098	0.6689
5	0.5334	0.8873	0.6012
6	0.466	0.8648	0.5389
7	0.4057	0.8423	0.4817
8	0.3519	0.8198	0.4292
9	0.304	0.7973	0.3813
10	0.2615	0.7748	0.3376
11	0.224	0.7523	0.2978
12	0.1915	0.7519	0.2546
13	0.1636	0.7519	0.2176
14	0.1399	0.7519	0.186
15	0.1195	0.7519	0.159
16	0.1022	0.7519	0.1359

$p_{SL} = 101325$ Pas
$T_{SL} = 288.2$ K
$r_{SL} = 1.225$ kg/m^3

Table A.3.2 English Units

Altitude (kft)	P/P_{SL}	T/T_{SL}	$\rho\rho_{SL}$
0	1.0	1.0	1.0
2	0.9298	0.9863	0.9428
4	0.8637	0.9725	0.8881
6	0.8014	0.9588	0.8359
8	0.7429	0.945	0.7861
10	0.6878	0.9313	0.7386
12	0.6362	0.9175	0.6933
14	0.5877	0.9038	0.6502
16	0.5422	0.8901	0.6092
18	0.4997	0.8763	0.5702
20	0.4599	0.8626	0.5332
22	0.4227	0.8489	0.498
24	0.388	0.8352	0.4646
26	0.3557	0.8215	0.433
28	0.3256	0.8077	0.4031
30	0.2975	0.794	0.3747
32	0.2715	0.7803	0.348
34	0.2474	0.7666	0.3227
36	0.225	0.7529	0.2988

Altitude (kft)	P/P_{SL}	T/T_{SL}	$\rho\rho_{SL}$
38	0.2044	0.7519	0.2719
40	0.1858	0.7519	0.2471
42	0.1688	0.7519	0.2245
44	0.1534	0.7519	0.204
46	0.1394	0.7519	0.1854
48	0.1267	0.7519	0.1685
50	0.1151	0.7519	0.1531

$p_{SL} = 2116$ lbf/ft^3
$T_{SL} = 518.7$ R
$r_{SL} = 0.07647$ lbm/ft^3

A.4 Critical State Properties of Gases

		Molar Mass	P_{crit} MPa	T_{crit} K	P_{crit} atm	T_{crit} R	v_{crit} cm^3/gm-mol	Z_{crit}
Air		28.996	3.77	133.0	37.21	239.0	88.3	0.300
Ammonia	NH_3	17.031	11.28	405.5	111.32	729.8	72.5	0.243
Argon	Ar	39.950	4.86	151.0	47.96	272.0	75.2	0.291
Carbon Dioxide	CO_2	44.010	7.39	304.2	72.93	547.5	94.0	0.293
Carbon Monoxide	CO	28.010	3.50	133.0	34.54	240.0	93.1	0.294
Chlorine	Cl_2	70.906	7.99	416.9	78.87	750.4	124.0	0.276
Helium	He	4.003	0.23	5.3	2.27	9.5	57.5	0.303
Hydrogen	H_2	2.016	1.30	33.3	12.83	59.9	65.0	0.304
Iodine	I2	253.809	11.70	819.0	115.47	1474.2	167.0	0.287
Krypton	Kr	83.800	5.50	209.4	54.28	376.9	92.2	0.291
Lithium	Li	6.941	67.00	3223.0	661.24	5801.4	120.0	0.300
Mercury	Hg	200.590	150.97	1763.2	1490.0	3173.7	29.1	0.300
Methane	CH_4	16.043	4.64	191.1	45.79	343.9	99.0	0.290
Neon	Ne	20.180	2.73	44.5	26.94	227.1	41.7	0.307
Nitrogen	N_2	28.013	3.39	126.2	33.46	227.1	90.1	0.291
Oxygen	O_2	31.999	5.08	154.8	50.14	278.6	73.4	0.288
Potassium	K	39.098	16.42	2105.9	162.02	3790.7	320.0	0.300
Propane	C_3H_8	44.097	4.26	370.0	42.04	665.9	200.0	0.277
Sodium	Na	22.990	25.60	2503.9	252.65	4507.0	244.0	0.300
Water	H_2O	18.015	22.10	647.4	218.11	1165.3	56.0	0.230
Water (Heavy)	D_2O	20.023	21.72	644.7	214.36	1160.4	54.9	0.222
Xenon	Xe	131.290	5.84	289.8	57.65	521.6	118.8	0.290

A.5 Constants for Van Der Waals Equation of State

Substance		Molar Mass	a kPa-m^6/ kmol^2	b m^3/kmol	a atm-ft^6/ lbmol^2	b ft^3/lbmol
Air		28.996	136.83	0.03666	345.21	0.58621
Ammonia	NH_3	17.031	425.09	0.03736	1075.80	0.59826
Argon	Ar	39.95	136.81	0.03229	346.85	0.51752
Carbon Dioxide	CO_2	44.0098	365.16	0.04278	924.18	0.68507
Carbon Monoxide	CO	28.0104	147.38	0.03949	374.96	0.63407
Chlorine	Cl_2	70.906	634.26	0.05422	1605.62	0.86835
Helium	He	4.0026	3.56	0.02395	8.94	0.38193
Hydrogen	H_2	2.0158	24.87	0.02662	62.88	0.42607
Iodine	I2	253.809	1671.81	0.07275	4232.15	1.16510
Krypton	Kr	83.8	232.51	0.03957	588.59	0.63372
Lithium	Li	6.941	4521.16	0.04999	11445.24	0.80067
Mercury	Hg	200.59	600.48	0.01214	1520.11	0.19439
Methane	CH_4	16.043	229.51	0.04280	580.74	0.68534
Neon	Ne	20.1797	21.15	0.01694	430.43	0.76922
Nitrogen	N_2	28.0134	137.00	0.03869	346.63	0.61946
Oxygen	O_2	31.9988	137.56	0.03167	348.12	0.50712
Potassium	K	39.0983	7877.53	0.13331	19942.70	2.13513
Propane	C_3H_8	44.097	937.13	0.09026	2371.61	1.44542
Sodium	Na	22.9898	7141.57	0.10165	18078.79	1.62795
Water	H_2O	18.0152	553.04	0.03044	1399.97	0.48757
Water (Heavy)	D_2O	20.023	557.95	0.03084	1412.43	0.49400
Xenon	Xe	131.29	419.20	0.05156	1061.20	0.82572

$R = 8314 \ J/kmol/K = 1545 \ ft\text{-}lbf/lbmol/R = 0.73023 \ atm\text{-}ft^3/lbmol/R$

A.6 Constants for Redlich-Kwong Equation of State

Substance		Molar Mass	a kPa-m^6-K$^{1/2}$/ kmol^2	b m^3/kmol	a atm-ft^6-R$^{1/2}$/ lbmol^2	b ft^3/lbmol
Air		28.996	1598.9	0.02541	5407.8	0.40631
Ammonia	NH_3	17.031	8673.7	0.02589	29448.7	0.41466
Argon	Ar	39.95	1703.5	0.02238	5796.3	0.35870
Carbon Dioxide	CO_2	44.0098	6453.4	0.02965	21912.0	0.47483
Carbon Monoxide	CO	28.0104	1722.3	0.02737	5886.1	0.43949
Chlorine	Cl_2	70.906	13122.4	0.03758	44568.3	0.60187

Substance		Molar Mass	a kPa-m^6-K$^{1/2}$/ kmol^2	b m^3/kmol	a atm-ft^6-R$^{1/2}$/ lbmol^2	b ft^3/lbmol
Helium	He	4.0026	8.3	0.01660	27.9	0.26473
Hydrogen	H$_2$	2.0158	145.4	0.01845	493.2	0.29531
Iodine	I2	253.809	48479.7	0.05042	164653.6	0.80756
Krypton	Kr	83.8	3409.3	0.02743	11579.2	0.43925
Lithium	Li	6.941	260082.9	0.03465	883330.7	0.55496
Mercury	Hg	200.59	25549.3	0.00841	86774.1	0.13473
Methane	CH$_4$	16.043	3214.9	0.02967	10912.6	0.47502
Neon	Ne	20.1797	143.0	0.01174	6572.7	0.53316
Nitrogen	N$_2$	28.0134	1559.5	0.02682	5293.1	0.42936
Oxygen	O$_2$	31.9988	1734.2	0.02195	5887.8	0.35150
Potassium	K	39.0983	366303.6	0.09240	1244158.2	1.47990
Propane	C$_3$H$_8$	44.097	18265.5	0.06256	62012.5	1.00185
Sodium	Na	22.9898	362104.2	0.07045	1229829.9	1.12836
Water	H$_2$O	18.0152	14258.5	0.02110	48424.8	0.33795
Water (Heavy)	D$_2$O	20.023	14354.4	0.02138	48752.6	0.34240
Xenon	Xe	131.29	7230.7	0.03574	24558.0	0.57232

$R = 8314$ J/kmol/K $= 1545$ ft-lbf/lbmol/R $= 0.73023$ atm-ft^3/lbmol/R

A.7 Constants for the Peng-Robinson Equation of State

Substance		Molar Mass	a kPa-m^6/ kmol^2	b m^3/kmol	a atm-ft^6/ lbmol^2	b ft^3/lbmol	ω
Air		28996	148.29	0.02282	374.15	0.36484	0.032
Ammonia	NH$_3$	17.031	460.72	0.02325	1165.97	0.37234	0.252
Argon	Ar	39.95	148.28	0.02010	375.92	0.32209	−0.004
Carbon Dioxide	CO$_2$	44.0098	395.76	0.02662	1001.65	0.42636	0.225
Carbon Monoxide	CO	28.0104	159.73	0.02458	406.39	0.39462	0.049
Chlorine	Cl$_2$	70.906	687.42	0.03374	1740.20	0.54044	0.073
Helium	He	4.0026	3.86	0.01490	9.69	0.23770	0
Hydrogen	H$_2$	2.0158	26.96	0.01657	68.16	0.26517	−0.215
Iodine	I$_2$	253.809	1811.93	0.04528	4586.88	0.72512	0.07
Krypton	Kr	83.8	252.00	0.02463	637.92	0.39441	0.001
Lithium	Li	6.941	4900.11	0.03111	12404.55	0.49831	−0.153
Mercury	Hg	200.59	650.81	0.00755	1647.52	0.12098	−0.139

		Molar	a	b	a	b	ω
Methane	CH_4	16.043	248.75	0.02664	629.41	0.42654	0.008
Neon	Ne	20.1797	22.93	0.01054	466.51	0.47874	−0.041
Nitrogen	N_2	28.0134	148.48	0.02408	375.69	0.38553	0.04
Oxygen	O_2	31.9988	149.09	0.01971	377.30	0.31562	0.021
Potassium	K	39.0983	8537.81	0.08297	21614.24	1.32884	−0.031
Propane	C_3H_8	44.097	1015.67	0.05618	2570.39	0.89958	0.152
Sodium	Na	22.9898	7740.16	0.06326	19594.10	1.01318	−0.076
Water	H_2O	18.0152	599.40	0.01895	1517.31	0.30345	0.344
Water(Heavy)	D_2O	20.023	604.71	0.01920	1530.82	0.30745	0.361
Xenon	Xe	131.29	454.34	0.03209	1150.15	0.51390	0.012

$R = 8314\ J/kmol/K = 1545\ ft\text{-}lbf/lbmol/R = 0.73023\ atm\text{-}ft^3/lbmol/R$

A.8 Thermophysical Properties of Solids

	Melting			(Cp in j/kg/K)		(k in W/m/K)					
	Point	Density									
Material	(K)		T(K)	200	300	400	600	800	1000	1200	1500
Aluminum (Pure)	933	2702	k	237	237	240	231	218			
			C_p	798	903	949	1033	1146			
Aluminum (2024-T6)	775	2770	k	163	177	186	186				
			C_p	787	875	925	1042				
Beryllium	1550	1850	k	301	200	161	126	106	90.8	78.7	
			C_p	1114	1825	2191	2604	2823	3018	3227	3519
Bismuth	545	9780	k	9.69	7.86	7.04					
			C_p	120	122	127					
Boron	2573	2500	k	55.5	27	16.8	10.6	9.6	9.85		
			C_p	600	1107	1463	1892	2160	2338		
Cadmium	594	8650	k	99.3	96.8	94.7					
			C_p	222	231	242					
Chromium	2118	7160	k	111	93.7	90.9	80.7	71.3	65.4	61.9	57.2
			C_p	384	449	484	542	581	616	682	779
Copper (Pure)	1358	8933	k	413	401	393	379	366	352	339	
			C_p	356	385	397	417	433	451	480	
Iron (Pure)	1810	7870	k	94	80.2	69.5	54.7	43.3	32.8	28.3	32.1
			C_p	384	447	490	574	680	975	609	654
Carbon Steel		7854	k		60.5	56.7	48	39.2	31.3		

Material	Melting Point (K)	Density	T(K)	200	300	400	600	800	1000	1200	1500
			C_p		434	487	559	685	1169		
302 Stainless		8055	k		15.1	17.3	20	22.8	25.4		
			C_p		480	512	559	585	606		
304 Stainless	1670	7900	k	12.6	14.9	16.6	19.8	22.6	25.4	28	31.7
			C_p	402	477	515	557	582	611	640	682
316 Stainless		8238	k		13.4	15.2	18.3	21.3	24.2		
			C_p		468	504	550	576	602		
347 Stainless		7978	k		14.2	15.8	18.9	21.9	24.7		
			C_p		480	513	559	585	606		
Lead	601	11340	k	36.7	35.3	34	31.4				
			C_p	125	129	132	142				
Lithium	454	534	k		84.8						
			C_p		3580						
Magnesium	923	1740	k	159	156	153	149	146			
			C_p	934	1024	1074	1170	1267			
Molybdenum	2894	10240	k	143	138	134	126	118	112	105	98
			C_p	224	251	261	275	285	295	308	330
Nickel (Pure)	1728	8900	k	107	90.7	80.2	65.6	67.6	71.8	76.2	82.6
			C_p	383	444	485	592	530	562	594	616
Inconel X-750	1665	8510	k	10.3	11.7	13.5	17	20.5	24	27.6	33
			C_p	372	439	473	510	546	626		
Niobium	2741	8570	k	52.6	53.7	55.2	58.2	61.3	64.4	67.5	72.1
			C_p	249	265	274	283	292	301	310	324
Palladium	1827	12020	k	71.6	71.8	73.6	79.7	86.9	94.2	102	110
			C_p	227	244	251	261	271	281	291	307
Platinum (Pure)	2045	21450	k	72.6	71.6	71.8	73.2	75.6	78.7	82.6	89.5
			C_p	125	133	136	141	146	152	157	165
60%Pt-40%Rh	1800	16630	k		47	52	59	65	69	73	76
			C_p		162						

	Melting			(Cp in j/ kg/K)	(k in W/m/K)						
	Point	Density									
Material	(K)		T(K)	200	300	400	600	800	1000	1200	1500
Plutonium	913	19860	k		25.1						
			C_p		104.4						
Potassium	337	862	k		102.5						
			C_p		757.1						
Rhenium	3453	21100	k	51	47.9	46.1	44.2	44.1	44.6	45.7	47.8
			C_p	127	136	139	145	151	156	162	171
Rhodium	2236	12450	k	154	150	146	136	127	121	116	110
			C_p	220	243	253	274	293	311	327	349
Silver	1235	10500	k	430	429	425	412	396	379	361	
			C_p	225	235	239	250	2623	277	292	
Sodium	371	986	k		142						
			C_p		1084						
Tantalum	3269	16600	k	57.5	57.5	57.8	58.6	59.4	60.2	61	62.2
			C_p	133	140	144	146	149	152	155	160
Thorium	2023	11700	k	54.6	54	54.5	55.8	56.9	56.9	58.7	
			C_p	112	118	124	134	145	156	167	
Tin	505	7310	k	73.3	66.6	62.2					
			C_p	215	227	243					
Titanium	1953	4500	k	24.5	21.9	20.4	19.4	19.7	20.7	22	24.5
			C_p	465	522	551	591	633	675	620	686
Tungsten	3660	19300	k	186	174	159	137	125	118	113	107
			C_p	122	132	137	142	145	148	152	157
Uranium	1406	19070	k	25.1	27.6	29.6	34	38.8	43.9	49	
			C_p	108	116	125	146	176	180	161	
Vanadium	2192	6100	k	31.3	30.7	31.3	33.3	35.7	38.2	40.8	44.6
			C_p	430	489	515	540	563	597	645	714
Zinc	693	7140	k	118	116	111	103				
			C_p	367	389	402	436				
Zirconium	2125	6570	k	25.2	22.7	21.6	20.7	21.6	23.7	26	28.8
			C_p	264	278	300	322	342	362	344	344
Beryllium Oxide	2725	3000	k		272	196	111	70	47	33	21.5
			C_p		1030	1350	1690	1865	1975	2055	2145
Carbon	1500	1950	k	1.18	1.6	1.89	2.19	2.37	2.53	2.84	3.48
Amorphous			C_p		509						
Graphite	2273	2210	k		5.7	16.8	9.23	4.09	2.68	2.01	1.6
Pyrolytic			C_p	411	709	992	1406	1650	1793	1890	1974

Material	Melting Point (K)	Density	T(K)	200	300	400	600	800	1000	1200	1500
					(Cp in j/kg/K)			(k in W/m/K)				
Silicon Carbide	3100	3160	k		490				87	58	30
			C_p		675	880	1050	1135	1195	1243	1310
Silicon Dioxide	1883	2220	k	1.14	1.38	1,51	1.75	2.17	2.87	4	
			C_p		745	905	1040	1105	1155	1195	
Silicon Nitride	2173	2400	k		16	13.9	11.3	9.88	8.76	8	7.16
			C_p	578	691	778	937	1063	1155	1226	1306
Thorium Dioxide	3573	9110	k		13	10.2	6.6	4.7	3.68	3.12	2.73
			C_p		235	255	274	285	295	303	315
Uranium Dioxide	3138	10980	k	13.1	10.05	8.17	5.95	4.67	3.849	3.27	2.67
			C_p	278	277.4	277.3	277	277	277.2	277	277
Plutonium Dioxide	2673	11460	k	16.3	11.17	8.489	5.73	4.33	3.477	2.91	2.33
			C_p		276.1						
Concrete		2300	k		1.4						
			C_p		880						
Glass		2500	k		1.4						
			C_p		750						
Ice	273	920	k	2.03	1.88						
			C_p	1945	2040						
Paraffin		900	k		0.24						
			C_p		2890						
Paper		930	k		0.18						
			C_p		1340						
Sand		1515	k		0.27						
			C_p		800						
Soil		2050	k		0.52						
			C_p		1840						

A.9 Thermophysical Properties of Liquids

	Temp	273	280	300	320	340	360	400
Engine oil	ρ (kg/m^3)	899.1	895.3	884.1	871.8	859.9	847.8	825.1
	C_p (J/kg/K)	1796	1827	1909	1993	2076	2161	2337
	μ (N-s/m^2)	3.85	2.17	0.486	0.141	0.00531	0.00252	0.00087
	k (W/m/K)	0.147	0.144	0.145	0.143	0.139	0.138	0.134

	Temp	273	300	350	400	450	500	600
Water (liquid)	ρ (kg/m³)	1000.0	997.0	973.7	937.2	890.5	831.3	648.9
T_m=273	C_p (J/kg/K)	4217.0	4179.0	4195.0	4256.0	4400.0	4660.0	7000.0
T_b=373	μ (N-s/m²)	1.75E-03	8.55E-04	3.43E-04	2.17E-04	1.52E-04	1.18E-04	8.10E-05
	k (W/m/K)	0.569	0.613	0.668	0.688	0.678	0.642	0.497
	Temp	273	300	350	400	450	500	600
Water (vapor)	ρ (kg/m³)	0.0048	0.0256	0.2600	1.3680	4.8077	13.05	72.99
T_m=273	C_p (J/kg/K)	1854	1872	1954	2158	2560	3270	8750
T_b=373	μ (N-s/m²)	8.02E-06	9.09E-06	1.11E-05	1.31E-05	1.49E-05	1.66E-05	2.27E-05
	k (W/m/K)	0.0182	0.0196	0.023	0.0272	0.0331	0.0423	0.103
Heavy	Temp	273	353					
Water (liquid)	ρ (kg/m³)	1105.4	1078.2					
T_m=277	C_p (J/kg/K)	4207.7	4178.4					
T_b=376	μ (N-s/m²)							
	k (W/m/K)	0.5931	0.632					
Heavy	Temp	313.2	300	350	400	450	500	600
Water (vapor)	ρ (kg/m³)	0.0058	0.0250	0.2717	1.4771	5.2966	14.75	83.96
T_m=277	C_p (J/kg/K)	1694.8	1712	1743.4	1779.4	1817.1	1856.4	1938.9
T_b=376	μ (N-s/m²)							
	k (W/m/K)	0.0187	0.0202	0.0237	0.02884	0.0309	0.0371	0.0482
	Temp	473	673	873	1073	1273		
Lithium	ρ (kg/m³)	508.9	491.2	475.1	457.5	441.4		
T_m=452 K	C_p (kJ/kg/K)	5861.5	4563.6	3810	3056.4	2302.7		
T_b=1590 K	μ (N-s/m²)	5.65E-04	4.56E-04	4.56E-04	4.56E-04	4.56E-04		
	k (W/m/K)	46.33	39.49	25.93	11.12	10.38		
	Temp	477.8	588.9	700.0	811.1	922.2	1033.3	1154.8
Sodium	ρ (kg/m³)	904.4	878.1	851.6	824.9	798.1	771.1	741.6
T_m=371	C_p (J/kg/K)	1.338	1.300	1.274	1.259	1.255	1.263	1.284
T_b=1156 K	μ (N-s/m²)	4.52E-04	3.33E-04	2.66E-04	2.26E-04	1.97E-04	1.74E-04	1.57E-04
	k (W/m/K)	81.52	75.81	70.27	65.08	60.23	55.56	50.88
	Temp	600	800	1000	1200	1392.6	1400	1800
		588.9	700.0	811.1	922.2	1029.2	1033.3	1255.6
Potassium	ρ (kg/m³)	771.9	745.8	719.1	692.0	665.4	664.2	607.7
T_m=337	C_p (J/kg/K)	0.783	0.791	0.804	0.825	0.854	0.854	0.934
T_b=1033 K	μ (N-s/m²)	2.54E-04	2.06E-04	1.71E-04	1.47E-04	1.31E-04	1.31E-04	1.09E-04
	k (W/m/K)	43.61	39.98	36.69	33.75	30.81	30.81	25.10
	Temp	366	644	977				
NaK (45/50)	ρ (kg/m³)	887.4	821.7	740.1				
T_m=292 K	C_p (J/kg/K)	1130	1055	1043				

T_b=1098 K	μ (N-s/m²)	5.79E-04	2.36E-04	1.61E-04					
	k (W/m/K)	25.6	27.5	28.9					
	Temp	*366*	*672*	*1033*					
NaK(22/78)	ρ (kg/m³)	849	775.3	690.4					
T_m=262 K	C_p (J/kg/K)	946	879	883					
T_b=1057 K	μ (N-s/m²)	4.92E-04	2.07E-04	1.46E-04					
	k (W/m/K)	24.4	26.7	29.4					
	Temp	*644*	*755*	*977*					
Lead	ρ (kg/m³)	10540	10412	10140					
T_m=600 K	C_p (J/kg/K)	159	155	151					
T_b=2010 K	μ (N-s/m²)	2.39E-03	1.93E-03	1.37E-03					
	k (W/m/K)	16.1	15.6	14.9					
	Temp	*273*	*300*	*350*	*400*	*450*	*500*	*600*	
Mercury	ρ (kg/m³)	13595	13529	13407	13287	13167	13048	12809	
T_m=234 K	C_p (J/kg/K)	140.4	139.3	137.7	136.5	135.7	135.3	135.5	
T_b=630 K	μ (N-s/m²)	1.69E-03	1.52E-03	1.31E-03	1.17E-03	1.08E-03	1.01E-03	9.11E-04	
	k (W/m/K)	8.18	8.54	9.18	9.8	10.4	10.95	11.95	
	Temp	*589*	*811*	*1033*					
Bismuth	ρ (kg/m³)	10011	9739	9467					
T_m=544 K	C_p (J/kg/K)	144.4	154.5	164.5					
T_b=1750 K	μ (N-s/m²)	1.62E-03	1.10E-03	7.90E-04					
	k (W/m/K)	16.4	15.6	15.6					
	Temp	*422*	*644*	*922*					
PbBi(44.5/55.5)	ρ (kg/m³)	10524	10236	9835					
T_m=398 K	C_p (J/kg/K)	147	147	147					
T_b=1943 K	μ (N-s/m²)	1.85E-03	1.53E-03	1.15E-03					
	k (W/m/K)	9.05	11.86	15.3788					

A.10 Thermophysical Properties of Gases

	Temp (K)	*250*	*350*	*500*	*650*	*800*	*1000*	*1200*	*1400*	*1600*	*1800*
Air	μ (N-s/m²)	1.60E-05	2.08E-05	2.70E-05	3.23E-05	3.70E-05	4.24E-05	4.73E-05	5.30E-05	5.84E-05	6.37E-05
	k (W/m/K)	0.0223	0.0300	0.0407	0.0497	0.0573	0.0667	0.0763	0.0910	0.1060	0.1200
	Temp (K)	*280*	*300*	*350*	*400*	*450*	*500*	*550*	*600*	*650*	*700*

Carbon	μ (N-s/m²)	1.40E-05	1.49E-05	1.69E-05	1.90E-03	2.10E-05	2.31E-05	2.51E-05	2.70E-05	2.88E-05	3.05E-05
Dioxide	k (W/m/K)	0.0152	0.0166	0.0205	0.0243	0.0283	0.0325	0.0366	0.0407	0.0445	0.0481
	Temp (K)	250	300	350	400	450	500	550	600	650	700
Carbon	μ (N-s/m²)	1.52E-05	1.75E-05	1.98E-05	2.18E-05	2.37E-05	2.54E-05	2.71E-05	2.86E-05	3.01E-05	3.15E-05
Monox-ide	k (W/m/K)	0.0214	0.0250	0.0285	0.0318	0.0350	0.0381	0.0411	0.0440	0.0470	0.0500
	Temp (K)	250	300	350	400	450	500	600	700	800	900
Helium	μ (N-s/m²)	1.70E-05	1.99E-05	2.21E-05	2.43E-05	2.63E-05	2.83E-05	3.20E-05	3.50E-05	3.82E-05	4.14E-05
	k (W/m/K)	0.1335	0.1520	0.1700	0.1870	0.2040	0.2200	0.2520	0.2780	0.3040	0.3300
	Temp (K)	250	300	350	400	500	600	800	1000	1200	1400
Hydro-gen	μ (N-s/m²)	7.89E-06	8.96E-06	9.88E-06	1.08E-05	1.26E-05	1.42E-05	1.72E-05	2.01E-05	2.26E-05	2.51E-05
	k (W/m/K)	0.1570	0.1830	0.2040	0.2260	0.2660	0.3050	0.3780	0.4480	0.5280	0.6100
	Temp (K)	250	300	350	400	500	600	700	800	900	1100
Nitrogen	μ (N-s/m²)	1.55E-05	1.78E-05	2.00E-05	2.22E-05	2.58E-05	2.91E-05	3.21E-05	3.49E-05	3.75E-05	4.23E-05
	k (W/m/K)	0.0222	0.0259	0.0293	0.0327	0.0389	0.0446	0.4990	0.0548	0.0597	0.0700
	Temp (K)	250	300	350	400	500	600	700	800	900	1100
Oxygen	μ (N-s/m²)	1.79E-05	2.07E-05	2.34E-05	2.58E-05	3.03E-05	3.44E-05	3.81E-05	4.15E-05	4.47E-05	5.06E-05
	k (W/m/K)	0.0226	0.0268	0.0296	0.0330	0.0412	0.0473	0.0528	0.0589	0.0649	0.0758
	Temp (K)	380	400	450	500	550	600	650	700	750	800
Water	μ (N-s/m²)	1.27E-05	1.34E-05	1.53E-05	1.70E-05	1.88E-05	2.07E-05	2.25E-05	2.43E-05	2.60E-05	2.79E-05
	k (W/m/K)	0.0246	0.0261	0.0299	0.0339	0.0379	0.0422	0.0464	0.0505	0.0549	0.0592

A.11 Ideal Gas Heat Capacities for Selected Gases

	τ	C_0^*	C_1	C_2	C_3	C_4	Range (°R)	Range (K)	Max Error (%)
Air	0.0140	9.0761	−3.4743	1.7804	−0.3134	0.0188	100–6200	56–3444	0.77
100% Air+CP(CH₁)	0.0400	9.5147	−3.9901	2.1908	−0.3898	0.0229	300–4000	167–2222	0.32
100% Air+CP(CH₂)	0.0400	9.7159	−4.0842	2.1394	−0.3635	0.0202	300–4000	167–2222	0.35

	τ	C_0^*	C_1	C_2	C_3	C_4	Range (°R)	Range (K)	Max Error (%)
100% Air+CP(CH₃)	0.0400	9.8608	−4.1581	2.1088	−0.3464	0.0184	300–4000	167–2222	0.37
Carbon Dioxide	1.0000	−5.8291	12.0893	−2.7155	0.2747	−0.0102	300–5800	167–3222	0.18
Carbon Monoxide	0.0750	12.1297	−7.2191	3.4222	−0.6162	0.0387	300–5800	167–3222	0.36
Hydrogen	0.5000	12.2786	−4.7650	1.4175	−0.1407	0.0037	300–5800	167–3222	0.48
Nitrogen	0.1000	12.2568	−7.0276	3.1766	−0.5465	0.0328	300–6400	167–3556	0.38
Oxygen	0.0005	8.5200	−3.1865	1.9424	−0.3855	0.0261	300–6400	167–3556	0.98
Water	0.1500	15.8257	−10.0486	4.3395	−0.6669	0.0355	300–6400	167–3556	0.28
Hydroxl Ion	0.0025	11.4039	−4.7972	1.7247	−0.2252	0.0099	540–6300	300–3500	0.62
Nitrous Oxide	7.5500	−21.2671	28.5044	−9.0836	1.2789	−0.0676	540–6300	300–3500	0.71
Nitrogen Dioxide	1.5000	−16.0807	21.1941	−5.8523	0.7382	−0.0356	540–6300	300–3500	0.81
Methane	2.8500	−21.3124	17.0304	−1.4849	−0.0505	0.0043	540–6300	300–3500	0.20

The Ideal Gas constant pressure specific heat is given by

$$\theta = \left(\frac{T(K)}{100}\right) = \left(\frac{T(°R)}{180}\right) \quad C_p = \sum_{i=0}^{i=4} C_i(\theta)^{1/2} = \frac{kcal}{kgmole-K} = \frac{Btu}{lbm-°R} \quad C_0 = C_0^*\left(\frac{\theta^2}{\theta^2+\tau}\right)$$

A.12 Enthalpy of Formation and Enthalpy of Vaporization

				25°C (77°F), 1 atm			
		h_f	g_f	$s°$	h_f	g_f	$s°$
		kj/kmol	kj/kmol	kj/kmol/K	Btu/lbmol	Btu/lbmol	Btu/lbmol/R
Carbon	C(s)	0	0	5.74	0	0	1.371
Hydrogen	H₂(g)	0	0	130.68	0	0	31.215
Nitrogen	N₂(g)	0	0	191.61	0	0	45.768
Oxygen	O₂(g)	0	0	205.04	0	0	48.976
Carbon Monoxide	CO(g)	−110,530	−137,150	197.65	−47523	−58968	47.211
Carbon Dioxide	CO₂(g)	−393,520	−394,360	213.8	−169195	−169556	51.069

				25°C (77°F), 1 atm			
		h_f	g_f	$s°$	h_f	g_f	$s°$
		kj/kmol	kj/kmol	kj/kmol/K	Btu/lbmol	Btu/lbmol	Btu/lbmol/R
Steam	$H_2O(g)$	−241,820	−228,590	188.83	−103971	−98283	45.104
Water	$H_2O(l)$	−285,830	−237,180	69.92	−122893	−101976	16.701
Ammonia	$NH_3(g)$	−46,190	−16,590	192.33	−19860	−7133	45.940
Methane	$CH_4(g)$	−74,850	−50,790	186.16	−32182	−21837	44.467
Propane	$C_3H_8(g)$	−103,850	−23,490	269.91	−44651	−10100	64.471
n-Octane	$C_8H_{18}(g)$	−208,450	16,530	466.73	−89624	7107	111.484
n-Octane	$C_8H_{18}(l)$	−249,950	6,610	360.79	−107467	2842	86.179
Methyl Alcohol	$CH_3OH(g)$	−200,670	−162,000	239.7	−86279	−69652	57.255
Methyl Alcohol	$CH_3OH(l)$	−277,690	−166,360	126.8	−119394	−71527	30.288
Monatomic Qxygen	$O(g)$	249,190	231,770	161.06	107140	99650	38.471
Monatomic Hydrogen	$H(g)$	218,000	203,290	114.72	93730	87405	27.402
Monatomic Nitrogen	$N(g)$	472,650	455,510	153.3	203217	195848	36.618
Hydroxyl Ion	$OH(g)$	39,460	34,280	183.7	16966	14739	43.879
Water		44,010	35.3	18922	8.43		
Propane		15,060	122.2	6475	29.18		
n-Octane		41,460	254.7	17826	60.84		
Methyl Alcohol		37,900	83.5	16295	19.95		
Sodium		100819	29.9	43347	7.14		
Potassium		83934	30.6	36088	7.31		
Mercury		61303	27.9	26358	6.66		

Appendix 13 Gas Property Tables for Selected Gases

A.13.1 Air Properties (SI Units)

Temperature		h	u	Pr	Vr	s_0	Gamma
K	C	kg/kgmol	kj/kgmol			kj/kgmol/K	
100	−173.1	1604.8	1142.6	0.03	6018.5	23.8878	1.400
125	−148.1	2331.0	1663.3	0.07	1916.6	30.3695	1.401
150	−123.1	3056.4	2182.4	0.12	1217.2	35.6598	1.402
175	−98.1	3781.8	2701.0	0.21	829.17	40.1327	1.401

Temperature		h	u	Pr	Vr	s_o	Gamma
K	C	kg/kgmol	kj/kgmol			kj/kgmol/K	
200	−73.1	4507.5	3219.7	0.34	594.50	44.0087	1.401
225	−48.1	5233.5	3738.5	0.51	443.22	47.4293	1.401
250	−23.1	5959.9	4257.7	0.73	340.76	50.4908	1.401
275	1.9	6686.9	4777.2	1.02	268.59	53.2620	1.400
300	26.9	7414.4	5297.2	1.39	216.07	55.7943	1.400
325	51.9	8142.9	5818.2	1.84	176.81	58.1266	1.399
350	76.9	8872.6	6340.2	2.38	146.79	60.2895	1.398
375	101.9	9603.7	6863.8	3.04	123.39	62.3073	1.396
400	126.9	10336.8	7389.2	3.82	104.82	64.1997	1.395
425	151.9	11072.0	7916.7	4.73	89.874	65.9826	1.393
450	176.9	11809.8	8446.8	5.79	77.685	67.6695	1.391
475	201.9	12550.5	8979.8	7.02	67.631	69.2711	1.389
500	226.9	13294.2	9515.8	8.44	59.253	70.7971	1.387
525	251.9	14041.4	10055.3	10.06	52.205	72.2555	1.384
550	276.9	14792.3	10598.4	11.90	46.231	73.6525	1.381
575	301.9	15547.0	11145.4	13.98	41.128	74.9945	1.379
600	326.9	16305.7	11696.3	16.33	36.742	76.2859	1.376
625	351.9	17068.5	12251.4	18.97	32.948	77.5314	1.373
650	376.9	17835.5	12810.7	21.92	29.648	78.7349	1.370
675	401.9	18606.9	13374.1	25.22	26.765	79.8992	1.367
700	426.9	19382.6	13942.1	28.89	24.233	81.0276	1.365
725	451.9	20162.6	14514.4	32.95	22.002	82.1225	1.362
750	476.9	20946.8	15090.8	37.45	20.027	83.1861	1.359
775	501.9	21735.5	15671.7	42.41	18.274	84.2203	1.357
800	526.8	22528.4	16256.8	47.87	16.711	85.2274	1.354
825	551.8	23325.6	16846.1	53.87	15.315	86.2087	1.351
850	576.8	24126.8	17439.7	60.44	14.064	87.1653	1.349
875	601.8	24932.1	18037.2	67.62	12.940	88.0990	1.347
900	626.8	25741.4	18638.5	75.46	11.926	89.0111	1.344
925	651.8	26554.5	19243.9	84.00	11.012	89.9020	1.342
950	676.8	27371.3	19852.8	93.28	10.184	90.7733	1.340
975	701.8	28191.9	20465.7	103.36	9.433	91.6261	1.338
1000	726.8	29016.1	21082.2	114.27	8.751	92.4606	1.336
1025	751.8	29843.6	21701.8	126.08	8.130	93.2783	1.334
1050	776.8	30674.4	22324.6	138.83	7.563	94.0789	1.333
1075	801.8	31508.7	22951.0	152.57	7.046	94.8638	1.331
1100	826.8	32346.0	23580.6	167.37	6.572	95.6337	1.329
1125	851.8	33186.2	24212.8	183.29	6.138	96.3890	1.328
1150	876.8	34029.3	24848.2	200.39	5.739	97.1304	1.326
1175	901.8	34875.2	25486.4	218.70	5.373	97.8575	1.325

Temperature		h	u	Pr	Vr	s_o	Gamma
K	C	kg/kgmol	kj/kgmol			kj/kgmol/K	
1200	926.8	35723.9	26127.4	238.35	5.035	98.5727	1.324
1225	951.8	36574.9	26770.7	259.35	4.723	99.2747	1.322
1250	976.8	37428.9	27416.6	281.80	4.436	99.9647	1.321
1275	1001.8	38285.6	28065.2	305.76	4.170	100.6433	1.320
1300	1026.8	39143.5	28715.8	331.30	3.924	101.3101	1.319
1325	1051.8	40004.7	29369.0	358.49	3.696	101.9659	1.318
1350	1076.8	40868.0	30024.6	387.42	3.485	102.6112	1.317
1375	1101.8	41733.0	30681.8	418.17	3.288	103.2462	1.316
1400	1126.8	42600.6	31341.7	450.85	3.105	103.8718	1.315
1425	1151.8	43470.4	32003.0	485.52	2.935	104.4876	1.314
1450	1176.8	44340.9	32665.9	522.22	2.777	105.0935	1.313
1475	1201.8	45214.2	33331.4	561.10	2.629	105.6904	1.312
1500	1226.8	46089.4	33999.0	602.25	2.491	106.2789	1.311
1525	1251.8	46966.6	34668.5	645.68	2.362	106.8578	1.310
1550	1276.8	47845.5	35339.7	691.69	2.241	107.4300	1.309
1575	1301.8	48726.2	36011.9	740.16	2.128	107.9931	1.309
1600	1326.8	49607.8	36685.8	791.40	2.022	108.5495	1.308
1625	1351.8	50491.7	37362.0	845.22	1.923	109.0966	1.307
1650	1376.8	51376.2	38038.9	902.11	1.829	109.6381	1.306
1675	1401.8	52262.5	38717.5	961.77	1.742	110.1705	1.306
1700	1426.8	53152.0	39397.5	1024.76	1.659	110.6979	1.305
1725	1451.8	54042.3	40080.2	1090.79	1.581	111.2171	1.304
1750	1476.8	54932.8	40763.0	1160.12	1.508	111.7294	1.304
1775	1501.8	55825.3	41449.4	1233.09	1.439	112.2365	1.303
1800	1526.8	56720.1	42134.9	1309.56	1.375	112.7367	1.303
1825	1551.8	57613.3	42820.4	1389.69	1.313	113.2305	1.302
1850	1576.8	58509.4	43508.9	1473.67	1.255	113.7182	1.302
1875	1601.8	59409.7	44201.5	1561.58	1.201	114.2000	1.301
1900	1626.8	60306.9	44891.0	1653.65	1.149	114.6762	1.300
1925	1651.8	61207.5	45582.2	1749.90	1.100	115.1466	1.300
1950	1676.8	62110.5	46279.2	1850.59	1.054	115.6117	1.299
1975	1701.8	63012.8	46972.2	1955.91	1.010	116.0718	1.299
2000	1726.8	63915.9	47669.2	2066.18	0.968	116.5278	1.298
2025	1751.8	64819.8	48365.4	2181.02	0.928	116.9775	1.298
2050	1776.8	65726.4	49062.8	2300.86	0.891	117.4222	1.298
2075	1801.8	66632.6	49762.9	2425.62	0.855	117.8612	1.297
2100	1826.8	67543.5	50464.5	2555.79	0.822	118.2958	1.297
2125	1851.8	68450.4	51165.3	2691.76	0.789	118.7267	1.296
2150	1876.8	69360.0	51865.6	2833.21	0.759	119.1525	1.296
2175	1901.8	70272.1	52568.3	2979.99	0.730	119.5725	1.295

Temperature		h	u	Pr	Vr	s_o	Gamma
K	C	kg/kgmol	kj/kgmol			kj/kgmol/K	
2200	1926.8	71182.2	53272.4	3133.78	0.702	119.9908	1.295
2225	1951.8	72097.2	53978.1	3293.20	0.676	120.4033	1.295
2250	1976.8	73011.9	54686.7	3458.58	0.651	120.8107	1.294
2275	2001.8	73923.4	55388.9	3630.89	0.627	121.2149	1.294
2300	2026.8	74842.5	56098.7	3809.81	0.604	121.6148	1.294
2325	2051.9	75754.3	56804.4	3995.67	0.582	122.0108	1.293
2350	2076.9	76674.1	57518.2	4189.73	0.561	122.4051	1.293
2375	2101.9	77592.9	58227.6	4389.97	0.541	122.7932	1.293
2400	2126.9	78507.2	58935.8	4597.72	0.522	123.1776	1.292
2425	2151.9	79426.7	59646.1	4813.26	0.504	123.5585	1.292
2450	2176.9	80349.7	60363.0	5037.23	0.486	123.9366	1.292
2475	2201.9	81273.4	61077.4	5270.24	0.470	124.3125	1.291
2500	2226.9	82192.5	61790.4	5509.22	0.454	124.6812	1.291
2525	2251.9	83114.8	62506.7	5757.86	0.439	125.0482	1.291
2550	2276.9	84041.1	63223.6	6015.53	0.424	125.4122	1.290
2575	2301.9	84969.4	63942.5	6280.78	0.410	125.7709	1.290
2600	2326.9	85886.2	64650.1	6557.81	0.396	126.1298	1.290
2625	2351.9	86807.6	65368.9	6844.26	0.384	126.4852	1.289
2650	2376.9	87741.2	66093.0	7139.73	0.371	126.8366	1.289
2675	2401.9	88664.3	66806.7	7447.82	0.359	127.1878	1.289
2700	2426.9	89596.2	67529.3	7761.68	0.348	127.5310	1.289
2725	2451.9	90518.9	68249.2	8089.94	0.337	127.8753	1.288
2750	2476.9	91450.9	68965.3	8423.94	0.326	128.2117	1.288
2775	2501.9	92381.9	69687.0	8769.42	0.316	128.5458	1.288
2800	2526.9	93311.9	70414.3	9131.02	0.307	128.8818	1.288
2825	2551.9	94248.4	71141.4	9502.36	0.297	129.2132	1.287
2850	2576.9	95182.0	71865.7	9888.05	0.288	129.5439	1.287
2875	2601.9	96115.2	72589.6	10282.3	0.280	129.8690	1.287
2900	2626.9	97050.2	73321.8	10686.3	0.271	130.1894	1.286
2925	2651.9	97967.6	74029.8	11103.6	0.263	130.5078	1.286
2950	2676.9	98908.9	74761.9	11543.9	0.256	130.8311	1.286
2975	2701.9	99845.8	75489.4	11986.1	0.248	131.1437	1.286
3000	2726.9	100781.8	76216.1	12444.3	0.241	131.4555	1.285
3025	2751.9	101706.6	76944.6	12933.0	0.234	131.7758	1.285
3050	2776.9	102662.5	77678.1	13408.5	0.227	132.0760	1.285
3075	2801.9	103596.5	78402.8	13907.1	0.221	132.3795	1.285
3100	2826.9	104539.8	79149.9	14431.7	0.215	132.6873	1.284
3125	2851.9	105465.8	79866.5	14965.8	0.209	132.9895	1.284
3150	2876.9	106407.4	80611.9	15519.7	0.203	133.2916	1.284
3175	2901.9	107353.1	81348.2	16072.2	0.198	133.5824	1.284

Temperature		h	u	Pr	Vr	s_o	Gamma
K	C	kg/kgmol	kj/kgmol			kj/kgmol/K	
3200	2926.9	108280.3	82053.1	16655.2	0.192	133.8786	1.284
3225	2951.9	109232.0	82808.5	17257.4	0.187	134.1739	1.283
3250	2976.9	110178.3	83545.5	17879.8	0.182	134.4685	1.283
3275	3001.9	111123.2	84267.9	18497.0	0.177	134.7506	1.283
3300	3026.9	112069.4	85004.8	19152.7	0.172	135.0402	1.283
3325	3051.9	113008.3	85747.5	19806.8	0.168	135.3194	1.283
3350	3076.9	113938.4	86468.3	20500.1	0.163	135.6055	1.283
3375	3101.9	114883.0	87203.5	21213.6	0.159	135.8899	1.282
3400	3126.9	115838.3	87949.5	21938.6	0.155	136.1693	1.283

A.13.2 Air Properties (English Units)

Temperature		h	u	Pr	Vr	s_o	Gamma
R	F	Btu/lbmol	Btu/lbmol			Btu/lbmol/R	
100	−359.7	690.0	491.3	0.004	26215.8	5.7059	1.402
140	−319.7	968.4	692.7	0.012	11288.7	8.0473	1.399
180	−279.7	1246.6	892.9	0.030	6018.51	9.7953	1.400
220	−239.7	1524.1	1091.9	0.060	3648.55	11.1878	1.401
260	−199.7	1801.4	1290.4	0.108	2406.77	12.3457	1.402
300	−159.7	2078.6	1488.6	0.178	1685.34	13.3375	1.402
340	−119.7	2355.9	1686.8	0.276	1233.92	14.2051	1.401
380	−79.7	2633.3	1885.1	0.406	935.173	14.9765	1.401
420	−39.7	2910.8	2083.4	0.576	728.608	15.6709	1.401
460	0.3	3188.5	2281.8	0.792	580.625	16.3024	1.401
500	40.3	3466.4	2480.4	1.061	471.455	16.8816	1.400
540	80.3	3744.5	2679.2	1.388	388.917	17.4166	1.400
580	120.3	4022.8	2878.2	1.784	325.178	17.9140	1.399
620	160.3	4301.6	3077.7	2.254	275.068	18.3788	1.398
660	200.3	4580.9	3277.6	2.808	235.036	18.8153	1.397
700	240.3	4860.7	3478.1	3.455	202.608	19.2269	1.396
740	280.3	5141.3	3679.3	4.204	176.016	19.6167	1.394
780	320.3	5422.6	3881.3	5.066	153.972	19.9869	1.393
820	360.3	5704.9	4084.2	6.051	135.513	20.3398	1.391
860	400.3	5988.1	4288.0	7.171	119.925	20.6771	1.389
900	440.3	6272.5	4493.0	8.438	106.655	21.0003	1.387
940	480.3	6558.0	4699.1	9.866	95.277	21.3106	1.384

Temperature		h	u	Pr	Vr	s₀	Gamma
R	F	Btu/lbmol	Btu/lbmol			Btu/ lbmol/R	
980	520.3	6844.7	4906.4	11.468	85.459	21.6094	1.382
1020	560.3	7132.7	5115.0	13.257	76.939	21.8974	1.380
1060	600.3	7422.1	5325.0	15.252	69.500	22.1757	1.377
1100	640.3	7712.8	5536.3	17.466	62.978	22.4449	1.375
1140	680.3	8004.9	5749.1	19.918	57.234	22.7058	1.372
1180	720.3	8298.6	5963.3	22.626	52.152	22.9589	1.370
1220	760.3	8593.7	6178.9	25.609	47.640	23.2048	1.367
1260	800.3	8890.2	6396.0	28.886	43.620	23.4439	1.365
1300	840.3	9188.2	6614.7	32.479	40.025	23.6768	1.362
1340	880.3	9487.6	6834.7	36.411	36.802	23.9037	1.360
1380	920.3	9788.6	7056.3	40.703	33.904	24.1250	1.357
1420	960.3	10091.0	7279.2	45.382	31.290	24.3410	1.355
1460	1000.3	10394.8	7503.7	50.467	28.930	24.5520	1.353
1500	1040.3	10700.1	7729.5	55.993	26.789	24.7583	1.351
1540	1080.3	11006.8	7956.8	61.980	24.847	24.9600	1.349
1580	1120.3	11314.8	8185.4	68.460	23.079	25.1575	1.346
1620	1160.3	11624.2	8415.3	75.463	21.468	25.3509	1.344
1660	1200.3	11934.8	8646.5	83.016	19.996	25.5403	1.343
1700	1240.3	12246.8	8879.1	91.154	18.650	25.7260	1.341
1740	1280.3	12560.0	9112.8	99.909	17.416	25.9082	1.339
1780	1320.3	12874.3	9347.9	109.313	16.283	26.0868	1.337
1820	1360.3	13190.1	9584.1	119.404	15.242	26.2621	1.336
1860	1400.3	13506.8	9821.5	130.218	14.284	26.4343	1.334
1900	1440.3	13824.8	10059.9	141.791	13.400	26.6034	1.332
1940	1480.3	14143.7	10299.3	154.172	12.583	26.7696	1.331
1980	1520.3	14463.8	10540.2	167.374	11.830	26.9328	1.329
2020	1560.3	14784.7	10781.6	181.464	11.132	27.0933	1.328
2060	1600.3	15106.9	11024.3	196.483	10.484	27.2512	1.327
2100	1640.3	15429.8	11267.8	212.465	9.884	27.4065	1.325
2140	1680.3	15753.9	11512.5	229.464	9.326	27.5594	1.324
2180	1720.3	16078.8	11757.9	247.495	8.808	27.7096	1.323
2220	1760.3	16404.3	12004.1	266.670	8.325	27.8578	1.322
2260	1800.3	16731.0	12251.4	286.998	7.875	28.0036	1.321
2300	1840.3	17058.3	12499.3	308.512	7.455	28.1472	1.320
2340	1880.3	17386.4	12748.0	331.296	7.063	28.2887	1.319
2380	1920.3	17715.5	12997.7	355.367	6.697	28.4280	1.318
2420	1960.3	18045.4	13247.9	380.859	6.354	28.5655	1.317
2460	2000.3	18375.8	13498.9	407.717	6.034	28.7009	1.316

Temperature		h	u	Pr	Vr	s_o	Gamma
R	F	Btu/lbmol	Btu/lbmol			Btu/lbmol/R	
2500	2040.3	18706.8	13750.4	436.076	5.733	28.8344	1.315
2540	2080.3	19038.8	14003.3	465.990	5.451	28.9661	1.314
2580	2120.3	19371.5	14256.2	497.469	5.186	29.0960	1.313
2620	2160.3	19704.9	14510.2	530.594	4.938	29.2240	1.313
2660	2200.3	20038.5	14764.7	565.557	4.703	29.3507	1.312
2700	2240.3	20372.8	15019.5	602.250	4.483	29.4755	1.311
2740	2280.3	20708.4	15275.7	640.811	4.276	29.5988	1.310
2780	2320.3	21044.0	15531.8	681.229	4.081	29.7202	1.310
2820	2360.3	21380.2	15788.6	723.753	3.896	29.8405	1.309
2860	2400.3	21716.9	16045.9	768.283	3.723	29.9591	1.308
2900	2440.3	22054.3	16303.8	814.954	3.558	30.0762	1.308
2940	2480.3	22392.2	16562.5	863.846	3.403	30.1919	1.307
2980	2520.3	22730.5	16821.4	915.164	3.256	30.3065	1.306
3020	2560.3	23069.5	17080.2	968.642	3.118	30.4192	1.306
3060	2600.3	23409.4	17340.7	1024.76	2.986	30.5311	1.305
3100	2640.3	23749.1	17601.7	1083.25	2.862	30.6413	1.305
3140	2680.3	24090.0	17863.1	1144.37	2.744	30.7503	1.304
3180	2720.3	24430.2	18123.9	1208.42	2.632	30.8585	1.303
3220	2760.3	24772.7	18386.8	1275.04	2.525	30.9650	1.303
3260	2800.3	25113.9	18647.9	1344.65	2.424	31.0706	1.302
3300	2840.3	25456.2	18911.5	1417.230	2.328	31.1750	1.302
3340	2880.3	25799.0	19174.8	1492.974	2.237	31.2784	1.301
3380	2920.3	26141.9	19438.3	1571.546	2.151	31.3802	1.301
3420	2960.3	26485.7	19702.6	1653.649	2.068	31.4814	1.300
3460	3000.3	26829.7	19967.2	1739.133	1.989	31.5814	1.300
3500	3040.3	27175.2	20233.2	1828.024	1.915	31.6804	1.300
3540	3080.3	27519.7	20498.2	1920.853	1.843	31.7788	1.299
3580	3120.3	27863.3	20762.4	2016.791	1.775	31.8756	1.299
3620	3160.3	28210.6	21030.3	2116.627	1.710	31.9715	1.298
3660	3200.3	28555.4	21295.6	2220.483	1.648	32.0667	1.298
3700	3240.3	28902.8	21563.5	2327.995	1.589	32.1606	1.297
3740	3280.3	29248.9	21830.9	2439.306	1.533	32.2533	1.297
3780	3320.3	29597.1	22098.9	2555.792	1.479	32.3459	1.297
3820	3360.3	29943.8	22366.9	2675.917	1.428	32.4371	1.296
3860	3400.3	30290.4	22633.4	2801.559	1.378	32.5283	1.296
3900	3440.3	30640.2	22904.4	2930.566	1.331	32.6177	1.295
3940	3480.3	30986.3	23170.4	3064.678	1.286	32.7065	1.295
3980	3520.3	31337.5	23442.8	3203.413	1.242	32.7944	1.295

Temperature		h	u	Pr	Vr	s_o	Gamma
R	F	Btu/lbmol	Btu/lbmol			Btu/ lbmol/R	
4020	3560.3	31685.2	23710.3	3347.464	1.201	32.8818	1.295
4060	3600.3	32034.6	23981.0	3496.259	1.161	32.9682	1.294
4100	3640.3	32384.9	24251.2	3650.195	1.123	33.0537	1.294
4140	3680.3	32735.3	24521.4	3809.811	1.087	33.1387	1.294
4180	3720.3	33082.6	24791.3	3975.642	1.051	33.2233	1.293
4220	3760.3	33434.7	25063.3	4144.901	1.018	33.3061	1.293
4260	3800.3	33783.8	25333.7	4322.743	0.985	33.3895	1.293
4300	3840.3	34136.5	25606.2	4503.871	0.955	33.4711	1.292
4340	3880.3	34488.3	25877.8	4692.876	0.925	33.5527	1.292
4380	3920.3	34839.6	26150.3	4886.152	0.896	33.6328	1.292
4420	3960.3	35191.0	26421.6	5089.733	0.868	33.7139	1.291
4460	4000.3	35543.2	26695.1	5296.056	0.842	33.7928	1.291
4500	4040.3	35895.4	26968.5	5509.218	0.817	33.8712	1.291
4540	4080.3	36248.5	27241.5	5729.551	0.792	33.9491	1.291
4580	4120.3	36602.1	27513.4	5957.094	0.769	34.0264	1.290
4620	4160.3	36952.2	27787.6	6193.744	0.746	34.1038	1.290
4660	4200.3	37307.9	28061.8	6433.935	0.724	34.1793	1.290
4700	4240.3	37662.5	28337.6	6686.213	0.703	34.2557	1.290
4740	4280.3	38016.2	28612.6	6939.447	0.683	34.3295	1.289
4780	4320.3	38370.7	28885.5	7206.479	0.663	34.4045	1.289
4820	4360.3	38721.3	29157.4	7476.027	0.645	34.4774	1.289
4860	4400.3	39078.7	29436.0	7761.679	0.626	34.5519	1.289
4900	4440.3	39433.7	29712.2	8050.20	0.609	34.6244	1.288
4940	4480.3	39787.4	29987.2	8348.87	0.592	34.6967	1.288
4980	4520.3	40143.3	30264.4	8651.44	0.576	34.7674	1.288
5020	4560.3	40499.7	30539.2	8967.89	0.560	34.8387	1.288
5060	4600.3	40853.1	30813.9	9294.72	0.544	34.9098	1.287
5100	4640.3	41209.2	31091.2	9628.23	0.530	34.9798	1.287
5140	4680.3	41569.5	31369.9	9975.90	0.515	35.0503	1.287
5180	4720.3	41920.0	31641.7	10326.3	0.502	35.1188	1.287
5220	4760.3	42283.6	31926.5	10686.3	0.488	35.1869	1.286
5260	4800.3	42634.6	32198.8	11058.6	0.476	35.2549	1.286
5300	4840.3	42998.2	32480.8	11436.7	0.463	35.3216	1.286
5340	4880.3	43352.6	32753.6	11837.8	0.451	35.3901	1.286
5380	4920.3	43707.3	33029.6	12237.7	0.440	35.4561	1.285
5420	4960.3	44063.5	33307.1	12652.2	0.428	35.5222	1.285
5460	5000.3	44425.7	33590.6	13080.4	0.417	35.5883	1.285
5500	5040.3	44778.9	33864.9	13523.8	0.407	35.6545	1.285

Temperature		h	u	Pr	Vr	s_o	Gamma
R	F	Btu/lbmol	Btu/lbmol			Btu/ lbmol/R	
5540	5080.3	45140.7	34148.0	13970.3	0.397	35.7190	1.285
5580	5120.3	45503.7	34432.3	14431.7	0.387	35.7835	1.284
5620	5160.3	45860.1	34704.3	14902.8	0.377	35.8473	1.284
5660	5200.3	46219.4	34984.8	15386.9	0.368	35.9108	1.284
5700	5240.3	46574.9	35261.6	15895.4	0.359	35.9754	1.284
5740	5280.3	46937.0	35550.6	16402.2	0.350	36.0377	1.284
5780	5320.3	47297.3	35826.5	16926.3	0.341	36.1002	1.283
5820	5360.3	47652.4	36102.9	17458.6	0.333	36.1617	1.283
5860	5400.3	48016.0	36387.7	18015.0	0.325	36.2240	1.283
5900	5440.3	48376.2	36669.2	18567.9	0.318	36.2840	1.283
5940	5480.3	48741.1	36949.6	19152.7	0.310	36.3456	1.283
5980	5520.3	49098.1	37233.5	19748.3	0.303	36.4064	1.283
6020	5560.3	49454.6	37511.3	20352.3	0.296	36.4662	1.282
6060	5600.3	49816.3	37794.3	20969.2	0.289	36.5255	1.283
6100	5640.3	50180.0	38073.6	21615.6	0.282	36.5858	1.282
6140	5680.3	50540.0	38354.8	22238.5	0.276	36.6422	1.282
6180	5720.3	50901.3	38637.4	22929.3	0.270	36.7030	1.282

A.13.3 H_2O Properties (SI Units)

Temperature		h	u	Pr	Vr	s_o	Gamma
K	C	kJ/kgmol	kj/kgmol			kj/kgmol/K	
200	−73.1	5510.1	4124.4	0.290	1259.160	169.2769	1.332
225	−48.1	6344.3	4757.4	0.460	490.529	173.2069	1.332
250	−23.1	7179.1	5389.8	0.700	356.982	176.7249	1.331
275	1.9	8015.2	6022.5	1.030	267.626	179.9124	1.330
300	26.9	8853.5	6656.6	1.460	205.544	182.8300	1.329
325	51.9	9694.8	7293.2	2.020	161.051	185.5235	1.327
350	76.9	10539.8	7933.1	2.730	128.322	188.0283	1.325
375	101.9	11389.1	8576.9	3.620	103.711	190.3722	1.323
400	126.9	12243.3	9225.3	4.710	84.853	192.5772	1.320
425	151.9	13102.8	9878.8	6.060	70.166	194.6613	1.318
450	176.9	13967.9	10537.7	7.680	58.563	196.6392	1.315
475	201.9	14838.9	11202.3	9.640	49.284	198.5230	1.312
500	226.9	15716.2	11873.1	11.970	41.779	200.3229	1.309
525	251.9	16600.0	12550.2	14.730	35.649	202.0476	1.306

Temperature		h	u	Pr	Vr	s_o	Gamma
K	C	kJ/kgmol	kj/kgmol			kj/kgmol/K	
550	276.9	17490.4	13233.8	17.970	30.599	203.7044	1.303
575	301.9	18387.7	13924.2	21.780	26.404	205.2997	1.300
600	326.9	19291.9	14621.5	26.210	22.895	206.8390	1.297
625	351.9	20203.2	15325.8	31.340	19.941	208.3270	1.294
650	376.9	21121.8	16037.4	37.280	17.438	209.7682	1.291
675	401.9	22047.9	16756.2	44.100	15.306	211.1660	1.288
700	426.9	22981.3	17482.5	51.930	13.481	212.5239	1.285
725	451.9	23922.3	18216.3	60.870	11.911	213.8447	1.282
750	476.9	24870.9	18957.7	71.050	10.556	215.1311	1.279
775	501.9	25827.4	19706.8	82.620	9.380	216.3855	1.276
800	526.8	26791.6	20463.6	95.730	8.357	217.6099	1.273
825	551.8	27763.5	21228.2	110.550	7.463	218.8063	1.271
850	576.8	28743.5	22000.8	127.260	6.679	219.9765	1.268
875	601.8	29731.3	22781.2	146.060	5.991	221.1218	1.265
900	626.8	30727.1	23569.5	167.160	5.384	222.2440	1.262
925	651.8	31731.0	24365.9	190.810	4.848	223.3439	1.260
950	676.8	32742.7	25170.1	217.260	4.373	224.4232	1.257
975	701.8	33762.4	25982.4	246.790	3.951	225.4828	1.255
1000	726.8	34790.1	26802.5	279.710	3.575	226.5235	1.252
1025	751.8	35825.8	27630.7	316.320	3.240	227.5464	1.250
1050	776.8	36869.4	28466.6	357.010	2.941	228.5522	1.248
1075	801.8	37920.9	29310.7	402.140	2.673	229.5419	1.245
1100	826.8	38980.1	30162.4	452.140	2.433	230.5162	1.243
1125	851.8	40047.3	31022.0	507.430	2.217	231.4753	1.241
1150	876.8	41122.3	31889.3	568.520	2.023	232.4204	1.239
1175	901.8	42204.8	32764.1	635.900	1.848	233.3515	1.237
1200	926.8	43294.8	33646.7	710.130	1.690	234.2695	1.235
1225	951.8	44392.7	34536.8	791.850	1.547	235.1750	1.233
1250	976.8	45497.7	35434.4	881.630	1.418	236.0679	1.231
1275	1002	46610.2	36338.9	980.230	1.301	236.9493	1.229
1300	1027	47729.9	37251.2	1088.310	1.195	237.8188	1.227
1325	1052	48856.9	38170.4	1206.690	1.098	238.6773	1.225
1350	1077	49991.0	39096.9	1336.320	1.010	239.5256	1.224
1375	1102	51131.9	40030.1	1477.870	0.930	240.3626	1.222
1400	1127	52279.6	40969.9	1632.520	0.858	241.1900	1.220
1425	1152	53434.1	41916.8	1801.100	0.791	242.0070	1.219
1450	1177	54595.2	42870.1	1984.990	0.730	242.8153	1.217
1475	1202	55762.9	43830.1	2185.000	0.675	243.6134	1.216
1500	1227	56936.6	44796.6	2402.600	0.624	244.4027	1.214

Temperature		h	u	Pr	Vr	s_o	Gamma
K	C	kJ/kgmol	kj/kgmol			kj/kgmol/K	
1525	1252	58116.8	45768.6	2638.960	0.578	245.1828	1.213
1550	1277	59302.9	46747.5	2895.720	0.535	245.9547	1.212
1575	1302	60495.3	47732.2	3174.180	0.496	246.7180	1.210
1600	1327	61693.9	48722.7	3475.590	0.460	247.4722	1.209
1625	1352	62898.1	49719.2	3802.260	0.427	248.2190	1.208
1650	1377	64107.6	50721.0	4155.830	0.397	248.9582	1.207
1675	1402	65323.3	51729.0	4537.600	0.369	249.6889	1.206
1700	1427	66543.8	52741.4	4950.400	0.343	250.4128	1.205
1725	1452	67769.7	53760.0	5395.200	0.320	251.1281	1.204
1750	1477	69000.6	54783.2	5875.270	0.298	251.8368	1.203
1775	1502	70237.3	55811.5	6392.490	0.278	252.5382	1.202
1800	1527	71478.6	56845.0	6949.430	0.259	253.2327	1.201
1825	1552	72724.0	57883.6	7548.220	0.242	253.9198	1.200
1850	1577	73975.0	58926.9	8192.660	0.226	254.6010	1.199
1875	1602	75230.7	59974.9	8884.150	0.211	255.2746	1.198
1900	1627	76490.6	61026.3	9627.270	0.197	255.9425	1.197
1925	1652	77754.6	62083.5	10424.460	0.185	256.6039	1.196
1950	1677	79024.1	63144.5	11278.480	0.173	257.2585	1.195
1975	1702	80296.8	64209.4	12194.130	0.162	257.9074	1.195
2000	1727	81574.7	65279.6	13173.870	0.152	258.5499	1.194
2025	1752	82855.8	66353.0	14222.670	0.142	259.1868	1.193
2050	1777	84140.9	67430.4	15346.000	0.134	259.8188	1.193
2075	1802	85430.3	68512.2	16540.930	0.125	260.4421	1.192
2100	1827	86723.8	69598.0	17823.400	0.118	261.0630	1.191
2125	1852	88019.4	70685.9	19185.650	0.111	261.6753	1.191
2150	1877	89320.2	71778.9	20643.450	0.104	262.2841	1.190
2175	1902	90626.4	72875.9	22198.160	0.098	262.8878	1.189
2200	1927	91933.1	73976.5	23853.41	0.092	263.4857	1.189
2225	1952	93244.3	75078.4	25619.01	0.087	264.0793	1.188
2250	1977	94559.1	76185.5	27490.67	0.082	264.6656	1.188
2275	2002	95876.4	77295.1	29486.28	0.077	265.2482	1.187
2300	2027	97196.6	78407.7	31609.24	0.073	265.8262	1.186
2325	2052	98522.0	79526.9	33860.18	0.069	266.3981	1.186
2350	2077	99849.5	80646.8	36256.23	0.065	266.9665	1.185
2375	2102	101181.8	81769.7	38795.49	0.061	267.5293	1.185
2400	2127	102514.1	82894.4	41493.24	0.058	268.0882	1.184
2425	2152	103850.5	84023.1	44350.89	0.055	268.6419	1.184
2450	2177	105189.8	85154.6	47387.27	0.052	269.1924	1.183
2475	2202	106531.3	86288.5	50591.75	0.049	269.7364	1.183

Temperature		h	u	Pr	Vr	s_o	Gamma
K	C	kJ/kgmol	kj/kgmol			kj/kgmol/K	
2500	2227	107878.4	87427.9	53995.10	0.046	270.2777	1.183
2525	2252	109226.4	88568.2	57590.81	0.044	270.8137	1.182
2550	2277	110576.9	89711.0	61398.02	0.042	271.3459	1.182
2575	2302	111929.5	90855.9	65430.73	0.039	271.8747	1.181
2600	2327	113285.2	92005.5	69692.74	0.037	272.3994	1.181
2625	2352	114646.5	93157.5	74183.61	0.035	272.9185	1.180
2650	2377	116004.6	94309.6	78938.94	0.034	273.4351	1.180
2675	2402	117370.5	95466.2	83955.33	0.032	273.9473	1.180
2700	2427	118734.1	96623.8	89240.35	0.030	274.4548	1.179
2725	2452	120104.6	97784.9	94835.10	0.029	274.9603	1.179
2750	2477	121475.6	98946.5	100725.1	0.027	275.4613	1.178
2775	2502	122851.6	100113.2	106929.9	0.026	275.9583	1.178
2800	2527	124226.5	101282.1	113469.0	0.025	276.4517	1.178
2825	2552	125603.7	102449.9	120392.1	0.023	276.9441	1.177
2850	2577	126984.2	103624.4	127621.4	0.022	277.4289	1.177
2875	2602	128364.0	104798.1	135249.5	0.021	277.9115	1.177
2900	2627	129749.0	105973.8	143271.5	0.020	278.3906	1.176
2925	2652	131137.9	107153.4	151752.6	0.019	278.8687	1.176
2950	2677	132529.4	108338.8	160597.6	0.018	279.3397	1.176
2975	2702	133916.9	109520.3	169925.2	0.018	279.8090	1.176
3000	2727	135309.9	110704.0	179739.8	0.017	280.2758	1.175
3025	2752	136707.7	111892.5	190023.6	0.016	280.7384	1.175
3050	2777	138101.4	113080.1	200807.2	0.015	281.1973	1.175
3075	2802	139504.2	114270.3	212141.3	0.014	281.6538	1.174
3100	2827	140902.4	115465.7	224116.0	0.014	282.1103	1.174
3125	2852	142306.1	116666.6	236620.6	0.013	282.5617	1.174
3150	2877	143714.6	117859.2	249705.6	0.013	283.0092	1.173
3175	2902	145122.6	119057.9	263395.2	0.012	283.4529	1.173
3200	2927	146534.1	120260.1	277801.660	0.012	283.8956	1.173

A.13.4 H_2O Properties (English Units)

Temperature		h	u	Pr	Vr	s_o	Gamma
R	F	Btu/lbmol	Btu/lbmol			Btu/lbmol/R	
300	−159.7	2369.1	1773.3	0.138	2180.3	40.4340	1.332
340	−119.7	2687.7	2016.1	0.227	1495.8	41.4308	1.332
380	−79.7	3006.4	2258.2	0.355	1069.9	42.3172	1.332

| Temperature | | h | u | Pr | Vr | s_o | Gamma |
R	F	Btu/lbmol	Btu/lbmol			Btu/lbmol/R	
420	−39.7	3325.3	2500.0	0.531	791.26	43.1150	1.332
460	0.3	3644.4	2741.7	0.765	601.31	43.8408	1.331
500	40.3	3964.1	2983.5	1.070	467.29	44.5071	1.330
540	80.3	4284.5	3226.0	1.460	369.98	45.1236	1.329
580	120.3	4606.0	3469.2	1.949	297.60	45.6979	1.327
620	160.3	4928.7	3713.5	2.555	242.62	46.2359	1.326
660	200.3	5252.8	3959.1	3.298	200.12	46.7425	1.324
700	240.3	5578.5	4206.2	4.198	166.75	47.2216	1.321
740	280.3	5905.9	4455.0	5.279	140.19	47.6765	1.319
780	320.3	6235.2	4705.5	6.566	118.80	48.1098	1.317
820	360.3	6566.5	4957.9	8.088	101.38	48.5240	1.314
860	400.3	6899.8	5212.3	9.877	87.07	48.9208	1.312
900	440.3	7235.2	5468.8	11.968	75.20	49.3020	1.309
940	480.3	7572.8	5727.4	14.397	65.29	49.6690	1.306
980	520.3	7912.7	5988.3	17.207	56.95	50.0231	1.304
1020	560.3	8254.8	6251.4	20.444	49.89	50.3653	1.301
1060	600.3	8599.4	6516.8	24.155	43.88	50.6966	1.298
1100	640.3	8946.3	6784.7	28.397	38.74	51.0179	1.296
1140	680.3	9295.7	7054.9	33.228	34.31	51.3299	1.293
1180	720.3	9647.5	7327.6	38.712	30.48	51.6332	1.290
1220	760.3	10001.9	7602.8	44.919	27.16	51.9285	1.288
1260	800.3	10358.8	7880.6	51.925	24.27	52.2164	1.285
1300	840.3	10718.3	8160.8	59.814	21.73	52.4972	1.282
1340	880.3	11080.3	8443.7	68.674	19.51	52.7716	1.280
1380	920.3	11445.0	8729.1	78.604	17.56	53.0397	1.277
1420	960.3	11812.3	9017.2	89.707	15.83	53.3021	1.275
1460	1000.3	12182.4	9308.0	102.10	14.30	53.5591	1.272
1500	1040.3	12555.0	9601.4	115.90	12.94	53.8109	1.270
1540	1080.3	12930.4	9897.5	131.25	11.73	54.0578	1.267
1580	1120.3	13308.4	10196.2	148.29	10.66	54.3002	1.265
1620	1160.3	13689.1	10497.7	167.16	9.69	54.5382	1.262
1660	1200.3	14072.6	10801.9	188.05	8.83	54.7719	1.260
1700	1240.3	14458.7	11108.7	211.13	8.05	55.0019	1.258
1740	1280.3	14847.6	11418.3	236.59	7.35	55.2280	1.256
1780	1320.3	15239.2	11730.6	264.64	6.73	55.4504	1.253
1820	1360.3	15633.5	12045.6	295.50	6.16	55.6695	1.251
1860	1400.3	16030.5	12363.2	329.42	5.65	55.8852	1.249
1900	1440.3	16430.2	12683.6	366.64	5.182	56.0979	1.247
1940	1480.3	16832.6	13006.6	407.45	4.761	56.3074	1.245

Temperature		h	u	Pr	Vr	s_o	Gamma
R	F	Btu/lbmol	Btu/lbmol			Btu/lbmol/R	
1980	1520.3	17237.5	13332.3	452.14	4.379	56.5141	1.243
2020	1560.3	17645.2	13660.6	501.01	4.032	56.7179	1.241
2060	1600.3	18055.5	13991.6	554.41	3.716	56.9190	1.239
2100	1640.3	18468.5	14325.1	612.69	3.428	57.1175	1.237
2140	1680.3	18883.9	14661.4	676.25	3.165	57.3135	1.235
2180	1720.3	19302.0	15000.1	745.49	2.924	57.5071	1.234
2220	1760.3	19722.6	15341.4	820.84	2.705	57.6983	1.232
2260	1800.3	20145.8	15685.2	902.78	2.503	57.8873	1.230
2300	1840.3	20571.4	16031.4	991.74	2.319	58.0739	1.229
2340	1880.3	20999.5	16380.1	1088.31	2.150	58.2584	1.227
2380	1920.3	21430.1	16731.3	1193.04	1.995	58.4409	1.225
2420	1960.3	21863.0	17084.8	1306.48	1.852	58.6213	1.224
2460	2000.3	22298.3	17440.8	1429.28	1.721	58.7997	1.222
2500	2040.3	22735.9	17799.0	1562.15	1.600	58.9762	1.221
2540	2080.3	23175.9	18159.7	1705.67	1.489	59.1507	1.220
2580	2120.3	23618.1	18522.4	1860.70	1.387	59.3235	1.218
2620	2160.3	24062.5	18887.5	2027.89	1.292	59.4944	1.217
2660	2200.3	24509.1	19254.7	2208.25	1.205	59.6636	1.216
2700	2240.3	24958.0	19624.3	2402.60	1.124	59.8311	1.214
2740	2280.3	25408.8	19995.7	2611.83	1.049	59.9969	1.213
2780	2320.3	25861.9	20369.3	2836.81	0.980	60.1610	1.212
2820	2360.3	26316.9	20745.1	3078.89	0.916	60.3236	1.211
2860	2400.3	26774.0	21122.7	3338.68	0.857	60.4844	1.210
2900	2440.3	27233.1	21502.4	3617.71	0.802	60.6438	1.209
2940	2480.3	27694.1	21884.1	3917.13	0.751	60.8017	1.208
2980	2520.3	28157.1	22267.6	4238.07	0.703	60.9581	1.207
3020	2560.3	28621.8	22652.9	4581.75	0.659	61.1130	1.206
3060	2600.3	29088.6	23040.2	4950.40	0.618	61.2666	1.205
3100	2640.3	29556.9	23429.2	5344.50	0.580	61.4188	1.204
3140	2680.3	30027.3	23820.2	5765.36	0.545	61.5693	1.203
3180	2720.3	30498.9	24212.7	6216.06	0.512	61.7188	1.202
3220	2760.3	30972.8	24607.1	6696.83	0.481	61.8667	1.201
3260	2800.3	31447.9	25002.8	7210.55	0.452	62.0135	1.200
3300	2840.3	31925.1	25400.5	7758.22	0.425	62.1589	1.199
3340	2880.3	32403.6	25799.5	8342.94	0.400	62.3031	1.199
3380	2920.3	32883.8	26200.3	8964.83	0.377	62.4459	1.198
3420	2960.3	33365.2	26602.3	9627.27	0.355	62.5875	1.197
3460	3000.3	33848.5	27006.1	10332.8	0.335	62.7279	1.196
3500	3040.3	34333.3	27411.5	11083.4	0.316	62.8672	1.196

Temperature		h	u	Pr	Vr	s_o	Gamma
R	F	Btu/lbmol	Btu/lbmol			Btu/lbmol/R	
3540	3080.3	34819.3	27818.0	11881.8	0.298	63.0053	1.195
3580	3120.3	35307.1	28226.7	12730.8	0.281	63.1424	1.194
3620	3160.3	35795.5	28635.7	13632.1	0.266	63.2782	1.194
3660	3200.3	36286.0	29046.7	14587.9	0.251	63.4128	1.193
3700	3240.3	36777.7	29459.3	15603.8	0.237	63.5465	1.192
3740	3280.3	37270.8	29872.9	16681.3	0.224	63.6791	1.192
3780	3320.3	37765.0	30287.7	17823.4	0.212	63.8106	1.191
3820	3360.3	38260.4	30703.7	19031.2	0.201	63.9408	1.191
3860	3400.3	38757.5	31121.3	20314.7	0.190	64.0704	1.190
3900	3440.3	39256.3	31540.6	21672.0	0.180	64.1988	1.189
3940	3480.3	39754.2	31959.8	23104.2	0.171	64.3259	1.189
3980	3520.3	40254.5	32380.6	24626.0	0.162	64.4526	1.188
4020	3560.3	40756.8	32802.8	26228.5	0.153	64.5778	1.188
4060	3600.3	41259.2	33225.7	27925.1	0.145	64.7022	1.187
4100	3640.3	41763.0	33650.8	29712.9	0.138	64.8255	1.187
4140	3680.3	42267.9	34075.5	31609.3	0.131	64.9483	1.186
4180	3720.3	42774.2	34502.4	33603.3	0.124	65.0698	1.186
4220	3760.3	43281.6	34930.3	35710.0	0.118	65.1906	1.185
4260	3800.3	43789.6	35358.9	37931.6	0.112	65.3104	1.185
4300	3840.3	44299.2	35789.0	40277.3	0.107	65.4296	1.185
4340	3880.3	44809.7	36220.1	42746.5	0.102	65.5477	1.184
4380	3920.3	45320.6	36652.2	45340.8	0.097	65.6647	1.184
4420	3960.3	45832.4	37084.5	48086.0	0.092	65.7815	1.183
4460	4000.3	46346.1	37518.8	50963.2	0.088	65.8969	1.183
4500	4040.3	46860.5	37953.7	53994.9	0.083	66.0116	1.183
4540	4080.3	47375.0	38388.8	57190.8	0.079	66.1258	1.182
4580	4120.3	47891.9	38826.2	60542.9	0.076	66.2389	1.182
4620	4160.3	48409.0	39263.8	64063.8	0.072	66.3512	1.181
4660	4200.3	48927.8	39702.5	67771.4	0.069	66.4629	1.181
4700	4240.3	49445.7	40141.6	71653.6	0.066	66.5735	1.181
4740	4280.3	49964.5	40581.7	75740.5	0.063	66.6837	1.180
4780	4320.3	50485.9	41023.0	80033.8	0.060	66.7932	1.180
4820	4360.3	51006.9	41465.2	84536.6	0.057	66.9019	1.180
4860	4400.3	51527.9	41907.5	89240.3	0.054	67.0094	1.179
4900	4440.3	52051.3	42350.8	94201.3	0.052	67.1168	1.179
4940	4480.3	52575.6	42794.9	99387.8	0.050	67.2233	1.179
4980	4520.3	53100.7	43241.2	104828	0.048	67.3291	1.178
5020	4560.3	53625.6	43687.4	110520	0.045	67.4341	1.178
5060	4600.3	54151.9	44133.5	116483	0.043	67.5385	1.178

Temperature		h	u	Pr	Vr	s_o	Gamma
R	F	Btu/lbmol	Btu/lbmol			Btu/lbmol/R	
5100	4640.3	54679.0	44581.9	122741	0.042	67.6424	1.177
5140	4680.3	55207.7	45030.4	129266	0.040	67.7452	1.177
5180	4720.3	55735.8	45478.4	136123	0.038	67.8479	1.177
5220	4760.3	56263.8	45927.6	143271	0.036	67.9495	1.176
5260	4800.3	56795.0	46378.6	150767	0.035	68.0508	1.176
5300	4840.3	57325.7	46830.5	158623	0.033	68.1516	1.176
5340	4880.3	57857.0	47281.7	166780	0.032	68.2512	1.176
5380	4920.3	58388.1	47734.1	175311	0.031	68.3503	1.175
5420	4960.3	58921.9	48189.1	184290	0.029	68.4495	1.175
5460	5000.3	59454.2	48641.3	193579	0.028	68.5471	1.175
5500	5040.3	59989.7	49098.0	203309	0.027	68.6445	1.175
5540	5080.3	60523.7	49551.9	213526	0.026	68.7419	1.174
5580	5120.3	61059.2	50008.6	224116	0.025	68.8380	1.174
5620	5160.3	61597.8	50465.0	235132	0.024	68.9333	1.174
5660	5200.3	62135.0	50924.1	246738	0.023	69.0290	1.173
5700	5240.3	62670.5	51380.8	258763	0.022	69.1235	1.173
5740	5280.3	63210.0	51838.8	271347	0.021	69.2178	1.173
5780	5320.3	63746.5	52299.4	284416	0.020	69.3112	1.173

A.13.5 CO_2 Properties (SI Units)

Temperature		h	u	Pr	Vr	s_o	Gamma
K	C	kJ/kgmol	kJ/kgmol			kJ/kgmol/K	
K	C	kJ/kgmol	kJ/kgmol			kJ/kgmol/K	
200	−73.1	4898.9	3513.4	0.28	1292.15	194.0903	1.345
225	−48.1	5723.6	4168.1	0.44	506.17	197.9744	1.329
250	−23.1	6579.3	4847.3	0.69	364.53	201.5795	1.313
275	1.9	7465.7	5552.2	1.03	267.07	204.9580	1.300
300	26.9	8382.0	6283.1	1.51	198.55	208.1462	1.288
325	51.9	9326.8	7039.4	2.17	149.50	211.1705	1.277
350	76.9	10298.7	7820.2	3.07	113.86	214.0510	1.267
375	101.9	11296.2	8624.6	4.28	87.607	216.8035	1.259
400	126.9	12318.0	9451.5	5.88	68.045	219.4409	1.252
425	151.9	13362.7	10299.9	7.97	53.308	221.9741	1.245
450	176.9	14429.0	11168.8	10.69	42.099	224.4118	1.239
475	201.9	15516.0	12057.2	14.18	33.493	226.7625	1.234
500	226.9	16622.5	12964.2	18.63	26.832	229.0325	1.229

Temperature		h	u	Pr	Vr	s_o	Gamma
K	C	kJ/kgmol	kJ/kgmol			kJ/kgmol/K	
525	251.9	17747.4	13889.0	24.27	21.635	231.2279	1.224
550	276.9	18890.0	14830.6	31.34	17.551	233.3538	1.220
575	301.9	20049.4	15788.5	40.15	14.320	235.4150	1.217
600	326.9	21224.9	16762.0	51.08	11.746	237.4160	1.213
625	351.9	22415.5	17750.2	64.54	9.684	239.3600	1.210
650	376.9	23620.8	18752.7	81.02	8.023	241.2508	1.207
675	401.9	24840.1	19768.7	101.10	6.677	243.0914	1.204
700	426.9	26072.5	20797.6	125.43	5.581	244.8847	1.202
725	451.9	27317.8	21839.1	154.78	4.684	246.6325	1.199
750	476.9	28575.4	22892.7	190.02	3.947	248.3377	1.197
775	501.9	29844.5	23957.4	232.14	3.339	250.0022	1.195
800	526.8	31124.9	25033.3	282.27	2.834	251.6279	1.193
825	551.8	32415.8	26119.5	341.74	2.414	253.2172	1.191
850	576.8	33717.0	27215.8	411.95	2.063	254.7706	1.189
875	601.8	35027.9	28321.6	494.59	1.769	256.2906	1.188
900	626.8	36348.2	29436.6	591.50	1.522	257.7783	1.186
925	651.8	37677.6	30560.6	704.80	1.312	259.2352	1.185
950	676.8	39015.2	31692.8	836.76	1.135	260.6620	1.183
975	701.8	40361.3	32833.2	990.10	0.985	262.0610	1.182
1000	726.8	41715.3	33981.6	1167.6	0.856	263.4322	1.181
1025	751.8	43076.4	35137.1	1372.6	0.747	264.7767	1.180
1050	776.8	44445.2	36299.4	1608.6	0.653	266.0959	1.179
1075	801.8	45820.7	37468.8	1879.7	0.572	267.3905	1.178
1100	826.8	47202.7	38644.7	2190.0	0.502	268.6610	1.177
1125	851.8	48591.4	39827.4	2544.8	0.442	269.9092	1.176
1150	876.8	49985.1	41015.0	2949.0	0.390	271.1346	1.175
1175	901.8	51385.5	42209.0	3408.9	0.345	272.3395	1.174
1200	926.8	52791.4	43408.4	3930.4	0.305	273.5230	1.173
1225	951.8	54201.5	44612.9	4521.0	0.271	274.6868	1.172
1250	976.8	55618.1	45822.6	5188.2	0.241	275.8313	1.172
1275	1001.8	57039.4	47037.0	5940.2	0.215	276.9566	1.171
1300	1026.8	58464.9	48255.7	6786.6	0.192	278.0639	1.170
1325	1051.8	59895.8	49479.7	7737.8	0.171	279.1545	1.170
1350	1076.8	61329.7	50707.6	8803.1	0.153	280.2268	1.169
1375	1101.8	62768.5	51939.5	9994.9	0.138	281.2824	1.169
1400	1126.8	64211.8	53175.9	11326.2	0.124	282.3220	1.168
1425	1151.8	65658.4	54415.7	12812.8	0.111	283.3473	1.168
1450	1176.8	67108.3	55658.6	14466.2	0.100	284.3563	1.167
1475	1201.8	68563.2	56906.7	16304.6	0.090	285.3509	1.167
1500	1226.8	70020.9	58157.6	18344.6	0.082	286.3310	1.166

Temperature		h	u	Pr	Vr	s_o	Gamma
K	C	kJ/kgmol	kJ/kgmol			kJ/kgmol/K	
1525	1251.8	71481.7	59410.6	20600.9	0.074	287.2954	1.166
1550	1276.8	72945.4	60668.3	23101.9	0.067	288.2480	1.165
1575	1301.8	74412.2	61927.4	25868.1	0.061	289.1882	1.165
1600	1326.8	75883.0	63192.1	28916.1	0.055	290.1142	1.164
1625	1351.8	77355.4	64456.8	32270.5	0.050	291.0267	1.164
1650	1376.8	78831.3	65726.7	35969.4	0.046	291.9289	1.164
1675	1401.8	80312.5	67000.2	40024.4	0.042	292.8169	1.163
1700	1426.8	81791.3	68272.9	44496.2	0.038	293.6975	1.163
1725	1451.8	83278.7	69551.0	49369.3	0.035	294.5615	1.163
1750	1476.8	84762.2	70828.5	54730.5	0.032	295.4185	1.162
1775	1501.8	86252.2	72110.8	60579.4	0.029	296.2627	1.162
1800	1526.8	87742.6	73395.1	66965.8	0.027	297.0959	1.162
1825	1551.8	89235.2	74680.0	73954.8	0.025	297.9212	1.162
1850	1576.8	90731.9	75972.3	81574.9	0.023	298.7365	1.161
1875	1601.8	92231.5	77262.6	89849.9	0.021	299.5398	1.161
1900	1626.8	93730.0	78555.0	98864.0	0.019	300.3346	1.161
1925	1651.8	95230.1	79845.8	108640.8	0.018	301.1186	1.160
1950	1676.8	96736.7	81146.4	119300.4	0.016	301.8968	1.160
1975	1701.8	98244.5	82444.8	130839.1	0.015	302.6643	1.160
2000	1726.8	99751.9	83746.2	143305.3	0.014	303.4209	1.160
2025	1751.8	101259.1	85044.1	156882.2	0.013	304.1735	1.160
2050	1776.8	102767.9	86346.8	171483.8	0.012	304.9133	1.159
2075	1801.8	104283.4	87653.0	187278.1	0.011	305.6458	1.159
2100	1826.8	105797.3	88960.8	204390.5	0.010	306.3727	1.159
2125	1851.8	107314.4	90271.8	222790.0	0.010	307.0894	1.159
2150	1876.8	108833.1	91581.2	242630.1	0.009	307.7986	1.159
2175	1901.8	110353.8	92895.8	264062.5	0.008	308.5023	1.158
2200	1926.8	111871.4	94204.1	287095.1	0.008	309.1976	1.158
2225	1951.8	113391.6	95521.6	311872.0	0.007	309.8858	1.158
2250	1976.8	114915.7	96836.3	338455.4	0.007	310.5658	1.158
2275	2001.8	116442.1	98153.4	367187.4	0.006	311.2432	1.158
2300	2026.8	117966.3	99471.5	397703.9	0.006	311.9070	1.158
2325	2051.9	119491.0	100793.5	430530.6	0.005	312.5663	1.157
2350	2076.9	121023.1	102109.7	465834.2	0.005	313.2215	1.157
2375	2101.9	122555.9	103439.7	503490.9	0.005	313.8678	1.157
2400	2126.9	124083.5	104758.0	544061.8	0.004	314.5121	1.157
2425	2151.9	125613.1	106078.3	587250.6	0.004	315.1472	1.157
2450	2176.9	127156.1	107418.5	633421.1	0.004	315.7764	1.157
2475	2201.9	128681.3	108734.4	682687.7	0.004	316.3991	1.157
2500	2226.9	130225.1	110068.8	735216.5	0.003	317.0154	1.156

Temperature		h	u	Pr	Vr	s_o	Gamma
K	C	kJ/kgmol	kJ/kgmol			kJ/kgmol/K	
2525	2251.9	131764.7	111399.1	791136.3	0.003	317.6248	1.156
2550	2276.9	133298.1	112723.1	851306.1	0.003	318.2342	1.156
2575	2301.9	134830.4	114052.7	914842.4	0.003	318.8326	1.156
2600	2326.9	136372.1	115385.0	982925.9	0.003	319.4294	1.156
2625	2351.9	137910.9	116721.1	1054793.8	0.002	320.0161	1.156
2650	2376.9	139455.6	118056.4	1132149.1	0.002	320.6044	1.156
2675	2401.9	140996.3	119387.8	1213719.6	0.002	321.1828	1.156
2700	2426.9	142542.7	120731.5	1299797.3	0.002	321.7525	1.155
2725	2451.9	144091.2	122070.6	1393380.3	0.002	322.3305	1.155
2750	2476.9	145633.1	123403.2	1491186.5	0.002	322.8945	1.155
2775	2501.9	147180.6	124741.3	1594517.1	0.002	323.4515	1.155
2800	2526.9	148735.4	126086.8	1705346.0	0.002	324.0101	1.155
2825	2551.9	150272.8	127428.0	1821001.6	0.002	324.5557	1.155
2850	2576.9	151831.0	128776.9	1945136.3	0.001	325.1039	1.155
2875	2601.9	153381.2	130104.7	2077527.0	0.001	325.6513	1.155
2900	2626.9	154935.0	131462.2	2215282.5	0.001	326.1851	1.155
2925	2651.9	156479.4	132797.3	2363265.3	0.001	326.7227	1.155
2950	2676.9	158032.4	134141.0	2516307.0	0.001	327.2444	1.154
2975	2701.9	159594.8	135494.0	2681335.5	0.001	327.7725	1.154
3000	2726.9	161139.8	136842.8	2854293.8	0.001	328.2921	1.154
3025	2751.9	162703.6	138197.3	3035087.0	0.001	328.8027	1.154
3050	2776.9	164262.6	139533.9	3231907.8	0.001	329.3251	1.154
3075	2801.9	165809.9	140884.9	3433278.0	0.001	329.8276	1.154
3100	2826.9	167368.2	142233.9	3647235.0	0.001	330.3302	1.154
3125	2851.9	168937.7	143594.1	3874568.3	0.001	330.8329	1.154
3150	2876.9	170504.6	144951.7	4115300.8	0.001	331.3340	1.154
3175	2901.9	172053.7	146304.5	4366548.0	0.001	331.8267	1.154
3200	2926.9	173619.6	147648.0	4633662.0	0.001	332.3203	1.154

A.13.6 CO_2 Properties (English Units)

Temperature		h	u	Pr	Vr	s_o	Gamma
R	F	Btu/lbmol	Btu/lbmol			Btu/ lbmol/R	
300	−159.7	2106.3	1510.6	0.139	2152.4	46.3610	1.369
340	−119.7	2405.9	1749.9	0.223	1521.8	47.2980	1.353
380	−79.7	2715.2	1995.7	0.345	1103.0	48.1580	1.338
420	−39.7	3035.0	2249.4	0.515	814.94	48.9579	1.323
460	0.3	3365.3	2511.7	0.752	611.48	49.7090	1.310

Temperature		h	u	Pr	Vr	s_o	Gamma
R	F	Btu/lbmol	Btu/lbmol			Btu/ lbmol/R	
500	40.3	3706.0	2782.8	1.076	464.84	50.4190	1.298
540	80.3	4056.8	3062.7	1.511	357.39	51.0939	1.288
580	120.3	4417.3	3351.2	2.090	277.56	51.7378	1.278
620	160.3	4787.1	3648.1	2.850	217.53	52.3541	1.269
660	200.3	5165.6	3953.0	3.839	171.90	52.9458	1.262
700	240.3	5552.6	4265.8	5.114	136.89	53.5150	1.255
740	280.3	5947.6	4586.0	6.741	109.78	54.0636	1.249
780	320.3	6350.1	4913.3	8.802	88.62	54.5933	1.243
820	360.3	6759.9	5247.5	11.392	71.98	55.1057	1.238
860	400.3	7176.6	5588.2	14.626	58.80	55.6018	1.233
900	440.3	7599.8	5935.3	18.635	48.30	56.0828	1.229
940	480.3	8029.4	6288.3	23.574	39.87	56.5498	1.225
980	520.3	8465.0	6647.1	29.626	33.08	57.0036	1.221
1020	560.3	8906.3	7011.6	37.001	27.57	57.4450	1.218
1060	600.3	9353.2	7381.3	45.939	23.07	57.8746	1.215
1100	640.3	9805.3	7756.2	56.722	19.39	58.2933	1.212
1140	680.3	10262.6	8136.0	69.673	16.36	58.7017	1.209
1180	720.3	10724.8	8520.6	85.149	13.86	59.1001	1.206
1220	760.3	11191.6	8909.7	103.578	11.78	59.4891	1.204
1260	800.3	11662.9	9303.3	125.433	10.05	59.8693	1.202
1300	840.3	12138.6	9701.1	151.242	8.596	60.2409	1.199
1340	880.3	12618.4	10102.9	181.629	7.378	60.6045	1.197
1380	920.3	13102.2	10508.7	217.266	6.352	60.9602	1.195
1420	960.3	13589.9	10918.2	258.923	5.484	61.3086	1.194
1460	1000.3	14081.3	11331.3	307.466	4.748	61.6498	1.192
1500	1040.3	14576.3	11748.0	363.861	4.122	61.9842	1.190
1540	1080.3	15074.6	12168.0	429.178	3.588	62.3121	1.189
1580	1120.3	15576.2	12591.3	504.618	3.131	62.6336	1.187
1620	1160.3	16080.9	13017.6	591.502	2.739	62.9491	1.186
1660	1200.3	16588.8	13447.0	691.365	2.401	63.2589	1.185
1700	1240.3	17099.6	13879.2	805.734	2.110	63.5629	1.184
1740	1280.3	17613.1	14314.1	936.428	1.858	63.8614	1.182
1780	1320.3	18129.4	14751.8	1085.49	1.640	64.1548	1.181
1820	1360.3	18648.2	15192.0	1255.08	1.450	64.4430	1.180
1860	1400.3	19169.6	15634.8	1447.55	1.285	64.7264	1.179
1900	1440.3	19693.5	16079.8	1665.6	1.141	65.0049	1.178
1940	1480.3	20219.4	16527.1	1912.1	1.015	65.2791	1.177
1980	1520.3	20747.9	16976.7	2190.0	0.904	65.5486	1.177

Temperature		h	u	Pr	Vr	s_o	Gamma
R	F	Btu/lbmol	Btu/lbmol			Btu/ lbmol/R	
2020	1560.3	21278.2	17428.2	2503.0	0.807	65.8138	1.176
2060	1600.3	21810.7	17882.0	2854.7	0.722	66.0749	1.175
2100	1640.3	22345.2	18337.8	3248.9	0.646	66.3318	1.174
2140	1680.3	22881.5	18795.1	3690.6	0.58	66.5849	1.173
2180	1720.3	23420.2	19254.7	4183.8	0.521	66.8340	1.173
2220	1760.3	23959.9	19715.7	4734.5	0.469	67.0796	1.172
2260	1800.3	24501.8	20178.5	5347.8	0.423	67.3215	1.172
2300	1840.3	25045.1	20643.0	6029.7	0.381	67.5598	1.171
2340	1880.3	25590.1	21108.9	6786.6	0.345	67.7946	1.170
2380	1920.3	26136.7	21576.7	7626.4	0.312	68.0263	1.170
2420	1960.3	26684.6	22045.6	8555.0	0.283	68.2545	1.169
2460	2000.3	27234.1	22515.9	9583.4	0.257	68.4799	1.169
2500	2040.3	27785.2	22988.0	10716.0	0.233	68.7017	1.168
2540	2080.3	28337.1	23461.1	11966.3	0.212	68.9209	1.168
2580	2120.3	28890.4	23935.0	13342.3	0.193	69.1370	1.167
2620	2160.3	29445.2	24411.0	14855.7	0.176	69.3504	1.167
2660	2200.3	30001.2	24887.9	16518.6	0.161	69.5611	1.166
2700	2240.3	30558.6	25366.2	18344.6	0.147	69.7693	1.166
2740	2280.3	31116.4	25845.2	20339.8	0.135	69.9743	1.166
2780	2320.3	31676.4	26325.8	22525.3	0.123	70.1770	1.165
2820	2360.3	32236.2	26806.9	24913.5	0.113	70.3771	1.165
2860	2400.3	32797.9	27289.1	27520.3	0.104	70.5747	1.165
2900	2440.3	33360.1	27772.5	30363.9	0.096	70.7700	1.164
2940	2480.3	33923.7	28256.7	33463.0	0.088	70.9630	1.164
2980	2520.3	34488.2	28742.4	36840.5	0.081	71.1540	1.164
3020	2560.3	35053.5	29228.3	40502.2	0.075	71.3421	1.163
3060	2600.3	35619.3	29715.4	44496.3	0.069	71.5289	1.163
3100	2640.3	36186.2	30202.8	48807.2	0.064	71.7125	1.163
3140	2680.3	36754.6	30691.8	53499.3	0.059	71.8948	1.162
3180	2720.3	37323.8	31182.2	58565.9	0.054	72.0745	1.162
3220	2760.3	37894.3	31673.2	64062.9	0.05	72.2526	1.162
3260	2800.3	38464.2	32163.7	70006.4	0.047	72.4288	1.162
3300	2840.3	39035.2	32655.3	76418.2	0.043	72.6029	1.161
3340	2880.3	39606.3	33147.6	83339.1	0.04	72.7750	1.161
3380	2920.3	40180.1	33642.6	90818.4	0.037	72.9457	1.161
3420	2960.3	40752.4	34136.2	98864.0	0.035	73.1143	1.161
3460	3000.3	41326.5	34630.1	107520.6	0.032	73.2809	1.161
3500	3040.3	41900.9	35125.8	116863.0	0.030	73.4464	1.160

Temperature		h	u	Pr	Vr	s_o	Gamma
R	F	Btu/lbmol	Btu/lbmol			Btu/ lbmol/R	
3540	3080.3	42476.5	35622.6	126858.2	0.028	73.6094	1.160
3580	3120.3	43051.5	36117.4	137653.7	0.026	73.7716	1.160
3620	3160.3	43629.7	36616.9	149204.5	0.024	73.9316	1.160
3660	3200.3	44205.8	37112.8	161585.4	0.023	74.0899	1.160
3700	3240.3	44783.2	37611.5	174927.8	0.021	74.2474	1.159
3740	3280.3	45361.9	38111.4	189169.0	0.020	74.4029	1.159
3780	3320.3	45940.8	38610.2	204391.3	0.018	74.5565	1.159
3820	3360.3	46519.2	39109.8	220690.0	0.017	74.7089	1.159
3860	3400.3	47100.7	39611.2	238098.2	0.016	74.8597	1.159
3900	3440.3	47680.2	40112.0	256730.5	0.015	75.0093	1.158
3940	3480.3	48261.0	40614.0	276623.2	0.014	75.1575	1.158
3980	3520.3	48843.2	41117.5	297891.2	0.013	75.3046	1.158
4020	3560.3	49423.9	41617.9	320487.7	0.013	75.4498	1.158
4060	3600.3	50007.8	42123.1	344603.6	0.012	75.5939	1.158
4100	3640.3	50588.8	42624.0	370381.8	0.011	75.7371	1.158
4140	3680.3	51172.8	43129.3	397704.7	0.010	75.8785	1.158
4180	3720.3	51756.2	43633.9	426811.9	0.010	76.0187	1.157
4220	3760.3	52341.1	44140.0	457729.2	0.009	76.1576	1.157
4260	3800.3	52925.3	44642.7	490688.2	0.009	76.2957	1.157
4300	3840.3	53509.1	45150.5	525804.1	0.008	76.4329	1.157
4340	3880.3	54095.4	45655.2	562836.0	0.008	76.5681	1.157
4380	3920.3	54680.1	46161.2	602004.0	0.007	76.7017	1.157
4420	3960.3	55268.8	46668.4	644330.8	0.007	76.8366	1.157
4460	4000.3	55853.4	47174.2	688207.0	0.006	76.9675	1.157
4500	4040.3	56443.5	47685.6	735215.0	0.006	77.0987	1.156
4540	4080.3	57030.7	48194.0	785035.6	0.006	77.2289	1.156
4580	4120.3	57616.4	48700.9	837485.5	0.005	77.3573	1.156
4620	4160.3	58205.2	49208.2	893302.4	0.005	77.4854	1.156
4660	4200.3	58793.7	49717.9	952144.0	0.005	77.6121	1.156
4700	4240.3	59383.1	50228.6	1014058.7	0.005	77.7372	1.156
4740	4280.3	59971.4	50738.1	1080090.8	0.004	77.8625	1.156
4780	4320.3	60560.5	51248.5	1149724.5	0.004	77.9866	1.156
4820	4360.3	61147.8	51757.1	1223372.8	0.004	78.1099	1.156
4860	4400.3	61739.5	52270.0	1299799.6	0.004	78.2302	1.155
4900	4440.3	62331.9	52783.6	1382923.1	0.004	78.3533	1.155
4940	4480.3	62921.9	53292.1	1468143.8	0.003	78.4720	1.155
4980	4520.3	63513.1	53807.4	1559496.6	0.003	78.5919	1.155
5020	4560.3	64102.9	54312.8	1655452.1	0.003	78.7105	1.155

Temperature		h	u	Pr	Vr	s_o	Gamma
R	F	Btu/lbmol	Btu/lbmol			Btu/lbmol/R	
5060	4600.3	64698.0	54829.1	1755773.1	0.003	78.8273	1.155
5100	4640.3	65286.3	55338.7	1862446	0.003	78.9445	1.155
5140	4680.3	65878.2	55851.9	1973642	0.003	79.0596	1.155
5180	4720.3	66474.8	56369.7	2091677	0.002	79.1750	1.155
5220	4760.3	67067.6	56883.7	2215282	0.002	79.2890	1.155
5260	4800.3	67664.1	57401.5	2345174	0.002	79.4021	1.154
5300	4840.3	68253.5	57912.1	2480459	0.002	79.5135	1.154
5340	4880.3	68847.6	58427.5	2625082	0.002	79.6260	1.154
5380	4920.3	69441.6	58937.1	2776344	0.002	79.7373	1.154
5420	4960.3	70032.3	59449.0	2934292	0.002	79.8472	1.154
5460	5000.3	70626.9	59964.9	3101107	0.002	79.9570	1.154
5500	5040.3	71225.6	60490.4	3274133	0.002	80.0648	1.154
5540	5080.3	71821.0	61001.5	3457955	0.002	80.1733	1.154
5580	5120.3	72413.3	61515.0	3647232	0.002	80.2791	1.154
5620	5160.3	73007.7	62030.7	3851144	0.001	80.3871	1.154
5660	5200.3	73615.6	62554.3	4063894	0.001	80.4939	1.154
5700	5240.3	74205.5	63071.0	4279592	0.001	80.5966	1.154
5740	5280.3	74808.4	63589.5	4513548	0.001	80.7023	1.154
5780	5320.3	75401.4	64103.8	4753843	0.001	80.8053	1.153

A.13.7 Nitrogen Properties (SI Units)

Temperature		h	u	Pr	Vr	s_o	Gamma
K	C	kJ/kgmol	kJ/kgmol			kJ/kgmol/K	
200	−73.1	4842.2	3456.7	0.34	1072.9	174.5226	1.399
225	−48.1	5570.4	3981.6	0.51	443.82	177.9535	1.399
250	−23.1	6298.6	4505.5	0.73	340.92	181.0223	1.400
275	1.9	7026.5	5028.6	1.02	268.58	183.7977	1.400
300	26.9	7754.4	5551.2	1.39	216.03	186.3310	1.400
325	51.9	8482.4	6073.4	1.84	176.82	188.6617	1.400
350	76.9	9210.7	6595.7	2.38	146.87	190.8208	1.399
375	101.9	9939.8	7118.5	3.04	123.53	192.8329	1.398
400	126.9	10670.1	7642.3	3.81	105.04	194.7179	1.397
425	151.9	11401.8	8167.4	4.71	90.150	196.4925	1.396
450	176.9	12135.5	8694.3	5.77	78.014	198.1698	1.394
475	201.9	12871.4	9223.4	6.99	68.000	199.7613	1.393

Temperature		h	u	Pr	Vr	s_o	Gamma
K	C	kJ/kgmol	kJ/kgmol			kJ/kgmol/K	
500	226.9	13609.9	9754.9	8.38	59.654	201.2764	1.391
525	251.9	14351.2	10289.2	9.97	52.632	202.7233	1.388
550	276.9	15095.7	10826.5	11.78	46.675	204.1086	1.386
575	301.9	15843.5	11367.2	13.83	41.584	205.4382	1.384
600	326.9	16594.9	11911.4	16.13	37.204	206.7174	1.381
625	351.9	17350.1	12459.2	18.71	33.413	207.9503	1.378
650	376.9	18109.0	13010.8	21.59	30.112	209.1411	1.376
675	401.9	18872.0	13566.4	24.79	27.225	210.2928	1.373
700	426.9	19639.0	14126.0	28.35	24.688	211.4086	1.370
725	451.9	20410.0	14689.6	32.30	22.448	212.4909	1.368
750	476.9	21185.3	15257.4	36.65	20.464	213.5421	1.365
775	501.9	21964.7	15829.3	41.44	18.700	214.5643	1.362
800	526.8	22748.2	16405.3	46.71	17.125	215.5594	1.360
825	551.8	23535.9	16985.5	52.49	15.717	216.5289	1.357
850	576.8	24327.7	17569.7	58.81	14.452	217.4743	1.355
875	601.8	25123.5	18158.0	65.72	13.314	218.3971	1.352
900	626.8	25923.4	18750.3	73.24	12.288	219.2984	1.350
925	651.8	26727.3	19346.6	81.43	11.359	220.1794	1.348
950	676.8	27534.9	19946.6	90.32	10.518	221.0410	1.345
975	701.8	28346.4	20550.5	99.96	9.754	221.8840	1.343
1000	726.8	29161.6	21158.1	110.40	9.058	222.7097	1.341
1025	751.8	29980.4	21769.3	121.68	8.424	223.5184	1.339
1050	776.8	30802.9	22384.0	133.85	7.844	224.3111	1.337
1075	801.8	31628.7	23002.2	146.97	7.314	225.0885	1.335
1100	826.8	32458.0	23623.8	161.09	6.828	225.8510	1.334
1125	851.8	33290.4	24248.6	176.26	6.383	226.5993	1.332
1150	876.8	34126.0	24876.5	192.54	5.973	227.3338	1.330
1175	901.8	34964.7	25507.5	210.00	5.595	228.0554	1.329
1200	926.8	35806.4	26141.4	228.69	5.247	228.7641	1.327
1225	951.8	36650.9	26778.3	248.67	4.926	229.4606	1.326
1250	976.8	37498.2	27417.9	270.02	4.629	230.1453	1.324
1275	1002	38348.2	28060.2	292.80	4.355	230.8186	1.323
1300	1027	39200.7	28704.9	317.08	4.100	231.4809	1.322
1325	1052	40055.8	29352.3	342.93	3.864	232.1324	1.321
1350	1077	40913.3	30002.1	370.42	3.645	232.7735	1.319
1375	1102	41773.1	30654.2	399.63	3.441	233.4045	1.318
1400	1127	42635.2	31308.5	430.64	3.251	234.0258	1.317
1425	1152	43499.3	31964.9	463.53	3.074	234.6377	1.316
1450	1177	44365.7	32623.6	498.38	2.909	235.2404	1.315

Temperature		h	u	Pr	Vr	s_o	Gamma
K	C	kJ/kgmol	kJ/kgmol			kJ/kgmol/K	
1475	1202	45234.0	33284.2	535.27	2.756	235.8341	1.314
1500	1227	46104.4	33946.7	574.30	2.612	236.4192	1.313
1525	1252	46976.4	34611.0	615.54	2.477	236.9958	1.312
1550	1277	47850.4	35277.3	659.11	2.352	237.5643	1.312
1575	1302	48726.0	35945.2	705.07	2.234	238.1247	1.311
1600	1327	49603.5	36614.8	753.54	2.123	238.6775	1.310
1625	1352	50482.5	37285.9	804.59	2.020	239.2224	1.309
1650	1377	51363.2	37958.9	858.34	1.922	239.7601	1.309
1675	1402	52245.1	38633.2	914.92	1.831	240.2907	1.308
1700	1427	53128.6	39308.8	974.39	1.745	240.8144	1.307
1725	1452	54013.4	39985.9	1036.90	1.664	241.3313	1.307
1750	1477	54899.8	40664.5	1102.50	1.587	241.8413	1.306
1775	1502	55787.6	41344.7	1171.36	1.515	242.3450	1.305
1800	1527	56676.6	42025.8	1243.57	1.447	242.8423	1.305
1825	1552	57566.8	42708.1	1319.26	1.383	243.3335	1.304
1850	1577	58458.1	43391.7	1398.50	1.323	243.8184	1.304
1875	1602	59350.7	44076.7	1481.52	1.266	244.2979	1.303
1900	1627	60244.3	44762.3	1568.34	1.211	244.7713	1.303
1925	1652	61139.2	45449.6	1659.15	1.160	245.2393	1.302
1950	1677	62035.0	46137.7	1754.00	1.112	245.7015	1.302
1975	1702	62932.4	46827.0	1853.15	1.066	246.1586	1.301
2000	1727	63830.5	47517.4	1956.68	1.022	246.6106	1.301
2025	1752	64729.4	48208.6	2064.67	0.981	247.0572	1.300
2050	1777	65629.4	48900.8	2177.38	0.942	247.4991	1.300
2075	1802	66530.4	49593.8	2294.82	0.904	247.9358	1.300
2100	1827	67432.2	50287.9	2417.25	0.869	248.3679	1.299
2125	1852	68335.2	50983.2	2544.70	0.835	248.7951	1.299
2150	1877	69238.7	51679.0	2677.57	0.803	249.2182	1.298
2175	1902	70143.8	52376.0	2815.68	0.772	249.6364	1.298
2200	1927	71049.1	53073.7	2959.36	0.743	250.0501	1.298
2225	1952	71955.4	53772.2	3108.89	0.716	250.4599	1.297
2250	1977	72862.1	54471.2	3264.18	0.689	250.8652	1.297
2275	2002	73770.2	55171.6	3425.64	0.664	251.2666	1.297
2300	2027	74679.0	55872.4	3593.33	0.640	251.6639	1.296
2325	2052	75588.3	56574.0	3767.45	0.617	252.0573	1.296
2350	2077	76498.6	57276.6	3948.23	0.595	252.4469	1.296
2375	2102	77409.8	57979.6	4135.53	0.574	252.8322	1.295
2400	2127	78321.5	58683.6	4329.93	0.554	253.2141	1.295
2425	2152	79233.6	59388.1	4531.69	0.535	253.5928	1.295

Temperature		h	u	Pr	Vr	s_o	Gamma
K	C	kJ/kgmol	kJ/kgmol			kJ/kgmol/K	
2450	2177	80147.2	60093.6	4740.30	0.517	253.9669	1.295
2475	2202	81061.3	60799.6	4956.64	0.499	254.3380	1.294
2500	2227	81974.9	61506.3	5180.90	0.483	254.7058	1.294
2525	2252	82889.7	62213.5	5412.57	0.467	255.0695	1.294
2550	2277	83806.5	62921.7	5653.13	0.451	255.4310	1.293
2575	2302	84722.2	63629.7	5901.50	0.436	255.7885	1.293
2600	2327	85639.7	64339.5	6158.56	0.422	256.1430	1.293
2625	2352	86557.4	65048.8	6424.15	0.409	256.4940	1.293
2650	2377	87474.6	65759.0	6698.76	0.396	256.8420	1.292
2675	2402	88394.1	66470.0	6982.77	0.383	257.1872	1.292
2700	2427	89313.6	67181.9	7276.13	0.371	257.5293	1.292
2725	2452	90232.9	67893.5	7579.28	0.360	257.8687	1.292
2750	2477	91153.4	68606.3	7891.81	0.348	258.2046	1.292
2775	2502	92074.3	69318.7	8214.82	0.338	258.5381	1.291
2800	2527	92995.6	70033.0	8547.82	0.328	258.8685	1.291
2825	2552	93917.5	70746.5	8891.44	0.318	259.1961	1.291
2850	2577	94840.0	71461.3	9246.23	0.308	259.5214	1.291
2875	2602	95762.3	72176.7	9612.03	0.299	259.8440	1.291
2900	2627	96685.2	72891.1	9988.31	0.290	260.1632	1.290
2925	2652	97609.5	73607.7	10376.35	0.282	260.4801	1.290
2950	2677	98533.2	74323.8	10777.49	0.274	260.7954	1.290
2975	2702	99457.2	75039.2	11188.71	0.266	261.1068	1.290
3000	2727	100382.7	75757.0	11613.83	0.258	261.4168	1.290
3025	2752	101307.1	76473.7	12049.96	0.251	261.7233	1.290
3050	2777	102233.1	77192.0	12500.42	0.244	262.0284	1.290
3075	2802	103158.5	77909.7	12963.57	0.237	262.3308	1.289
3100	2827	104084.9	78628.4	13439.74	0.231	262.6307	1.289
3125	2852	105011.0	79346.8	13930.07	0.224	262.9286	1.289
3150	2877	105937.8	80065.2	14434.00	0.218	263.2241	1.289
3175	2902	106865.5	80786.8	14951.77	0.212	263.5171	1.289
3200	2927	107793.2	81505.2	15484.30	0.207	263.8080	1.289
3225	2952	108720.8	82225.1	16030.68	0.201	264.0963	1.289
3250	2977	109649.0	82945.6	16593.23	0.196	264.3831	1.289
3275	3002	110577.2	83666.1	17171.93	0.191	264.6681	1.288
3300	3027	111504.7	84386.0	17764.55	0.186	264.9502	1.288
3325	3052	112434.6	85108.1	18374.79	0.181	265.2310	1.288
3350	3077	113363.7	85829.6	18999.02	0.176	265.5087	1.288
3375	3102	114292.6	86550.8	19643.30	0.172	265.7859	1.288
3400	3127	115224.0	87272.8	20301.83	0.167	266.0601	1.288

Temperature		h	u	Pr	Vr	s_o	Gamma
K	C	kJ/kgmol	kJ/kgmol			kJ/kgmol/K	
3425	3152	116154.8	87995.9	20979.58	0.163	266.3331	1.288
3450	3177	117084.1	88719.2	21673.51	0.159	266.6036	1.287
3475	3202	118014.7	89442.1	22386.46	0.155	266.8727	1.287
3500	3227	118947.9	90166.0	23118.61	0.151	267.1403	1.287
3525	3252	119880.8	90891.2	23867.95	0.148	267.4055	1.287
3550	3277	120813.0	91615.7	24637.25	0.144	267.6692	1.287

A.13.8 Nitrogen Properties (English Units)

Temperature		h	u	Pr	Vr	s_o	Gamma
R	F	Btu/lbmol	Btu/lbmol			Btu/lbmol/R	
300	−159.7	2081.9	1486.2	0.177	1692.8	41.6870	1.400
340	−119.7	2360.0	1687.3	0.275	1237.9	42.5571	1.400
380	−79.7	2638.2	1888.1	0.406	937.06	43.3308	1.399
420	−39.7	2916.5	2088.5	0.576	729.35	44.0272	1.399
460	0.3	3194.8	2288.7	0.792	580.81	44.6601	1.400
500	40.3	3473.0	2488.5	1.061	471.43	45.2400	1.400
540	80.3	3751.2	2688.2	1.389	388.86	45.7752	1.400
580	120.3	4029.4	2887.8	1.784	325.19	46.2722	1.400
620	160.3	4307.7	3087.4	2.253	275.18	46.7363	1.399
660	200.3	4586.3	3287.2	2.805	235.26	47.1717	1.399
700	240.3	4865.2	3487.2	3.449	202.95	47.5819	1.398
740	280.3	5144.5	3687.6	4.194	176.46	47.9700	1.397
780	320.3	5424.5	3888.5	5.048	154.50	48.3384	1.396
820	360.3	5705.1	4090.1	6.024	136.12	48.6893	1.394
860	400.3	5986.5	4292.4	7.131	120.60	49.0243	1.392
900	440.3	6268.8	4495.6	8.382	107.38	49.3451	1.391
940	480.3	6552.0	4699.8	9.788	96.040	49.6531	1.389
980	520.3	6836.4	4904.9	11.362	86.252	49.9493	1.387
1020	560.3	7121.8	5111.2	13.119	77.752	50.2348	1.384
1060	600.3	7408.4	5318.6	15.072	70.330	50.5104	1.382
1100	640.3	7696.3	5527.3	17.237	63.815	50.7770	1.380
1140	680.3	7985.5	5737.2	19.631	58.073	51.0352	1.378
1180	720.3	8275.9	5948.5	22.269	52.988	51.2856	1.375
1220	760.3	8567.8	6161.0	25.171	48.469	51.5288	1.373
1260	800.3	8861.0	6375.0	28.354	44.438	51.7653	1.370

Temperature		h	u	Pr	Vr	s_o	Gamma
R	F	Btu/lbmol	Btu/lbmol			Btu/lbmol/R	
1300	840.3	9155.6	6590.3	31.839	40.831	51.9955	1.368
1340	880.3	9451.6	6807.1	35.645	37.593	52.2198	1.366
1380	920.3	9749.0	7025.2	39.795	34.677	52.4385	1.363
1420	960.3	10047.9	7244.7	44.312	32.046	52.6520	1.361
1460	1000.3	10348.1	7465.6	49.217	29.664	52.8605	1.359
1500	1040.3	10649.8	7688.0	54.537	27.504	53.0643	1.356
1540	1080.3	10952.8	7911.7	60.297	25.540	53.2637	1.354
1580	1120.3	11257.2	8136.8	66.523	23.751	53.4588	1.352
1620	1160.3	11563.0	8363.2	73.244	22.118	53.6499	1.350
1660	1200.3	11870.1	8591.0	80.487	20.624	53.8372	1.348
1700	1240.3	12178.6	8820.1	88.285	19.256	54.0208	1.346
1740	1280.3	12488.3	9050.5	96.664	18.001	54.2009	1.344
1780	1320.3	12799.3	9282.2	105.660	16.846	54.3776	1.342
1820	1360.3	13111.6	9515.1	115.308	15.784	54.5511	1.340
1860	1400.3	13425.1	9749.2	125.636	14.805	54.7215	1.339
1900	1440.3	13739.7	9984.5	136.685	13.901	54.8889	1.337
1940	1480.3	14055.5	10221.0	148.490	13.065	55.0534	1.335
1980	1520.3	14372.5	10458.6	161.090	12.291	55.2151	1.334
2020	1560.3	14690.6	10697.3	174.521	11.575	55.3741	1.332
2060	1600.3	15009.8	10937.1	188.827	10.909	55.5306	1.331
2100	1640.3	15330.0	11177.9	204.048	10.292	55.6845	1.329
2140	1680.3	15651.2	11419.8	220.229	9.717	55.8361	1.328
2180	1720.3	15973.4	11662.6	237.408	9.183	55.9852	1.327
2220	1760.3	16296.6	11906.4	255.635	8.684	56.1321	1.325
2260	1800.3	16620.7	12151.2	274.959	8.219	56.2768	1.324
2300	1840.3	16945.7	12396.7	295.426	7.785	56.4194	1.323
2340	1880.3	17271.6	12643.3	317.078	7.380	56.5599	1.322
2380	1920.3	17598.4	12890.6	339.970	7.001	56.6983	1.321
2420	1960.3	17925.9	13138.8	364.158	6.645	56.8348	1.320
2460	2000.3	18254.3	13387.8	389.695	6.313	56.9694	1.319
2500	2040.3	18583.4	13637.5	416.625	6.001	57.1021	1.318
2540	2080.3	18913.3	13888.0	445.019	5.708	57.2330	1.317
2580	2120.3	19243.9	14139.2	474.922	5.432	57.3622	1.316
2620	2160.3	19575.2	14391.1	506.397	5.174	57.4896	1.315
2660	2200.3	19907.2	14643.7	539.497	4.931	57.6153	1.314
2700	2240.3	20239.9	14897.0	574.296	4.701	57.7395	1.313
2740	2280.3	20573.1	15150.8	610.851	4.486	57.8620	1.313
2780	2320.3	20907.0	15405.4	649.229	4.282	57.9830	1.312
2820	2360.3	21241.5	15660.4	689.473	4.090	58.1024	1.311

Temperature		h	u	Pr	Vr	s_o	Gamma
R	F	Btu/lbmol	Btu/lbmol			Btu/lbmol/R	
2860	2400.3	21576.5	15916.1	731.685	3.909	58.2204	1.310
2900	2440.3	21912.2	16172.2	775.912	3.738	58.3370	1.310
2940	2480.3	22248.3	16429.0	822.200	3.576	58.4520	1.309
2980	2520.3	22585.0	16686.3	870.690	3.423	58.5658	1.308
3020	2560.3	22922.3	16944.2	921.370	3.278	58.6782	1.308
3060	2600.3	23260.0	17202.4	974.393	3.140	58.7893	1.307
3100	2640.3	23598.0	17461.2	1029.80	3.010	58.8991	1.307
3140	2680.3	23936.7	17720.4	1087.65	2.887	59.0077	1.306
3180	2720.3	24275.9	17980.2	1148.04	2.770	59.1150	1.306
3220	2760.3	24615.5	18240.3	1211.08	2.659	59.2211	1.305
3260	2800.3	24955.4	18500.9	1276.77	2.553	59.3260	1.305
3300	2840.3	25295.9	18761.9	1345.29	2.453	59.4298	1.304
3340	2880.3	25636.7	19023.3	1416.60	2.358	59.5324	1.304
3380	2920.3	25977.8	19285.1	1490.97	2.267	59.6340	1.303
3420	2960.3	26319.4	19547.2	1568.34	2.181	59.7345	1.303
3460	3000.3	26661.4	19809.7	1648.83	2.098	59.8339	1.302
3500	3040.3	27003.8	20072.7	1732.57	2.020	59.9322	1.302
3540	3080.3	27346.5	20336.0	1819.63	1.945	60.0296	1.302
3580	3120.3	27689.5	20599.7	1910.11	1.874	60.1260	1.301
3620	3160.3	28032.8	20863.5	2004.12	1.806	60.2214	1.301
3660	3200.3	28376.6	21127.9	2101.77	1.741	60.3158	1.300
3700	3240.3	28720.7	21392.7	2203.02	1.680	60.4093	1.300
3740	3280.3	29065.2	21657.7	2308.13	1.620	60.5018	1.300
3780	3320.3	29409.8	21922.9	2417.25	1.564	60.5936	1.299
3820	3360.3	29754.8	22188.4	2530.38	1.510	60.6844	1.299
3860	3400.3	30100.3	22454.5	2647.55	1.458	60.7743	1.299
3900	3440.3	30445.8	22720.7	2769.04	1.408	60.8634	1.298
3940	3480.3	30791.8	22987.2	2894.83	1.361	60.9516	1.298
3980	3520.3	31137.9	23253.9	3025.20	1.316	61.0391	1.298
4020	3560.3	31484.6	23521.1	3159.98	1.272	61.1256	1.297
4060	3600.3	31831.1	23788.4	3299.54	1.230	61.2115	1.297
4100	3640.3	32178.1	24055.9	3443.92	1.191	61.2965	1.297
4140	3680.3	32525.6	24324.0	3593.33	1.152	61.3808	1.296
4180	3720.3	32873.0	24591.9	3747.78	1.115	61.4644	1.296
4220	3760.3	33220.9	24860.3	3907.33	1.080	61.5472	1.296
4260	3800.3	33569.1	25129.1	4072.25	1.046	61.6293	1.296
4300	3840.3	33917.3	25398.0	4242.58	1.014	61.7107	1.295
4340	3880.3	34266.0	25667.3	4418.55	0.982	61.7914	1.295
4380	3920.3	34614.7	25936.5	4600.30	0.952	61.8714	1.295

Temperature		h	u	Pr	Vr	s_o	Gamma
R	F	Btu/lbmol	Btu/lbmol			Btu/lbmol/R	
4420	3960.3	34963.6	26206.0	4787.81	0.923	61.9508	1.294
4460	4000.3	35313.2	26476.1	4981.34	0.895	62.0295	1.294
4500	4040.3	35662.5	26746.3	5180.90	0.869	62.1075	1.294
4540	4080.3	36012.3	27016.7	5386.68	0.843	62.1848	1.294
4580	4120.3	36362.4	27287.3	5598.82	0.818	62.2615	1.294
4620	4160.3	36712.3	27557.7	5817.90	0.794	62.3377	1.293
4660	4200.3	37062.9	27828.8	6043.10	0.771	62.4132	1.293
4700	4240.3	37413.4	28099.9	6275.62	0.749	62.4881	1.293
4740	4280.3	37764.0	28371.1	6514.80	0.728	62.5624	1.293
4780	4320.3	38115.1	28642.7	6761.13	0.707	62.6361	1.292
4820	4360.3	38466.5	28914.7	7015.17	0.687	62.7094	1.292
4860	4400.3	38817.8	29186.5	7276.14	0.668	62.7819	1.292
4900	4440.3	39169.1	29458.4	7544.44	0.649	62.8538	1.292
4940	4480.3	39520.6	29730.4	7821.43	0.632	62.9254	1.292
4980	4520.3	39873.0	30003.4	8105.91	0.614	62.9963	1.291
5020	4560.3	40224.8	30275.7	8398.19	0.598	63.0667	1.291
5060	4600.3	40577.1	30548.5	8699.26	0.582	63.1366	1.291
5100	4640.3	40929.5	30821.5	9008.12	0.566	63.2059	1.291
5140	4680.3	41281.6	31094.5	9326.59	0.551	63.2749	1.291
5180	4720.3	41634.7	31368.2	9653.38	0.537	63.3433	1.291
5220	4760.3	41987.2	31641.2	9988.32	0.523	63.4110	1.290
5260	4800.3	42340.2	31914.7	10333.02	0.509	63.4784	1.290
5300	4840.3	42693.8	32188.9	10686.94	0.496	63.5453	1.290
5340	4880.3	43046.5	32462.5	11050.74	0.483	63.6118	1.290
5380	4920.3	43399.9	32736.1	11423.45	0.471	63.6776	1.290
5420	4960.3	43753.4	33010.2	11806.61	0.459	63.7432	1.290
5460	5000.3	44107.0	33284.6	12198.95	0.448	63.8081	1.290
5500	5040.3	44460.7	33558.8	12602.40	0.436	63.8727	1.290
5540	5080.3	44814.5	33832.9	13015.39	0.426	63.9367	1.289
5580	5120.3	45168.7	34108.0	13439.72	0.415	64.0004	1.289
5620	5160.3	45522.8	34382.3	13874.76	0.405	64.0637	1.289
5660	5200.3	45876.8	34657.2	14319.60	0.395	64.1264	1.289
5700	5240.3	46231.2	34931.8	14777.22	0.386	64.1888	1.289
5740	5280.3	46586.3	35207.4	15245.71	0.376	64.2508	1.289
5780	5320.3	46940.0	35482.3	15725.32	0.368	64.3123	1.289
5820	5360.3	47294.6	35757.5	16217.27	0.359	64.3735	1.289
5860	5400.3	47648.8	36032.3	16721.14	0.350	64.4342	1.288
5900	5440.3	48004.0	36308.0	17236.29	0.342	64.4945	1.288
5940	5480.3	48358.9	36583.4	17764.58	0.334	64.5545	1.288

Temperature		h	u	Pr	Vr	s₀	Gamma
R	F	Btu/lbmol	Btu/lbmol			Btu/lbmol/R	
5980	5520.3	48714.6	36859.7	18306.96	0.327	64.6142	1.288
6020	5560.3	49069.8	37135.4	18859.23	0.319	64.6732	1.288
6060	5600.3	49424.7	37410.9	19427.11	0.312	64.7321	1.288
6100	5640.3	49780.2	37687.0	20007.33	0.305	64.7906	1.288
6140	5680.3	50135.4	37962.7	20600.29	0.298	64.8486	1.288
6180	5720.3	50491.1	38238.9	21209.51	0.291	64.9064	1.288
6220	5760.3	50847.7	38516.1	21830.38	0.285	64.9637	1.287
6260	5800.3	51202.9	38791.8	22466.66	0.279	65.0208	1.287
6300	5840.3	51559.1	39068.6	23118.64	0.273	65.0776	1.287
6340	5880.3	51915.2	39345.2	23782.05	0.267	65.1338	1.287
6380	5920.3	52271.8	39622.4	24462.81	0.261	65.1898	1.287

A.13.9 Oxygen Properties (SI Units)

Temperature		h	u	Pr	Vr	s₀	Gamma
K	C	kJ/kgmol	kJ/kgmol			kJ/kgmol/K	
200	−73.1	4844.7	3459.2	0.330	1074.6	188.0074	1.399
225	−48.1	5573.2	3979.9	0.510	444.49	191.4393	1.399
250	−23.1	6302.3	4501.1	0.730	341.27	194.5121	1.398
275	1.9	7032.8	5023.8	1.020	268.54	197.2970	1.397
300	26.9	7765.6	5548.8	1.390	215.56	199.8474	1.395
325	51.9	8501.6	6077.0	1.850	175.89	202.2038	1.392
350	76.9	9241.6	6609.2	2.410	145.49	204.3974	1.389
375	101.9	9986.3	7146.0	3.080	121.74	206.4524	1.385
400	126.9	10736.2	7688.1	3.890	102.89	208.3882	1.381
425	151.9	11491.8	8235.9	4.850	87.693	210.2205	1.377
450	176.9	12253.4	8789.6	5.980	75.307	211.9616	1.373
475	201.9	13021.2	9349.5	7.300	65.100	213.6220	1.369
500	226.9	13795.1	9915.7	8.830	56.611	215.2100	1.365
525	251.9	14575.5	10488.2	10.610	49.493	216.7328	1.361
550	276.9	15362.2	11067.0	12.650	43.479	218.1965	1.357
575	301.9	16155.0	11652.1	14.990	38.366	219.6063	1.353
600	326.9	16954.0	12243.2	17.650	33.992	220.9665	1.350
625	351.9	17759.0	12840.3	20.670	30.230	222.2808	1.346
650	376.9	18569.7	13443.2	24.090	26.980	223.5526	1.343
675	401.9	19386.0	14051.6	27.940	24.158	224.7849	1.340

Temperature		h	u	Pr	Vr	s_o	Gamma
K	C	kJ/kgmol	kJ/kgmol			kJ/kgmol/K	
700	426.9	20207.7	14665.5	32.260	21.697	225.9803	1.337
725	451.9	21034.5	15284.5	37.100	19.544	227.1409	1.334
750	476.9	21866.4	15908.5	42.490	17.653	228.2688	1.332
775	501.9	22702.9	16537.2	48.480	15.986	229.3661	1.329
800	526.8	23544.0	17170.5	55.130	14.512	230.4342	1.327
825	551.8	24389.3	17807.9	62.470	13.206	231.4744	1.325
850	576.8	25238.9	18449.7	70.580	12.043	232.4888	1.323
875	601.8	26092.4	19095.3	79.500	11.006	233.4786	1.321
900	626.8	26949.3	19744.4	89.300	10.079	234.4443	1.319
925	651.8	27810.0	20397.4	100.030	9.248	235.3876	1.318
950	676.8	28674.0	21053.5	111.750	8.501	236.3092	1.316
975	701.8	29541.2	21712.9	124.550	7.828	237.2105	1.314
1000	726.8	30411.7	22375.4	138.470	7.222	238.0915	1.313
1025	751.8	31284.8	23040.8	153.610	6.673	238.9542	1.312
1050	776.8	32160.9	23708.9	170.030	6.175	239.7984	1.310
1075	801.8	33039.5	24379.7	187.820	5.723	240.6257	1.309
1100	826.8	33921.0	25053.3	207.050	5.313	241.4361	1.308
1125	851.8	34804.5	25729.1	227.810	4.938	242.2303	1.307
1150	876.8	35690.4	26407.2	250.190	4.597	243.0094	1.306
1175	901.8	36578.8	27087.6	274.280	4.284	243.7737	1.305
1200	926.8	37469.3	27770.3	300.160	3.998	244.5233	1.304
1225	951.8	38362.2	28455.3	327.960	3.735	245.2596	1.303
1250	976.8	39256.8	29142.2	357.750	3.494	245.9826	1.302
1275	1002	40153.4	29830.7	389.660	3.272	246.6929	1.301
1300	1027	41051.8	30521.4	423.800	3.068	247.3910	1.300
1325	1052	41952.2	31214.1	460.240	2.879	248.0768	1.300
1350	1077	42854.5	31908.4	499.160	2.705	248.7517	1.299
1375	1102	43758.2	32604.3	540.600	2.543	249.4148	1.298
1400	1127	44664.2	33302.6	584.770	2.394	250.0678	1.297
1425	1152	45572.1	34002.5	631.780	2.256	250.7106	1.297
1450	1177	46480.9	34703.6	681.690	2.127	251.3426	1.296
1475	1202	47391.8	35406.7	734.760	2.007	251.9659	1.295
1500	1227	48304.3	36111.1	791.040	1.896	252.5795	1.295
1525	1252	49218.7	36817.5	850.640	1.793	253.1835	1.294
1550	1277	50133.9	37525.4	913.810	1.696	253.7789	1.293
1575	1302	51051.3	38234.7	980.670	1.606	254.3660	1.293
1600	1327	51970.3	38946.0	1051.430	1.522	254.9453	1.292
1625	1352	52890.2	39657.4	1126.040	1.443	255.5152	1.291
1650	1377	53811.9	40372.2	1205.050	1.369	256.0790	1.291

Temperature		h	u	Pr	Vr	s_o	Gamma
K	C	kJ/kgmol	kJ/kgmol			kJ/kgmol/K	
1675	1402	54735.2	41087.9	1288.250	1.300	256.6340	1.290
1700	1427	55660.0	41804.1	1375.960	1.236	257.1816	1.290
1725	1452	56586.7	42523.1	1468.570	1.175	257.7232	1.289
1750	1477	57513.6	43242.3	1565.810	1.118	258.2562	1.288
1775	1502	58444.0	43964.2	1668.390	1.064	258.7838	1.288
1800	1527	59374.3	44687.6	1776.380	1.013	259.3051	1.287
1825	1552	60306.7	45412.4	1889.640	0.966	259.8190	1.287
1850	1577	61239.9	46137.1	2008.650	0.921	260.3268	1.286
1875	1602	62175.3	46864.8	2133.750	0.879	260.8291	1.286
1900	1627	63111.9	47593.7	2264.990	0.839	261.3253	1.285
1925	1652	64050.0	48323.3	2402.790	0.801	261.8163	1.284
1950	1677	64989.3	49054.8	2547.020	0.766	262.3010	1.284
1975	1702	65929.9	49786.9	2697.750	0.732	262.7790	1.283
2000	1727	66870.6	50521.6	2856.260	0.700	263.2536	1.283
2025	1752	67815.3	51258.6	3021.750	0.670	263.7219	1.282
2050	1777	68759.6	51993.6	3195.660	0.641	264.1871	1.282
2075	1802	69706.0	52732.3	3376.570	0.615	264.6449	1.281
2100	1827	70653.6	53472.2	3565.950	0.589	265.0986	1.281
2125	1852	71602.0	54212.9	3763.990	0.565	265.5479	1.280
2150	1877	72552.5	54955.7	3970.630	0.541	265.9922	1.280
2175	1902	73505.1	55699.0	4186.520	0.520	266.4324	1.279
2200	1927	74456.5	56444.3	4412.320	0.499	266.8691	1.279
2225	1952	75410.8	57190.9	4646.560	0.479	267.2992	1.278
2250	1977	76366.4	57938.9	4892.130	0.460	267.7273	1.278
2275	2002	77324.9	58688.0	5147.180	0.442	268.1498	1.277
2300	2027	78282.5	59439.6	5413.640	0.425	268.5695	1.277
2325	2052	79242.9	60190.7	5690.300	0.409	268.9838	1.276
2350	2077	80202.2	60942.3	5978.470	0.393	269.3945	1.276
2375	2102	81165.1	61697.5	6278.510	0.378	269.8016	1.275
2400	2127	82128.8	62451.9	6591.310	0.364	270.2058	1.275
2425	2152	83092.6	63209.4	6915.330	0.351	270.6048	1.274
2450	2177	84060.2	63967.9	7253.860	0.338	271.0021	1.274
2475	2202	85028.5	64726.9	7605.280	0.325	271.3954	1.273
2500	2227	85997.5	65489.8	7969.600	0.314	271.7845	1.273
2525	2252	86964.4	66250.6	8348.850	0.302	272.1710	1.273
2550	2277	87935.4	67015.6	8740.540	0.292	272.5521	1.272
2575	2302	88909.4	67780.3	9149.560	0.281	272.9323	1.272
2600	2327	89883.7	68545.3	9573.050	0.272	273.3085	1.271
2625	2352	90857.1	69312.6	10011.66	0.262	273.6809	1.271

Temperature		h	u	Pr	Vr	s_o	Gamma
K	C	kJ/kgmol	kJ/kgmol			kJ/kgmol/K	
2650	2377	91831.8	70077.9	10468.70	0.253	274.0521	1.270
2675	2402	92811.6	70851.8	10942.12	0.244	274.4198	1.270
2700	2427	93788.0	71618.8	11430.69	0.236	274.7829	1.269
2725	2452	94769.9	72391.4	11939.27	0.228	275.1448	1.269
2750	2477	95754.4	73166.6	12466.23	0.221	275.5039	1.269
2775	2502	96733.4	73939.4	13008.47	0.213	275.8579	1.268
2800	2527	97720.0	74716.8	13570.96	0.206	276.2098	1.268
2825	2552	98698.0	75492.0	14159.12	0.200	276.5625	1.267
2850	2577	99688.5	76269.9	14760.84	0.193	276.9085	1.267
2875	2602	100674.1	77052.6	15385.88	0.187	277.2533	1.267
2900	2627	101660.0	77829.3	16033.14	0.181	277.5959	1.266
2925	2652	102649.4	78609.3	16702.24	0.175	277.9358	1.266
2950	2677	103642.2	79392.9	17392.24	0.170	278.2724	1.265
2975	2702	104632.1	80176.6	18109.35	0.164	278.6083	1.265
3000	2727	105629.1	80964.3	18842.33	0.159	278.9381	1.265
3025	2752	106618.5	81747.7	19605.69	0.154	279.2683	1.264
3050	2777	107617.3	82537.1	20394.44	0.150	279.5962	1.264
3075	2802	108613.8	83324.3	21208.77	0.145	279.9217	1.264
3100	2827	109604.7	84112.4	22050.11	0.141	280.2451	1.263
3125	2852	110603.4	84901.8	22912.96	0.136	280.5643	1.263
3150	2877	111600.9	85690.0	23813.95	0.132	280.8849	1.263
3175	2902	112598.4	86484.7	24740.19	0.128	281.2021	1.262
3200	2927	113599.7	87276.7	25688.12	0.125	281.5147	1.262
3225	2952	114600.5	88068.1	26669.83	0.121	281.8265	1.262
3250	2977	115604.2	88862.5	27674.83	0.117	282.1341	1.261
3275	3002	116609.5	89658.5	28729.51	0.114	282.4450	1.261
3300	3027	117612.1	90458.3	29793.74	0.111	282.7474	1.261
3325	3052	118627.2	91257.6	30908.52	0.108	283.0528	1.260
3350	3077	119629.9	92057.5	32048.53	0.105	283.3539	1.260
3375	3102	120639.1	92857.4	33222.66	0.102	283.6530	1.260
3400	3127	121644.6	93653.5	34440.44	0.099	283.9523	1.259
3425	3152	122653.0	94459.1	35676.52	0.096	284.2455	1.259
3450	3177	123666.0	95262.8	36961.71	0.093	284.5397	1.259
3475	3202	124674.9	96062.4	38297.42	0.091	284.8348	1.259
3500	3227	125690.8	96869.0	39663.05	0.088	285.1261	1.258
3525	3252	126703.0	97671.8	41036.38	0.086	285.4091	1.258
3550	3277	127718.6	98484.7	42482.84	0.084	285.6971	1.258

A.13.10 Oxygen Properties (English Units)

Temperature		h	u	Pr	Vr	s_o	Gamma
R	F	Btu/lbmol	Btu/lbmol			Btu/lbmol/R	
300	−159.7	2083.0	1487.3	0.177	1696.0	44.9080	1.400
340	−119.7	2361.2	1686.1	0.274	1240.0	45.7784	1.399
380	−79.7	2639.5	1885.0	0.405	938.55	46.5524	1.399
420	−39.7	2918.0	2084.0	0.575	730.37	47.2491	1.399
460	0.3	3196.7	2283.4	0.791	581.31	47.8831	1.398
500	40.3	3476.1	2483.2	1.061	471.30	48.4653	1.397
540	80.3	3756.2	2683.9	1.392	388.01	49.0043	1.395
580	120.3	4037.4	2885.7	1.792	323.61	49.5066	1.392
620	160.3	4319.9	3088.8	2.272	272.87	49.9777	1.390
660	200.3	4604.0	3293.5	2.841	232.28	50.4217	1.387
700	240.3	4889.8	3499.8	3.511	199.35	50.8421	1.383
740	280.3	5177.5	3708.1	4.294	172.33	51.2418	1.380
780	320.3	5467.1	3918.3	5.203	149.92	51.6229	1.376
820	360.3	5758.8	4130.6	6.252	131.17	51.9876	1.372
860	400.3	6052.6	4344.9	7.456	115.34	52.3375	1.369
900	440.3	6348.6	4561.5	8.832	101.90	52.6738	1.365
940	480.3	6646.7	4780.1	10.398	90.403	52.9979	1.361
980	520.3	6946.9	5001.0	12.171	80.517	53.3106	1.358
1020	560.3	7249.3	5223.9	14.174	71.965	53.6131	1.355
1060	600.3	7553.8	5448.9	16.425	64.536	53.9058	1.351
1100	640.3	7860.3	5676.0	18.949	58.052	54.1897	1.348
1140	680.3	8168.8	5905.1	21.768	52.371	54.4651	1.345
1180	720.3	8479.2	6136.0	24.909	47.372	54.7328	1.342
1220	760.3	8791.5	6368.9	28.397	42.962	54.9931	1.340
1260	800.3	9105.7	6603.7	32.262	39.055	55.2465	1.337
1300	840.3	9421.6	6840.1	36.531	35.586	55.4932	1.335
1340	880.3	9739.2	7078.3	41.237	32.495	55.7339	1.332
1380	920.3	10058.4	7318.1	46.411	29.734	55.9686	1.330
1420	960.3	10379.2	7559.4	52.088	27.262	56.1977	1.328
1460	1000.3	10701.4	7802.3	58.302	25.042	56.4216	1.326
1500	1040.3	11025.2	8046.6	65.092	23.044	56.6403	1.324
1540	1080.3	11350.2	8292.2	72.494	21.243	56.8542	1.322
1580	1120.3	11676.6	8539.1	80.549	19.615	57.0634	1.321
1620	1160.3	12004.2	8787.4	89.298	18.142	57.2682	1.319
1660	1200.3	12333.1	9036.8	98.787	16.804	57.4687	1.318
1700	1240.3	12663.1	9287.3	109.057	15.588	57.6651	1.316

Temperature		h	u	Pr	Vr	s_o	Gamma
R	F	Btu/lbmol	Btu/lbmol			Btu/lbmol/R	
1740	1280.3	12994.2	9539.1	120.159	14.481	57.8577	1.315
1780	1320.3	13326.4	9791.8	132.140	13.471	58.0464	1.314
1820	1360.3	13659.6	10045.5	145.054	12.547	58.2316	1.312
1860	1400.3	13993.7	10300.2	158.939	11.703	58.4131	1.311
1900	1440.3	14328.8	10555.9	173.867	10.928	58.5914	1.310
1940	1480.3	14664.9	10812.5	189.878	10.217	58.7663	1.309
1980	1520.3	15001.7	11069.9	207.053	9.563	58.9383	1.308
2020	1560.3	15339.4	11328.1	225.424	8.961	59.1071	1.307
2060	1600.3	15677.9	11587.3	245.068	8.406	59.2730	1.306
2100	1640.3	16017.1	11847.0	266.042	7.893	59.4361	1.305
2140	1680.3	16357.1	12107.6	288.437	7.419	59.5966	1.304
2180	1720.3	16697.9	12368.9	312.265	6.981	59.7542	1.304
2220	1760.3	17039.2	12630.9	337.650	6.575	59.9094	1.303
2260	1800.3	17381.4	12893.6	364.668	6.197	60.0623	1.302
2300	1840.3	17724.2	13157.0	393.349	5.847	60.2126	1.301
2340	1880.3	18067.6	13420.9	423.798	5.522	60.3607	1.300
2380	1920.3	18411.9	13685.7	456.060	5.219	60.5064	1.300
2420	1960.3	18756.6	13951.0	490.289	4.936	60.6501	1.299
2460	2000.3	19101.8	14216.9	526.500	4.672	60.7916	1.298
2500	2040.3	19447.6	14483.3	564.786	4.426	60.9310	1.298
2540	2080.3	19794.1	14750.4	605.319	4.196	61.0686	1.297
2580	2120.3	20141.4	15017.9	648.093	3.981	61.2042	1.296
2620	2160.3	20489.0	15286.3	693.189	3.780	61.3378	1.296
2660	2200.3	20837.0	15554.9	740.793	3.591	61.4697	1.295
2700	2240.3	21185.9	15824.3	791.037	3.413	61.6000	1.295
2740	2280.3	21535.4	16094.4	843.836	3.247	61.7283	1.294
2780	2320.3	21885.1	16364.7	899.451	3.091	61.8551	1.293
2820	2360.3	22235.5	16635.4	957.983	2.944	61.9803	1.293
2860	2400.3	22586.5	16907.2	1019.44	2.805	62.1038	1.292
2900	2440.3	22938.1	17179.4	1084.06	2.675	62.2258	1.292
2940	2480.3	23290.0	17451.8	1151.88	2.552	62.3463	1.291
2980	2520.3	23642.0	17724.4	1223.12	2.436	62.4655	1.291
3020	2560.3	23994.9	17997.9	1297.84	2.327	62.5832	1.290
3060	2600.3	24348.5	18272.0	1375.96	2.224	62.6993	1.290
3100	2640.3	24702.3	18546.3	1458.02	2.126	62.8143	1.289
3140	2680.3	25057.2	18821.8	1543.78	2.034	62.9278	1.289
3180	2720.3	25412.4	19097.5	1633.68	1.947	63.0402	1.288
3220	2760.3	25767.9	19373.6	1727.83	1.864	63.1515	1.287
3260	2800.3	26123.7	19649.9	1826.07	1.785	63.2613	1.287
3300	2840.3	26480.3	19927.1	1928.68	1.711	63.3699	1.286

Temperature		h	u	Pr	Vr	s_o	Gamma
R	F	Btu/lbmol	Btu/lbmol			Btu/lbmol/R	
3340	2880.3	26836.9	20204.6	2036.13	1.640	63.4776	1.286
3380	2920.3	27194.4	20482.6	2147.96	1.574	63.5837	1.285
3420	2960.3	27552.4	20761.2	2264.99	1.510	63.6891	1.285
3460	3000.3	27910.7	21039.7	2387.13	1.449	63.7934	1.285
3500	3040.3	28269.7	21319.2	2514.14	1.392	63.8963	1.284
3540	3080.3	28629.2	21599.6	2646.86	1.337	63.9985	1.284
3580	3120.3	28988.8	21879.8	2785.34	1.285	64.0997	1.283
3620	3160.3	29348.9	22160.4	2929.13	1.236	64.1997	1.283
3660	3200.3	29710.1	22442.2	3078.65	1.189	64.2986	1.282
3700	3240.3	30071.2	22723.8	3235.02	1.144	64.3970	1.282
3740	3280.3	30433.1	23006.3	3397.46	1.101	64.4942	1.281
3780	3320.3	30795.0	23288.7	3565.94	1.060	64.5904	1.281
3820	3360.3	31157.5	23571.8	3741.25	1.021	64.6857	1.280
3860	3400.3	31520.6	23855.4	3924.16	0.984	64.7805	1.280
3900	3440.3	31884.3	24139.6	4114.02	0.948	64.8743	1.279
3940	3480.3	32247.9	24424.7	4310.84	0.914	64.9671	1.279
3980	3520.3	32612.7	24709.1	4515.11	0.881	65.0590	1.278
4020	3560.3	32978.3	24995.2	4727.45	0.850	65.1503	1.278
4060	3600.3	33343.0	25280.5	4947.04	0.821	65.2404	1.278
4100	3640.3	33708.5	25566.6	5175.92	0.792	65.3303	1.277
4140	3680.3	34075.1	25854.4	5413.64	0.765	65.4194	1.277
4180	3720.3	34441.8	26141.0	5659.10	0.739	65.5075	1.276
4220	3760.3	34809.4	26429.1	5914.01	0.714	65.5950	1.276
4260	3800.3	35176.7	26717.0	6177.31	0.690	65.6815	1.275
4300	3840.3	35545.1	27005.9	6450.56	0.667	65.7674	1.275
4340	3880.3	35913.5	27294.8	6734.85	0.644	65.8531	1.275
4380	3920.3	36282.0	27584.6	7026.65	0.623	65.9373	1.274
4420	3960.3	36650.8	27874.7	7330.08	0.603	66.0213	1.274
4460	4000.3	37021.3	28163.6	7644.56	0.583	66.1047	1.273
4500	4040.3	37392.1	28455.7	7969.59	0.565	66.1874	1.273
4540	4080.3	37762.7	28746.1	8305.63	0.547	66.2694	1.273
4580	4120.3	38133.7	29038.3	8653.25	0.529	66.3508	1.272
4620	4160.3	38504.2	29330.1	9013.26	0.513	66.4318	1.272
4660	4200.3	38876.5	29623.7	9383.72	0.497	66.5117	1.271
4700	4240.3	39249.3	29916.3	9767.59	0.481	66.5914	1.271
4740	4280.3	39622.2	30210.4	10162.9	0.466	66.6702	1.271
4780	4320.3	39995.4	30503.5	10573.0	0.452	66.7487	1.270
4820	4360.3	40369.8	30797.7	10995.9	0.438	66.8266	1.270
4860	4400.3	40741.7	31090.9	11430.7	0.425	66.9036	1.269
4900	4440.3	41117.8	31386.8	11879.5	0.412	66.9801	1.269

Temperature		h	u	Pr	Vr	s_o	Gamma
R	F	Btu/lbmol	Btu/lbmol			Btu/lbmol/R	
4940	4480.3	41492.4	31682.7	12343.5	0.400	67.0562	1.269
4980	4520.3	41867.4	31977.5	12823.6	0.388	67.1319	1.268
5020	4560.3	42244.0	32275.3	13321.3	0.377	67.2076	1.268
5060	4600.3	42619.0	32570.2	13831.8	0.366	67.2822	1.268
5100	4640.3	42995.1	32867.5	14356.2	0.355	67.3561	1.267
5140	4680.3	43373.6	33165.5	14897.9	0.345	67.4297	1.267
5180	4720.3	43750.6	33464.1	15455.8	0.335	67.5027	1.267
5220	4760.3	44126.3	33761.1	16033.1	0.326	67.5755	1.266
5260	4800.3	44505.8	34060.4	16624.9	0.316	67.6475	1.266
5300	4840.3	44883.6	34359.5	17240.2	0.307	67.7197	1.265
5340	4880.3	45262.6	34658.3	17867.5	0.299	67.7906	1.265
5380	4920.3	45641.6	34958.6	18513.8	0.291	67.8612	1.265
5420	4960.3	46019.8	35258.0	19178.5	0.283	67.9312	1.265
5460	5000.3	46400.1	35556.8	19868.9	0.275	68.0015	1.264
5500	5040.3	46779.6	35860.3	20578.2	0.267	68.0711	1.264
5540	5080.3	47162.7	36159.1	21300.1	0.260	68.1396	1.264
5580	5120.3	47542.1	36462.5	22050.1	0.253	68.2083	1.263
5620	5160.3	47924.1	36762.9	22821.5	0.246	68.2766	1.263
5660	5200.3	48306.8	37066.9	23609.7	0.240	68.3440	1.263
5700	5240.3	48689.4	37370.7	24424.0	0.233	68.4114	1.262
5740	5280.3	49071.1	37670.9	25258.6	0.227	68.4781	1.262
5780	5320.3	49452.4	37973.4	26122.2	0.221	68.5449	1.262
5820	5360.3	49836.4	38278.6	27005.5	0.216	68.6109	1.262
5860	5400.3	50219.4	38583.0	27911.8	0.210	68.6765	1.261
5900	5440.3	50604.0	38886.0	28843.0	0.205	68.7416	1.261
5940	5480.3	50984.9	39190.9	29793.6	0.199	68.8060	1.261
5980	5520.3	51369.0	39496.3	30780.5	0.194	68.8707	1.260
6020	5560.3	51755.9	39801.7	31791.3	0.189	68.9349	1.260
6060	5600.3	52142.4	40106.5	32828.2	0.185	68.9986	1.260
6100	5640.3	52526.9	40412.4	33889.5	0.180	69.0618	1.260
6140	5680.3	52910.4	40717.1	34992.4	0.175	69.1254	1.259
6180	5720.3	53295.7	41023.6	36107.9	0.171	69.1877	1.259
6220	5760.3	53684.3	41333.4	37269.0	0.167	69.2506	1.259
6260	5800.3	54071.3	41638.9	38424.6	0.163	69.3112	1.259
6300	5840.3	54458.4	41947.2	39662.9	0.159	69.3742	1.258
6340	5880.3	54843.1	42253.2	40889.6	0.155	69.4347	1.258
6380	5920.3	55229.9	42561.3	42160.0	0.151	69.4955	1.258

A.13.11 Hydrogen Properties (SI Units)

Temperature		h	u	Pr	Vr	s$_0$	Gamma
K	C	kJ/kgmol	kJ/kgmol			kj/kgmol/K	
200	−73.1	4805.7	3420.1	0.350	1029.2	114.446	1.438
225	−48.1	5495.8	3924.5	0.520	435.08	117.697	1.425
250	−23.1	6198.6	4437.6	0.740	338.55	120.659	1.416
275	1.9	6910.5	4956.6	1.020	268.69	123.373	1.409
300	26.9	7628.8	5479.7	1.380	216.99	125.873	1.405
325	51.9	8351.6	6005.5	1.830	177.96	128.187	1.402
350	76.9	9077.5	6532.9	2.370	147.95	130.338	1.400
375	101.9	9805.5	7061.1	3.010	124.49	132.347	1.399
400	126.9	10534.8	7589.8	3.780	105.88	134.230	1.398
425	151.9	11265.0	8118.5	4.670	90.912	136.001	1.398
450	176.9	11995.8	8647.1	5.720	78.734	137.672	1.397
475	201.9	12727.0	9175.5	6.910	68.713	139.253	1.397
500	226.9	13458.4	9703.7	8.280	60.384	140.754	1.397
525	251.9	14190.1	10231.7	9.830	53.397	142.182	1.397
550	276.9	14922.0	10759.6	11.580	47.487	143.544	1.396
575	301.9	15654.3	11287.4	13.550	42.449	144.846	1.396
600	326.9	16386.9	11815.4	15.740	38.124	146.093	1.396
625	351.9	17120.1	12343.6	18.180	34.387	147.290	1.395
650	376.9	17853.8	12872.2	20.870	31.138	148.441	1.395
675	401.9	18588.3	13401.3	23.850	28.299	149.550	1.394
700	426.9	19323.5	13931.1	27.130	25.804	150.619	1.394
725	451.9	20059.8	14461.6	30.720	23.601	151.653	1.393
750	476.9	20797.1	14993.2	34.640	21.649	152.653	1.392
775	501.9	21535.7	15525.8	38.930	19.910	153.621	1.391
800	526.8	22275.6	16059.6	43.580	18.356	154.561	1.390
825	551.8	23016.8	16594.8	48.640	16.962	155.473	1.389
850	576.8	23759.7	17131.3	54.120	15.707	156.360	1.388
875	601.8	24504.3	17669.5	60.040	14.575	157.224	1.387
900	626.8	25250.5	18209.4	66.430	13.549	158.065	1.385
925	651.8	25998.7	18751.0	73.310	12.617	158.884	1.384
950	676.8	26748.8	19294.5	80.720	11.769	159.685	1.383
975	701.8	27500.9	19840.0	88.670	10.995	160.466	1.381
1000	726.8	28255.2	20387.5	97.210	10.287	161.230	1.380
1025	751.8	29011.6	20937.1	106.340	9.638	161.977	1.378
1050	776.8	29770.2	21489.0	116.120	9.042	162.708	1.377
1075	801.8	30531.1	22043.0	126.570	8.493	163.425	1.375

Temperature		h	u	Pr	Vr	s_o	Gamma
K	C	kJ/kgmol	kJ/kgmol			kj/kgmol/K	
1100	826.8	31294.6	22599.6	137.720	7.987	164.126	1.373
1125	851.8	32060.3	23158.3	149.620	7.519	164.815	1.372
1150	876.8	32828.6	23719.8	162.280	7.087	165.490	1.370
1175	901.8	33599.4	24283.3	175.750	6.686	166.153	1.368
1200	926.8	34372.8	24849.8	190.080	6.313	166.8049	1.367
1225	951.8	35148.7	25418.6	205.290	5.967	167.4448	1.365
1250	976.8	35927.5	25990.4	221.430	5.645	168.0741	1.363
1275	1001.8	36708.8	26564.4	238.530	5.345	168.6927	1.362
1300	1026.8	37493.0	27141.3	256.670	5.065	169.3019	1.360
1325	1051.8	38279.6	27720.8	275.850	4.803	169.9012	1.358
1350	1076.8	39068.9	28303.0	296.150	4.559	170.4915	1.357
1375	1101.8	39861.4	28888.0	317.590	4.329	171.0727	1.355
1400	1126.8	40656.2	29475.7	340.260	4.115	171.6458	1.353
1425	1151.8	41453.9	30066.2	364.180	3.913	172.2106	1.352
1450	1176.8	42254.3	30659.3	389.400	3.724	172.7673	1.350
1475	1201.8	43057.5	31254.8	415.990	3.546	173.3165	1.348
1500	1226.8	43863.2	31853.6	444.010	3.378	173.8584	1.347
1525	1251.8	44671.7	32454.5	473.490	3.221	174.3928	1.345
1550	1276.8	45483.5	33058.9	504.510	3.072	174.9204	1.344
1575	1301.8	46297.5	33665.3	537.150	2.932	175.4416	1.342
1600	1326.8	47113.8	34274.7	571.450	2.800	175.9562	1.341
1625	1351.8	47933.5	34886.7	607.450	2.675	176.4641	1.339
1650	1376.8	48755.8	35501.8	645.290	2.557	176.9665	1.338
1675	1401.8	49580.5	36118.8	684.920	2.446	177.4620	1.336
1700	1426.8	50407.3	36738.3	726.580	2.340	177.9529	1.335
1725	1451.8	51237.3	37361.0	770.150	2.240	178.4370	1.333
1750	1476.8	52069.7	37985.7	815.840	2.145	178.9162	1.332
1775	1501.8	52903.8	38612.9	863.700	2.055	179.3901	1.331
1800	1526.8	53740.8	39241.4	913.720	1.970	179.8581	1.330
1825	1551.8	54581.4	39875.1	966.120	1.889	180.3218	1.328
1850	1576.8	55424.0	40510.0	1020.85	1.812	180.7799	1.327
1875	1601.8	56267.7	41146.0	1078.02	1.739	181.2329	1.326
1900	1626.8	57114.5	41785.2	1137.85	1.670	181.6819	1.325
1925	1651.8	57963.9	42427.7	1200.31	1.604	182.1262	1.324
1950	1676.8	58816.2	43072.4	1265.35	1.541	182.5649	1.322
1975	1701.8	59669.1	43717.5	1333.32	1.481	183.0000	1.321
2000	1726.8	60524.9	44365.6	1404.24	1.424	183.4308	1.320
2025	1751.8	61382.1	45016.0	1478.15	1.370	183.8572	1.319
2050	1776.8	62242.3	45667.7	1555.15	1.318	184.2794	1.318

Temperature		h	u	Pr	Vr	s$_o$	Gamma
K	C	kJ/kgmol	kJ/kgmol			kj/kgmol/K	
2075	1801.8	63105.5	46323.1	1635.33	1.269	184.6973	1.317
2100	1826.8	63969.2	46979.2	1718.79	1.222	185.1112	1.316
2125	1851.8	64836.1	47638.4	1805.64	1.177	185.5210	1.315
2150	1876.8	65704.2	48300.4	1896.32	1.134	185.9284	1.314
2175	1901.8	66575.0	48963.5	1990.34	1.093	186.3307	1.313
2200	1926.8	67446.9	49627.8	2088.08	1.054	186.729	1.312
2225	1951.8	68324.0	50295.5	2189.74	1.016	187.124	1.311
2250	1976.8	69197.7	50963.2	2295.26	0.980	187.516	1.310
2275	2001.8	70077.9	51634.0	2405.30	0.946	187.905	1.310
2300	2026.8	70956.1	52304.6	2519.14	0.913	188.290	1.309
2325	2051.9	71839.3	52980.1	2637.34	0.882	188.671	1.308
2350	2076.9	72722.6	53657.3	2759.89	0.851	189.048	1.307
2375	2101.9	73608.8	54335.9	2887.56	0.822	189.424	1.306
2400	2126.9	74496.5	55015.8	3019.41	0.795	189.795	1.305
2425	2151.9	75383.7	55697.0	3156.42	0.768	190.164	1.305
2450	2176.9	76275.0	56378.9	3298.22	0.743	190.530	1.304
2475	2201.9	77168.5	57063.2	3444.71	0.718	190.891	1.303
2500	2226.9	78065.4	57754.0	3597.46	0.695	191.252	1.302
2525	2251.9	78958.7	58441.2	3755.16	0.672	191.608	1.302
2550	2276.9	79858.3	59131.5	3918.30	0.651	191.962	1.301
2575	2301.9	80757.3	59824.4	4087.42	0.630	192.313	1.300
2600	2326.9	81660.9	60518.7	4262.98	0.610	192.663	1.299
2625	2351.9	82562.7	61211.2	4443.41	0.591	193.008	1.299
2650	2376.9	83468.5	61911.0	4630.75	0.572	193.351	1.298
2675	2401.9	84374.3	62607.4	4824.38	0.554	193.691	1.297
2700	2426.9	85280.6	63307.7	5023.60	0.537	194.028	1.297
2725	2451.9	86190.2	64007.9	5230.92	0.521	194.364	1.296
2750	2476.9	87100.1	64711.8	5444.85	0.505	194.697	1.295
2775	2501.9	88012.9	65418.5	5666.24	0.490	195.029	1.295
2800	2526.9	88928.4	66124.7	5893.76	0.475	195.356	1.294
2825	2551.9	89842.6	66829.6	6130.21	0.461	195.683	1.293
2850	2576.9	90761.3	67545.5	6370.52	0.447	196.003	1.293
2875	2601.9	91678.1	68252.9	6621.13	0.434	196.323	1.292
2900	2626.9	92599.3	68964.8	6879.94	0.422	196.642	1.291
2925	2651.9	93523.4	69679.6	7146.94	0.409	196.959	1.291
2950	2676.9	94447.0	70393.8	7424.87	0.397	197.276	1.290
2975	2701.9	95365.3	71109.4	7707.28	0.386	197.586	1.290
3000	2726.9	96299.9	71828.1	8001.78	0.375	197.898	1.289
3025	2751.9	97225.9	72551.3	8304.25	0.364	198.206	1.289

Temperature		h	u	Pr	Vr	s_o	Gamma
K	C	kJ/kgmol	kJ/kgmol			kj/kgmol/K	
3050	2776.9	98155.4	73271.5	8614.29	0.354	198.511	1.288
3075	2801.9	99082.3	73989.1	8934.31	0.344	198.814	1.287
3100	2826.9	100015.5	74712.9	9264.16	0.335	199.116	1.287
3125	2851.9	100948.4	75443.1	9604.56	0.325	199.416	1.286
3150	2876.9	101874.6	76160.0	9951.58	0.317	199.711	1.286
3175	2901.9	102824.7	76900.7	10313.1	0.308	200.008	1.285
3200	2926.9	103754.9	77628.2	10687.4	0.299	200.3040	1.284

A.13.12 Hydrogen Properties (English Units)

Temperature		h	u	Pr	Vr	s_o	Gamma
R	F	Btu/lbmol	Btu/lbmol			Btu/lbmol/R	
300	−159.7	2066.2	1470.5	0.195	1541.3	27.3370	1.467
340	−119.7	2319.9	1656.3	0.290	1171.3	28.1307	1.446
380	−79.7	2580.6	1847.0	0.418	908.85	28.8554	1.432
420	−39.7	2846.5	2041.2	0.584	718.63	29.5205	1.421
460	0.3	3116.2	2238.0	0.796	577.94	30.1338	1.414
500	40.3	3388.7	2436.7	1.060	471.90	30.7019	1.409
540	80.3	3663.4	2636.7	1.383	390.58	31.2303	1.405
580	120.3	3939.5	2837.6	1.772	327.23	31.7236	1.403
620	160.3	4216.8	3039.0	2.237	277.16	32.1859	1.401
660	200.3	4494.8	3240.8	2.784	237.05	32.6204	1.399
700	240.3	4773.4	3442.8	3.422	204.54	33.0302	1.398
740	280.3	5052.3	3644.9	4.160	177.90	33.4177	1.398
780	320.3	5331.5	3847.0	5.005	155.84	33.7851	1.397
820	360.3	5610.8	4049.0	5.968	137.41	34.1344	1.397
860	400.3	5890.3	4250.9	7.056	121.88	34.4671	1.397
900	440.3	6169.8	4452.8	8.280	108.69	34.7848	1.397
940	480.3	6449.4	4654.5	9.650	97.410	35.0888	1.397
980	520.3	6729.1	4856.3	11.175	87.693	35.3802	1.397
1020	560.3	7009.0	5058.0	12.867	79.276	35.6601	1.396
1060	600.3	7288.9	5259.8	14.734	71.940	35.9293	1.396
1100	640.3	7569.0	5461.6	16.790	65.514	36.1887	1.396
1140	680.3	7849.2	5663.5	19.045	59.857	36.4389	1.395
1180	720.3	8129.7	5865.6	21.512	54.854	36.6807	1.395
1220	760.3	8410.5	6067.8	24.202	50.410	36.9147	1.394
1260	800.3	8691.5	6270.3	27.128	46.447	37.1414	1.394

Temperature		h	u	Pr	Vr	s_o	Gamma
R	F	Btu/lbmol	Btu/lbmol			Btu/lbmol/R	
1300	840.3	8972.9	6473.1	30.303	42.899	37.3612	1.393
1340	880.3	9254.6	6676.1	33.742	39.713	37.5747	1.392
1380	920.3	9536.7	6879.6	37.457	36.842	37.7821	1.391
1420	960.3	9819.3	7083.4	41.465	34.246	37.9839	1.391
1460	1000.3	10102.3	7287.7	45.779	31.892	38.1805	1.390
1500	1040.3	10385.8	7492.4	50.417	29.752	38.3721	1.389
1540	1080.3	10669.9	7697.6	55.392	27.802	38.5590	1.388
1580	1120.3	10954.6	7903.4	60.724	26.019	38.7415	1.387
1620	1160.3	11239.8	8109.8	66.427	24.388	38.9198	1.385
1660	1200.3	11525.7	8316.7	72.521	22.890	39.0941	1.384
1700	1240.3	11812.3	8524.3	79.026	21.512	39.2647	1.383
1740	1280.3	12099.5	8732.6	85.961	20.242	39.4317	1.382
1780	1320.3	12387.4	8941.5	93.341	19.070	39.5953	1.380
1820	1360.3	12676.1	9151.2	101.193	17.985	39.7557	1.379
1860	1400.3	12965.5	9361.5	109.533	16.981	39.9129	1.378
1900	1440.3	13255.7	9572.7	118.388	16.049	40.0673	1.376
1940	1480.3	13546.7	9784.6	127.776	15.183	40.2189	1.375
1980	1520.3	13838.5	9997.4	137.723	14.377	40.3677	1.373
2020	1560.3	14131.1	10210.8	148.254	13.625	40.5141	1.372
2060	1600.3	14424.6	10425.3	159.396	12.924	40.6580	1.370
2100	1640.3	14718.8	10640.4	171.165	12.269	40.7994	1.369
2140	1680.3	15014.1	10856.5	183.598	11.656	40.9387	1.367
2180	1720.3	15310.2	11073.4	196.724	11.082	41.0758	1.366
2220	1760.3	15607.1	11291.2	210.553	10.544	41.2107	1.364
2260	1800.3	15905.1	11510.0	225.140	10.038	41.3437	1.363
2300	1840.3	16203.8	11729.5	240.496	9.564	41.4747	1.361
2340	1880.3	16503.5	11950.1	256.667	9.117	41.6040	1.360
2380	1920.3	16804.0	12171.5	273.665	8.697	41.7313	1.358
2420	1960.3	17105.6	12393.8	291.537	8.301	41.8569	1.357
2460	2000.3	17408.0	12617.1	310.317	7.927	41.9809	1.355
2500	2040.3	17711.4	12841.2	330.038	7.575	42.1033	1.354
2540	2080.3	18015.7	13066.4	350.727	7.242	42.2240	1.352
2580	2120.3	18321.1	13292.4	372.431	6.927	42.3433	1.351
2620	2160.3	18627.3	13519.5	395.188	6.630	42.4610	1.350
2660	2200.3	18934.6	13747.4	419.029	6.348	42.5774	1.348
2700	2240.3	19242.4	13976.2	444.012	6.081	42.6924	1.347
2740	2280.3	19551.5	14206.0	470.143	5.828	42.8059	1.345
2780	2320.3	19861.4	14436.7	497.506	5.588	42.9183	1.344
2820	2360.3	20172.2	14668.2	526.107	5.360	43.0293	1.343
2860	2400.3	20483.9	14900.6	555.982	5.144	43.1389	1.341

Temperature		h	u	Pr	Vr	s_o	Gamma
R	F	Btu/lbmol	Btu/lbmol			Btu/lbmol/R	
2900	2440.3	20796.5	15133.9	587.240	4.938	43.2476	1.340
2940	2480.3	21110.2	15368.3	619.862	4.743	43.3549	1.339
2980	2520.3	21424.6	15603.4	653.922	4.557	43.4611	1.337
3020	2560.3	21739.9	15839.4	689.489	4.380	43.5663	1.336
3060	2600.3	22056.0	16076.3	726.583	4.211	43.6704	1.335
3100	2640.3	22373.2	16314.2	765.241	4.051	43.7733	1.334
3140	2680.3	22690.9	16552.8	805.452	3.898	43.8750	1.332
3180	2720.3	23009.8	16792.3	847.464	3.752	43.9760	1.331
3220	2760.3	23329.4	17032.4	891.139	3.613	44.0758	1.330
3260	2800.3	23650.2	17273.7	936.736	3.480	44.1749	1.329
3300	2840.3	23971.1	17515.6	983.986	3.354	44.2726	1.328
3340	2880.3	24293.4	17758.7	1033.37	3.232	44.3698	1.327
3380	2920.3	24616.2	18002.0	1084.55	3.116	44.4658	1.326
3420	2960.3	24939.8	18246.3	1137.85	3.006	44.5611	1.325
3460	3000.3	25264.2	18491.2	1193.24	2.900	44.6555	1.324
3500	3040.3	25589.6	18737.5	1250.77	2.798	44.7490	1.323
3540	3080.3	25916.0	18984.8	1310.49	2.701	44.8416	1.322
3580	3120.3	26242.3	19231.6	1372.46	2.608	44.9334	1.321
3620	3160.3	26570.1	19479.9	1436.67	2.520	45.0242	1.320
3660	3200.3	26898.3	19729.1	1503.57	2.434	45.1146	1.319
3700	3240.3	27227.4	19978.7	1572.68	2.353	45.2038	1.318
3740	3280.3	27557.4	20229.3	1644.49	2.274	45.2925	1.317
3780	3320.3	27887.0	20479.4	1718.79	2.199	45.3802	1.316
3820	3360.3	28218.9	20731.8	1795.98	2.127	45.4675	1.315
3860	3400.3	28550.0	20983.5	1875.66	2.058	45.5537	1.314
3900	3440.3	28882.4	21237.2	1958.71	1.991	45.6397	1.314
3940	3480.3	29215.3	21490.6	2044.45	1.927	45.7248	1.313
3980	3520.3	29549.5	21745.3	2132.74	1.866	45.8087	1.312
4020	3560.3	29883.8	22000.2	2224.60	1.807	45.8925	1.311
4060	3600.3	30219.1	22256.8	2319.38	1.750	45.9753	1.310
4100	3640.3	30555.0	22513.2	2417.69	1.696	46.0578	1.310
4140	3680.3	30891.1	22769.1	2519.14	1.643	46.1394	1.309
4180	3720.3	31227.8	23027.1	2623.96	1.593	46.2204	1.308
4220	3760.3	31566.3	23286.1	2731.96	1.545	46.3005	1.307
4260	3800.3	31904.5	23544.9	2844.50	1.498	46.3806	1.307
4300	3840.3	32243.1	23804.0	2959.95	1.453	46.4596	1.306
4340	3880.3	32582.4	24064.6	3079.53	1.409	46.5383	1.305
4380	3920.3	32922.2	24325.6	3203.30	1.367	46.6165	1.304
4420	3960.3	33263.4	24586.7	3330.60	1.327	46.6939	1.304
4460	4000.3	33604.6	24849.2	3461.81	1.288	46.7706	1.303

Temperature		h	u	Pr	Vr	s_o	Gamma
R	F	Btu/lbmol	Btu/lbmol			Btu/lbmol/R	
4500	4040.3	33947.7	25112.1	3597.46	1.251	46.8470	1.302
4540	4080.3	34288.8	25374.5	3737.92	1.215	46.9230	1.302
4580	4120.3	34632.8	25638.3	3882.32	1.180	46.9983	1.301
4620	4160.3	34976.3	25901.6	4030.11	1.146	47.0725	1.300
4660	4200.3	35318.8	26166.8	4183.69	1.114	47.1468	1.300
4700	4240.3	35664.8	26432.6	4342.71	1.082	47.2208	1.299
4740	4280.3	36010.7	26698.3	4504.54	1.052	47.2935	1.298
4780	4320.3	36354.3	26963.2	4672.92	1.023	47.3664	1.298
4820	4360.3	36702.3	27231.0	4846.26	0.995	47.4387	1.297
4860	4400.3	37049.9	27499.9	5023.59	0.967	47.5101	1.297
4900	4440.3	37397.7	27767.5	5207.93	0.941	47.5816	1.296
4940	4480.3	37745.1	28036.2	5396.59	0.915	47.6523	1.295
4980	4520.3	38094.5	28305.4	5591.04	0.891	47.7226	1.295
5020	4560.3	38442.8	28575.0	5790.58	0.867	47.7922	1.294
5060	4600.3	38793.1	28846.5	5998.17	0.844	47.8622	1.294
5100	4640.3	39141.9	29116.6	6207.91	0.822	47.9304	1.293
5140	4680.3	39495.4	29388.5	6425.75	0.800	47.9989	1.293
5180	4720.3	39846.4	29660.7	6649.98	0.779	48.0670	1.292
5220	4760.3	40196.6	29932.2	6879.93	0.759	48.1345	1.291
5260	4800.3	40549.5	30203.6	7117.84	0.739	48.2021	1.291
5300	4840.3	40902.4	30477.7	7361.83	0.720	48.2690	1.290
5340	4880.3	41256.2	30752.8	7610.55	0.702	48.3350	1.290
5380	4920.3	41610.1	31027.9	7868.25	0.684	48.4011	1.289
5420	4960.3	41960.6	31299.7	8132.56	0.666	48.4667	1.289
5460	5000.3	42316.4	31576.7	8406.08	0.650	48.5324	1.288
5500	5040.3	42674.0	31852.7	8681.65	0.634	48.5965	1.288
5540	5080.3	43026.8	32129.6	8971.08	0.618	48.6616	1.287
5580	5120.3	43385.2	32403.6	9264.15	0.602	48.7254	1.287
5620	5160.3	43743.0	32685.5	9566.16	0.587	48.7891	1.286
5660	5200.3	44099.9	32960.9	9875.62	0.573	48.8523	1.286
5700	5240.3	44455.9	33238.0	10197.0	0.559	48.9160	1.285
5740	5280.3	44814.2	33517.6	10519.0	0.546	48.9777	1.285
5780	5320.3	45171.6	33796.3	10854.5	0.532	49.0400	1.284

Appendix 14 Thermodynamic Properties for Water

A.14.1 The Saturation Temperature vs. Pressure (SI Units)

Press (kPa)	Temp (K)	Specific voulme (m**3/kg)		Internal energy (kJ/kg)			Enthalpy (kJ/kg)			Entropy (kJ/kg/K)		
		vf	vg	uf	ufg	ug	hf	hfg	hg	sf	sfg	g
0.60	273.2	0.001000	205.9975	0.00	2374.9	2374.9	0.00	2500.9	2500.9	0.00000	9.15549	9.15549
0.80	276.9	0.001000	159.6461	15.81	2364.3	2380.1	15.81	2492.0	2507.8	0.05748	8.99925	9.05672
1.00	280.1	0.001000	129.1833	29.30	2355.2	2384.5	29.30	2484.4	2513.7	0.10591	8.86902	8.97493
1.20	282.8	0.001000	108.6740	40.57	2347.6	2388.2	40.57	2478.0	2518.6	0.14595	8.76236	8.90831
1.40	285.1	0.001001	93.9033	50.28	2341.1	2391.4	50.28	2472.5	2522.8	0.18016	8.76236	8.85214
1.60	287.2	0.001001	82.7463	58.83	2335.3	2394.2	58.84	2467.7	2526.6	0.21005	8.59355	8.80360
1.80	289.0	0.001001	74.0143	66.49	2330.2	2396.7	66.49	2463.4	2529.9	0.23663	8.52424	8.76087
2.00	290.6	0.001001	66.9896	73.43	2325.5	2398.9	73.43	2459.5	2532.9	0.26058	8.46214	8.72272
2.50	294.2	0.001002	54.2421	88.43	2315.4	2403.8	88.43	2451.0	2539.4	0.31186	8.33030	8.64215
3.00	297.2	0.001003	45.6550	100.99	2306.9	2407.9	100.99	2443.9	2544.9	0.35433	8.22223	8.57656
3.50	299.8	0.001003	39.4678	111.83	2299.6	2411.4	111.84	2437.7	2549.6	0.39066	8.13060	8.52126
4.00	302.1	0.001004	34.7925	121.40	2293.1	2414.5	121.40	2432.3	2553.7	0.42245	8.05104	8.47349
5.00	306.0	0.001005	28.1863	137.76	2282.1	2419.8	137.76	2423.0	2560.8	0.47625	7.91766	8.39391
6.00	309.3	0.001006	23.7342	151.49	2272.8	2424.3	151.49	2415.2	2566.7	0.52087	7.80827	8.32915
7.00	312.2	0.001007	20.5252	163.36	2264.7	2428.1	163.37	2408.4	2571.8	0.55908	7.71549	8.27456
8.00	314.7	0.001008	18.0994	173.84	2257.6	2431.4	173.85	2402.4	2576.2	0.59253	7.63488	8.22741
9.00	316.9	0.001009	16.1997	183.25	2251.2	2434.5	183.26	2397.0	2580.3	0.62233	7.56359	8.18592
10.00	319.0	0.001010	14.6706	191.80	2245.4	2437.2	191.81	2392.1	2583.9	0.64922	7.49968	8.14889

Press (kPa)	Temp (K)	Specific voulme (m**3/kg)		Internal energy (kJ/kg)			Enthalpy (kJ/kg)			Entropy (kJ/kg/K)		
		vf	vg	uf	ufg	ug	hf	hfg	hg	sf	sfg	g
15.00	327.1	0.001014	10.0204	225.92	2222.1	2448.0	225.94	2372.4	2598.3	0.75484	7.25228	8.00712
20.00	333.2	0.001017	7.6482	251.38	2204.6	2456.0	251.40	2357.5	2608.9	0.83195	7.07528	7.90724
25.00	338.1	0.001020	6.2034	271.90	2190.5	2462.4	271.93	2345.5	2617.4	0.89309	6.93708	7.83016
30.00	342.2	0.001022	5.2286	289.20	2178.5	2467.7	289.23	2335.3	2624.6	0.94394	6.82351	7.76745
40.00	349.0	0.001026	3.9931	317.53	2158.8	2476.3	317.57	2318.5	2636.1	1.02590	6.64307	7.66897
50.00	354.5	0.001030	3.2401	340.42	2142.8	2483.2	340.48	2304.7	2645.2	1.09101	6.50196	7.59296
60.00	359.1	0.001033	2.7318	359.77	2129.2	2488.9	359.84	2293.0	2652.9	1.14524	6.38586	7.53110
70.00	363.1	0.001036	2.3649	376.61	2117.3	2493.9	376.68	2282.7	2659.4	1.19186	6.28709	7.47895
80.00	366.6	0.001038	2.0872	391.56	2106.6	2498.2	391.64	2273.5	2665.2	1.23283	6.20106	7.43389
90.00	369.8	0.001041	1.8695	405.03	2097.0	2502.1	405.13	2265.2	2670.3	1.26944	6.12479	7.39423
101.33	373.1	0.001043	1.6733	418.88	2087.1	2506.0	418.99	2256.5	2675.5	1.30672	6.04766	7.35439
120.00	377.9	0.001047	1.4284	439.17	2072.5	2511.6	439.30	2243.8	2683.1	1.36075	5.93688	7.29763
140.00	382.4	0.001051	1.2366	458.22	2058.6	2516.9	458.37	2231.6	2690.0	1.41085	5.83517	7.24602
160.00	386.4	0.001054	1.0914	475.17	2046.2	2521.4	475.34	2220.7	2696.0	1.45494	5.74643	7.20137
180.00	390.1	0.001058	0.9775	490.48	2035.0	2525.5	490.67	2210.7	2701.4	1.49437	5.66765	7.16203
200.00	393.4	0.001061	0.8857	504.47	2024.6	2529.1	504.68	2201.6	2706.2	1.53010	5.59676	7.12686
300.00	406.7	0.001073	0.6058	561.13	1982.0	2543.2	561.46	2163.4	2724.9	1.67176	5.31980	6.99157
400.00	416.8	0.001084	0.4624	604.29	1948.8	2553.1	604.72	2133.3	2738.1	1.77660	5.11882	6.89542
500.00	425.0	0.001093	0.3748	639.64	1921.1	2560.7	640.19	2107.9	2748.1	1.86060	4.95998	6.82058
600.00	432.0	0.001101	0.3156	669.84	1897.0	2566.8	670.50	2085.6	2756.1	1.93110	4.82807	6.75917
700.00	438.1	0.001108	0.2728	696.37	1875.4	2571.8	697.14	2065.6	2762.7	1.99208	4.71490	6.70698
800.00	443.6	0.001115	0.2403	720.13	1855.9	2576.0	721.02	2047.3	2768.3	2.04599	4.61555	6.66154
900.00	448.5	0.001121	0.2149	741.72	1837.9	2579.7	742.72	2030.3	2773.0	2.09440	4.52683	6.62124

Press (kPa)	Temp (K)	Specific voulme (m**3/kg)		Internal energy (kJ/kg)			Enthalpy (kJ/kg)			Entropy (kJ/kg/K)		
		vf	vg	uf	ufg	ug	hf	hfg	hg	sf	sfg	g
1000.00	453.0	0.001127	0.1943	761.56	1821.2	2582.8	762.68	2014.4	2777.1	2.13843	4.44655	6.58498
1200.00	461.1	0.001139	0.1632	797.13	1790.7	2587.9	798.50	1985.3	2783.8	2.21630	4.30539	6.52169
1400.00	468.2	0.001149	0.1408	828.52	1763.3	2591.8	830.13	1958.8	2788.9	2.28388	4.18364	6.46752
1600.00	474.5	0.001159	0.1237	856.76	1738.2	2594.9	858.61	1934.3	2792.9	2.34381	4.07621	6.42002
1800.00	480.3	0.001168	0.1104	882.51	1714.8	2597.3	884.61	1911.4	2796.0	2.39779	3.97980	6.37760
2000.00	485.5	0.001177	0.0996	906.27	1693.0	2599.2	908.62	1889.8	2798.4	2.44702	3.89214	6.33916
2500.00	497.1	0.001197	0.0799	958.99	1643.2	2602.2	961.98	1840.1	2802.0	2.55443	3.70155	6.25597
3000.00	507.0	0.001217	0.0667	1004.72	1598.6	2603.3	1008.37	1794.9	2803.3	2.64562	3.54017	6.18579
4000.00	523.5	0.001253	0.0498	1082.42	1519.4	2601.8	1087.43	1713.5	2800.9	2.79665	3.27306	6.06971
5000.00	537.1	0.001286	0.0394	1148.07	1448.9	2597.0	1154.50	1639.7	2794.2	2.92075	3.05296	5.97370
6000.00	548.7	0.001319	0.0324	1205.82	1384.1	2589.9	1213.73	1570.8	2784.6	3.02744	2.86263	5.89007
7000.00	559.0	0.001352	0.0274	1257.97	1322.9	2580.9	1267.44	1505.1	2772.6	3.12199	2.69264	5.81463
8000.00	568.2	0.001385	0.0235	1306.00	1264.4	2570.4	1317.08	1441.5	2758.6	3.20765	2.53720	5.74485
9000.00	576.5	0.001418	0.0205	1350.89	1207.6	2558.4	1363.65	1379.2	2742.9	3.28657	2.39244	5.67901
10000.00	584.1	0.001453	0.0180	1393.34	1151.8	2545.1	1407.87	1317.6	2725.5	3.36029	2.25560	5.61589
11000.00	591.2	0.001489	0.0160	1433.90	1096.6	2530.5	1450.28	1256.1	2706.4	3.42995	2.12458	5.55453
12000.00	597.8	0.001526	0.0143	1473.01	1041.3	2514.4	1491.33	1194.3	2685.6	3.49646	1.99766	5.49412
13000.00	604.0	0.001566	0.0128	1511.04	985.6	2496.7	1531.40	1131.5	2662.9	3.56058	1.87331	5.43388
14000.00	609.8	0.001610	0.0115	1548.34	928.9	2477.3	1570.88	1067.2	2638.1	3.62300	1.75005	5.37305
15000.00	615.3	0.001657	0.0103	1585.30	870.5	2455.8	1610.15	1000.7	2610.9	3.68445	1.62636	5.31080
16000.00	620.5	0.001710	0.0093	1622.32	809.6	2431.9	1649.67	931.1	2580.8	3.74568	1.50059	5.24627
17000.00	625.4	0.001769	0.0084	1659.98	745.2	2405.1	1690.05	857.4	2547.4	3.80770	1.37081	5.17850

Press (kPa)	Temp (K)	Specific voulme (m**3/kg)		Internal energy (kJ/kg)			Enthalpy (kJ/kg)			Entropy (kJ/kg/K)		
		vf	vg	uf	ufg	ug	hf	hfg	hg	sf	sfg	g
18000.00	630.1	0.001840	0.0075	1698.93	675.6	2374.6	1732.04	777.5	2509.5	3.87170	1.23384	5.10554
19000.00	634.6	0.001925	0.0036	1740.32	279.6	2019.9	1776.91	310.7	2087.6	3.93967	0.48827	4.42795
20000.00	638.9	0.002039	0.0034	1786.35	235.7	2022.1	1827.12	263.1	2090.2	4.01541	0.41112	4.42653
21000.00	643.0	0.002212	0.0033	1842.97	180.8	2023.8	1889.43	203.0	2092.4	4.10930	0.31555	4.42485
22064.00	647.1	0.003209	0.0032	2034.47	0.0	2034.5	2105.27	0.0	2105.3	4.43941	0.00000	4.43941

A.14.2 The Saturation Pressure vs. Temperature (SI Units)

Temp (K)	Press (kPa)	Specific voulme (m**3/kg)		Internal Energy (kJ/kg)			Ethalphy (kJ/kg)			Entropy (kJ/kg/K)		
		vf	vg	uf	ufg	ug	hf	hfg	hg	sf	sfg	sg
273.1	0.61	0.001000	206.1397	0.00	2374.9	2374.9	0.00	2500.9	2500.9	0.00000	9.15591	9.15591
275.0	0.70	0.001000	181.6044	7.76	2369.7	2377.4	7.76	2496.5	2504.3	0.02831	9.07831	9.10662
280.0	0.99	0.001000	130.1941	28.79	2355.5	2384.3	28.80	2484.7	2513.5	0.10412	8.87382	8.97794
285.0	1.39	0.001001	94.6073	49.78	2341.4	2391.2	49.78	2472.8	2522.6	0.17840	8.67661	8.85501
290.0	1.92	0.001001	69.6305	70.73	2327.3	2398.1	70.73	2461.0	2531.7	0.25128	8.48623	8.73751
295.0	2.62	0.001002	51.8694	91.66	2313.2	2404.9	91.66	2449.2	2540.8	0.32283	8.30229	8.62511
300.0	3.54	0.001003	39.0821	112.57	2299.1	2411.7	112.57	2437.3	2549.9	0.39312	8.12441	8.51754
305.0	4.72	0.001005	29.7669	133.47	2285.0	2418.4	133.48	2425.4	2558.9	0.46222	7.95228	8.41451
310.0	6.23	0.001007	22.9051	154.37	2270.8	2425.2	154.38	2413.5	2567.9	0.53018	7.78559	8.31577

Temp (K)	Press (kPa)	Specific voulme (m**3/kg)		Internal Energy (kJ/kg)			Ethalphy (kJ/kg)			Entropy (kJ/kg/K)		
		vf	vg	uf	ufg	ug	hf	hfg	hg	sf	sfg	sg
315.0	8.14	0.001009	17.7969	175.26	2256.6	2431.9	175.27	2401.6	2576.8	0.59704	7.62405	8.22110
320.0	10.55	0.001011	13.9557	196.16	2242.4	2438.6	196.17	2389.6	2585.7	0.66285	7.46741	8.13026
325.0	13.53	0.001013	11.0396	217.06	2228.1	2445.2	217.07	2377.5	2594.6	0.72765	7.31540	8.04305
330.0	17.21	0.001015	8.8056	237.96	2213.8	2451.8	237.98	2365.4	2603.3	0.79148	7.16779	7.95928
335.0	21.72	0.001018	7.0792	258.87	2199.4	2458.3	258.89	2353.2	2612.1	0.85438	7.02436	7.87875
340.0	27.19	0.001021	5.7341	279.80	2185.0	2464.8	279.82	2340.9	2620.7	0.91638	6.88491	7.80128
345.0	33.78	0.001024	4.6778	300.73	2170.5	2471.2	300.77	2328.5	2629.3	0.97751	6.74921	7.72673
350.0	41.68	0.001027	3.8420	321.69	2155.9	2477.6	321.73	2316.0	2637.7	1.03781	6.61710	7.65492
355.0	51.08	0.001030	3.1760	342.66	2141.2	2483.9	342.71	2303.4	2646.1	1.09731	6.48839	7.58570
360.0	62.19	0.001034	2.6415	363.66	2126.4	2490.1	363.72	2290.7	2654.4	1.15604	6.36289	7.51894
365.0	75.26	0.001037	2.2099	384.68	2111.5	2496.2	384.75	2277.8	2662.5	1.21403	6.24046	7.45449
370.0	90.54	0.001041	1.8591	405.72	2096.5	2502.3	405.81	2264.8	2670.6	1.27129	6.12094	7.39223
375.0	108.30	0.001045	1.5723	426.79	2081.4	2508.2	426.91	2251.6	2678.5	1.32787	6.00417	7.33204
380.0	128.85	0.001049	1.3365	447.90	2066.1	2514.0	448.03	2238.2	2686.3	1.38378	5.89001	7.27379
385.0	152.52	0.001053	1.1415	469.04	2050.7	2519.8	469.20	2224.7	2693.9	1.43904	5.77834	7.21738
390.0	179.64	0.001058	0.9794	490.21	2035.2	2525.4	490.40	2210.9	2701.3	1.49369	5.66901	7.16270
395.0	210.59	0.001062	0.8440	511.43	2019.5	2530.9	511.65	2197.0	2708.6	1.54775	5.56190	7.10964
400.0	245.75	0.001067	0.7303	532.68	2003.5	2536.2	532.95	2182.8	2715.7	1.60122	5.45690	7.05812
405.0	285.55	0.001072	0.6345	553.99	1987.5	2541.4	554.29	2168.3	2722.6	1.65415	5.35389	7.00804
410.0	330.42	0.001077	0.5533	575.33	1971.2	2546.5	575.69	2153.6	2729.3	1.70654	5.25277	6.95931
415.0	380.82	0.001082	0.4842	596.73	1954.7	2551.4	597.15	2138.7	2735.8	1.75843	5.15343	6.91186
420.0	437.24	0.001087	0.4253	618.19	1938.0	2556.1	618.66	2123.4	2742.1	1.80982	5.05578	6.86560

Temp (K)	Press (kPa)	Specific voulme (m**3/kg)		Internal Energy (kJ/kg)			Ethalphy (kJ/kg)			Entropy (kJ/kg/K)		
		vf	vg	uf	ufg	ug	hf	hfg	hg	sf	sfg	sg
425.0	500.18	0.001093	0.3747	639.70	1921.0	2560.7	640.24	2107.9	2748.1	1.86074	4.95972	6.82046
430.0	570.18	0.001098	0.3311	661.27	1903.8	2565.1	661.90	2092.0	2753.9	1.91121	4.86516	6.77637
435.0	647.77	0.001104	0.2935	682.91	1886.4	2569.3	683.62	2075.8	2759.4	1.96124	4.77202	6.73327
440.0	733.55	0.001110	0.2609	704.61	1868.7	2573.3	705.43	2059.3	2764.7	2.01086	4.68022	6.69108
445.0	828.10	0.001117	0.2326	726.39	1850.7	2577.1	727.32	2042.4	2769.7	2.06009	4.58966	6.64975
455.0	1046.02	0.001130	0.1862	770.18	1813.9	2584.1	771.37	2007.4	2778.8	2.15744	4.41199	6.56943
460.0	1170.68	0.001137	0.1672	792.21	1795.0	2587.2	793.54	1989.4	2782.9	2.20560	4.32472	6.53032
465.0	1306.72	0.001144	0.1504	814.33	1775.8	2590.1	815.82	1970.8	2786.7	2.25345	4.23839	6.49183
470.0	1454.84	0.001152	0.1356	836.55	1756.2	2592.7	838.22	1951.9	2790.1	2.30099	4.15292	6.45392
475.0	1615.75	0.001159	0.1226	858.87	1736.3	2595.1	860.74	1932.4	2793.2	2.34826	4.06826	6.41652
480.0	1790.19	0.001167	0.1110	881.30	1715.9	2597.2	883.39	1912.5	2795.9	2.39527	3.98431	6.37958
485.0	1978.94	0.001176	0.1006	903.85	1695.2	2599.0	906.18	1892.0	2798.2	2.44204	3.90101	6.34305
490.0	2182.77	0.001185	0.0914	926.53	1674.0	2600.6	929.11	1871.0	2800.1	2.48859	3.81829	6.30688
495.0	2402.48	0.001194	0.0832	949.34	1652.4	2601.8	952.20	1849.3	2801.5	2.53495	3.73606	6.27101
500.0	2638.90	0.001203	0.0758	972.29	1630.3	2602.6	975.46	1827.1	2802.6	2.58113	3.65426	6.23539
505.0	2892.85	0.001213	0.0691	995.40	1607.8	2603.2	998.90	1804.3	2803.2	2.62717	3.57281	6.19997
510.0	3165.22	0.001223	0.0632	1018.66	1584.7	2603.3	1022.53	1780.7	2803.3	2.67307	3.49163	6.16470
515.0	3456.86	0.001233	0.0578	1042.10	1561.0	2603.1	1046.37	1756.5	2802.8	2.71888	3.41064	6.12952
520.0	3768.70	0.001245	0.0529	1065.73	1536.8	2602.5	1070.42	1731.5	2801.9	2.76461	3.32977	6.09437
525.0	4101.65	0.001256	0.0485	1089.55	1511.9	2601.4	1094.70	1705.7	2800.4	2.81029	3.24892	6.05921
530.0	4456.65	0.001268	0.0445	1113.58	1486.4	2599.9	1119.23	1679.1	2798.3	2.85595	3.16802	6.02396
535.0	4834.69	0.001281	0.0409	1137.84	1460.1	2598.0	1144.04	1651.5	2795.6	2.90162	3.08696	5.98857

Temp (K)	Press (kPa)	Specific voulme (m**3/kg)		Internal Energy (kJ/kg)			Ethalphy (kJ/kg)			Entropy (kJ/kg/K)		
		vf	vg	uf	ufg	ug	hf	hfg	hg	sf	sfg	sg
540.0	5236.75	0.001294	0.0376	1162.35	1433.1	2595.5	1169.13	1623.1	2792.2	2.94734	3.00564	5.95298
545.0	5663.85	0.001308	0.0345	1187.12	1405.4	2592.5	1194.53	1593.6	2788.1	2.99314	2.92397	5.91711
550.0	6117.05	0.001323	0.0318	1212.18	1376.7	2588.9	1220.27	1563.0	2783.3	3.03906	2.84182	5.88087
555.0	6597.43	0.001339	0.0292	1237.54	1347.2	2584.7	1246.37	1531.3	2777.6	3.08515	2.75905	5.84420
560.0	7106.12	0.001355	0.0269	1263.25	1316.6	2579.9	1272.88	1498.3	2771.2	3.13146	2.67553	5.80700
565.0	7644.26	0.001373	0.0248	1289.32	1285.0	2574.3	1299.82	1464.0	2763.8	3.17805	2.59110	5.76915
570.0	8213.06	0.001392	0.0228	1315.80	1252.2	2568.0	1327.23	1428.2	2755.4	3.22497	2.50557	5.73054
575.0	8813.76	0.001412	0.0210	1342.73	1218.0	2560.8	1355.17	1390.8	2745.9	3.27231	2.41873	5.69103
580.0	9447.69	0.001433	0.0193	1370.16	1182.5	2552.7	1383.70	1351.6	2735.3	3.32014	2.33033	5.65048
585.0	10116.21	0.001457	0.0178	1398.14	1145.4	2543.5	1412.88	1310.5	2723.3	3.36858	2.24011	5.60869
590.0	10820.77	0.001482	0.0163	1426.75	1106.4	2533.2	1442.79	1267.1	2709.9	3.41772	2.14771	5.56544
595.0	11562.92	0.001510	0.0150	1456.07	1065.5	2521.6	1473.53	1221.4	2694.9	3.46773	2.05273	5.52046
600.0	12344.30	0.001540	0.0137	1486.21	1022.2	2508.5	1505.22	1172.8	2678.0	3.51877	1.95463	5.47340
605.0	13166.69	0.001573	0.0126	1517.30	976.3	2493.6	1538.01	1120.9	2658.9	3.57108	1.85272	5.42380
610.0	14032.02	0.001611	0.0114	1549.53	927.1	2476.6	1572.14	1065.1	2637.3	3.62498	1.74610	5.37108
615.0	14942.40	0.001654	0.0104	1583.17	873.9	2457.1	1607.89	1004.6	2612.5	3.68092	1.63352	5.31444
620.0	15900.20	0.001704	0.0094	1618.61	815.8	2434.4	1645.70	938.2	2583.9	3.73955	1.51329	5.25284
630.0	17969.08	0.001837	0.0075	1697.70	677.9	2375.6	1730.71	780.1	2510.8	3.86968	1.23822	5.10789
640.0	20265.91	0.002076	0.0034	1799.92	222.7	2022.7	1842.00	248.9	2090.9	4.03783	0.38844	4.42627
647.1	22063.97	0.003209	0.0032	2034.52	0.0	2034.5	2105.33	0.0	2105.3	4.43950	0.00000	4.43950

A.14.3 Superheated Steam Table (SI Units)

Press = 0.010 MPa, Tsat = 319.0 K

T(K)	325.	350.	400.	450.	500.	550.	600.	650.	700.	800.	900.	1000.
v-m³/kg	14.9541	16.1208	18.4417	20.7557	23.0669	25.3768	27.6860	29.9947	32.3032	36.9196	41.5356	46.1514
u-kj/kg	2446.0	2482.1	2554.3	2627.2	2701.3	2776.8	2853.8	2932.4	3012.6	3177.9	3350.3	3529.6
h-kj/kg	2595.6	2643.3	2738.7	2834.7	2932.0	3030.6	3130.7	3232.3	3335.6	3547.1	3765.6	3991.2
s-kj/kg/K	8.18513	8.32676	8.58137	8.80763	9.01251	9.20046	9.37463	9.53734	9.69038	9.97277	10.23002	10.46758

Press = 0.015 MPa, Tsat = 327.1 K

T(K) =	350.	400.	450.	500.	550.	600.	650.	700.	800.	900.	1000.	1100.
v-m³/kg	10.7362	12.2880	13.8328	15.3748	16.9155	18.4555	19.9950	21.5343	24.6123	27.6899	30.7672	33.8444
u-kj/kg	2481.4	2553.9	2627.0	2701.2	2776.7	2853.7	2932.3	3012.5	3177.9	3350.2	3529.6	3716.1
h-kj/kg	2642.5	2738.2	2834.5	2931.8	3030.4	3130.6	3232.2	3335.5	3547.1	3765.6	3991.1	4223.8
s-kj/kg/K	8.13764	8.39336	8.62004	8.82509	9.01314	9.18737	9.35011	9.50318	9.78559	10.04286	10.28043	10.50212

Press = 0.025 MPa, Tsat = 338.1 K

T(K) =	350.	400.	450.	500.	550.	600.	650.	700.	800.	900.	1000.	1100.
v-m³/kg	6.4284	7.3650	8.2944	9.2211	10.1464	11.0710	11.9952	12.9191	14.7664	16.6132	18.4599	20.3063
u-kj/kg	2480.0	2553.2	2626.6	2700.9	2776.5	2853.6	2932.2	3012.4	3177.8	3350.2	3529.6	3716.1
h-kj/kg	2640.7	2737.3	2833.9	2931.4	3030.2	3130.4	3232.1	3335.4	3547.0	3765.5	3991.1	4223.7
s-kj/kg/K	7.89785	8.15585	8.38335	8.58878	8.77701	8.95134	9.11416	9.26727	9.54974	9.80704	10.04463	10.26633

Press = 0.040 MPa, Tsat = 349.0 K

T(K) =.	350.	400.	450.	500.	550.	600.	650.	700.	800.	900.	1000.	1100.
v-m³/kg	4.0050	4.5957	5.1791	5.7596	6.3388	6.9173	7.4953	8.0731	9.2281	10.3826	11.5370	12.6912
u-kj/kg	2477.8	2552.1	2625.9	2700.5	2776.2	2853.3	2932.0	3012.3	3177.7	3350.1	3529.5	3716.0
h-kj/kg	2638.0	2736.0	2833.1	2930.8	3029.8	3130.0	3231.8	3335.2	3546.9	3765.4	3991.0	4223.7

s-kj/kg/K	7.67465	7.93627	8.16504	8.37102	8.55954	8.73403	8.89695	9.05013	9.33269	9.59003	9.82764	10.04936
Press=0.060 MPa	Tsat=359.1 K											
T(K)=.	400.	450.	500.	550.	600.	650.	700.	800.	900.	1000.	1100.	1200.
v-m³/kg	3.05719	3.44835	3.83661	4.22350	4.60967	4.99541	5.38086	6.15124	6.92120	7.69092	8.46050	9.22998
u-kj/kg	2550.7	2625.1	2699.9	2775.8	2853.0	2931.8	3012.0	3177.6	3350.0	3529.4	3716.0	3909.4
h-kj/kg	2734.1	2832.0	2930.1	3029.2	3129.6	3231.5	3334.9	3546.7	3765.3	3990.9	4223.6	4463.2
s-kj/kg/K	7.74553	7.97603	8.18276	8.37166	8.54637	8.70943	8.86270	9.14537	9.40277	9.64043	9.86217	10.07059
Press=0.101 MPa	Tsat=373.1 K											
T(K)=	400.	450.	500.	550.	600.	650.	700.	800.	900.	1000.	1100.	1200.
v-m³/kg	1.80206	2.03654	2.26798	2.49803	2.72735	2.95623	3.18482	3.64148	4.09772	4.55372	5.00959	5.46535
u-kj/kg	2547.7	2623.3	2698.7	2774.9	2852.4	2931.2	3011.6	3177.3	3349.8	3529.3	3715.8	3909.3
h-kj/kg	2730.3	2829.7	2928.5	3028.1	3128.7	3230.8	3334.3	3546.3	3765.0	3990.7	4223.4	4463.0
s-kj/kg/K	7.49608	7.73027	7.93859	8.12828	8.30345	8.46679	8.62025	8.90315	9.16068	9.39842	9.62020	9.82866
Press=0.200 MPa	Tsat=393.4 K											
T(K)=	400.	450.	500.	550.	600.	650.	700.	800.	900.	1000.	1100.	1200.
v-m³/kg	0.90251	1.02507	1.14427	1.26200	1.37897	1.49549	1.61172	1.84366	2.07517	2.30645	2.53758	2.76862
u-kj/kg	2540.0	2619.0	2695.9	2772.9	2850.8	2930.0	3010.6	3176.6	3349.2	3528.8	3715.5	3909.0
h-kj/kg	2720.5	2824.0	2924.8	3025.3	3126.6	3229.1	3332.9	3545.3	3764.3	3990.1	4223.0	4462.7
s-kj/kg/K	7.16292	7.40676	7.61907	7.81074	7.98701	8.15104	8.30496	8.58842	8.84625	9.08417	9.30606	9.51459
Press=0.300 MPa	Tsat=406.7 K											
T(K)=	450.	500.	550.	600.	650.	700.	800.	900.	1000.	1100.	1200.	1300.
v-m³/kg	0.67872	0.75960	0.83891	0.91743	0.99550	1.07327	1.22829	1.38288	1.53724	1.69145	1.84557	1.99962
u-kj/kg	2614.4	2693.0	2770.8	2849.2	2928.7	3009.6	3175.8	3348.7	3528.4	3715.1	3908.7	4108.8
h-kj/kg	2818.1	2920.8	3022.5	3124.5	3227.4	3331.5	3544.3	3763.5	3989.6	4222.6	4462.4	4708.7
s-kj/kg/K	7.20939	7.42604	7.61979	7.79722	7.96195	8.11634	8.40036	8.65851	8.89661	9.11862	9.32722	9.52438

Press = 0.400 MPa   Tsat = 416.8 K

T(K)=	450.	500.	550.	600.	650.	700.	800.	900.	1000.	1100.	1200.	1300.
v-m³/kg	0.50545	0.56722	0.62735	0.68666	0.74550	0.80405	0.92061	1.03674	1.15263	1.26838	1.38404	1.49963
u-kj/kg	2609.7	2690.0	2768.7	2847.6	2927.4	3008.5	3175.1	3348.1	3528.0	3714.8	3908.4	4108.6
h-kj/kg	2811.9	2916.9	3019.6	3122.3	3225.6	3330.1	3543.3	3762.8	3989.0	4222.2	4462.0	4708.5
s-kj/kg/K	7.06594	7.28723	7.48316	7.66177	7.82722	7.98208	8.26667	8.52513	8.76341	8.98553	9.19421	9.39142

Press = 0.500 MPa   Tsat = 425.0 K

T(K)=	450.	500.	550.	600.	650.	700.	800.	900.	1000.	1100.	1200.	1300.
v-m³/kg	0.40140	0.45177	0.50040	0.54819	0.59550	0.64251	0.73599	0.82905	0.92187	1.01455	1.10712	1.19964
u-kj/kg	2604.8	2686.9	2766.6	2846.0	2926.2	3007.5	3174.3	3347.6	3527.6	3714.5	3908.1	4108.4
h-kj/kg	2805.5	2912.8	3016.8	3120.1	3223.9	3328.7	3542.3	3762.1	3988.5	4221.7	4461.7	4708.2
s-kj/kg/K	6.95182	7.17807	7.37626	7.55608	7.72226	7.87760	8.16276	8.42153	8.66000	8.88223	9.09098	9.28825

Press = 0.600 MPa   Tsat = 432.0 K

T(K)=	450.	500.	550.	600.	650.	700.	800.	900.	1000.	1100.	1200.	1300.
v-m³/kg	0.33195	0.37477	0.41576	0.45587	0.49549	0.53482	0.61292	0.69059	0.76803	0.84532	0.92251	0.99965
u-kj/kg	2599.7	2683.8	2764.4	2844.4	2924.9	3006.4	3173.6	3347.0	3527.1	3714.1	3907.8	4108.1
h-kj/kg	2798.8	2908.7	3013.9	3117.9	3222.2	3327.3	3541.4	3761.4	3988.0	4221.3	4461.3	4707.9
s-kj/kg/K	6.85603	7.08760	7.28815	7.46921	7.63613	7.79195	8.07769	8.33677	8.57542	8.79777	9.00659	9.20392

Press = 0.700 MPa   Tsat = 438.1 K

T(K)=	450.	500.	550.	600.	650.	700.	800.	900.	1000.	1100.	1200.	1300.
v-m³/kg	0.28228	0.31976	0.35529	0.38992	0.42406	0.45789	0.52501	0.59170	0.65814	0.72444	0.79065	0.85679
u-kj/kg	2594.3	2680.6	2762.2	2842.7	2923.6	3005.4	3172.9	3346.5	3526.7	3713.8	3907.5	4107.9
h-kj/kg	2791.9	2904.4	3010.9	3115.7	3220.4	3325.9	3540.4	3760.7	3987.4	4220.9	4461.0	4707.6
s-kj/kg/K	6.77268	7.00998	7.21299	7.39533	7.56300	7.71931	8.00562	8.26502	8.50385	8.72631	8.93521	9.13258

Press = 0.800 MPa   Tsat = 443.6 K

Press = 0.900 MPa (Tsat = 448.5 K)

T(K)=	450.	500.	550.	600.	650.	700.	800.	900.	1000.	1100.	1200.	1300.
v-m³/kg	0.24496	0.27847	0.30993	0.34046	0.37049	0.40020	0.45908	0.51752	0.57572	0.63379	0.69175	0.74965
u-kj/kg	2588.7	2677.4	2760.0	2841.1	2922.3	3004.3	3172.1	3345.9	3526.3	3713.4	3907.3	4107.6
h-kj/kg	2784.7	2900.1	3007.9	3113.5	3218.7	3324.5	3539.4	3759.9	3986.9	4220.4	4460.7	4707.4
s-kj/kg/K	6.69816	6.94173	7.14729	7.33094	7.49938	7.65617	7.94306	8.20278	8.44180	8.66437	8.87334	9.07077

Press = 1.000 MPa (Tsat = 453.0 K)

T(K)=	450.	500.	550.	600.	650.	700.	800.	900.	1000.	1100.	1200.	1300.
v-m³/kg	0.21586	0.24634	0.27465	0.30199	0.32881	0.35533	0.40779	0.45983	0.51162	0.56327	0.61483	0.66632
u-kj/kg	2582.7	2674.0	2757.8	2839.4	2921.0	3003.3	3171.4	3345.4	3525.9	3713.1	3907.0	4107.4
h-kj/kg	2777.0	2895.7	3004.9	3111.2	3216.9	3323.1	3538.4	3759.2	3986.3	4220.0	4460.3	4707.1
s-kj/kg/K	6.63002	6.88060	7.08881	7.27380	7.44301	7.60031	7.88778	8.14781	8.38701	8.60970	8.81874	9.01622

Press = 1.200 MPa (Tsat = 461.1 K)

T(K)=	500.	550.	600.	650.	700.	800.	900.	1000.	1100.	1200.	1300.	1400.
v-m³/kg	0.22063	0.24641	0.27121	0.29547	0.31943	0.36677	0.41368	0.46034	0.50687	0.55329	0.59966	0.64598
u-kj/kg	2670.6	2755.5	2837.8	2919.7	3002.2	3170.6	3344.8	3525.4	3712.7	3906.7	4107.1	4313.9
h-kj/kg	2891.3	3001.9	3109.0	3215.2	3321.6	3537.4	3758.5	3985.8	4219.6	4460.0	4706.8	4959.9
s-kj/kg/K	6.82505	7.03601	7.22237	7.39237	7.55017	7.83822	8.09857	8.33796	8.56076	8.76988	8.96741	9.15493

Press = 1.400 MPa (Tsat = 468.2 K)

T(K)=	500.	550.	600.	650.	700.	800.	900.	1000.	1100.	1200.	1300.	1400.
v-m³/kg	0.18201	0.20405	0.22503	0.24546	0.26557	0.30523	0.34445	0.38342	0.42225	0.46099	0.49966	0.53829
u-kj/kg	2663.7	2750.9	2834.4	2917.1	3000.1	3169.2	3343.7	3524.6	3712.0	3906.1	4106.7	4313.5
h-kj/kg	2882.1	2995.7	3104.4	3211.6	3318.8	3535.4	3757.1	3984.7	4218.7	4459.3	4706.2	4959.4
s-kj/kg/K	6.72659	6.94336	7.13256	7.30417	7.46298	7.75221	8.01319	8.25296	8.47598	8.68525	8.88289	9.07048

T(K)=	500.	550.	600.	650.	700.	800.	900.	1000.	1100.	1200.	1300.	1400.

v-m³/kg	0.15437	0.17376	0.19203	0.20974	0.22711	0.26127	0.29500	0.32848	0.36181	0.39506	0.42823	0.46137
u-kj/kg	2656.4	2746.2	2831.0	2914.4	2998.0	3167.7	3342.6	3523.7	3711.3	3905.5	4106.2	4313.1
h-kj/kg	2872.5	2989.4	3099.8	3208.1	3315.9	3533.4	3755.6	3983.6	4217.9	4458.6	4705.7	4959.0
s-kj/kg/K	6.64047	6.86348	7.05567	7.22894	7.38878	7.67920	7.94082	8.18096	8.40421	8.61362	8.81137	8.99904
Press=1.600 MPa	Tsat=474.5 K											
T(K)=	500.	550.	600.	650.	700.	800.	900.	1000.	1100.	1200.	1300.	1400.
v-m³/kg	0.13360	0.15103	0.16728	0.18294	0.19825	0.22830	0.25791	0.28727	0.31649	0.34561	0.37466	0.40368
u-kj/kg	2648.8	2741.3	2827.5	2911.8	2995.8	3166.2	3341.5	3522.9	3710.7	3905.0	4105.7	4312.6
h-kj/kg	2862.6	2983.0	3095.2	3204.5	3313.0	3531.5	3754.2	3982.5	4217.0	4457.9	4705.1	4958.5
s-kj/kg/K	6.56317	6.79289	6.98820	7.16319	7.32408	7.61571	7.87796	8.11847	8.34195	8.55151	8.74936	8.93711
Press=1.800 MPa	Tsat=480.3 K											
T(K)=	500.	550.	600.	650.	700.	800.	900.	1000.	1100.	1200.	1300.	1400.
v-m³/kg	0.11740	0.13334	0.14802	0.16209	0.17581	0.20266	0.22906	0.25522	0.28123	0.30715	0.33300	0.35881
u-kj/kg	2640.9	2736.4	2824.0	2909.1	2993.7	3164.7	3340.4	3522.0	3710.0	3904.4	4105.2	4312.2
h-kj/kg	2852.2	2976.4	3090.5	3200.9	3310.1	3529.5	3752.7	3981.4	4216.2	4457.2	4704.6	4958.1
s-kj/kg/K	6.49238	6.72934	6.92792	7.10466	7.26663	7.55947	7.82237	8.06326	8.28696	8.49667	8.69463	8.38246
Press=2.000 MPa	Tsat=485.5 K											
T(K)=.	500.	550.	600.	650.	700.	800.	900.	1000.	1100.	1200.	1300.	1400.
v-m³/kg	0.10439	0.11917	0.13261	0.14541	0.15786	0.18215	0.20599	0.22958	0.25303	0.27638	0.29967	0.32291
u-kj/kg	2632.6	2731.3	2820.5	2906.4	2991.5	3163.2	3339.3	3521.2	3709.3	3903.8	4104.7	4311.8
h-kj/kg	2841.4	2969.7	3085.7	3197.2	3307.2	3527.5	3751.3	3980.3	4215.3	4456.6	4704.0	4957.6
s-kj/kg/K	6.42645	6.67128	6.87328	7.05184	7.21490	7.50897	7.77251	8.01377	8.23771	8.44757	8.64563	8.33353
Press=2.500 MPa	Tsat=497.1 K											
T(K)=	500.	550.	600.	650.	700.	800.	900.	1000.	1100.	1200.	1300.	1400.
v-m³/kg	0.08080	0.09361	0.10484	0.11538	0.12553	0.14522	0.16445	0.18342	0.20226	0.22099	0.23967	0.25830

u-kj/kg	2609.7	2718.1	2811.4	2899.5	2986.1	3159.4	3336.5	3519.0	3707.5	3902.4	4103.5	4310.8
h-kj/kg	2811.7	2952.2	3073.5	3188.0	3299.9	3522.5	3747.6	3977.6	4213.2	4454.8	4702.7	4956.5
s-kj/kg/K	6.27535	6.54361	6.75484	6.93818	7.10407	7.40126	7.66643	7.90864	8.13314	8.34338	8.54170	8.72980
Press=3.000 MPa	Tsat=507.0 K											
T(K)=	550.	600.	650.	700.	800.	900.	1000.	1100.	1200.	1300.	1400.	1500.
v-m³/kg	0.07650	0.08631	0.09534	0.10397	0.12060	0.13675	0.15265	0.16841	0.18407	0.19967	0.21522	0.23074
u-kj/kg	2704.1	2801.9	2892.5	2980.5	3155.6	3333.7	3516.8	3705.8	3900.9	4102.3	4309.7	4523.0
h-kj/kg	2933.7	3060.9	3178.5	3292.5	3517.4	3744.0	3974.8	4211.0	4453.1	4701.3	4955.4	5215.2
s-kj/kg/K	6.43310	6.65462	6.84310	7.01196	7.31236	7.57918	7.82233	8.04742	8.25802	8.45662	8.64491	8.82414
Press=3.500 MPa	Tsat=515.7 K											
T(K)=	550.	600.	650.	700.	800.	900.	1000.	1100.	1200.	1300.	1400.	1500.
v-m³/kg	0.06422	0.07304	0.08102	0.08857	0.10301	0.11697	0.13068	0.14424	0.15770	0.17110	0.18445	0.19777
u-kj/kg	2689.3	2792.2	2885.4	2974.9	3151.8	3330.9	3514.7	3704.1	3899.5	4101.0	4308.7	4522.1
h-kj/kg	2914.0	3047.8	3168.9	3284.9	3512.4	3740.3	3972.1	4208.9	4451.4	4699.9	4954.3	5214.3
s-kj/kg/K	6.33371	6.56677	6.76077	6.93274	7.23644	7.50492	7.74903	7.97469	8.18567	8.38453	8.57302	8.75240
Press=4.000 MPa	Tsat=523.5 K											
T(K)=	550.	600.	650.	700.	800.	900.	1000.	1100.	1200.	1300.	1400.	1500.
v-m³/kg	0.05494	0.06306	0.07027	0.07701	0.08982	0.10213	0.11419	0.12610	0.13792	0.14967	0.16138	0.17305
u-kj/kg	2673.4	2782.0	2878.1	2969.3	3148.0	3328.1	3512.5	3702.3	3898.0	4099.8	4307.6	4521.2
h-kj/kg	2893.1	3034.3	3159.1	3277.3	3507.3	3736.7	3969.3	4206.7	4449.7	4698.5	4953.1	5213.4
s-kj/kg/K	6.24171	6.48774	6.68767	6.86291	7.17000	7.44017	7.68524	7.91147	8.12284	8.32196	8.51064	8.69017
Press=4.500 MPa	Tsat=530.6 K											
T(K)=	550.	600.	650.	700.	800.	900.	1000.	1100.	1200.	1300.	1400.	1500.
v-m³/kg	0.04765	0.05528	0.06189	0.06802	0.07955	0.09059	0.10137	0.11200	0.12254	0.13300	0.14343	0.15382
u-kj/kg	2656.3	2771.5	2870.6	2963.5	3144.1	3325.3	3510.4	3700.6	3896.6	4098.6	4306.6	4520.3

h-kj/kg	2870.7	3020.3	3149.1	3269.6	3502.1	3733.0	3966.5	4204.6	4448.0	4697.1	4952.0	5212.5
s-kj/kg/K	6.15444	6.41525	6.62155	6.80022	7.11079	7.38267	7.62871	7.85553	8.06727	8.26666	8.45554	8 63521
Press=5.000 MPa	Tsat=537.1 K											
T(K)=	550.	600.	650.	700.	800.	900.	1000.	1100.	1200.	1300.	1400.	1500.
v-m³/kg	0.04175	0.04904	0.05519	0.06082	0.07134	0.08136	0.09112	0.10072	0.11023	0.11967	0.12907	0 13843
u-kj/kg	2637.7	2760.6	2862.9	2957.7	3140.3	3322.5	3508.2	3698.8	3895.1	4097.4	4305.5	4519.4
h-kj/kg	2846.5	3005.8	3138.9	3261.8	3497.0	3729.3	3963.8	4202.4	4446.2	4695.7	4950.9	5211.6
s-kj/kg/K	6.06984	6.34772	6.56087	6.74313	7.05728	7.33089	7.57790	7.80531	8.01743	8.21709	8.40617	8 58598
Press=6.000 MPa	Tsat=548.7 K											
T(K)=.	550.	600.	650.	700.	800.	900.	1000.	1100.	1200.	1300.	1400.	1500.
v-m³/kg	0.03267	0.03961	0.04510	0.05001	0.05902	0.06751	0.07573	0.08380	0.09177	0.09967	0.10753	0.11536
u-kj/kg	2594.6	2737.5	2847.1	2945.8	3132.5	3316.8	3503.8	3695.3	3892.2	4094.9	4303.5	4517.6
h-kj/kg	2790.6	2975.2	3117.7	3245.9	3486.6	3721.9	3958.2	4198.1	4442.8	4693.0	4948.7	5209.7
s-kj/kg/K	5.90115	6.22319	6.45166	6.64164	6.96328	7.24041	7.48939	7.71798	7.93087	8.13107	8.32053	8.50064
Press=7.000 MPa	Tsat=559.0 K											
T(K)=	600.	650.	700.	800.	900.	1000.	1100.	1200.	1300.	1400.	1500.	1600.
v-m³/kg	0.03280	0.03787	0.04227	0.05022	0.05762	0.06474	0.07171	0.07858	0.08539	0.09215	0.09888	0.10558
u-kj/kg	2712.3	2830.5	2933.5	3124.5	3311.1	3499.4	3691.8	3889.3	4092.5	4301.4	4515.8	4735.4
h-kj/kg	2941.9	3095.6	3229.5	3476.1	3714.4	3952.6	4193.8	4439.4	4690.2	4946.4	5207.9	5474.5
s-kj/kg/K	6.10772	6.35409	6.55258	6.88213	7.16288	7.41385	7.64363	7.85730	8.05804	8.24789	8.42829	8.50030
Press=8.000 MPa	Tsat=568.2 K											
T(K)=	600.	650.	700.	800.	900.	1000.	1100.	1200.	1300.	1400.	1500.	1600.
v-m³/kg	0.02761	0.03241	0.03646	0.04361	0.05020	0.05650	0.06264	0.06869	0.07468	0.08061	0.08652	0.09239
u-kj/kg	2684.6	2813.1	2920.9	3116.5	3305.3	3495.0	3688.3	3886.4	4090.0	4299.3	4514.0	4733.8
h-kj/kg	2905.5	3072.4	3212.6	3465.4	3706.9	3947.0	4189.4	4435.9	4687.4	4944.2	5206.1	5473.0

s-kj/kg/K	5.99692	6.26455	6.47242	6.81035	7.09482	7.34781	7.57880	7.79324	7.99453	8.18477	8.36547	8.53770
Press=9.000 MPa	Tsat=576.5 K											
T(K)=	600.	650.	700.	800.	900.	1000.	1100.	1200.	1300.	1400.	1500.	1600.
v-m³/kg	0.02349	0.02815	0.03193	0.03847	0.04442	0.05009	0.05559	0.06100	0.06634	0.07164	0.07690	0.08214
u-kj/kg	2653.8	2794.8	2907.9	3108.4	3299.5	3490.6	3684.7	3883.4	4087.5	4297.2	4512.2	4732.3
h-kj/kg	2865.2	3048.1	3195.2	3454.6	3699.3	3941.4	4185.1	4432.5	4684.6	4941.9	5204.3	5471.5
s-kj/kg/K	5.88729	6.18066	6.39887	6.74568	7.03398	7.28902	7.52122	7.73645	7.93828	8.12892	8.30991	8.48236
Press=10.000 MPa	Tsat=584.1 K											
T(K)=	600.	650.	700.	800.	900.	1000.	1100.	1200.	1300.	1400.	1500.	1600.
v-m³/kg	0.02009	0.02471	0.02829	0.03436	0.03980	0.04496	0.04995	0.05485	0.05968	0.06446	0.06921	0.07393
u-kj/kg	2618.9	2775.4	2894.4	3100.1	3293.7	3486.2	3681.2	3880.5	4085.1	4295.1	4510.4	4730.7
h-kj/kg	2819.8	3022.5	3177.3	3443.7	3691.7	3935.8	4180.7	4429.0	4681.9	4939.7	5202.5	5470.0
s-kj/kg/K	5.77538	6.10069	6.33038	6.68660	6.97883	7.23595	7.46938	7.68539	7.88777	8.07881	8.26009	8.43276
Press=12.000 MPa	Tsat=597.8 K											
T(K)=	600.	650.	700.	750.	800.	850.	900.	950.	1000.	1050.		
v-m³/kg	0.01459	0.01947	0.02280	0.02562	0.02818	0.03058	0.03287	0.03510	0.03727	0.03940		
u-kj/kg	2528.8	2733.1	2866.2	2979.0	3083.3	3183.5	3281.9	3379.6	3477.2	3575.3		
h-kj/kg	2703.9	2966.8	3139.8	3286.5	3421.4	3550.4	3676.4	3800.7	3924.4	4048.1		
s-kj/kg/K	5.52472	5.94753	6.20430	6.40683	6.58105	6.73752	6.88148	7.01595	7.14285	7.26349		
Press=14.000 MPa	Tsat=609.8 K											
T(K)=		650.	700.	750.	800.	850.	900.	950.	1000.	1050.		
v-m³/kg		0.01563	0.01885	0.02145	0.02376	0.02589	0.02792	0.02987	0.03177	0.03363		
u-kj/kg		2684.9	2835.9	2957.0	3066.0	3169.3	3269.9	3369.2	3468.2	3567.3		
h-kj/kg		2903.7	3099.8	3257.3	3398.6	3531.8	3660.8	3787.5	3913.0	4038.1		
s-kj/kg/K		5.79671	6.08799	6.30540	6.48786	6.64947	6.79693	6.93391	7.06267	7.18475		

Press=16.000 MPa	Tsat=620.5 K								
T(K)=	650.	700.	750.	800.	850.	900.	950.	1000.	1050.
v-m³/kg	0.01263	0.01586	0.01831	0.02044	0.02238	0.02421	0.02596	0.02765	0.02930
u-kj/kg	2628.3	2803.4	2933.9	3048.2	3154.8	3257.8	3358.8	3459.1	3559.3
h-kj/kg	2830.4	3057.1	3226.9	3375.1	3512.9	3645.1	3774.1	3901.5	4028.1
s-kj/kg/K	5.64059	5.97752	6.21205	6.40350	6.57059	6.72171	6.86128	6.99195	7.11548

Press=18.000 MPa	Tsat=630.1 K								
T(K)=.	650.	700.	750.	800.	850.	900.	950.	1000.	1050.
v-m³/kg	0.01014	0.01350	0.01586	0.01785	0.01964	0.02132	0.02291	0.02445	0.02594
u-kj/kg	2559.0	2768.3	2909.9	3029.8	3140.1	3245.4	3348.3	3449.9	3551.2
h-kj/kg	2741.5	3011.2	3195.3	3351.1	3493.7	3629.2	3760.7	3890.0	4018.1
s-kj/kg/K	5.46893	5.87012	6.12446	6.32572	6.49863	6.65356	6.79580	6.92843	7.05344

Press=20.000 MPa	Tsat=638.9 K								
T(K)=	650.	700.	750.	800.	850.	900.	950.	1000.	1050.
v-m³/kg	0.00790	0.01158	0.01388	0.01577	0.01745	0.01901	0.02047	0.02188	0.02325
u-kj/kg	2466.8	2730.1	2884.7	3011.0	3125.0	3232.9	3337.6	3440.7	3543.0
h-kj/kg	2624.9	2961.6	3162.4	3326.5	3474.1	3613.1	3747.1	3878.4	4008.0
s-kj/kg/K	5.26183	5.76355	6.04101	6.25298	6.43206	6.59096	6.73595	6.87059	6.99709

Press=22.000 MPa	Tsat=646.9 K								
T(K)=	650.	700.	750.	800.	850.	900.	950.	1000.	1050.
v-m³/kg	0.00147	0.00997	0.01226	0.01407	0.01566	0.01711	0.01848	0.01979	0.02105
u-kj/kg	1596.4	2688.3	2858.4	2991.6	3109.7	3220.3	3326.9	3431.4	3534.9
h-kj/kg	1698.6	2907.8	3128.1	3301.2	3454.2	3596.8	3733.5	3866.7	3997.9
s-kj/kg/K	3.69088	5.65591	5.96051	6.18419	6.36978	6.53282	6.68065	6.81732	6.94536

Press = 24.000 MPa

T(K) =	650.	700.	750.	800.	850.	900.	950.	1000.	1050.
v-m³/kg	0.00146	0.00861	0.01090	0.01265	0.01416	0.01554	0.01682	0.01804	0.01921
u-kj/kg	1590.2	2642.3	2830.8	2971.7	3094.1	3207.5	3316.1	3422.0	3526.6
h-kj/kg	1695.8	2848.9	3092.4	3275.4	3434.0	3580.4	3719.7	3855.0	3987.8
s-kj/kg/K	3.68017	5.54534	5.88208	6.11853	6.31101	6.47834	6.62908	6.76784	6.89743

Press = 26.000 MPa

T(K) =	650.	700.	750.	800.	850.	900.	950.	1000.	1050.
v-m³/kg	0.00145	0.00742	0.00974	0.01145	0.01290	0.01420	0.01542	0.01656	0.01766
u-kj/kg	1584.4	2591.2	2801.9	2951.2	3078.2	3194.5	3305.1	3412.6	3518.4
h-kj/kg	1693.3	2784.1	3055.2	3248.9	3413.6	3563.8	3705.9	3843.2	3977.7
s-kj/kg/K	3.67008	5.42987	5.80500	6.05541	6.25513	6.42692	6.58065	6.72153	6.85270

Press = 28.000 MPa

T(K) =	650.	700.	750.	800.	850.	900.	950.	1000.	1050.
v-m³/kg	0.00144	0.00637	0.00874	0.01042	0.01182	0.01306	0.01421	0.01530	0.01634
u-kj/kg	1578.9	2533.7	2771.6	2930.2	3062.0	3181.4	3294.1	3403.1	3510.1
h-kj/kg	1691.1	2712.0	3016.4	3221.9	3392.8	3547.1	3692.0	3831.5	3967.5
s-kj/kg/K	3.66050	5.30724	5.72875	5.99434	6.20169	6.37808	6.53487	6.67792	6.81068

Press = 30.000 MPa

T(K) =	650.	700.	750.	800.	850.	900.	950.	1000.	1050.
v-m³/kg	0.00143	0.00543	0.00787	0.00952	0.01088	0.01207	0.01317	0.01420	0.01519
u-kj/kg	1573.7	2468.6	2739.9	2908.7	3045.6	3168.1	3283.0	3393.6	3501.8
h-kj/kg	1689.1	2631.5	2976.2	3194.4	3371.8	3530.2	3678.1	3819.7	3957.4
s-kj/kg/K	3.65138	5.17540	5.65290	5.93499	6.15032	6.33146	6.49139	6.63664	6.77102

Press = 35.000 MPa

T(K)=	650.	700.	750.	800.	850.	900.	950.	1000.	1050.
v-m³/kg	0.00141	0.00141	0.00613	0.00773	0.00900	0.01010	0.01109	0.01201	0.01289
u-kj/kg	1561.7	1701.7	2654.4	2852.5	3003.4	3134.3	3254.9	3369.6	3480.9
h-kj/kg	1685.0	1899.7	2868.9	3123.2	3318.4	3487.6	3643.0	3790.1	3932.0
s-kj/kg/K	3.63027	3.83785	5.46347	5.79237	6.02927	6.22286	6.39089	6.54179	6.68027

Press = 40.000 MPa

T(K)=	650.	700.	750.	800.	850.	900.	950.	1000.	1050.
v-m³/kg	0.00139	0.00139	0.00483	0.00640	0.00760	0.00862	0.00953	0.01038	0.01117
u-kj/kg	1550.8	1691.1	2560.8	2793.3	2959.9	3099.8	3226.4	3345.4	3459.9
h-kj/kg	1681.8	1897.7	2754.1	3049.3	3263.9	3444.6	3607.7	3760.4	3906.6
s-kj/kg/K	3.61114	3.81901	5.27412	5.65605	5.91650	6.12322	6.29967	6.45636	6.59905

A.14.4 H_2O Compressed Liquid Table (SI units)

Temperature = 273.1 K	Psat = 0.001 Mpa									
Press(MPa)=	1	2	5	8	10	15	20	25	30	50
v-m³/kg	0.001	0.001	0.001	0.001	0.001	0.00099	0.00099	0.00099	0.00099	0.00098
u-kj/kg	0	0	0	0.1	0.1	0.2	0.2	0.3	0.3	0.3
h-kj/kg	1	2	5	8.1	10.1	15.1	20	25	29.9	49.1
s-kj/kg/K	0	0	0.00014	0.00027	0.00034	0.00045	0.00047	0.00041	0.00028	0
Temperature = 280.0 K	Psat = 0.001 MPa									
Press(MPa)=	1	2	5	8	10	15	20	25	30	50

	P=1	P=2	P=5	P=8	P=10	P=15	P=20	P=25	P=30	P=50
v-m³/kg	0.001	0.001	0.001	0.001	0.001	0.00099	0.00099	0.00099	0.00099	0.00098
u-kj/kg	28.8	28.8	28.7	28.7	28.7	28.6	28.5	28.4	28.3	27.9
h-kj/kg	29.8	30.8	33.7	36.7	38.6	43.5	48.3	53.1	57.9	76.7
s-kj/kg/K	0.10407	0.10402	0.10386	0.10368	0.10354	0.10315	0.1027	0.10219	0.10162	0.09879
Temperature = 300.0 K (Psat = 0.004 Mpa)										
Press(MPa) =	1	2	5	8	10	15	20	25	30	50
v-m³/kg	0.001	0.001	0.001	0.001	0.001	0.001	0.00099	0.00099	0.00099	0.00098
u-kj/kg	112.5	112.4	112.2	111.9	111.8	111.4	111	110.6	110.2	108.6
h-kj/kg	113.5	114.4	117.2	119.9	121.7	126.3	130.8	135.4	139.9	157.8
s-kj/kg/K	0.39285	0.39257	0.39174	0.39089	0.39033	0.3889	0.38744	0.38597	0.38448	0.37831
Temperature = 320.0 K (Psat = 0.011 MPa)										
Press(MPa) =	1	2	5	8	10	15	20	25	30	50
v-m³/kg	0.00101	0.00101	0.00101	0.00101	0.00101	0.001	0.001	0.001	0.001	0.00099
u-kj/kg	196	195.9	195.5	195	194.8	194.1	193.4	192.8	192.1	189.7
h-kj/kg	197	197.9	200.5	203.1	204.8	209.2	213.5	217.8	222.1	239.2
s-kj/kg/K	0.66242	0.66198	0.66066	0.65934	0.65847	0.65628	0.65409	0.65191	0.64973	0.64104
Temperature = 340.0 K (Psat = 0.027 MPa)										
Press(MPa) =	1	2	5	8	10	15	20	25	30	50
v-m³/kg	0.00102	0.00102	0.00102	0.00102	0.00102	0.00101	0.00101	0.00101	0.00101	0.001
u-kj/kg	279.6	279.4	278.8	278.3	277.9	276.9	276	275.1	274.2	270.9
h-kj/kg	280.6	281.5	283.9	286.4	288	292.2	296.3	300.4	304.5	320.9
s-kj/kg/K	0.91582	0.91524	0.91352	0.91181	0.91067	0.90785	0.90504	0.90226	0.89949	0.88862
Temperature = 360.0 K (Psat = 0.062 MPa)										
Press(MPa) =	1	2	5	8	10	15	20	25	30	50
v-m³/kg	0.00103	0.00103	0.00103	0.00103	0.00103	0.00103	0.00102	0.00102	0.00102	0.00101
u-kj/kg	363.4	363.2	362.4	361.7	361.2	360	358.8	357.7	356.5	352.2

h-kj/kg	402.8	387.1	383.2	379.3	375.4	371.5	369.9	367.6	365.2	364.5
s-kj/kg/K	1.12283	1.13572	1.13902	1.14236	1.14573	1.14914	1.15052	1.15259	1.15468	1.15538

Temperature = 380.0 K, Psat = 0.129 MPa

Press(MPa)=	50	30	25	20	15	10	8	5	2	1
v-m³/kg	0.00102	0.00103	0.00104	0.00104	0.00104	0.00104	0.00104	0.00105	0.00105	0.00105
u-kj/kg	433.8	439.1	440.5	441.9	443.4	444.9	445.5	446.4	447.3	447.6
h-kj/kg	485.1	470.1	466.4	462.7	459	455.3	453.8	451.6	449.4	448.7
s-kj/kg/K	1.34526	1.36011	1.36394	1.36782	1.37175	1.37574	1.37735	1.37978	1.38224	1.38306

Temperature = 400.0 K, Psat = 0.246 MPa

Press(MPa)=	50	30	25	20	15	10	8	5	2	1
v-m³/kg	0.00104	0.00105	0.00105	0.00106	0.00106	0.00106	0.00106	0.00106	0.00107	0.00107
u-kj/kg	515.8	522.1	523.8	525.5	527.3	529	529.8	530.9	532	532.4
h-kj/kg	567.8	553.6	550.1	546.6	543.1	539.7	538.3	536.2	534.1	533.5
s-kj/kg/K	1.5573	1.57417	1.57854	1.58298	1.58749	1.59207	1.59392	1.59672	1.59955	1.60051

Temperature = 420.0 K, Psat = 0.437 MPa

Press(MPa)=	50	30	25	20	15	10	8	5	2	1
v-m³/kg	0.00106	0.00107	0.00107	0.00107	0.00108	0.00108	0.00108	0.00108	0.00109	0.00109
u-kj/kg	598.1	605.6	607.6	609.6	611.7	613.9	614.8	616.1	617.5	617.9
h-kj/kg	650.9	637.7	634.4	631.1	627.9	624.7	623.4	621.5	619.6	619
s-kj/kg/K	1.76019	1.7792	1.78415	1.78919	1.79432	1.79954	1.80166	1.80487	1.80811	1.8092

Temperature = 440.0 K, Psat = 0.734 MPa

Press(MPa)=	50	30	25	20	15	10	8	5	2	1
v-m³/kg	0.00108	0.00109	0.00109	0.0011	0.0011	0.0011	0.0011	0.00111	0.00111	0.00111
u-kj/kg	680.9	689.8	692.1	694.6	697.1	699.6	700.7	702.3	703.9	704.5
h-kj/kg	734.7	722.4	719.5	716.5	713.6	710.7	709.5	707.8	706.1	705.6
s-kj/kg/K	1.95505	1.9764	1.98199	1.9877	1.99352	1.99947	2.00189	2.00556	2.00928	2.01053

Temperature = 460.0 K, Psat = 1.171 MPa

Press(MPa) =	1	2	5	8	10	15	20	25	30	50
v-m³/kg	0.19849	0.00114	0.00113	0.00113	0.00113	0.00113	0.00112	0.00112	0.00111	0.0011
u-kj/kg	2597	791.7	789.7	787.8	786.5	783.4	780.5	777.6	774.8	764.3
h-kj/kg	2795.5	793.9	795.4	796.8	797.8	800.3	802.9	805.5	808.2	819.2
s-kj/kg/K	6.62522	2.20441	2.20012	2.19591	2.19314	2.18633	2.1797	2.17322	2.16689	2.1429

Temperature = 480.0 K, Psat = 1.790 MPa

Press(MPa) =	1	2	5	8	10	15	20	25	30	50
v-m³/kg	0.20981	0.00117	0.00116	0.00116	0.00116	0.00115	0.00115	0.00114	0.00114	0.00112
u-kj/kg	2634.9	881.1	878.7	876.4	874.9	871.2	867.6	864.2	860.9	848.5
h-kj/kg	2844.7	883.5	884.6	885.7	886.5	888.5	890.6	892.8	895.1	904.7
s-kj/kg/K	6.73004	2.39491	2.38993	2.38504	2.38184	2.374	2.36639	2.35899	2.35179	2.32474

Temperature = 500.0 K, Psat = 2.639 MPa

Press(MPa) =	1	2	5	8	10	15	20	25	30	50
v-m³/kg	0.22063	0.10439	0.0012	0.0012	0.00119	0.00119	0.00118	0.00118	0.00117	0.00115
u-kj/kg	2670.6	2632.6	970	967.1	965.3	960.8	956.5	952.3	948.4	933.8
h-kj/kg	2891.3	2841.4	976	976.7	977.2	978.6	980.1	981.7	983.5	991.3
s-kj/kg/K	6.82505	6.42645	2.57651	2.57075	2.56699	2.55784	2.54901	2.54049	2.53223	2.50153

Temperature = 520.0 K, Psat = 3.769 MPa

Press(MPa) =	1	2	5	8	10	15	20	25	30	50
v-m³/kg	0.23111	0.11055	0.00124	0.00124	0.00123	0.00123	0.00122	0.00121	0.00121	0.00118
u-kj/kg	2705.1	2674	1064.2	1060.7	1058.3	1052.8	1047.5	1042.5	1037.7	1020.3
h-kj/kg	2936.2	2895.1	1070.4	1070.6	1070.7	1071.2	1071.9	1072.8	1073.8	1079.4
s-kj/kg/K	6.91324	6.53188	2.7617	2.75477	2.75027	2.73938	2.72897	2.719	2.70941	2.67426

Temperature = 540.0 K, Psat = 5.237 MPa

Press(MPa) =	1	2	5	8	10	15	20	25	30	50
v-m³/kg				0.00129	0.00128	0.00127	0.00126	0.00125	0.00125	0.00122
u-kj/kg				1158.1	1155.1	1148.1	1141.5	1135.2	1129.3	1108.5
h-kj/kg				1168.4	1168	1167.2	1166.7	1166.6	1166.7	1169.3
s-kj/kg/K				2.93939	2.93382	2.9205	2.90793	2.89602	2.88469	2.84393

Temperature = 560.0 K, Psat = 7.106 MPa

Press(MPa) =	1	2	5	8	10	15	20	25	30	50
v-m³/kg				0.00135	0.00135	0.00133	0.00132	0.00131	0.00129	0.00126
u-kj/kg				1261.4	1257.4	1248	1239.4	1231.5	1224.1	1198.7
h-kj/kg				1272.2	1270.8	1268	1265.8	1264.2	1263	1261.6
s-kj/kg/K				3.12812	3.12086	3.1038	3.08808	3.07344	3.0597	3.01163

Temperature = 580.0 K, Psat = 9.45 MPa

Press(MPa) =	1	2	5	8	10	15	20	25	30	50
v-m³/kg					0.00143	0.00141	0.00139	0.00137	0.00135	0.00131
u-kj/kg					1368.5	1355.1	1343.3	1332.8	1323.2	1291.5
h-kj/kg					1382.9	1376.2	1371.1	1367	1363.8	1356.7
s-kj/kg/K					3.31731	3.29363	3.27272	3.25386	3.23661	3.17855

Temperature = 600.0 K, Psat = 12.34 MPa

Press(MPa) =	1	2	5	8	10	15	20	25	30	50
v-m³/kg						0.00152	0.00148	0.00145	0.00143	0.00136
u-kj/kg						1474.8	1456.6	1441.5	1428.3	1387.5
h-kj/kg						1497.5	1486.3	1477.8	1471.2	1455.6
s-kj/kg/K						3.49917	3.46794	3.44163	3.41862	3.34608

Temperature=620.0 K	Psat= 15.9 Mpa									
Press(MPa)=	1	2	5	8	10	15	20	25	30	50
v-m3/kg							0.00163	0.00157	0.00153	0.00143
u-kj/kg							1588.6	1562.9	1542.9	1487.6
h-kj/kg							1621.2	1602.3	1588.9	1559.1
s-kj/kg/K							3.68902	3.64556	3.6115	3.5159

Temperature=640.0 K	Psat= 20.27 Mpa									
Press(MPa)=	1	2	5	8	10	15	20	25	30	50
v-m3/kg								0.01322	0.01322	0.01322
u-kj/kg								2616.4	2616.4	2616.4
h-kj/kg								2815.1	2815.1	2815.1
s-kj/kg/K							5.63646	5.63646	5.63646	5.63646

A.14.5 The Saturation Temperature vs. Pressure (English Units)

Press (psi)	Temp (R)	Specific volume (ft**3/lbm)		Intrenal energy (Btu/lbm)			Enthalpy (Btu/lbm)			Entropy (Btu/lbm/R)		
		vf	vg	uf	ufg	ug	hf	hfg	hg	sf	sfg	sg
0.1	494.7	0.0161	2951.1577	3.01	1021.2	1024.2	3.02	1075.8	1078.8	0.00611	2.17479	2.18090
0.2	512.8	0.0161	1529.1317	21.25	1008.9	1030.2	21.25	1065.5	1086.8	0.04231	2.07785	2.12017
0.3	524.1	0.0161	1041.6294	32.60	1001.3	1033.9	32.60	1059.1	1091.7	0.06421	2.02075	2.08496
0.4	532.5	0.0161	793.5160	41.00	995.6	1036.6	41.00	1054.4	1095.4	0.08010	1.98002	2.06013
0.6	544.8	0.0161	541.0339	53.35	987.3	1040.6	53.36	1047.4	1100.7	0.10305	1.92229	2.02534
0.8	554.0	0.0161	412.4360	62.52	981.1	1043.6	62.52	1042.1	1104.7	0.11973	1.88108	2.00081

Press (psi)	Temp (R)	Specific volume (ft**3/lbm)		Internal energy (Btu/lbm)			Enthalpy (Btu/lbm)			Entropy (Btu/lbm/R)		
		vf	vg	uf	ufg	ug	hf	hfg	hg	sf	sfg	sg
1.0	561.4	0.0162	334.2068	69.88	976.1	1046.0	69.88	1037.9	1107.8	0.13292	1.84897	1.98188
1.5	575.3	0.0162	228.1557	83.83	966.6	1050.4	83.83	1029.9	1113.8	0.15746	1.79023	1.94769
2.5	594.1	0.0163	141.1594	102.58	953.8	1056.4	102.59	1019.1	1121.7	0.18954	1.71547	1.90501
3.0	601.1	0.0163	118.9527	109.62	949.0	1058.6	109.63	1015.0	1124.6	0.20132	1.68857	1.88989
3.5	607.2	0.0164	102.9342	115.73	944.7	1060.5	115.74	1011.4	1127.1	0.21143	1.66571	1.87715
4.0	612.6	0.0164	90.8181	121.13	941.0	1062.1	121.15	1008.2	1129.4	0.22030	1.64584	1.86614
5.0	621.9	0.0164	73.6775	130.43	934.6	1065.0	130.44	1002.7	1133.2	0.23536	1.61246	1.84781
6.0	629.7	0.0165	62.1094	138.27	929.1	1067.4	138.29	998.0	1136.3	0.24789	1.58501	1.83290
7.0	636.5	0.0165	53.7612	145.08	924.3	1069.4	145.11	994.0	1139.1	0.25865	1.56168	1.82033
8.0	642.5	0.0166	47.4442	151.13	920.1	1071.2	151.15	990.3	1141.5	0.26811	1.54137	1.80947
10.0	652.8	0.0166	38.5033	161.54	912.7	1074.3	161.57	984.0	1145.5	0.28418	1.50721	1.79139
12.5	663.6	0.0167	31.2495	172.36	905.0	1077.4	172.40	977.3	1149.7	0.30062	1.47276	1.77337
14.7	671.6	0.0167	26.8597	180.48	899.2	1079.7	180.52	972.2	1152.8	0.31278	1.44757	1.76035
20.0	687.6	0.0169	20.1340	196.61	887.6	1084.2	196.67	962.0	1158.7	0.33651	1.39911	1.73562
25.0	699.7	0.0170	16.3406	208.88	878.6	1087.5	208.96	954.1	1163.1	0.35421	1.36356	1.71777
30.0	710.0	0.0170	13.7771	219.30	870.9	1090.2	219.40	947.3	1166.6	0.36900	1.33420	1.70320
35.0	718.9	0.0171	11.9251	228.41	864.1	1092.5	228.52	941.2	1169.7	0.38175	1.30914	1.69089
40.0	726.9	0.0172	10.5224	236.53	857.9	1094.5	236.66	935.7	1172.4	0.39298	1.28724	1.68022
45.0	734.1	0.0172	9.4220	243.88	852.3	1096.2	244.02	930.7	1174.7	0.40304	1.26777	1.67081
50.0	740.7	0.0173	8.5350	250.61	847.2	1097.8	250.77	926.0	1176.8	0.41217	1.25022	1.66239
55.0	746.7	0.0174	7.8041	256.82	842.4	1099.2	257.00	921.6	1178.6	0.42052	1.23424	1.65476
60.0	752.4	0.0174	7.1912	262.61	837.9	1100.5	262.80	917.5	1180.3	0.42824	1.21955	1.64779

Press (psi)	Temp (R)	Specific volume (ft**3/lbm)		Intrenal energy (Btu/lbm)			Enthalpy (Btu/lbm)			Entropy (Btu/lbm/R)		
		vf	vg	uf	ufg	ug	hf	hfg	hg	sf	sfg	sg
65.0	757.6	0.0175	6.6696	268.02	833.7	1101.7	268.24	913.7	1181.9	0.43542	1.20595	1.64137
70.0	762.6	0.0175	6.2201	273.13	829.6	1102.8	273.35	910.0	1183.3	0.44213	1.19329	1.63542
75.0	767.3	0.0176	5.8286	277.95	825.8	1103.8	278.19	906.5	1184.7	0.44844	1.18143	1.62987
80.0	771.7	0.0176	5.4844	282.53	822.2	1104.7	282.79	903.1	1185.9	0.45439	1.17028	1.62467
85.0	775.9	0.0176	5.1794	286.89	818.7	1105.6	287.17	899.9	1187.0	0.46003	1.15975	1.61977
90.0	779.9	0.0177	4.9071	291.06	815.3	1106.4	291.36	896.8	1188.1	0.46539	1.14977	1.61515
95.0	783.8	0.0177	4.6626	295.05	812.1	1107.1	295.37	893.7	1189.1	0.47049	1.14028	1.61078
100.0	787.5	0.0178	4.4417	298.89	809.0	1107.9	299.22	890.8	1190.1	0.47538	1.13124	1.60661
110.0	794.5	0.0178	4.0581	306.14	803.0	1109.2	306.50	885.3	1191.8	0.48454	1.11432	1.59886
120.0	800.9	0.0179	3.7365	312.90	797.4	1110.3	313.29	880.0	1193.3	0.49302	1.09874	1.59176
130.0	807.0	0.0180	3.4626	319.24	792.1	1111.4	319.67	875.0	1194.7	0.50091	1.08429	1.58520
140.0	812.7	0.0181	3.2267	325.22	787.1	1112.3	325.69	870.2	1195.9	0.50830	1.07080	1.57910
150.0	818.1	0.0181	3.0211	330.89	782.3	1113.2	331.40	865.7	1197.1	0.51525	1.05814	1.57340
160.0	823.2	0.0182	2.8404	336.28	777.7	1114.0	336.82	861.3	1198.1	0.52183	1.04622	1.56805
170.0	828.1	0.0183	2.6802	341.43	773.3	1114.7	342.00	857.0	1199.0	0.52806	1.03494	1.56300
180.0	832.8	0.0183	2.5373	346.35	769.0	1115.4	346.96	852.9	1199.9	0.53399	1.02423	1.55822
190.0	837.2	0.0184	2.4088	351.07	764.9	1116.0	351.72	849.0	1200.7	0.53965	1.01403	1.55367
200.0	841.5	0.0184	2.2928	355.62	760.9	1116.5	356.30	845.1	1201.4	0.54506	1.00429	1.54935
250.0	860.7	0.0187	1.8478	376.11	742.6	1118.7	376.97	827.2	1204.2	0.56915	0.96114	1.53029
300.0	877.0	0.0189	1.5467	393.80	726.3	1120.1	394.85	811.1	1206.0	0.58953	0.92485	1.51438
350.0	891.4	0.0192	1.3290	409.48	711.5	1121.0	410.72	796.3	1207.1	0.60728	0.89335	1.50063
400.0	904.3	0.0194	1.1640	423.66	697.8	1121.5	425.09	782.6	1207.6	0.62308	0.86537	1.48845

Press (psi)	Temp (R)	Specific volume (ft**3/lbm)		Intrenal energy (Btu/lbm)			Enthalpy (Btu/lbm)			Entropy (Btu/lbm/R)		
		vf	vg	uf	ufg	ug	hf	hfg	hg	sf	sfg	sg
450.0	916.0	0.0196	1.0345	436.64	685.0	1121.6	438.27	769.5	1207.8	0.63737	0.84010	1.47747
500.0	926.7	0.0198	0.9301	448.67	672.9	1121.6	450.50	757.1	1207.6	0.65045	0.81698	1.46743
600.0	945.9	0.0202	0.7718	470.48	650.3	1120.8	472.72	733.7	1206.5	0.67378	0.77569	1.44947
700.0	962.8	0.0206	0.6572	490.01	629.3	1119.4	492.68	711.8	1204.5	0.69430	0.73931	1.43361
800.0	977.9	0.0209	0.5704	507.83	609.6	1117.4	510.93	691.0	1201.9	0.71271	0.70654	1.41925
900.0	991.7	0.0213	0.5021	524.31	590.8	1115.1	527.86	670.9	1198.7	0.72950	0.67650	1.40600
1000.0	1004.3	0.0216	0.4470	539.73	572.7	1112.4	543.73	651.4	1195.1	0.74501	0.64861	1.39362
1100.0	1016.0	0.0220	0.4015	554.27	555.1	1109.4	558.75	632.4	1191.1	0.75948	0.62242	1.38190
1200.0	1026.9	0.0224	0.3632	568.10	538.0	1106.1	573.07	613.7	1186.8	0.77310	0.59760	1.37069
1300.0	1037.2	0.0227	0.3306	581.34	521.2	1102.5	586.81	595.2	1182.0	0.78600	0.57389	1.35990
1400.0	1046.8	0.0231	0.3023	594.07	504.6	1098.6	600.06	576.9	1177.0	0.79831	0.55110	1.34941
1500.0	1055.9	0.0235	0.2776	606.38	488.1	1094.5	612.91	558.6	1171.5	0.81012	0.52904	1.33916
1600.0	1064.6	0.0239	0.2558	618.34	471.7	1090.0	625.42	540.4	1165.8	0.82151	0.50757	1.32907
1800.0	1080.7	0.0248	0.2189	641.42	438.9	1080.3	649.67	503.5	1153.2	0.84327	0.46589	1.30916
2000.0	1095.5	0.0257	0.1886	663.74	405.4	1069.2	673.25	465.7	1139.0	0.86408	0.42510	1.28918
2250.0	1112.4	0.0270	0.1573	691.19	361.7	1052.9	702.44	416.0	1118.4	0.88942	0.37393	1.26335
2500.0	1127.8	0.0287	0.1309	719.11	314.2	1033.3	732.37	361.5	1093.9	0.91498	0.32054	1.23552
2750.0	1142.0	0.0308	0.0573	749.09	121.2	870.3	764.78	134.6	899.4	0.94233	0.11756	1.05990
3000.0	1155.1	0.0345	0.0531	785.51	86.3	871.8	804.64	96.7	901.3	0.97571	0.08362	1.05933
3200.0	1164.8	0.0517	0.0517	877.11	0.0	877.1	907.70	0.0	907.7	1.06317	0.00000	1.06317

A.14.6 The Saturation Pressure vs. Temperature (English Units)

Press (psi)	Temp (R)	Specific volume (ft**3/lbm)		Intrenal energy (Btu/lbm)			Enthalpy (Btu/lbm)			Entropy (Btu/lbm/R)		
		vf	vg	uf	ufg	ug	hf	hfg	hg	sf	sfg	sg
491.7	0.09	0.0161	3308.9548	0.00	1023.2	1023.2	0.00	1077.5	1077.5	0.00000	2.19152	2.19152
500.0	0.12	0.0161	2418.8743	8.38	1017.6	1026.0	8.38	1072.8	1081.2	0.01690	2.14557	2.16247
510.0	0.18	0.0161	1686.7268	18.44	1010.8	1029.3	18.44	1067.1	1085.6	0.03681	2.09238	2.12920
520.0	0.26	0.0161	1195.2240	28.47	1004.1	1032.5	28.47	1061.5	1089.9	0.05630	2.04126	2.09755
530.0	0.37	0.0161	859.8069	38.49	997.3	1035.8	38.49	1055.8	1094.3	0.07538	1.99205	2.06744
540.0	0.51	0.0161	627.3451	48.50	990.6	1039.1	48.50	1050.1	1098.6	0.09410	1.94466	2.03876
550.0	0.71	0.0161	463.8765	58.51	983.8	1042.3	58.51	1044.4	1102.9	0.11246	1.89896	2.01142
560.0	0.96	0.0162	347.3375	68.51	977.0	1045.5	68.51	1038.7	1107.2	0.13048	1.85486	1.98535
570.0	1.29	0.0162	263.1730	78.51	970.2	1048.7	78.52	1033.0	1111.5	0.14819	1.81227	1.96045
580.0	1.71	0.0162	201.6415	88.52	963.4	1051.9	88.52	1027.2	1115.8	0.16558	1.77109	1.93668
590.0	2.25	0.0163	156.1334	98.52	956.6	1055.1	98.53	1021.4	1120.0	0.18269	1.73126	1.91395
600.0	2.92	0.0163	122.1055	108.53	949.7	1058.2	108.54	1015.6	1124.2	0.19951	1.69269	1.89220
610.0	3.75	0.0164	96.3962	118.55	942.8	1061.3	118.56	1009.7	1128.3	0.21606	1.65531	1.87137
620.0	4.78	0.0164	76.7798	128.57	935.9	1064.4	128.58	1003.8	1132.4	0.23236	1.61906	1.85142
630.0	6.05	0.0165	61.6720	138.60	928.9	1067.5	138.62	997.8	1136.5	0.24841	1.58387	1.83228
640.0	7.57	0.0165	49.9327	148.64	921.8	1070.5	148.66	991.8	1140.5	0.26422	1.54969	1.81391
650.0	9.42	0.0166	40.7337	158.69	914.8	1073.4	158.72	985.7	1144.4	0.27981	1.51645	1.79626
660.0	11.62	0.0167	33.4671	168.76	907.6	1076.4	168.79	979.5	1148.3	0.29518	1.48411	1.77928
670.0	14.23	0.0167	27.6832	178.84	900.4	1079.2	178.88	973.3	1152.1	0.31034	1.45261	1.76294

Press (psi)	Temp (R)	Specific volume (ft**3/lbm))		Intrenal energy (Btu/lbm)			Enthalpy (Btu/lbm)			Entropy (Btu/lbm/R)		
		vf	vg	uf	ufg	ug	hf	hfg	hg	sf	sfg	sg
680.0	17.31	0.0168	23.0457	188.93	893.1	1082.1	188.99	966.9	1155.9	0.32529	1.42190	1.74720
690.0	20.92	0.0169	19.3017	199.05	885.8	1084.8	199.11	960.5	1159.6	0.34006	1.39195	1.73201
700.0	25.13	0.0170	16.2589	209.18	878.4	1087.5	209.26	953.9	1163.2	0.35464	1.36270	1.71734
710.0	30.02	0.0170	13.7705	219.33	870.8	1090.2	219.43	947.2	1166.7	0.36904	1.33412	1.70316
720.0	35.64	0.0171	11.7232	229.51	863.2	1092.7	229.62	940.4	1170.1	0.38327	1.30616	1.68943
730.0	42.10	0.0172	10.0291	239.71	855.5	1095.2	239.84	933.5	1173.4	0.39734	1.27880	1.67613
740.0	49.48	0.0173	8.6197	249.93	847.7	1097.6	250.09	926.5	1176.6	0.41125	1.25198	1.66323
750.0	57.86	0.0174	7.4409	260.18	839.8	1100.0	260.37	919.3	1179.6	0.42501	1.22569	1.65070
760.0	67.35	0.0175	6.4501	270.46	831.7	1102.2	270.68	911.9	1182.6	0.43863	1.19989	1.63852
770.0	78.05	0.0176	5.6133	280.77	823.6	1104.3	281.03	904.4	1185.4	0.45211	1.17454	1.62665
780.0	90.07	0.0177	4.9034	291.12	815.3	1106.4	291.41	896.7	1188.1	0.46546	1.14963	1.61509
790.0	103.52	0.0178	4.2985	301.50	806.8	1108.3	301.84	888.8	1190.7	0.47869	1.12511	1.60380
800.0	118.52	0.0179	3.7809	311.92	798.2	1110.2	312.31	880.8	1193.1	0.49180	1.10098	1.59277
810.0	135.18	0.0180	3.3362	322.38	789.5	1111.9	322.83	872.5	1195.4	0.50480	1.07719	1.58198
820.0	153.64	0.0181	2.9527	332.89	780.6	1113.5	333.40	864.1	1197.5	0.51769	1.05372	1.57141
830.0	174.03	0.0183	2.6207	343.44	771.5	1115.0	344.03	855.4	1199.4	0.53048	1.03055	1.56104
840.0	196.49	0.0184	2.3323	354.04	762.3	1116.3	354.71	846.4	1201.1	0.54319	1.00766	1.55085
850.0	221.14	0.0185	2.0809	364.69	752.9	1117.6	365.45	837.3	1202.7	0.55580	0.98501	1.54082
860.0	248.15	0.0187	1.8611	375.41	743.2	1118.6	376.26	827.8	1204.1	0.56834	0.96259	1.53093
870.0	277.66	0.0188	1.6683	386.18	733.4	1119.6	387.15	818.1	1205.3	0.58080	0.94038	1.52118
880.0	309.82	0.0190	1.4986	397.02	723.3	1120.3	398.11	808.1	1206.2	0.59320	0.91833	1.51153
890.0	344.79	0.0191	1.3488	407.93	713.0	1120.9	409.15	797.8	1207.0	0.60553	0.89644	1.50198

Press	Temp	Specific volume (ft**3/lbm)		Intrenal energy (Btu/lbm)			Enthalpy (Btu/lbm)			Entropy (Btu/lbm/R)		
(psi)	(R)	vf	vg	uf	ufg	ug	hf	hfg	hg	sf	sfg	sg
900.0	382.74	0.0193	1.2163	418.91	702.4	1121.3	420.28	787.2	1207.5	0.61782	0.87468	1.49250
910.0	423.83	0.0195	1.0986	429.98	691.6	1121.6	431.50	776.3	1207.8	0.63006	0.85302	1.48309
920.0	468.23	0.0197	0.9940	441.13	680.5	1121.6	442.83	764.9	1207.8	0.64227	0.83144	1.47371
930.0	516.12	0.0199	0.9006	452.37	669.1	1121.5	454.27	753.2	1207.5	0.65444	0.80992	1.46436
940.0	567.68	0.0201	0.8171	463.72	657.4	1121.1	465.83	741.1	1206.9	0.66660	0.78841	1.45501
950.0	623.10	0.0203	0.7422	475.17	645.3	1120.5	477.51	728.6	1206.1	0.67874	0.76691	1.44565
960.0	682.56	0.0205	0.6749	486.74	632.9	1119.6	489.33	715.6	1204.9	0.69089	0.74537	1.43625
970.0	746.26	0.0207	0.6143	498.44	620.1	1118.5	501.30	702.1	1203.4	0.70304	0.72376	1.42680
980.0	814.41	0.0210	0.5595	510.28	606.9	1117.1	513.44	688.0	1201.5	0.71522	0.70206	1.41728
990.0	887.21	0.0212	0.5100	522.27	593.2	1115.4	525.75	673.4	1199.2	0.72743	0.68022	1.40765
1000.0	964.87	0.0215	0.4651	534.42	579.0	1113.4	538.26	658.2	1196.5	0.73969	0.65819	1.39789
1010.0	1047.63	0.0218	0.4243	546.75	564.3	1111.0	550.98	642.3	1193.3	0.75202	0.63595	1.38796
1020.0	1135.71	0.0221	0.3871	559.29	549.0	1108.3	563.94	625.7	1189.6	0.76443	0.61341	1.37784
1030.0	1229.35	0.0225	0.3531	572.04	533.0	1105.1	577.16	608.3	1185.4	0.77695	0.59054	1.36749
1040.0	1328.81	0.0229	0.3220	585.05	516.4	1101.4	590.67	589.9	1180.6	0.78960	0.56724	1.35685
1050.0	1434.35	0.0233	0.2935	598.34	498.9	1097.2	604.52	570.6	1175.1	0.80242	0.54344	1.34587
1060.0	1546.25	0.0237	0.2672	611.95	480.5	1092.5	618.73	550.2	1168.9	0.81544	0.51904	1.33448
1070.0	1664.83	0.0242	0.2429	625.93	461.1	1087.0	633.38	528.5	1161.9	0.82869	0.49390	1.32260
1080.0	1790.40	0.0247	0.2205	640.33	440.4	1080.8	648.52	505.3	1153.8	0.84225	0.46786	1.31011
1090.0	1923.31	0.0253	0.1995	655.24	418.3	1073.6	664.26	480.4	1144.6	0.85619	0.44069	1.29688
1100.0	2063.95	0.0260	0.1800	670.78	394.5	1065.3	680.71	453.3	1134.0	0.87060	0.41209	1.28270
1110.0	2212.74	0.0268	0.1616	687.10	368.4	1055.5	698.08	423.6	1121.7	0.88566	0.38163	1.26730

Press (psi)	Temp (R)	Specific volume (ft**3/lbm)		Internal energy (Btu/lbm)			Enthalpy (Btu/lbm)			Entropy (Btu/lbm/R)		
		vf	vg	uf	ufg	ug	hf	hfg	hg	sf	sfg	sg
1120.0	2370.19	0.0278	0.1441	704.47	339.5	1044.0	716.64	390.6	1107.2	0.90160	0.34870	1.25030
1130.0	2536.83	0.0289	0.1274	723.34	306.7	1030.0	736.92	352.9	1089.8	0.91885	0.31230	1.23114
1140.0	2713.33	0.0305	0.0580	744.45	125.6	870.1	759.76	139.4	899.2	0.93811	0.12195	1.06006
1150.0	2900.52	0.0327	0.0547	769.61	101.6	871.2	787.17	113.4	900.5	0.96110	0.09844	1.05953
1160.0	3099.69	0.0371	0.0517	805.97	65.9	871.9	827.28	74.2	901.5	0.99470	0.06398	1.05868
1164.8	3200.12	0.0515	0.0515	876.57	0.0	876.6	907.08	0.0	907.1	1.06264	0.00000	1.06264

A.14.7 Superheated Steam Table (English Units)

Press= 1.0 psi, Tsat= 561.4 R

T(R) =	600.	650.	700.	750.	800.	850.	900.	1000.	1100.	1200.	1300.	1400.
v-ft3/lbm	357.745	387.798	417.779	447.720	477.638	507.541	537.435	597.207	656.965	716.715	776.461	836.203
u-Btu/lbm	1059.3	1076.5	1093.7	1111.0	1128.5	1146.1	1163.9	1200.1	1237.0	1274.9	1313.6	1353.1
h-Btu/lbm	1125.5	1148.2	1170.9	1193.8	1216.8	1240.0	1263.3	1310.5	1358.6	1407.4	1457.2	1507.8
s-Btu/lbm/R	2.01229	2.04866	2.08238	2.11392	2.14360	2.17168	2.19835	2.24811	2.29388	2.33640	2.37620	2.41372

Press= 2.5 psi, Tsat= 594.1 R,

T(R) =	600.	650.	700.	750.	800.	850.	900.	1000.	1100.	1200.	1300.	1400.
v-ft3/lbm	142.701	154.841	166.904	178.926	190.924	202.909	214.883	238.816	262.735	286.646	310.553	334.456
u-Btu/lbm	1058.5	1076.0	1093.3	1110.7	1128.3	1145.9	1163.8	1200.0	1237.0	1274.8	1313.5	1353.1
h-Btu/lbm	1124.5	1147.5	1170.5	1193.5	1216.5	1239.8	1263.1	1310.4	1358.5	1407.4	1457.1	1507.7

s-Btu/lbm/R	1.90967	1.94663	1.98065	2.01236	2.04214	2.07028	2.09699	2.14680	2.19260	2.23513	2.27495	2.31247
Press=5.0	psi,	Tsat=621.9	R									
T(R)=650.	650.	750.	800.	850.	900.	1000.	1100.	1200.	1300.	1400.	1500.	
v-ft3/lbm	77.185	83.277	89.327	95.353	101.364	107.366	119.352	131.325	143.290	155.250	167.206	179.161
u-Btu/lbm	1075.1	1092.7	1110.3	1127.9	1145.7	1163.6	1199.8	1236.9	1274.7	1313.4	1353.0	1393.6
h-Btu/lbm	1146.4	1169.7	1192.9	1216.1	1239.4	1262.8	1310.2	1358.3	1407.2	1457.0	1507.7	1559.2
s-Btu/lbm/R	1.86870	1.90322	1.93521	1.96516	1.99340	2.02019	2.07008	2.11592	2.15848	2.19832	2.23586	2.27143
Press=7.5	psi,	Tsat=639.6	R									
T(R)=650.	650.	750.	800.	850.	900.	1000.	1100.	1200.	1300.	1400.	1500.	
v-ft3/lbm	51.297	55.400	59.460	63.495	67.516	71.527	79.531	87.521	95.504	103.482	111.457	119.429
u-Btu/lbm	1074.2	1092.1	1109.9	1127.6	1145.4	1163.4	1199.7	1236.8	1274.6	1313.4	1353.0	1393.5
h-Btu/lbm	1145.3	1169.0	1192.4	1215.7	1239.1	1262.6	1310.0	1358.2	1407.1	1456.9	1507.6	1559.2
s-Btu/lbm/R	1.82250	1.85755	1.88983	1.91995	1.94831	1.97516	2.02514	2.07103	2.11362	2.15347	2.19103	2.22661
Press=10.0 psi		Tsat=652.8 R										
T(R)=	700.	750.	800.	850.	900.	1000.	1100.	1200.	1300.	1400.	1500.	1600.
v-ft3/lbm	41.4602	44.5258	47.5662	50.5914	53.6069	59.6202	65.6198	71.6115	77.5984	83.5819	89.5631	95.5426
u-Btu/lbm	1091.5	1109.4	1127.3	1145.2	1163.1	1199.5	1236.7	1274.6	1313.3	1352.9	1393.5	1434.9
h-Btu/lbm	1168.2	1191.8	1215.3	1238.7	1262.3	1309.8	1358.0	1407.0	1456.8	1507.5	1559.1	1611.6
s-Btu/lbm/R	1.82489	1.85746	1.88776	1.91622	1.94315	1.99321	2.03915	2.08177	2.12164	2.15920	2.19480	2.22869
Press=12.5 psi		Tsat=663.6 R										
T(R)=	700.	750.	800.	850.	900.	1000.	1100.	1200.	1300.	1400.	1500.	1600.
v-ft3/lbm	33.0955	35.5648	38.0085	40.4367	42.8550	47.6737	52.4787	57.2759	62.0681	66.8570	71.6435	76.4284

u-Btu/lbm	1090.9	1109.0	1126.9	1144.9	1162.9	1199.4	1236.5	1274.5	1313.2	1352.9	1393.4	1434.9
h-Btu/lbm	1167.4	1191.2	1214.8	1238.4	1262.0	1309.6	1357.9	1406.9	1456.7	1507.4	1559.0	1611.6
s-Btu/lbm/R	1.79933	1.83221	1.86269	1.89126	1.91827	1.96841	2.01440	2.05704	2.09693	2.13451	2.17011	2.20401

Press = 14.7 psi   Tsat = 671.6 R

T(R)=.	700.	750.	800.	850.	900.	1000.	1100.	1200.	1300.	1400.	1500.	1600.
v-ft3/lbm	28.0952	30.2085	32.2956	34.3669	36.4283	40.5332	44.6241	48.7072	52.7854	56.8603	60.9328	65.0036
u-Btu/lbm	1090.3	1108.6	1126.7	1144.7	1162.7	1199.3	1236.4	1274.4	1313.2	1352.8	1393.4	1434.8
h-Btu/lbm	1166.7	1190.7	1214.4	1238.1	1261.8	1309.4	1357.7	1406.8	1456.6	1507.4	1559.0	1611.5
s-Btu/lbm/R	1.78065	1.81380	1.84444	1.87311	1.90018	1.95040	1.99643	2.03910	2.07901	2.11659	2.15220	2.18611

Press = 20.0 psi   Tsat = 687.6 R

T(R)=	700.	750.	800.	850.	900.	1000.	1100.	1200.	1300.	1400.	1500.	1600.
v-ft3/lbm	20.5449	22.1217	23.6710	25.2040	26.7268	29.7540	32.7671	35.7724	38.7726	41.7696	44.7642	47.7572
u-Btu/lbm	1088.9	1107.6	1125.9	1144.1	1162.3	1198.9	1236.2	1274.2	1313.0	1352.7	1393.3	1434.8
h-Btu/lbm	1164.9	1189.4	1213.5	1237.3	1261.2	1309.0	1357.4	1406.5	1456.4	1507.2	1558.8	1611.4
s-Btu/lbm/R	1.74461	1.77846	1.80949	1.83841	1.86564	1.91605	1.96217	2.00490	2.04485	2.08246	2.11809	2.15201

Press = 40.0 psi   Tsat = 726.9 R

T(R)=	750.	800.	850.	900.	1000.	1100.	1200.	1300.	1400.	1500.	1600.	1700.
v-ft3/lbm	10.9125	11.7197	12.5083	13.2855	14.8204	16.3406	17.8527	19.3597	20.8634	22.3648	23.8644	25.3628
u-Btu/lbm	1103.8	1123.1	1142.0	1160.6	1197.8	1235.3	1273.5	1312.4	1352.2	1392.9	1434.4	1476.9
h-Btu/lbm	1184.5	1209.8	1234.5	1258.9	1307.4	1356.2	1405.6	1455.7	1506.6	1558.3	1611.0	1664.5
s-Btu/lbm/R	1.69668	1.72936	1.75928	1.78714	1.83827	1.88478	1.92774	1.96784	2.00555	2.04126	2.07523	2.10768

Press = 60.0 psi   Tsat = 752.4 R

T(R)=	800.	850.	900.	1000.	1100.	1200.	1300.	1400.	1500.	1600.	1700.	1800.
v-ft3/lbm	7.7325	8.2745	8.8041	9.8422	10.8650	11.8794	12.8887	13.8947	14.8983	15.9002	16.9008	17.9005
u-Btu/lbm	1120.1	1139.7	1158.8	1196.5	1234.4	1272.8	1311.9	1351.8	1392.5	1434.1	1476.6	1520.0
h-Btu/lbm	1205.9	1231.5	1256.5	1305.8	1355.0	1404.6	1454.9	1505.9	1557.8	1610.5	1664.1	1718.6
s-Btu/lbm/R	1.68079	1.71180	1.74036	1.79226	1.83917	1.88236	1.92261	1.96043	1.99620	2.03022	2.06272	2.09388

Press = 80.0 psi,   Tsat = 771.7 R

T(R)=	800.	850.	900.	1000.	1100.	1200.	1300.	1400.	1500.	1600.	1700.	1800.
v-ft3/lbm	5.7361	6.1561	6.5625	7.3528	8.1271	8.8927	9.6532	10.4103	11.1650	11.9180	12.6698	13.4206
u-Btu/lbm	1117.0	1137.4	1157.0	1195.3	1233.5	1272.1	1311.3	1351.3	1392.1	1433.7	1476.3	1519.8
h-Btu/lbm	1201.8	1228.4	1254.1	1304.1	1353.7	1403.7	1454.1	1505.3	1557.3	1610.1	1663.8	1718.3
s-Btu/lbm/R	1.64497	1.67722	1.70652	1.75924	1.80656	1.84999	1.89039	1.92831	1.96416	1.99823	2.03077	2.06196

Press = 100.0 psi   Tsat = 787.5 R

T(R)=	800.	850.	900.	1000.	1100.	1200.	1300.	1400.	1500.	1600.	1700.	1800.
v-ft3/lbm	4.5358	4.8839	5.2169	5.8589	6.4842	7.1007	7.7119	8.3196	8.9250	9.5287	10.1312	10.7327
u-Btu/lbm	1113.6	1134.9	1155.1	1194.1	1232.6	1271.4	1310.7	1350.8	1391.7	1433.4	1476.0	1519.5
h-Btu/lbm	1197.5	1225.2	1251.6	1302.4	1352.5	1402.7	1453.4	1504.7	1556.7	1609.6	1663.4	1718.0
s-Btu/lbm/R	1.61596	1.64965	1.67975	1.73333	1.78107	1.82474	1.86529	1.90332	1.93924	1.97337	2.00595	2.03717

Press = 120.0 psi   Tsat = 800.9 R

T(R)=	850.	900.	1000.	1100.	1200.	1300.	1400.	1500.	1600.	1700.	1800.	1900.
v-ft3/lbm	4.0347	4.3192	4.8628	5.3889	5.9059	6.4176	6.9259	7.4317	7.9359	8.4388	8.9407	9.4420
u-Btu/lbm	1132.4	1153.2	1192.8	1231.7	1270.7	1310.2	1350.3	1391.3	1433.1	1475.7	1519.2	1563.7
h-Btu/lbm	1221.9	1249.0	1300.7	1351.3	1401.7	1452.6	1504.0	1556.2	1609.2	1663.0	1717.7	1773.2

s-Btu/lbm/R	1.62645	1.65742	1.71191	1.76009	1.80401	1.84471	1.88284	1.91884	1.95302	1.98564	2.01689	2.04692
Press=140.0 psi,	Tsat=812.7 R											
T(R)=	850.	900.	1000.	1100.	1200.	1300.	1400.	1500.	1600.	1700.	1800.	1900.
v-ft3/lbm	3.4272	3.6775	4.1511	4.6064	5.0525	5.4931	5.9303	6.3651	6.7981	7.2299	7.6608	8.0909
u-Btu/lbm	1129.7	1151.2	1191.5	1230.7	1269.9	1309.6	1349.9	1390.9	1432.7	1475.4	1519.0	1563.5
h-Btu/lbm	1218.5	1246.4	1299.0	1350.0	1400.8	1451.8	1503.4	1555.7	1608.8	1662.6	1717.4	1772.9
s-Btu/lbm/R	1.60622	1.63813	1.69358	1.74221	1.78638	1.82724	1.86547	1.90154	1.93578	1.96844	1.99972	2.02977
Press=160.0psi,	Tsat=823.2R											
T(R)=	850.	900.	1000.	1100.	1200.	1300.	1400.	1500.	1600.	1700.	1800.	1900.
v-ft3/lbm	2.9707	3.1958	3.6172	4.0195	4.4124	4.7998	5.1836	5.5651	5.9448	6.3233	6.7008	7.0776
u-Btu/lbm	1127.0	1149.1	1190.2	1229.8	1269.2	1309.0	1349.4	1390.5	1432.4	1475.1	1518.7	1563.2
h-Btu/lbm	1214.9	1243.7	1297.2	1348.7	1399.8	1451.0	1502.8	1555.2	1608.3	1662.2	1717.0	1772.7
s-Btu/lbm/R	1.58812	1.62105	1.67750	1.72659	1.77102	1.81204	1.85038	1.88653	1.92082	1.95351	1.98482	2.01490
Press=180.0 psi	Tsat=832.8 R											
T(R)=	850.	900.	1000.	1100.	1200.	1300.	1400.	1500.	1600.	1700.	1800.	1900.
v-ft3/lbm	2.6148	2.8207	3.2017	3.5630	3.9145	4.2605	4.6029	4.9428	5.2811	5.6181	5.9541	6.2895
u-Btu/lbm	1124.1	1147.0	1188.9	1228.8	1268.5	1308.4	1348.9	1390.1	1432.1	1474.8	1518.5	1563.0
h-Btu/lbm	1211.1	1240.9	1295.4	1347.5	1398.8	1450.3	1502.1	1554.6	1607.9	1661.9	1716.7	1772.4
s-Btu/lbm/R	1.57158	1.60564	1.66314	1.71272	1.75740	1.79858	1.83703	1.87325	1.90759	1.94033	1.97167	2.00177
Press=200.0 psi,	Tsat=841.5 R											
T(R)=	850.	900.	1000.	1100.	1200.	1300.	1400.	1500.	1600.	1700.	1800.	1900.
v-ft3/lbm	2.3291	2.5202	2.8693	3.1977	3.5162	3.8290	4.1382	4.4451	4.7501	5.0540	5.3568	5.6590

u–Btu/lbm	1121.0	1144.8	1187.5	1227.9	1267.8	1307.8	1348.4	1389.7	1431.7	1474.6	1518.2	1562.8
h–Btu/lbm	1207.2	1238.1	1293.6	1346.2	1397.8	1449.5	1501.5	1554.1	1607.4	1661.5	1716.4	1772.1
s–Btu/lbm/R	1.55619	1.59152	1.65013	1.70020	1.74515	1.78649	1.82505	1.86134	1.89574	1.92852	1.95989	1.99002

Press = 250.0 psi,   Tsat = 860.7 R

T(R) =	900.	1000.	1100.	1200.	1300.	1400.	1500.	1600.	1700.	1800.	1900.	2000.
v–ft3/lbm	1.9779	2.2703	2.5400	2.7991	3.0523	3.3019	3.5490	3.7944	4.0385	4.2816	4.5241	4.7660
u–Btu/lbm	1139.1	1184.1	1225.5	1265.9	1306.4	1347.2	1388.7	1430.9	1473.8	1517.6	1562.2	1607.7
h–Btu/lbm	1230.6	1289.0	1342.9	1395.4	1447.5	1499.9	1552.8	1606.3	1660.5	1715.6	1771.4	1828.1
s–Btu/lbm/R	1.56033	1.62195	1.67333	1.71896	1.76072	1.79954	1.83603	1.87056	1.90344	1.93489	1.96507	1.99414

Press = 300.0 psi,   Tsat = 877.0 R

T(R) =	900.	1000.	1100.	1200.	1300.	1400.	1500.	1600.	1700.	1800.	1900.	2000.
v–ft3/lbm	1.6144	1.8704	2.1013	2.3209	2.5345	2.7443	2.9517	3.1572	3.3615	3.5649	3.7675	3.9696
u–Btu/lbm	1133.0	1180.5	1223.0	1264.1	1304.9	1346.0	1387.7	1430.0	1473.1	1517.0	1561.7	1607.2
h–Btu/lbm	1222.6	1284.2	1339.6	1392.9	1445.5	1498.3	1551.5	1605.2	1659.6	1714.8	1770.7	1827.5
s–Btu/lbm/R	1.53305	1.59813	1.65091	1.69726	1.73944	1.77854	1.81522	1.84989	1.88287	1.91439	1.94463	1.97375

Press = 350.0 psi,   Tsat = 891.4 R

T(R) =	900.	1000.	1100.	1200.	1300.	1400.	1500.	1600.	1700.	1800.	1900.	2000.
v–ft3/lbm	1.3527	1.5842	1.7877	1.9793	2.1646	2.3460	2.5250	2.7021	2.8780	3.0529	3.2270	3.4007
u–Btu/lbm	1126.2	1176.7	1220.5	1262.2	1303.4	1344.8	1386.7	1429.2	1472.4	1516.3	1561.1	1606.7
h–Btu/lbm	1213.8	1279.3	1336.2	1390.3	1443.6	1496.7	1550.1	1604.1	1658.7	1713.9	1770.0	1826.8
s–Btu/lbm/R	1.50812	1.57726	1.63155	1.67865	1.72127	1.76065	1.79752	1.83233	1.86541	1.89701	1.92731	1.95647

Press=400.0 psi,	Tsat=904.3 R											
T(R) =	1000.	1100.	1200.	1300.	1400.	1500.	1600.	1700.	1800.	1900.	2000.	2100.
v-ft3/lbm	1.3691	1.5523	1.7230	1.8871	2.0473	2.2049	2.3608	2.5153	2.6689	2.8217	2.9741	3.1259
u-Btu/lbm	1172.8	1217.9	1260.3	1302.0	1343.6	1385.7	1428.3	1471.6	1515.7	1560.5	1606.2	1652.7
h-Btu/lbm	1274.1	1332.7	1387.8	1441.6	1495.1	1548.8	1603.0	1657.7	1713.1	1769.3	1826.2	1884.0
s-Btu/ lbm/R	1.55851	1.61441	1.66230	1.70536	1.74503	1.78209	1.81704	1.85022	1.88190	1.91226	1.94146	1.96963

Press=450.0 psi,	Tsat=916.0 R											
T(R) =	1000.	1100.	1200.	1300.	1400.	1500.	1600.	1700.	1800.	1900.	2000.	2100.
v-ft3/lbm	1.2012	1.3691	1.5236	1.6712	1.8149	1.9560	2.0953	2.2332	2.3702	2.5065	2.6422	2.7775
u-Btu/lbm	1168.8	1215.3	1258.4	1300.5	1342.4	1384.7	1427.5	1470.9	1515.0	1560.0	1605.7	1652.3
h-Btu/lbm	1268.8	1329.2	1385.2	1439.5	1493.5	1547.5	1601.8	1656.8	1712.3	1768.6	1825.6	1883.4
s-Btu/ lbm/R	1.54131	1.59895	1.64767	1.69119	1.73115	1.76840	1.80349	1.83678	1.86853	1.89895	1.92820	1.95640

Press=500.0 psi	Tsat=926.7 R											
T(R) =	1000.	1100.	1200.	1300.	1400.	1500.	1600.	1700.	1800.	1900.	2000.	2100.
v-ft3/lbm	1.0665	1.2223	1.3640	1.4985	1.6290	1.7568	1.8828	2.0075	2.1313	2.2543	2.3767	2.4987
u-Btu/lbm	1164.6	1212.6	1256.4	1298.9	1341.2	1383.7	1426.6	1470.2	1514.4	1559.4	1605.2	1651.8
h-Btu/lbm	1263.2	1325.6	1382.6	1437.5	1491.8	1546.1	1600.7	1655.8	1711.5	1767.9	1825.0	1882.9
s-Btu/ lbm/R	1.52530	1.58481	1.63438	1.67838	1.71863	1.75608	1.79132	1.82471	1.85654	1.88702	1.91631	1.94454

Press=550.0 psi	Tsat=936.7 R											
T(R) =	1000.	1100.	1200.	1300.	1400.	1500.	1600.	1700.	1800.	1900.	2000.	2100.
v-ft3/lbm	0.9559	1.1021	1.2333	1.3572	1.4769	1.5939	1.7091	1.8229	1.9358	2.0479	2.1595	2.2707
u-Btu/lbm	1160.2	1209.8	1254.5	1297.4	1340.0	1382.6	1425.7	1469.4	1513.8	1558.8	1604.7	1651.3
h-Btu/lbm	1257.5	1321.9	1379.9	1435.5	1490.2	1544.8	1599.6	1654.8	1710.7	1767.2	1824.4	1882.3

s-Btu/ lbm/R	1.51020	1.57171	1.62218	1.66666	1.70722	1.74487	1.78025	1.81374	1.84565	1.87619	1.90552	1.93380
Press=600.0 psi	Tsat=945.9 R											
T(R)=	1000.	1100.	1200.	1300.	1400.	1500.	1600.	1700.	1800.	1900.	2000.	2100.
v-ft3/lbm	0.8632	1.0017	1.1243	1.2394	1.3501	1.4581	1.5642	1.6690	1.7729	1.8760	1.9785	2.0806
u-Btu/lbm	1155.7	1207.0	1252.5	1295.9	1338.7	1381.6	1424.9	1468.7	1513.1	1558.3	1604.2	1650.9
h-Btu/lbm	1251.5	1318.2	1377.2	1433.4	1488.5	1543.4	1598.5	1653.9	1709.8	1766.4	1823.7	1881.8
s-Btu/ lbm/R	1.49578	1.55946	1.61087	1.65586	1.69672	1.73458	1.77010	1.80369	1.83568	1.86628	1.89566	1.92397
Press=700.0 psi	Tsat=962.8 R											
T(R)=	1000.	1100.	1200.	1300.	1400.	1500.	1600.	1700.	1800.	1900.	2000.	2100.
v-ft3/lbm	0.7162	0.8437	0.9530	1.0541	1.1508	1.2447	1.3366	1.4272	1.5169	1.6057	1.6941	1.7820
u-Btu/lbm	1145.9	1201.2	1248.4	1292.8	1336.2	1379.6	1423.2	1467.2	1511.8	1557.1	1603.2	1650.0
h-Btu/lbm	1238.6	1310.4	1371.8	1429.3	1485.2	1540.7	1596.2	1652.0	1708.2	1765.0	1822.5	1880.7
s-Btu/ lbm/R	1.46837	1.53696	1.59037	1.63641	1.67790	1.71617	1.75198	1.78579	1.81793	1.84865	1.87812	1.90650
Press=800.0 psi,	Tsat=977.9 R											
T(R)=	1000.	1100.	1200.	1300.	1400.	1500.	1600.	1700.	1800.	1900.	2000.	2100.
v-ft3/lbm	0.6041	0.7246	0.8243	0.9151	1.0013	1.0846	1.1659	1.2459	1.3249	1.4031	1.4807	1.5580
u-Btu/lbm	1134.9	1195.1	1244.2	1289.6	1333.7	1377.5	1421.4	1465.7	1510.5	1556.0	1602.1	1649.0
h-Btu/lbm	1224.3	1302.3	1366.2	1425.0	1481.9	1538.0	1593.9	1650.0	1706.6	1763.6	1821.2	1879.6
s-Btu/ lbm/R	1.44194	1.51644	1.57205	1.61919	1.66134	1.70004	1.73614	1.77016	1.80246	1.83330	1.86286	1.89131
Press=900.0 psi	Tsat=991.7 R											
T(R)=	1000.	1100.	1200.	1300.	1400.	1500.	1600.	1700.	1800.	1900.	2000.	2100.
v-ft3/lbm	0.5148	0.6315	0.7239	0.8069	0.8850	0.9600	1.0331	1.1048	1.1755	1.2454	1.3148	1.3838

u-Btu/lbm	1122.5	1188.7	1239.9	1286.4	1331.2	1375.4	1419.7	1464.2	1509.2	1554.9	1601.1	1648.1
h-Btu/lbm	1208.2	1293.8	1360.4	1420.7	1478.5	1535.2	1591.6	1648.1	1704.9	1762.2	1820.0	1878.5
s-Btu/lbm/R	1.41552	1.49735	1.55536	1.60367	1.64649	1.68563	1.72203	1.75627	1.78873	1.81969	1.84934	1.87787
Press=1000.0 psi,		Tsat=1004.3 R										
T(R)=	1100.	1200.	1300.	1400.	1500.	1600.	1700.	1800.	1900.	2000.	2100.	2200.
v-ft3/lbm	0.5565	0.6435	0.7203	0.7919	0.8604	0.9269	0.9920	1.0560	1.1193	1.1821	1.2444	1.3063
u-Btu/lbm	1181.9	1235.5	1283.2	1328.6	1373.3	1417.9	1462.7	1507.9	1553.7	1600.1	1647.2	1695.1
h-Btu/lbm	1284.9	1354.5	1416.4	1475.1	1532.5	1589.3	1646.2	1703.2	1760.7	1818.7	1877.4	1936.7
s-Btu/lbm/R	1.47927	1.53992	1.58948	1.63300	1.67258	1.70929	1.74374	1.77636	1.80744	1.83719	1.86579	1.89338
Press=1100.0 psi,		Tsat=1016.0 R										
T(R)=	1100	1200.	1300.	1400.	1500.	1600.	1700.	1800.	1900.	2000.	2100.	2200.
v-ft3/lbm	0.4947	0.5776	0.6493	0.7157	0.7788	0.8399	0.8996	0.9583	1.0162	1.0735	1.1304	1.1869
u-Btu/lbm	1174.8	1230.9	1279.8	1326.1	1371.2	1416.1	1461.2	1506.6	1552.6	1599.1	1646.3	1694.2
h-Btu/lbm	1275.5	1348.4	1411.9	1471.6	1529.7	1587.0	1644.2	1701.6	1759.3	1817.5	1876.3	1935.7
s-Btu/lbm/R	1.46191	1.52547	1.57634	1.62060	1.66064	1.69765	1.73233	1.76511	1.79631	1.82615	1.85482	1.88247
Press=1200.0 psi,		Tsat=1026.9 R										
T(R)=	1100	1200.	1300.	1400.	1500.	1600.	1700.	1800.	1900.	2000.	2100.	2200.
v-ft3/lbm	0.4428	0.5224	0.5901	0.6521	0.7109	0.7675	0.8227	0.8768	0.9302	0.9830	1.0353	1.0873
u-Btu/lbm	1167.3	1226.2	1276.5	1323.4	1369.1	1414.4	1459.7	1505.3	1551.4	1598.1	1645.4	1693.4
h-Btu/lbm	1265.6	1342.2	1407.4	1468.2	1526.9	1584.7	1642.3	1699.9	1757.9	1816.2	1875.1	1934.7
s-Btu/lbm/R	1.44502	1.51180	1.56408	1.60909	1.64959	1.68692	1.72182	1.75477	1.78609	1.81602	1.84477	1.87248

Press = 1300.0 psi, Tsat = 1037.2 R

T(R) =	1100.	1200.	1300.	1400.	1500.	1600.	1700.	1800.	1900.	2000.	2100.	2200.
v-ft3/lbm	0.3983	0.4756	0.5400	0.5983	0.6533	0.7062	0.7576	0.8079	0.8575	0.9064	0.9549	1.0031
u-Btu/lbm	1159.3	1221.4	1273.0	1320.8	1367.0	1412.6	1458.2	1504.0	1550.3	1597.0	1644.4	1692.6
h-Btu/lbm	1255.0	1335.8	1402.9	1464.6	1524.0	1582.4	1640.3	1698.2	1756.4	1815.0	1874.0	1933.7
s-Btu/lbm/R	1.42838	1.49877	1.55252	1.59832	1.63931	1.67696	1.71208	1.74519	1.77664	1.80667	1.83549	1.86326

Press = 1400.0 psi, Tsat = 1046.8 R

T(R) =	1100	1200.	1300.	1400.	1500.	1600.	1700.	1800.	1900.	2000.	2100.	2200.
v-ft3/lbm	0.3596	0.4354	0.4969	0.5522	0.6040	0.6536	0.7017	0.7488	0.7951	0.8408	0.8860	0.9309
u-Btu/lbm	1150.7	1216.4	1269.5	1318.1	1364.8	1410.8	1456.6	1502.7	1549.1	1596.0	1643.5	1691.7
h-Btu/lbm	1243.8	1329.2	1398.2	1461.1	1521.2	1580.0	1638.3	1696.6	1755.0	1813.7	1872.9	1932.8
s-Btu/lbm/R	1.41181	1.48626	1.54158	1.58819	1.62967	1.66764	1.70300	1.73627	1.76785	1.79796	1.82686	1.85470

Press = 1500.0 psi, Tsat = 1055.9 R

T(R) =	1100	1200.	1300.	1400.	1500.	1600.	1700.	1800.	1900.
v-ft3/lbm	0.3255	0.4003	0.4596	0.5122	0.5613	0.6081	0.6534	0.6976	0.7411
u-Btu/lbm	1141.5	1211.3	1266.0	1315.4	1362.6	1409.0	1455.1	1501.4	1547.9
h-Btu/lbm	1231.7	1322.4	1393.5	1457.5	1518.3	1577.7	1636.4	1694.9	1753.5
s-Btu/lbm/R	1.39510	1.47415	1.53114	1.57860	1.62058	1.65889	1.69447	1.72792	1.75962

Press = 1600.0 psi, Tsat = 1064.6 R

T(R) =	1100	1200.	1300.	1400.	1500.	1600.	1700.	1800.	1900.
v-ft3/lbm	0.2950	0.3695	0.4268	0.4772	0.5238	0.5682	0.6110	0.6528	0.6938
u-Btu/lbm	1131.4	1206.0	1262.4	1312.7	1360.4	1407.2	1453.6	1500.0	1546.8
h-Btu/lbm	1218.7	1315.3	1388.7	1453.9	1515.5	1575.3	1634.4	1693.2	1752.1

s-Btu/ lbm/R	1.37804	1.46238	1.52113	1.56948	1.61198	1.65062	1.68643	1.72005	1.75187
Press = 1700.0 psi	Tsat = 1072.9 R								
T(R) =	1100.	1200.	1300.	1400.	1500.	1600.	1700.	1800.	1900.
v-ft3/lbm	0.2673	0.3422	0.3979	0.4462	0.4908	0.5330	0.5737	0.6133	0.6521
u-Btu/lbm	1120.4	1200.5	1258.7	1309.9	1358.2	1405.4	1452.0	1498.7	1545.6
h-Btu/lbm	1204.5	1308.1	1383.8	1450.2	1512.6	1572.9	1632.4	1691.5	1750.6
s-Btu/ lbm/R	1.36037	1.45087	1.51150	1.56076	1.60379	1.64277	1.67883	1.71261	1.74456
Press = 1800.0 psi	Tsat = 1080.7 R								
T(R) =	1100.	1200.	1300.	1400.	1500.	1600.	1700.	1800.	1900.
v-ft3/lbm	0.2418	0.3178	0.3721	0.4187	0.4614	0.5017	0.5405	0.5781	0.6150
u-Btu/lbm	1108.2	1194.8	1254.9	1307.1	1356.0	1403.5	1450.5	1497.4	1544.4
h-Btu/lbm	1188.7	1300.6	1378.8	1446.5	1509.6	1570.6	1630.4	1689.8	1749.2
s-Btu/ lbm/R	1.34174	1.43955	1.50219	1.55240	1.59598	1.63531	1.67160	1.70555	1.73764
Press = 1900.0 psi	Tsat = 1088.3 R								
T(R) =	1100.	1200.	1300.	1400.	1500.	1600.	1700.	1800.	1900.
v-ft3/lbm	0.2179	0.2958	0.3490	0.3941	0.4351	0.4738	0.5108	0.5467	0.5818
u-Btu/lbm	1094.2	1189.0	1251.1	1304.3	1353.8	1401.7	1448.9	1496.0	1543.3
h-Btu/lbm	1170.8	1292.9	1373.7	1442.7	1506.7	1568.2	1628.4	1688.1	1747.7
s-Btu/ lbm/R	1.32165	1.42838	1.49317	1.54436	1.58849	1.62817	1.66471	1.69884	1.73105
Press = 2000.0 psi	Tsat = 1095.5 R								
T(R) =.	1100.	1200.	1300.	1400.	1500.	1600.	1700.	1800.	1900.
v-ft3/lbm	0.1949	0.2758	0.3281	0.3719	0.4115	0.4486	0.4840	0.5184	0.5519

u-Btu/lbm	1077.8	1182.8	1247.2	1301.4	1351.5	1399.8	1447.4	1494.7	1542.1
h-Btu/lbm	1149.9	1284.9	1368.5	1439.0	1503.7	1565.8	1626.4	1686.4	1746.2
s-Btu/lbm/R	1.29918	1.41730	1.48438	1.53660	1.58130	1.62134	1.65812	1.69242	1.72476

Press = 2500.0 psi   Tsat = 1127.8 R

T(R) =	1200	1300	1400	1500	1600	1700	1800	1900.
v-ft3/lbm	0.1980	0.2483	0.2873	0.3214	0.3528	0.3824	0.4109	0.4385
u-Btu/lbm	1148.2	1226.4	1286.5	1340.0	1390.4	1439.5	1487.9	1536.2
h-Btu/lbm	1239.7	1341.3	1419.4	1488.6	1553.6	1616.3	1677.9	1738.9
s-Btu/lbm/R	1.36160	1.44310	1.50107	1.54884	1.59077	1.62881	1.66401	1.69701

Press = 3000.0 psi   Tsat = 1155.1 R

T(R) =	1200	1300.	1400.	1500.	1600.	1700.	1800.	1900.
v-ft3/lbm	0.1421	0.1943	0.2307	0.2613	0.2889	0.3147	0.3392	0.3629
u-Btu/lbm	1103.1	1203.5	1270.8	1328.0	1380.8	1431.5	1481.1	1530.2
h-Btu/lbm	1181.9	1311.3	1398.8	1473.0	1541.1	1606.1	1669.3	1731.6
s-Btu/lbm/R	1.30046	1.40439	1.46935	1.52058	1.56455	1.60393	1.64007	1.67375

Press = 3500.0 psi

T(R) =	1200	1300.	1400.	1500.	1600.	1700.	1800.	1900.
v-ft3/lbm	0.0244	0.1550	0.1900	0.2183	0.2433	0.2663	0.2881	0.3089
u-Btu/lbm	720.0	1177.7	1254.1	1315.6	1370.9	1423.3	1474.1	1524.2
h-Btu/lbm	768.0	1278.1	1377.1	1456.9	1528.4	1595.7	1660.6	1724.2
s-Btu/lbm/R	0.91193	1.36649	1.44004	1.49513	1.54131	1.58210	1.61920	1.65358

Press = 4000.0 psi								
T(R) =	1200	1300.	1400.	1500.	1600.	1700.	1800.	1900.
v–ft3/lbm	0.0240	0.1249	0.1594	0.1860	0.2091	0.2301	0.2497	0.2685
u–Btu/lbm	714.7	1148.8	1236.4	1302.7	1360.8	1415.0	1467.1	1518.2
h–Btu/lbm	765.3	1241.1	1354.3	1440.3	1515.5	1585.2	1651.9	1716.8
s–Btu/ lbm/R	0.90685	1.32818	1.41226	1.47167	1.52020	1.56247	1.60058	1.63567
Press = 4500.0 psi								
T(R) =	1200	1300.	1400.	1500.	1600.	1700.	1800.	1900.
v–ft3/lbm	0.0237	0.1009	0.1355	0.1609	0.1825	0.2019	0.2200	0.2370
u–Btu/lbm	709.9	1116.0	1217.7	1289.4	1350.5	1406.6	1460.0	1512.1
h–Btu/lbm	763.2	1200.0	1330.4	1423.4	1502.4	1574.6	1643.1	1709.3
s–Btu/ lbm/R	0.90232	1.28852	1.38550	1.44969	1.50072	1.54453	1.58367	1.61949
Press = 5000.0 psi								
T(R) =	1200	1300.	1400.	1500.	1600.	1700.	1800.	1900.
v–ft3/lbm	0.0234	0.0815	0.1163	0.1409	0.1613	0.1794	0.1962	0.2119
u–Btu/lbm	705.6	1079.1	1198.0	1275.8	1340.0	1398.1	1452.9	1505.9
h–Btu/lbm	761.3	1154.4	1305.5	1406.0	1489.1	1564.0	1634.3	1701.9
s–Btu/ lbm/R	0.89820	1.24703	1.35941	1.42885	1.48252	1.52792	1.56811	1.60468

A.14.8 H_2O Compressed Liquid Tables (English Units)

Temperature = 500.0 R		Psat = 0.1 psi								
Press(psi)=	250	500	1000	1500	2000	2500	3000	3500	4000	5000
v-ft3/lbm	0.016	0.016	0.016	0.016	0.016	0.0159	0.0159	0.0159	0.0159	0.0158
u-Btu/lbm	8.4	8.4	8.4	8.4	8.3	8.3	8.3	8.3	8.3	8.3
h-Btu/lbm	9.1	9.9	11.3	12.8	14.3	15.7	17.1	18.6	20	22.9
s-Btu/lbm/R	0.0169	0.01689	0.01687	0.01684	0.0168	0.01675	0.0167	0.01664	0.01657	0.01641

Temperature = 550.0 R		Psat = 0.7 psi								
Press(psi)=	250	500	1000	1500	2000	2500	3000	3500	4000	5000
v-ft3/lbm	0.0161	0.0161	0.0161	0.0161	0.016	0.016	0.016	0.016	0.016	0.0159
u-Btu/lbm	58.4	58.4	58.2	58.1	57.9	57.8	57.6	57.5	57.4	57.1
h-Btu/lbm	59.2	59.9	61.2	62.5	63.9	65.2	66.5	67.8	69.2	71.8
s-Btu/lbm/R	0.11232	0.11219	0.11192	0.11165	0.11137	0.1111	0.11082	0.11054	0.11026	0.1097

Temperature = 600.0 R		Psat = 2.9 psi								
Press(psi)=	250	500	1000	1500	2000	2500	3000	3500	4000	5000
v-ft3/lbm	0.0163	0.0163	0.0163	0.0163	0.0162	0.0162	0.0162	0.0162	0.0161	0.0161
u-Btu/lbm	108.4	108.3	108	107.8	107.5	107.3	107	106.8	106.5	106.1
h-Btu/lbm	109.2	109.8	111	112.3	113.5	114.8	116	117.2	118.5	120.9
s-Btu/lbm/R	0.19929	0.19907	0.19864	0.1982	0.19777	0.19734	0.19691	0.19648	0.19606	0.19521

Temperature = 650.0 R		Psat = 9.4 psi								
Press(psi)=	250	500	1000	1500	2000	2500	3000	3500	4000	5000
v-ft3/lbm	0.0166	0.0166	0.0166	0.0165	0.0165	0.0165	0.0165	0.0164	0.0164	0.0164
u-Btu/lbm	158.5	158.3	158	157.6	157.2	156.9	156.5	156.2	155.8	155.2
h-Btu/lbm	159.3	159.9	161	162.2	163.3	164.5	165.6	166.8	168	170.3

s-Btu/lbm/R	0.27953	0.27924	0.27866	0.27809	0.27752	0.27695	0.27639	0.27584	0.27529	0.27419
Temperature=700.0 R		Psat=25.1 psi								
Press(psi)=	250	500	1000	1500	2000	2500	3000	3500	4000	5000
v-ft3/lbm	0.017	0.0169	0.0169	0.0169	0.0169	0.0168	0.0168	0.0168	0.0167	0.0167
u-Btu/lbm	209	208.7	208.2	207.7	207.2	206.8	206.3	205.8	205.4	204.5
h-Btu/lbm	209.7	210.3	211.3	212.4	213.5	214.5	215.6	216.7	217.8	219.9
s-Btu/lbm/R	0.35431	0.35395	0.35323	0.35252	0.35182	0.35113	0.35044	0.34976	0.34908	0.34775
Temperature=750.0 R		Psat=57.9 psi								
Press(psi)=	250	500	1000	1500	2000	2500	3000	3500	4000	5000
v-ft3/lbm	0.0174	0.0174	0.0173	0.0173	0.0173	0.0172	0.0172	0.0172	0.0171	0.0171
u-Btu/lbm	259.9	259.6	259	258.3	257.7	257.1	256.5	255.9	255.3	254.2
h-Btu/lbm	260.7	261.2	262.2	263.1	264.1	265.1	266	267	268	270
s-Btu/lbm/R	0.42467	0.42423	0.42337	0.42251	0.42167	0.42083	0.42001	0.41919	0.41839	0.41681
Temperature=800.0 R		Psat=118.5 psi								
Press(psi)=	250	500	1000	1500	2000	2500	3000	3500	4000	5000
v-ft3/lbm	0.0179	0.0179	0.0178	0.0178	0.0178	0.0177	0.0177	0.0176	0.0176	0.0175
u-Btu/lbm	311.7	311.3	310.4	309.6	308.8	308.1	307.3	306.5	305.8	304.4
h-Btu/lbm	312.5	312.9	313.7	314.6	315.4	316.2	317.1	318	318.8	320.6
s-Btu/lbm/R	0.49152	0.49099	0.48994	0.48891	0.4879	0.4869	0.48592	0.48495	0.484	0.48212
Temperature=850.0 R		Psat=221.1 psi								
Press(psi)=	250	500	1000	1500	2000	2500	3000	3500	4000	5000
v-ft3/lbm	0.0185	0.0185	0.0185	0.0184	0.0184	0.0183	0.0183	0.0182	0.0182	0.0181
u-Btu/lbm	364.6	364.1	363	362	360.9	359.9	358.9	358	357.1	355.2
h-Btu/lbm	365.5	365.8	366.4	367.1	367.7	368.4	369.1	369.8	370.5	372
s-Btu/lbm/R	0.55573	0.55508	0.5538	0.55255	0.55132	0.55012	0.54894	0.54778	0.54664	0.54442

Temperature = 900.0 R   Psat = 382.7 psi

Press(psi)=	250	500	1000	1500	2000	2500	3000	3500	4000	5000
v-ft3/lbm		0.0193	0.0192	0.0192	0.0191	0.019	0.019	0.0189	0.0188	0.0187
u-Btu/lbm		418.6	417.1	415.8	414.4	413.1	411.8	410.6	409.4	407.1
h-Btu/lbm		420.4	420.7	421.1	421.5	421.9	422.4	422.9	423.4	424.4
s-Btu/lbm/R		0.61744	0.61584	0.61428	0.61277	0.61129	0.60985	0.60844	0.60706	0.6044

Temperature = 950.0 R   Psat = 623.1 psi

Press(psi)=	250	500	1000	1500	2000	2500	3000	3500	4000	5000
v-ft3/lbm			0.0202	0.0201	0.02	0.0199	0.0198	0.0197	0.0197	0.0195
u-Btu/lbm			473.7	471.8	470	468.2	466.6	464.9	463.4	460.4
h-Btu/lbm			477.4	477.4	477.4	477.4	477.6	477.7	477.9	478.4
s-Btu/lbm/R			0.67717	0.67515	0.67321	0.67133	0.66951	0.66775	0.66604	0.66277

Temperature = 1000.0 R   Psat = 964.9 psi

Press(psi)=	250	500	1000	1500	2000	2500	3000	3500	4000	5000
v-ft3/lbm			0.0215	0.0214	0.0212	0.0211	0.0209	0.0208	0.0207	0.0205
u-Btu/lbm			534.2	531.4	528.8	526.3	524	521.8	519.6	515.7
h-Btu/lbm			538.2	537.4	536.7	536.1	535.6	535.2	534.9	534.6
s-Btu/lbm/R			0.73949	0.73666	0.73399	0.73146	0.72905	0.72675	0.72454	0.72038

Temperature = 1050.0 R   Psat = 1434.3 psi

Press(psi)=	250	500	1000	1500	2000	2500	3000	3500	4000	5000
v-ft3/lbm				0.0232	0.0229	0.0227	0.0224	0.0222	0.0221	0.0217
u-Btu/lbm				597.7	593.4	589.5	585.9	582.6	579.5	573.9
h-Btu/lbm				604.2	601.9	599.9	598.3	597	595.8	594
s-Btu/lbm/R				0.80183	0.7976	0.79376	0.79024	0.78696	0.7839	0.7783

Temperature = 1100.0 R   Psat = 2063.9 psi

Press(psi) =	250	500	1000	1500	2000	2500	3000	3500	4000	5000
v-ft3/lbm						0.0254	0.0249	0.0244	0.024	0.0234
u-Btu/lbm						663.3	656.4	650.6	645.4	636.7
h-Btu/lbm						675.1	670.2	666.4	663.2	658.4
s-Btu/lbm/R						0.86361	0.85706	0.85149	0.8466	0.83821

Appendix 15 Thermodynamic Property Tables for Carbon Dioxide

A.15.1 CO_2 Saturation Temperature Table (SI Units)

T(K)	P(Mpa)	Volume (m**3/kg)			Energy (kJ/kg)			Enthalpy (kJ/kg)			Entropy (kJ/kg/K)		
		vf	vfg	vg	uf	ufg	ug	hf	hfg	hg	sf	sfg	sg
216.5	0.5173	0.000847	0.071931	0.072778	0.00	314.66	314.22	0.00	351.87	351.87	0.00000	1.62497	1.62497
220.0	0.6000	0.000857	0.062281	0.063137	6.49	308.75	315.24	7.00	346.12	353.12	0.03175	1.57326	1.60501
225.0	0.7365	0.000871	0.050940	0.051811	16.88	299.69	316.57	17.52	337.21	354.73	0.07848	1.49872	1.57720
230.0	0.8949	0.000885	0.041984	0.042869	27.28	290.46	317.73	28.07	328.03	356.09	0.12425	1.42620	1.55046
235.0	1.0769	0.000901	0.034828	0.035729	37.55	281.17	318.72	38.52	318.68	357.19	0.16850	1.35607	1.52457
240.0	1.2849	0.000918	0.029049	0.029967	47.72	271.77	319.49	48.90	309.10	358.00	0.21142	1.28791	1.49934
245.0	1.5211	0.000936	0.024333	0.025269	57.89	262.15	320.04	59.32	299.16	358.48	0.25348	1.22106	1.47454
250.0	1.7875	0.000956	0.020449	0.021405	68.15	252.18	320.33	69.85	288.74	358.59	0.29503	1.15495	1.44998
255.0	2.0866	0.000977	0.017222	0.018199	78.52	241.80	320.32	80.56	277.73	358.30	0.33630	1.08914	1.42545
260.0	2.4208	0.001001	0.014516	0.015516	89.06	230.91	319.97	91.49	266.05	357.53	0.37744	1.02326	1.40070
265.0	2.7924	0.001027	0.012227	0.013254	99.79	219.43	319.22	102.66	253.57	356.23	0.41856	0.95688	1.37544

T(K)	P(Mpa)	Volume (m**3/kg)			Energy (kJ/kg)			Enthalpy (kJ/kg)			Entropy (kJ/kg/K)		
		vf	vfg	vg	uf	ufg	ug	hf	hfg	hg	sf	sfg	sg
270.0	3.2043	0.001056	0.010272	0.011328	110.77	207.22	317.99	114.16	240.13	354.29	0.45995	0.88937	1.34933
275.0	3.6592	0.001090	0.008586	0.009676	122.13	194.04	316.17	126.12	225.46	351.57	0.50204	0.81985	1.32188
280.0	4.1602	0.001129	0.007113	0.008242	134.02	179.57	313.59	138.72	209.16	347.88	0.54545	0.74701	1.29246
285.0	4.7106	0.001177	0.005805	0.006982	146.67	163.35	310.01	152.21	190.69	342.90	0.59096	0.66910	1.26006
290.0	5.3143	0.001239	0.004616	0.005854	160.36	144.62	304.98	166.94	169.15	336.09	0.63966	0.58327	1.22294
295.0	5.9765	0.001324	0.003484	0.004808	175.75	121.81	297.55	183.66	142.62	326.29	0.69391	0.48347	1.17739
300.0	6.7058	0.001471	0.002278	0.003749	195.40	39.57	234.56	205.06	104.89	309.95	0.76240	0.34960	1.11200
304.2	7.3834	0.002155	0.000000	0.002155	241.37	0.0	241.37	257.31	0.00	257.31	0.93116	0.00000	0.93116

A.15.2 CO_2 Saturation Pressure Table (SI Units)

T(K)	P(Mpa)	Volume (m**3/kg)			Energy (kJ/kg)			Enthalpy (kJ/kg)			Entropy (kJ/kg/K)		
	vf	vfg	vg	uf	ufg	ug	hf	hfg	hg	sf	sfg	sg	vf
0.5173	216.5	0.000847	0.071926	0.072774	0.00	314.66	314.23	0.00	351.51	351.87	0.00001	1.62495	1.62496
0.6000	220.0	0.000857	0.062281	0.063137	6.49	308.75	315.24	7.00	346.12	353.12	0.03175	1.57326	1.60501
0.7000	223.7	0.000867	0.053559	0.054425	14.23	302.01	316.24	14.84	339.50	354.34	0.06669	1.51745	1.58413
0.8000	227.1	0.000877	0.046942	0.047819	21.24	295.84	317.08	21.94	333.39	355.33	0.09780	1.46810	1.56590
0.9000	230.2	0.000886	0.041744	0.042630	27.59	290.18	317.77	28.38	327.75	356.13	0.12561	1.42406	1.54967
1.0000	233.0	0.000895	0.037547	0.038442	33.39	284.95	318.34	34.29	322.50	356.78	0.15071	1.38428	1.53499
1.5000	244.6	0.000935	0.024695	0.025630	57.04	262.97	320.00	58.44	300.01	358.45	0.24998	1.22664	1.47662
2.0000	253.6	0.000971	0.018062	0.019033	75.62	244.73	320.36	77.56	280.86	358.42	0.32484	1.10744	1.43229
2.5000	261.1	0.001006	0.013974	0.014981	91.43	228.41	319.84	93.94	263.35	357.29	0.38657	1.00857	1.39514

T(K)	P(Mpa)	Volume (m**3/kg)			Energy (kJ/kg)			Enthalpy (kJ/kg)			Entropy (kJ/kg/K)		
	vf	vfg	vg	uf	ufg	ug	hf	hfg	hg	sf	sfg	sg	vf
3.0000	267.6	0.001041	0.011180	0.012221	105.43	213.22	318.65	108.55	246.76	355.32	0.43989	0.92220	1.36208
3.5000	273.3	0.001078	0.009131	0.010209	118.23	198.63	316.86	122.00	230.59	352.59	0.48766	0.84371	1.33137
4.0000	278.5	0.001116	0.007549	0.008665	130.27	184.22	314.48	134.73	214.41	349.15	0.53182	0.77001	1.30183
4.5000	283.1	0.001158	0.006275	0.007433	141.86	169.62	311.48	147.07	197.86	344.93	0.57374	0.69880	1.27254
5.0000	287.5	0.001205	0.005210	0.006415	153.23	154.53	307.77	159.26	180.58	339.84	0.61437	0.62821	1.24258
5.5000	291.5	0.001260	0.004285	0.005545	164.60	138.54	303.14	171.53	162.11	333.64	0.65465	0.55620	1.21085
6.0000	295.2	0.001328	0.003445	0.004773	176.32	120.92	297.23	184.28	141.59	325.87	0.69592	0.47968	1.17560
6.5000	298.6	0.001420	0.002199	0.003619	189.34	77.22	266.56	198.56	90.42	288.95	0.74169	0.30631	1.04800
7.0000	301.9	0.001571	0.000955	0.002526	205.65	14.70	220.35	216.64	39.24	255.88	0.79940	0.13294	0.93234
7.3834	304.2	0.002155	0.000000	0.002155	241.37	0.00	241.37	257.31	0.00	257.31	0.93116	0.00000	0.93116

A.15.3 Superheated CO_2 Table (SI Units)

P=0.5173 MPa	T(K) 216.5	220.	250.	300	350	400	500	600	700	800	900	1000
v,m**3/kg	0.07277	0.07429	0.08691	0.10672	0.12586	0.14469	0.18190	0.21881	0.25559	0.29229	0.32895	0.36557
u,kJ/kg	314.23	316.52	336.19	369.47	404.41	441.28	520.67	606.76	698.42	794.67	894.67	997.77
h,kJ/kg	351.87	354.95	381.15	424.68	469.52	516.13	614.77	719.95	830.64	945.88	1064.84	1186.88
s,kJ/kg/K	1.62496	1.63907	1.75074	1.90943	2.04760	2.17203	2.39184	2.58344	2.75397	2.90778	3.04786	3.17642

P=0.6000MPa	T(K) 220.0	220.0	250.	300.	350.	400.	500.	600.	700.	800.	900.	1000
v,m**3/kg	0.06314	0.00000	0.07429	0.09161	0.10824	0.12456	0.15673	0.18861	0.22036	0.25202	0.28365	0.31524
u,kJ/kg	315.24	0.00	335.31	368.92	404.01	440.97	520.47	606.60	698.30	794.57	894.58	997.69
h,kJ/kg	353.12	0.00	379.89	423.89	468.96	515.71	614.50	719.77	830.51	945.78	1064.77	1186.83

s,kJ/kg/K	1.60501	0.00000	1.71914	1.87954	2.01844	2.14324	2.36341	2.55516	2.72578	2.87964	3.01975	3.14832
P=0.7000 MPa	T(K) 223.7	250.	300.	350.	400.	500.	600.	700.	800.	900.	1000.	1200.
v,m**3/kg	0.05443	0.06301	0.07811	0.09250	0.10656	0.13424	0.16162	0.18887	0.21604	0.24316	0.27026	0.32439
u,kJ/kg	316.24	334.23	368.24	403.52	440.60	520.21	606.41	698.15	794.44	894.48	997.60	1211.21
h,kJ/kg	354.34	378.34	422.91	468.27	515.19	614.18	719.55	830.36	945.67	1064.69	1186.78	1438.28
s,kJ/kg/K	1.58413	1.68559	1.84813	1.98793	2.11318	2.33378	2.52573	2.69644	2.85036	2.99051	3.11911	3.34826
P=0.8000 MPa	T(K) 227.1	250.	300.	350.	400.	500.	600.	700.	800.	900.	1000.	1200.
v,m**3/kg	0.04782	0.05454	0.06798	0.08069	0.09307	0.11737	0.14138	0.16526	0.18905	0.21280	0.23652	0.28390
u,kJ/kg	317.08	333.12	367.55	403.04	440.22	519.96	606.23	698.00	794.32	894.37	997.50	1211.13
h,kJ/kg	355.33	376.75	421.94	467.59	514.68	613.86	719.33	830.21	945.56	1064.61	1186.72	1438.26
s,kJ/kg/K	1.56590	1.65582	1.82060	1.96130	2.08702	2.30805	2.50019	2.67100	2.82498	2.96516	3.09379	3.32297
P=0.9000 MPa	T(K) 23.02	250.	300.	350.	400.	500.	600.	700.	800.	900.	1000.	1200.
v,m**3/kg	0.04263	0.04794	0.06010	0.07150	0.08258	0.10425	0.12564	0.14689	0.16806	0.18919	0.21028	0.25242
u,kJ/kg	317.77	331.99	366.86	402.55	439.85	519.71	606.04	697.85	794.20	894.27	997.41	1211.06
h,kJ/kg	356.13	375.13	420.95	466.90	514.17	613.54	719.11	830.05	945.45	1064.54	1186.67	1438.23
s,kJ/kg/K	1.54967	1.62888	1.79601	1.93764	2.06383	2.28530	2.47762	2.64854	2.80257	2.94280	3.07145	3.30066
P=1.0000 MPa	T(K) 233.0	250.	300.	350.	400.	500.	600.	700.	800.	900.	1000.	1200.
v,m**3/kg	0.03844	0.04265	0.05379	0.06416	0.07418	0.09376	0.11305	0.13220	0.15127	0.17030	0.18929	0.22723
u,kJ/kg	318.34	330.83	366.16	402.05	439.47	519.46	605.85	697.70	794.07	894.16	997.32	1210.99
h,kJ/kg	356.78	373.47	419.95	466.21	513.65	613.22	718.90	829.90	945.34	1064.46	1186.61	1438.21
s,kJ/kg/K	1.53499	1.60415	1.77374	1.91632	2.04298	2.26489	2.45741	2.62842	2.78251	2.92277	3.05145	3.28069
P=1.5000 MPa	T(K) 244.6	250.	300.	350.	400.	500.	600.	700.	800.	900.	1000.	1200.
v,m**3/kg	0.02563	0.02665	0.03485	0.04211	0.04899	0.06228	0.07527	0.08812	0.10090	0.11362	0.12632	0.15166
u,kJ/kg	320.00	324.48	362.54	399.55	437.58	518.19	604.91	696.95	793.45	893.63	996.86	1210.62
h,kJ/kg	358.45	364.46	414.82	462.71	511.06	611.61	717.81	829.13	944.80	1064.07	1186.34	1438.11

s,kJ/kg/K	1.47662	1.50091	1.68488	1.83254	1.96164	2.18577	2.37925	2.55076	2.70515	2.84559	2.97439	3.20379
P=2.0000 MPa	T(K) 253.6	300.	350.	400.	500.	600.	700.	800.	900.	1000.	1200.	1400.
v,m**3/kg	0.01903	0.02535	0.03107	0.03640	0.04654	0.05638	0.06608	0.07571	0.08529	0.09484	0.11388	0.13287
u,kJ/kg	320.36	358.72	396.99	435.65	516.92	603.97	696.21	792.83	893.10	996.40	1210.25	1431.83
h,kJ/kg	358.42	409.41	459.13	508.45	610.01	716.73	828.37	944.25	1063.68	1186.07	1438.01	1697.57
s,kJ/kg/K	1.43229	1.61745	1.77081	1.90250	2.12891	2.32335	2.49535	2.65003	2.79066	2.91959	3.14914	3.34913
P=2.5000 MPa	T(K) 261.1	300.	350.	400.	500.	600.	700.	800.	900.	1000.	1200.	1400.
v,m**3/kg	0.01498	0.01962	0.02445	0.02884	0.03710	0.04505	0.05286	0.06060	0.06829	0.07595	0.09121	0.10642
u,kJ/kg	319.84	354.65	394.35	433.71	515.65	603.03	695.46	792.22	892.58	995.94	1209.89	1431.53
h,kJ/kg	357.29	403.69	455.48	505.81	608.40	715.66	827.62	943.71	1063.30	1185.81	1437.91	1697.58
s,kJ/kg/K	1.39514	1.56125	1.72105	1.85548	2.08423	2.27964	2.45215	2.60711	2.74793	2.87698	3.10668	3.30675
P=3.0000 MPa	T(K) 267.6	300.	350.	400.	500.	600.	700.	800.	900.	1000.	1200.	1400.
v,m**3/kg	0.01222	0.01577	0.02003	0.02381	0.03081	0.03750	0.04405	0.05050	0.05695	0.06335	0.07609	0.08879
u,kJ/kg	318.65	350.28	391.64	431.73	514.38	602.09	694.72	791.60	892.06	995.49	1209.53	1431.22
h,kJ/kg	355.32	397.58	451.73	503.15	606.80	714.59	826.87	943.18	1062.92	1185.55	1437.81	1697.60
s,kJ/kg/K	1.36208	1.51160	1.67877	1.81611	2.04727	2.24365	2.41666	2.57182	2.71291	2.84208	3.07193	3.27210
P=3.5000 MPa	T(K) 273.3	300.	350.	400.	500.	600.	700.	800.	900.	1000.	1200.	1400.
v,m**3/kg	0.01021	0.01299	0.01687	0.02021	0.02632	0.03211	0.03776	0.04333	0.04886	0.05436	0.06530	0.07620
u,kJ/kg	316.86	345.55	388.85	429.73	513.10	601.15	693.98	790.99	891.53	995.03	1209.17	1430.92
h,kJ/kg	352.59	391.01	447.90	500.47	605.21	713.52	826.13	942.65	1062.55	1185.29	1437.72	1697.61
s,kJ/kg/K	1.33137	1.46584	1.64157	1.78201	2.01563	2.21299	2.38650	2.54204	2.68321	2.81251	3.04251	3.24276
P=4.0000 MPa	T(K) 278.5	300.	350.	400.	500.	600.	700.	800.	900.	1000.	1200.	1400.
v,m**3/kg	0.00867	0.01087	0.01450	0.01751	0.02295	0.02806	0.03304	0.03794	0.04279	0.04761	0.05720	0.06675
u,kJ/kg	314.48	340.37	385.98	427.71	511.82	600.21	693.24	790.38	891.02	994.58	1208.81	1430.62
h,kJ/kg	349.15	383.86	443.96	497.76	603.61	712.47	825.39	942.13	1062.18	1185.03	1437.62	1697.63

s,kJ/kg/K	1.30183	1.42218	1.60799	1.75173	1.98788	2.18624	2.36023	2.51606	2.65742	2.78684	3.01699	3.21732
P=4.5000 MPa	T(K) 283.1	300.	350.	400.	500.	600.	700.	800.	900.	1000.	1200.	1400
v,m**3/kg	0.00743	0.00919	0.01265	0.01541	0.02033	0.02492	0.02937	0.03374	0.03807	0.04237	0.05091	0.05940
u,kJ/kg	311.48	334.61	383.01	425.66	510.53	599.27	692.50	789.77	890.50	994.13	1208.45	1430.33
h,kJ/kg	344.93	375.94	439.92	495.02	602.02	711.41	824.66	941.61	1061.81	1184.78	1437.53	1697.65
s,kJ/kg/K	1.27254	1.37913	1.57711	1.72436	1.96311	2.16246	2.33695	2.49307	2.63460	2.76414	2.99444	3.19486
P=5.0000 MPa	T(K) 287.5	300.	350.	400.	500.	600.	700.	800.	900.	1000.	1200.	1400
v,m**3/kg	0.00641	0.00779	0.01116	0.01374	0.01824	0.02241	0.02643	0.03039	0.03429	0.03817	0.04587	0.05353
u,kJ/kg	307.77	328.03	379.95	423.57	509.24	598.34	691.77	789.17	889.98	993.68	1208.09	1430.03
h,kJ/kg	339.84	366.98	435.76	492.26	600.43	710.37	823.93	941.10	1061.45	1184.53	1437.45	1697.67
s,kJ/kg/K	1.24258	1.33514	1.54826	1.69927	1.94068	2.14103	2.31602	2.47242	2.61414	2.74380	2.97424	3.17474
P=5.5000 MPa	T(K) 291.5	300.	350.	400.	500.	600.	700.	800.	900.	1000.	1200.	1400
v,m**3/kg	0.00554	0.00659	0.00995	0.01236	0.01653	0.02035	0.02403	0.02764	0.03120	0.03474	0.04175	0.04872
u,kJ/kg	303.14	320.24	376.78	421.47	507.95	597.40	691.03	788.56	889.47	993.23	1207.73	1429.73
h,kJ/kg	333.64	356.46	431.48	489.47	598.84	709.33	823.21	940.59	1061.09	1184.29	1437.36	1697.69
s,kJ/kg/K	1.21085	1.28813	1.52098	1.67602	1.92015	2.12151	2.29699	2.45368	2.59557	2.72535	2.95594	3.15653
P=6.0000 MPa	T(K) 295.2	300.	350.	400.	500.	600.	700.	800.	900.	1000.	1200.	1400
v,m**3/kg	0.00477	0.00548	0.00893	0.01122	0.01510	0.01864	0.02203	0.02535	0.02863	0.03188	0.03832	0.04471
u,kJ/kg	297.23	310.38	373.50	419.33	506.66	596.47	690.30	787.96	888.95	992.78	1207.38	1429.44
h,kJ/kg	325.87	343.29	427.06	486.66	597.26	708.29	822.50	940.08	1060.73	1184.04	1437.28	1697.71
s,kJ/kg/K	1.17560	1.23416	1.49490	1.65426	1.90119	2.10356	2.27954	2.43650	2.57858	2.70848	2.93921	3.13988
P=6.5000 MPa	T(K) 298.6	300.	350.	400.	500.	600.	700.	800.	900.	1000.	1200.	1400
v,m**3/kg	0.00142	0.00435	0.00806	0.01025	0.01389	0.01719	0.02034	0.02342	0.02645	0.02946	0.03541	0.04132
u,kJ/kg	189.33	295.63	370.09	417.16	505.36	595.54	689.57	787.36	888.44	992.34	1207.02	1429.14
h,kJ/kg	198.55	323.90	422.50	483.81	595.68	707.26	821.79	939.58	1060.38	1183.80	1437.19	1697.73

s,kJ/kg/K	0.74165	1.16131	1.46975	1.63375	1.88354	2.08694	2.26341	2.42065	2.56290	2.69292	2.92380	3.12455
P=7.0000 MPa	T(K) 301.9	350.	400.	500.	600.	700.	800.	900.	1000.	1200.	1400.	1600.
v,m**3/kg	0.00157	0.00732	0.00943	0.01286	0.01595	0.01889	0.02176	0.02459	0.02738	0.03292	0.03842	0.04389
u,kJ/kg	205.74	366.55	414.96	504.06	594.61	688.84	786.76	887.93	991.90	1206.67	1428.85	1657.11
h,kJ/kg	216.75	417.79	480.95	594.10	706.24	821.08	939.09	1060.03	1183.57	1437.11	1697.76	1964.31
s,kJ/kg/K	0.79975	1.44529	1.61429	1.86703	2.07144	2.24840	2.40593	2.54835	2.67849	2.90951	3.11034	3.28826
P=7.3834 MPa	T(K) 304.2	350.	400.	500.	600.	700.	800.	900.	1000.	1200.	1400.	1600.
v,m**3/kg	0.00221	0.00682	0.00887	0.01217	0.01511	0.01791	0.02064	0.02333	0.02598	0.03124	0.03645	0.04164
u,kJ/kg	243.42	363.73	413.26	503.07	593.89	688.29	786.30	887.55	991.56	1206.40	1428.62	1656.92
h,kJ/kg	260.02	414.06	478.23	592.89	705.46	820.55	938.71	1059.77	1183.57	1437.05	1697.78	1964.39
s,kJ/kg/K	0.94013	1.42690	1.5999	1.85503	2.06022	2.23756	2.39530	2.53786	2.66808	2.89922	3.10011	3.27807
P=7.500 MPa		350.	400.	500.	600.	700.	800.	900.	1000.	1200.	1400.	1600.
v,m**3/kg	0.00166	0.00667	0.00871	0.01197	0.01487	0.01764	0.02032	0.02297	0.02558	0.03076	0.03590	0.04101
u,kJ/kg	214.30	362.86	412.74	502.76	593.68	688.12	786.16	887.43	991.46	1206.32	1428.56	1656.86
h,kJ/kg	226.75	412.90	478.05	592.53	705.23	820.38	938.59	1059.69	1183.33	1437.03	1697.79	1964.41
s,kJ/kg/K	0.83007	1.42135	1.59573	1.85148	2.05691	2.23437	2.39217	2.53477	2.66502	2.89619	3.09710	3.27507
P=8.000 MPa		350.	400.	500.	600.	700.	800.	900.	1000.	1200.	1400.	1600.
v,m**3/kg	0.00148	0.00610	0.00808	0.01119	0.01393	0.01654	0.01907	0.02155	0.02401	0.02887	0.03369	0.03849
u,kJ/kg	200.44	359.01	410.48	501.46	592.75	687.39	785.57	886.92	991.02	1205.97	1428.27	1656.61
h,kJ/kg	212.29	407.83	475.13	590.96	704.22	819.69	938.11	1059.35	1183.10	1436.96	1697.81	1964.51
s,kJ/kg/K	0.78001	1.39775	1.57794	1.83678	2.04324	2.22118	2.37926	2.52203	2.65240	2.88371	3.08470	3.26273
P=9.000 MPa		350.	400.	500.	600.	700.	800.	900.	1000.	1200.	1400.	1600.
v,m**3/kg	0.00137	0.00515	0.00704	0.00989	0.01237	0.01471	0.01697	0.01920	0.02139	0.02573	0.03002	0.03429
u,kJ/kg	189.73	350.78	405.88	498.85	590.90	685.95	784.39	885.92	990.14	1205.27	1427.69	1656.11

h,kJ/kg	202.07	397.10	469.21	587.83	702.21	818.32	937.15	1058.68	1182.65	1436.81	1697.87	1964.72
s,kJ/kg/K	0.74177	1.35107	1.54429	1.80950	2.01803	2.19696	2.35559	2.49870	2.62930	2.86089	3.06204	3.24016
P=10.000 MPa		350.	400.	500.	600.	700.	800.	900.	1000.	1200.	1400.	1600
v,m**3/kg	0.00132	0.00437	0.00620	0.00885	0.01112	0.01325	0.01530	0.01731	0.01929	0.02321	0.02708	0.03093
u,kJ/kg	183.35	341.76	401.15	496.24	589.05	684.52	783.21	884.92	989.27	1204.58	1427.11	1655.62
h,kJ/kg	196.51	385.51	463.19	584.73	700.24	816.97	936.20	1058.02	1182.21	1436.67	1697.94	1964.93
s,kJ/kg/K	0.71907	1.30438	1.51273	1.78460	1.99520	2.17510	2.33428	2.47773	2.60855	2.84042	3.04173	3.21995
P=12.500 MPa		350.	400.	500.	600.	700.	800.	900.	1000.	1200.	1400.	1600
v,m**3/kg	0.00124	0.00300	0.00472	0.00699	0.00888	0.01062	0.01229	0.01392	0.01552	0.01868	0.02180	0.02489
u,kJ/kg	173.03	315.85	388.84	489.68	584.47	680.97	780.30	882.45	987.12	1202.87	1425.69	1654.40
h,kJ/kg	188.49	353.29	447.78	577.07	695.41	813.69	933.93	1056.45	1181.16	1436.36	1698.12	1965.48
s,kJ/kg/K	0.68224	1.18642	1.44045	1.73002	1.94585	2.12816	2.28867	2.43295	2.56433	2.79688	2.99857	3.17705
P=15.000 MPa		350.	400.	500.	600.	700.	800.	900.	1000.	1200.	1400.	1600
v,m**3/kg	0.00119	0.00222	0.00374	0.00576	0.00739	0.00887	0.01029	0.01166	0.01301	0.01566	0.01827	0.02086
u,kJ/kg	165.95	289.68	376.00	483.13	579.93	677.47	777.44	880.02	985.01	1201.19	1424.28	1653.19
h,kJ/kg	183.79	322.92	432.17	569.58	690.74	810.55	931.75	1054.96	1180.18	1436.10	1698.34	1966.05
s,kJ/kg/K	0.65685	1.08136	1.37520	1.68334	1.90438	2.08905	2.25087	2.39596	2.52788	2.76108	2.96315	3.14185
P=20.000 MPa		350.	400.	500.	600.	700.	800.	900.	1000.	1200.	1400.	1600
v,m**3/kg	0.00113	0.00163	0.00262	0.00426	0.00554	0.00670	0.00779	0.00885	0.00988	0.01189	0.01387	0.01582
u,kJ/kg	155.78	255.86	350.57	470.16	571.04	670.64	771.86	875.29	980.90	1197.91	1421.54	1650.83
h,kJ/kg	178.36	288.43	403.03	555.30	681.94	804.67	927.73	1052.23	1178.42	1435.70	1698.87	1967.26
s,kJ/kg/K	0.61997	0.95621	1.26338	1.60545	1.83658	2.02578	2.19009	2.33671	2.46965	2.70411	2.90689	3.08604

Appendix 16 Thermodynamic Property Tables for Sodium

A.16.1 Sodium Temperature Saturation Table (SI units)

T(K)	P(MPa)	Volume (m**3/kg)			Energy (kJ/kg)			Enthalpy (kJ kg)			Entropy (kJ/kg)		
		vf	vfg	vg	uf	ufg	ug	h	hfg	hg	sf	sfg	sg
800.0	0.001	0.001211	298.0341	298.0353	0.0	4034.4	4034.3	0.0	4315.6	4315.5	0.0000	5.3945	5.3944
850.0	0.002	0.001229	127.4088	127.4100	66.4	3970.8	4037.2	66.4	4267.0	4333.4	0.0804	5.0200	5.1004
900.0	0.005	0.001247	60.0689	60.0702	132.7	3904.9	4037.6	132.7	4215.5	4348.2	0.1563	4.6839	4.8401
950.0	0.011	0.001266	30.7359	30.7372	200.2	3836.0	4036.2	200.2	4160.4	4360.6	0.2292	4.3793	4.6085
1000.0	0.020	0.001286	16.8575	16.8587	268.5	3765.2	4033.7	268.5	4102.8	4371.3	0.2991	4.1028	4.4019
1050.0	0.036	0.001306	9.8135	9.8148	336.6	3694.3	4030.9	336.7	4044.6	4381.2	0.3653	3.8520	4.2172
1100.0	0.060	0.001327	6.0137	6.0150	405.1	3623.4	4028.5	405.2	3986.0	4391.2	0.4285	3.6236	4.0521
1154.6	0.101	0.001351	3.7069	3.7083	479.9	3547.6	4027.5	480.0	3923.2	4403.2	0.4939	3.3979	3.8917
1200.0	0.150	0.001372	2.5685	2.5699	542.4	3486.0	4028.4	542.6	3872.2	4414.9	0.5457	3.2269	3.7725
1250.0	0.224	0.001395	1.7722	1.7736	611.0	3421.1	4032.2	611.3	3818.8	4430.1	0.5996	3.0550	3.6547
1300.0	0.325	0.001419	1.2599	1.2613	682.9	3356.1	4039.0	683.3	3765.0	4448.4	0.6528	2.8962	3.5490
1400.0	0.625	0.001469	0.6885	0.6899	831.8	3232.2	4064.0	832.7	3662.8	4495.5	0.7528	2.6163	3.3691
1500.0	1.101	0.001523	0.4091	0.4106	998.1	3107.4	4105.5	999.7	3557.9	4557.7	0.8494	2.3719	3.2213
1600.0	1.802	0.001581	0.2590	0.2606	1194.0	2970.8	4164.8	1196.8	3437.4	4634.2	0.9463	2.1484	3.0947
1700.0	2.776	0.001642	0.1713	0.1729	1440.8	2804.2	4245.0	1445.4	3279.7	4725.1	1.0487	1.9292	2.9780

A.16.2 Sodium Pressure Saturation Table (SI units)

P(MPa)	T(K)	Volume (m**3/kg)			Energy (kJ/kg)			Enthalpy (kJ kg)			Entropy (kJ/kg)		
		vf	vfg	vg	uf	ufg	ug	hf	hfg	hg	sf	sfg	sg
0.009	941.1	0.001263	34.4471	34.4483	188.2	3848.4	4036.6	188.2	4170.3	4358.5	0.2165	4.4313	4.6479
0.020	999.9	0.001286	16.8773	16.8786	268.2	3765.5	4033.7	268.2	4103.1	4371.3	0.2988	4.1035	4.4023
0.040	1060.5	0.001311	8.8223	8.8236	351.0	3679.4	4030.3	351.0	4032.3	4383.3	0.3788	3.8024	4.1811
0.060	1099.5	0.001327	6.0411	6.0425	404.3	3624.2	4028.6	404.4	3986.7	4391.1	0.4278	3.6258	4.0536
0.080	1129.1	0.001340	4.6196	4.6209	444.9	3582.8	4027.7	445.0	3952.4	4397.4	0.4637	3.5006	3.9643
0.100	1153.1	0.001351	3.7528	3.7541	477.9	3549.6	4027.5	478.0	3924.9	4402.9	0.4921	3.4036	3.8958
0.101	1154.6	0.001351	3.7070	3.7084	480.0	3547.5	4027.5	480.1	3923.1	4403.2	0.4939	3.3978	3.8917
0.200	1235.2	0.001388	1.9714	1.9728	591.3	3439.4	4030.7	591.5	3833.7	4425.3	0.5844	3.1038	3.6882
0.300	1289.0	0.001413	1.3548	1.3562	667.1	3370.0	4037.2	667.6	3776.5	4444.0	0.6414	2.9298	3.5712
0.400	1330.2	0.001434	1.0390	1.0404	726.6	3318.3	4044.9	727.2	3733.9	4461.1	0.6836	2.8070	3.4905
0.500	1364.1	0.001451	0.8459	0.8474	777.0	3276.3	4053.2	777.7	3699.2	4476.9	0.7175	2.7118	3.4293
0.750	1430.6	0.001485	0.5826	0.5841	880.3	3194.6	4074.9	881.4	3631.6	4513.0	0.7825	2.5385	3.3210
1.000	1481.9	0.001513	0.4471	0.4486	966.2	3130.5	4096.8	967.7	3577.6	4545.3	0.8320	2.4142	3.2461
2.000	1623.0	0.001594	0.2348	0.2364	1245.1	2936.1	4181.1	1248.3	3405.6	4653.8	0.9691	2.0983	3.0675
3.000	1719.4	0.001655	0.1586	0.1603	1497.8	2766.0	4263.7	1502.7	3241.9	4744.6	1.0698	1.8855	2.9553

A.16.3 Superheated Sodium Table (SI Units)

	T(K)	950	1000	1050	1100	1150	1200	1250	1300	1350	1400	1450
P=0.0093 MPa	T(K) 941.1											
v,m**3/kg	34.44834	34.9420	37.4751	39.7787	41.9451	44.0527	46.0856	48.0955	50.0854	52.0603	54.0195	55.9839
u,kJ/kg	4036.59	4053.95	4131.43	4187.49	4231.91	4269.71	4303.53	4334.91	4364.73	4393.52	4421.62	4449.24
h,kJ/kg	4358.54	4380.48	4481.67	4559.26	4623.98	4681.33	4734.23	4784.41	4832.83	4880.08	4926.55	4972.47
s,kJ/kg/K	4.64786	4.6711	4.7751	4.8509	4.9111	4.9622	5.0072	5.0481	5.0861	5.1218	5.1555	5.1878
P=0.0500 MPa	T(K)1081.6	1100.	1150.	1200.	1250.	1300.	1350.	1400.	1450.	1500.	1550.	1600.
v,m**3/kg	7.16339	7.3656	7.8738	8.3370	8.7728	9.1880	9.5898	9.9813	10.3661	10.7461	11.1220	11.4955
u,kJ/kg	4029.30	4067.50	4152.25	4217.57	4270.57	4315.56	4355.23	4391.29	4424.85	4456.64	4487.15	4516.73
h,kJ/kg	4387.47	4435.77	4545.94	4634.47	4709.21	4774.96	4834.71	4890.36	4943.17	4993.95	5043.27	5091.51
s,kJ/kg/K	4.11076	4.1553	4.2536	4.3291	4.3902	4.4418	4.4870	4.5275	4.5645	4.5990	4.6313	4.6619
P=0.1000 MPa	T(K)1153.1	1200.	1250.	1300.	1350.	1400.	1450.	1500.	1550.	1600.	1700.	1800.
v,m**3/kg	3.75411	4.0095	4.2583	4.4909	4.7099	4.9198	5.1238	5.3224	5.5175	5.7113	6.0886	6.4619
u,kJ/kg	4027.47	4118.13	4194.78	4257.01	4309.31	4354.73	4395.35	4432.52	4467.19	4500.04	4561.93	4620.60
h,kJ/kg	4402.87	4519.07	4620.66	4706.08	4780.30	4846.74	4907.72	4964.76	5018.94	5071.02	5170.79	5266.81
s,kJ/kg/K	3.89575	3.9956	4.0790	4.1462	4.2023	4.2507	4.2936	4.3322	4.3678	4.4009	4.4614	4.5162
P=0.1013 MPa	T(K)1154.6	1200.	1250.	1300.	1350.	1400.	1450.	1500.	1550.	1600.	1700.	1800.
v,m**3/kg	3.70836	3.9531	4.2086	4.4293	4.6460	4.8546	5.0551	5.2515	5.4449	5.6342	6.0083	6.3769
u,kJ/kg	4027.46	4115.61	4193.14	4255.49	4308.11	4353.79	4394.57	4431.88	4466.67	4499.59	4561.62	4620.37
h,kJ/kg	4403.21	4516.16	4618.74	4704.29	4778.88	4845.62	4906.79	4963.99	5018.31	5070.48	5170.40	5266.51
s,kJ/kg/K	3.89173	3.9887	4.0736	4.1403	4.1967	4.2453	4.2883	4.3271	4.3627	4.3958	4.4564	4.5114
P=0.2000 Mpa	T(K)1235.2	1250.	1300.	1350.	1400.	1450.	1500.	1550.	1600.	1700.	1800.	
v,m**3/kg	1.97278	2.0149	2.1492	2.2746	2.3923	2.5043	2.6118	2.7161	2.8176	3.0153	3.2077	
u,kJ/kg	4030.70	4060.32	4148.92	4222.58	4284.70	4338.27	4385.53	4428.13	4467.23	4538.19	4602.85	

h,kJ/kg	5244.38	5141.23	5030.76	4971.33	4907.90	4839.12	4763.15	4677.48	4578.76	4463.26	4425.26
s,kJ/kg/K	4.2557	4.1968	4.1297	4.0920	4.0503	4.0036	3.9501	3.8874	3.8120	3.7197	3.68816
P=0.3000 Mpa		1800.	1700.	1600.	1550.	1500.	1450.	1400.	1350.	1300.	T(K)1289.0
v,m**3/kg		2.1230	1.9910	1.8541	1.7830	1.7094	1.6330	1.5524	1.4673	1.3768	1.35620
u,kJ/kg		4585.34	4514.89	4435.34	4390.39	4340.57	4284.39	4219.93	4145.04	4057.97	4037.18
h,kJ/kg		5222.25	5112.20	4991.58	4925.30	4853.41	4774.27	4685.67	4585.23	4471.01	4444.04
s,kJ/kg/K		4.0994	4.0364	3.9632	3.9210	3.8736	3.8197	3.7568	3.6823	3.5931	3.57121
P=0.4000 Mpa			1800.	1700.	1600.	1550.	1500.	1450.	1400.	1350.	T(K)1330.2
v,m**3/kg			1.5807	1.4792	1.3729	1.3172	1.2593	1.1987	1.1349	1.0678	1.04040
u,kJ/kg			4568.10	4492.10	4404.49	4354.22	4298.05	4234.45	4161.92	4079.76	4044.92
h,kJ/kg			5200.44	5083.78	4953.64	4881.10	4801.77	4713.95	4615.92	4506.88	4461.08
s,kJ/kg/K			3.9858	3.9190	3.8398	3.7935	3.7410	3.6806	3.6100	3.5271	3.49055
P=0.5000 MPA				1800.	1700.	1600.	1550.	1500.	1450.	1400.	T(K)1364.1
v,m**3/kg				1.2557	1.1723	1.0845	1.0382	0.9901	0.9396	0.8867	0.84736
u,kJ/kg				4551.15	4469.86	4374.83	4319.85	4258.40	4189.33	4112.29	4053.23
h,kJ/kg				5178.99	5056.02	4917.08	4838.98	4753.43	4659.12	4555.62	4476.92
s,kJ/kg/K				3.8957	3.8252	3.7404	3.6903	3.6332	3.5674	3.4912	3.42932
P=0.6000 MPa				1800.	1700.	1600.	1550.	1500.	1450.	1400.	T(K)1393.2
v,m**3/kg				1.0390	0.9679	0.8926	0.8529	0.8113	0.7679	0.7229	0.71675
u,kJ/kg				4534.51	4448.24	4346.48	4287.59	4222.09	4149.78	4072.44	4061.83
h,kJ/kg				5157.89	5028.97	4882.03	4799.27	4708.87	4610.55	4506.21	4491.87
s,kJ/kg/K				3.8205	3.7464	3.6564	3.6030	3.5419	3.4718	3.3918	3.38018
P=0.7000 MPa					1800.	1700.	1600.	1550.	1500.	1450.	T(K)1418.8
v,m**3/kg					0.8842	0.8220	0.7558	0.7208	0.6843	0.6464	0.62230
u,kJ/kg					4518.19	4427.27	4319.61	4257.59	4189.55	4116.64	4070.56

h,kJ/kg	4506.11	4569.10	4668.54	4762.13	4848.67	5002.68	5137.18
s,kJ/kg/K	3.33927	3.3893	3.4624	3.5268	3.5833	3.6782	3.7556
P=0.8000 Mpa	T(K)1441.8	1450.	1500.	1550.	1600.	1700.	1800.
v,m**3/kg	0.55049	0.5561	0.5895	0.6221	0.6535	0.7128	0.7683
u,kJ/kg	4079.29	4090.60	4161.22	4230.19	4294.39	4407.05	4502.25
h,kJ/kg	4519.68	4535.45	4632.88	4727.89	4817.18	4977.26	5116.91
s,kJ/kg/K	3.30409	3.3172	3.3922	3.4592	3.5184	3.6178	3.6984
P=0.9000 MPa	T(K)1462.7	1500.	1550.	1600.	1700.	1800.	
v,m**3/kg	0.49413	0.5164	0.5458	0.5741	0.6280	0.6782	
u,kJ/kg	4088.04	4137.55	4205.62	4270.94	4387.58	4486.67	
h,kJ/kg	4532.75	4602.32	4696.79	4787.66	4952.69	5097.06	
s,kJ/kg/K	3.27341	3.3295	3.3985	3.4599	3.5633	3.6470	
P=1.0000 MPa	T(K)1481.9	1500.	1550.	1600.	1700.	1800.	
v,m**3/kg	0.44859	0.4583	0.4849	0.5109	0.5601	0.6062	
u,kJ/kg	4096.75	4118.89	4184.13	4249.42	4368.93	4471.50	
h,kJ/kg	4545.34	4577.20	4669.06	4760.28	4929.03	5077.67	
s,kJ/kg/K	3.24613	3.2730	3.3433	3.4066	3.5136	3.6002	
P=2.0000 MPa	T(K)1623.0	1700.	1800.				
v,m**3/kg	0.23636	0.2574	0.2834				
u,kJ/kg	4181.14	4241.43	4347.27				
h,kJ/kg	4653.85	4756.31	4914.13				
s,kJ/kg/K	3.06745	3.1640	3.2716				

A.16.4 Sodium Temperature Saturation Table (English Units)

T(R)	P(psi)	Volume (ft**3/lbm)			Energy (Btu/lbm)			Enthalpy (Btu/lbm)			Entropy (Btu/lbm)		
		vf	vfg	vg	uf	ufg	ug	hf	hfg	hg	sf	sfg	sg
1440.0	0.1	0.01944	4784.0732	4784.0928	−0.1	8412.9	8412.8	−0.1	8999.3	8999.2	2.77752	3.47195	6.24947
1500.0	0.3	0.01963	2683.4402	2683.4597	91.3	8326.1	8417.4	91.3	8933.5	9024.8	2.70908	3.30871	6.01779
1600.0	0.6	0.01996	1131.0094	1131.0293	245.7	8174.2	8419.9	245.7	8815.3	9061.0	2.61041	3.06085	5.67126
1700.0	1.4	0.02029	527.3803	527.4006	436.5	7979.5	8416.0	436.6	8652.6	9089.1	2.53727	2.82764	5.36491
1800.0	2.9	0.02065	270.5976	270.6182	559.9	7851.7	8411.6	559.9	8555.6	9115.5	2.45895	2.64062	5.09957
1900.0	5.5	0.02101	148.8262	148.8472	717.8	7687.2	8405.0	717.9	8420.5	9138.5	2.40120	2.46215	4.86335
2000.0	9.8	0.02139	87.1279	87.1493	876.3	7523.6	8400.0	876.5	8285.3	9161.8	2.35325	2.30148	4.65473
2100.0	16.4	0.02178	53.8157	53.8375	1035.4	7363.3	8398.6	1035.7	8152.4	9188.1	2.31376	2.15673	4.47049
2200.0	26.2	0.02218	34.8169	34.8391	1195.4	7207.8	8403.2	1195.9	8023.8	9219.7	2.28164	2.02621	4.30784
2300.0	40.1	0.02260	23.4509	23.4735	1357.7	7057.5	8415.3	1358.6	7899.8	9258.4	2.25603	1.90818	4.16421
2400.0	59.2	0.02304	16.3607	16.3838	1523.8	6912.6	8436.4	1525.0	7780.7	9305.7	2.23610	1.80108	4.03717
2500.0	84.7	0.02349	11.7652	11.7887	1698.8	6768.5	8467.3	1700.6	7661.5	9362.1	2.22161	1.70256	3.92417
2600.0	117.8	0.02396	8.6896	8.7136	1882.8	6626.0	8508.8	1885.3	7543.0	9428.3	2.21143	1.61176	3.82319
2800.0	211.7	0.02495	5.0603	5.0853	2298.4	6326.7	8625.1	2303.1	7286.1	9589.2	2.20253	1.44566	3.64818
3000.0	350.7	0.02602	3.1440	3.1700	2819.0	5971.7	8790.7	2827.1	6959.5	9786.6	2.20581	1.28880	3.49461

A.16.5 Sodium Pressure Saturation Table (English Units)

P(psi)	T(R)	Volume (ft**3/lbm)			Energy (Btu/lbm)			Enthalpy (Btu/lbm)			Entropy (Btu/lbm)		
		vf	vfg	vg	uf	ufg	ug	hf	hfg	hg	sf	sfg	sg
1.4	1694.0	0.02027	552.9115	552.9318	392.7	8024.8	8417.5	392.7	8696.2	9088.9	0.25091	5.13362	5.38454
2.5	1778.1	0.02057	311.3963	311.4169	525.6	7887.4	8412.9	525.6	8584.7	9110.3	0.32730	4.82810	5.15540
5.0	1884.3	0.02095	162.7399	162.7609	693.0	7713.0	8406.0	693.0	8441.9	9134.9	0.41842	4.48007	4.89849
7.5	1952.7	0.02121	111.4133	111.4345	801.4	7600.7	8402.0	801.4	8349.2	9150.6	0.47446	4.27565	4.75011
10.0	2004.5	0.02140	85.1776	85.1990	883.8	7516.0	8399.8	883.8	8279.1	9162.9	0.51564	4.13034	4.64598
12.5	2046.6	0.02157	69.1843	69.2059	950.6	7448.2	8398.7	950.6	8223.0	9173.6	0.54812	4.01794	4.56606
14.7	2078.3	0.02169	59.5053	59.5270	1001.2	7397.3	8398.5	1001.2	8180.9	9182.1	0.57219	3.93639	4.50858
15.0	2082.4	0.02171	58.3823	58.4040	1007.5	7391.0	8398.5	1007.5	8175.6	9183.2	0.57520	3.92615	4.50135
17.5	2113.6	0.02183	50.5836	50.6055	1057.4	7341.4	8398.9	1057.4	8134.7	9192.1	0.59845	3.84866	4.44711
20.0	2141.5	0.02194	44.6815	44.7034	1101.8	7297.9	8399.7	1101.8	8098.7	9200.5	0.61875	3.78177	4.40052
25.0	2189.8	0.02214	36.3213	36.3435	1179.6	7222.8	8402.4	1179.6	8036.6	9216.2	0.65350	3.66995	4.32345
30.0	2231.0	0.02231	30.6741	30.6964	1246.1	7160.0	8406.0	1246.1	7984.8	9230.9	0.68238	3.57897	4.26135
45.0	2328.6	0.02272	21.0823	21.1050	1405.9	7014.4	8420.3	1405.9	7865.1	9271.0	0.74868	3.37755	4.12623
60.0	2403.4	0.02305	16.1681	16.1911	1531.9	6905.3	8437.2	1531.9	7775.5	9307.4	0.79783	3.23519	4.03302
75.0	2464.9	0.02333	13.1648	13.1881	1638.8	6816.4	8455.2	1638.8	7702.3	9341.2	0.83749	3.12481	3.96230
100.0	2549.2	0.02372	10.1038	10.1275	1790.1	6696.3	8486.4	1790.1	7603.4	9393.5	0.89046	2.98265	3.87311
200.0	2779.2	0.02485	5.3337	5.3586	2256.0	6354.9	8610.9	2256.0	7314.8	9570.8	1.03310	2.63194	3.66503
300.0	2935.0	0.02567	3.6502	3.6758	2640.2	6090.6	8730.7	2640.2	7078.3	9718.5	1.13168	2.41166	3.54334

A.16.6 Superheated Sodium Table (English Units)

	T(R)1694.0	1700.	1800.	1900.	2000.	2100.	2200.	2300.	2400.	2500.	2600.	2700.
P= 1.4 psi	T(R)1694.0	1700.	1800.	1900.	2000.	2100.	2200.	2300.	2400.	2500.	2600.	2700.
v,ft**3/lbm	552.9318	555.9260	601.5211	642.4615	680.8981	717.9449	754.1944	789.7718	825.1837	860.1784	895.1399	929.9736
u,Btu/lbm	8417.5	8431.5	8615.3	8743.4	8843.3	8927.9	9003.8	9074.5	9142.0	9207.5	9271.7	9334.9
h,Btu/lbm	9088.9	9106.5	9345.6	9523.5	9670.0	9799.6	9919.4	10033.4	10143.8	10252.0	10358.5	10464.1
s,Btu/lbm/R	5.3845	5.3949	5.5319	5.6282	5.7034	5.7667	5.8224	5.8731	5.9201	5.9642	6.0060	6.0459
P= 2.5 psi	T(R)1778.1	1800.	1900.	2000.	2100.	2200.	2300.	2400.	2500.	2600.	2700.	2800.
v,ft**3/lbm	311.4169	317.4174	342.4559	365.0533	386.2528	406.6337	426.4664	446.0121	465.3169	484.4529	503.5057	522.4739
u,Btu/lbm	8412.9	8464.1	8647.7	8780.0	8884.3	8972.8	9051.9	9125.1	9194.5	9261.5	9326.9	9391.0
h,Btu/lbm	9110.3	9174.9	9414.5	9597.4	9749.3	9883.4	10006.9	10123.8	10236.5	10346.3	10454.3	10561.0
s,Btu/lbm/R	5.1554	5.1916	5.3214	5.4153	5.4895	5.5519	5.6068	5.6566	5.7026	5.7457	5.7864	5.8252
P= 5.0 psi	T(R)1884.3	1900.	2000.	2100.	2200.	2300.	2400.	2500.	2600.	2700.	2800.	2900.
v,ft**3/lbm	162.7609	164.9819	178.0859	189.8740	200.8626	211.3520	221.5073	231.4522	241.2410	250.9281	260.5435	270.0991
u,Btu/lbm	8406.0	8444.5	8644.1	8790.5	8905.9	9002.9	9088.3	9166.3	9239.4	9309.2	9376.8	9442.7
h,Btu/lbm	9134.9	9183.3	9441.7	9640.8	9805.5	9949.4	10080.3	10202.8	10319.8	10433.0	10543.6	10652.3
s,Btu/lbm/R	4.8985	4.9242	5.0571	5.1545	5.2312	5.2952	5.3509	5.4009	5.4468	5.4895	5.5298	5.5679
P= 7.5 psi	T(R)1952.7	2000.	2100.	2200.	2300.	2400.	2500.	2600.	2700.	2800.	2900.	3000.
v,ft**3/lbm	111.4345	115.8719	124.4683	132.3047	139.6526	146.6837	153.5009	160.1333	166.7368	173.2347	179.6686	186.0625
u,Btu/lbm	8402.0	8513.1	8698.8	8840.1	8954.4	9051.8	9138.2	9217.4	9291.7	9362.6	9431.1	9497.8
h,Btu/lbm	9150.6	9291.4	9534.9	9728.9	9892.6	10037.2	10169.4	10293.4	10411.8	10526.3	10638.1	10747.8
s,Btu/lbm/R	4.7501	4.8217	4.9410	5.0314	5.1043	5.1658	5.2198	5.2684	5.3132	5.3548	5.3940	5.4312
P= 10.0 psi	T(R)2004.5	2100.	2200.	2300.	2400.	2500.	2600.	2700.	2800.	2900.	3000.	
v,ft**3/lbm	85.1990	91.8104	98.0478	103.8186	109.2796	114.5324	119.6472	124.6441	129.5703	134.5207	139.2899	
u,Btu/lbm	8399.8	8609.7	8775.6	8906.7	9015.7	9110.4	9195.6	9274.3	9348.5	9419.5	9488.2	

h,Btu/lbm	9162.9	9432.0	9653.8	9836.6	9994.5	10136.3	10267.2	10390.7	10509.1	10623.8	10735.8
s,Btu/lbm/R	4.6460	4.7781	4.8816	4.9630	5.0303	5.0882	5.1396	5.1862	5.2293	5.2697	5.3075
P=12.5 psi	T(R)2046.6	2100.	2200.	2300.	2400.	2500.	2600.	2700.	2800.	2900.	3000.
v,ft**3/lbm	69.2059	72.2589	77.5131	82.3293	86.8434	91.1549	95.3236	99.3923	103.3869	107.3276	111.2298
u,Btu/lbm	8398.7	8523.8	8712.5	8859.6	8980.0	9082.8	9173.9	9257.0	9334.5	9408.0	9478.6
h,Btu/lbm	9173.6	9332.9	9580.3	9781.4	9952.3	10103.4	10241.2	10369.8	10492.0	10609.6	10723.9
s,Btu/lbm/R	4.5661	4.6436	4.7594	4.8491	4.9220	4.9837	5.0378	5.0863	5.1308	5.1721	5.2108
P=14.7 psi	T(R)2078.3	2100.	2200.	2300.	2400.	2500.	2600.	2700.	2800.	2900.	3000.
v,ft**3/lbm	59.5270	60.6072	65.2546	69.4941	73.4376	77.1835	80.7923	84.2977	87.7346	91.1159	94.4552
u,Btu/lbm	8398.5	8451.6	8658.4	8819.0	8949.0	9058.8	9155.0	9241.8	9322.2	9397.9	9470.2
h,Btu/lbm	9182.1	9249.3	9517.4	9733.7	9915.7	10074.8	10218.4	10351.5	10477.0	10597.2	10713.5
s,Btu/lbm/R	4.5086	4.5412	4.6670	4.7637	4.8413	4.9063	4.9627	5.0129	5.0586	5.1008	5.1402
P=15.0 psi	T(R)2082.4	2100.	2200.	2300.	2400.	2500.	2600.	2700.	2800.	2900.	3000.
v,ft**3/lbm	58.4040	59.2735	63.8431	68.0124	71.8927	75.5716	79.1147	82.5565	85.9165	89.2437	92.5195
u,Btu/lbm	8398.5	8441.9	8651.0	8813.4	8944.8	9055.5	9152.4	9239.7	9320.5	9396.5	9469.0
h,Btu/lbm	9183.2	9238.1	9508.8	9727.2	9910.7	10070.9	10215.3	10348.9	10474.9	10595.5	10712.1
s,Btu/lbm/R	4.5013	4.5280	4.6551	4.7526	4.8309	4.8964	4.9531	5.0036	5.0494	5.0917	5.1312
P=20.0 psi	T(R)2141.5	2200.	2300.	2400.	2500.	2600.	2700.	2800.	2900.	3000.	
v,ft**3/lbm	44.7034	46.8068	50.1409	53.2172	56.1055	58.8591	61.5165	64.1042	66.6435	69.1359	
u,Btu/lbm	8399.7	8534.1	8723.8	8875.8	9001.7	9109.8	9205.6	9292.7	9373.6	9450.0	
h,Btu/lbm	9200.5	9372.6	9622.1	9829.1	10006.7	10164.2	10307.6	10441.0	10567.4	10688.5	
s,Btu/lbm/R	4.4005	4.4811	4.5931	4.6817	4.7544	4.8162	4.8704	4.9190	4.9633	5.0044	
P=25.0 psi	T(R)2189.8	2200.	2300.	2400.	2500.	2600.	2700.	2800.	2900.	3000.	
v,ft**3/lbm	36.3435	36.6427	39.4482	42.0280	44.4280	46.7121	48.8968	51.0140	53.0731	55.1066	
u,Btu/lbm	8402.4	8426.7	8638.6	8808.9	8949.0	9067.9	9171.9	9265.2	9350.9	9431.1	

h,Btu/lbm	9216.2	9247.2	9521.9	9750.0	9944.0	10113.9	10266.8	10407.5	10539.5	10665.0
s,Btu/lbm/R	4.3235	4.3379	4.4623	4.5603	4.6398	4.7067	4.7645	4.8157	4.8620	4.9046
P=30.0 psi	T(R)2231.0	2300.	2400.	2500.	2600.	2700.	2800.	2900.	3000.	
v,ft**3/lbm	30.6964	32.3466	34.5823	36.6599	38.6280	40.4872	42.2885	44.0402	45.7548	
u,Btu/lbm	8406.0	8558.4	8744.6	8897.8	9026.9	9138.6	9238.0	9328.4	9412.3	
h,Btu/lbm	9230.9	9427.6	9673.9	9882.9	10064.6	10226.5	10374.3	10511.8	10641.7	
s,Btu/lbm/R	4.2614	4.3505	4.4569	4.5429	4.6146	4.6758	4.7296	4.7779	4.8220	
P=45.0 psi	T(R)2328.6	2400.	2500.	2600.	2700.	2800.	2900.	3000.		
v,ft**3/lbm	21.1050	22.2378	23.7421	25.1384	26.4846	27.7569	28.9821	30.1735		
u,Btu/lbm	8420.3	8570.1	8753.6	8908.8	9042.1	9158.4	9262.3	9356.8		
h,Btu/lbm	9271.0	9466.5	9710.6	9922.5	10109.5	10277.2	10430.4	10572.9		
s,Btu/lbm/R	4.1262	4.2126	4.3147	4.3988	4.4701	4.5313	4.5852	4.6336		
P=60.0 psi	T(R)2403.4	2500.	2600.	2700.	2800.	2900.	3000.			
v,ft**3/lbm	16.1911	17.3290	18.4446	19.4989	20.5009	21.4601	22.3860			
u,Btu/lbm	8437.2	8627.0	8800.6	8951.1	9082.2	9198.3	9302.7			
h,Btu/lbm	9307.4	9558.3	9791.9	9999.0	10184.0	10351.6	10505.8			
s,Btu/lbm/R	4.0330	4.1415	4.2360	4.3155	4.3834	4.4426	4.4951			
P=75.0 psi	T(R)2464.9	2500.	2600.	2700.	2800.	2900.	3000.			
v,ft**3/lbm	13.1881	13.4599	14.4440	15.3240	16.1567	16.9534	17.7190			
u,Btu/lbm	8455.2	8520.2	8703.9	8866.7	9009.9	9136.7	9250.3			
h,Btu/lbm	9341.2	9428.0	9674.2	9896.0	10095.3	10275.6	10440.6			
s,Btu/lbm/R	3.9623	3.9992	4.1034	4.1899	4.2637	4.3277	4.3840			
P=100.0 psi	T(R)2549.2	2600.	2700.	2800.	2900.	3000.				
v,ft**3/lbm	10.1275	10.4854	11.1702	11.8280	12.4567	13.0588				

u,Btu/lbm	8486.4	8573.8	8743.2	8899.6	9040.4	9166.8
h,Btu/lbm	9393.5	9512.9	9743.8	9959.1	10156.1	10336.5
s,Btu/lbm/R	3.8731	3.9260	4.0204	4.1022	4.1732	4.2353
P=200.0 psi	T(R)2779.2	2800.	2900.	3000.		
v,ft**3/lbm	5.3586	5.4301	5.7768	6.1134		
u,Btu/lbm	8610.9	8633.3	8760.4	8898.8		
h,Btu/lbm	9570.8	9606.2	9795.2	9994.0		
s,Btu/lbm/R	3.6650	3.6835	3.7690	3.8462		
P=300.0 psi	T(R)2935.0	3000.				
v,ft**3/lbm	3.6758	3.8323				
u,Btu/lbm	8730.7	8776.9				
h,Btu/lbm	9718.5	9806.6				
s,Btu/lbm/R	3.5433	3.5974				

Index

© Springer International Publishing Switzerland 2015   721
B. Zohuri, P. McDaniel, *Thermodynamics In Nuclear Power Plant Systems,*
DOI 10.1007/978-3-319-13419-2

CPI Antony Rowe
Chippenham, UK
2017-01-23 22:33